NEW BOOKS

PLEASE STAMP DATE DUE, BOTH BELOW AND ON CARD

DATE DUE	DATE DUE	DATE DUE	DATE DUE
PLEASE DO NOT BORROW UNTIL	MAY 2 1991		SEP 2 5 2000
SEP 3 0	JUN 1 3 1997		
DEC 1 4 1992			
JUN 0 3			
MAR 3 1 1997			
JUN 1 5 1998			
DEC 1 4 2001			
APR 0 8 2003			

GL-15

PROGRESS IN BRAIN RESEARCH

VOLUME 85

THE PREFRONTAL CORTEX
Its Structure, Function and Pathology

Recent volumes in PROGRESS IN BRAIN RESEARCH

Volume 60: The Neurohypophysis: Structure, Function and Control, by B.A. Cross and G. Leng (Eds.) – 1983
Volume 61: Sex Differences in the Brain: The Relation Between Structure and Function, by G.J. de Vries, J.P.C. De Bruin, H.B.M. Uylings and M.A. Corner (Eds.) – 1984
Volume 62: Brain Ischemia: Quantitative EEG and Imaging Techniques, by G. Pfurtscheller, E.J. Jonkman and F.H. Lopes de Silva (Eds.) – 1984
Volume 63: Molecular Mechanisms of Ischemic Brain Damage, by K. Kogure, K.-A. Hossmann, B.K. Siesjö and F.A. Welsh (Eds.) – 1985
Volume 64: The Oculomotor and Skeletal Motor Systems: Differences and Similarities, by H.-J. Freund, U. Büttner, B. Cohen and J. Noth (Eds.) – 1986
Volume 65: Psychiatric Disorders: Neurotransmitters and Neuropeptides, by J.M. Van Ree and S. Matthysse (Eds.) – 1986
Volume 66: Peptides and Neurological Disease, by P.C. Emson, M.N. Rossor and M. Tohyama (Eds.) – 1986
Volume 67: Visceral Sensation, by F. Cervero and J.F.B. Morrison (Eds.) – 1986
Volume 68: Coexistence of Neuronal Messengers: A New Principle in Chemical Transmission, by T. Hökfelt, K. Fuxe and B. Pernow (Eds.) – 1986
Volume 69: Phosphoproteins in Neuronal Function, by W.H. Gispen and A. Routtenberg (Eds.) – 1986
Volume 70: Aging of the Brain and Alzheimer's Disease, by D.F. Swaab, E. Fliers, M. Mirmiran, W.A. van Gool and F. van Haaren (Eds.) – 1986
Volume 71: Neural Regeneration, by F.J. Seil, E. Herbert and B.M. Carlson (Eds.) – 1987
Volume 72: Neuropeptides and Brain Function, by E.R. de Kloet, V.M. Wiegant and D. de Wied (Eds.) – 1987
Volume 73: Biochemical Basis of Functional Neuroteratology, by G.J. Boer, M.G.P. Feenstra, M. Mirmiran, D.F. Swaab and F. van Haaren (Eds.) – 1988
Volume 74: Transduction and Cellular Mechanisms in Sensory Receptors, by W. Hamann and A. Iggo (Eds.) – 1988
Volume 75: Vision within Extrageniculo-striate Systems, by T.B. Hicks and G. Benedek (Eds.) – 1988
Volume 76: Vestibulospinal Control of Posture and Locomotion, by O. Pompeiano and J.H.J. Allum (Eds.) – 1988
Volume 77: Pain Modulation, by H.L. Fields and J.-M. Besson (Eds.) – 1988
Volume 78: Transplantation into the Mammalian CNS, by D.M. Gash and J.R. Sladek, Jr. (Eds.) – 1988
Volume 79: Nicotinic Receptors in the CNS: Their Role in Synaptic Transmission, by A. Nordberg, K. Fuxe, B. Holmstedt and A. Sundwall (Eds.) – 1989
Volume 80: Afferent Control of Posture and Locomotion, by J.H.J. Allum and M. Hulliger (Eds.) – 1989
Volume 81: The Central Neural Organization of Cardiovascular Control, by J. Ciriello, M.M. Caverson and C. Polosa (Eds.) – 1989
Volume 82: Neural Transplantation: From Molecular Basis to Clinical Applications, by S.B. Dunnet and S.-J. Richards (Eds.) – 1990
Volume 83: Understanding the Brain through the Hippocampus, by J. Storm-Mathisen, J. Zimmer and O.P. Ottersen (Eds.) – 1990
Volume 84: Cholinergic Neurotransmission: Functional and Clinical Aspects, by S.-M. Aquilonius and P.-G. Gillberg (Eds.) – 1990

PROGRESS IN BRAIN RESEARCH

VOLUME 85

THE PREFRONTAL CORTEX

Its Structure, Function and Pathology

Proceedings of the 16th International Summer School of Brain Research, held at the Royal Tropical Institute and the Royal Netherlands Academy of Sciences, Amsterdam (The Netherlands), from 28 August to 1 September 1989

EDITED BY

H.B.M. UYLINGS, C.G. VAN EDEN, J.P.C. DE BRUIN,
M.A. CORNER and M.G.P. FEENSTRA

Netherlands Institute for Brain Research, Meibergdreef 33, 1105 AZ Amsterdam ZO (The Netherlands)

ELSEVIER
AMSTERDAM – NEW YORK – OXFORD
1990

© 1990, Elsevier Science Publishers B.V. (Biomedical Division)

All rights reserved. No part of this publication may be reproduced, stored in a retrieval system or transmitted in any form or by any means, electronic, mechanical, photocopying, recording or otherwise without the prior written permission of the publisher, Elsevier Science Publishers B.V. (Biomedical Division), P.O. Box 1527, 1000 BM Amsterdam (The Netherlands).

No responsibility is assumed by the Publisher for any injury and/or damage to persons or property as a matter of products liability, negligence or otherwise, or from any use or operation of any methods, products, instructions or ideas contained in the material herein. Because of the rapid advances in the medical sciences, the Publisher recommends that independent verification of diagnoses and drug dosages should be made.

Special regulations for readers in the U.S.A.:
This publication has been registered with the Copyright Clearance Center Inc. (CCC), Salem, Massachusetts. Information can be obtained from the CCC about conditions under which the photocopying of parts of this publication may be made in the USA. All other copyright questions, including photocopying outside of the USA, should be referred to the copyright owner, Elsevier Science Publishers B.V. (Biomedical Division) unless otherwise specified.

ISBN 0-444-81124-9 (volume)
ISBN 0-444-80104-9 (series)

This book is printed on acid-free paper

Published by:
Elsevier Science Publishers B.V. (Biomedical Division)
P.O. Box 211
1000 AE Amsterdam
(The Netherlands)

Sole distributors for the USA and Canada:
Elsevier Science Publishing Company, Inc.
655 Avenue of the Americas
New York, NY 10010
(U.S.A.)

Library of Congress Cataloging-in-Publication Data

International Summer School of Brain Research (16th : 1989 : Royal
   Tropical Institute and Royal Netherlands Academy of Sciences)
      The prefrontal cortex : its structure, function, and pathology :
   proceedings of the 16th International Summer School of Brain
   Research, held at the Royal Tropical Institute and the Royal
   Netherlands Academy of Sciences Amsterdam, The Netherlands, from 28
   August to 1 September 1989 / edited by H.B.M. Uylings . . . [et al.].
         p.   cm. -- (Progress in brain research ; v. 85)
      Includes bibliographical references.
      Includes index.
         ISBN 0-444-81124-9 (alk. paper)
      1. Prefrontal cortex--Congresses.   2. Prefrontal cortex-
   -Pathophysiology--Congresses.   I. Uylings. H. B. M. (Harry B. M.)
   II. Title.   III. Series.
      [DNLM: 1. Frontal Lobe--anatomy & histology--congresses.
   2. Frontal Lobe--physiopathology--congresses.   W1 PR667J v. 85 / WL
   307 I615p 1989]
   QP376.P7   vol. 85
   [QP383. 17]
   612.8'2 s--dc20
   [612.8'25]
   DNLM/DLC
   for Library of Congress   90-14128
                                                                           CIP

Printed in The Netherlands

List of Contributors

G.E. Alexander, Dept. of Neurology, Johns Hopkins School of Medicine, N. Wolfe Street, Baltimore, MD 21205, USA

H.W. Berendse, Dept. of Anatomy and Embryology, Vrije Universiteit, van der Boechorststraat 7, 1081 BT Amsterdam, The Netherlands

K.F. Berman, Dept. of Nuclear Medicine, NIH Building 10, Room 1C401, 9000 Rockville Pike, Bethesda, MD 20892, USA

M. Bertolucci-D'Angio, Mental Retardation Research Center, UCLA, 760 Westwood Plaza, Los Angeles, CA 90024, USA

S.L. Buchanan, Neuroscience Lab, Wm. Jennings Bryan Dorn Veterans' Hospital, Columbia, SC 29201, USA

M. Cobo, Dept. of Physiology, Faculty of Medicine, University Complutense de Madrid, 28040 Madrid, Spain

M.A. Corner, Netherlands Institute for Brain Research, Meibergdreef 33, 1105 AZ Amsterdam ZO, The Netherlands

M.D. Crutcher, Dept. of Neurology, Johns Hopkins School of Medicine, N. Wolfe Street, Baltimore, MD 21205, USA

J.P.C. De Bruin, Netherlands Institute for Brain Research, Meibergdreef 33, 1105 AZ Amsterdam ZO, The Netherlands

M.R. DeLong, Dept. of Neurology, Johns Hopkins School of Medicine, N. Wolfe Street, Baltimore, MD 21205, USA

A.Y. Deutch, Dept. of Psychiatry, Yale University School of Medicine, Connecticut Mental Health Center, 34 Park Street, New Haven, CT 06508, USA

P. Driscoll, Eidgenössische Technische Hochschule Zürich, Institut für Verhaltenswissenschaft, Laboratorium für Vergleichende Physiologie und Verhaltensbiologie, Turnerstrasse 1, EHT-Zentrum, CH-8092 Zürich, Switzerland

S.B. Dunnett, Dept. of Exp. Psychology, University of Cambridge, Downing Street, Cambridge CB2 3EB, Great Britain

M.G.P. Feenstra, Netherlands Institute for Brain Research, Meibergdreef 33, 1105 AZ Amsterdam ZO, The Netherlands

J.M. Fuster, Dept. of Psychiatry, UCLA School of Medicine, 760 Westwood Plaza, Los Angeles, CA 90024, USA

M. Gabriel, Dept. of Physiology, University of Illinois, 603 East Daniel Street, Champaign, IL 61820, USA

A. Gevins, EEG Systems Laboratory, 51 Federal St., Suite 401, San Francisco, CA 94107, USA

R. Gibb, Dept. of Psychology, The University of Lethbridge, 401 University Drive, Lethbridge, Alberta, Canada, T1K 3M4

C.M. Gibbs, Neuroscience Lab, Wm. Jennings Bryan Dorn Veterans' Hospital, Columbia, SC 29201, USA

J. Glowinski, INSERM U 114, Collège de France, Chaire de Neuropharmacologie, 11 Place Marcelin-Berthelot, 75231 Paris Cedex 05, France

R. Godbout, INSERM U 114, Collège de France, Chaire de Neuropharmacologie, 11 Place Marcelin-Berthelot, 75231 Paris Cedex 05, France

P.S. Goldman-Rakic, Section of Neuroanatomy, Yale University School of Medicine, C303 SHM, PO Box 3333, New Haven, CT 06510-8001, USA

H.J. Groenewegen, Dept. of Anatomy and Embryology, Vrije Universiteit, van der Boechorststraat 7, 1081 BT Amsterdam, The Netherlands

E.R. John, Brain Research Labs, Dept. of Psychiatry, y, NYU Medical Center, 550 First Ave, New York, NY 10016, USA; Nathan S. Kline Institute for Psychiatric Research, Orangeburg, NY

M. Judas, Section of Neuroanatomy, Dept. of Anatomy, Medical Faculty, University of Zagreb, Salata 11, PP. 286, 41001 Zagreb, Yugoslavia

J.H. Kaas, Dept. of Psychology, Vanderbilt University, 134 Wesley Hall, Nashville, TN 37240, USA

A. Kalsbeek, Netherlands Institute for Brain Research, Meibergdreef 33, 1105 AZ Amsterdam ZO, The Netherlands

S.E. Karpiak, Div. of Neuroscience, Coll. of Physicians and Surgeons, Columbia University, 722 West 168th Street, New York, NY 10032, USA

B. Kolb, Dept. of Psychology, The University of Lethbridge, 401 University Drive, Lethbridge, Alberta, Canada, T1K 3M4

I. Kostovic, Section of Neuroanatomy, Dept. of Anatomy, Medical Faculty, University of Zagreb, Salata 11, PP. 286, 41001 Zagreb, Yugoslavia

J.M. Kros, Dept. of Pathology, Erasmus University, Dr. Molewaterplein 40, 3015 GD Rotterdam, The Netherlands

A.H.M. Lohman, Dept. of Anatomy and Embryology, Vrije Universiteit, van der Boechorststraat 7, 1081 BT Amsterdam, The Netherlands

S.P. Mahadik, Div. of Neuroscience, Coll. of Physicians and Surgeons, Columbia University, 722 West 168th Street, New York, NY 10032, USA

J. Mantz, INSERM U 114, Collège de France, Chaire de Neuropharmacologie, 11 Place Marcelin-Berthelot, 75231 Paris Cedex 05, France

F. Mora, Dept. of Physiology, Faculty of Medicine, University Complutense de Madrid, 28040 MADRID, Spain

L. Mrzljak, Section of Neuroanatomy, Dept. of Anatomy, Medical Faculty, University of Zagreb, Salata 11, PP. 286, 41001 Zagreb, Yugoslavia

E.J. Neafsey, Dept. of Anatomy, Loyola University Stritch, School of Medicine, 2160 South First Ave, Maywood, IL 60153, USA

D.N. Pandya, Edith Nourse Rogers Memorial Veterans Hospital, 200 Springs Road, Bedford, MA 01730, USA

J.G. Parnavelas, Dept. of Anatomy & Dev. Biology, University College London, Gower Street, London WC1E 6BT, Great Britain

D.A. Powell, Neuroscience Lab, Wm. Jennings Bryan Dorn Veterans' Hospital, Columbia, SC 29201, USA

L.S. Prichep, Brain Research Labs, Dept. of Psychiatry, NYU Medical Center, 550 First Ave, New York, NY 10016, USA; Nathan S. Kline Institute for Psychiatric Research, Orangeburg, NY

R.H. Roth, Dept. of Pharmacology, Yale University School of Medicine, Connecticut Mental Health Center, 34 Park Street, New Haven, CT 06508, USA

B. Scatton, Synthélabo Recherche (LERS), Biology Department, 31 Avenue Paul Vaillant-Couturier, 92220 Bagneux, France

A. Serrano, Synthélabo Recherche (LERS), Biology Department, 31 Avenue Paul Vaillant-Couturier, 92220 Bagneux, France

A.-M. Thierry, INSERM U 114, Collège de France, Chaire de Neuropharmacologie, 11 Place Marcelin-Berthelot, 75231 Paris Cedex 05, France

H.B.M. Uylings, Netherlands Institute for Brain Research, Meibergdreef 33, 1105 AZ Amsterdam ZO, The Netherlands

E.S. Valenstein, Neuroscience Lab Bldg, The University of Michigan, 1103 East Huron, Ann Arbor, MI 48104, USA

C.G. Van Eden, Netherlands Institute for Brain Research, Meibergdreef 33, 1105 AZ Amsterdam ZO, The Netherlands

D.R. Weinberger, Dept. of Nuclear Medicine, NIH Building 10, Room 1C401, 9000 Rockville Pike, Bethesda, MD 20892, USA

J.G. Wolters, Dept. of Anatomy and Embryology, Vrije Universiteit, van der Boechorststraat 7, 1081 BT Amsterdam, The Netherlands

E.H. Yeterian, Dept. of Psychology, Colby College, Waterville, ME 04901, USA

Preface

The prefrontal cortex has been regarded to be involved in a variety of functions: as the cortex for "higher psychic activity", as the cortex of convergence of cognition and emotion, as being involved in schizophrenia and depression, and as the visceral cortex. The prefrontal cortex is thus on the one hand "attributed the highest integrative faculties of the human mind", whereas others emphasized "the surprising paucity of cognitive deficits" following frontal lesions (Editorial Mesulam, Ann. Neurol., 19, 1986). The prefrontal cortex is therefore an intriguing topic to deal with, given the many different concepts for this part of the cortex.

Over the past decade many new data on the prefrontal cortex have been obtained especially in primates and rodents, thanks to progress in research techniques and an increased interest in this topic. Therefore, the 16th International Summer school of Brain Research, held in Amsterdam in August 1989 (the year of the 80th anniversary of the Netherlands Institute for Brain Research) was entirely devoted to the topic of Prefrontal Cortex. This book is one of the products of this stimulating Summer School. Unfortunately, Professors B. Milner, M. Mishkin and P. Roland were, for several reasons unable to prepare a chapter on their important studies presented during the Summer School.

At first sight, the term prefrontal cortex appears to be illogical: "How can any cortex be in front of the frontal cortex?". This term is still widely used by neurobiologists, since, at present there is no appropriate alternative term. Among alternative terms, the "frontal granular cortex" has been used. This term, however, is not appropriate for many reasons. The cytoarchitectonic criterion implied in the name is not sufficient to define the prefrontal cortex; not only during early development, since the entire frontal lobe of primates *including* the prefrontal cortex is granular, but also in adult animals. There are also *agranular* cortical areas which belong to the prefrontal cortex. The historical origin of the term "prefrontal" has been described elsewhere (I. Divac, IBRO News, 16 (2), 2, 1988). In the classical cortical maps of the human brain we can see that the meaning of the term "prefrontal cortical area" changed during the first decades of this century. Brodmann's area as prefrontalis (conform Smith, 1907) indicates only Brodmann's areas 11 and 12. Also the map of Campbell (1905) names a small rostral cortical area as prefrontal, but this area is different from Brodmann's. Since Walker (1940), the term *prefrontal* cortex defines what we presently consider as prefrontal cortex in primates. On the other hand, even today the borders of the cortical area regarded as prefrontal cortex are under discussion for different mammalian species. In connection with this, the first section describes the present knowledge on the organization of the prefrontal cortex, after the two introductory chapters by Kaas and Parnavelas.

In the second section, developmental and plasticity aspects in rodent and human prefrontal cortex are considered. The third section deals rather extensively with the functional aspects characteristic for the prefrontal cortex in primates, rats and rabbits. In the last section, some topics on dysfunction of prefrontal cortex in rat and human are reviewed, including a historical review on psychosurgery.

This book will therefore be of interest for neuroscientists, neurologists and psychiatrists.

Amsterdam, February 1990

Harry B.M. Uylings
Corbert G. van Eden
Jan P.C. de Bruin
Michael A. Corner
Matthijs G.P. Feenstra

Acknowledgements

The Sixteenth International Summer School of Brain Research was organized under the auspices of:

The Royal Netherlands Academy of Sciences
UNESCO/IBRO

Financial support was also obtained from:

The Dr. Saal van Zwanenberg Foundation
The Netherlands Society for the Advancement of Natural Sciences, Medicine and Surgery
Organon International B.V.
Bayer Nederland B.V.

We wish to express our gratitude to T. Eikelboom and A.A.M. Janssen for taking care of the innumerable organizational details for the Summer School as well as the secretarial work required to compile the book.

Contents

List of Contributors v
Preface vii
Acknowledgements ix

Section I – Organization of Prefrontal Cortical Systems

1. How sensory cortex is subdivided in mammals : Implications for studies of prefrontal cortex
   J.H. Kaas (Nashville, TN, USA) 3

2. Neurotransmitters in the cerebral cortex
   J.G. Parnavelas (London, UK) 13

3. Qualitative and quantitative comparison of the prefrontal cortex in rat and in primates, including humans
   H.B.M. Uylings and C.G. van Eden (Amsterdam, The Netherlands) 31

4. Prefrontal cortex in relation to other cortical areas in rhesus monkey: Architecture and connections
   D.N. Pandya and E.H. Yeterian (Bedford, MA, Waterville, ME and Boston, MA, USA) 63

5. The anatomical relationship of the prefrontal cortex with the striatopallidal system, the thalamus and the amygdala: evidence for a parallel organization
   H.J. Groenewegen, H.W. Berendse, J.G. Wolters and A.H.M. Lohman (Amsterdam, The Netherlands) 95

6. Basal ganglia-thalamocortical circuits: Parallel substrates for motor, oculomotor, "prefrontal" and "limbic" functions
   G.E. Alexander, M.D. Crutcher and M.R. DeLong (Baltimore, MD, USA) 119

7. Prefrontal cortical control of the autonomic nervous system: Anatomical and physiological observations
   E.J. Neafsey (Maywood, IL, USA) 147

Section II — Development and Plasticity in Prefrontal Cortex

8. The development of the rat prefrontal cortex
   Its size and development of connections with thalamus, spinal cord and other cortical areas
   C.G. van Eden, J.M. Kros and H.B.M. Uylings (Amsterdam, The Netherlands) 169

9. Neuronal development in human prefrontal cortex in prenatal and postnatal stages
   L. Mrzljak, H.B.M. Uylings, C.G. van Eden and M. Judáš (Zagreb, Yugoslavia and Amsterdam, The Netherlands) 185

10. Structural and histochemical reorganization of the human prefrontal cortex during perinatal and postnatal life
    I. Kostović (Zagreb, Yugoslavia) 223

11. Anatomical correlates of behavioural change after neonatal prefrontal lesions in rats
    B. Kolb and R. Gibb (Alberta, Canada) 241

12. Age-dependent effects of lesioning the mesocortical dopamine system upon prefrontal cortex morphometry and PFC-related behaviors
    A. Kalsbeek, J.P.C. De Bruin, M.G.P. Feenstra and H.B.M. Uylings (Amsterdam, The Netherlands) 257

13. Is it possible to repair the damaged prefrontal cortex by neural tissue transplantation?
    S.B. Dunnett (Cambridge, UK) 285

14. Enhanced cortical maturation: Gangliosides in CNS plasticity
    S.E. Karpiak and S.P. Mahadik (New York, NY, USA) 299

Section III — Functional Aspects of Prefrontal Cortex

15. Behavioral electrophysiology of the prefrontal cortex of the primate
    J.M. Fuster (Los Angeles, CA, USA) 313

16. Cellular and circuit basis of working memory in prefrontal cortex of nonhuman primates
    P.S. Goldman-Rakic (New Haven, CT, USA) 325

17. Distributed neuroelectric patterns of human neocortex during simple cognitive tasks
    A. Gevins (San Francisco, CA, USA) 337

18. Influence of the ascending monoaminergic systems on the activity of the rat prefrontal cortex
    A.-M. Thierry, R. Godbout, J. Mantz and J. Glowinski (Paris, France) 357

19. The determinants of stress-induced activation of the prefrontal cortical dopamine system
    A.Y. Deutch and R.H. Roth (New Haven, CT, USA) 367

20. Involvement of mesocorticolimbic dopaminergic systems in emotional states
    M. Bertolucci-D'Angio, A. Serrano, P. Driscoll and B. Scatton (Bagneux, France) 405

21. The neurobiological basis of prefrontal cortex self-stimulation: A review and an integrative hypothesis
    F. Mora and M. Cobo (Madrid, Spain) 419

22. Role of the prefrontal – thalamic axis in classical conditioning
    D.A. Powell, S.L. Buchanan and C.M. Gibbs (Columbia, SC, USA) 433

23. Functions of anterior and posterior cingulate cortex during avoidance learning in rabbits
    M. Gabriel (Urbana, IL, USA) 467

24. Social behaviour and the prefrontal cortex
    J.P.C. de Bruin (Amsterdam, The Netherlands) 485

Section IV – Pathology of Prefrontal Cortex

25. Animal models for human PFC-related disorders
    B. Kolb (Lethbridge, Canada) 501

26. The prefrontal cortex in schizophrenia and other neuropsychiatric diseases: in vivo physiological correlates of cognitive deficits
    K.F. Berman and D.R. Weinberger (Washington, DC, USA) ... 521

27. The prefrontal area and psychosurgery
    E.S. Valenstein (Ann Arbor, MI, USA) 539

28. Neurometric studies of aging and cognitive impairment
      E.R. John and L.S. Prichep (New York, NY and Orangeburg, NY, USA) 555

Subject Index 567

SECTION I

Organization of Prefrontal Cortical Systems

CHAPTER 1

How sensory cortex is subdivided in mammals: Implications for studies of prefrontal cortex

Jon H. Kaas

Department of Psychology, Vanderbilt University, Nashville, TN 37240, USA

Introduction

A general approach toward understanding brain functions is to determine how the brain is subdivided into systems, and then study parts of systems and how they interact. Thus it is initially important to determine with some accuracy what the parts are and how they are interconnected. Prefrontal cortex is an especially intriguing part of the brain because it is obviously expanded in humans, and this expansion undoubtedly accounts for many of our unique mental and behavioral abilities. There is widespread agreement that prefrontal cortex consists of a number of functionally distinct subdivisions, but opinions vary on how this cortex is subdivided in any particular mammalian species, and on how subdivisions of prefrontal cortex compare in various species. This disagreement implies uncertainty, and it would be valuable to obtain accurate and compelling information that could lead to a consensus. However, the issues of how to subdivide cortex and compare species are not specific to prefrontal cortex. Experimental and comparative studies of sensory-perceptual cortex will have the advantage of subdivisions that are denoted by the presence of systematic representations of receptor surfaces, and can provide guidelines, general principles, and basic conclusions that apply to prefrontal cortex. This paper considers conclusions stemming from studies of sensory cortex, and then briefly discusses several proposed subdivisions of the frontal lobe in respect to these conclusions.

The premise of subdividing cortex

A basic premise of Brodmann (1909), Smith (1907), Von Economo (1929), and other early investigators was that cortex is divided into a patchwork of areas or fields, the "organs" of the brain, that perform distinct functions and yet are functionally interrelated. They also believed that because different functions are based on structural variations in cortex, areas must differ in histological appearance, and these differences can be used to distinguish fields. The borders between at least some areas were considered to be sharp, and mammalian species with large brains and more complex behavior, especially humans, were thought to have more areas. All mammals had a number of areas in common, and the number of areas increased in evolution by existing areas (or in some sense, composite areas) subdividing by "differentiating". While these and other early investigators largely agreed on these few principles, they were limited by the techniques of that time to examine brain sections stained for cell bodies or myelinated fibers in order to discover distinctions

---

Correspondence: Jon H. Kaas, Ph.D., 301 A & S Psychology Building, Department of Psychology, Vanderbilt University, Nashville, TN 37240, USA. Tel.: 615-322-6029.

denoting the extents and boundaries of possible subdivisions, and they came to quite different conclusions on how brains are subdivided. Some later investigators, notably Lashley (e.g. Lashley and Clark (1946)) and Von Bonin and Bailey (1961), questioned the assumption that species differ in number of subdivisions and even the validity of the architectonic method. However, it is now clear that cortex is divided into a number of functionally distinct subdivisions or areas, borders are often if not always sharp, species differ in numbers of areas, and areas do differ in architectonic appearances (see Kaas, 1987a, for review). In addition, we are now more aware of the difficulties and potential for error in subdividing cortex. These difficulties both explain why investigators have so often differed in conclusions, and indicate that even current proposals should be evaluated with great caution.

The difficulties of obtaining a comparative understanding of how cortex is subdivided

There are several major reasons why it has been difficult to obtain an accurate portrayal of how cortex is subdivided into areas in different mammals. First and foremost is the enormous scope of the problem. Brains vary greatly in such obvious features as size, shape, and patterns of fissures, and there has been a long time for the evolution of these and other variations. Mammals evolved from therapsid reptiles some 250 million years ago (M.Y.A.), and have formed a number of continuing branches or lines of evolution (see Kaas, 1987b). Prototherian mammals (monotremes) diverged from therian mammals some 200 M.Y.A., metatherian (marsupials) mammals from eutherian (placental) mammals about 150 M.Y.A., and edentates from other eutherian mammals about 120 M.Y.A. The remaining eutherian mammals later divided into 17 or so major orders, largely about 65 M.Y.A. (see Fig. 5 in Kaas, 1987b). Within a given order, especially in primates, the range of obvious brain differences can be considerable. In addition, the rate of brain change can be quite rapid. For example, the evolution of modern human brains from those of one-third the size in Australopithecus occurred within the relatively short time of 3.5 million years.

A second difficulty is that brain subdivisions are usually not obvious. While the specialized functions of some subdivisions of the brain may be based largely on differences in internal organization, other areas may be structurally similar because they mediate basically the same computations, differing largely in what inputs serve as the basis for the computations (see Sur et al., 1988). Whether this is the case or not, many subdivisions of cortex are highly similar in cytoarchitecture, and very difficult to reliably distinguish by differences in appearance using traditional stains. A related problem is that many and possibly most areas of cortex are functionally and structurally heterogeneous (e.g., Livingstone and Hubel, 1988). As a clear example, in all mammals primary visual cortex, V-I or area 17, has a large part activated by both eyes and a small part activated only by the contralateral eye. The binocular portion is notably thicker and is often more conspicuously laminated than the monocular portion. Such structural differences within fields can be mistaken for differences between fields. Thus, Brodmann (1909) mistook the monocular portion of area 17 in squirrels for area 18 (Fig. 1), and the error persists in most descriptions of visual cortex in rodents (see Kaas et al., 1989, for review).

Another difficulty is that brain areas are capable of change in all aspects including appearance. Area 17 or striate cortex was the first recognized area of cortex, and it is perhaps the most distinctive in appearance. Yet, area 17 varies from being rather indistinctly laminated in such mammals as hedgehogs (see Kaas, 1987a) to being conspicuously laminated in tarsiers (e.g., Hassler, 1966). If it were not for the existence of extant species with an area 17 of intermediate degrees of lamination and of a similar position, together with more current and compelling comparative evidence on connections and neural response properties, claims that area 17 is the same area (homologous, stemming

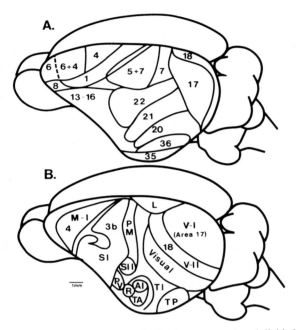

Fig. 1. Different proposals for how cortex is subdivided in rodents (squirrels). (A) Subdivisions of cortex according to the cytoarchitectonic studies of Brodmann (1909). (B) Subdivisions described in recent studies using microelectrode recordings, connection patterns, and architectonics. See Kaas et al. (1989), Luethke et al. (1988) and Krubitzer et al. (1986) for details and references. Note that there is little correspondence between modern and classical views, other than in the parntial overlap of fields designated as primary visual (area 17) and primary motor (area 4) cortex. Brodmann recognized no primary somatosensory or auditory fields in rodents. Modified from Kaas (1989b).

from a common ancestor) in the two mammals would be incredulous. In addition, area 17 has become more laminated in several lines of descent, and ocular dominance columns have evolved independently in area 17 of cats and several groups of primates. Thus, current resemblances in different species may not reflect the ancestral state. It is also possible that a cortical area would change over time to come to resemble another field, creating the likelihood of misidentification. For instance, the primary auditory area, A-I, and the rostral auditory area, R, closely resemble each other in architectonic appearance (see Luethke et al., 1988), and they probably have been confused

or, more likely, combined in architectonic studies. However, A-I and R contain separate maps of the cochlea, and have somewhat different connections.

An experimental approach

A logical conclusion is that cortical areas are most reliably defined by multiple criteria (e.g., Campbell and Hodos, 1970; Kaas, 1987a). Thus, the weaknesses and potential for error inherent in one method may be compensated by strengths of another method. A functionally distinct area should have unifying characteristics that distinguish all parts of the area from other areas. Such characteristics may include a host of architectonic and histochemical features reflecting structural specialization, a unique pattern of connections with other parts of the brain as a station in a processing system, a population of neurons with response properties that are distinct from populations in other fields, and specific behavioral defects as a result of deactivation. In sensory systems, areas often contain a systematic representation of a receptor surface (skin, retina, cochlea). Since areas need not be homogeneous in structure or function, subregions and modules within an area may be misidentified as an area. For example, modules in area 17 have different inputs, outputs, histological appearance, and neuron properties (e.g., Livingstone and Hubel, 1988). Nevertheless, these modules clearly are repeating, interacting parts of a larger area. However, there may be regions of cortex where it becomes difficult and perhaps pointless to distinguish modules from areas. In any case, we need to consider many types of evidence in attempts to subdivide cortex, and to weigh information carefully with regard to reliability and power.

In addition to defining areas within a species, it is critically important to evaluate the probability that a given area in one species is or is not homologous (the same area) with an area in another species. Areas are judged to be homologous when the number of observed similarities is so great, and the observed differences so explainable, that the possibility of resemblance due to common inheritance outweighs the alternative of resemblance due to convergent evolution. A problem, however, is that without a better understanding of the mechanisms of cortical development and evolution (see Kaas, 1988), it is difficult to evaluate the probability of resemblances due to convergence.

Two examples of experimentally revealed areas

Given the marked differences in published schemes for subdividing cortex, it sometimes seems as if there are as many ways of subdividing brains as there are investigators. Yet, there has been considerable progress. Many subdivisions have now been defined with considerable certainty, and many traditional concepts about brain organization have been revised (see Merzenich and Kaas, 1980). Two of many possible examples of clear progress follow.

A number of years ago, Allman and Kaas (1971) discovered a systematic representation of the visual field in a densely myelinated oval of cortex in the upper temporal lobe of owl monkeys. This representation, which we called the middle temporal visual area (MT) (Fig. 2), was subsequently shown to be present in all primates investigated, and to be further characterized by a number of features including inputs from areas 17 and 18, neurons that are highly selective for direction of stimulus movement but nonselective for color, and outputs that relay to posterior parietal cortex (see Kaas, 1989a, for review). Inactivation experiments suggest a role in visual tracking and attention. While MT is now a firmly established and extensively studied subdivision of visual cortex in primates, it was not part of any previous proposal for subdividing cortex (however, Kuypers et al. (1965) had demonstrated projections from area 17 to the location of MT). Instead, MT was included as part or parts of one or more larger fields in other proposals. For example, MT roughly overlaps the dorsal third of "area 21" of Brod-

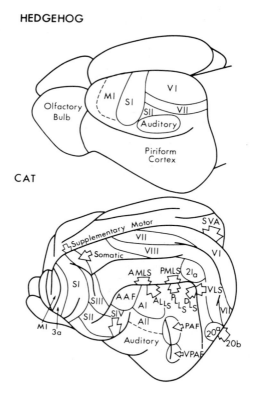

Fig. 2. Proposed subdivisions of neocortex in a mammal with little neocortex (hedgehog) and expanded neocortex (cat). See Fig. 1B for squirrels and Fig. 3 for owl monkeys. Primary motor (M-I), primary and secondary somatic (S-I and S-II), primary auditory (A-I) and primary and secondary visual (V-I and V-II) are present in all these mammals. Other terms for areas reflect different proposals, uncertainties about homologies, and the probable acquisition of different areas in separate lines of descent. In cats, a number of auditory and visual fields have been named by position, such as the posterior auditory field (PAF). Areas 21a, 20a and 20b are after Brodmann. See Kaas (1982) for details.

mann (1909) and the ventral half of area PFG of Peden and Von Bonin (1947) in marmoset monkeys.

Another notable example of progress has been in understanding the organization of cortex in the region of the primary somatosensory area, S-I, in rodents. Welker (1971) made an important early advance when she correlated a microelectrode map of body surface receptors with the surface-view distribution of cortex characterized by densely packed granule cells in layer IV. More recently, Dawson and Killackey (1987) have used a reaction for the mitochondrial enzyme, succinic dehydrogenase, to display a thalamocortical input pattern that precisely indicates the extent and somatotopic organization of S-I. Before these and other related investigations, there was much confusion about how cortex in the S-I region of rodents was subdivided, and many different proposals resulted. Brodmann (1909), for instance, included parts of areas 1, 5 and 7, and other fields in the region in S-I, while failing to recognize an area 3 (Fig. 1).

Conclusions supported by modern approaches

As a result of modern approaches, comprehensive proposals have been published by a number of investigators on how cortex is divided in monkeys, cats, rodents, and a number of other mammals (Figs. 1 – 3). Many proposed subdivisions are well supported by experimental evidence, others are more tentative, and we can expect much further progress in the near future (see Kaas, 1987a, b, 1989b; Krubitzer et al., 1986; Luethke et al., 1988, for reviews). Nevertheless, a number of clear conclusions are justified. Mammals such as hedgehogs and opossums with little neocortex have only a few sensory and motor areas, and these areas occupy most of cortex. These areas include primary and secondary somatosensory areas, S-I and S-II, primary and secondary visual fields, V-I and V-II, primary auditory cortex, and, in eutherian mammals at least, primary motor cortex. Mammals with expanded neocortex have these basic sensory and motor fields inherited from a common ancestor, and have additional sensory fields. For example, monkeys appear to have as many as 20 or more visual areas (see Kaas, 1989b). Advanced mammals have not only larger brains, but more subdivisions. The increase in number has occurred independently in several lines of descent, and thus some of the areas in different lines are not homologous. Furthermore, if the mechanisms for increasing the numbers of fields are limited (see

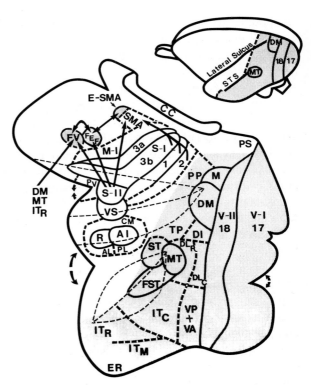

Fig. 3. Some connections of the supplementary motor area (SMA), the frontal eye field (FEF) and the frontal visual area (FV) in New World monkeys. Cortical areas are superimposed on an outline of cortex that has been separated from the brain and flattened. CC, corpus callosum; 3a, 3b, 1, 2, S-II, PV, and VS are somatosensory areas; CM, AL, PL, R, and AI are auditory areas. Shaded areas are visual or visuomotor. See Kaas (1989b) for abbreviations. Modified from Kaas (1989b).

Kaas, 1987b, 1989c), comparable areas may have been independently acquired in different progressive species. Advances in the complexity and variety of behavior seem to depend on increases in the number of areas, and increases in the number of areas is a way of avoiding constraints on modifying existing areas (see Kaas, 1989b).

Motor and visuomotor areas in the frontal lobe

Unlike posterior and temporal regions of neocortex, it is difficult to apply sensory mapping methods to frontal cortex because sensory input is rather indirect, activation is often unreliable in anesthetized preparations, and activation patterns are likely to be complex. However, it is useful to determine regions of input from sensory-perceptual cortex into the frontal lobe. In addition, microstimulation procedures have the potential for revealing subdivisions of the caudal portion of the frontal lobe, and connection patterns of these subdivisions may help reveal valid subdivisions of prefrontal cortex. These approaches will help determine what subdivisions of the frontal lobe are common to all or most mammals, and what areas are common to members of specific lines of evolution, such as primates.

Because the primary motor area, M-I, has been identified in a wide range of mammals (Fig. 2), a reasonable premise is that M-I predated mammals and it now exists as a homologous area in all mammals. M-I is typically characterized by a systematic representation of body movements proceeding from the hindfoot to face in a mediolateral direction across cortex, a position just rostral to S-I, or a somatosensory fringe (area 3a) bordering S-I, a cytoarchitectonic specialization emphasizing layer V pyramidal cells and de-emphasizing layer IV granule cells (hence, the term, agranular cortex is applied), and a pattern of connections that includes inputs from subdivisions of the somatosensory cortex and inputs relayed in the ventrolateral thalamus from the cerebellum.

While there is good evidence that M-I exists in all primates, and even in all eutherian mammals, there is the early claim of Lende (1963) that M-I and S-I are coextensive and form an "amalgam" in marsupials. Since this "amalgam" had all of the major characteristics of S-I, an alternative interpretation is that the cortical area is S-I and that M-I either does not exist or has not been found. Evidence for at least a partially separate M-I rostral to S-I in other studies in marsupials (e.g., Haight and Neylon, 1979; Bohringer and Rowe, 1977) and in monotremes (Ulinsky, 1984) suggests either that "M-I" has evolved independently a number of times in mammals, perhaps by "differentiating" from S-I as suggested by the claim of partial overlap of S-I and M-I in rats (see Wise and Donoghue, 1986), or that M-I evolved early in mammals or premammals and has been retained in most living mammals, but not always discovered.

All primates also appear to have a supplementary motor area (SMA) and a frontal eye field (FEF), and an oval of cortex rostrolateral to the FEF that we call the frontal visual area (FV) (Kaas and Krubitzer, 1988). The SMA is in medial cortex adjoining the hindlimb representation of M-I (see Gould et al., 1986). The motor map in SMA proceeds from hindlimb to face in a caudorostral direction, with as much as the rostral fourth of the area devoted to eye movements (hence, this part of SMA is sometimes referred to as the supplementary eye field (E-SMA)). Stimulation experiments have also provided good evidence for SMA in carnivores (Górska, 1974; Jameson et al., 1968) suggesting that SMA evolved early and is widespread in mammals. However, as of yet, there is no convincing evidence for SMA in other mammals (see Gould et al., 1986). A premotor field in the relative position of SMA appears to exist in rodents (see Li et al., 1990, for review), but the identity of this field remains uncertain.

The FEF is another area that may be common to many or most mammals, since eye movements have been elicited by electrical stimulation of frontal cortex in a number of mammalian species (e.g., see Kolb, 1984). However, the demonstration that eye movements can be elicited from part of frontal cortex does not conclusively indicate the presence of a field homologous with the FEF of primates, since eye movements can be elicited from several regions of the frontal lobe in primates, including the E-SMA. Thus, evidence on other features of visuomotor fields is needed for further comparisons.

An intriguing possibility, given that a part of the SMA is devoted to eye movements, is that the FEF is the eye movement portion of M-I. In this case, all mammals with M-I should then have a FEF. The FEF has the proper topographic relationship with the rest of the motor map in M-I of New

World monkeys and prosimians (Gould et al., 1986; Kaas and Krubitzer, 1988), but in Old World monkeys, a premotor area separates the FEF from M-I (see Wise, 1985). Possibly a later evolving premotor field displaced the FEF from M-I in more advanced primates.

The FV is a small oval of cortex just ventral to the FEF of primates (Kaas and Krubitzer, 1988). FV has dense interconnections with the FEF and the E-SMA, but higher levels of current are needed to elicit eye movements from the FV. While the FEF and especially the FV have inputs from a number of higher-order visual areas, the E-SMA appears to be relatively free of such direct visual influences. The second somatosensory area, S-II, also has connections with both the FEF and the FV in some primates. The SMA relates mainly to subdivisions of somatosensory cortex. The FEF and the SMA are interconnected with more rostral parts of the frontal lobe (e.g., Huerta et al., 1987), and some of this cortex also has input from higher-order auditory fields (Morel et al., 1989). Thus, connection patterns of these more established fields (Fig. 3) may help reveal other functional subdivisions of the frontal lobe.

Summary

A problem for studies of all cortex, including prefrontal cortex, is how to relate results to proposed subdivisions of the brain and how to compare results across species. Investigators have been highly dependent on traditional architectonic maps of neocortex. However, over the last 20 years, other methods, especially the microelectrode mapping method, have been applied to regions of sensory cortex, where orderly representations of receptor surfaces have provided strong evidence for new proposals of how cortex is subdivided. As a result, we can more reliably indicate functional subdivisions of cortex and identify valid homologies across species.

Obtaining accurate maps of functional subdivisions of prefrontal cortex in a range of mammalian species is an important goal in that such maps are basic to interpreting other data. While microelectrode mapping methods may be more difficult to apply to prefrontal than sensory cortex, a range of useful anatomical and histochemical methods are available. Studies of sensory cortex suggest that all mammals have a few sensory areas in common, and that additional areas have evolved in some lines. One approach toward a better comparative understanding of frontal cortex might be to investigate the possibility that primary motor cortex, the supplementary motor area, and the frontal eye field are areas that exist in most or all mammals, and use these fields, when they can be clearly demonstrated, as reference areas for further studies on how frontal and prefrontal cortex is subdivided.

References

Allman, J.M. and Kaas, J.H. (1971) A representation of the visual field in the caudal third of the middle temporal gyrus of the owl monkey *(Aotus trivirgatus). Brain Res.,* 31: 85 – 105.

Bohringer, R.C. and Rowe, M.J. (1977) The organization of the sensory and motor areas of cerebral cortex in the platypus *(Ornithorhynchus anatinus). J. Comp. Neurol.,* 174: 1 – 14.

Brodmann, K. (1909) *Vergleichende Lokalisationslehre der Grosshirnrinde,* Barth, Leipzig.

Campbell, C.B.G. and Hodos, W. (1970) The concept of homology and the evolution of the nervous system. *Brain Behav. Evol.,* 3: 353 – 367.

Dawson, D.R. and Killackey, H.P. (1987) The organization and mutability of the forepaw and hindpaw representations in the somatosensory cortex of the neonatal rat. *J. Comp. Neurol.,* 256: 246 – 256.

Gould, J.H., III, Cusick, C.G., Pons, T.P. and Kaas, J.H. (1986) The relationship of corpus callosum connections to electrical stimulation maps of motor, supplementary motor, and frontal eye fields in owl monkeys. *J. Comp. Neurol.,* 247: 297 – 325.

Górska, T. (1974) Functional organization of cortical motor areas in adult dogs and puppies. *Acta Neurobiol. Exp.,* 34: 171 – 203.

Haight, J.R. and Neylon, L. (1979) The organization of neocortical projections from the ventrolateral thalamic nucleus in the bush-tailed possum, *Trichosurus vulpecula,* and the problem of motor and sensory convergence within the mammalian brain. *J. Anat.,* 129: 673 – 694.

Hassler, R. (1966) Comparative anatomy of the central visual systems in day- and night-active primates. In R. Hassler and

H. Stephan (Eds.), *Evolution of the Forebrain,* Thieme Verlag, Stuttgart, pp. 419–434.

Huerta, M.F., Krubitzer, L.A. and Kaas, J.H. (1987) The frontal eye field as defined by intracortical microstimulation in squirrel monkeys, owl monkeys, and macaque monkeys. II. Cortical connections. *J. Comp. Neurol.,* 265: 332–361.

Jameson, H.P., Arumugasamy, N. and Hardin, W.B. (1968) The supplementary motor area in the raccoon. *Brain Res.,* 11: 628–637.

Kaas, J.H. (1982) The segregation of function in the nervous system. Why do sensory systems have so many subdivisions? In W.P. Neff (Ed.), *Contributions to Sensory Psysiology, Vol. 7,* Academic Press, New York, pp. 201–240.

Kaas, J.H. (1987a) The organization of neocortex in mammals: Implications for theories of brain function. *Annu. Rev. Psychol.,* 38: 124–151.

Kaas, J.H. (1987b) The organization and evolution of neocortex. In S.P. Wise (Ed.), *Higher Brain Functions,* Wiley, New York, pp. 347–378.

Kaas, J.H. (1988) Development of cortical sensory maps. In. P. Rakic and W. Singer (Eds.), *Neurobiology of Neocortex,* Wiley, New York, pp. 101–113.

Kaas, J.H. (1989a) Changing concepts of visual cortex organization in primates. In J.W. Brown (Ed.), *Neuropsychology of Visual Perception,* Lawrence Erlbaum, Hillsdale, NJ, pp. 3–32.

Kaas, J.H. (1989b) Why does the brain have so many visual areas? *J. Cogn. Neurosci.,* 1: 121–135.

Kaas, J.H. (1989c) The evolution of complex sensory systems in mammals. *J. Exp. Biol.,* 146: 165–176.

Kaas, J.H. and Krubitzer, L.A. (1988) Subdivisions of visuomotor and visual cortex in the frontal lobe of primates: The frontal eye field and the target of the middle temporal area. *Soc. Neurosci. Abstr.,* 14: 820.

Kaas, J.H., Krubitzer, L.A. and Johanson, K.L. (1989) Cortical connections of areas 17 (V-I) and 18 (V-II) of squirrels. *J. Comp. Neurol.,* 281: 426–446.

Kolb, B. (1984) Functions of the frontal cortex of the rat: A comparative review. *Brain Res. Rev.,* 8: 65–98.

Krubitzer, L.A., Sesma, M.A. and Kaas, J.H. (1986) Microelectrode maps, myeloarchitecture, and cortical connections of three somatotopically organized representations of the body surface in parietal cortex of squirrels. *J. Comp. Neurol.,* 250: 403–430.

Kuypers, H.G.J.M., Szwarcbart, M.K., Mishkin, M. and Rosvold, H.E. (1965) Occipito-temporal cortico-cortical connections in the rhesus monkey. *Exp. Neurol.,* 11: 245–262.

Lashley, K.S. and Clark, G. (1946) The cytoarchitecture of the cerebral cortex of Ateles: A critical examination of architectonic studies. *J. Comp. Neurol.,* 85: 223–305.

Lende, R.A. (1963) Cerebral cortex: A sensorimotor amalgam in the marsupialia. *Science,* 141: 730–732.

Li, X.-G, Florence, S.L. and Kaas, J.H. (1990) Areal distributions of cortical neurons projecting to different levels of the caudal brain stem and spinal cord in rats. *Somatosens. Motor Res.,* in press.

Livingstone, M.S. and Hubel, P.H. (1988) Segregation of form, color, movements, and depth: Anatomy, physiology, and perception. *Science,* 240: 740–749.

Luethke, L.E., Krubitzer, L.A. and Kaas, J.H. (1988) Cortical connections of electrophysiologically and architectonically defined subdivisions of auditory cortex in squirrels. *J. Comp. Neurol.,* 268: 181–203.

Merzenich, M.M. and Kaas, J.H. (1980) Principles of organization of sensory-perceptual systems in mammals. *Prog. Psychobiol. Physiol. Psychol.,* 9: 1–42.

Morel, A.E., Krubitzer, K.A. and Kaas, J.H. (1989) Connections of auditory cortex in owl monkeys. *Soc. Neurosci. Abstr.,* 15: 111.

Peden, J.K. and Von Bonin, G. (1947) The neocortex of Hapale. *J. Comp. Neurol.,* 86: 37–64.

Smith, G.E. (1907) A new topographic survey of human cerebral cortex, being an account of the distribution of the anatomically distinct cortical areas and their relationship to the cerebral sulci. *J. Anat.,* 41: 237–254.

Sur, M., Garraghty, P.E. and Roe, A.W. (1988) Experimentally induced visual projections into auditory thalamus and cortex. *Science,* 242: 1437–1441.

Ulinski, P.S. (1984) Thalamic projections to the somatosensory cortex in echidna *(Tachyglossus aculeatus). J. Comp. Neurol.,* 229: 153–170.

Von Bonin, G. and Bailey, P. (1961) Pattern of the cerebral isocortex. In H. Hofer, A.H. Schultz and D. Starch (Eds.), *Primatologia; Handbook of primatology, Vol. 10,* Karger, Basel, pp. 1–42.

Von Economo, C. (1929) *The Cytoarchitectonics of the Human Cortex,* Oxford University Press, Oxford.

Welker, C. (1971) Microelectrode delineation of fine grain somatotopic organization of SmI cerebral neocortex in albino rat. *Brain Res.,* 26: 259–275.

Wise, S.P. (1985) The primate premotor cortex: Past, present, and preparatory. *Annu. Rev. Neurosci.,* 8: 1–19.

Wise, S.P. and Donoghue, J.P. (1986) Motor cortex of rodents. In. E.G. Jones and A. Peters (Eds.), *Cerebral Cortex. Vol. 5. Sensory-Motor Areas and Aspects of Cortical Connectivity,* Plenum Press, New York, pp. 243–270.

Discussion

E.J. Neafsey: A comment on Dr. Fuster's question – Dr. Michael Merzenich has shown that motor cortex stimulation maps in owl monkeys trained to make movements reveal a much larger representation of the trained movement than seen in untrained animals.

J.H. Kaas: All sensory and motor representations appear to be mutable. Changes have been most clearly documented in deactivation or lesion experiments, that reduce or eliminate some neural activity, with the result that normally active parts of the system expand. There is also some evidence that increasing the activity in part of a representation, by use of electrical stimulation, will produce an expansion. All these results are consistent with the theory that topographic features of representations are maintained by relatively stable activity patterns.

E.J. Neafsey: As the number of maps of sensory fields has proliferated, what has happened to "association" cortex?

J.H. Kaas: The term "association cortex" has many implications, but if one means cortex where neurons are activated equally or nearly so by inputs from more than one modality, then there seems to be very little association cortex in most mammals. Instead, most cortex appears to largely relate to one type of sensory input or another. An exception may be parts of prefrontal cortex, but prefrontal cortex is not extensive in most mammals.

CHAPTER 2

Neurotransmitters in the cerebral cortex

John G. Parnavelas

Department of Anatomy and Developmental Biology, University College London, London UK

Introduction

The object of this review is to summarize information about the localization of putative neurotransmitter substances in the cerebral cortex. I shall focus on the more recent neurochemical, histochemical and immunohistochemical studies, because a number of comprehensive reviews published in the past 10 years (e.g. Emson, 1979; Emson and Lindvall, 1979; Parnavelas and McDonald, 1983) have provided extensive accounts of the older literature. Most studies in this field until now have focused on the neocortex of the rat. For this reason, I shall concentrate on the findings in this species and only briefly mention observations on the cat, monkey and human cerebral cortex. Various neurotransmitters and particularly dopamine in the prefrontal cortex will be dealt with, sometimes in considerable detail, in a number of subsequent chapters.

Monoamines

Noradrenaline

It has been known since the early 1960s that the entire cerebral cortex receives a widespread projection from a small number of noradrenaline (NA)-containing neurons located in the locus coeruleus (LC) (Dahlström and Fuxe, 1964). More recent studies, which utilized the retrograde transport of horseradish peroxidase (HRP), have suggested that the projections of the LC to various cortical areas are topographically organized (Waterhouse et al., 1983). The principles of organization of the LC neurons and of their pathways of projection to the neocortex and other areas of the CNS have been reviewed extensively (Lindvall and Björklund, 1974, 1984; Moore and Bloom, 1979; Moore and Card, 1984).

The noradrenergic innervation of the neocortex was first detected by the histofluorescence technique of Falck and Hillarp (Andén et al., 1966; Fuxe et al., 1968) and was confirmed later by the glyoxylic acid fluorescence method (Levitt and Moore, 1978). The latter technique as well as more recent immunohistochemical methods with antibodies to dopamine-β-hydroxylase (Morrison et al., 1978) or NA (Papadopoulos et al., 1989a) have provided a detailed description of the organization of the noradrenergic innervation in a number of cortical areas in the rat and other species. Biochemical (Levitt and Moore, 1978; Palkovits et al., 1979; Reader, 1981) and histochemical (Levitt and Moore, 1978) investigations have shown that the density of innervation varies in different regions of the cortex. Levitt and Moore (1978) have measured the highest NA concentration in the cingulate and frontal areas followed by the visual, auditory and sensorimotor cortices, respectively. The density of innervation of the various layers of the neocortex also varies from one area to another. For example,

---

Correspondence to: John G. Parnavelas, Department of Anatomy and Developmental Biology, University College London, Gower Street, London WC1E 6BT, UK.

the outer half of the rat visual cortex is more densely innervated than the infragranular layers (Morrison et al., 1978; Papadopoulos et al., 1989a) which is the reverse of the pattern observed in the somatosensory and motor cortices (Morrison et al., 1978; Molliver et al., 1982). More detailed measurements of NA in the rat visual cortex, using high pressure liquid chromatography, have shown that layer I contains the highest concentration of the monoamine while layer V the lowest (Parnavelas et al., 1985).

The intracortical path of noradrenergic axons conforms to a pattern which is preserved, for the most part, in different cytoarchitectonic areas and in various species (Lindvall and Björklund, 1984). The general features of this pattern include predominantly long, tangential axons in layers I and VI, radially oriented fibers in layers II and III, and tortuous and oblique axons in layers IV and V. More recent detailed observations of immunohistochemically stained sections of visual cortex permitted tracing of individual fibers over quite long distances and the establishment of continuity between fibers running at various depths of field (Papadopoulos et al., 1989a). These observations have revealed the presence of noradrenergic axons that follow a sinuous course, with a periodicity of 60 – 80 μm, between the upper part of layer II and deep layer IV as they proceed in the mediolateral direction (Fig. 1).

The primate and human cerebral cortices exhibit far greater regional and laminar differences in the density and pattern of innervation than the rat neocortex (Levitt et al., 1984; Morrison et al., 1984; Gaspar et al., 1989; Lewis and Morrison, 1989). Biochemical (Brown et al., 1979) and histochemical studies (Levitt et al., 1984) in monkey and immunohistochemical studies in human cerebral cortex (Gaspar et al., 1989) have shown a characteristic density gradient, with the densest innervation in sensorimotor areas which decreases both caudally and rostrally. There are also considerable differences in the laminar pattern of innervation between cortical areas (Levitt et al., 1984; Morrison et al., 1984; Gaspar et al., 1989).

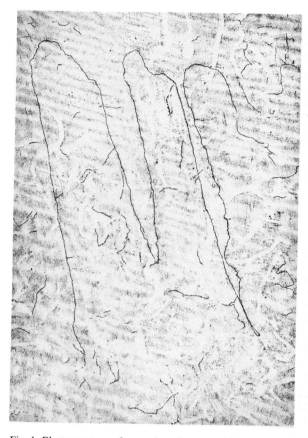

Fig. 1. Photomontage of anoradrenaline-containing axon in the rat visual cortex showing several oscillations between the upper portion of layer II (top) and deep layer IV (bottom) before leaving the plane of section. × 350. (From Papadopoulos et al., 1989a, with permission.)

It should be said that the pattern of innervation in primary visual cortex differs fundamentally from all other cortical areas in terms of laminar specialization and monoamine complementarity. Morrison and colleagues (Morrison et al., 1982; Morrison and Foote, 1986) have shown that within the monkey primary visual cortex, layers III, V and VI receive a moderately dense noradrenergic projection whereas layer IV is largely devoid of noradrenergic fibers. It appears as though the NA projection is directed predominantly at the layers containing the cells that project out of the visual cortex. Precisely at the border of areas 17 and 18 there is an abrupt shift in the innervation pattern. Layer IV in area 18 contains substantially more fibers than the corresponding layer in area 17, and the overall density of NA innervation increases from area 17 to area 18. This abrupt change in density and fiber distribution occurs at all borders of area 17 and surrounding cortical areas.

The way in which noradrenergic axons communicate with neuronal elements in the cerebral cortex and the issue of whether these fibers form conventional synapses have been the source of controversy. Descarries and colleagues, using high resolution autoradiography following topical application of [^{3}H]NA onto the rat frontoparietal cortex, reported that NA-containing axon terminals rarely formed synaptic contacts (see Beaudet and Descarries, 1978, for review). When quantitative analysis was performed in series of a maximum of 5 consecutive sections, the incidence of terminals engaged in synaptic contacts was estimated to be 18%. This observation led these authors to propose the nonsynaptic hypothesis for the action of monoamines in the cortex. However, there is now ample evidence in the literature, based on a variety of methods including that used by Descarries and colleagues (see Parnavelas et at., 1985), which supports the frequent incidence of noradrenergic terminals in the cortex and elsewhere in the brain with specialized junctional appositions.

Recently, investigators have used electron microscopic immunocytochemistry in the rat visual and frontoparietal cortices combined with serial section analysis to examine further the ultrastructural features of monoaminergic terminals (Papadopoulos et al., 1989a). This approach avoids some of the problems inherent in the methods applied previously (i.e. steep diffusion gradients after topical application of labeled NA and lack of extensive serial section analysis; see Molliver et al., 1982; Parnavelas et al., 1985, for discussion). These studies have shown that nearly all labeled vesicle-containing axon terminals or varicosities in these cortical areas form conventional synapses with dendritic spines, dendritic shafts of various diameters or somata of pyramidal and nonpyramidal neurons. Although these results do not exclude the possibility that NA may be released to some degree from nonsynaptic varicosities, they do strongly suggest that release through conventional synapses is likely to represent a major mode of action.

Dopamine

Evidence for the dopaminergic innervation of the cerebral cortex was first presented in the biochemical studies of Thierry et al. (1973, 1974). Subsequent anatomical studies confirmed that dopamine (DA)-containing neurons located in the ventral midbrain tegmentum (VTA) innervate restricted cortical fields (see Lindvall and Björklund, 1984, for review). These fields include the following cortical areas: (1) anteromedial prefrontal; (2) anterior cingulate; (3) suprarhinal and perirhinal; and (4) pyriform and entorhinal. Lesion experiments have shown a clear topographical organization of the VTA projections to different cortical regions (Kalsbeek et al., 1987). For some time, it was generally assumed that other cortical areas lack DA innervation. However, a number of biochemical studies have suggested the presence of lower but significant amounts of endogenous DA in other areas of the rat cerebral cortex (Palkovits et al., 1979; Reader, 1981). Subsequent immunohistochemical studies have provided further evidence for the existence of additional dopaminergic terminal fields in the rat retrosplenial, sensorimotor and occipital cortices (Berger et al., 1985). The latter observations have been confirmed recently by Descarries et al. (1987), who utilized autoradiography to visualize terminals in cortical slices incubated with tritiated DA, and by DA immunohistochemistry (Phillipson et al., 1987; Papadopoulos et al., 1989b). In addition to significant differences in the density of DA innervation between cortical areas, there are pronounced differences in the laminar organization of the dopaminergic fibers in the various cortical regions (Lindvall and Björklund, 1984; Berger et al., 1985; Descarries et al., 1987; Van Eden et al., 1987; Papadopoulos et al., 1989b). For example, in the anterior cingulate cortex the innervation is restricted to the supragranular layers (Lindvall and Björklund, 1984). In the prefrontal cortex, DA fibers are found throughout layers II – VI with the highest density in the infragranular layers (Lindvall and Björklund, 1984; Van Eden et al., 1987), and in the primary visual cortex the few dopaminergic axons are concentrated in layer VI (Phillipson et al., 1987; Papadopoulos et al., 1989b).

The DA innervation of the cerebral cortex shows major differences between rodents and primates, characterized by expanded cortical targets and by a highly differentiated laminar distribution in the latter (see Berger et al., 1988, for discussion). Earlier biochemical studies suggested that the concentration of DA in primate cortex was highest in the prefrontal and temporal areas and decreased along the fronto-occipital axis (Brown et al., 1979). However, more recent immunohistochemical studies with antisera directed against tyrosine hydroxylase (TH) have shown that there is regional heterogeneity in DA innervation density, which does not follow a rostro-caudal gradient (Lewis et al., 1987; Berger et al., 1988). For example, the primary motor cortex is more densely innervated by TH-positive fibers than any frontal region rostral to it. Although the density of innervation decreases immediately caudal to the motor cortex (primary somatosensory cortex) it increases again further caudally, showing a preference for somatosensory association over primary somatosensory cortex (Lewis et al., 1987; Berger et al., 1988) (Fig. 2). Lewis et al. (1987) summarize their findings in primate neocortex by suggesting that dopaminergic fibers preferentially innervate motor over sensory cortical regions, sensory association over primary sensory regions, and auditory association over visual association regions. In addition, the laminar pattern of innervation in a

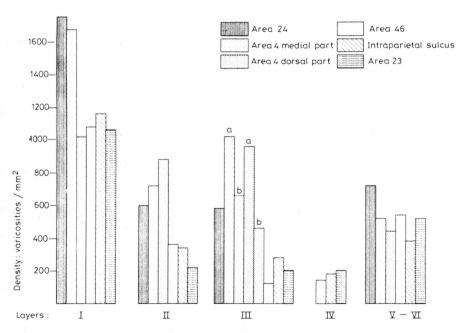

Fig. 2. Laminar density of dopamine innervation for 6 different cortical areas in the macaque brain. Layer I exhibits the highest density in all 6 areas. Layers II-IV of the granular cortices (areas 23, 46, intraparietal sulcus) contain the lowest density (28, 18 and 15% of layer I, respectively). (From Berger et al., 1988, with permission.)

given cortical area is correlated with its fiber density. Thus, sparsely innervated regions show labeled fibers only in layer I and sometimes layer VI. In regions of intermediate density, labeled fibers are located primarily in layers I, III, V and VI, whereas in densely innervated regions dopaminergic axons are present throughout the cortical thickness. Immunohistochemical studies in humans, also with an antibody to TH, have shown very similar innervation patterns (Gaspar et al., 1989).

Dopamine-containing varicosities or axon terminals in the rat prefrontal (Van Eden et al., 1987) or visual cortex (Papadopoulos et al., 1989b) examined in single ultrathin sections were seen at times to form synaptic contacts with dendritic profiles. When varicosities in the visual cortex were examined in uninterrupted series of serial sections, they almost invariably appeared to form synaptic contacts with postsynaptic targets (Papadopoulos et al., 1989b). Synapses were usually of the asymmetrical variety, and the postsynaptic targets were chiefly dendritic shafts of a wide range of diameters, but also dendritic spines. Seguela et al. (1988) also reported a high incidence of synaptic contacts in two different cortical areas. In an analysis of serial sections, these authors noted that 93% of DA varicosities formed synaptic contacts in the anteromedial frontal cortex and 56% in the suprarhinal cortical region. These observations suggest that regional differences may exist in the fine structural relationships of the DA inputs to different regions of the cerebral cortex.

Serotonin

The serotonin (5-HT) innervation of the cerebral cortex arises in the midbrain raphe region, and more specifically within the dorsal (DR) and median raphe (MR) nuclei (Dählstrom and Fuxe, 1964). The existence of a direct projection was first suggested by lesion studies (Ungerstedt, 1971) and confirmed later by anterograde tracing techniques (Conrad et al., 1974; Bobillier et al., 1975; Azmitia and Segal, 1978). There is general agreement that the projections of the DR are predominantly ipsilateral whereas the sparse projections of the MR are bilateral (Jacobs et al., 1978; Köhler and Steinbusch, 1982; Waterhouse et al., 1986). Waterhouse et al. (1986) examined the topographical distribution of DR and MR neurons projecting to different cortical areas of the rat by using HRP histochemistry or double-labeling techniques. The major finding to arise from these studies is that the rostro-caudally aligned motor, somatosensory and visual cortical areas each receive afferents from anatomically discrete portions of the DR while the motor portion of the frontal cortex receives additional input from cells scattered within the MR. The use of computer-assisted image reconstruction in this study facilitated the visualization of spatial relationships between groups of retrogradely labeled neurons within the DR. Thus, it was revealed that cells projecting to the motor and somatosensory areas are positioned successively more dorsal and rostral within the nucleus than those projecting to visual areas 17 and 18. Furthermore, double labeling experiments showed that individual DR neurons give rise to collateral branches to functionally related areas of the cerebral cortex.

A number of studies have shown that not all neurons within the raphe nuclei are serotoninergic. The proportion of cells which contain 5-HT varies within the subdivisions of the raphe nuclei (Köhler and Steinbusch, 1982; O'Hearn and Molliver, 1984). More recent experiments by Kosofsky and Molliver (1987) confirmed that most of the raphe projections to the cortex are indeed serotoninergic and estimated that 5-HT is not contained in 5 – 20% of DR axons and about 33% of MR fibers. Based on these observations, these authors have postulated that adjacent raphe neurons may project in parallel to the same cortical region, where each may release different neurotransmitters.

The serotoninergic innervation of the rat neocortex was initially examined with the formaldehyde-induced histofluorescence method (Fuxe, 1965). Immunohistochemistry with antibodies to 5-HT (Steinbusch et al., 1978) allowed visualization of serotoninergic axons with much increased sensitivity, and revealed a rich network of fibers throughout all cortical layers. Such immunohistochemical studies also showed the existence of some regional differences in the density and laminar distribution of 5-HT axons in the rat cerebral cortex (Lidov et al., 1980; Steinbusch, 1981).

High pressure liquid chromatography measurements in individual layers of the visual cortex have shown the presence of 5-HT in all layers with the concentration decreasing progressively with distance from the cortical surface (Parnavelas et al., 1985). A comparable picture has emerged from observations of immunohistochemically stained sections through the same cortical area (Papadopoulos et al., 1987). This immunohistochemical study has shown a dense plexus of 5-HT fibers in all cortical layers. Most axons were fine and tortuous bearing numerous small varicosities but some thick fibers with large, spherical or elongated varicosities were also present. Thus, morphologically distinct groups of axons project to the visual cortex in agreement with the previous observations of Lidov et al. (1980) and with the recent findings of Kosofsky and Molliver (1987) which suggest that the cerebral cortex is innervated by two 5-HT axon types. This latter study has shown that the axons of the DR neurons are very fine and typically show small varicosities, whereas the cortical axons which arise from the MR have large spherical varicosities and show variations in axonal diameter. Although the DR and MR projections overlap in most cortical areas, there is a topographic distribution of the two fiber types with the MR fibers innervating preferentially the hippocampus and other limbic areas of the cortex. It is interesting that the two axon types exhibit striking differences in vulnerability to the neurotoxic effects of substituted amphetamines (Mamounas and Molliver, 1988). The recent description (Mulligan and Törk, 1988) of two morphologically distinct serotoninergic axonal networks in the cat also supports the notion that there exist two parallel 5-HT projection systems to the cortex.

Work in primates has shown regional variations in terms of the density and laminar distribution of the 5-HT axons (Brown et al., 1979; Takeuchi and Sano, 1983; Morrison et al., 1984; Morrison and Foote, 1986; Berger et al., 1988). These studies suggest a greater degree of organization of the serotoninergic innervation in animals with more highly developed cortices as well as a close relationship between cortical fields and their patterns of innervation. Thus, the 5-HT innervation of the primary visual cortex of primates is the most dense of all neocortical regions. The same area receives a relatively low density of DA and NA innervation (Morrison and Foote, 1986; Berger et al., 1988). The NA and 5-HT systems exhibit a high degree of laminar specialization in this area and appear to be distributed in a complementary fashion: layers V and VI receive a moderately dense NA projection and a sparse 5-HT projection whereas layers IVa and IVc receive a very dense 5-HT projection and are largely devoid of NA axons (Morrison et al., 1982). These patterns of innervation have prompted these authors to speculate that the two transmitter systems may have different cellular targets and may be involved in different aspects of cortical information processing.

Earlier attempts to examine the incidence of specialized 5-HT contacts in the cerebral cortex have resulted in conflicting findings (Descarries et al., 1975; Takeuchi and Sano, 1984; Molliver et al., 1982; Parnavelas et al., 1985). However, recently published studies which employed immunocytochemical methods combined with extensive serial section analysis in rat visual and frontoparietal cortices (Papadopoulos et al., 1987) and monkey primary visual cortex (De Lima et al., 1988), have shown that the great majority of 5-HT-containing axon terminals and varicosities form conventional synapses. These findings provide further support for the notion that 5-HT provides a highly organized innervation of the cerebral cortex.

Acetylcholine

There are two sources of cholinergic innervation of all areas of the rat cerebral cortex: projections from the basal forebrain and intrinsic cortical neurons (Eckenstein et al., 1988; Lysakowski et al., 1989). The projections from the basal forebrain, the major source of cholinergic input to the cortex, have been known since the early histochemical studies of Shute and Lewis (1967).

The basal forebrain comprises a band of large neurons which extends from the medial septal nucleus through the diagonal band of Broca to the nucleus basalis of Meynert (NB); the majority are cholinergic (Mesulam et al., 1983). Mesulam et al. (1983), based on observations in monkey and rat, have proposed a classification scheme that provides a common nomenclature for the forebrain and upper brainstem cholinergic neuronal groups. According to this scheme, cholinergic neurons within the basal forebrain are divided into 4 subgroups (Ch1 – Ch4). Of these, the Ch4 subgroup contains the neurons of the NB, the substantia innominata, laterally situated neurons of the nucleus of the vertical limb of the diagonal band of Broca (NVL), as well as medially situated cells of the nucleus of the horizontal limb of the diagonal band of Broca (NHL). These neuronal populations are grouped into a single sector on the basis that they all project to the cortical mantle and the amygdala. Of the other sectors, Ch1 and Ch2 are contained within the medial septal nucleus and medial portion of the NVL respectively and project to the hippocampus. As for the Ch3 sector, it is contained mostly within the lateral portion of the NHL and provides the major cholinergic innervation to the olfactory bulb. These observations are in good agreement with more recent tracing studies in the rat which analyzed the topographical organization of the basal forebrain projections (Saper, 1984; Eckenstein et al., 1988). By tracing the course of choline acetyltransferase (ChAT)-stained axons from cells in the basal forebrain to the cortex, Eckenstein et al. (1988) described three different subsystems originating from specific nuclei in the basal forebrain and innervating different cortical areas. The three areas of origin of these projection systems are: (1) cholinergic nuclei within the septum, (2) the lateral portion of the NHL, and (3) the NB; these areas innervate respectively (a) the cingulate and retrosplenial cortex, (b) entorhinal and olfactory cortex and (c) all other cortical areas. Saper (1984), using autoradiographic tracing techniques, described discrete pathways between subdivisions of the basal forebrain and specific areas of the cortex. He reported that the basal forebrain projections to the cerebral cortex are organized into medial and lateral pathways. The medial pathway originates primarily in the medial septal nucleus, nucleus of the diagonal band of Broca, and medial substantia innominata. These fibers innervate the cingulate and retrosplenial areas as well as adjacent regions of the neocortex. In contrast, the lateral pathway arises within the caudal components of the basal forebrain including the NB. These fibers sweep laterally through the substantia innominata and ventral part of the putamen to enter the external capsule, from which they distribute to the cortex. The occipital cortex represents the most distal extent of the projections with the two pathways overlapping in the middle of area 17; areas 18 and 18a are the targets of only one of the pathways. This observation has been confirmed more recently by the HRP study of Carey and Rieck (1987).

The density of the cholinergic innervation appears similar throughout most of the cortex, with the exception of olfactory and entorhinal cortices where considerably higher densities were observed (Eckenstein et al., 1988). Cholinergic terminals show a marked laminar distribution in many cortical areas. The laminar organization appears to be due mostly to the projections from the basal forebrain, since lesions of these projections have led to the disappearance of the lamination. Eckenstein et al. (1988) suggested that most cortical regions show only slight variations from a basic laminar pattern, although each area appears to exhibit its own unique characteristics. Details of the innervation pattern of most areas of the rat cerebral cortex may be found in the comprehensive studies of Eckenstein et al. (1988) and Lysakowski et al. (1989). The basic pattern is characterized by a moderately high density of terminals in layers I–III, low density in layer IV, high density in layer V, and varying decrease of density in layer VI. Studies on the density and distribution of ChAT-positive axons in the cerebral cortex of primates are limited. A ChAT immunohistochemical study in the auditory cortex of the monkey (Campbell et al., 1987) has shown a fairly dense innervation of layers I–IV and marked paucity of fibers in layers V and VI.

The intrinsic cholinergic neurons in the rat cerebral cortex, most of which also contain vasoactive intestinal polypeptide (VIP) (Eckenstein and Baughman, 1984; Eckenstein et al., 1988), were first visualized with ChAT immunohistochemistry by Eckenstein and Thoenen (1983) and subsequently examined in considerable detail in a number of cortical areas (Houser et al., 1985; Parnavelas et al., 1986; Eckenstein et al., 1988). These studies have shown that ChAT-labeled cells are scattered in layers II–VI with the majority present in layers II and III. They are exclusively non-pyramidal cells, with bipolar cells predominating in layers II–IV and a smaller number of multipolar cells in layers V and VI. Bipolar cells are oriented vertically, although an occasional cell with horizontally oriented dendrites is observed in layer VI. Observations with the electron microscope (Parnavelas et al., 1986) have shown that both bipolar and multipolar ChAT-labeled neurons show common features. Their nucleus occupies a large part of the cell body and shows an irregular nuclear envelope. They typically possess a small amount of perinuclear cytoplasm, which contains a sparse complement of organelles and receives a small number of asymmetrical and symmetrical axosomatic synaptic contacts. It is interesting to note that the cortices of higher mammals lack intrinsic cholinergic neurons. Such neurons appear to be absent from the cortices of the ferret (Henderson, 1987) and monkey (Mesulam et al., 1984), and reports about their presence in the cat are equivocal. Some authors (Kimura et al., 1981; Vincent and Reiner, 1987) failed to observe cholinergic cells in the cerebral cortex of this animal, while others (Stichel et al., 1987) have reported the presence of a very small number of faintly stained neurons. These observations suggest that intrinsic cholinergic neurons only play a minor, if any, role in the cortex of some mammals.

Electron microscopic investigation of ChAT-labeled axonal segments and varicosities in the rat visual cortex has shown that only a small proportion form synapses with postsynaptic targets in a thin section. However, serial section analysis of a number of labeled axon terminals and vesicle-containing varicosities has shown most of them in synaptic association with neuronal profiles (Parnavelas et al., 1986). The great majority of synapses in the visual (Parnavelas et al., 1986) as well as in the motor and somatosensory areas (Houser et al., 1985) are symmetrical although a small number of asymmetrical synapses are also present. Postsynaptic targets are usually dendrites of a wide range of diameters but sometimes also the perikarya of unlabeled and, on very few occasions, of ChAT-labeled nonpyramidal neurons (Parnavelas et al., 1986).

Amino acids

Excitatory amino acids

It has been known for some time that L-glutamate (Glu) and L-aspartate (Asp) exert a potent excitatory action when applied iontophoretically to single neurons in the CNS including the cerebral cortex. A large body of evidence derived from biochemical, pharmacological and anatomical studies suggests a neurotransmitter role for both amino acids in corticofugal systems (see Fagg and Foster, 1983; Streit, 1984, for reviews). These studies have documented decreases in endogenous transmitter content, high affinity uptake, and release following lesions of the origins of the corticofugal pathways. In addition, selective retrograde labeling of cortical pyramidal cells was shown following injection of D-[^{3}H]Asp into a number of subcortical areas in rats and cats (Baughman and Gilbert, 1981; Matute and Streit, 1985). However, it has been difficult to identify Glu- and Asp-containing neurons histologically since both Glu and Asp are key metabolites in the brain.

A major advance in the attempt to identify excitatory amino acid-containing neurons in the CNS was the first visualization of Glu in individual neurons by immunohistochemistry, using antibodies raised against this amino acid coupled with glutaraldehyde to bovine serum albumin (Storm-Mathisen et al., 1983). Since this breakthrough by Stom-Mathisen and colleagues, efforts in other laboratories have resulted in the production of antibodies, directed against the excitatory amino acids conjugated to protein carriers, which demonstrate Glu or Asp in perfusion fixed tissue (Campistron et al., 1986; Hepler et al., 1988).

Using such antibodies investigators have begun to study the morphology, distribution and connections of transmitter amino acid-containing neurons in the CNS including the neocortex. In a recent light and electron microscopic immunocytochemical study in the rat visual cortex (Dori et al., 1989), we observed that Glu-labeled neurons are pyramidal cells distributed in layers II–VI, with an increased concentration in layers II and III. Aspartate immunoreactivity was localized chiefly to pyramidal neurons in layers II–VI, although approximately 10% of immunolabeled cells were nonpyramidal neurons scattered throughout the cortex. Cell body measurements revealed that, for both groups of neurons, layer V contained the largest labeled neurons, whereas layers IV and VI contained the smallest. Furthermore, in every layer, Asp-stained neurons were larger than Glu-positive cells. As for the axons of Glu- and Asp-labeled neurons, they were seen to form asymmetrical synapses, which are presumably excitatory in nature, primarily with dendritic spines.

Although we have shown some differences in the distribution, size and morphology of Glu- and Asp-labeled neurons, we noted frequently Glu- and Asp- immunoreactive cells showing overlapping morphologies. We did not examine the possible coexistence of the two amino acids in individual neurons in the visual cortex, but Conti et al. (1987a) have shown that Glu and Asp are colocalized in a small percentage of immunoreactive neurons (about 10% of the entire population of cells stained for Glu and Asp). The above description of the distribution and morphology of Glu-positive cells in the rat visual cortex resembles closely the description by Conti et al. (1987b) of stained cells in the somatosensory cortex of the rat and monkey, although these authors also reported, based entirely on light microscopic preparations, the presence of a small number of nonpyramidal neurons. However, our results and those of Conti and co-workers differ significantly from the findings of Donoghue et al. (1985) who examined the distribution of glutaminase immunoreactivity in the somatosensory cortex of rats, guinea pigs and monkeys. They reported that in all 3 species, glutaminase-positive neurons are almost exclusively present in the infragranular layers. A possible explanation for this discrepancy is that Glu is synthesized by more than one enzymatic pathway.

A double-labeling technique, which combines the immunohistochemical localization of Glu or Asp with the presence of retrogradely transported HRP, has been used recently to examine the Glu- and Asp-containing neurons in the cortex of the rat, cat and monkey which give rise to subcortical and corticocortical projections (Conti et al., 1988; Dinopoulos et al., 1989). Such studies in the rat (Dinopoulos et al., 1989, and in preparation) have shown that in the various corticofugal pathways differing proportions of neurons are Glu- or Asp-immunoreactive (Table I). These findings suggest that cortical projection neurons exhibit a previously unsuspected neurochemical heterogeneity and may explain the reported differential involvement of these neurons in neurodegenerative disorders such as Huntington's disease and Alzheimer's dementia (Greenamyre, 1986).

TABLE I

Glu and Asp in corticofugal and corticocortical projections

Cortical area	Injection site	Double labeled/single labeled	
		Glutamate	Aspartate
V	P	42	50
SS	P	56	57
SS	C/P	40	57
M	C/P	53	59
V	cVC	38	49

Note: Percentages of HRP- and glutamate- or aspartate-positive neurons (double labeled) versus HRP-positive cells (single labeled) in various cortical areas (V, visual; SS, somatosensory; M, motor) following HRP injections in subcortical and cortical target sites (P, pons; C/P, caudate/putamen; cVC, contralateral visual cortex).

GABA

Gamma-aminobutyric acid (GABA) is a major inhibitory neurotransmitter in the cerebral cortex (see Curtis and Johnston, 1974; Krnjević, 1981; Sillito, 1984, for reviews). GABA-mediated transmission is known to contribute to very specific response properties of cortical neurons (see Sillito, 1984, for review).

GABA within the cortex is contained in intrinsic neurons (Emson and Lindvall, 1979), although recent evidence in rat and cat suggests that some corticopetal neurons in the basal forebrain contain GABA as well as acetylcholine (Fisher et al., 1988, 1989). Ribak (1978) was the first to localize glutamic acid decarboxylase (GAD) in neurons and axon terminals in the cerebral cortex. In his preparations of rat visual cortex, labeled cells were distributed fairly evenly throughout the cortical thickness and displayed multipolar or bitufted dendritic forms. In light microscopic preparations, numerous dark punctate structures were also seen both around cell bodies and in the intervening neuropil. Electron microscopy revealed that these structures were immunoreactive axon terminals forming symmetrical synaptic contacts with the cell bodies and dendrites of pyramidal and nonpyramidal neurons, as well as with some axon initial segments.

In a more recent immunohistochemical analysis, also in the visual cortex of the rat, Lin et al. (1986) reported that the distribution of GAD-positive cells is not uniform throughout the cortical layers but described a prominent band of stained cells in layer IV. These authors also described increased concentrations of GAD-immunoreactive puncta in layer IV as well as in the superficial portion of layer I and in layer VI. In this study, GAD-labeled cells formed 15% of the neuronal population and, in agreement with Ribak (1978), showed dendritic features typical of multipolar and bitufted nonpyramidal neurons. Another study (Meinecke and Peters, 1987), using an antibody against GABA, has also shown that about 15% of neurons in the rat visual cortex are GABAergic. These authors were able to demonstrate that, with the exception of one-fifth of the bipolar cells, all nonpyramidal neurons in this area are GABAergic.

The density, distribution and morphology of GABAergic neurons have been examined in a variety of cortical areas in a number of species. Hendry et al. (1987) have reported that the density and proportion of GABA-containing neurons relative to the total neuronal population is similar in most of 10 structurally and functionally distinct areas examined in the monkey cerebral cortex. Exception is provided by area 17 which contains 50% more GABA-immunoreactive neurons than other areas but also contains more than twice the total number of neurons. Thus, in most cortical areas GABA-containing cells form approximately 25% of the neuronal population but in area 17 they make up less than 20% of the total. This estimate for area 17 correlates closely with estimates of GABAergic neurons in the cat visual cortex (Gabbott and Somogyi, 1986) but is higher than other reports in the monkey visual (Fitzpatrick et al., 1987) and sensorimotor cortices (Houser et al., 1983).

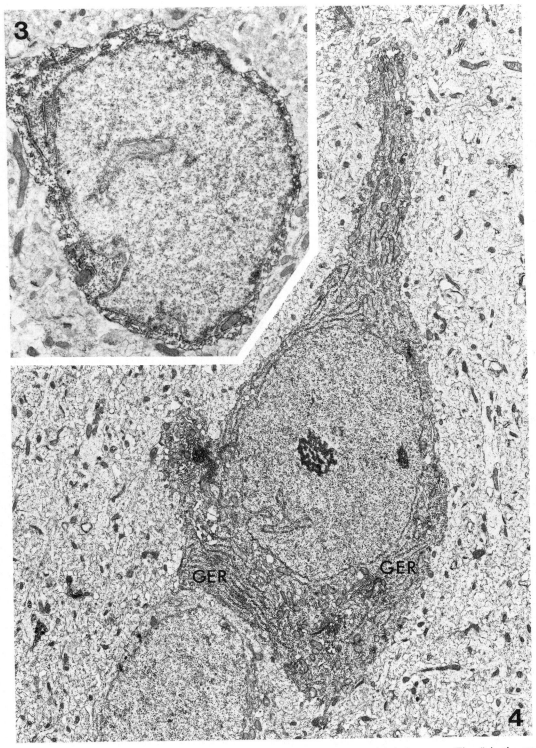

Figs. 3 and 4. Vasoactive intestinal polypeptide (Fig. 3) and neuropeptide Y-labeled neurons (Fig. 4) in the rat visual cortex. The neuropeptide Y-containing neuron shows abundant cytoplasm packed with cisternae of granular endoplasmic veticulum (GER). Fig. 3, × 9300; Fig. 4, × 7800.

Neuropeptides

A number of biologically active peptides have been identified in the cerebral cortex since the mid 1970s (Emson, 1979; Emson and Lindvall, 1979; Krieger, 1983). Although their functional role remains unclear (see Kelly, 1982), the development of radioimmunoassay and immunohistochemical techniques have enabled investigators to study their precise anatomical localization. Using such methods, some peptides have been localized in cortical neurons in every mammalian species examined. They include somatostatin (somatotropin release inhibiting factor (SRIF)), vasoactive intestinal polypeptide (VIP), cholecystokinin (CCK) and neuropeptide Y (NPY). The distribution, morphology, synaptic organization and development of these peptide-containing neurons in the rat and other species have been reviewed extensively in recent years (Parnavelas and McDonald, 1983; Emson and Hunt, 1984; Jones and Hendry, 1986; Parnavelas, 1986; Parnavelas et al., 1988) and, therefore, I shall only outline some salient features.

As I mentioned earlier with respect to the visual cortex, evidence suggests that virtually all nonpyramidal neurons contain the neurotransmitter GABA. These cells may be divided into subpopulations on the basis of a coexisting peptide. These subpopulations may be distinguished not only on the basis of their peptide content but also by their cytological and synaptic features. For example, NPY-containing neurons consistently show a prominent nucleolus, abundant cytoplasm which contains numerous strands of granular endoplasmic reticulum organized in parallel arrays, and axons which tend to contact medium sized dendrites (Fig. 4). It should be said that these features are common to all SRIF neurons irrespective of laminar position, soma size or dendritic form. VIP neurons, on the other hand, typically show a thin rim of cytoplasm containing a sparse complement of organelles surrounding a relatively large nucleus (Fig. 3). Their axons contact predominantly the dendrites and cell bodies of other nonpyramidal neurons including some which are VIP positive. In addition to the distinguishing ultrastructural features found in adult neurons, peptidergic cells in the cortex show distinct developmental profiles. For example, the peak of neurogenesis for SRIF neurons is at embryonic days 16 – 17 (E16 – E17), for NPY neurons at E17 and for VIP cells at E19 (Cavanagh and Parnavelas, 1988, 1989, 1990). These birthdates are reflected in the first detectable appearances of these neurons: SRIF cells appear in the cortex before birth, NPY neurons at about the time of birth and VIP cells towards the end of the first postnatal week. Furthermore, the development of peptide-containing subpopulations of cortical neurons is altered differentially following sensory manipulation early in life (Jeffery and Parnavelas, 1987; Parnavelas et al., 1990).

There is now considerable evidence which shows the presence of corticotropin-releasing factor (CRF) immunoreactivity throughout the rat neocortex (Merchenthaler et al., 1982; Swanson et al., 1983; Hökfelt et al., 1987). It appears from the brief illustrations in these studies that some of the cells display features typical of bipolar neurons and resemble VIP- and/or ChAT-positive neurons. However the evidence for the presence of opioid peptide-immunoreactive neurons in the neocortex of the rat is less compelling (Khachaturian et al., 1982; Cuello, 1983; McGinty et al., 1984; Petrusz et al., 1985). There are also isolated reports illustrating the presence of substance P-labeled cell bodies in the rat neocortex (Ljungdall et al., 1978; Penny et al., 1986), a finding which has not been substantiated by other authors, although such neurons are widely present in the cortices of "higher" mammals (Jones et al., 1988). A number of other peptides have also been measured by radioimmunoassay and immunohistochemistry in the rat neocortex (see Parnavelas and McDonald, 1983; Palkovits and Brownstein, 1985). However, these studies either have not been widely confirmed or they have not provided a description of the neurons involved. Undoubtedly, more peptides will eventually be found in the cortex. However, this field is in desperate need for information about the role of these substances in cortical function.

Acknowledgements

My thanks to Anna Alongi for typing this manuscript and to Stephen Davies for his always helpful comments. I am grateful to Drs. B. Berger and P. Gaspar for providing Fig. 2.

References

Andén, N.-E., Dahlström, A., Fuxe, K., Larsson, K., Olson, L. and Ungerstedt, U. (1966). Ascending monoamine neurons to the telencephalon and diencephalon. *Acta Physiol. Scand.*, 67: 313–326.

Azmitia, E.C. and Segal, M. (1978) An autoradiographic analysis of the differential ascending projections of the dorsal and median raphe nuclei in the rat. *J. Comp. Neurol.*, 179: 641–668.

Baughman, R.W. and Gilbert, C.D. (1981) Aspartate and glutamate as possible neurotransmitters in the visual cortex. *J. Neurosci.*, 1: 427–439.

Beaudet, A. and Descarries, L. (1978) The monoamine innervation of rat cerebral cortex: synaptic and nonsynaptic axon terminals. *Neuroscience*, 3: 851–860.

Berger, B., Verney, C., Alvarez, C., Vigny, A. and Helle, K.B. (1985) New dopaminergic terminal fields in the motor, visual (area 18b) and retrosplenial cortex in the young and adult rat. Immunocytochemical and catecholamine histochemical analyses. *Neuroscience*, 15: 983–998.

Berger, B., Trottier, S., Verney, C., Gaspar, P. and Alvarez, C (1988) Regional and laminar distribution of the dopamine and serotonin innervation in the macaque cerebral cortex: a radioautographic study. *J. Comp. Neurol.*, 273: 99–119.

Bobillier, P., Petitjean, F., Salvert, D., Ligier, M. and Seguin, S. (1975) Differential projections of the nucleus raphe dorsalis and nucleus raphe centralis as revealed by autoradiography. *Brain Res.*, 85: 205–210.

Brown, R.M., Crane, A.M. and Goldman, P.S. (1979) Regional distribution of monoamines in the cerebral cortex and subcortical structures of the rhesus monkey: concentrations and in vivo synthesis rates. *Brain Res.*, 168: 133–150.

Campbell, M.J., Lewis, D.A., Foote, S.L. and Morrison, J.J. (1987) Distribution of choline acetyltransferase-, serotonin-, dopamine-β-hydroxylase-, tyrosine hydroxylase-immunoreactive fibers in monkey primary auditory cortex. *J. Comp. Neurol.*, 261: 209–220.

Campistron, G., Buijs, R.M. and Geffard, M. (1986) Specific antibodies against aspartate and their immunocytochemical application in the rat brain. *Brain Res.*, 365: 179–184.

Carey, R.G. and Rieck, R.W. (1987) Topographic projections to the visual cortex from the basal forebrain in the rat. *Brain Res.*, 424: 205–215.

Cavanagh, M.E. and Parnavelas, J.G. (1988) Development of somatostatin immunoreactive neurons in the rat occipital cortex: a combined immunocytochemical-autoradiographic study. *J. Comp. Neurol.*, 268: 1–12.

Cavanagh, M.E. and Parnavelas, J.G. (1989) Development of vasoactive-intestinal-polypeptide-immunoreactive neurons in the rat occipital cortex: a combined immunohistochemical-autoradiographic study. *J. Comp. Neurol.*, 284: 637–645.

Cavanagh, M.E. and Parnavelas, J.G. (1990) Development of neuropeptide Y (NPY) immunoreactive neurons in the rat occipital cortex: a combined immunohistochemical-autoradiographic study. *J. Comp. Neurol.*, 297: 553–563.

Conrad, L.C.A., Leonard, C.M. and Pfaff, D.W. (1974) Connections of the median and dorsal raphe nuclei in the rat: An autoradiographic and degeneration study. *J. Comp. Neurol.*, 156: 179–206.

Conti, F., Rustioni, A. and Petrusz, P. (1987a) Co-localization of glutamate and aspartate immunoreactivity in neurons of the rat somatic sensory cortex. In T.P. Hicks, D. Lodge and H. McLennan (Eds.), *Excitatory Amino Acid Transmission*, Alan R. Liss, New York, pp. 169–172.

Conti, F., Rustioni, A., Petrusz, P. and Towle, A.C. (1987b) Glutamate-positive neurons in the somatic sensory cortex of rats and monkeys. *J. Neurosci.*, 7: 1887–1901.

Conti, F., Fabri, M. and Manzoni, T. (1988) Glutamate-positive cortico-cortical neurons in the somatic sensory areas I and II of cats. *J. Neurosci.*, 8: 2948–2960.

Cuello, A.C. (1983) Central distribution of opiod peptides. *Br. Med. Bull.*, 39: 11–16.

Curtis, D.R. and Johnston, G.A.R. (1974) Amino acid transmitters in the mammalian central nervous system. *Ergebn. Physiol.*, 69: 97–188.

Dahlström, A. and Fuxe, K. (1964) Evidence for the existence of monoamine-containing neurons in the central nervous system. I. Demonstration of monoamines in the cell bodies of brain stem neurons. *Acta Physiol. Scand.*, 62: Suppl. 232: 1–55.

De Lima, A.D., Bloom, F.E. and Morrison, J.H. (1988) Synaptic organization of serotonin-immunoreactive fibers in primary visual cortex of the macaque monkey. *J. Comp. Neurol.*, 274: 280–294.

Descarries, L., Beaudet, A. and Watkins, K.C. (1975) Serotonin nerve terminals in adult rat neocortex. *Brain Res.*, 100: 563–588.

Descarries, L., Lemay, B., Doucet, G. and Berger, B. (1987) Regional and laminar density of the dopamine innervation in adult rat cerebral cortex. *Neuroscience*, 21: 807–824.

Dinopoulos, A., Dori, I., Davies, S.W. and Parnavelas, J.G. (1989) Neurochemical heterogeneity among corticofugal and callosal projections. *Exp. Neurol.*, 105: 36–44.

Donoghue, J.P., Wenthold, R.J. and Altschuler, R.A. (1985)

Localization of glutaminase-like and aspartate aminotransferase-like immunoreactivity in neurons of cerebral neocortex. *J. Neurosci.*, 5: 2597–2608.

Dori, I., Petrou, M. and Parnavelas, J.G. (1989) Excitatory amino acid-containing neurons in the rat visual cortex: a light and electron microscopic immunocytochemical study. *J. Comp. Neurol.*, 290: 169–184.

Eckenstein, F. and Baughman, R.W. (1984) Two types of cholinergic innervation in cortex, one co-localized with vasoactive intestinal polypeptide. *Nature*, 309: 153–155.

Eckenstein, F. and Thoenen, H. (1983) Cholinergic neurons in the rat cerebral cortex demonstrated by immunohistochemical localization of choline acetyltransferase. *Neurosci. Lett.*, 36: 211–215.

Eckenstein, F.P., Baughman, R.W. and Quinn, J. (1988) An anatomical study of cholinergic innervation in rat cerebral cortex. *Neuroscience*, 25: 457–474.

Emson, P.C. (1979) Peptides as neurotransmitter candidates in the mammalian CNS. *Progr. Neurobiol.*, 13: 61–116.

Emson, P.C. and Hunt, S.P. (1984) Peptide-containing neurons in the visual cortex. In E.G. Jones and A. Peters (Eds.) *Cerebral Cortex. Vol. 2. Functional Properties of Cortical Cells*, Plenum Press, New York, pp. 145–169.

Emson, P.C. and Lindvall, O. (1979) Distribution of putative neurotransmitters in the neocortex. *Neuroscience*, 4: 1–30.

Fagg, G.E. and Foster, A.C. (1983) Amino acid neurotransmitters and their pathways in the mammalian central nervous system, *Neuroscience*, 9: 701–719.

Fisher, R.S. and Levine, M.S. (1989) Transmitter cosynthesis by corticopetal basal forebrain neurons. *Brain Res.*, 491: 163–168.

Fisher, R.S., Buchwald, N.A., Hull, C.D. and Levine, M.S. (1988) GABAergic basal forebrain neurons project to the neocortex: the localization of glutamic acid decarboxylase and choline acetyltransferase in feline corticopetal neurons. *J. Comp. Neurol.*, 272: 489–502.

Fitzpatrick, D., Lund, J.S., Schmechel, D.E. and Towles, A.C. (1987) Distribution of GABAergic neurons and axon terminals in the macaque striate cortex. *J. Comp. Neurol.*, 264: 73–91.

Fuxe, K. (1965) Evidence for the existence of monoamine neurons in the central nervous system. IV. The distribution of monoamine terminals in the central nervous system. *Acta Physiol. Scand.*, 64, Suppl. 247: 39–85.

Fuxe, K., Hamberger, B. and Hökfelt, T. (1968) Distribution of noradrenaline nerve terminals in cortical areas of the rat. *Brain Res.*, 8: 125–131.

Gabbott, P.L.A. and Somogyi, P. (1986) Quantitative distribution of GABA-immunoreactive neurons on the visual cortex (area 17) of the cat. *Exp. Brain Res.*, 61: 323–331.

Gaspar, P., Berger, B., Febvret, A., Vigny, A. and Henry, J.P. (1989) Catacholamine innervation of the human cerebral cortex as revealed by comparative immunohistochemistry of tyrosine hydroxylase and dopamine-beta-hydroxylase. *J. Comp. Neurol.*, 279: 249–271.

Greenamyre, J.T. (1986) The role of glutamate in neurotransmission and in neurological disease. *Arch. Neurol.*, 43: 1058–1063.

Henderson, Z. (1987) Cholinergic innervation of ferret visual system. *Neuroscience*, 20: 503–518.

Hendry, S.H.C., Schwark, H.D., Jones, E.G. and Yan, J. (1987) Numbers and proportions of GABA-immunoreactive neurons in different areas of monkey cerebral cortex. *J. Neurosci.*, 7: 1503–1519.

Hepler, J.R., Toomim, C.S., McCarthy, K.D., Conti, F., Battaglia, G., Rustioni, A. and Petrusz, P. (1988) Characterization of antisera to glutamate and aspartate. *J. Histochem. Cytochem.*, 36: 13–22.

Hökfelt, T., Fahrenkrug, J., Ju, G., Ceccatelli, S., Tsuruo, Y., Meister, B., Mutt, V., Rundgren, M., Brodin, E., Terenius, L., Hulting, A.-L., Werner, S., Björklund, H. and Vale, W. (1987) Analysis of peptide histidine-isoleucine/vasoactive intestinal polypeptide-immunoreactive neurons in the central nervous system with special reference to their relation to corticotropin releasing factor- and enkephalin-like immunoreactivities in the paraventricular hypothalamic nucleus. *Neuroscience*, 23: 827–857.

Houser, C.R., Hendry, S.H.C., Jones, E.G. and Vaughn, J.E. (1983) Morphological diversity of immunocytochemically identified GABA neurons in the monkey sensory-motor cortex. *J. Neurocytol.*, 12: 617–638.

Houser, C.R., Crawford, G.D., Slavaterra, P.M. and Vaughn, J.E. (1985) Immunocytochemical localization of choline acetyltransferase in rat cerebral cortex: A study of cholinergic neurons and synapses. *J. Comp. Neurol.*, 234: 17–34.

Jacobs, B.L., Foote, S.L. and Bloom, F.E. (1978) Differential projections of neurons within the dorsal raphe nucleus of the rat: A horseradish peroxidase (HRP) study. *Brain Res.*, 147: 149–153.

Jeffery, G. and Parnavelas, J.G. (1987) Early visual deafferentation of the cortex results in an asymmetry of somatostatin labelled cells. *Exp. Brain. Res.*, 67: 651–655.

Jones, E.G. and Hendry, S.H.C. (1986) Peptide-containing neurons of the primate cerebral cortex. In J.B. Martin and J.D. Barchas (Eds.), *Neuropeptides in Neurologic and Psychiatric Disease*, Raven Press, New York, pp. 163–178.

Jones, E.G., DeFelipe, J. and Hendry, S.H.C. (1988) A study of tachykinin-immunoreactive neurons in the monkey cerebral cortex. *J. Neurosci.*, 8: 1208–1224.

Kalsbeek, A., Buijs, R.M. Hofman, M.A., Matthijsen, M.A.H., Pool, C.W. and Uylings, H.B.M. (1987) Effects of neonatal thermal lesioning of the mesocortical dopaminergic projection on the development of the rat prefrontal cortex. *Dev. Brain Res.*, 32: 123–132.

Kelly, J.S. (1982) Electrophysiology of peptides in the central nervous system. *Br. Med. Bull.*, 38: 283–290.

Khachaturian, H., Watson, S.J., Lewis, M.E., Coy, D., Goldstein, A. and Akil, H. (1982) Dynorphin immunocytochemistry in the rat central nervous system. *Peptides,* 3: 941–954.

Kimura, H., McGeer, P.L., Peng, J.H. and McGeer, E.G. (1981) The central cholinergic system studied by choline acetyltransferase immunohistochemistry in the cat. *J. Comp. Neurol.,* 200: 151–201.

Köhler, C. and Steinbusch, H. (1982) Identification of serotonin and non-serotonin-containing neurons of the midbrain raphe projecting to the entorhinal area and the hippocampal formation. A combined immunohistochemical and fluorescent retrograde tracing study in the rat brain. *Neuroscience,* 7: 951–975.

Kosofsky, B.E. and Molliver, M.E. (1987) The serotonergic innervation of cerebral cortex: Different classes of axon terminals arise from dorsal and median raphe nuclei. *Synapse,* 1: 153–168.

Krieger, D.T. (1983) Brain peptides: What, where and why? *Science,* 222: 975–985.

Krnjević, K. (1981) Transmitters in motor systems. In V.B. Brooks (Ed.), *Handbook of Physiology, Section 1; The Nervous System, Vol. II, Motor Control, Part 1,* American Physiological Society, Bethesda, MD, pp. 107–154.

Levitt, P. and Moore, R.Y. (1978) Noradrenaline neuron innervation of the neocortex in the rat. *Brain Res.,* 139: 219–231.

Levitt, P., Rakic, P. and Goldman-Rakic, P. (1984) Region-specific distribution of catecholamine afferents in primate cerebral cortex: a fluorescence histochemical analysis. *J. Comp. Neurol.,* 227: 23–26.

Lewis, D. and Morrison, J.H. (1989) Noradrenergic innervation of monkey prefrontal cortex: a dopamine-β-hydroxylase immunohistochemical study. *J. Comp. Neurol.,* 282: 317–330.

Lewis, D.A., Campbell, M.J., Foote, S.L., Goldstein, M. and Morrison, J.H. (1987) The distribution of tyrosine hydroxylase-immunoreactive fibers in primate neocortex is widespread but regionally specific. *J. Neurosci.,* 7: 279–290.

Lidov, H.G.W., Grzanna, R. and Molliver, M.E. (1980) The serotonin innervation of the cerebral cortex in the rat – an immunohistochemical analysis. *Neuroscience,* 5: 207–227.

Lin, C.-S., Lu, S.M. and Schmechel, D.E. (1986) Glutamic acid decarboxylase and somatostatin immunoreactivities in rat visual cortex. *J. Comp. Neurol.,* 244: 369–383.

Lindvall, O. and Björklund, A. (1974) The organization of the ascending catecholamine neuron system in the rat brain as revealed by the glyoxylic acid fluorescence method. *Acta Physiol. Scand.,* Suppl., 412: 1–48.

Lindvall, O. and Björklund, A. (1984) General organization of cortical monoamine systems. In L. Descarries, T.R. Reader and H.H. Jasper (Eds.), *Monoamine Innervation of Cerebral Cortex,* Alan R. Liss, New York, pp, 9–40.

Ljungdahl, A., Hökfelt, T. and Nilsson, G. (1978) Distribution of substance P-like immunoreactivity in the central nervous system of the rat. I. Cell bodies and nerve terminals. *Neuroscience,* 3: 861–943.

Lysakowski, A., Wainer, B.H., Bruce, G. and Hersh, L.B. (1989) An atlas of the regional and laminar distribution of choline acetyltransferase immunoreactivity in rat cerebral cortex. *Neuroscience,* 28: 291–336.

Mamounas, L.A. and Molliver, M.E. (1988) Evidence for dual serotonergic projections to neocortex: axons from the dorsal and median raphe nuclei are differentially vulnerable to the neurotoxin *p*-chloroamphetamine (PCA). *Exp. Neurol.,* 102: 23–36.

Matute, C. and Streit, P. (1985) Selective retrograde labeling with D-[^{3}H]-aspartate in afferents to the mammalian superior colliculus. *J. Comp. Neurol.,* 241: 34–49.

McGinty, J.F., Van der Kooy, D. and Bloom, F.E. (1984) The distribution and morphology of opioid peptide immunoreactive neurons in the cerebral cortex of rats. *J. Neurosci.,* 4: 1104–1117.

Meinecke, D.L. and Peters, A. (1987) GABA immunoreactive neurons in rat visual cortex. *J. Comp. Neurol.,* 261: 388–404.

Merchenthaler, I., Vigh, S., Petrusz, P. and Schally, A.V. (1982) Immunocytochemical localization of corticotropin-releasing factor (CRF) in the rat brain. *Am. J. Anat.,* 165: 385–396.

Mesulam, M.-M., Mufson, E.J., Wainer, B.H. and Levey, A.I. (1983) Central cholinergic pathways in the rat: an overview based on an alternative nomenclature (Ch1–Ch6). *Neuroscience,* 10: 1185–1201.

Mesulam, M.-M., Mufson, E.J., Levey, A.I. and Wainer, B.H. (1984) Atlas of cholinergic neurons in the forebrain and upper brainstem of the macaque based on monoclonal choline acetyltransferase immunohistochemistry and acetycholinesterase histochemistry. *Neuroscience,* 12: 669–686.

Molliver, M.E., Grzanna, R., Lidov, H.G.W., Morrison, J.H. and Olschowka, J.A. (1982) Monoamine systems in the cerebral cortex. In V. Chan-Palay and S.L. Palay (Eds.), *Cytochemical Methods in Neuroanatomy,* Alan R. Liss, New York, pp. 255–277.

Moore, R.Y. and Bloom, F.E. (1979) Central catecholamine neuron systems: anatomy and physiology of the norepinephrine and epinephrine systems. *Annu. Rev. Neurosci.,* 2: 113–168.

Moore, R.Y. and Card, J.P. (1984) Noradrenaline-containing neuron systems. In A. Björklund and T. Hökfelt (Eds.), *Handbook of Chemical Neuroanatomy. Vol. 2. Classical Transmitters in the CNS, Part I.* Elsevier, Amsterdam, pp. 123–156.

Morrison, J.H. and Foote, S.L. (1986) Noradrenergic and serotonergic innervation of cortical, thalamic and tectal visual structures in old and new world monkeys. *J. Comp. Neurol.,* 243: 117–138.

Morrison, J.H., Grzanna, R., Molliver, M.E. and Coyle, J.T. (1978) The distribution and orientation of noradrenergic fibers in neocortex of the rat: an immunofluorescence study.

J. Comp. Neurol., 181: 17–40.

Morrison, J.H., Foote, S.L., Molliver, M.E., Bloom, F.E. and Lidov, H.G.W. (1982) Noradrenergic and serotonergic fibers innervate complementary layers in monkey primary visual cortex: An immunohistochemical study. Proc. Natl. Acad. Sci. (U.S.A.), 79: 2401–2405.

Morrison, J.H., Foote, S.L. and Bloom F.E. (1984) Regional, laminar, developmental and functional characteristics of noradrenaline and serotonin innervation patterns in monkey cortex. In L. Descarries, T.A. Reader and H.H. Jasper (Eds.), Monoamine Innervation of Cerebral Cortex, Alan R. Liss, New York, pp. 61–75.

Mulligan, K.A. and Törk, I. (1988) Serotonergic innervation of the cat cerebral cortex. J. Comp. Neurol., 270: 86–110.

O'Hearn, E. and Molliver, M.E. (1984) Organization of raphe-cortical projections in rat: A quantitative retrograde study. Brain Res. Bull., 13: 709–726.

Palkovits, M. and Brownstein, M.J. (1985) Distribution of neuropeptides in the central nervous system using biochemical micromethods. In A. Björklund and T. Hökfelt (Eds.), Handbook of Chemical Neuroanatomy. Vol. 4. GABA and Neuropeptides in the CNS, Part I, Elsevier, Amsterdam, pp. 1–71.

Palkovits, M., Záborsky, L., Brownstein, M.J., Fekete, M.I.K., Herman, J.P. and Kanyicska, B. (1979) Distribution of norepinephrine and dopamine in cerebral cortical areas of the rat. Brain Res. Bull., 4: 593–601.

Papadopoulos, G.C., Parnavelas, J.G. and Buijs, R.M. (1987) Light and electron microscopic immunocytochemical analysis of the serotonin innervation of the rat visual cortex. J. Neurocytol., 16: 883–892.

Papadopoulos, G.C., Parnavelas, J.G. and Buijs, R.M. (1989a) Light and electron microscopic immunocytochemical analysis of the noradrenaline innervation of the rat visual cortex. J. Neurocytol., 18: 1–10.

Papadopoulos, G.C., Parnavelas, J.G. and Buijs, R.M. (1989b) Light and electron microscopic immunocytochemical analysis of the dopamine innervation of the rat visual cortex. J. Neurocytol., 18: 303–310.

Parnavelas, J.G. (1986) Morphology and distribution of peptide-containing neurones in the cerebral cortex. In P.C. Emson, M.N. Rossor and M. Tohyama (Eds.), Peptides in Neurological Disease. Progress in Brain Research, Vol. 66, Elsevier, Amsterdam, pp. 119–134.

Parnavelas, J.G. and McDonald, J.K. (1983) The cerebral cortex. In P.C. Emson (Ed.), Chemical Neuroanatomy, Raven Press, New York, pp. 505–549.

Parnavelas, J.G., Moises, H.C. and Speciale, S.G. (1985) The monoaminergic innervation of the rat visual cortex. Proc. R. Soc. Lond. B, 223: 319–329.

Parnavelas, J.G., Kelly, W., Franke, E. and Eckenstein, F. (1986) Cholinergic neurons and fibres in the rat visual cortex. J. Neurocytol., 15: 329–336.

Parnavelas, J.G., Papadopoulos, G.C. and Cavanagh, M.E. (1988) Changes in neurotransmitters during development. In A. Peters and E.G. Jones (Eds.), Cerebral Cortex. Vol. 7. Development and Maturation of Cerebral Cortex, Plenum Press, New York, pp. 177–209.

Parnavelas, J.G., Jeffery, G., Cope, J. and Davies, S.W. (1990) Early removal of mystacial vibrissae in rats results in an increase of somatostatin labelled cells in the somatosensory cortex. Exp. Brain Res., in press.

Penny, G.R., Afsharpour, S. and Kitai, S.T. (1986) Substance P-immunoreactive neurons in the neocortex of the rat: a subset of the glutamic acid decarboxylase immunoreactive neurons. Neurosci. Lett., 65: 53–59.

Petrusz, P., Merchenthaler, I. and Maderdrut, J.L. (1985) Distribution of enkephalin-containing neurons in the central nervous system. In A. Björklund and T. Hökfelt (Eds.), Handbook of Chemical Neuroanatomy. Vol. 4. GABA and Neuropeptides in the CNS, Part I, Elsevier, Amsterdam, pp. 273–334.

Phillipson, O.T., Kilpatrick, I.C. and Jones, M.W. (1987) Dopaminergic innervation of the primary visual cortex in the rat, and some correlations with human cortex. Brain Res. Bull., 18: 621–633.

Reader, T.A. (1981) Distribution of catecholamines and serotonin in the rat cerebral cortex: Absolute levels and relative proportions. J. Neur. Transm., 50: 13–27.

Ribak, C.E. (1978) Aspinous and sparsely-spinous stellate neurons in the visual cortex of rats contain glutamic acid decarboxylase. J. Neurocytol., 7: 461–478.

Saper, C.B. (1984) Organization of cerebral cortical afferent systems in the rat. II. Magnocellular basal nucleus. J. Comp. Neurol., 222: 313–342.

Seguela, P., Watkins, K.C. and Descarries, L. (1988) Ultrastructural features of dopamine axon terminals in the anteromedial and the suprarhinal cortex of adult rat. Brain Res., 442: 1–12.

Shute, C.C.D. and Lewis, P.R. (1967) The ascending cholinergic reticular system: Neocortical, olfactory and subcortical projections. Brain, 90: 497–520.

Sillito, A.M. (1984) Functional considerations of the operation of GABAergic inhibitory processes in the visual cortex. In E.G. Jones and A. Peters (Eds.), Cerebral Cortex. Vol. 2. Functional Properties of Cortical Cells, Plenum Press, New York, pp. 91–117.

Steinbusch, H.W.M. (1981) Distribution of serotonin-immunoreactivity in the central nervous system of the rat — cell bodies and terminals. Neuroscience, 6: 557–618.

Steinbusch, H.W.M., Verhofstad, A.A.J. and Joosten, H.W.J. (1978) Localization of serotonin in the central nervous system by immunohistochemistry: Description of a specific and sensitive technique and some applications. Neuroscience, 3: 811–819.

Stichel, C.C., De Lima, A.D. and Singer, W. (1987) A search for choline-acetyltransferase-like immunoreactivity in neurons of cat striate cortex. Brain Res., 405: 395–399.

Storm-Mathisen, J., Leknes, A.K., Bore, A.T., Vaaland, J.L., Edminson, P., Haug, F.-M.S. and Ottersen, O.P. (1983) First visualization of glutamate and GABA in neurones by immunocytochemistry. *Nature*, 301: 517–520.

Streit, P. (1984) Glutamate and aspartate as transmitter candidates for systems of the cerebral cortex. In E.G. Jones and A. Peters (Eds.), *Cerebral Cortex. Vol. 2. Functional Properties of Cortical Cells*, Plenum Press, New York, pp. 119–143.

Swanson, L.W., Sawchenko, P.E., Rivier, J. and Vale, W.W. (1983) Organization of ovine corticotropin-releasing factor immunoreactive cells and fibers in the rat brain: an immunohistochemical study. *Neuroendocrinology*, 36: 165–186.

Takeuchi, Y. and Sano, Y. (1983) Immunohistochemical demonstration of serotonin nerve fibers in the neocortex of the monkey *(Macaca fuscata)*. *Anat. Embryol.*, 166: 155–168.

Takeuchi, Y. and Sano, Y. (1984) Serotonin nerve fibers in the primary visual cortex of the monkey. Quantitative and immunoelectronmicroscopical analysis. *Anat. Embryol.*, 169: 1–8.

Thierry, A.M., Blanc, G., Sobel, A., Stinus, L. and Glowinski, J. (1973) Dopaminergic terminals in the rat cortex. *Science*, 182: 499–501.

Thierry, A.M., Hirsch, J.C., Tassin, J.P., Blanc, G. and Glowinski, J. (1974) Presence of dopaminergic terminals and absence of dopaminergic cell bodies in the cerebral cortex of the cat. *Brain Res.*, 79: 77–88.

Ungerstedt, U. (1971) Stereotaxic mapping of the monoamine pathways in the rat brain. *Acta Physiol. Scand.*, Suppl., 367: 1–48.

Van Eden, C.G. Hoorneman, E.M.D., Buijs, R.M., Matthijssen, M.A.H., Geffard, M. and Uylings, H.B.M. (1987) Immunocytochemical localization of dopamine in the prefrontal cortex of the rat at the light and electron microscopical level. *Neuroscience*, 22: 849–862.

Vincent, S.R. and Reiner, P.B. (1987) The immunohistochemical localization of choline acetyltransferase in the cat brain. *Brain Res. Bull.*, 18: 371–415.

Waterhouse, B.D., Lin, C.-S., Burne, R.A. and Woodward, D.J. (1983) The distribution of neocortical projection neurons in the locus coeruleus. *J. Comp. Neurol.*, 217: 418–431.

Waterhouse, B.D., Mihailoff, G.A. Baack, J.C. and Woodward, D.J. (1986) Topographical distribution of dorsal and median raphe neurons projecting to motor, sensorimotor and visual cortical areas in the rat. *J. Comp. Neurol.*, 249: 460–476.

Discussion

A.Y. Deutch: You showed 50% of corticostriatal neurons retrogradely labeled were glutamate-immunoreactive. If you want both aspartate *and* glutamate, what is the composition of corticostriatal pathway?

J. Parnavelas: Table I suggests that Glu- and Asp-containing neurons could account for the entire corticostriatal pathway. It also suggests that some axons contain both amino acids.

M.A. Corner: Does distribution of individual neurons (monoaminergic, cholinergic, etc.) approximate that the whole pool, or are there specific subclasses with more selective terminations?

J. Parnavelas: It is now widely believed that some of the nonspecific afferents to the cortex (e.g. monoaminergic, cholinergic) are topographically organized. Thus, there is ample experimental evidence which suggests that a cortical area or even a portion of an area receives projections from distinct and separate neuronal subpopulations. This type of analysis has not been carried out at the single cell level.

P.E. Roland: Do pyramidal cell axon terminals contain Glu/Asp – and have these terminals been identified to belong to the perikarya cells?

J.E. Parnavelas: There are Glu- and Asp-positive axon terminals in the cortex. One assumes that they belong to cortical pyramidal neurons although such terminals have not been traced to the cells of origin.

J.E. Pisetsky: (1) Do transmitters work synergetically? (2) Does GABA inhibit other neurotransmitters in Alzheimer's?

J. Parnavelas: (1) There is evidence that some transmitters act synergetically in the cortex. (2) GABA is known to inhibit cortical pyramidal neurons which contain Glu or Asp. These excitatory amino acids are effected in Alzheimer's dementia.

J. Cavada: Which are the targets of the monoamine synaptic contacts? Is there anything known about the cell types involved?

J. Parnavelas: We do not know which cells in the cortex are contacted by monoamine-containing axon terminals. We have ultrastructural evidence that both pyramidal and nonpyramidal neurons are contacted by noradrenergic axons.

G.J. Boer: You shortly mentioned some features in the ontogeny of neuropeptide-containing cortical cells. If you talk about birth date of these various types are you then referring to birth for gene expression of the neuropeptide or to genuine cell birth? And secondly, what is the time course of cell number increase in relation to the degeneration you mentioned?

J. Parnavelas: I am referring to the date of birth of these neurons. Our work with a number of peptides in the rat visual cortex has shown that peptide-containing neurons gradually increase in number during the first 3 postnatal weeks after which time their numbers decline considerably.

CHAPTER 3

Qualitative and quantitative comparison of the prefrontal cortex in rat and in primates, including humans

Harry B.M. Uylings and Corbert G. van Eden

Netherlands Institute for Brain Research, Meibergdreef 33, 1105 AZ Amsterdam, The Netherlands

Introduction

Since Brodmann's studies (1909, 1912) it has been generally accepted that the phylogenetic development of the prefrontal cortex (PFC) is "highest" in primates, and reaches its apogee in man. Until the publication of Rose and Woolsey's classical paper (1948a), it was mainly Brodmann's cytoarchitectonic criterion that was used to delineate the PFC in the various mammalian species. Brodmann's criterion for defining the PFC was the presence of a granular layer IV in the cortical areas rostral to the motor and premotor areas. As a consequence, the belief was held until 1948 that many mammals "below" primates, such as insectivores, marsupials, the rat and many other rodents, did not have a PFC. Since Rose and Woolsey (1948a) proposed that the PFC should be defined as the "essential cortical projection area of the mediodorsal nucleus of the thalamus" (MD), all mammals have been considered to have a PFC.

Recently, more accurate tracing techniques have revealed new data on the connectivity pattern of the PFC, especially in primates and rat. For (the interpretation of) experimental studies it is important to know to what extent the prefrontal cortices

in rat and primates are homologous. After formulating the criteria for inferring homologies of particular brain regions, this chapter will deal with the comparison of rats and primates, mainly on the basis of recently obtained anatomical data. For a functional comparison of the prefrontal cortices of rat and primates we refer to the chapter by Kolb (this volume).

Brodmann's data (1912) on the relative size (surface area) of the PFC in different mammalian species are still being reproduced by current researchers (e.g. Passingham, 1973; Markowitsch, 1988). Since other criteria are now used to delineate the PFC, we will attempt to determine and compare the relative volumes of the PFC in rat, New World and Old World monkeys, anthropoid ape, and man.

Criteria for inferring homologies of brain regions

Before Brodmann (1909, 1912), the pattern of cortical sulci was used to determine whether brain areas in gyrencephalic mammals were homologous. This approach was strongly disputed by Brodmann (e.g. 1912), who emphasized that cytoarchitectonic characteristics should be applied instead. Later Walker (1940) and Rose and Woolsey (1948a), among others, suggested that it was better to rely on the physiological and neural connectivity

---

Correspondence: H.B.M. Uylings, Netherlands Institute for Brain Research, Meibergdreef 33, 1105 AZ Amsterdam, The Netherlands.

characteristics of the different brain regions than to consider only its architecture in detecting homologous areas. A list of criteria for inferring the homology of brain regions has been drawn up by Campbell and Hodos (1970). On the basis of this list, the following criteria are regarded as important for the homology of pertinent brain regions: (1) their pattern and relative density of nerve connections; (2) their functional (i.e. electrophysiological and behavioral) properties resulting from (electrical) stimulation, lesions etc.; (3) the distribution of different immunocytochemical and neuroactive "markers"; (4) their embryologic development; (5) for closely related species, their cytoarchitectonic characteristics.

The more concordant the characteristics are, the more likely it is that brain regions are homologous. However, in addition to the long list of characteristics, the evolution of more specialized cortical areas from a smaller number of cortical fields should also be considered. This may impede the firm establishment of the homology of brain regions between different species. Since Rose and Woolsey (1948a), the thalamocortical projection area of the mediodorsal nucleus of the thalamus has generally been applied as the definition of the PFC (e.g. Leonard, 1969; Markowitsch and Pritzel, 1981; Benjamin and Golden, 1985; Reep, 1984). In this chapter we will follow the first of the above-mentioned homology criteria in comparing the rat PFC with the primate PFC, and will consider the connections with the thalamus, corticocortical connections and the connections with other subcortical nuclei.

Neural connections between prefrontal cortex and thalamus

Thalamocortical connections are considered to be of importance for the delineation of cortical areas, since one of their functions may be the regulation of cortical differentiation and specialization (e.g. Rakic, 1988; O'Leary, 1989). When Rose and Woolsey conducted their studies (1948a,b), the knowledge about thalamocortical connections was restricted by the limitations of the techniques then available. At that time the MD was assumed to be the only thalamic nucleus with thalamocortical connections to the PFC, and therefore proposed to be the so-called "definition nucleus". Later, with improved silver degeneration staining techniques (Leonard, 1969; Domesick, 1972) and more refined anterograde and retrograde axoplasmic staining techniques (Kievit and Kuypers, 1977; Krettek and Price, 1977a; Divać et al., 1978; Beckstead, 1979; Groenewegen, 1988; Giguere and Goldman-Rakic, 1988), it became evident that the mediodorsal nucleus is not the only thalamic nucleus to project to the PFC (see for rat Table I). In addition, it has been noted recently, e.g. in macaque monkey, cat and sheep (e.g. Akert and Hartmann-Von Monakov, 1980; Dinopoulos et al., 1985; Russchen et al., 1987; Giguere and Goldman-Rakic, 1988; Morán and Reinoso-Suárez, 1988) that some of the mediodorsal nucleus projections reach a few cortical areas outside the PFC, such as the premotor, temporal and parietal cortices.

Given the fact that cortical regions receive projections from several thalamic nuclei, the class of thalamic nuclei with specific cortical projections is considered to be important for the delineation of cortical areas. This class of nuclei has cortical projections which terminate mainly in the middle cortical layers within restricted cortical regions, which in turn reciprocate these projections (Caviness and Frost, 1980; Herkenham, 1980, 1986; Jones, 1985). This group of thalamic nuclei includes the relay nuclei, such as the dorsal lateral geniculate nucleus, and the ventral posterior nucleus, for the primary visual cortex, and the primary somatosensory cortex, respectively. The mediodorsal nucleus also has specific cortical projections which are reciprocal and terminate in the middle cortical layers. Nauta (1961, 1962), and Leonard (1969), among others, regarded the reciprocity of the MD cortical projections as an important criterion for PFC delineation. The concept of specific cortical projections, however, does not lead to an unambiguous delineation. For example, the submedial nucleus (nucleus gelatinosus) has specific cortical connections with a part of the rat frontal cortex (Price and Slotnick, 1983) that also receives specific cortical projections from the MD (see below). In addition, the thalamic nuclei with specific projections have only one major ascending input (e.g. Herkenham, 1986), but Groenewegen (1988) showed that the medial dorsal nucleus receives afferents from many different sources. Therefore, a better approach to delineating the PFC might be to characterize as prefrontal cortical areas all those cortical regions of which the reciprocal connections with the MD show a

TABLE I

Connections between rat prefrontal cortex and thalamus

Thalamus nuclei	Fr2	$AC_{(d)}$	IL/PL/MO	AI_d	AI_v
Mediodorsal (MD)					
paralamellar	↑↓				
lateral		↑↓			
central					↑↓
ventral/medial			↑↓	↑↓	(↑↓)
Midline					
medioventral (reuniens)	↑↓	↑↓	↑↓	↓	(↑)↓
rhomboid	·↓	↓	↓	·↓	↓
paratenial		·↓	↑↓		
Intralaminar					
rostral group					
cent. medial, cent. lat	↑↓	↑↓	↑↓	↑(↓)	↑(↓)
paracentral	·↓				
caudal group					
parafascicular	↑↓	↑↓	↑		
Anterior nuclei					
anteromedial	↑↓	·↓	·		
lateral dorsal	↓	↓			
Ventral medial complex					
submedial (= gelatinosus)	↓	↓	↓	↓	↑↓(VLO)
ventral medial (principal)	↑↓	↑↓	↑↓	↑↓	↑↓
basal ventral medial[a]	·↓	·↓	·↓	↑↓	·↓
ventral anterior – ventral lateral complex	·	·			(·)
Lateral posterior	↑↓	(↑)↓			
Reticular	↓	↓	↓	↓	↓

↑: thalamocortical; ↓: corticothalamic connections; small arrow indicates minor projections for one thalamic nucleus; arrows in parentheses indicate that the connection is not firmly established yet.
For abbreviations see list at the end of this chapter.
[a] = medial ventrobasal nucleus = parvocellular part of ventroposterior medial nucleus.
Note that a complete comparison of relative densities of nerve projections is possible within one study, but proves difficult when different studies with different (thalamic) injections and techniques are used.

relatively higher density than those with any other thalamic nucleus. The density and laminar location of the different thalamic projections within the different prefrontal cortical areas can only be revealed with anterograde tracers. The number of labeled neurons after a retrograde tracer injection is, of course, not an acceptable measure for the relative density of terminations.

In conclusion, the reciprocity of projections of the mediodorsal nucleus of the thalamus can no longer be used exclusively in defining the PFC. It is better to include only those cortical areas in the PFC for which the reciprocal connections with the mediodorsal nucleus are stronger than are the connections with other thalamic nuclei. The present knowledge of mammals is not yet sufficient for that, but as far as the available data permit we will review these aspects in rat and monkey.

Connections between prefrontal cortex and thalamus in the rat

Since Rose and Woolsey (1948a), the techniques for tracing nerve connections have improved enormously. As a result, the part of the rat cerebral cortex regarded as the PFC increased (see Fig. 1; Leonard, 1969; Krettek and Price, 1977a; Van Eden, 1986; Groenewegen, 1988; Verwer, personal communication). For a summary of the thalamocortical and corticocortical connections with different PFC areas, we refer to Table I. The present view is that the rat prefrontal cortex has its main reciprocal thalamic connections with the MD (e.g. Groenewegen, 1988). Fig. 2 indicates that reciprocal MD connections lead to the inclusion in the rat PFC of the following cortical regions: the *medial PFC,* i.e. Fr2, the dorsal and ventral anterior cingulate cortices (ACd and ACv, respectively), the prelimbic cortex (PL) and the infralimbic and medial orbital cortices (IL and MO, respectively); the *lateral PFC* (frequently called orbital or sulcal PFC) consists of the dorsal and ventral agranular insular cortices (AId and AIv, respectively), and the lateral orbital cortex (LO), which was included in the AIv delineated by Van Eden

and Uylings (1985). According to Leonard (1969), Groenewegen (1988) and Groenewegen et al. (this volume) the medial and lateral prefrontal cortices are not two separate cortical regions, but form an uninterrupted area around the rostral pole. The small ventral orbital cortex (VO) and ventral

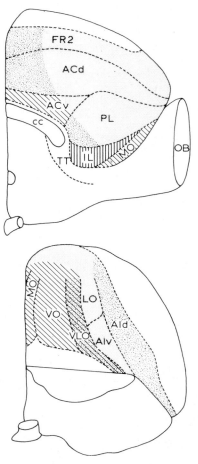

Fig. 1. The extent of what is regarded as prefrontal cortex has increased with the improvement of anatomical techniques. Fine dots indicate the PFC found by Leonard (1969). The heavy dots indicate the extension proposed by Krettek and Price (1977a), the oblique line the extension proposed by Groenewegen (1988), and the vertical lines the extension proposed by Verwer (personal communication, 1989). For abbreviations denoting the various cortical areas see list at the end of this chapter. The nomenclature of Krettek and Price (1977a) has been followed, with the exception of the neutral term frontal area 2 (Fr2) (Zilles, 1985; Schober, 1986).

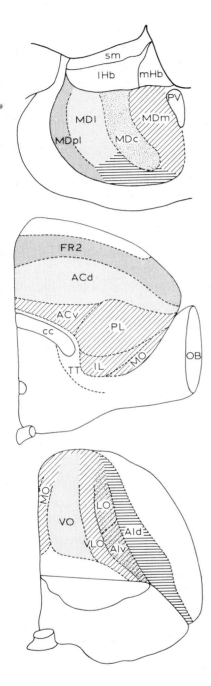

Fig. 2. The reciprocal connections between parts of the mediodorsal nucleus and prefrontal cortical areas in the rat (adapted from Groenewegen, 1988). For abbreviations see list at the end of this chapter. The neutral term frontal area 2 (Fr2) (Zilles, 1985; Schober, 1986) is used instead of the frequently used term medial precentral area (PrCm) (Krettek and Price, 1977a; Van

lateral orbital cortex (VLO) can be called the *ventral PFC*.

Concerning the criteria of highest relative density of the MD reciprocal connections, the following can be said for the various cortical areas which are illustrated in Fig. 2. At the time of Rose and Woolsey (1948b) the anterior cingulate cortex was excluded from the PFC, because the only thalamocortical projection known was the one from the anterior medial nucleus of the thalamus. However, Domesick (1972), Krettek and Price (1977a) and Vogt et al. (1981) have reported that the MD has more projections to the anterior cingulate cortex, and also to Fr2, than other thalamic nuclei examined. The projections of the thalamic anterior nuclei appear to be lighter in the medial PFC and to terminate mainly in layers I and VI (e.g. Krettek and Price, 1977a). The relative density of the projections of the different thalamic nuclei to Fr2 (= PrCm = AGm) has not been studied extensively. The main studies reported are based upon retrograde HRP tracing (e.g. Donoghue and Parham, 1983). The conclusion of these studies (Donoghue and Parham, 1983; Reep et al., 1984) is that many more cells project to Fr2 from the MD than from other thalamic nuclei, e.g. the ventral lateral nucleus (VL) (see also Table I).

The VL projection to Fr2 has been taken by Donoghue and Parham, among others, to be an argument for regarding this area, too, as part of

Eden and Uylings, 1985), because the rat cortex has no central sulcus. Fr2 is also preferred to Donoghue and Wise's term medial agranular cortex (AGm) (1982) since their lateral agranular cortex (AGl) is not really agranular. Fr2 is called area 8 in Vogt et al. (1981). The indicated Fr2 area is narrower in the rostral tip and the border is drawn less caudally than in Zilles' Fr2 (1985). Our Fr2 borders are based upon the cytoarchitectural characteristics and the MD connections described by Van Eden and Uylings (1985), Groenewegen (1988) and last section of this chapter. Vogt et al. (1981) call the infralimbic cortex (IL) area 25, the prelimbic cortex (PL) area 32, and the ventral and dorsal anterior cingulate cortices (ACv, ACd) areas 24a and 24b, respectively. The AIv and AId caudal border with the posterior agranular insular cortex (AI) is more frontal in Van Eden and Uylings (1985) and Groenewegen (1988) than in Zilles (1985) and Krushel and Van der Kooy (1988).

the sensorimotor cortex (see below). The ventral anterior nucleus (VA), however, is usually included in the ventral lateral nucleus of the rat, due to its difficult cytoarchitectonic delineation (Jones, 1985). A part of the VL cells, projecting to the cortical Fr2 (Donoghue and Parham, 1983; Reep et al., 1984), probably belongs to the ventral anterior nucleus described by Jones (1985, pp. 55–56). Herkenham, therefore, used the term "VA–VL complex" in his 1986 review. In this paper Herkenham also showed that a medial part of this thalamic complex, the paralaminar VL, (i) receives cerebellar afferents, (ii) projects outside the primary motor cortex to frontal, parietal and occipital areas, including Fr2 and the anterior cingulate cortex, and (iii) terminates mainly in layer I.

The posterior part of Fr2, at least, also has clear reciprocal connections with the lateral posterior complex (LP) (Sukekawa, 1988a). The pulvinar can not be distinguished in the rat (Jones, 1985, p. 541), but the rat LP shows features which are "comparable" to those of the primate pulvinar (Herkenham, 1986, p. 428).

So far, we know of no thalamic nuclei with a reciprocal connection with the lateral PFC stronger than that of the MD. According to Krushel and Van der Kooy (1988), the MD has stronger reciprocal connections with AId than with the basal ventral medial nucleus (VMb).

In the literature the VMb is frequently mentioned as medial ventrobasal nucleus, mTVB (e.g. Krushel and Van der Kooy, (1988) or parvocellular part of ventroposterior medial nucleus, VPMpc (e.g. Kosar et al., 1986); see for discussion of nomenclature Jones (1985, p. 417).

Besides having strong connections with MD, the cortical areas AIv and VLO also receive specific cortical projections from the submedial nucleus (i.e. the nucleus gelatinosus; Price and Slotnick, 1983). At present it is unclear whether or not the VLO has a stronger reciprocal connection with MD or submedial nucleus (cf. Price and Slotnick, 1983; Groenewegen, 1988). The projections of the submedial nucleus currently form no argument for excluding the VLO from the PFC. The submedial nucleus, like the central segment of the MD, receives an olfactory input (Price and Slotnick, 1983; Groenewegen et al., this volume). For the ventral orbital cortex (VO), too, more data are required to firmly establish its "membership" in the PFC (Groenewegen, personal communication).

The MD projections terminate mainly in the middle cortical layer III and also in the superficial layer I in the rat PFC. Outside the agranular rat PFC, hardly any MD terminations are found in the neocortex. From Table I it appears that the principal ventral medial nucleus (VMp, nomenclature of Jones, 1985, p. 418) also has reciprocal connections with the PFC areas. These projections are less dense than the MD projections. Furthermore, the VMp projects to cortical layer I, and most densely to the frontal part, but also to almost the entire neocortex (Herkenham, 1979).

In conclusion, when the criterion of heaviest reciprocal connections with MD is applied to the present data on thalamic connections, the cortical areas displayed in Fig. 2 will be regarded as PFC areas. More precise data on thalamic connections are desirable and could change our views about the extension of the rat PFC (e.g. for VLO and VO).

Connections between the prefrontal cortex and thalamic nuclei in primates

Since the study by Kievit and Kuypers (1977) it has been accepted that in the primate brain the MD is not the only thalamic nucleus that projects to the PFC, but that the ventral anterior (VA) and medial pulvinar nuclei do the same, in addition to a smaller contribution from the intralaminar and midline nuclei (e.g. Giguere and Goldman-Rakic, 1988). As in the rat, the fact that the different frontal cortical areas have reciprocal connections with other thalamic nuclei appears to entail that the presence of reciprocal connections between MD and cortical areas is an insufficient criterion for defining the primate PFC. Mainly on the basis of cytoarchitectonics, Walker (1940) has described the areas of the macaque PFC. In addition to the complete granular frontal region of the brain, i.e. the traditional PFC, he also includes the agranular anterior cingulate cortex, area 24, and area 25 without explaining why. Walker's map is still widely used for the delineation of the macaque PFC, although the anterior cingulate area is not generally regarded as a part of the PFC. For the parcellation of this region we prefer the map of Vogt et al. (1987) and that of Barbas and Pandya (1989) (see Fig. 3). We will therefore consider the thalamic connections with the cortical areas indicated in Fig. 3 and some other cortical regions with reciprocal connections with MD.

As is the case in the rat (see Fig. 2), the primate MD consists of several segments with reciprocal connections with particular frontal cortical areas (e.g. Kievit and Kuypers, 1977; Akert and Hartmann-Von Monakov, 1980; Goldman-Rakic and Porrino, 1985; Giguere and Goldman-Rakic, 1988; Pandya and Yeterian, this volume). The medial magnocellular part of the MD (MDmc) has reciprocal connections with the orbitofrontal cortex (areas 11 – 14), with area 25, and with the ventromedial part of cortical area 24. Reciprocal connections also exist between the lateral parvocellular part of the MD (MDpc) and the dorsolateral and dorsomedial areas 46, 9, 32, 24 and 6m (6m, the supplementary motor area (SMA), is the rostromedial part of the premotor area 6; see Schell and Strick, 1984); between the paralamellar, multiform part of the MD (MDmf) and the prearcuate frontal eye field (FEF), area 8, i.e. Walker's areas 8a and 45; between the anterior region of MD and area 10; and between the posterior densocellular part of MD (MDdc) and areas 9, 46 and SMA (see for a recent report on this subject Giguere and Goldman-Rakic, 1988; and for dorsal-ventral subdivision of MD Pandya and Yeterian, this volume).

The granular prefrontal areas 8 – 14 and 46 have in common that they all receive a substantial afferent projection from MD, VA and medial pulvinar (e.g. Kievit and Kuypers, 1977; Goldman-Rakic and Porrino, 1985). Giguere and Goldman-Rakic (1988) demonstrated a very dense terminal projection of the MD in layer IV and the deep part of layer III (not in layer I) of these cortical areas. However, data allowing relative comparison of the density of terminal projection from MD, VA and medial pulvinar are lacking at present. Traditionally, the cingulate cortex and other paralimbic cortical areas are thought to have the most "closely related" thalamic connections with the so-called limbic nuclei of the thalamus, i.e. the anterior nuclei: anteroventral (AV), anteromedial (AM), anterodorsal (AD) and the lateral dorsal nucleus

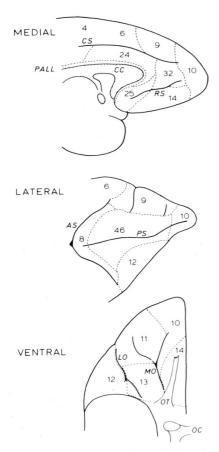

Fig. 3. Cytoarchitectonic parcellation of the macaque prefrontal cortex. The lateral and ventral views are adapted from Walker (1940). The medial view is adapted from Vogt et al. (1987) and Barbas and Pandya (1988). As, arcuate sulcus; CC, corpus callosum; CS, cingulate sulcus; LO, lateral orbital sulcus; MO, medial orbital sulcus; OC, optic chiasm; OT, olfactory tract; PALL, limbic periallocortex; PS, principal sulcus; RS, rostral sulcus.

(LD) (Yeterian and Pandya, 1988). The primate anterior cingulate cortex (area 24), however, receives its main thalamic input from the MD, and from the intralaminar and midline nuclei, and, to a limited extent, from AM and VA (Vogt et al., 1979; Baleydier and Mauguière, 1980; Vogt et al., 1987). The MD projections terminate mainly in the middle layer of this cortex, i.e. in deep layer III, like in the granular PFC, although less densely (Giguere and Goldman-Rakic, 1988). Most of the corticothalamic projections from the anterior cingulate cortex terminate in the MD, in MDmc or MDpc (Yeterian and Pandya, 1988). Corticothalamic terminations are also found in VA and AM, and in the intralaminar and midline nuclei, but not in the medial pulvinar (Baleydier and Mauguière, 1980; Yeterian and Pandya, 1988). On the other hand, the posterior cingulate cortex (area 23) receives its main thalamic projections from the anterior nuclei, AV, AM, AD and LD, and from the medial pulvinar (Vogt et al., 1979; Baleydier and Mauguière, 1980), while its corticothalamic efferents terminate preferentially in the medial pulvinar, partly in LD and AV, but not at all in MD (Baleydier and Mauguière, 1980; Yeterian and Pandya, 1988). Therefore, in contrast with the traditional view of the cingulate cortex, the MD, and not the anterior thalamic nuclei, appears to be the thalamic nucleus with the main thalamic reciprocal connections with the primate anterior cingulate cortex (area 24).

Walker's areas 8a and 45 are regarded by Huerta et al. (1986) and Stanton et al. (1989) as one single functional area 8, the prearcuate frontal eye field (FEF). From this region the saccadic eye movements can be elicited with the lowest currents, below 50 μA. This region thus participates in visuomotor function and is also characterized by its direct input from the deeper layers VI – VII of the superior colliculus (e.g. Huerta et al., 1986). The clear cytoarchitectonic features of the prearcuate FEF are its large pyramidal cells in layer V and its transition from the dense granular cortex on the prearcuate gyral convexity to the agranular premotor cortex in the posterior bank of the arcuate sulcus (Stanton et al., 1989). Huerta et al. (1986) have shown in more detail than Akert and Hartmann-Von Monakov (1980) that the heaviest reciprocal thalamic connections of FEF are those with the MD (especially MDmf). There are also dense reciprocal connections with VA, though less so with medial pulvinar and intralaminar nuclei, in the Old World and New World monkey. In addition, FEF receives some thalamic input from the rostral part of the ventrolateral (VL) nucleus (Akert and Hartmann-Von Monakov, 1980; Huerta et al., 1986). We therefore regard FEF as a PFC area.

The supplementary motor area (SMA) is located in the medial wall of the premotor area 6, (e.g. Schell and Strick, 1984). Both the premotor area and SMA have reciprocal connections with MD (Akert and Hartmann-Von Monakov, 1980; Giguere and Goldman-Rakic, 1988; Matelli et al., 1989) but they are less abundant than those with VL (Schell and Strick, 1984), and with VA (e.g. Akert and Hartmann-Von Monakov, 1980; Jürgens, 1984). The MD terminals end in the deep part of layer III in the SMA, but they are much less dense than those in the granular PFC, and also less dense than those in the anterior cingulate cortex. The number of cortical SMA cells with reciprocal connections with the MD is considerably lower than that in anterior cingulate cortex and granular PFC (Giguere and Goldman-Rakic, 1988). The SMA and the caudal premotor area (area F4 of Matelli et al., 1989) have especially strong connections with the oral and the medial part of VL, VLo and VLm (nomenclature of Olszewski, 1952; according to the nomenclature of Jones (1985) this is the anterior part of VL, VLa); see Schell and Strick (1984), Goldman-Rakic and Porrino (1985) and Matelli et al. (1989). The postarcuate premotor area (APA; Schell and Strick, 1984; comparable to areas F5 and F2 in Matelli et al., 1989) receives its major input from the posterior VL nucleus (nomenclature of Jones, 1985), mainly from its medial part, which Olszewski (1952) calls area X. In view of these reciprocal connections of these premotor areas, they are not included in the PFC.

A reciprocal connection, though only a limited one, also appears to exist between the MD and the primary motor area 4. Akert and Hartmann-Von Monakov (1980) have shown the projection from area 4 to MDmf, while Schell and Strick (1984), Goldman-Rakic and Porrino (1985), Ilinsky et al. (1985), and Leichnetz (1986) have found that some projections run to area 4 from the most ventrolateral part of the caudal MDmf, next to the intralaminar nuclei, central lateral and centrum medianum, and from a cluster of cells in the MDdc. MD projections to area 4 have not been found by Jones et al. (1979) and Matelli et al. (1989). However, the great majority of the reciprocal thalamic connections of primary motor area 4 are with the oral part of the ventral posterior lateral nucleus, VPLo (Akert and Hartmann-Von Monakov, 1980; Schell and Strick, 1984).

According to Jones (1985, p. 384) and Ilinsky and Kultas-Ilinsky (1987), this nucleus VPLo (according to Olszewski, 1952) is a part of a larger common nucleus, the posterior nucleus of the ventral lateral complex, VLp (see also Schell and Strick, 1984). VLo also projects to the primary motor area 4 (Goldman-Rakic and Porrino, 1985; Leichnetz, 1986; Matelli et al., 1989). No connections have been reported with VA, although VLo (Olszewski's nomenclature) is called densocellular part of VA by Ilinsky and Kultas-Ilinsky (1987). Jones (1985) regards VLo as a part of VL, but calls this the anterior part, VLa.

In the macaque insula an anterior and a posterior part can be distinguished, on the basis of its architecture (e.g. Mesulam and Mufson, 1985) and its thalamic connections (Mufson and Mesulam, 1984). The anterior insula has agranular and dysgranular subfields, and the posterior insula has a granular cortex (Mesulam and Mufson, 1985). As the rat agranular insula cortex, the macaque anterior insular has extensive reciprocal connections with MD (especially MDmc). However, these are not predominant in the primate, and the anterior insula is, therefore, not included in the PFC. The preferential reciprocal thalamic connections are those with the basal ventral medial nucleus (VMb) (according to the nomenclature of Jones, 1985; i.e. the parvocellular part of the ventral posterior medial nucleus (VPMpc) in the nomenclature of Olszewski, 1952). In the monkey, the VMb receives an important direct ipsilateral projection from the solitary tract nucleus (NTS), which is the first central relay nucleus for gustatory input (Beckstead et al., 1980). In addition, the anterior insula has some reciprocal connections with the medial pulvinar, the ventral posterior inferior nucleus (VPI) and some intralaminar nuclei (Mufson and Mesulam, 1984). On the other hand, the posterior insula predominantly has reciprocal connections with the medial pulvinar, the suprageniculate nucleus and, also, the VPI. A limited reciprocal connection appears to exist with MDpc. Together with the corticocortical connections, these connections between insula and thalamus lead to the conclusion that the anterior insula is more related to olfactory, gustatory, and visceral functions, whereas the posterior insula is more related to somesthetic, auditory and skeletomotor functions (Mesulam and Mufson, 1985).

Connections of the mediodorsal nucleus in rats and primates

In previous sections we related the PFC to the MD, in both rat and primate. However, is the rat MD really comparable with the primate MD? In order to answer this question we have compared the connectivity patterns of rat and primate MDs. Table II gives a summary of the present data on reciprocal cortical connections with the different parts of the mediodorsal nucleus in rat and primate. It shows that the topographies of these cortical connections in rat and primate are quite similar. The same applies to the subcortical afferents to the different parts of the mediodorsal nucleus in rat and primate (Table III). In fact, the connectivity pattern is more complex than would appear from Tables II and III. The projection field does not always terminate homogeneously within the different parts of MD or the other brain regions indicated in

TABLE II

Cortical reciprocal connections with mediodorsal nucleus

Rat		Primate	
Cortex area	MD region	Cortex area	MD region
ACd	MDl	AC	MDpc, (MDdc)
ACv, PL, IL, MO	MDm	areas 25, 32, ventral part area 24	MDmc
AI	MDm, MDv, MDc	AI	MDmc, ventral part MDpc
VLO	MDm	orbital areas 11–14	MDmc
VO	MDl		
Fr2	MDpl	FEF	MDmf
		area 9	MDpc
		SMA	MDpc, MDdc
		APA, superior arcuate part	dorsal part MDpc
		APA, inferior arcuate part	ventral part MDpc, ventrocaudal part MDmc
		area 4	lateral part of both MDmf and MDdc
		area 46, dorsal to PS	dorsal part MDpc
		area 46, ventral to PS	ventral part MDpc
		anterior temporal isocortex	MDpc
Ent.[a,b]	MDm	ento-/perirhinal cortex[a]	MDmc
		temporal periallocortex and proisocortex[a]	MDmc
Pir.[a,b]	MDm, MDc	(temporal) pyriform cortex[a]	MDmc
subiculum[a]	MDm	subiculum[a]	MDmc

For legend see Table III.

Tables II and III. For example, there is a dorsoventral differentiation in the primate MD for the reciprocal connections with different cortical fields (see Fig. 12 of Pandya and Yeterian, this volume; Goldman-Rakic and Porrino, 1985), and a characteristic pattern exists in the connections with the basal ganglia in the rat (see Fig. 6 of Groenewegen et al., this volume). In rat and primate MD there is also a rostrocaudal differentiation for several terminal fields. In both primate and rat, the rostral part of the medial MD (MDm) receives projections from more amygdaloid neurons and fewer ventral pallidal neurons than the caudal part (Aggleton and Mishkin, 1984; Russchen et al., 1987; Groenewegen, 1988). In rat it also receives more projections than the caudal part from entorhinal, prepiriform and preoptic neurons (Groenewegen, 1988). In addition, some terminal fields within a part of MD end in patches (e.g. the amygdaloid fibres in primate (Russchen et al., 1987)), and some efferent MD fibers originate in clusters of neurons (e.g. Goldman-Rakic and Porrino, 1985). Furthermore, topographic similarity of the connections in rat and primate does not necessarily imply a similarity of the relative "weight" of the different projections. Some examples are described in the preceding section on the thalamocortical projections. Yet another example is that both in primate (Nauta, 1961; Porrino et al., 1981; Aggleton and Mishkin, 1984; Russchen et al., 1987) and in rat (Krettek and Price, 1977b; Groenewegen, 1988) different amygdaloid nuclei project to the medial part of the MD, while the reciprocal connection is largely limited to the basolateral

TABLE III

Subcortical afferents to thalamic mediodorsal nucleus

Rat		Primate	
MD segment	Subcortical field	MD segment	Subcortical field
MDm	Ventral pallidum[c], nucleus of diagonal band of Broca, nucleus accumbens, bed nucleus of stria terminalis, dorsomedial and lateral hypothalamus, several nuclei of amygdala[b], ventral tegmental area, pars compacta of substantia nigra, parabrachial nuclei, nucleus of solitary tract, medial and lateral preoptic area[b], also prepiriform cortex[b] and entorhinal cortex[b]	MDmc	Several nuclei of amygdala[b], (olfactory bulb), olfactory cortex, ventral pallidum[c], bed nucleus of the stria terminalis, anteromedial part of substantia nigra, nucleus of diagonal band of Broca, internal segment of globus pallidus, periaqueductal gray
MDc	Olfactory tubercle, prepiriform cortex, ventral endopiriform nucleus, lateral preoptic area, lateral hypothalamus, nucleus of diagonal band of Broca		
MDv	Ventral striatum[c], ventral pallidum[c], pars reticulata of substantia nigra, ventral tegmental area		
MDl	Pars reticulata of substantia nigra, interpeduncular nucleus, dorsolateral tegmental nucleus, medial part of globus pallidus, and lateral hypothalamus	MDmf, pc	Lateral anterior part of substantia nigra (mf), superior colliculus (mf), medial vestibular nuclei, mesencephalic tegmental fields, cervical and lumbar spinal cord (pc), periaqueductal gray (pc)
MDpl	Superior colliculus, pars reticulata of substantia nigra, dorsolateral tegmental nucleus, pretectal nuclei, fastigial nucleus of cerebellum	MDdc	Dentate nucleus of cerebellum, lateral anterior part of substantia nigra, cervical and lumbar spinal cord
all	Reticular thalamic nucleus, locus coeruleus, raphe nucleus, rostral part of central gray, reticular formation		

For abbreviations see list at the end of this chapter.
Rat data based primarily upon Groenewegen (1988) and Groenewegen et al. (this volume). Primate data based primarily upon Akert and Hartmann-Von Monakov (1980), Mantyh (1983a,b), Schell and Strick (1984), Goldman-Rakic and Porrino (1985), Ilinsky et al. (1985), Leichnetz (1986), Ilinsky and Kultas-Ilinsky (1987), Russchen et al. (1987), Cornwall and Phillipson (1988), Giguere and Goldman-Rakic (1988), Gower (1989), Matelli et al. (1989) and Pandya and Yeterian (this volume).
[a] Thalamocortical MD projections not (yet) found.
[b] More neurons project to rostral MDm.
[c] More neurons project to caudal part.

nucleus of the amygdala (Nauta, 1962, Van Vulpen and Verwer, 1989). On the other hand, the amygdala projections to the medial magnocellular MD are rather prominent in monkey, but much less so in rat (e.g. Russchen et al., 1987; Groenewegen, 1988).

Another minor difference is the differential acetylcholinesterase (AChE) staining of the MD. In the adult primate, a few cells from the nuclei of the diagonal band of Broca and from the nucleus basalis of Meynert project to MDmc (Russchen et al., 1987). These nuclei of the substantia innominata contain a vast number of cholinergic neurons (Mesulam et al., 1983a), whereas the MDmc-projecting neurons are probably non-cholinergic, given the negative AChE staining of the entire MD from early postnatal development onward (Kostović and Goldman-Rakic, 1983).

Before birth the MD displays a transient positive AChE staining (Kostović and Goldman-Rakic, 1983). It might be, therefore, that the adult non-cholinergic neurons were transiently cholinergic during early life, or that only the afferent fibers to MD of cholinergic neurons are transiently present, in addition to temporary cholinergic neurons in MD.

On the other hand, in the adult rat many cells from the nuclei of the diagonal band of Broca project to MD, predominantly to the medial part, MDm (Groenewegen, 1988). In the rat, these Broca band nuclei contain a very large number of cholinergic neurons (e.g. Mesulam et al., 1983b). The neurons of Broca's diagonal band that project to MDm, however, are probably non-cholinergic in rat, too, since the adult MDm shows almost no AChE staining (Groenewegen, 1988). However, the lateral and ventral segments of MD in the rat show a moderate AChE staining, probably caused by the projection from the dorsolateral tegmental nucleus in the brainstem (Groenewegen, 1988). The moderate dopaminergic innervation of only the most medial zone of MDm in the rat derives from the dopaminergic neurons in pars compacta of the substantia nigra (Groenewegen, 1988). Unfortunately, however, no data on the dopaminergic innervation of the primate MD are known to us. We do not know, therefore, whether or not the projection neurons from the periaqueductal gray and from the substantia nigra (Mantyh, 1983b; Björklund and Lindvall, 1984) are dopaminergic.

In conclusion, Tables II and III show a largely similar topography of the MD connections in rat and primate, and suggest that the medial and central part of the rat, MD, MDm and MDc, resemble the medial magnocellular part of the primate MD; and that the lateral and paralaminar part of the rat MD are comparable with the parvocellular, paralaminar multiform and densocellular part of the primate MD.

Corticocortical connections of prefrontal cortex

Prefrontal corticocortical connections in the rat

The rat prefrontal corticocortical connections show some specific topographical organization (Sesack et al., 1989; Reep et al., 1990; Van Eden et al., 1990). Until now, this topographical organization has not been studied in such detail as that of the primate cortex (see section below). In the rat, most of the prefrontal corticocortical connections are reciprocal. The study by Van Eden and Lamme especially shows that the medial PFC can be divided into 3 subfields with regard to these connections, viz. a ventral subfield consisting of the prelimbic and infralimbic cortices, a shoulder subfield (Fr2 and ACd), and a rostral subfield (see Fig. 4). The first of these, i.e. the prelimbic and infralimbic cortices, has reciprocal connections with the perirhinal and entorhinal cortex and non-recurrent connections with the temporal half of the hippocampal CA_1 region (see Fig. 4). These are all limbic cortices. The second subfield (Fr2 and ACd) has reciprocal connections mainly with the secondary visual cortices and also with the primary visual cortex and the retrosplenial cortex (Fig. 4). The third subfield, the rostral part of the medial PFC (ACd, Fr2), has reciprocal connections with motor, mixed somatosensory-motor and somatosensory association cortices (Donoghue and

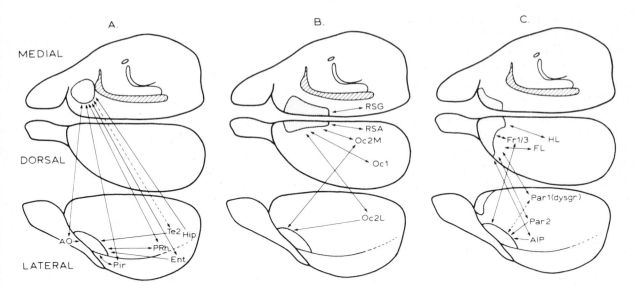

Fig. 4. This diagram of cortico-cortical connections of the rat prefrontal cortex is mainly based upon Van Eden et al. (1990), and also on the studies of Swanson (1981), Vogt and Miller (1983), Swanson and Köhler (1986), Audinat et al. (1988), Sukekawa (1988b) and Sesack et al., (1989). See for abbreviations the list at the end of this chapter.

Parham, 1983; Reep et al., 1990; Van Eden et al., 1990) (Fig. 4C). (Fr1 is called AGl by Donoghue and Parham (1983).) The presence of reciprocal connections with Fr1 is also a property of the primate premotor cortex (see next section below) (Fig. 4). The distinction between the shoulder and the ventral subregions of the medial PFC parallels the distinction in the reciprocal projections from the medial and lateral parts of the MD (see Fig. 2), e.g. the lateral part of the MD (MDl) and the paralaminar part of the MD (MDpl) project to the shoulder region. In this respect, it is of interest that the MDpl receives afferents from the deep layers of the superior colliculus (see rat frontal eye field, below). Our initial results also indicate some specific topographical organization for the lateral and ventral PFC.

Primate prefrontal corticocortical connections

Since this topic will be dealt with extensively in the chapter by Pandya and Yeterian (this volume), we will only summarize their main conclusions. They describe how the entire cortex has evolved in 2 bands from 2 moieties, i.e. paleocortical and archicortical. This evolution is cytoarchitectonically apparent in a paleo(ventro)fugal and an archi(medio)fugal trend of neocortical differentiation in 2 separate belts for each of the different cortical subsystems (e.g prefrontal, motor, somatosensory etc.). This theory implies that a higher level of differentiation corresponds with an increased thickness of the supragranular layers, a more pronounced granular layer IV and a decreased thickness of the infragranular layers (see Figs 5 and 7 in Pandya and Yeterian, this volume). This applies to both paleofugal and archifugal trends. The principal sulcus region (Fig. 3), where the two trends meet, is the area of highest differentiation. The authors show that according to this dual trend of cortex differentiation, corticocortical connections are organized in such a way that cortical areas that have a similar level of differentiation will tend to be reciprocally connected. This appears to be the case both for connections within the PFC and for those between the PFC and other cortical areas. PFC areas that are interconnected with other cortical areas belong to the same trend, either paleofugal or archifugal (see e.g. Fig. 17 in Pandya and Yeterian, this volume), whereas areas that are interconnected within the PFC belong to different trends (see Fig. 11 in Pandya and Yeterian, this volume). According to this theory, which is supported by data of tracing studies, prefrontal cortical areas generally have reciprocal connections with the premotor region, somatosensory association region, visual association region, auditory association region and paralimbic region, which regions are hierarchically organized within each trend (i.e. dorsal or ventral). Thus, the highly developed frontal eye field (area 8) and other premotor areas are connected with first-order association areas; the lateral PFC with second-order sensory association areas, the orbital and medial PFC with third-order sensory association areas; and the rostral PFC with third-order somatosensory association areas (Pandya and Yeterian, this volume). In primates, the FEF is not directly interconnected with the primary motor cortex, but has extensive connections with

TABLE IV

Rat prefrontal cortical connections

	Fr2	AC$_{(d)}$	PL/IL/MO	AI$_d$	AI$_v$
Forebrain (see also Table II and Fig. 1)					
Caudoputamen (dorsal striatum)	↓	↓	↓	↓	
Ventral striatum (nucleus accumbens)	↓	↓	↓	↓	↓
Ventral pallidum	˙(↓)	˙↓	↑↓	↑↓	↓
Lateral septum			↓	↓	
Medial septum (Ch 1)		↑↓	↑↓		
Vertical band of Broca (Ch 2)	↑	↑↓	↑↓	↑	↑
Horizontal band of Broca (Ch 3, 4)	˙(↓)	↑↓	↑↓	˙	˙
Nucleus basalis (Ch 4)	↑↓	˙↓	˙↓	↑↓	↑↓
Claustrum	↓	↓	↓	↓	↓
Amygdala					
basolateral	(↓)↓	˙↓	↑↓	↑↓˙	˙
basomedial	˙	˙	↑	↑	
central					˙
amygdalohippocampal			↑		
periamygdaloid cortex			↑	↑	
cortical			˙	˙	
Endopiriform nucleus	↑		↑	↑	
Hippocampus (CA1)	↑	↑	↑		
Subiculum			↑↓		
Olfactory bulb			↓	(˙)	(˙)
Interbrain (diencephalon) (see also Table I)					
Subthalamic nucleus	↓	↓	↓		
Lateral preoptic nucleus	(↓)	↓	↓		
Lateral hypothalamus	↑↓	↑↓	↑↓	↓	↑
Habenula nuclei		↓	↓		
Limbic mammillary nuclei		↓	↓		
Supra- and tubero-mammillary nuclei	↑	↑	↑↓		

See for legend Table V.

premotor areas, such as the SMA and the postarcuate premotor area (APA), which, in turn, have direct reciprocal connections with the primary motor cortex (e.g. Pandya and Kuypers, 1969; Schell and Strick, 1984; Huerta et al., 1987).

As was indicated in the previous section, it is not

TABLE V

Brainstem-prefrontal cortical connections in rat

Midbrain/pons	Fr2	$AC_{(d)}$	PL/IL/MO	AI_d	AI_v
Prerubral field (H field of Forel)	↑(↓)	(↓)	(↓)		
Retrorubral field (A8)	(↓)	(↓)	(↓)		
Substantia nigra, pars compacta (A9)	(↓)	↑↓	↓		
Ventral tegmental area (Tsai) (A10)	↑(↓)	↑↓	↑↓	↑↓	↑↓
Interpeduncular nucleus			↓		
Ventral/dorsal tegmental area (Gudden)	↓	↓	↓	↓	↓
Periaqueductal gray	↓	↓	↓	↓	↓
Nucleus Edinger–Westphal	↓	↓			
Pretectum	↓	↓			
Superior colliculus	↓	↓	↓		
Pedunculopontine tegmental nucleus (Ch 5)	↓	↓	↑↓		
Laterodorsal tegmental nucleus (Ch 6)	↓	↓	↑↓		
Parabrachial nucleus	↑	↑	↑↓	↑↓	↑↓
Locus coeruleus (A6)	↑↓	↑↓	↑↓	↑↓	↑↓
Nucleus subcoeruleus (A7)	↓	↓	↓		
Lateral paragigantocellular nucleus (A5)	↓	↓	↓		
Reticular formation	↓	↓	↓	↓	↓
Cuneiform nucleus		↓	↓		
B9 (in Lemniscus medialis)	(·)	(·)	(·)↓	(·)	(·)
Dorsal raphe nucleus (B7)	↑↓	↑↓	(↑)↓	(↑)	(↑)
Median raphe nucleus[a] (B8)	↑↓	↑↓	↑↓	(·)	(·)
Nucleus raphe magnus	↓	↓	↓		
Nucleus raphe pallidus	↓	↓	↓		
Nucleus raphe obscuris	↓	↓	↓		
Pontine nuclei	↓	↓	↓		
Solitary tract nucleus (A2, C2)	(↓)	↓	↓	↓	↓
Trigeminal nucleus	↓	↓	↓	↓	↓
Spinal cord	↓	↓	↓	↓	

[a] = central superior nucleus = nucleus Bechterew.
↑: corticopetal; ↓: corticofugal connections; small arrow indicates minor projections; arrows in parentheses indicate that the connection is not firmly established yet.
For abbreviations see list at the end of this chapter.
Ch1–6, groups of cholinergic neurons (Mesulam et al., 1983); A2–A7, noradrenergic cell groups (Dahlström and Fuxe, 1964); A8–A10, dopaminergic cell groups (Dahlström and Fuxe, 1964); C2, an adrenergic cell group (Hökfelt et al., 1974; Howe et al., 1980); B7–9, serotonergic cell groups (Dahlström and Fuxe, 1964).
Tables IV and V are based upon Swanson (1981), Mesulam et al. (1983b), Milner et al. (1983) O'Hearn and Molliver (1984), Saper (1984), Afsharpour (1985), Van Eden (1985), Hardy (1986), Neafsey et al. (1986b), Satoh and Fibiger (1986), Luiten et al. (1987), McDonald (1987), Miller (1987), Ruggiero et al. (1987), Terreberry and Neafsey (1987), Sesack et al. (1989), Groenewegen et al. (this volume), Gaykema et al. (1990a, b).

known at present whether a similar hierarchical organization of both the cortex and corticocortical connections exists in the rat. The rat data that are currently available do not contradict this hierarchical theory, but much more information is certainly required before definitive conclusions can be drawn.

Other connections of prefrontal cortex

Connections with the rat PFC

In Tables IV and V the afferent and efferent connections of the rat PFC with (sub)cortical regions are summarized (excluding the isocortical areas and the thalamus). The tables do not allow specification of particular topographical distributions of the terminations within the regions indicated. Specific topography of the distributions has been described for several brain regions. For the basal ganglia the reader is referred to Groenewegen et al. (this volume); for the amygdala to e.g. McDonald (1987) and Groenewegen et al. (this volume); for the ventral tegmental area (VTA) to e.g. Kalsbeek et al. (1987); for the periaqueductal gray to e.g. Hardy (1986) and Sesack et al. (1989); for the dorsal raphe nucleus to O'Hearn and Molliver (1984); and for the solitary tract nucleus to Ruggiero et al. (1987) and Terreberry and Neafsey (1987).

Groenewegen et al. (this volume) review the involvement of different parts of the rat PFC in parallel circuits with particular parts of the basal ganglia, the thalamus and the amygdala. They currently distinguish 4 parallel circuits, viz. the "prelimbic circuit" (see their Fig. 11), the "dorsal agranular insular circuit" (see their Fig. 12), the "dorsal anterior cingulate circuit" and the "ventral agranular insular circuit". No data are yet available on the involvement of the ventral PFC and the frontal area 2 (Fr2) in anatomical circuits which involve the basal ganglia. As the connectivity data of Tables I – V point out, the different areas of the rat PFC are involved in many different circuits of e.g. motor systems, visual systems, olfactory system, sensorimotor system and visceral/autonomic systems. The extensive connectivity between the different PFC areas forms the basis of the integration of this information at the level of the prefrontal cortex.

Data on neurotransmitter systems together with data on corticocortical connections also show the unique (integrating and/or gating) position of the PFC. This is most evident for the dopamine system (see e.g. Kalsbeek et al., this volume). The neocortical projection of the major mesocortical dopamine system from the midbrain ventral tegmental area (VTA, A10) in particular terminates in the rat PFC. In the supragenual AC, this projection coincides with the dopaminergic projection from the pars compacta of the substantia nigra, although this projection is restricted to other layers, viz. layers III and II (see Kalsbeek et al., this volume). Until now, no other cortical projection areas for this substantia nigra nucleus have been described in the rat. The PL and lateral PFC have direct reciprocal connections with the VTA and the pars compacta of the substantia nigra. A second important indirect pathway to the VTA originates in ACd and PL. Fibers of this pathway originate in ACd and PL and make synaptic contacts with neurons in the medial part of the lateral habenula (Phillipson and Griffith, 1980; Greatrex and Phillipson, 1982). The projections from these cells run directly and indirectly, via the interpeduncular nucleus, to the pars compacta of substantia nigra and to the VTA (Thierry et al., 1983; Christoph et al., 1986; Nishikawa et al., 1986). This important second pathway is under influence of neurons in the serotonergic dorsal and median raphe nuclei in the brainstem and neurons in the cholinergic basal forebrain nuclei (Nishikawa et al., 1986; Satoh and Fibiger, 1986). Another example of multiple origin of a particular neurotransmitter innervation in the PFC is the cholinergic innervation. The entire neocortex, and also the lateral PFC and a large part of the medial PFC, receive a cholinergic projection from the nucleus basalis of the basal forebrain nuclei (e.g. Mesulam et al., 1983b; Luiten et al., 1987), but both the medial and lateral PFC also receive cholinergic projections from other cholinergic neurons, i.e. the ventral part of the vertical and part of the horizontal limb of Broca's band (Gaykema et al., 1990a). In addition, the medial PFC, especially the prelimbic region, appears to be the only cortical area with a clear cholinergic innervation from the midbrain laterodorsal tegmental nucleus, Ch6, and, to a lesser extent, from the pedunculopontine tegmental nucleus (Ch5) (Satoh and Fibiger, 1986).

The unique gating position of the PFC becomes especially clear with regard to its efferent connections with those cholinergic and monoaminergic nuclei that have a widespread efferent termination in the entire neocortex. The cholinergic nucleus basalis has afferent terminations in the entire neocortex, but according to Gaykema and Van Eden (in preparation, a PHA-L study) only the lateral PFC (AId,v) has a massive reciprocal connection with this nucleus. In addition, PL and IL reciprocate the projections from the nuclei of the diagonal band of Broca (Gaykema et al., 1990b). Apart from the small projection from the perirhinal cortex (Gaykema and Van Eden, in preparation), no direct recurrent projections from any of the 15 neocortical areas examined were found. This is in accordance with primate data (see section below), but seems to be at variance with the suggestion of Saper (1984) and Lemann and Saper (1985). The noradrenergic fibers from the locus coeruleus and the serotonergic fibers from the dorsal and median raphe nuclei are widely distributed over almost the entire cortex, including the PFC (see Parnavelas, this volume). As we have described above for the dopaminergic and cholinergic recurrent projections, the PFC is also the only cortical area which projects back to the locus coeruleus (mainly from AI: Cedarbaum and Aghajanian, 1978), and to the raphe nuclei (mainly from PL, ACd and AId,v: Aghajanian and Wang, 1977).

When we consider Tables I – V, it becomes evident that areas resembling the frontal eye field and premotor cortex can be distinguished in the rat PFC. For a review of the electrophysiological data on this aspect we refer to Neafsey et al. (1986a). Given its particular thalamic and corticocortical connections, the frontal eye field can be located in the caudal half of Fr2 and ACd (see Fig. 2). Both areas are connected with the visual cortex (see Fig. 4), and project to the superior colliculus and other brainstem nuclei related with eye movements (e.g. Sesack et al., 1989; Van Eden et al., 1990). In addition, it receives projections from that part of the MD which receives afferents from the superior colliculus (see Table II). On the other hand, ACd also appears to be involved in the limbic circuitry, in view of its connections with limbic regions in the forebrain (Fig. 4, Tables III and IV) and the interbrain (e.g. the mammillary nuclei; Table IV). Both the prelimbic region (including the infralimbic cortex) and the agranular insular region have quite extensive connections with limbic structures and with the autonomic and visceral centers, such as the solitary tract nucleus as well as several regions within the periaqueductal gray (Neafsey et al., 1986b; Ruggiero et al., 1987; Terreberry and Neafsey, 1987). We will not elaborate on these autonomic and visceral relations, since they are reviewed extensively in Neafsey (this volume).

Connections of the primate PFC

Tables VI and VII summarize the subcortical connections as far as they are known to us at present. As was the case in the previous tables, only connections as such are indicated, and not the presence or absence of topographical heterogeneity in the projection pattern. Topographical differentiation is known for instance within the caudate nucleus (e.g. Goldman and Nauta, 1977; Arikuni and Kubota, 1986), the amygdala (e.g. Price et al., 1987), the substantia innominata (Mesulam et al., 1983a; Russchen et al., 1985), the subthalamic nucleus (e.g. Alexander et al., this volume) and the periaqueductal gray.

In comparing Tables IV and VI, and Tables V and VII, it appears that there are no essential differences between rat and primate with regard to the PFC connectivity pattern with the areas

TABLE VI

Primate prefrontal cortical connections

	Cortical areas					
	10	8	46	9	24(25, 32)	11 – 14
Caudate nucleus		↓	↓	↓	↓	↓
Putamen			↓	↓	↓	
Ventral striatum					↓	↓
Lateral septum					↑	
Dorsal septum		↓			↑	
Ventral band of Broca (Ch 2)						
Horizontal band of Broca (Ch 3)					(↓)	↑(↓)
Nucleus basalis (Ch 4)	↑↓	↑↓	↑↓	↑↓	↑↓	↑↓
Claustrum		↑↓	↓	↓	↓	↓
Amygdala						
basolateral nuclei[a]			↑↓		↑↓	↑↓[b]
periamygdaloid cortex					·	·
Hippocampus						
subiculum				↑		
presubiculum				↑↓	(↑)↓	
Subthalamic nucleus	↓	↓	↓	↓	↓	↓
Lateral hypothalamus		↓	(↓)	(↓)		(·)

↑: corticopetal; ↓: corticofugal connections; small arrow indicates minor projections.
[a] especially the magnocellular basal nucleus, but also accessory basal nucleus.
[b] With the exception of area 11.
Ch1 – 6, groups of cholinergic neurons (Mesulam et al., 1983a).
For abbreviations see list at the end of this chapter. The cortical area numbers are displayed in Fig. 3.
Based mainly on Kievit and Kuypers (1975), Tanaka and Goldman (1976), Goldman and Nauta (1977), Künzle (1978), Takagi (1979), Baleydier and Mauguière (1980), Mufson and Mesulam (1982), Mesulam et al. (1983a), Amaral and Price (1984), Goldman-Rakic et al. (1984), Russchen et al. (1985), Leichnetz (1986), Kitt et al. (1987), Price et al. (1987).

studied. The information on the midbrain, however, is more detailed in the rat than in the primate. Therefore, a comparison of the different PFC areas in the rat and primate solely on the basis of the connections mentioned in Tables IV – VII will be rather imcomplete, with the exception of the frontal eye field. Both in rat and in primate this field is connected with pretectum and superior colliculus. In the primate the projections to the superior colliculus arise not only from the FEF, but also from the posterior part of area 46 and area 9 (see Fig. 3). These PFC neurons with projections to the superior colliculus coincide with PFC neurons that receive visual and auditory information and appear to be important for regulating visual attention, and for coding the information about an object's location in space, using eye fixation, eye movements and auditory information (Suzuki, 1985). The chapters by Alexander et al. and Groenewegen et al. (this volume) show a resemblance between primate and rat regarding circuits in which particular areas of the PFC are involved. However, the data on primate basal ganglion connections are not yet sufficiently detailed to permit a good comparison with the data presented by Groenewegen et al. (this volume).

There are many similarities in the topography of the amygdala connections. A similar amygdala

TABLE VII

Primate brainstem-prefrontal cortical connections

	Cortical areas					
	10	8	46	9	24(25, 32)	11 – 14
Retrorubral field (A8)	↑	(↓)	↑(↓)	↑(↓)	↑	·
SNc (A9)		↓	↑	↑↓	↑	
Ventral tegmental area (A10)	↑	↑	↑	↑	↑	↑
Periaqueductal gray	·	↓	↑(↓)	↓		↑
Pretectum		↓				
Superior colliculus		↓	↓	↓		
Pedunculopontine tegm. n. (Ch 5)	↑		↑	↑	(*)	(*)
Laterodorsal tegm. n. (Ch 6)	↑		↑		↑	↑
Parabrachial nucleus	↑		↑	↑	↑	—
Locus coeruleus (A6)	↑	↑	↑↓	↑↓	↑	↑
Reticular formation		↓		(↓)		
Dorsal raphe nucleus (B7)	↑	↑	↑↓	↑↓	↑	↑
Median raphe nucleus (B8)	↑	↑	↑↓	↓	↑	↑
Pontine nucleus		↓	↓	↓		

↑: corticopetal; ↓: corticofugal connections; small arrow indicates minor projections.
C1 – 6, groups of cholinergic neurons (Mesulam et al., 1983a); A2 – A7, noradrenergic cell groups (Dahlström and Fuxe, 1964); A8 – A10, dopaminergic cell groups (Dahlström and Fuxe, 1964);
B7 – 9, serotonergic cell groups (Dahlström and Fuxe, 1964).
For abbreviations see list at the end of this chapter.
Based mainly on Künzle (1978), Porrino and Goldman-Rakic (1982), Arnsten and Goldman-Rakic (1984); Huerta et al. (1986), Leichnetz (1986).

nucleus appears to be involved in these connections in both rat and primate, viz. the basolateral complex. The main cortical projections of the amygdala in the rat are to the medial and lateral PFC, the (anterior) insular and perirhinal cortices, and the primary olfactory cortex. This roughly resembles the connection pattern of the primate amygdala to the orbital and medial PFC, to a large part of the insula and to part of the temporal lobe (Price et al., 1987). In both rodents and primates, it is in the amygdala that the main afferent projection to the nucleus basalis originates.

In the monkey, too, the unique gating position of the PFC is apparent in the neocortical efferents to cholinergic and monoaminergic nuclei that have a distribution of terminals over the entire neocortex. As is the case in rat, the primate cholinergic nucleus basalis projects to the entire cortex (e.g. Russchen et al., 1985), but only the medial and orbital PFC and the rostral part of the temporal cortex have reciprocal connections with the substantia innominata (the nucleus basalis, and part of the horizontal limb of the diagonal band of Broca). It is in precisely these cortical areas that the majority of amygdala efferents to the neocortex terminate, which corroborates the rat data discussed in the previous section.

It is notable that the primate substantia innominata receives projections from many cells of the substantia nigra pars compacta, the ventral tegmental area and the retrorubral area (which contain the A9, A10 and A8 dopaminergic groups, respectively). It also receives projections from the serotonergic dorsal and median raphe nuclei, from the parabrachial nucleus (i.e. a major integrative center for visceral information) and from the central gray. Finally, it has reciprocal connections with the cholinergic pedunculopontine nucleus (Ch5) (Russchen et al., 1985).

In the rat, the noradrenergic distribution in the cerebral cortex is more or less uniform, as is the case for the serotonergic distribution. In the rhesus monkey cerebral cortex, the noradrenaline content is moderately high in the frontal lobe, reaches its maximum in the somatosensory cortex, and decreases to relatively low levels in the occipital cortex (Levitt et al., 1984; Lewis and Morrison, 1989), while the serotonergic innervation is densest in visual cortical areas, and decreases only slightly into rostral directions (Levitt et al., 1984; Berger et al., 1988). As in the rat, neocortical recurrent projections to the noradrenergic locus coeruleus region and the serotonergic dorsal and median raphe nuclei have only been found to originate in the PFC, i.e. the dorsomedial and dorsolateral PFC areas 9 and 46 (Arnsten and Goldman-Rakic, 1984).

The cortical expansion of some neurotransmitter systems in the primate is visible in Table VII. It appears that the cholinergic pedunculopontine and laterodorsal tegmental nuclei in the brainstem project to more (PFC) cortical areas than in the rat (Table V). The same appears to apply to the dopaminergic substantia nigra pars compacta and the dopaminergic retrorubral field (Table VII). Therefore, the fact that the dopaminergic distribution in the primate cerebral cortex is more widespread than in the rat (Berger et al., 1988; Gaspar et al., 1989), could also be caused by the extension of the substantia nigra and retrorubral projections. The highest dopamine content in primate cortex is found in the orbital and dorsolateral prefrontal cortex (Levitt et al., 1984). In addition, the highest density of dopaminergic afferents is found in the anterior cingulate (area 24), the primary motor and premotor areas (4 and 6) (Lewis et al., 1987; Berger et al., 1988).

Anatomical conclusions on rat and primate prefrontal cortex

From the previous sections it is evident that the presence of reciprocal connections with the mediodorsal nucleus of the thalamus by itself is not sufficient for defining a "homologous" PFC in rat and primate. For both groups it seems that those cortical areas with thalamic reciprocal connections which are densest with the MD can be regarded as PFC areas. Whether or not this criterion can be used to define the PFC in all mammalian species is not clear. In rat and primate, it coincides with the differential characteristics of corticocortical and other subcortical connections (see section concerned above), and with the evolutionary concepts concerning the cytoarchitecture of the different prefrontal areas (described in detail in Pandya and Yeterian, this volume, and summarized in the above-mentioned section). In the same way, the primate anterior cingulate cortex can be regarded as a PFC subfield. The present anatomical data further imply that in the rat the rostral part of Fr2 and the anterior cingulate area incorporate primate premotor characteristics. Taking into consideration also the functions in which this region is involved (described by Kolb, 1984, and this volume), the conclusion seems justified that premotor areas in the rat are not segregated from the prefrontal cortical areas. Moreover, the corticocortical connections of the rat rostral PFC (see Fig. 4) also encompass some similar connections shown in rhesus monkey between area 46, and the parietal and somatosensory cortices (see for rhesus monkey e.g. Selemon and Goldman-Rakic, 1988; Cavada and Goldman-Rakic, 1989; Preuss and Goldman-Rakic, 1989). These data do not warrant denoting the rat Fr2 as a pure premotor region distinct from the prefrontal cortex in the rat and analogous to the one in monkey (e.g. Passingham et al., 1988). Similarly, the caudal dorsal shoulder region (including Fr2) in the rat PFC certainly incorporates the frontal eye field characteristics, but, in view of both the functional aspects (e.g. Neafsey et al., 1986b) and the anatomical circuitry, also incorporates anterior cingulate characteristics (see e.g. Tables I – VII). Our tentative conclusion is that characteristics of the rat Fr2 area and some of the rat AC become more specialized in segregated primate areas 9, 46,

8, 6 and, possibly, 10, while the so-called cingulate characteristics of the rat AC have their equivalent in primate area 24. Insufficient data are available at present to draw definitive conclusions about the "homologies" of rat PL. The counterparts of some aspects (e.g. olfactory ones) of the rat PL and IL areas are probably found rather in the primate orbitofrontal areas (e.g. area 13, ventral part of area 12, see e.g. Potter and Nauta, 1979; Takagi, 1979), while other cingulate-related aspects may be represented in primate areas 25 and 32 (see Fig. 3 and Tables). Insular aspects of the rat agranular insular cortex (see Tables I – V; Neafsey, this volume) are certainly more specialized in the primate agranular and dysgranular insula (see e.g. Mesulam and Mufson, 1985). Some characteristics of the rat AI might nevertheless still correspond with those of the primate orbitofrontal PFC.

It can be concluded, therefore, that due to the fact that many anatomical and functional aspects are not segregated very well (see also Kaas, 1987, and this volume), there is an element of subjectivity in indicating different areas in the rat cortex. Another good example of an intermingling of different anatomical and functional aspects in the rat cortex is given by the detailed work of Donoghue et al. (1979) and Donoghue and Wise (1982). Donoghue et al. (1979) have shown that the so-called defining thalamic projections for the motor cortex and the somatosensory cortex, respectively, partially overlap in the rat (see their Fig. 12). Donoghue and Wise (1982) have also found in their microstimulation experiments that there is a partial functional overlap of the motor cortex and the somatosensory cortex (see their Fig. 9). This does not necessarily indicate that the overlapping motor or somatosensory cortices are inferior ones, but demonstrates that motor and somatosensory characteristics are partially intermingled in the rat cortex, while being segregated into specialized areas in primate cortex. A similar segregation appears to occur in the PFC. On the basis of the present data on thalamic, corticocortical and other subcortical connections, we consider the rat PFC to be made up of the areas displayed in Fig. 2, and the primate PFC of areas 8 – 14, 24, 25, 32 and 46, which are displayed in Fig. 3.

Cat prefrontal cortex

In previous sections we compared the rodent PFC with the primate PFC, since ample anatomical data are available for both groups of animals. Except for the fine studies of Reinoso-Suárez and coworkers (e.g. Cavada, 1984; Cavada and Reinoso-Suárez, 1985; Martínez-Moreno et al., 1987) relatively few studies have been done on the cat PFC (e.g. Markowitsch et al., 1978; Irle et al., 1984). The data available show that the anatomy of the cat PFC system differs in some respects from that in both rats and primates. For example, according to Robertson and Kaitz (1981) and Musil and Olson (1988), the cat AC receives its thalamic input mainly from the anteromedial nucleus (AM) and less so from the MD. In the cat no projection from the entorhinal cortex to the neocortex (Witter and Groenewegen, 1986), or from the amygdala to the MD (e.g. Russchen et al., 1987) has been found, in contrast to what is the case in rat and primate. Further research on the cat prefrontal cortical system will be of interest for establishing the differences and similarities. With regard to the prefrontal cortical system, the cat does not seem to be "in line" with the rat and primate PFC.

Quantitative comparison of prefrontal cortex in higher primates and rat

Since Brodmann's publication (1912) on the relative surface area of the granular prefrontal cortex (called regio frontalis) in different mammalian species (see Table VIII for a summary of his data), both the concept of the prefrontal cortex has changed (see above) and morphometric methods have been enormously improved (see Uylings et al., 1986a). In the previous section we described the PFC subareas in the rat (see e.g. Fig. 2) and primate brain (see e.g. Fig. 3). In the primate brain the granular frontal lobe areas plus the agranular anterior cingulate area and area 25 have been

TABLE VIII

% Surface area of regio frontalis[a] of total cerebral cortex according to Brodmann (1912)

Human	29[b]
Chimpanzee (anthropoid ape)	17
Macaque (*M. maurus*)	11
Marmoset	9
Dog	7
Cat	3
Rabbit	2
Rat	0

[a] Excluding regio precentralis/cingularis (Brodmann, 1909, p. 128).
[b] According to Sarkisov (1955) 23.5%, or 25% when areas 24 and 25 are included as well.

defined as the PFC (see Fig. 3). We have determined the relative volume of the PFC gray matter in 6 rats (both female and male, ranging from day 30 to day 90); in a marmoset, a New World monkey; in a pigtail monkey *(Macaca nemestrina)*, an Old World monkey; in an orang-utan, an anthropoid ape; and in 3 men (44, 70 and 76 years old, respectively, in whom no neurological disease had been detected).

The preparations of the rat brains are from the collection of the Netherlands Institute for Brain Research (NIBR), the preparations of the marmoset brain are from the collection of the Department of Anatomy of the Vrije Universiteit, Amsterdam (courtesy of Dr. M. Witter and Prof. H. Groenewegen), the preparations of the orang-utan were made by Zilles and Rehkämper (1988) from a brain of the C.U. Ariëns Kappers collection of the NIBR, the preparations of the brain of the 70-year-old man are from the collection of Prof. Zilles (Dept. of Anatomy, Cologne) and the preparations of the two other human brains were obtained from the division of neuropathology of the Erasmus University, Rotterdam (courtesy of Dr. J.M. Kros).

For the delineation of the rat PFC in the Nissl-stained sections, the description of Van Eden and Uylings (1985) was used, with the description of Krettek and Price (1977a) added for the ventral PFC areas. At the rostral tip, the lateral border of Fr2 in the coronal sections was characterized by transition of layer II to layer I. The layer II cells formed a more serrated boundary with layer I in the region lateral to Fr2. These cytoarchitectonic borders parallel the border of the MD projections. The delineation of the marmoset, macaque, orang-utan and human PFCs was determined with the aid of the descriptions of Von Bonin and Bailey (1947); Peden and Von Bonin (1947); Bailey et al. (1950); Sarkisov (1955); Braak (1980); Stephan et al. (1980); Vogt et al. (1987); Barbas and Pandya (1989). The volume was determined with Cavalieri's estimator, i.e. by systematic sampling and cross-sectional area measurements of the prefrontal cortex and isocortex, i.e. neocortex and mesocortex (e.g. Uylings et al., 1986b). The area measurements were obtained with a superimposed grid test system, in the microscope or on standardized photographs (as illustrated in Fig. 5 in Gundersen et al., 1988). For estimating the volume of the PFC relative to that of the isocortex, the ratio is used of the number of grid points covering the PFC and that of those covering the entire isocortex (see for further description Uylings et al., 1986b, and e.g., Gundersen et al., 1988). This is a ratio of 2 dependent variables, viz. PFC volume and entire isocortex volume, for which reason the jackknife procedure was applied to obtain a least-biased estimate of the mean value of the ratios in rat and man, along with a confidence interval (Cochran, 1977, sections 6.15 – 6.17).

Fig 5 shows our estimate of the relative PFC volumes in higher primates and rats. Although we need to extend the number of specimens, especially for the different monkey species examined, the available data enable us to present a first comparison. The increase of the relative PFC volume in the different primate species examined, reaching 30% in the human brain, confirms the general trend indicated for primates in Brodmann's table (Table VIII). On the other hand, the comparatively high value found for the relative PFC volume in the rat (24% of total isocortex) is, at first sight, conspicuous and unexpected. Thus the question arises of whether the rat really is a special(ized) "prefrontal" animal. As has been explained elsewhere (e.g. Uylings et al., 1986b, 1987), ratios are often inappropriate measures for making comparisons. They may actually be used only if a linear isometric relation between the two variables exists (i.e. if the slope of the regression line for these variables is 1.00) *and* if the regression line passes through the origin of the two axes. For a better interpretation of the data on relative volume, we display the regression analysis of the absolute data in Figs. 6 – 8. We have applied the standard major axis method (Hofman et al., 1986) to data based on our own measurements and on the literature (including those of Frahm et al., 1982; Zilles and Rehkämper, 1988). Fig. 6 shows logarithmic values of the volume of PFC gray matter vs. the volume of the isocortical gray matter, which appear highly correlated ($r = 0.995$). The slope of the regression

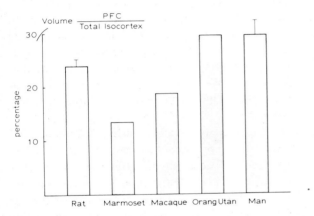

Fig. 5. A histogram of the ratio of prefrontal cortex volume and total isocortex volume in rat and several higher primate species. The isocortex is the neocortex and mesocortex.

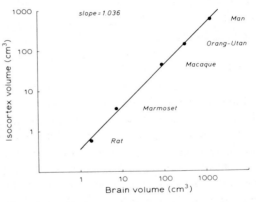

Fig. 7. The bivariate graph of isocortex volume and total brain volume in the species studied, together with the regression line, which resulted from the regression analysis (see text for further explanation).

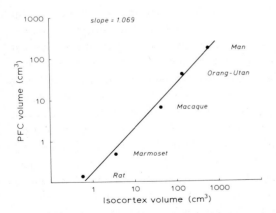

Fig. 6. The bivariate graph of PFC volume and isocortex volume in the species studied, together with the regression line, which resulted from the regression analysis (see text for further explanation).

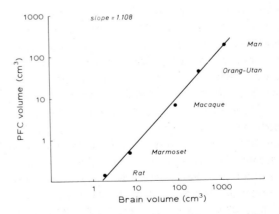

Fig. 8. The bivariate graph of PFC volume and total brain volume in the species studied, together with the regression line, which resulted from the regression analysis (see text for further explanation).

line is 1.07, just failing to be significantly above 1.00, i.e. the value for isometric increase. From Fig. 6, it follows that the use of ratios is not allowed in comparing the relative PFC size, since the regression line does not pass through the intersection of the two axes. The relations between logarithmic values of isocortical gray volume and entire brain volume displayed in Fig. 7 indicate a different regression relation between PFC volume and the entire brain volume. In Fig. 7 the values are very strongly correlated ($r = 0.999$), while the slope of the regression line is 1.04 and does not deviate significantly from the 1.00 slope of an isometric line. Fig. 8 shows the regression relation for the logarithmic values of PFC gray matter volume vs. brain volume. They are very strongly correlated ($r = 0.999$) and the slope of this regression line, 1.11, is significantly higher than 1.00. This implies a disproportionally large increase in PFC volume with respect to total brain volume, which means that humans have the largest relative PFC volume. Since the number of species studied is rather limited, we have to compare our findings with data reported in the literature. This is possible only for the relation between total brain volume and isocortical gray volume. Our isometric regression relation (Fig. 7) is in agreement with the findings of Harman (1947) and Schlenska (1974), but at variance with the relation between total brain volume and neocortical gray volume found by Frahm et al. (1982). The latter emphasize the use of bivariate regression, preferably within a family or subfamily. In this context, a more particular question is whether the PFC bivariate point of the rat deviates significantly from the regression relation of the group of higher primates. For the primate group, the regression of the logarithmic values of isocortex volume and brain volume is represented by an almost isometric line (slope is 0.981 and $r = 0.9998$), which is in agreement with the results of both Schlenska (1974) and Frahm et al. (1982). The logarithmic values of PFC volume and brain volume within the primate group are also strongly correlated ($r = 0.9996$, see Fig. 8) and the bivariate point for the rat lies within the confidence interval of the primate line, and does not deviate significantly from this primate regression line (for statistics used see Uylings et al., 1986b, 1987). In this regard, the volume of the rat PFC relative to brain volume is not special, and the above-mentioned conclusion of a disproportionally large increase in PFC volume with respect to total brain volume remains valid. On the other hand, the bivariate point for the rat deviates significantly from the primates' regression relation of the logarithmic values of PFC and isocortex volume. This means that we cannot arrive at a clear statement for the PFC size in relation to isocortex volume in the rat. It is evident we need data on more non-human primate species to be more certain that the rat is significantly different in this respect. We also need more data on the relation of the PFC and isocortex volume in mammalian orders "lower" than primates. In view of the difficulties we already have in defining the PFC in the rat (see previous sections), it is not feasible at present to delineate the PFC in many other mammals "lower" than primates. For the time being, therefore, the question of the size of the rat PFC relative to its total brain size seems to be answered. After *allometric scaling* it appears that the rat PFC does not deviate from the primate PFC with regard to its proportion of the total brain volume. However, the question of rat PFC size relative to its isocortical gray volume is still open. The answer to this latter question is probably confounded by the overlapping of cortical fields in the rat, as explained in previous sections. In these sections we have suggested that, with regard to the anatomical connectivity wiring diagram and their functions, cortical areas are much less segregated in rat than in higher primates. Therefore, the prefrontal cortex can be relatively larger in rat, although its absolute size is far less developed. This is illustrated by Fig. 9, in which each of the 3 open squares are proportionally larger in the left diagram than in the right one, although the absolute size of the open squares is increased in the right diagram (up to 3.4 times as large).

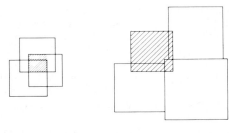

Lower mammals   Higher mammals

Fig. 9. A diagram of cortical "evolution", including both increase in cortical size and further segregation of cortical fields (see text). The area indicated with oblique lines is completely overlapped in the left figure, which might lead to the unjustified impression that a new cortical area appears in the right figure.

In conclusion, comparing the different connectivity patterns and the relative PFC size in rat and primates, the evolution of the PFC is most probably a result both of an increase in cortical size and of segregation and specialization of the cortical regions. This process of expansion and segregation is not unique to cortical "evolutionary" development, since similar processes also occur in cortical ontogenetic development, be it on a different scale. An illustrative example is the segregation and extension of the cortical ocular dominance columns after birth (e.g. Hubel et al., 1977; Hitchcock and Hickey, 1980), which can coincide with elimination of neuronal structures that have become exuberant during development (see e.g. Huttenlocher et al., 1982; and for review Ebbeson, 1984; Finlay et al., 1987; Fawcett, 1988; Innocenti, 1988; Uylings et al., 1990).

Acknowledgements

The authors are indebted to Mr. A. Janssen for correcting the English and styling and typing the manuscript, to Mr. G. van der Meulen for his assistance in photographing preparations etc. and to Mr. H. Stoffels for the artwork.

List of abbreviations

ACd	dorsal anterior cingulate cortex
AChE	acetylcholinesterase
ACv	ventral anterior cingulate cortex
AD	anterodorsal nucleus of the thalamus
AGm	medial agranular cortex = Fr2
AGl	lateral agranular cortex = Fr1
AId	dorsal agranular insular cortex
AIp	posterior agranular insular cortex
AIv	ventral agranular insular cortex
AM	anteromedial nucleus of the thalamus
AO	anterior olfactory nucleus
APA	(post)arcuate premotor area
AV	anteroventral nucleus of the thalamus
A2-A7	noradrenergic cell groups (Dahlström and Fuxe, 1964)
A8-A10	dopaminergic cell groups (Dahlström and Fuxe, 1964)
A10	midbrain ventral tegmental area
B7-9	serotonergic cell groups (Dahlström and Fuxe, 1964)
CA_1	cornu ammonis field 1
Ch1-6	groups of cholinergic neurons (Mesulam et al., 1983a,b)
C2	an adrenergic cell group (Hökfelt et al., 1974; Howe et al., 1980)
Ent.	entorhinal cortex
FEF	prearcuate frontal eye field
FL	forelimb area (Zilles, 1985)
Fr1	primary motor cortex (Zilles, 1985)
Fr2	frontal area 2 = PrCm = AGm
Fr3	frontal area 3 (Zilles, 1985)
F2	premotor area as defined by Matelli et al. (1989)
F5	premotor area as defined by Matelli et al. (1989)
Hip	hippocampus
HL	hindlimb area (Zilles, 1985)
IL	infralimbic cortex
LD	lateral dorsal nucleus
LHb	lateral habenula
LO	lateral orbital cortex
LP	lateral posterior complex of the thalamus
MD	mediodorsal nucleus of the thalamus
MDc	central part of the rat MD
MDdc	posterior densocellular part of MD
MDl	lateral part of the rat MD
MDm	medial part of the rat MD
MDmc	medial magnocellular part of the MD
MDmf	paralamellar, multiform part of the MD
MDpc	lateral parvocellular part of the MD
MDpl	paralaminar part of the rat MD
MDv	ventral part of the rat MD
mHb	medial habenala
MO	medial orbital cortex
mTVB	medial ventrobasal nucleus = VMb
NTS	solitary tract nucleus
Oc 1	primary visual cortex (Zilles, 1985)
Oc 2L	lateral part occipital area 2 (Zilles, 1985)
Oc 2M	medial part occipital area 2 (Zilles, 1985)

Par1 (dysgr)	dysgranular part within parietal area 1 (Zilles, 1985)
Par2	parietal area 2, supplementary somatosensory cortex (Zilles, 1985)
PFC	prefrontal cortex
PHA-L	Phaseolus vulgaris leucoagglutinin
Pir.	prepyriform cortex, primary olfactory cortex (Zilles, 1985)
PL	prelimbic cortex
PrCm	medial precentral area = Fr2
PRh	perirhinal cortex (Zilles, 1985)
PS	principal sulcus
PV	paraventricular nucleus of the (epi)thalamus
RSA	agranular retrosplenial cortex (Zilles, 1985)
RSG	granular retrosplenial cortex (Zilles, 1985)
sm	stria medullaris
SMA	supplementary motor area
SNc	substantia nigra pars compacta
Te2	temporal area 2 (Zilles, 1985)
VA	ventral anterior nucleus of the thalamus
VAdc	densocellular part of VA (Ilinsky and Kultas-Ilinsky, 1987) = VLa
VL	ventral lateral nucleus of the thalamus
VLa	anterior part of VL (Jones, 1985)
VLm	medial part of VL (Olszewski, 1952) = VLa
VLo	oral part of VL (Olszewski, 1952) = VAdc = VLa
VLO	ventral lateral orbital cortex
VLp	posterior nucleus of the ventral lateral complex (Jones, 1985)
VMb	basal ventral medial nucleus (Jones, 1985)
VMp	"principal" ventral medial nucleus (Jones, 1985)
VO	ventral orbital cortex
VPI	ventral posterior inferior nucleus of the thalamus
VPLo	oral part of ventral posterior lateral nucleus (Olszewski, 1952) = VLp
VPMpc	parvocellular part of ventroposterior medial nucleus = VMb
VTA	ventral tegmental area
WGA-HRP	wheat germ agglutinin-horseradish peroxidase
X	thalamic area X (Olszewski, 1952) = VLp
6m	rostromedial part of area 6

References

Afsharpour, S. (1985) Topographical projections of the cerebral cortex to the subthalamic nucleus. *J. Comp. Neurol.*, 236: 14 – 28.

Aggleton, J.P. and Mishkin, M. (1984) Projections of the amygdala to the thalamus in the cynomolgus monkey. *J. Comp. Neurol.*, 222: 56 – 68.

Aghajanian, G.K. and Wang, R.Y. (1977) Habenular and other midbrain raphe afferents demonstrated by a modified retrograde tracing technique. *Brain Res.*, 306: 9 – 18.

Akert, K. and Hartmann-Von Monakov, K. (1980) Relationships of precentral, premotor, and prefrontal cortex to the mediodorsal and intralaminar nuclei of the monkey thalamus. *Acta Neurol. Exp.*, 40: 7 – 25.

Alexander, G.E., Crutcher, M.D. and DeLong, M.R. (1990) Basal ganglia-thalamocortical circuits: parallel substrates for motor oculomotor, "prefrontal" and "limbic" functions. *This volume*, Ch. 6.

Amaral, D.G. and Price, J.L. (1984) Amygdalo-cortical projections in the monkey *(Macaca fascicularis)*. *J. Comp. Neurol.*, 230: 465 – 496.

Arikuni, T. and Kubota, K. (1986) The organization of prefronto-caudate projections and their laminar origin in the macaque monkey: a retrograde study using HRP-gel. *J. Comp. Neurol.*, 244: 492 – 510.

Arnsten, A.F.T. and Goldman-Rakic, P.S. (1984) Selective prefrontal cortical projections to the region of the locus coeruleus and raphe nuclei in the rhesus monkey. *Brain Res.*, 306: 9 – 18.

Audinat, E., Condé, F. and Crépel, F. (1988) Cortico-cortical connections of the limbic cortex of the rat. *Exp. Brain Res.*, 69: 439 – 443.

Bailey, P., Bonin, G. von and McCullough, W.S. (1950) *The Isocortex of the Chimpanzee*, University of Illinois Press, Urbana, IL, 440 pp.

Baleydier, C. and Mauguière, F. (1980) The duality of the cingulate gyrus in monkey. Neuroanatomical study and functional hypothesis. *Brain*, 103: 525 – 554.

Barbas, H. and Pandya, D.N. (1989) Architecture and intrinsic connections of the prefrontal cortex in the rhesus monkey. *J. Comp. Neurol.*, 286: 353 – 375.

Beckstead, R.M. (1979) An autoradiographic examination of corticocortical and subcortical projections of the mediodorsal-projection (prefrontal) cortex in the rat. *J. Comp. Neurol.*, 184: 43 – 62.

Beckstead, R.M., Morse, J.R. and Norgren, R. (1980) The nucleus of the solitary tract in the monkey: projections to the thalamus and brain stem nuclei. *J. Comp. Neurol.*, 190: 259 – 282.

Benjamin, R.M. and Golden, G.T. (1985) Extent and organization of opossum prefrontal cortex defined by anterograde and retrograde transport methods. *J. Comp. Neurol.*, 238: 77 – 91.

Berger, B., Trottier, S., Verney, C., Gaspar, P. and Alvarez, C. (1988) Regional and laminar distribution of the dopamine and serotonin innervation in the macaque cerebral cortex: a radioautographic study. *J. Comp. Neurol.*, 273: 99 – 119.

Björklund, A. and Lindvall, O. (1984) Dopamine containing systems in the CNS. In A. Björklund and T. Hökfelt (Eds.), *Classical Transmitters in the CNS, Handbook of Chemical Neuroanatomy*, Vol. 2, Part I, Elsevier, Amsterdam, pp. 55 – 122.

Braak, H. (1980) *Architectonics of the Human Telencephalic*

Cortex, Springer, Berlin, 147 pp.
Brodmann, K. (1909) *Vergleichende Lokalisationslehre der Grosshirnrinde,* Barth-Verlag, Leipzig, 324 pp.
Brodmann, K. (1912) Neue Ergebnisse über die vergleichende histologische Lokalisation der Grosshirnrinde mit besondere Berücksichtigung des Stirnhirns. *Suppl. Anat Anz.,* 41: 157–216.
Campbell, C.B.G. and Hodos, W. (1970) The concept of homology and the evolution of the nervous system. *Brain Behav. Evol.,* 3: 353–367.
Cavada, C. (1984) Transcortical sensory pathways to the prefrontal cortex with special attention to the olfactory and visual modalities. In F. Reinoso-Suárez and C. Ajmone-Marsan (Eds.), *Cortical Integration,* Raven Press, New York, pp. 317–328.
Cavada, C. and Goldman-Rakic, P.S. (1989) Postorior parietal cortex in rhesus monkey. II. Evidence for segregated corticocortical networks linking sensory and limbic areas with the frontal lobe. *J. Comp. Neurol.,* 287: 422–445.
Cavada, C. and Reinoso-Suárez, F. (1985) Topographical organization of the cortical afferent connections of the prefrontal cortex in the cat. *J. Comp. Neurol.,* 242: 293–324.
Caviness, V.S. and Frost, D.O. (1980) Tangential organization of thalamic projections to the neocortex in the mouse. *J. Comp. Neurol.,* 194: 335–367.
Cedarbaum, J.M. and Aghajanian, G.K. (1978) Afferent projections to the rat locus coeruleus as determined by a retrograde tracing technique. *J. Comp. Neurol.,* 178: 1–16.
Christoph, G.R., Leonzio, R.J. and Wilcox, K.S. (1986) Stimulation of the lateral habenula inhibits dopamine-containing neurons in the substantia nigra and ventral tegmental area. *J. Neurosci.,* 6: 613–619.
Cochran, W.G. (1977) *Sampling Techniques,* 3rd Edn., Wiley, New York, 428 pp.
Cornwall, J. and Phillipson, O.T. (1988) Mediodorsal and reticular thalamic nuclei receive collateral axons from prefrontal cortex and laterodorsal tegmental nucleus in the rat. *Neurosci. Lett.,* 88: 121–126.
Dahlström, A. and Fuxe, K. (1964) Evidence for the existence of monoamine-containing neurons in the central nervous system. I. Demonstration of monoamines in the cell bodies of brain stem neurons. *Acta Physiol. Scand.,* Suppl. 62: 1–55.
Dinopoulos, A., Karamanlidis, A.N., Papadopoulos, G., Antonopoulos, J. and Michaloudi, H. (1985) Thalamic projections to motor, prefrontal and somatosensory cortex in the sheep studied by means of the horseradish peroxidase retrograde transport method. *J. Comp. Neurol.,* 241: 63–81.
Divać, I., Kosmal, A., Björklund, A. and Lindvall, O. (1978) Subcortical projections to the prefrontal cortex in the rat as revealed by the horseradish peroxidase technique. *Neuroscience,* 3: 785–796.

Domesick, V.B. (1972) Thalamic relationships of the medial cortex in the rat. *Brain Behav. Evol.,* 6: 457–483.
Donoghue, J.P. and Parham, C. (1983) Afferent connections of the lateral agranular field of the rat motor cortex. *J. Comp. Neurol.,* 217: 390–404.
Donoghue, J.P. and Wise, S.P. (1982) The motor cortex of the rat: cytoarchitecture and microstimulation mapping. *J. Comp. Neurol.,* 212: 76–88.
Donoghue, J.P., Kerman, K.L. and Ebner, F.F. (1979) Evidence for two organizational plans within the somatic sensory-motor cortex of the rat. *J. Comp. Neurol.,* 183: 647–664.
Ebbeson, S.O.E. (1984) Evoluation and ontogeny of neural circuits. *Behav. Brain Sci.,* 7: 321–366.
Fawcett, J.W. (1988) Retinotopic maps, cell death, and electrical activity in the retinotectal and retinocollicular projections. In J.G. Parnavelas, C.D. Stern and R.V. Stirling (Eds.), *The making of the Nervous System,* Oxford University Press, Oxford, pp. 319–339.
Finlay, B.L., Wikler, K.C. and Sengelaub, D.R. (1987) Regressive events in brain development and scenerados for vertebrate brain evolution. *Brain Behav. Evol.* 301: 102–117.
Frahm, H.D., Stephan, H. and Stephan, M. (1982) Comparison of brain structure in insectivora and primates. I. Neocortex. *J. Hirnforsch.,* 23: 375–389.
Gaspar, P., Berger, B., Febvret, A., Vigny, A. and Henry, J.P. (1989) Catecholamine innervation of the human cerebral cortex as revealed by comparative immunohistochemistry of tyrosine hydroxylase and dopamine-beta-hydroxylase. *J. Comp. Neurol.,* 279: 249–271.
Gaykema, R.P.A., Luiten, P.G.M., Nyakas, C. and Traber, J. (1990a) Cortical projection patterns of the medial septum–diagonal band complex. *J. Comp.Neurol.,* 293: 103–124.
Gaykema, R.P.A., Van Weeghel, R., Hersh, L.B. and Luiten, P.G.M. (1990b) Prefrontal cortical projections to the cholinergic neurons in the basal forebrain. *J. Comp. Neurol.,* in press.
Giguere, M. and Goldman-Rakic, P.S. (1988) Mediodorsal nucleus: areal, laminar and tangential distribution of afferents and efferents in the frontal lobe of rhesus monkeys. *J. Comp. Neurol.,* 277: 195–213.
Goldman, P.S. and Nauta, W.J.H. (1977) An intricately patterned prefrontocaudate projection in the rhesus monkey. *J. Comp. Neurol.,* 171: 369–386.
Goldman-Rakic, P.S. and Porrino, L.J. (1985) The primate mediodorsal (MD) nucleus and its projection to the frontal lobe. *J. Comp. Neurol.,* 242: 535–560.
Goldman-Rakic, P.S., Selemon, L.D. and Schwartz, M.L. (1984) Dual pathways connecting the dorsolateral prefrontal cortex with the hippocampal formation and parahippocampal cortex in the rhesus monkey. *Neuroscience,* 12: 719–743.
Gower, E.C. (1989) Efferent projections from limbic cortex of

the temporal pole to the magnocellular medial dorsal nucleus in the rhesus monkey. *J. Comp. Neurol.,* 280: 343–358.

Greatrex, R.M. and Phillipson, O.T. (1982) Demonstration of synaptic input from prefrontal cortex to the habenula in the rat. *Brain Res.,* 238: 192–197.

Groenewegen, H.J. (1988) Organization of the afferent connections of the mediodorsal thalamic nucleus in the rat, related to the mediodorsal-prefrontal topography. *Neuroscience,* 24: 379–431.

Groenewegen, H.J. Berendse, H.W., Wolters, J.G. and Lohman, A.H.M. (1990) The anatomical relationship of the prefrontal cortex with the striatopallidal system, the thalamus and the amygdala: evidence for a parallel organization. *This Volume,* Ch. 5.

Gundersen, H.J.G., Bendtsen, T.F., Korbo, L., Marcussen, N., Møller, A., Nielsen, K., Nyengaard, J.R., Pakkenberg, B., Sørensen, F.B., Vesterby, A. and West, M.J. (1988) Some new, simple and efficient stereological methods and their use in pathological research and diagnosis. *Acta Pathol. Microbiol. Scand., C, Immunol.,* 96: 379–394.

Hardy, S.G.P. (1986) Projections to the midbrain from the medial versus lateral prefrontal cortices of the rat. *Neurosci. Lett.,* 63: 159–164.

Harman, P.J. (1947) On the significance of fissuration of the isocortex. *J. Comp. Neurol.,* 87: 161–168.

Herkenham, M. (1979) The afferent and efferent connections of the ventromedial thalamic nucleus in the rat. *J. Comp. Neurol.,* 183: 487–518.

Herkenham, M. (1980) Laminar organization of thalamic projections to the rat neocortex. *Science,* 207: 532–535.

Herkenham (1986) New perspectives on the organization and evolution of nonspecific thalamocortical projections. In E.G. Jones and A. Peters (Eds.), *Sensory-Motor Areas and Aspects of Cortical Connectivity. Cerebral Cortex, Vol. 5,* Plenum Press, New York, pp. 403–445.

Hitchcock, P.F. and Hickey, T.L. (1980) Ocular dominance columns: evidence for their presence in humans. *Brain Res.,* 182: 176–179.

Hofman, M.A., Laan, A.C. and Uylings, H.B.M. (1986) Bivariate linear models in neurobiology: problems of concept and methodology. *J. Neurosci. Methods,* 18: 103–114.

Hökfelt, T., Fuxe, K., Goldstein, M. and Johansson, O. (1974) Immunohistochemical evidence for the existence of adrenaline neurons in the rat brain. *Brain Res.,* 66: 235–251.

Howe, P.R.C., Costa, M., Furness, J.B. and Chalmers, J.P. (1980) Simultaneous demonstration of phenylethanolamine N-methyltransferase immunofluorescent and catecholamine nerve cell bodies in the rat medulla oblongata. *Neuroscience,* 5: 2229–2238.

Hubel, D.H., Wiesel, T.N. and LeVay, S. (1977) Plasticity of ocular dominance columns in monkey striate cortex. *Phil. Trans. R. Soc. B,* 278: 377–409.

Huerta, M.F., Krubitzer, L.A. and Kaas, J.H. (1986) Frontal eye field as defined by intracortical microstimulation in squirrel monkeys, owl monkeys, and macaque monkeys. I. Subcortical connections. *J. Comp. Neurol.,* 253: 415–439.

Huerta, M.F., Krubitzer, L.A. and Kaas, J.H. (1987) Frontal eye field as defined by intracortical microstimulation in squirrel monkeys, owl monkeys, and macaque monkeys. II. Cortical connections. *J. Comp. Neurol.,* 265: 332–361.

Huttenlocher, P.R., Courten, C. de, Garey, L.J. and Van der Loos, H. (1982) Synaptogenesis in human visual cortex – evidence for synapse elimination during normal development. *Neurosci. Lett.,* 33: 247–252.

Ilinsky, I.A. and Kultas-Ilinsky, K. (1987) Sagittal cytoarchitectonic maps of the *Macaca mulatta* thalamus with a revised nomenclature of the motor-related nuclei validated by observations on their connectivity. *J. Comp. Neurol.,* 262: 331–364.

Ilinsky, I.A., Jouandet, M.L. and Goldman-Rakic, P.S. (1985) Organization of the nigrothalamocortical system in the rhesus monkey. *J. Comp. Neurol.,* 236: 315–330.

Innocenti, G.M. (1988) Loss of axonal projections in the development of the mammalian brain. In J.G. Parnavelas, C.D. Stern and R.V. Stirling (Eds.), *The Making of the Nervous System,* Oxford University Press, Oxford, pp. 319–339.

Irle, E., Markowitsch, H.J. and Streicher, M. (1984) Cortical and subcortical, including sensory-related, afferents to the thalamic mediodorsal nucleus of the cat. *J. Hirnforsch.,* 25: 29–51.

Jones, E.G. (1985) *The Thalamus,* Plenum Press, New York, 935 pp.

Jones, E.G., Wise, S.P. and Coulter, J.D. (1979) Differential thalamic relationships of sensory-motor and parietal cortical fields in monkeys. *J. Comp. Neurol.,* 183: 833–882.

Jürgens, U. (1984) The efferent and afferent connections of the supplementary motor area. *Brain Res.,* 300: 63–81.

Kaas, J.H. (1987) The organization and evolution of neocortex. In S.P. Wise (Ed.), *Higher Brain Functions: Recent Explorations of the Brain's Emergent Properties,* John Wiley, New York, pp. 347–378.

Kaas, J.H. (1990) How sensory cortex is subdivided in mammals: implications for studies in prefrontal cortex. *This Volume,* Ch. 1.

Kalsbeek, A., Buijs, R.M., Hofman, M.A., Matthijssen, M.A.H., Pool, C.W. and Uylings, H.B.M. (1987) Effects of neonatal thermal lesioning of the mesocortical dopaminergic projection on the development of the rat prefrontal cortex. *Dev. Brain Res.,* 32: 123–132.

Kalsbeek, A., De Bruin, J.P.C., Feenstra, M.G.P. and Uylings H.B.M. (1990) Age-dependent effects of lesioning the mesocortical dopamine system upon prefrontal cortex morphometry and PFC-related behaviors. *This Volume,* Ch. 12.

Kievit, J. and Kuypers, H.G.J.M. (1977) Organization of the thalamocortical connections to the frontal lobe in the rhesus monkey. *Exp. Brain Res.,* 29: 299–322.

Kitt, C.A., Mitchell, S.J., DeLong, M.R., Wainer, B.H. and

Price, D.L. (1987) Fiber pathways of basal forebrain cholinergic neurons in monkeys. *Brain Res.*, 406: 192–206.

Kolb, B. (1984) Functions of the frontal cortex of the rat: a comparative review. *Brain Res. Rev.*, 8: 65–98.

Kolb, B. (1990) Animal models for human PFC-related disorders. *This Volume*, Ch.

Kosar, E., Grill, H.J. and Norgren, R. (1986) Gustatory cortex in the rat. II. Thalamocortical projections. *Brain Res.*, 379: 342–352.

Kostović, I. and Goldman-Rakic, P.S. (1983) Transient cholinesterase staining in the mediodorsal nucleus of the thalamus and its connections in the developing human and monkey brain. *J. Comp. Neurol.*, 219: 431–447.

Krettek, J.E. and Price, J.L. (1977a) The cortical projections of the mediodorsal nucleus and adjacent thalamic nuclei in the rat. *J. Comp. Neurol.*, 171: 157–192.

Krettek, J.E. and Price, J.L. (1977b) Projections from the amygdaloid complex to the cerebral cortex and thalamus in the rat and cat. *J. Comp. Neurol.*, 172: 687–722.

Krushel, L.A. and Van der Kooy, D (1988) Visceral cortex: integration of the mucosal senses with limbic information in the rat agranular insular cortex. *J. Comp. Neurol.*, 270: 39–54.

Künzle, H. (1978) An autoradiographic analysis of the efferent connections from premotor and adjacent prefrontal regions (areas 6 and 9) in *Macaca fascicularis*. *Brain Behav. Evol.*, 15: 185–234.

Leichnetz, G.R. (1986) Afferent and efferent connections of the dorsolateral precentral gyrus (area 4, hand/arm region) in the macaque monkey, with comparisons to area 8. *J. Comp. Neurol.*, 254: 460–492.

Lemann, W. and Saper, C.B. (1985) Evidence for a cortical projection to the magnocellular basal nucleus in the rat: an electron microscopic axonal transport study. *Brain Res.*, 334: 339–343.

Leonard, C.M. (1969) The prefrontal cortex of the rat. I. Cortical projection of the mediodorsal nucleus. II. Efferent connections. *Brain Res.*, 12: 321–343.

Levitt, P., Rakic, P. and Goldman-Rakic, P. (1984) Comparative assessment of monoamine afferents in mammalian cerebral cortex. In L. Descarries. T.A. Reader and H.H. Jasper (Eds.), *Monoamine Innervation of Cerebral Cortex*, Alan R. Liss, New York, pp. 41–59.

Lewis, D.A. and Morrison, J.H. (1989) Noradrenergic innervation of monkey prefrontal cortex: a dopamine-β-hydroxylase immunocytochemical study. *J. Comp. Neurol.*, 282: 317–330.

Lewis, D.A., Campbell, M.J., Foote, S.L., Goldstein, M. and Morrison, J.H. (1987) The distribution of tyrosine hydroxylase-immunoreactive fibers in primate neocortex is widespread but regionally specific. *J. Neurosci.*, 7: 279–290.

Luiten, P.G.M., Gaykema, R.P.A., Traber, J. and Spencer, D.G. Jr. (1987) Cortical projection patterns of magnocellular basal nucleus subdivisions as revealed by anterogradely transported *Phaseolus vulgaris* leucoagglutinin. *Brain Res.*, 413: 229–250.

Mantyh, P.W. (1983a) Connections of midbrain periaqueductal gray in the monkey. I. Ascending efferent projections. *J. Neurophysiol.*, 49: 567–581.

Mantyh, P.W. (1983b) The spinothalamic tract in the primate: a re-examination using wheatgerm agglutinin conjugated to horseradish peroxidase. *Neuroscience*, 9: 847–862.

Markowitsch, H.J. (1988) Anatomical and functional organization of the primate prefrontal cortical system. In H.D. Steklis and J. Erwin (Eds.), *Neurosciences, Comparative Primate Biology, Vol. 4*, Alan R. Liss, New York, pp. 99–153.

Markowitsch, H.J. and Pritzel, M. (1981) Prefrontal cortex of the guinea pig *(Cavia porcellus)* defined as cortical projection area of the thalamic mediodorsal nucleus. *Brain Behav. Evol.*, 18: 80–95.

Markowitsch, H.J., Pritzel, M. and Divac, I. (1978) The prefrontal cortex of the cat: Anatomical subdivisions based on retrograde labeling of cells in the mediodorsal thalamic nucleus. *Exp. Brain Res.*, 32: 335–344.

Martínez-Moreno, E., Llamas, A., Avendaño, C., Renes, E. and Reinoso-Suárez, F. (1987) General plan of the thalamic projections to the prefrontal cortex in the cat. *Brain Res.*, 407: 17–26.

Matelli, M., Luppino, G., Fogassi, L. and Rizzolatti, G. (1989) Thalamic input to inferior area 6 and area 4 in the macaque monkey. *J. Comp. Neurol.*, 280: 468–488.

McDonald, A.J. (1987) Organization of amygdaloid projections to the mediodorsal thalamus and prefrontal cortex: a fluorescence retrograde transport study in the rat. *J. Comp. Neurol.*, 262: 46–58.

Mesulam, M.-M. and Mufson, E.J. (1985) The insula of Reil in man and monkey. Architectonics, connectivity, and function. In A. Peters and E.G. Jones (Eds.), *Cerebral Cortex, Vol. 4*, Plenum Press, New York, pp. 179–226.

Mesulam, M.-M., Mufson, E.J., Levey, A.I. and Wainer, B.H. (1983a) Cholinergic innervation of cortex by the basal forebrain: cytochemistry and cortical connections of the septal area, diagonal band nuclei, nucleus basalis (substantia innominata), and hypothalamus in the rhesus monkey. *J. Comp. Neurol.*, 214: 170–197.

Mesulam, M.-M., Mufson, E.J., Wainer, B.H. and Levey, A.I. (1983b) Central cholinergic pathways in the rat: an overview based on an alternative nomenclature (Ch1–Ch6). *Neuroscience*, 10: 1185–1201.

Miller, M.W. (1987) The origin of corticospinal projection neurons in rat. *Exp. Brain Res.*, 67: 339–351.

Milner, T.A., Loy, R. and Amaral, D.G. (1983) An anatomical study of the development of the septo-hippocampal projection in the rat. *Develop. Brain Res.*, 8: 343–371.

Morán, M.A. and Reinoso-Suárez (1988) Topographical organization of the thalamic afferent connections to the motor cortex in the cat. *J. Comp. Neurol.*, 270: 64–85.

Mufson, E.J. and Mesulam, M.-M. (1982) Insula of the Old World monkey. II. Afferent cortical input and comments on the claustrum. *J. Comp. Neurol.*, 212: 23–37.

Mufson, E.J. and Mesulam, M.-M. (1984) Thalamic connections of the insula in the rhesus monkey and comments on the paralimbic connectivity of the medial pulvinar nucleus. *J. Comp. Neurol.*, 227: 109–120.

Musil, S.Y. and Olson, C.R. (1988) Organization of cortical and subcortical projections to anterior cingulate cortex in the cat. *J. Comp Neurol.*, 272: 203–218.

Nauta, W.J.H. (1961) Fiber degeneration following lesions of the amygdaloid complex in the monkey. *J. Anat.*, 95: 515–531.

Nauta, W.J.H. (1962) Neural associations of the amygdaloid complex in the monkey, *Brain*, 85: 505–520.

Neafsey, E.J. (1990) Prefrontal cortical control of the autonomic nervous system: anatomical and physiological observations. *This Volume*, Ch. 7.

Neafsey, E.J., Bold, E.L., Haas, G., Hurley-Gius, K.M., Quirk, G., Sievert, C.F. and Terreberry, R.R. (1986a) The organization of the rat motor cortex: a microstimulation mapping study. *Brain Res. Rev.*, 11: 77–96.

Neafsey, E.J., Hurley-Gius, K.M. and Arvanitis, D. (1986b) The topographical organization of neurons in the rat medial frontal, insular and olfactory cortex projecting to the solitary nucleus, olfactory bulb, periaqueductal gray and superior colliculus. *Brain Res.*, 377: 261–270.

Nishikawa, T., Fage, D. and Scatton, B. (1986) Evidence for, and nature of, the tonic inhibitory influence of habenulointerpeduncular pathways upon cerebral dopaminergic transmission in the rat. *Brain Res.*, 373: 324–336.

O'Hearn, E. and Molliver, M.E. (1984) Organization of the raphe-cortical projections in rat: A quantitative retrograde study. *Brain Res. Bull.*, 13: 709–726.

O'Leary, D.D.M. (1989) Do cortical areas emerge from a protocortex? *Trends Neurosci.*, 12: 400–406.

Olszewski, J. (1952) *The Thalamus of the Macaca mulatta. An Atlas for Use with the Stereotaxic Instrument*, Karger, New York, 93 pp.

Pandya, D.N. and Kuypers, H.G.J.M. (1969) Cortico-cortical connections in the rhesus monkey. *Brain Res.*, 13: 13–36.

Pandya, D.N. and Yeterian, E.H. (1990) Prefrontal cortex in relation to other cortical areas in rhesus monkey: architecture and connections. *This Volume*, Ch. 4.

Parnavelas, J.G. (1990) Neurotransmitters in the cerebral cortex. *This Volume*, Ch. 2.

Passingham, R.E. (1973) Anatomical differences between the neocortex of man and other primates. *Brain Behav. Evol.*, 7: 337–359.

Passingham, R.E., Myers, C., Rawlins, N., Lightfoot, V. and Fearn, S. (1988) Premotor cortex in the rat. *Behav. Neurosci.*, 102: 101–109.

Peden, J.K. and Von Bonin, G. (1947) The neocortex of hapale. *J. Comp. Neurol.*, 86: 37–63.

Phillipson, O.T. and Griffith, A.C. (1980) The neurones of origin for the mesohabenular dopamine pathway. *Brain Res.*, 197: 213–218.

Porrino, L.J. and Goldman-Rakic, P.S. (1982) Brainstem innervation of prefrontal and anterior cingulate cortex in the rhesus monkey revealed by retrograde transport of HRP. *J. Comp. Neurol.*, 205: 63–76.

Porrino, L.J., Crane, A.M. and Goldman-Rakic, P.S. (1981) Direct and indirect pathways from the amygdala to the frontal lobe in rhesus monkeys. *J. Comp. Neurol.*, 198: 121–136.

Potter, H. and Nauta, W.J.H. (1979) A note on the problem of olfactory associations of the orbitofrontal cortex in the monkey. *Neuroscience*, 4: 361–367.

Preuss, T.M. and Goldman-Rakic, P.S. (1989) Connections of the ventral granular frontal cortex of macaques with perisylvian premotor and somatosensory areas: anatomical evidence for somatic representation in primate frontal association cortex. *J. Comp. Neurol.*, 282: 293–316.

Price, J.L. and Slotnick, B.M. (1983) Dual olfactory representation in the rat thalamus: an anatomical and electrophysiological study. *J. Comp. Neurol.*, 215: 63–77.

Price, J.L., Russchen, F.T. and Amaral, D.G. (1987) The limbic region. II. The amygdaloid complex. In A. Björklund, T. Hökfelt and L.W. Swanson (Eds.), *Integrated Systems of the CNS, Part I. Handbook of Chemical Neuroanatomy, Vol. 5*, Elsevier, Amsterdam, pp. 279–388.

Rakic, P. (1988) Specification of cerebral cortical areas. *Science*, 241: 170–176.

Reep, R.L. (1984) Relationship between prefrontal and limbic cortex: a comparative anatomical review. *Brain Behav. Evol.*, 25: 5–80.

Reep, R.L., Corwin, J.V., Hashimoto, A. and Watson, R.T. (1984) Afferent connections of medial precentral cortex in the rat. *Neurosci. Lett.*, 44: 247–252.

Reep, R.L., Goodwin, G.S. and Corwin, J.V. (1990) Topographic organization in the corticocortical connections of medial agranular cortex in rats. *J. Comp. Neurol.*, 294: 262–280.

Robertson, R.T. and Kaitz, S.S. (1981) Thalamic connections with limbic cortex. I. Thalamocortical projections. *J. Comp. Neurol.*, 195: 501–525.

Rose, J.E. and Woolsey, C.N. (1948a) The orbitofrontal cortex and its connections with the mediodorsal nucleus in rabbit, sheep and cat. *Res. Publ. Ass. Nerv. Ment. Dis.*, 27: 210–232.

Rose, J.E. and Woolsey, C.N. (1948b) Structure and relations of limbic cortex and anterior thalamic nuclei in rabbit and cat. *J. Comp. Neurol.*, 89: 279–347.

Ruggiero, D.A., Mraovitch, S., Granata, A.R., Anwar, M. and Reis, D.J. (1987) A role of insular cortex in cardiovascular function. *J. Comp. Neurol.*, 257: 189–207.

Russchen, F.T., Amaral, D.G. and Price, J.L. (1985) The afferent connections of the substantia innominata in the monkey, *Macaca fascicularis*. *J. Comp. Neurol.*, 242: 1–27.

Russchen, F.T., Amaral, D.G. and Price, J.L. (1987) The afferent input to the magnocellular division of the mediodorsal thalamic nucleus in the monkey, *Macaca fascicularis*. *J. Comp. Neurol.*, 256: 175–210.

Saper, C.B. (1984) Organization of cerebral cortical afferent systems in the rat. I. Magnocellular basal nucleus. *J. Comp. Neurol.*, 222: 313–342.

Sarkisov, S.A. (Ed.) (1955) *Atlas of the Cytoarchitectonics of the Human Cerebral Cortex,* Medgiz, Moscow.

Satoh, K. and Fibiger, H.C. (1986) Cholinergic neurons of the laterodorsal tegmental nucleus: efferent and afferent connections. *J. Comp. Neurol.*, 253: 277–302.

Schell, G.R. and Strick, P.L. (1984) The origin of thalamic inputs to the arcuate premotor and supplementary motor areas. *J. Neurosci.*, 4: 539–560.

Schlenska, G. (1974) Volumen- und Oberflächenmessungen an Gehirnen verschiedener Säugetiere im Vergleich zu einem errechneten Modell. *J. Hirnforsch.*, 15: 401–408.

Schober, W. (1986) The rat cortex in stereotaxic coordinates. *J. Hirnforsch.*, 27: 121–143.

Selemon, L.D. and Goldman-Rakic, P.S. (1988) Common cortical and subcortical targets of the dorsolateral prefrontal and posterior parietal cortices in the rhesus monkey: evidence for a distributed neural network subserving spatially guided behavior. *J. Neurosci.*, 8: 4049–4068.

Sesack, S.R., Deutch, A.Y., Roth, R.H. and Bunney, B.S. (1989) Topographical organization of the efferent projections of the medial prefrontal cortex in the rat: An anterograde tract-tracing study with *Phaseolus vulgaris* leucoagglutinin. *J. Comp. Neurol.*, 290: 213–242.

Stanton, G.B., Deng, S.-Y., Goldberg, M.E. and McMullen, N.T. (1989) Cytoarchitectural characteristics of the frontal eye fields in macaque monkeys. *J. Comp. Neurol.*, 282: 415–427.

Stephan, H., Baron, G. and Schwerdtfeger, W.K. (1980) *The Brain of the Common Marmoset (Callithrix jacchus). A Stereotaxic Atlas,* Springer, Berlin.

Sukekawa, K. (1988a) Reciprocal connections between medial prefrontal cortex and lateral posterior nucleus in rats. *Brain Behav. Evol.*, 32: 246–251.

Sukekawa, K. (1988b) Interconnections of the visual cortex with the frontal cortex in the rat. *J. Hirnforsch.*, 29: 83–93.

Suzuki, H. (1985) Distribution and organization of visual and auditory neurons in monkey prefrontal cortex. *Vision Res.*, 25: 465–469.

Swanson, L.W. (1981) A direct projection from Ammon's horn to prefrontal cortex in the rat. *Brain Res.*, 217: 150–154.

Swanson, L.W. and Köhler, C. (1986) Anatomical evidence for direct projections from the entorhinal area to the entire cortical mantle in the rat. *J. Neurosci.*, 6: 3010–3023.

Takagi, S.F. (1979) Dual systems for sensory olfactory processing in higher primates. *Trends Neurosci.*, 2: 313–315.

Tanaka, D. and Goldman, P.S. (1976) Silver degeneration and autoradiographic evidence for a projection from the principal sulcus to the septum in the rhesus monkey. *Brain Res.*, 103: 535–540.

Terreberry, R.R. and Neafsey, E.J. (1987) The rat medial frontal cortex projects directly to autonomic regions of the brainstem. *Brain Res. Bull.*, 19: 639–649.

Thierry, A.-M., Chevalier, G., Ferron, A. and Glowinski, J. (1983) Diencephalic and mesencephalic efferents of the medial prefrontal cortex in the rat. Electrophysiological evidence for the existence of branched axons. *Exp. Brain Res.*, 50: 275–282.

Uylings, H.B.M., Verwer, R.W.H. and Van Pelt, J. (Eds.) (1986a) *Morphometry and Stereology in Neurosciences.* Special issue of *J. Neurosci. Methods*, 18: 1–242.

Uylings, H.B.M., Van Eden, C.G. and Hofman, M.A. (1986b) Morphometry of size/volume variables and comparison of their bivariate relations in the nervous system under different conditions. *J. Neurosci. Methods,* 18: 19–37.

Uylings, H.B.M., Hofman, M.A. and Matthijssen, M.A.H. (1987) Comparison of bivariate linear relations in biological allometry research. *Acta Stereol.*, 6: 467–472.

Uylings, H.B.M., Van Eden, C.G., Parnavelas, J.G. and Kalsbeek, A. (1990) The prenatal and postnatal development of rat cerebral cortex. In B. Kolb and R.C. Tees (Eds.), *The Cerebral Cortex of the Rat,* MIT Press, Cambridge, MA, pp. 35–76.

Van Eden, C.G. (1986) Development of connections between the mediodorsal nucleus of the thalamus and the prefrontal cortex in the rat. *J. Comp. Neurol.*, 244: 349–359.

Van Eden, C.G. and Uylings, H.B.M. (1985) Cytoarchitectonic development of the prefrontal cortex in the rat. *J. Comp. Neurol.*, 241: 253–267.

Van Eden, C.G., Lamme, V.A.F., and Uylings, H.B.M. (1990) Heterotopic cortical afferents to the medial prefrontal cortex in the rat. A combined retrograde and anterograde tracer study. *Neuroscience*, submitted.

Van Vulpen, E.H.S. and Verwer, R.W.H. (1989) Organization of projections from the mediodorsal nucleus of the thalamus to the basolateral complex of the amygdala in the rat. *Brain Res.*, 500: 389–394.

Vogt, B.A. and Miller, M.W. (1983) Cortical connections between rat cingulate cortex and visual, motor, and postsubicular cortices. *J. Comp. Neurol.*, 216; 192–210.

Vogt, B.A., Rosene, D.L. and Pandya, D.N. (1979) Thalamic and cortical afferents differentiate anterior from posterior cingulate cortex in the monkey. *Science*, 204: 205–207.

Vogt, B.A., Rosene, D.I. and Peters, A.P. (1981) Synaptic termination of thalamic and callosal afferents in cingulate cortex of the rat. *J. Comp. Neurol.*, 201: 265–283.

Vogt, B.A., Pandya, D.N. and Rosene, D,L. (1987) Cingulate cortex of the rhesus monkey. I. Cytoarchitecture and thalamic afferents. *J. Comp. Neurol.*, 262: 256–270.

Von Bonin, G. and Bailey, P. (1947) *The Neocortex of Macaca*

mulatta, University of Illinois Press, Urbana, IL, 163 pp.
Walker, A.E. (1940) A cytoarchitectural study of the prefrontal area of the macaque monkey. *J. Comp. Neurol.,* 73: 59 – 86.
Wise, S.P. and Donoghue, J.P. (1986) Motor cortex of rodents. In E.G. Jones and A. Peters (Eds.), *Sensory-Motor Areas and Aspects of Cortical Connectivity. Cerebral Cortex, Vol. 5,* Plenum Press, New York, pp. 243 – 270.
Yeterian, E.H. and Pandya, D.N. (1988) Corticothalamic connections of paralimbic regions in the rhesus monkey. *J. Comp. Neurol.,* 269: 130 – 146.
Zilles, K. (1985) *The Cortex of the Rat. A Stereotaxic Atlas,* Springer, Berlin, 121 pp.
Zilles, K. and Rehkämper, G. (1988) The brain, with special reference to the telencephalon. In J.H. Schwartz (Ed.), *Orang-Utan Biology,* Oxford University Press, Oxford, pp. 157 – 176.

Discussion

P. Goldman-Rakic: It seems that you used different criteria to define prefrontal cortex in the different species.

H.B.M. Uylings: If that were true, a fair comparison of the PFC in different mammalian species (in this case the rat and the primate species) would be impossible indeed. However, my approach was to use a similar set of criteria for defining the PFC in rat and primates, viz. the anatomical set of connections with the thalamus, cortico-cortical connections and connections with other brain regions.

B. Kolb: The cat seems to have a small PFC by all criteria.

H.B.M. Uylings: The present data do indeed suggest this. However, I think that many more data, both anatomical and functional, are needed to come to definite conclusions with regard to the cat PFC.

G.E. Alexander: To what extent is the apparent constancy of PFC propositional volume attributable to inclusion of supplementary motor area (SMA) within your definition of PFC?

H.B.M. Uylings: SMA was not included in primates' PFC but it *was* in rat's. Due to the lack of segregation some aspects of the primate SMA have their counterpart within the rat PFC.

A.Y. Deutch: The rat frontal area 2 (Fr2) is suggested to be homologous to the frontal eye-field and the premotor area. Are these different functional regions spatially distinct or intermingled?

H.B.M. Uylings: Our studies on the corticocortical and PFC connections with the brainstem (Van Eden et al., 1990) together with the electrophysiological data obtained by microstimulation (see review Neafsey et al., 1986a, cited in our chapter) suggest that the part of the rat PFC which includes premotor characteristics is located in the rostral tip of Fr2 and anterior cingulate cortex (AC), and the part of PFC which includes the frontal eye-field characteristics is located in the caudal half of Fr2 and ACd. There is only a small overlap (see Fig. 4).

E.J. Neafsey: The volume of PFC area in rat and monkey appears to be 24%, and in the macaque 19% (see your Fig. 5). How many PFC subfields can be distinguished in each of the species?

H.B.M. Uylings: On the basis of cytoarchitectonics only, we currently distinguish 10 or 11 subfields in the rat (see Fig. 2) and 11 subfields in the rhesus monkey, i.e., when no distinction is made in the monkey between dorsal and ventral part in areas 46 and 8 (see Fig. 3 and, e.g., Pandya and Yeterian, this volume). As I stated in my presentation, there is some unclarity regarding a few subfields (in particular in the rat) as to whether or not they should be regarded as parts of the PFC. A major problem in determining the number of subfields, however, is presented by the difficulties inherent in cytoarchitectonic parcellation. Comparing the number of subfields on the basis of connectivity patterns and functional criteria does not solve this question either, since, as I explained before, that method also leads to problems due to a partial overlap of different cortical "fields" in the rat cortex.

J. Kaas: If fields are thought to overlap in some species, such as in rats, would estimates of the amount of cortex devoted to a region of areas not tend to overestimate the size of the region?

H.B.M. Uylings: This is one of the reasons why I showed Fig. 9.

R.W.H. Verwer: Do you think that tracer studies will be able to give a trustworthy quantification of strength?

H.B.M. Uylings: I do agree with you if you imply that a completely reliable quantification is impossible. What is possible, however, is to obtain an impression of anterograde tracing with the sensitive PHA-L: with this tracer it can be determined whether or not a particular brain region receives denser and/or more widespread axonal terminations from one or the other brain region with a similar number of labeled neurons.

CHAPTER 4

Prefrontal cortex in relation to other cortical areas in rhesus monkey: Architecture and connections

Deepak N. Pandya[1,3,4] and Edward H. Yeterian[1,2]

[1] *Edith Nourse Rogers Memorial Veterans Hospital, 200 Springs Road, Bedford, MA 01730,* [2] *Department of Psychology, Colby College Waterville, ME 04901,* [3] *Departments of Anatomy and Neurology, Boston University School of Medicine, Boston, MA 02118, and* [4] *Harvard Neurological Unit, Beth Israel Hospital, Boston, MA 02215, USA.*

Introduction

The frontal lobes have long been regarded as crucial for a variety of complex functions, including drive and motivation, planning and sequencing in time and space, orientation and attention, provisional or representational memory, emotionality, personality, behavioral adjustment, and inhibition of behavior (Fulton and Jacobsen, 1935; Goldstein, 1939; Pribram, 1960; Nauta, 1964, 1973; Teuber, 1972; Luria, 1973a,b; Milner, 1982; Milner and Petrides, 1984; Damasio, 1985; Lhermitte, 1986; Lhermitte et al., 1986; Stuss and Benson, 1986; Goldman-Rakic, 1987; Fuster, 1989). All of these presumed frontal lobe functions have been derived from clinical as well as experimental observations. Since the late 19th century, the frontal lobes have been regarded as involved in higher cognitive functions. For example, Ferrier (1876), based on his lesion-behavior studies in dogs and monkeys, viewed the frontal lobes as critical for attention. Thus, following cortical ablation, Ferrier found that his monkeys had a decreased interest in their surroundings, and that they appeared to be apathetic. Similarly, Hitzig (1874) and Bianchi (1895) emphasized the involvement of the frontal lobes in higher intellectual functions. Bianchi described his experimental observations in monkeys as follows: "No sensory defects are noticeable, on the closest investigation. But discrimination, and the higher co-ordinations of sensory factors on which the more complex psychical manifestations depend, and which are necessary to the preservation of the individual and the species, are reduced to mere rudiments" (p. 518). In other words, there is disintegration of "personality" as such in the monkeys. Furthermore, Bianchi hypothesized that "the frontal lobes are the seat of co-ordination and fusion of the incoming and outgoing products of the several sensory and motor areas of the cortex.... The frontal lobes would thus sum up into series the products of the sensory-motor regions as well as the emotive states which accompany all perceptions..." (p. 521).

In this century, many investigators have elaborated further upon the functions of the frontal lobe, and their observations have had a significant impact on our current thinking. Several investigators (e.g., Luria, Milner) have attempted to establish structure-function relationships for specific subareas of the frontal lobe. In contrast, others (e.g., Nauta, Teuber) have tended to view the frontal lobe in terms of integrative principles, emphasizing the complex linkages with other cor-

Correspondence: Deepak N. Pandya, M.D., E.N. Rogers Memorial Veterans Hospital, 200 Springs Road, Bedford, MA 01730, USA.

tical and subcortical structures. Whether or not one examines the frontal lobe from the viewpoint of localization of function, or integrated systems, knowledge of its structural features is crucial to understanding its functional roles.

This chapter examines architecture and connections of the frontal lobe. The frontal lobe has connections with several different cortical and subcortical regions. We will focus on the role of cortical connections, and only briefly discuss thalamic connections. It is reasonable to assume that, over the course of evolution, the adaptation of organisms to increasingly complex conditions is reflected in concomitant morphological features of the brain in general, and of the frontal lobe in particular. It has been well established that during evolution the frontal lobe has undergone marked increases in relative size and complexity, reaching its highest levels in primates (e.g., Reep, 1985). These changes are expressed, in part, in terms of cellular architecture and connectional organization.

Early in this century, a great deal of interest was shown in understanding the basic cellular make-up of the cerebral cortex in primates. In particular, parcellation of the cerebral cortex including the frontal lobe was carried out in the human brain (Fig. 1) by Brodmann (1908), and in the monkey (Fig. 2) by Brodmann (1909), Vogt and Vogt (1919) and Von Bonin and Bailey (1947). With specific regard to the frontal lobe in the monkey, Walker's (1940) subdivisions have remained a mainstay for interpreting the morphology of frontal cortical regions (Fig. 3). These cytoarchitectonic maps have provided an invaluable framework, as well as a consistent nomenclature, for a wide range of neuroscientists.

Architecture and evolution of the frontal lobe

The existence of several discrete architectonic areas in the frontal lobe raises an important question: Is there some organizational principle which accounts for, and interrelates, different cytoarchitectonic regions of the frontal lobe? Given that the anatomical elaboration of the frontal lobe is a con-

comitant of primate evolution, is it possible to find systematic architectonic changes which reflect evolutionary progression? Indeed, Dart (1934) and Abbie (1940, 1942) proposed a dual origin of the cerebral cortex based on architectonic studies of reptilian and marsupial brains, respectively. Sanides (1969, 1972), using comparative architectonic methods, provided evidence for a dual origin of the cerebral cortex in different mammalian species, including insectivores, prosimians, and primates.

Briefly, the concept of the dual origin of the cerebral cortex begins with the proposition that there are two prime moieties from which all cortical regions have evolved — the archicortical (or hippocampal) and the paleocortical (or olfactory).

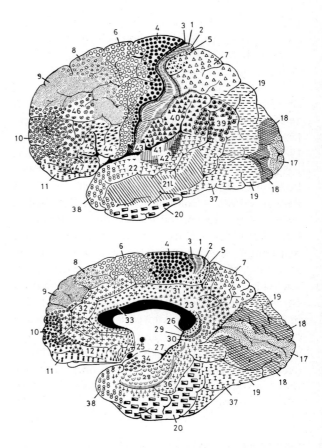

Fig. 1. Diagram of the human cerebral hemisphere showing the architectonic parcellation according to Brodmann (1908).

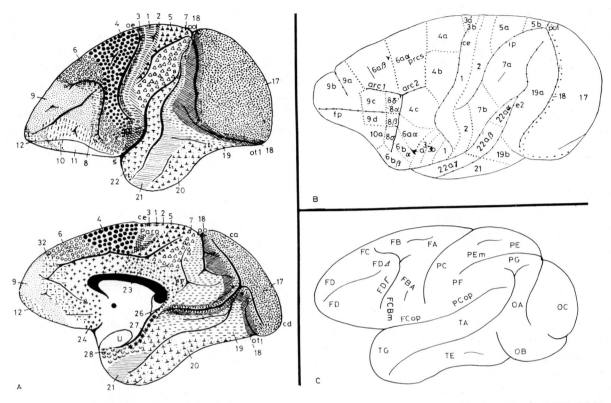

Fig. 2. Diagrams of 3 monkey cerebral hemispheres showing architectonic parcellations according to (A) Brodmann (1909), (B) Vogt and Vogt (1919), and (C) Von Bonin and Bailey (1947).

Each moiety is tied to one of the two basic issues common to all behaving organisms, the questions of "what" and "where". Even at the unicellular level, an organism which moves through a medium must deal with what is around it as well as where it is in the environment. These two basic issues can be related conceptually and more broadly to the two prime moieties – the issue of what, i.e., sensory processing, to the paleocortical moiety, and the issue of where, i.e., spatial processing crucial for effecting behavior, to the archicortical moiety. The following is a simplified summary, based on architectonic observations, of the way in which the cerebral cortex appears to have evolved from these two moieties.

As mentioned above, the cortex can be viewed as having evolved from two prime moieties, paleocortex and archicortex (Fig. 4A). Further differentiation from these moieties gives rise to a stage which is termed periallocortex (PALL). The next architectonic step is toward proisocortex (Pro-iso), or nearly 6-layered cortex, which leads to the isocortices (Fig. 4B). Thus, caudally further steps in the progression of isocortex emanate from two structures, namely the parinsular (rostral in-

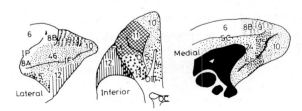

Fig. 3. Diagrams showing the architectonic parcellation of the prefrontal cortex according to Walker (1940).

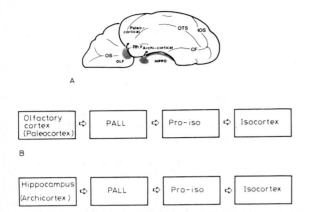

Fig. 4. (A) Diagram of the basal surface of the cerebral cortex in the rhesus monkey showing the location of the two primordial moieties: olfactory cortex (paleocortical) and hippocampus (archicortical). (B) Flow diagram depicting the cortical architectonic sequences from the two primordial moieties. Abbreviations: PALL, periallocortex; Pro-iso, proisocortex.

sula and temporal pole) and the paralimbic (cingulate) cortices, which are related to the paleo- and archicortical moieties, respectively (Fig. 5). Parinsular areas, which show progressive emphasis on granular cells, differentiate into second somatosensory (SII), second visual (MT or VII), and second auditory (AII) areas. Further development from these areas leads to a stage in each sensory modality from which primary sensory areas originate. These primary areas derived from parinsular regions are related to the face, head and neck areas of SI, the central visual representation of VI, and to the auditory modality. Likewise, paralimbic areas which show a predominant emphasis on pyramidal cells, give rise to the second sensory area (SSA) on the medial surface, as well as the prostriate area (Pro st) on the ventral surface of the cerebral hemisphere. In turn these areas would differentiate into the primary sensory regions relating to the trunk and limbs in SI, and to areas having the peripheral visual field representation (Pandya and Yeterian, 1985).

Just as there are systematic architectonic elaborations within post-Rolandic cortical regions, there are similar progressions within the frontal lobe (Pandya and Yeterian, 1985). Thus, the paleocortical moiety of the olfactory tubercle gives rise to periallocortex and orbital proisocortex. From the bilaminated-appearing proisocortex of the orbital surface, the ventral and lateral prefrontal isocortices would evolve (Fig. 6A). This architectonic differentiation is evident as a progressive emphasis from infragranular layers in the proisocortical region to supragranular layers in isocortical regions. Similarly, from the archicor-

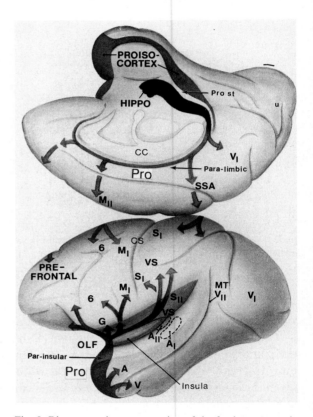

Fig. 5. Diagrammatic representation of the further progression of the two cortical evolutionary trends shown in Fig. 4, culminating in pre- and post-Rolandic sensory and motor cortical areas (Pandya and Yeterian, 1985). Abbreviations: A_I, primary auditory area; A_{II}, second auditory area; CC, corpus callosum; CS, central sulcus; G, gustatory area; HIPPO, hippocampus; M_I, motor cortex; M_{II}, supplementary motor area; OLF, olfactory cortex; Pro, proisocortex; Pro st, prostriate region; S_I, primary somatosensory area; S_{II}, second somatosensory area; SSA, supplementary sensory area; V_I, primary visual area; V_{II} (MT), second visual area; VS, vestibular area.

tical moiety on the medial surface, one can identify surrounding proisocortical areas, namely areas 24, 25 and 32 (Fig. 6B). From these proisocortical regions, one can trace the different cortical areas on the medial and dorsolateral surfaces of the prefrontal cortex. Like the ventral prefrontal regions, these areas show progressive laminar differentiation with emphasis on infragranular layers in medial regions, and towards supragranular layers on the lateral surface (Fig. 7). In other words, there are two distinct architectonic trends in the prefrontal cortex; the ventral trend comprised of areas Pro, 13, 12, 14, 11, 10, 46 and 8, and the dorsal trend consisting of areas Pro (24, 25, and 32), 9, 10, 46 and 8. Moreover, the architectonic areas of the frontal lobe can be viewed as organized in a radial manner, as successive tiers, with gradual architectonic changes within both cortical trends (Barbas and Pandya, 1989) (Fig. 8).

In order to evaluate the significance of the pro-

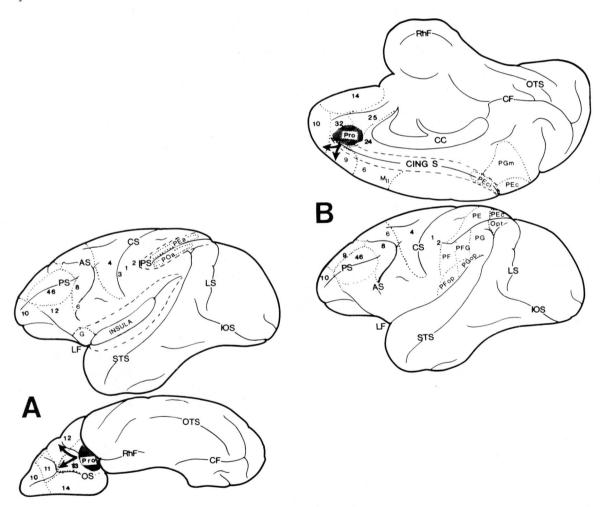

Fig. 6. (A) Progressive architectonic steps from orbital proisocortex leading to ventral area 8. (B) Similar steps from medial proisocortex leading to dorsal area 8. Abbreviations for sulci in this and subsequent figures: AS, arcuate sulcus; CC, corpus callosum; CF, calcarine fissure; CING S, cingulate sulcus; CS, central sulcus, IOS, inferior occipital sulcus; IPS, intraparietal sulcus; LF, lateral fissure; LS, lunate sulcus; OS, orbital sulcus; OTS, occipitotemporal sulcus; POMS, parieto-occipito-medial sulcus; PS, principal sulcus; Rh F, rhinal fissure; STS, superior temporal sulcus.

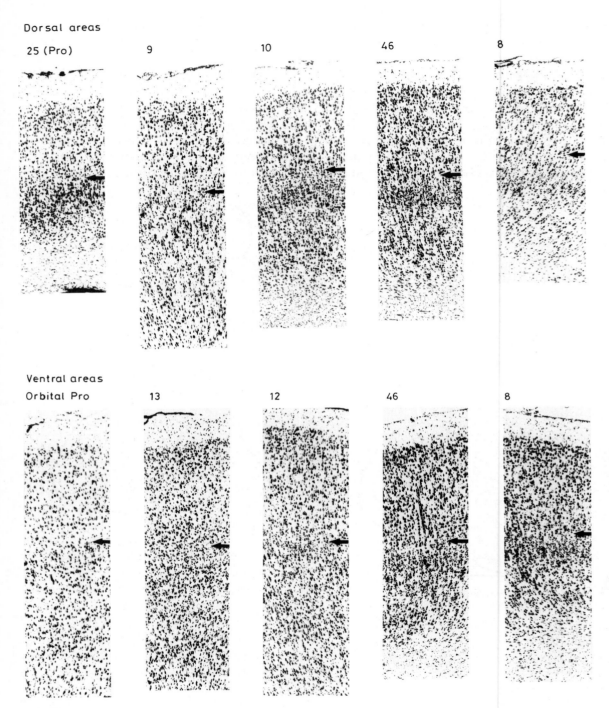

Fig. 7. Bright-field photomicrographs of representative frontal cortical regions belonging to the two architectonic trends (Siwek, 1989). The upper panel depicts a sequence of regions within the dorsal or archicortical trend. The lower panel depicts a sequence of regions within the ventral or paleocortical trend. Note that in both of these trends the medial proisocortex (area 25) and the orbital proisocortex have a predominance of infragranular layers and less developed supragranular layers. In both trends, cortical areas show progressive differentiation and development of the supragranular layers, with the greatest degree of differentiation in area 8. Arrows indicate the border between infra- and supragranular cortical layers.

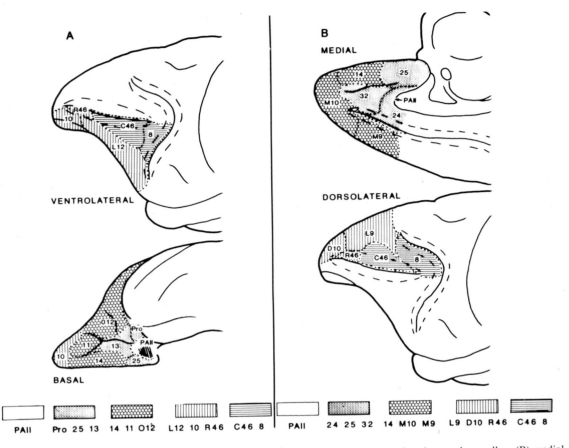

Fig. 8. Diagrams showing architectonic areas arranged in stages within (A) the orbital and ventral as well as (B) medial and dorsolateral prefrontal cortices, i.e., within the paleocortical and archicortical trends, respectively. Within each trend the axis of differentiation proceeds in a direction from the least differentiated periallocortex (PAll) to area 8, which represents the most highly differentiated cortex (Barbas and Pandya, 1989).

gressive architectonic trends in the prefrontal cortex, we have examined relationships between architectonic and connectional organization. The intrinsic connections of the prefrontal cortex appear to be organized in a manner consistent with dual architectonic trends (Pandya and Yeterian, 1985; Barbas and Pandya, 1989).

Intrinsic and thalamic connections of the frontal lobe

Each architectonic area within the two frontal lobe trends has bidirectional connections with more differentiated regions on the one hand, and with less differentiated regions on the other (Barbas and Pandya, 1989). Thus, the ventral proisocortical area projects mainly to areas 13, 12, 11 and 14. Area 13 in turn projects back to the proisocortical area as well as to areas 12, 11 and 14. Area 12 projects to areas 11, 13 and 14, and to ventral areas 10 and 46. Area 46 sends its connections to areas 12 and 10 rostroventrally and to ventral area 8 caudally. Area 8 projects to ventral area 46 rostrally, to dorsal area 8, and to area 6 caudally (Fig. 9). Similarly, area 32, a medial proisocortical region, projects to areas 25 and 14 ventrally, to area 24 caudally, and to areas 9 and 10 dorsally. Area 9 in turn has connections with proisocortex (areas 32 and 24) medially and dorsal area 10 and area 46 laterally. Area 46 projects to areas 9 and 10 on the one hand and to dorsal area 8 and area 6 on the other. In addition, dorsal area 8 projects to dorsal area 46 and to area 9 rostrally, to dorsal area 6 caudally, and to area 8 ventrally (Fig. 10).

It is noteworthy that these two trends are interconnected at certain levels (Fig. 11). Thus, the ventral and medial proisocortical areas are interconnected, as are areas 9 and 12 (Pandya et al., 1971). The dorsal and ventral portions of area 8 are also interconnected. These observations indicate that within each trend, a given cortical region projects to architectonically less differentiated areas as well as to regions with a more developed cortical laminar organization.

The dual organization of the frontal lobe is also reflected in the pattern of frontothalamic connections. For instance, based on the connectional studies of a number of different investigators (Akert, 1964; Kievit and Kuypers, 1977; Goldman-Rakic and Porrino, 1985; Giguere and Goldman-Rakic, 1988; Siwek, 1989), it appears that prefrontal connections with specific subdivisions of the mediodorsal (MD) nucleus (Olszewski, 1952) follow the mediolateral and rostrocaudal topography of the frontal lobe (Fig. 12). Thus, the medial prefrontal and orbitofrontal proisocortical areas project to and receive afferents from the most medial portion of MD (magnocellular divi-

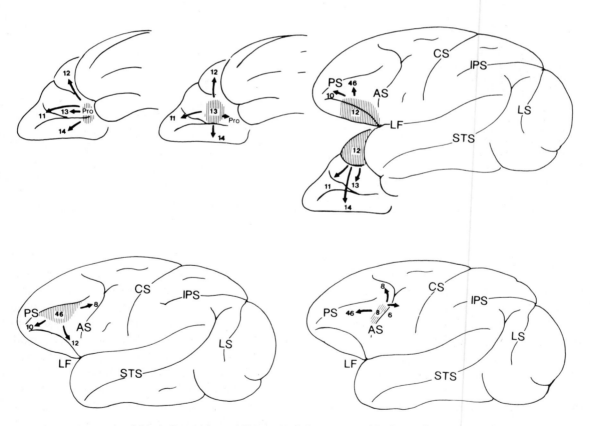

Intrinsic Connections of Paleocortical Trend areas of Prefrontal Cortex

Fig. 9. Intrinsic connections of subregions of the ventral (paleocortical) trend of the prefrontal cortex, from orbital proisocortex to ventral area 8.

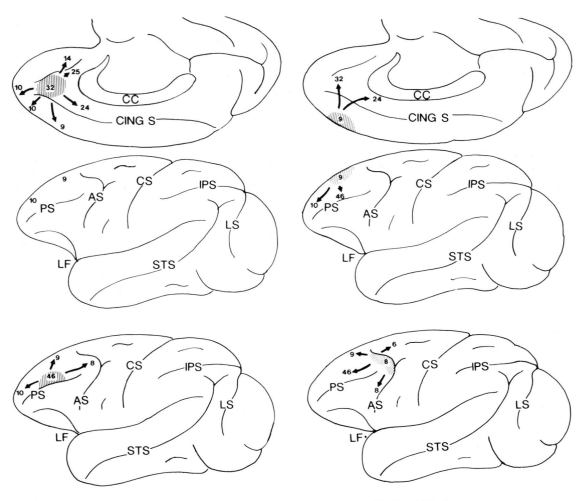

Intrinsic Connections of Archicortical Trend areas of Prefrontal Cortex

Fig. 10. Intrinsic connections of subregions of the dorsal (archicortical) trend of the prefrontal cortex, from medial proisocortex to dorsal area 8.

sion), maintaining their respective dorsal and ventral topographies in this nucleus. The lateral prefrontal region, above and below the principal sulcus, is connected with the middle part of MD (parvocellular division), again showing a dorsoventral topography. Finally, dorsal area 8 and ventra area 8 within the concavity of the arcuate sulcus project to and receive afferents from, respectively, dorsal and ventral portions of the most lateral part of MD (multiformis division).

These frontal lobe projections to and from MD correspond to the patterns of cytoarchitectonic differentiation of the prefrontal cortex, and are consistent with the concept of dual origins of the frontal lobe (Siwek, 1989).

It appears that the dual architectonic trends in the prefrontal cortex can be related systematically to intrinsic prefrontal connectivity, as well as to the thalamic connectivity of these regions. This concept can be extended to premotor and motor

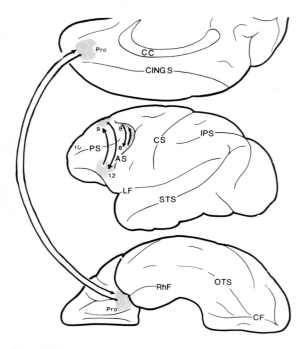

Fig. 11. Diagram showing the interconnections between dorsal and ventral cytoarchitectonic trends, at proisocortical, prefrontal (areas 9 and 12), and premotor (area 8) levels (Pandya and Barbas, 1985; Pandya and Barnes, 1987).

systems of intrinsic or local connections that parallel the architectonic sequences (Fig. 14). The rostral cingulate gyrus (area 24) projects to the supplementary motor area and to dorsal area 6. The supplementary motor area in turn projects to dorsal area 6 and area 4 on the one hand and to area 24 on the other (Damasio and Van Hoesen, 1980). The dorsal portion of the premotor cortex (area 6), particularly its caudal subdivision, projects to dorsal area 4 (MI) as well as to MII and area 24 (Barbas and Pandya, 1987). A similar situation exists for ventral regions. Thus, the insular cortex projects to area ProM in the frontal operculum and to ventral area 6 (Mesulam and Mufson, 1982). Area ProM sends projections back

regions of the frontal lobe (Fig. 13). Thus, from the hippocampal moiety can be traced the proisocortex of the cingulate gyrus (area 24), whereas the olfactory region leads to insular proisocortex. The subsequent architectonic steps continue from these two proisocortices to the premotor and precentral cortices. The dorsal trend leads from the cingulate gyrus to the supplementary motor area, MII, and to the dorsal portion of the premotor cortex (area 6) as well as to dorsal motor cortex (area 4). The ventral trend, in contrast, progresses from insular proisocortex to area ProM (proisocortical motor area, analogous to MII of the dorsal trend) and to the ventral portion of the premotor cortex as well as to ventral motor cortex (Pandya and Barbas, 1985; Barbas and Pandya, 1987).

The different architectonic regions of the dorsal and ventral premotor and motor trends have

Summary of thalamocortical projections

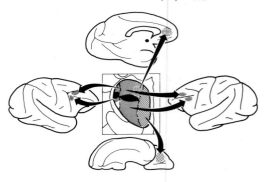

Summary of corticothalamic projections

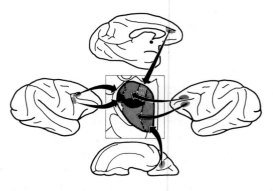

Fig. 12. Summary diagram showing the topographic distribution of both the thalamocortical and corticothalamic connections between the mediodorsal nucleus and the prefrontal cortex (Siwek, 1989).

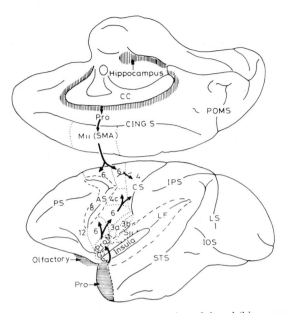

Fig. 13. Diagrammatic representation of dorsal (hippocampal-cingulate) and ventral (olfactory-insular) architectonic trends showing progressive architectonic steps leading to dorsal and ventral sectors of premotor (area 6) and motor (area 4) cortices in the rhesus monkey. Note that the dorsal (SMA) and ventral (Pro M) supplementary motor areas are interposed between their respective proisocortices (Pro) and the premotor regions (area 6) (Pandya and Barbas, 1985).

to the rostral insula and also projects to ventral area 6. Ventral area 6, in turn, projects back to area ProM and to the ventral precentral motor area. Many of these connections within each trend are reciprocal; and there are also interconnections between areas in the dorsal and ventral pathways.

Thus, there appears to be a dual organization of motor-related cortices. The dorsal region, related to the hippocampus and rostral cingulate gyrus, includes the supplementary motor area (MII), dorsal premotor cortex (area 6), and dorsal motor cortex (MI, area 4). The ventral division, related to the rostral insula, includes area ProM in the frontal operculum, the ventral premotor region (area 6) and ventral motor cortex (MI, area 4).

Traditionally, the functions of the prefrontal cortex have been viewed from two major perspectives. On the one hand, a number of complex processes (e.g., frontal lobe syndrome) have been ascribed to the prefrontal cortex as a whole. On the other hand several investigators have viewed the prefrontal cortex as consisting of distinct functional zones (e.g., Rosenkilde, 1979; Damasio, 1985). As will be discussed below, it may be useful

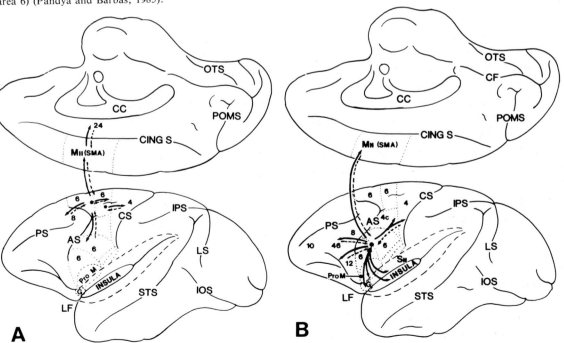

Fig. 14. Diagrams showing intrinsic (intrafrontal) connections of (A) dorsal and (B) ventral premotor areas (Pandya and Barbas, 1985).

to consider the dual architectonic trends and their related connectional features in attempting to understand prefrontal function.

It has been customary to focus on motor, premotor and supplementary motor areas as well as immediately adjoining regions when studying the cortical components of the motor system. It may, however, be more meaningful to include all related areas from which cortical motor areas have evolved sequentially. A consideration of dorsal and ventral evolutionary pathways may provide a valuable context for understanding certain basic aspects of motor function, even though the older regions of these pathways (cingulate and insular cortices) do not appear to have a motor role in the strictest sense. It is possible that these regions do influence motor behavior at a broader, more fundamental level. Thus, the cortical components of the dorsal motor pathway may be engaged in providing a spatial context for motility by virtue of their connections with the dorsal parietal lobe and with the medial occipital lobe. In contrast, the ventral motor pathway may provide emotional input which influences motor activity, since it is closely tied to insular, orbital and temporal polar cortices as well as to ventral parietal regions (Pandya and Yeterian, 1985). According to this view it is the combined contributions from different regions of these two motor pathways which result in fully integrated motor behavior (Pandya, 1987).

Corticocortical connections leading to the frontal lobe from the post-Rolandic sensory association areas (somatosensory, visual and auditory) and paralimbic regions (cingulate and parahippocampal gyri) are also organized according to dual trends. Connections from the frontal lobe back to the sensory association areas and paralimbic regions seem to be organized in a similar manner.

Architecture and frontal lobe connections of the somatosensory cortices

The physiological studies of Woolsey (1958) have shown a topographic somatic representation within the primary somatosensory area, SI, of the postcentral gyrus (Fig. 15A). In terms of architecture, the cortex of the postcentral gyrus includes areas 3, 1 and 2 (Fig. 15B), and it receives thalamic input from the ventroposterior nuclei (Powell and Mountcastle, 1959; Jones and Powell, 1970a; Jones et al., 1979). Further physiological studies have shown re-representations of somatosensory regions in the parietal lobe (Paul et al., 1975; Kaas et al., 1981; Merzenich et al., 1981). In terms of function, SI is involved in basic processing of somatic sensation, e.g., texture and angularity (Randolph and Semmes, 1974). In contrast, the somatosensory association areas are considered to occupy most of the posterior parietal cortex (areas 5 and 7 of Brodmann, and Vogt and Vogt; areas PE, PEm, PF, and PG of Von Bonin and Bailey), and are thought to be involved in more complex and integrative functions in the somatosensory sphere (Duffy and Burchfiel, 1971; Mountcastle et al., 1975; Sakata, 1975; Robinson and Goldberg, 1978; Lynch, 1980; Hyvärinen, 1982). Unlike the primary somatosensory areas, these association areas maintain connections with the associative thalamic nuclei, predominantly the lateral posterior and pulvinar nuclei, and also intralaminar and reticular nuclei (Trojanowski and Jacobson, 1975; Jones et al., 1979; Weber and Yin, 1984; Yeterian and Pandya, 1985).

An architectonic analysis indicates that the somatosensory areas have progressed from two different moieties, archicortical as well as paleocortical (Sanides, 1972). Thus, the superior parietal region is thought to have progressed from the hippocampal moiety, passing through the paralimbic-proisocortical stages (within the cingulate region) in successive steps to culminate in the trunk, tail, and extremity representations in the postcentral gyrus on the one hand and the association areas in the superior and medial parietal lobule on the other (Fig. 15C). Likewise, the face representation of the postcentral gyrus and its association areas in the inferior parietal lobule are thought to be derived from the paleocortical moieties in successive steps passing through the

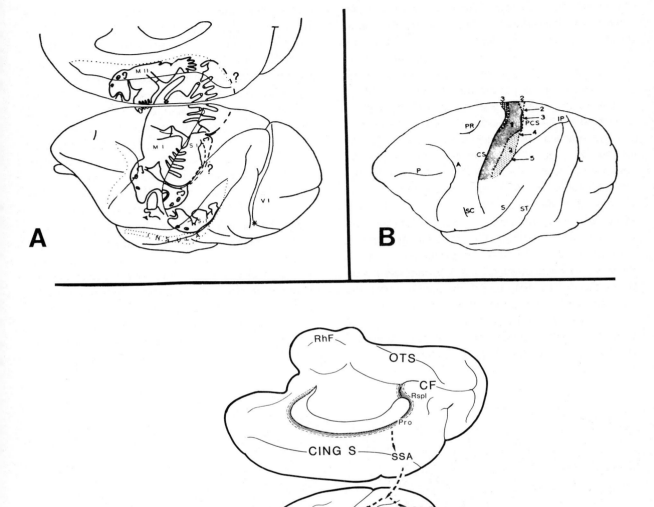

Fig. 15. (A) Diagram of primary sensorimotor (SI and MI), second sensory (SII), and supplementary motor (MII) representations in the rhesus monkey as described by Woolsey (1958) using physiological methods. (B) Architectonic boundaries of SI of the postcentral gyrus as described by Powell and Mountcastle (1959). (C) Two progressive architectonic trends leading to the primary somatosensory and association areas of the parietal lobe. The dorsal trend (archicortical) is shown by dashed lines and the ventral trend (paleocortical) by solid lines.

parinsular proisocortical stages in the frontal and pericentral opercula (Fig. 15C). These two trends seem to merge in the depths of the intraparietal sulcus.

It has been well established that the postcentral gyrus (area SI) is connected with the primary motor cortex, MI, as well as the supplementary motor cortex, MII, whereas the association areas of the parietal lobe are connected with the premotor, supplementary motor, and prefrontal regions (Pandya and Kuypers, 1969; Jones and Powell, 1969, 1970b; Chavis and Pandya, 1976; Barbas and Mesulam, 1981, 1985; Petrides and Pandya, 1984; Barbas, 1986; Selemon and Goldman-Rakic, 1988). The postcentral gyrus is flanked caudally by distinct architectonic areas which have been divided into 3 broad sectors on the basis of their differential connectivity with the frontal lobe (Chavis and Pandya, 1976). The first-order sensory association region SA1 of the superior parietal lobule (SPL) consists of areas PE and PEa (Figs. 16 and 17A), and projects to rostral area 4, to dorsal premotor cortex, area 6, and to the supplementary motor cortex, MII (Fig. 17C). Likewise, area PF, representing the SA1 for the inferior parietal lobule (IPL), projects to ventral premotor area 6 and to the precentral and frontal opercula (Fig. 17B). The next regions in the parietal lobe are designated SA2, or second somatosensory association regions. In the SPL, area SA2 consists of area PEc and projects to dorsal area 6 as well as to MII. However, the area 6 projections are considerably more rostral than those of area SA1. Area SA2 of the IPL consists of an area in the lower lip of the intraparietal sulcus. This area projects predominantly to ventral area 46, below the principal sulcus. The next parietal association sectors are designated SA3. The dorsal SA3 area is comprised of area PGm on the medial surface of the parietal lobe and projects to rostral area 6 in and above the upper limb of the arcuate sulcus, to adjacent area 8, and to dorsal areas 46 and 9. Unlike the areas SA1 and SA2 of the SPL, area SA3 lacks a projection to MII. Area

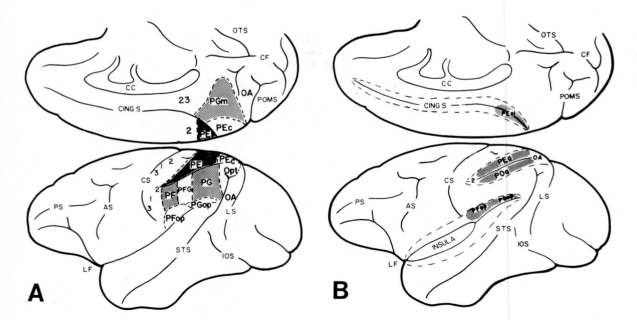

Fig. 16. Diagrams showing the architectonic parcellation of the posterior parietal cortex (Pandya and Seltzer, 1982). (A) Architectonic areas on the exposed lateral and medial surfaces. (B) Architectonic areas within the cingulate and intraparietal sulci, and the caudal Sylvian fissure.

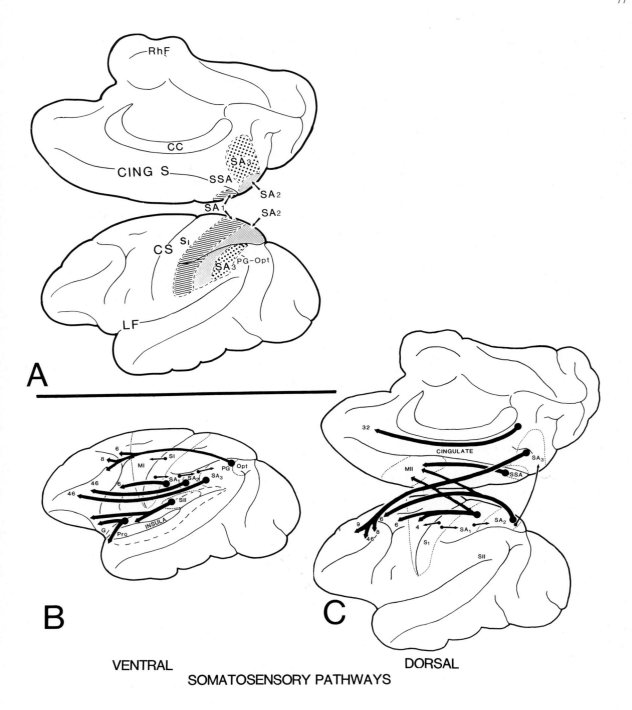

Fig. 17. (A) Diagrammatic representations of the 3 subdivisions, SA_1, SA_2, SA_3, of the somatosensory association cortex of the inferior and superior parietal lobules. (B) Diagram of the frontal lobe connections of the inferior parietal lobule (ventral trend). (C) Diagram of the frontal lobe connections of the superior parietal lobule (dorsal trend).

SA3 of the IPL consists of area PFG and rostral area PG and projects mainly to rostral area 46 in the ventral bank of the principal sulcus. The caudalmost portion of IPL consists of caudal area PG and area Opt. This area is situated at the junction of the parietal and occipital association regions and can be considered as equivalent to the angular gyrus, or area 39, of the human brain. Although located within the parietal lobe, this region is functionally as well as connectionally complex; it projects to area 46 and to dorsal areas 8 and 6.

On the basis of their cortical connectivity, the association areas of the SPL and IPL appear to be involved in somewhat different aspects of the somatosensory sphere (Pandya and Yeterian, 1985). Areas SA1, SA2, and SA3 of the SPL comprise a sequence of projection areas originating from the trunk and extremity representations of SI, ultimately leading to the cingulate gyrus and dorsal premotor, supplementary motor, as well as prefrontal regions (Fig. 17C). In contrast, areas SA1, SA2, and SA3 of the IPL form a series of projection regions which begin in the face, head, and neck representations of SI and ultimately relate to the cingulate and parahippocampal gyri, the ventral premotor and prefrontal regions, as well as the gustatory area in the frontal operculum (Pandya et al., 1980) (Fig. 17B). This differential arrangement coincides with the proposed dual nature of origin of somatosensory related regions (Sanides, 1972), and corresponds to the dual nature of frontal lobe phylogenetic development, where the principal sulcus represents a dividing line between upper and lower architectonic trends.

Architecture and frontal lobe connections of the visual cortices

The primary visual area lies within the occipital lobe and is related to a specific thalamic nucleus, the lateral geniculate body (Hubel and Wiesel, 1972; Winfield et al., 1975; Benevento and Yoshida, 1981; Yukie and Iwai, 1981; Bullier and Kennedy, 1983; Doty, 1983; Weller and Kaas, 1983). This region, the so-called striate cortex, has a characteristic architecture and has been designated area 17 by Brodmann (1909) and area OC by Von Bonin and Bailey (1947). Physiological studies have shown that it is involved in the analysis of visual input as it first reaches the cortical level (Hubel and Wiesel, 1968). Daniel and Whitteridge (1961) have shown that the midline of central vision is located at the very edge of the striate cortex (juxtastriate), whereas the peripheral visual field is represented in the remainder of the striate cortex. In recent years, on the basis of physiological as well as anatomical observations, several investigators have described multiple visual representations in surrounding peristriate belt (Zeki, 1978; Allman et al., 1981; Van Essen and Maunsell, 1983). The classical visual association region includes areas 18 and 19, the circumstriate belt which surrounds the striate cortex, and extends into the inferior temporal region, areas TE1, TE2, and TE3, or areas 20 and 21. Unlike the primary visual area (area 17), the thalamic connections of these regions are predominantly with associative nuclei, especially the lateral and the inferior pulvinar nucleus (Clark, 1936; Benevento and Rezak, 1976; Rosene and Pandya, unpublished observations). Recent anatomical and physiological studies have shown that the posterior parahippocampal region also contains visual association areas (Rosene and Pandya, 1983; Newsome et al., 1986). These association areas have been shown to be involved in sequential processing of visual information (Mishkin, 1972; Gross et al., 1981; Ungerleider and Mishkin, 1982; Mishkin et al., 1983).

Like the somatosensory system, the visual system appears to have evolved from both primordial moieties, i.e., paleocortical and archicortical. An analysis of architecture in primates carried out by Rosene and Pandya (1983) indicates that two distinct trends can be traced within the cortical visual system (Fig. 18). Both trends originate from the proisocortex of the temporal lobe. One trend progresses ventrolaterally with sequential modification of granularization through areas TE1,

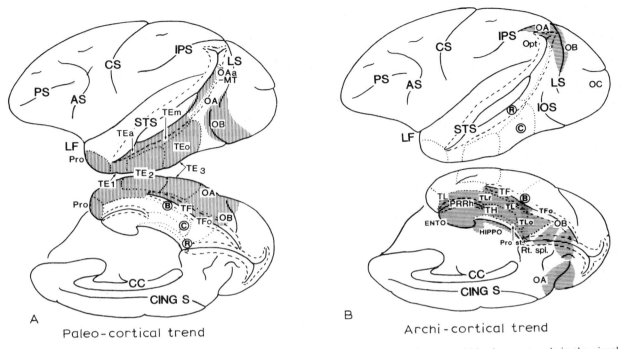

Fig. 18. Diagrammatic representations of the distribution of the different architectonic areas within the two trends in the visual system. (A) Paleocortical trend in inferotemporal and lateral occipital regions. (B) Archicortical trend in parahippocampal gyrus and medial, dorsal and ventral occipital areas (Pandya et al., 1988).

TE2, and TE3 in the inferotemporal region and area OAa (designated area VII, second visual area, and corresponding to area MT) as well as ventral areas OA, OB, and OC in the occipital lobe. Similarly, the other trend progresses ventromedially in successive steps through the parahippocampal gyrus (areas TH, TL, TF and the prostrate region, Pro st) and the occipital lobe regions (ventral and medial areas OA, OB, and OC). Although these observations need to be elaborated upon in further studies, it can be suggested that the differential function in terms of central and peripheral vision at the cortical level may be reflected in these two trends. Thus, in the ventrolateral trend, area OAa (area MT) in the superior temporal sulcus may represent the prokoniocortical step in the phylogenetic development of areas of central vision. In the ventromedial trend, the prostriate area can be considered as the prokoniocortical stage related primarily to areas of peripheral vision. In this schema for the visual cortices, prokoniocortical regions are those which are immediately adjacent and connected to the highly granularized primary sensory area.

The visual association areas of the lateral occipital and inferotemporal regions can be grouped into 3 rostrocaudal divisions (Fig. 19A) according to their frontal lobe projections (Chavis and Pandya, 1976). The lateral prestriate cortex, VA1, is shown to project to the premotor region (periarcuate cortex, area 8). Area VA2 of the caudal inferotemporal region, in contrast, sends projections to the premotor region (rostral area 8) and to the prearcuate area (area 46) below the principal sulcus. The rostral inferotemporal area VA3 (area 21) projects to the same frontal area as VA2, but in addition has a distinct connection to areas 12 and 11 on the orbital surface (Fig. 19B). Thus, visual association areas VA1, VA2, and VA3 show a sequential distribution of projections to the frontal lobe, beginning in the periarcuate region and progressing rostrally through the prearcuate and orbitofrontal regions (Kuypers et al., 1965; Chavis and Pandya, 1976; Barbas and Mesulam, 1981; Van Hoesen, 1982; Barbas, 1986, 1988). The medial and dorsal peristriate areas project to dorsal area 8. The areas rostral to the ventral peristriate belt project to dorsal area 46 and to area 12 (Rosene and Pandya, 1983; Barbas, 1988) (Fig. 19C).

The visual association areas are related to those frontal lobe regions which appear to occupy similar levels of architectural organization (Pandya and Yeterian, 1985). Thus, for example, area

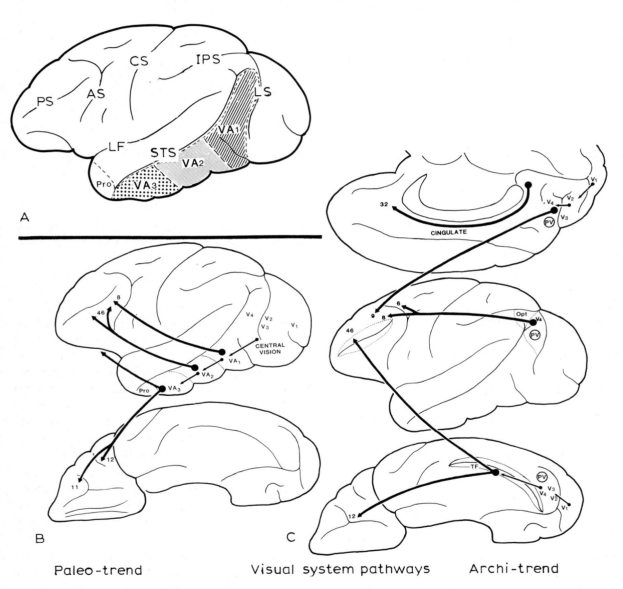

Fig. 19. Diagrammatic representations of (A) the 3 visual association regions of the inferotemporal cortex, VA_1, VA_2 and VA_3 relating to central vision, and (B) their long association connections to the frontal lobe. (C) Long association connections to the frontal lobe from areas subserving peripheral vision (PV) on the medial, dorsal and ventral surfaces of the cerebral hemisphere.

VA1, which has well-developed supragranular layers, is connected with area 8, which has an emphasis on the supragranular layers. In contrast, area VA3, which shows dysgranularity and an emphasis on infragranular layers, is related to areas 12 and 11, which have basically similar architectonic features. Area VA2, which has architectonic characteristics intermediate to those of areas VA1 and VA3, is related to those cortical regions with similar characteristics.

At the beginning of the description of visual association areas, it was suggested that this system might have evolved from 2 moieties, the temporal proisocortex, and the prostriate region of the medial parahippocampal gyrus. The connections described above pertain mainly to the trend which is related to the temporal proisocortex, and are related to the central visual field (Fig. 19B). Recent studies (Barbas and Mesulam, 1981, 1985; Rosene and Pandya, 1983; Barbas, 1988) have shown that unlike the connections of the polar proisocortical trend, the ventral, medial, and dorsolateral visual association areas are connected preferentially with the dorsal prefrontal regions (Fig. 19C). These preoccipital regions are related to the peripheral visual field. Thus, the dichotomous patterns of projections within the visually related areas support the concept of the dual nature of cortical visual systems with one trend (archicortical) relating to the peripheral visual field and the other (paleocortical) to central vision. This concept is supported by the physiological observations of Suzuki and Azuma (1983) and Suzuki (1985).

Architecture and frontal lobe connections of the auditory cortices

The primary auditory area in primates has been shown to be located in the supratemporal plane by its architectonic specialization, i.e., markedly granular cortex (koniocortex), its thalamic connections with a specific nucleus, the medial geniculate body (MGB), and its electrophysiological properties (Ferrier, 1876; Von Economo and Horn, 1930; Walker, 1938; Ades and Felder, 1942; Walzl and Woolsey, 1943; Kennedy, 1955; Merzenich and Brugge, 1973; Mesulam and Pandya, 1973; Pandya and Sanides, 1973; Brugge and Reale, 1985). The primary auditory area, AI, is flanked by the auditory association region in the superior temporal gyrus (STG) and by the second auditory area (AII) medially around the circular sulcus (Woolsey and Fairman, 1946) (Fig. 20A). The auditory association region has distinct architecture and has been designated area TA by Von Bonin and Bailey and area 22 by Brodmann. The thalamic connections of these regions are related more to the pulvinar nucleus than to the MGB (Walker, 1938; Mesulam and Pandya, 1973; Trojanowski and Jacobson, 1975; Pandya et al., 1986). Behavioral studies of the primary auditory area have implicated it in elementary auditory processing such as the analysis of frequency and amplitude (Walzl and Woolsey, 1943; Merzenich and Brugge, 1973; Brugge and Reale, 1985), whereas the association regions are considered to be involved in more integrative functions such as auditory pattern recognition and sound localization (Weiskrantz and Mishkin, 1958; Wegener, 1969; Leinonen et al., 1980; Hyvärinen, 1982). Despite morphological and functional observations, the precise limits of the primary auditory area and the auditory association regions are not fully delineated. Merzenich and Brugge (1973) have shown the existence of multiple auditory representations in the supratemporal plane and superior temporal gyrus (Fig. 20B). Likewise, the thalamic projections from the MGB are shown to extend quite rostrally in the superior temporal region (Mesulam and Pandya, 1973; Wegener, 1976).

More recent architectonic analyses of the superior temporal region have revealed a beltlike organization for the auditory areas, with a koniocortical core (areas Kam and Kalt, corresponding to area AI) surrounded by belt areas (Pandya and Sanides, 1973; Jones and Burton, 1976). The belt is composed of one prokoniocortical area (proA, corresponding to area AII) in a parinsular location, and of a caudal (paAc), a lateral (paAlt), and a rostral (paAr) parakoniocortical area (Fig. 20C). Parakoniocortical areas are somewhat less granularized than koniocortical regions, and immediately adjoin the auditory koniocortex caudally, laterally, and rostrally.

Besides outlining these opercular areas, Pandya and Sanides have demarcated additional areas in the superior temporal gyrus, areas Tpt, Ts3, Ts2, Ts1 and the proisocortical region (Pro) in the tem-

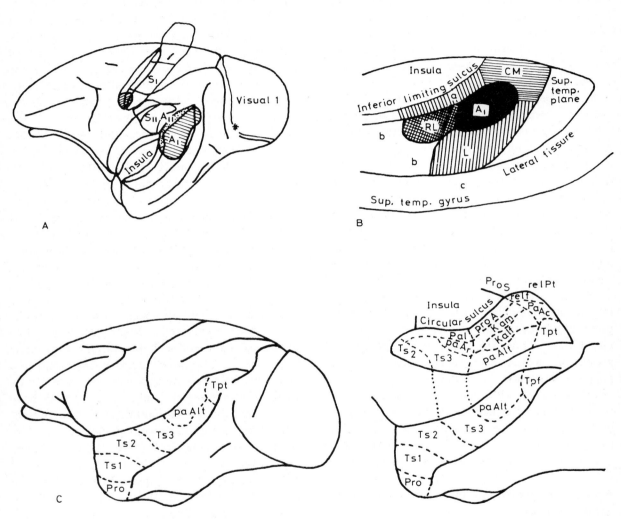

Fig. 20. (A) Diagram of the lateral surface of the cerebral hemisphere of the rhesus monkey showing the locations of primary and second auditory (AI and AII), somatosensory (SI and SII), and visual (VI) areas as described by Woolsey and Fairman (1946). Note that the Sylvian fissure is opened to expose the depths. (B) Schematic drawing of the supratemporal plane and superior temporal gyrus showing major subdivisions of auditory areas based upon the cytoarchitectonic and physiological observations of Merzenich and Brugge (1973). Abbreviations: AI, primary auditory field; RL, rostrolateral field; L, lateral field; CM, caudomedial field; a, field bordering AI medially and rostrally, buried in the circular sulcus; b, area of auditory cortex rostral and lateral to other fields; c, lateral surface of the superior temporal gyrus with additional auditory fields. (C) Architectonic parcellation of the superior temporal gyrus and supratemporal plane as described by Pandya and Sanides (1973) (from Pandya and Yeterian, 1985). Abbreviations: Kalt, lateral koniocortex; Kam, medial koniocortex; paAc, caudal parakoniocortex; paAlt, lateral parakoniocortex; paAr, rostral parakoniocortex; paI, parinsular cortex; proA, prokoniocortex; proS, somatic prokoniocortex (SII); reIpt, retroinsular parietal cortex; reIt, retroinsular temporal cortex; Tpt, temporoparietal cortex; Ts1, Ts2, Ts3, temporalis superior 1, 2, 3.

poral pole. The studies of Jones and Burton have revealed a somewhat similar parcellation of the superior temporal region, albeit using a different nomenclature.

Many investigators have studied the intracortical connections of the auditory association cortices in the rhesus monkey (Whitlock and Nauta, 1956; Hurst, 1959; Myers, 1967; Pandya and Kuypers, 1969; Pandya et al., 1969, 1971; Jones and Powell, 1970b; Trojanowski and Jacobson, 1975; Chavis and Pandya, 1976; Seltzer and Pandya, 1976, 1978; Jacobson and Trojanowski, 1977; Barbas and Mesulam, 1981, 1985; Amaral et al., 1983; Barbas, 1986). On the basis of long association connections the belt areas of the STG are divided into 3 major sectors in a caudorostral direction (Chavis and Pandya, 1976) (Fig. 21A). The first auditory association area (AA1) consists of architectonic regions Tpt, paAlt, and the caudal portion of Ts3. The second auditory association area (AA2) is comprised of the rostral portion of Ts3, and area Ts2, whereas the third auditory association area (AA3) includes the Ts1 and Pro regions. Each of these auditory association areas has a distinctive pattern of long association projections to the frontal lobe, parietotemporal region, and the paralimbic area (Fig. 21B). Thus, area AA1 projects mainly to the dorsal periarcuate cortex in the frontal lobe, in particular dorsal area 8 in the concavity of the arcuate sulcus. The frontal lobe projections from area AA2 are directed primarily to the prearcuate region, area 46 below the principal sulcus, and to the dorsal prefrontal cortex, areas 9 and 10. The third auditory association area, AA3, in the temporal polar region projects mainly to medial prefrontal and orbitofrontal cor-

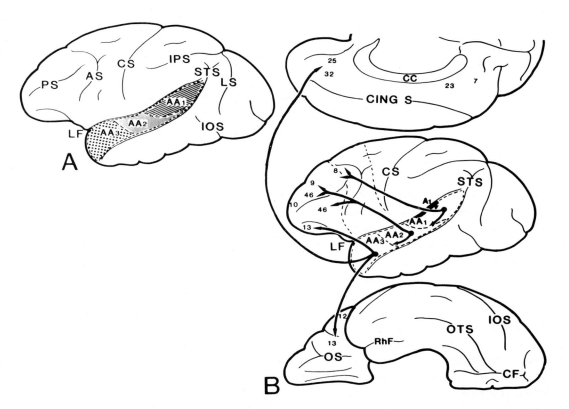

Fig. 21. Diagrams showing (A) the 3 divisions of the auditory association areas of the superior temporal regions, AA_1, AA_2 and AA_3, and (B) their long connections to the frontal lobe.

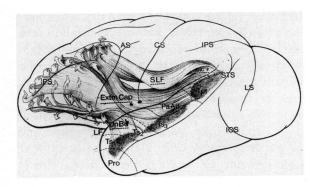

Fig. 22. Artist's schematic rendition of the trajectories of the temporofrontal fibers originating from the rostral (via the uncinate bundle, Un Bd), middle (via the extreme capsule, Extm Cap), and caudal (via the arcuate fasciculus, ARC F, and the superior longitudinal fasciculus, SLF) parts of the superior temporal gyrus (Petrides and Pandya, 1984).

periarcuate region (area 8), whereas area AA2 projects to the prefrontal region (area 46). Finally, area AA3 projects to the orbitofrontal and medial prefrontal regions (areas 12, 13, 25 and 32). As already shown, the distant projections of the auditory association areas to the frontal lobe are directed to specific regions, some of which are recipients of afferents from other sensory modalities. Even more striking is the fact that each of the auditory association regions projects specifically to those other association areas, in the frontal regions, which appear to have analogous architectonic features and to occupy comparable stages within their own architectonic lines (Petrides and Pandya, 1984; Pandya and Yeterian, 1985).

tices, i.e., areas 12, 13, 25 and 32. It should be pointed out that most of these connections between the auditory association areas and the frontal lobe are known to be reciprocal.

These temporofrontal connections are mediated via 3 distinct pathways (Petrides and Pandya, 1988). The fiber pathway from the rostral part of the superior temporal gyrus (Pro, Ts1 and Ts2) projects to the proisocortical areas of the orbital and medial frontal cortices, as well as to nearby orbital and medial areas, and travels to the frontal lobe as part of the uncinate fasciculus. The middle part of the superior temporal gyrus (areas Ts3 and paAlt) projects predominantly to the lateral frontal cortex and to the dorsal aspect of the medial frontal lobe. These temporofrontal fibers originating from the middle part of the superior temporal gyrus travel via the extreme capsule and lie dorsal to the fibers of the uncinate fasciculus. The posterior part of the superior temporal gyrus sends its fibers to the frontal lobe through the extreme capsule as well as the arcuate and superior longitudinal fasciculi (Fig. 22).

Each of the auditory association areas has specific connectional relationships with association areas of the frontal lobe (Pandya and Yeterian, 1985). Thus, area AA1 is connected with the

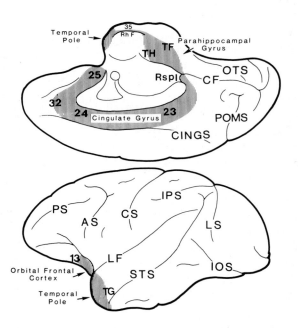

Fig. 23. Diagram showing the various paralimbic areas in the medial and orbitofrontal cortices of the cerebral hemisphere in the rhesus monkey (Yeterian and Pandya, 1988). Note the distribution of paralimbic areas on the orbital surface and the paleo-olfactory region (area 13); in the medial prefrontal cortex (areas 25 and 32); in the cingulate gyrus (areas 24, 23, and the retrosplenial area, Rspl); in the parahippocampal gyrus (areas TH and TF); in the perirhinal cortex (area 35); and in the temporal pole (area TG or 38).

Frontal lobe connections of paralimbic association areas

The paralimbic areas are situated between the sensory association areas and the limbic regions (Fig. 23) and have direct connections with limbic structures. These areas include the cingulate region (areas 23, 24, 25 and 32), the caudal orbitofrontal cortex (area 13), the temporal pole (area TG), and the parahippocampal region (areas TF and TH). Unlike the other association areas described above, which have well-developed 6-layered cortex, the paralimbic regions in general have architectonic features intermediate to those of isocortex and allocortex (Yakovlev et al., 1966; Sanides, 1972; Braak, 1980). In terms of cortical relationships the cingulate gyrus has reciprocal connections with the prefrontal association cortex and the insula (Nauta, 1964; Jones and Powell, 1970b; Pandya et al., 1971, 1981; Baleydier and Mauguière, 1980; Mesulam and Mufson, 1982; Mufson and Mesulam, 1982). The temporal pole receives input from second-order visual and auditory association areas and projects to medial frontal and orbitofrontal cortex (Jones and Powell, 1970b; Van Hoesen et al., 1972; Seltzer and Pandya, 1976; Morán et al., 1987). The orbitofrontal cortex is connected with other prefrontal areas on the one hand and with perirhinal, entorhinal, and parahippocampal regions on the other (Pandya and Kuypers, 1969; Jones and Powell, 1970b; Van Hoesen et al., 1972; Barbas and Pandya, 1987). Finally, the orbitofrontal and the parahippocampal regions are interconnected (Jones and Powell, 1970b; Seltzer and Pandya, 1976; Van Hoesen, 1982; Amaral et al., 1983) (Fig. 24).

There are also interconnections among these paralimbic regions. For example, the caudal orbitofrontal cortex and the temporal pole are reciprocally connected as are the cingulate gyrus and the parahippocampal region (Bailey et al., 1943; Pandya and Kuypers, 1969; Jones and Powell, 1970b; Pandya et al., 1971, 1981; Baleydier and Mauguière, 1980). Although the detailed efferent connections of the paralimbic

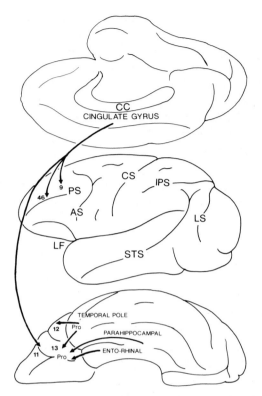

Fig. 24. Diagram showing frontal lobe connections from dorsal (cingulate-archicortical) and ventral (temporal polar-parahippocampal-paleocortical) paralimbic regions.

regions remain to be delineated, a clear dichotomy appears to exist (Fig. 24). Thus, area 23 of the cingulate gyrus is connected predominantly with the lateral prefrontal region, whereas area 24 is related mainly to the premotor region. In contrast, the temporal polar and parahippocampal regions are connected mainly with the caudal orbitofrontal, ventral prefrontal and temporal cortices. Therefore, the cingulate region ties in with the dorsal or hippocampal trend of the frontal lobe as described above, whereas the temporopolar region seems to be related to the ventral or paleocortical trend of the frontal lobe (Sanides, 1972; Galaburda, 1984; Pandya and Yeterian, 1985).

Efferent cortical connections of the frontal lobe

The long association connections of premotor regions, including areas 6 and 8, indicate that these areas are connected with the first-order sensory association regions (Fig. 25B). Thus, area 6 is topographically connected with the rostral parietal association region, SA1. Likewise, the premotor region in front of the arcuate sulcus, area 8, is connected with the first-order auditory association area, AA1. The connections from area 8 to the visual association area are limited (Jones and Powell, 1970b; Pandya and Vignolo, 1971; Künzle and Akert, 1977; Deacon et al., 1982; Godschalk et al., 1984).

The long connections of the prefrontal cortex (Fig. 25A) are directed to the posterior parietal cortex, to the temporal lobe (superior and inferior temporal regions including the superior temporal sulcus) and to the paralimbic areas (Nauta, 1964; Pandya and Kuypers, 1969; Jones and Powell, 1970b; Pandya et al., 1971; Selemon and Goldman-Rakic, 1988). The posterior parietal projections originate from caudal areas 46 and 9, and are directed to areas SA2 and SA3 of the IPL and SPL including the intraparietal sulcus in a

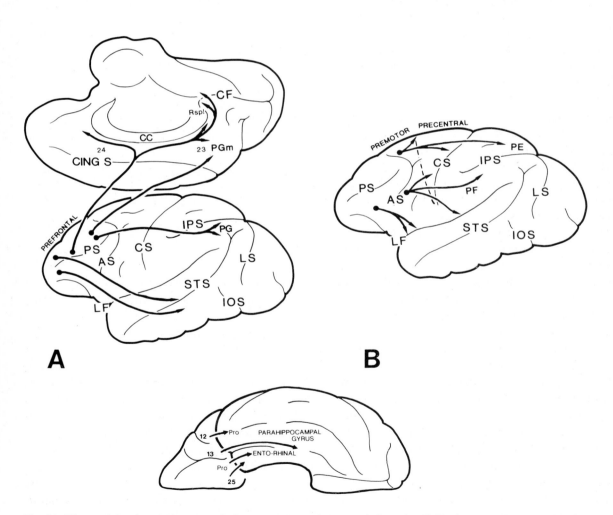

Fig. 25. Diagrams showing the long association connections of the (A) prefrontal and (B) premotor areas.

topographic manner. The temporal lobe connections from prefrontal regions are organized as follows: Dorsal areas 46 and 10 project to the AA2 region as well as to the adjacent superior temporal sulcus, whereas ventral areas 46 and 10 are connected with area VA2 and with the adjacent superior temporal sulcus. The orbitofrontal regions, i.e., areas 12 and 13, as well as medial prefrontal area 25, are connected with areas AA3 and VA3. Finally, the paralimbic connections originate from the dorsomedial and ventral prefrontal regions. Thus, the orbital proisocortical region and area 13 are connected with the temporal polar proisocortical area, as well as the perirhinal and entorhinal regions and the parahippocampal gyrus, areas TH and TL (Van Hoesen, 1982). The connections from dorsal areas 46, 9, and medial proisocortex (area 32) are directed mainly to the cingulate gyrus and to the retrosplenial cortex.

The long association connections of the prefrontal regions appear to be directed to those sensory association areas from which the frontal region receives afferent connections (Pandya and Yeterian, 1985; Selemon and Goldman-Rakic, 1988). That is, most of the connections between pre- and post-Rolandic association areas are reciprocal in nature. The frontal projections to these sensory association areas are organized such that those from the premotor regions are related to first-order association areas. Those from the lateral prefrontal cortex are connected with second-order sensory association areas, and those from the orbital and medial prefrontal regions are related to third-order sensory association areas. The third-order projections to somatosensory association areas originate from the rostral prefrontal region. The paralimbic connections of the frontal region indicate a dichotomy, such that the ventral trend regions, especially ventral proisocortex as well as surrounding areas, are related to the ventral temporal paralimbic regions. In contrast, the dorsal proisocortex and its immediately surrounding regions project to the cingulate gyrus and retrosplenial cortex. These connectional and architectural dichotomies of prefrontal regions were suggested originally by Nauta (1964) and Sanides (1972), and further emphasize the dual nature of cortical development.

Discussion

These observations indicate that the frontal lobe as well as the major divisions of the post-Rolandic cortex are organized systematically in terms of both architecture and connections. With regard to architecture it seems that the entire cortex has evolved from two moieties, namely the paleocortical and archicortical. In terms of the frontal lobe, the paleocortical trend begins in the orbital region and progresses towards the ventrolateral prefrontal cortex. Likewise, the archicortical trend stems from the medial proisocortical region and continues toward the dorsolateral prefrontal cortex. Within these trends, as one moves from the prime moiety outward, there is a progressive elaboration of supragranular layers and a decreased emphasis on infragranular layers. It is this elaboration of supragranular layers that allows the frontal lobe cortex to receive and to forward highly integrated information. Along with the observed elaboration in cortical architecture there are concomitant systematic patterns of corticocortical connectivity. It appears that this connectivity mirrors architectonic organization such that less differentiated frontal lobe cortical regions are closely related to less differentiated post-Rolandic areas and more differentiated frontal regions preferentially with

Paleocortical trend (Ventral trend)

Deficits: Emotional lability
 Personality changes
 Perseveration

Functions: Planning and sequencing
 mood

Archicortical trend (Dorsal trend)

Deficits: Behaviour initiation
 attention

Functions: Drive, motivation and will

Fig. 26. Summary of functional roles of the paleocortical (ventral) and archicortical (dorsal) trends of the frontal lobe.

more differentiated post-Rolandic areas (Pandya and Yeteran, 1985).

The dual architectonic origin of the frontal lobe may be the basis for the dichotomous functional roles ascribed to the frontal region (Fig. 26). Clinical observations following damage to the ventral frontal lobe, which stems from the paleocortical moiety, typically emphasize emotional lability, personality changes and perseveration, i.e., in part deficits in day-to-day function relating to planning, sequencing and mood. In contrast, observations of behavioral changes following damage to midline frontal regions, which arise from the archicortical moiety, often include deficits in behavioral initiation, attention and communication, i.e., in functions relating to drive, motivation and will (e.g., Stuss and Benson, 1986).

Within each of the frontal architectonic trends there appears to be increasing functional complexity as one proceeds towards the lateral convexity of the frontal lobe (Fig. 27). Thus, deficits resulting from damage to the lateral frontal region involve mainly attentional, temporal and integrative functions and are different from changes relating to emotion following orbitofrontal involvement and motivational changes following medial frontal involvement (e.g., Fuster, 1989).

It is of interest to note that the existence of dual architectonic trends in the frontal lobe is reflected in the patterns of connectivity between post-Rolandic and frontal regions. In the somatosensory system (Fig. 28), the inferior parietal association regions relating to head, neck and face project to the ventral frontal cortices, i.e., to the ventral

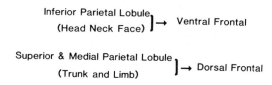

Fig. 28. Summary of the relationships between the inferior parietal, and superior and medial parietal, lobules and the ventral and dorsal frontal trends.

trend. In contrast, the superior and medial parietal association areas relating to trunk and limbs project to the dorsal and medial regions, i.e., to the dorsal trend of the frontal lobe (Barbas and Mesulam, 1981, 1985; Petrides and Pandya, 1984; Pandya and Yeterian, 1985). In the visual system, the association areas of the inferior temporal region related to central vision project to the ventral portion of the frontal lobe, i.e., to the ventral trend (Fig. 29). In contrast, the dorsomedial and ventral visual association areas relating to peripheral vision project to the dorsal and medial portions of the frontal lobe, i.e., to the dorsal trend (Rosene and Pandya, 1983).

With regard to limbic cortices, there also appears to be a dichotomous pattern of connections (Fig. 30). The ventral frontal cortices receive principal limbic input from the ventral temporal limbic areas (entorhinal and perirhinal cortices, and parahippocampal gyrus), whereas the dorsal and medial frontal regions receive input predominantly from the cingulate and retrosplenial cortices

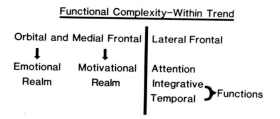

Fig. 27. Summary of functional roles of the orbital and medial, and lateral frontal trends of the frontal lobe.

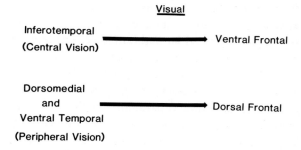

Fig. 29. Summary of relationships between the visual cortices and the ventral and dorsal frontal trends.

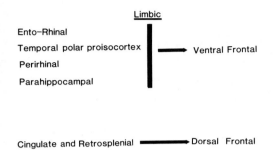

Fig. 30. Summary of relationships between paralimbic regions and the ventral and dorsal frontal trends.

(Baleydier and Mauguière, 1980; Pandya et al., 1981; Van Hoesen, 1982; Pandya and Yeterian, 1985).

These data are consistent with the notion that the dorsal and ventral trends within the frontal lobe have differential functions. Thus, based on connections from the face, head and neck regions of the somatosensory association cortex, the inferior temporal region relating to central vision and from ventral temporal paralimbic regions, the ventral frontal areas would appear to be closely tied to the synthesis of object – emotion relationships in a behavioral context. It is not unreasonable to view the information regarding object identity and significance conveyed from post-Rolandic areas to the frontal lobe as a key aspect of the emotional reactivity long ascribed to ventral frontal regions. Damage to the ventral frontal region has been associated with shallow mentation or emotion. The dorsal frontal trend, in light of its input from the trunk and limb regions of somatosensory association cortex, from visual association areas relating to peripheral vision as well as from medial paralimbic cortices appears to be involved in the integration of visospatial and motivational processes. In this regard it is reasonable to consider the coming together of trunk and limb information, peripheral visual and midline limbic influences as essential for the energizing and guidance of behavior in three-dimensional space. This notion receives support from clinical observations that damage to the cingulate gyrus leads to an akinetic state and likewise that supplementary motor area damage leads to hesitancy of speech and in initiating movement. Moreover, medial prefrontal damage leads to so-called apathetic states (e.g., Stuss and Benson, 1986).

Another significant observation regarding the post-Rolandic input to the frontal lobe is that the cortical areas at similar architectonic stages appear to be strongly interconnected. Additionally, specific frontal regions have distinct connections with the mediodorsal nucleus of the thalamus. The importance of these types of connectivity has been emphasized by several investigators (e.g., Nauta, 1964; Damasio, 1985; Stuss and Benson, 1986). It would appear that as cortical areas evolved progressively in the frontal and post-Rolandic regions, those which arose at the same time formed close connectional relationships. This would imply that frontal areas along with their related post-Rolandic as well as thalamic regions may work in concert to subserve common functions.

Moreover, there are systematic connections from the frontal lobe to the post-Rolandic sensory association and limbic cortices as well as to the thalamus. These connections are reciprocal to those just outlined. Thus, the ventral frontal regions project back to the inferior parietal lobule, the inferior temporal region, the superior temporal gyrus, ventral paralimbic cortices and ventral portions of the MD nucleus. In contrast, the dorsal frontal region projects back to the superior parietal lobule, the medial and ventral occipitotemporal regions, the superior temporal gyrus, medial paralimbic cortices and the dorsal portion of the MD nucleus. The existence of substantial reciprocal connections from the frontal lobe may be considered as a neural substrate for the concept of corollary discharge as proposed by Teuber (1972). These connections may have a significant role in the generation and maintenance of meaningful behavior. Nauta (1971, 1973) has proposed an interpretation of frontal lobe function based on anatomical considerations. He suggests that the unique feature of the neural circuitry of the frontal lobe is its reciprocal relationship with post-Rolandic sensory association areas and with the telencephalic limbic system. Nauta suggests further that the frontal cortex both monitors and modulates limbic mechanisms and that frontal lobe syndromes may involve the inability to integrate the "internal milieu" with cortical sensory information regarding the external environment. It seems that the concepts proposed by Teuber and Nauta, which emphasize the reciprocal relationships of the frontal lobe with other cortical association and paralimbic regions, are consistent with evolutionary architectonic and connectional concepts.

In conclusion, the evolutionary concept described in terms of cortical architecture and connections may lead to a better understanding of frontal lobe mechanisms as well as clinical disorders. The dual organizational trends with sequential progression from the rudimentary cortices toward highly developed frontal regions may provide a framework for further experimental studies. This cortical evolutionary approach may allow the generation of systematic hypotheses aimed at revealing the contribution of specific architectonic regions and their related connections. In other words, architectonic and connectional data may allow for a stepwise approach for future physiological and behavioral studies of the frontal lobe. Finally, the general concept of structural and functional duality at the cortical level has a parallel in the work of Mishkin and his collegues in the visual system (Mishkin et al., 1983). Based on anatomical, physiological and behavioral studies, they have proposed that there are two cortical visual systems; a ventral system engaged in object analysis, and a dorsal system involved in visuospatial function. The functional contributions of the frontal lobe may be organized along similar lines, with the ventral frontal lobe dealing with ramifications emanating from the basic question of "what" and the dorsal frontal lobe with the ramifications of "where".

Acknowledgments

We are very grateful to Mr. Michael Schorr for excellent technical assistance.

This study is supported by the Veterans Administration, Edith Nourse Rogers Memorial Veterans Hospital, Bedford, MA; by NIH grant 16841; and by Colby College Social Science grant 01 2216.

References

Abbie, A.A. (1940) Cortical lamination in the Monotremata. *J. Comp. Neurol.*, 72: 428–467.

Abbie, A.A. (1942) Cortical lamination in a polyprodont marsupial, *Perameles nasuta. J. Comp. Neurol.*, 76: 509–536.

Ades, H.W. and Felder, R. (1942) The acoustic area of the monkey *(Macaca mulatta). J. Neurophysiol.*, 5: 49–54.

Akert, K. (1964) Comparative anatomy of frontal cortex and thalamo-frontal connections. In J.M. Warren and K. Akert, (Eds.), *The Frontal Granular Cortex and Behavior*, McGraw-Hill, New York, pp. 372–396.

Allman, J.M., Baker, J.F., Newsome, W.T. and Petersen, S.E. (1981) Visual topography and function: Cortical visual areas in the owl monkey. In C.N. Woolsey (Ed.), *Cortical Sensory Organization. Vol. 2. Multiple Visual Areas*, Humana Press, Clifton, NJ, pp. 171–185.

Amaral, D.G., Insausti, R. and Cowan, W.M. (1983) Evidence for a direct projection from the superior temporal gyrus to the entorhinal cortex in the monkey. *Brain Res.*, 275: 263–277.

Bailey, P., Von Bonin, G., Garol, H.W. and McCulloch, W.S. (1943) Long association fibers in cerebral hemispheres of monkey and chimpanzee. *J. Neurophysiol.*, 6: 129–134.

Baleydier, C. and Mauguière, F. (1980) The duality of the cingulate gyrus in monkey. *Brain*, 103: 525–554.

Barbas, H. (1986) Pattern in the laminar organization of corticocortical connections. *J. Comp. Neurol.*, 252: 415–422.

Barbas, H. (1988) Anatomic organization of basoventral and mediodorsal visual recipient prefrontal regions in the rhesus monkey. *J. Comp. Neurol.*, 276: 313–342.

Barbas, H. and Mesulam, M.-M. (1981) Organization of afferent input to subdivisions of area 8 in the rhesus monkey. *J. Comp. Neurol.*, 200: 407–432.

Barbas, H. and Mesulam, M.-M. (1985) Cortical afferent input to the principalis region of the rhesus monkey. *Neuroscience*, 15: 619–637.

Barbas, H. and Pandya, D.N. (1987) Architecture and frontal

cortical connections of the premotor cortex (area 6) in the rhesus monkey. *J. Comp. Neurol.,* 256: 211–228.

Barbas, H. and Pandya, D.N. (1989) Architecture and intrinsic connections of the prefrontal cortex in the rhesus monkey. *J. Comp. Neurol.,* 286: 353–375.

Benevento, L.A. and Rezak, M. (1976) The cortical projections of the inferior pulvinar and adjacent lateral pulvinar in the rhesus monkey *(Macaca mulatta)*: An autoradiographic study. *Brain Res.,* 108: 1–24.

Benevento, L.A. and Yoshida, K. (1981) The afferent and efferent organization of the lateral geniculo-prestriate pathways in the macaque monkey. *J. Comp. Neurol.,* 203: 455–474.

Bianchi, L. (1895) The function of the frontal lobes. *Brain,* 18: 497–522.

Braak, H. (1980) *Architectonics of the Human Telencephalic Cortex,* Springer, Berlin.

Brodmann, K. (1908) Beitrage zur histologischen Lokalisation der Grosshirnrinde, VI. Mitteilung: Die Cortexgliederung des Menschen. *J. Psychol. Neurol.,* 10: 231–246.

Brodmann, K. (1909) *Vergleichende Lokalisationlehre der Grosshirnrinde in ihren Prinzipien dargestellt auf Grund des Zellenbaues,* Barth, Leipzig.

Brugge, J.F. and Reale, R.A. (1985) Auditory cortex. In A. Peters and E.G. Jones (Eds.), *Cerebral Cortex. Vol. 4. Association and Auditory Cortices,* Plenum, New York, pp. 229–271.

Bullier, J. and Kennedy, H. (1983) Projection of the lateral geniculate nucleus onto cortical area V2 in the macaque monkey. *Exp. Brain Res.,* 53: 168–172.

Chavis, D.A. and Pandya, D.N. (1976) Further observation on corticofrontal connections in the rhesus monkey. *Brain Res.,* 117: 369–386.

Clark, W.E.L. (1936) The thalamic connections of the temporal lobe of the brain in the monkey. *J Anat.,* 70: 447–464.

Damasio, A. (1985) The frontal lobes. In K.M. Heilman and E. Valenstein (Eds.), *Clinical Neuropsychology, 2nd Edn.,* Oxford University Press, New York, pp. 339–375.

Damasio, A.R. and Van Hoesen, G.W. (1980) Structure and function of the supplementary motor area. *Neurology,* 30: 359.

Daniel, P.M. and Whitteridge, D. (1961) The representation of the visual field on the cerebral cortex in monkeys. *J. Physiol.,* 159: 203–221.

Dart, R.A. (1934) The dual structure of the neopallium: Its history and significance. *J Anat.,* 69: 3–19.

Deacon, T.W., Rosenberg, P., Eckert, M.K. and Shank, C.E. (1982) Afferent connections of the primate inferior arcuate cortex. *Anat. Rec.,* 8: 933.

Doty, R.W. (1983) Nongeniculate afferents to striate cortex in macaques. *J. Comp. Neurol.,* 218: 159–173.

Duffy, F.H. and Burchfiel, J.L. (1971) Somatosensory system: Organizational hierarchy from single units in monkey area 5. *Science,* 172: 273–275.

Ferrier, D. (1876) *The Functions of the Brain,* Smith, Elder, London.

Fulton, J.F. and Jacobson, C.F. (1935) The functions of the frontal lobes: A comparative study in monkeys, chimpanzees and man. *Adv. Mod. Biol. (Moscow),* 4: 113–123.

Fuster, J.M. (1989) *The Prefrontal Cortex,* 2nd Edn., Raven Press, New York.

Galaburda, A.M. (1984) Anatomy of language; lessons from comparative anatomy. In D. Kaplan, A.R. Lecours and J. Marshall (Eds.), *Neurolinguistics,* MIT Press, Cambridge, MA, pp. 398–415.

Giguere, M. and Goldman-Rakic, P.S. (1988) Mediodorsal nucleus: Areal, laminar, and tangential distribution of afferents and efferents in the frontal lobe of rhesus monkeys. *J. Comp. Neurol.,* 277: 195–213.

Godschalk, M., Lemon, R., Nijs, H.G.T. and Kuypers, H.G.J.M. (1984) Cortical afferents and efferents of monkey postarcuate area: An anatomical and electrophysiological study. *Exp. Brain Res.,* 56: 410–424.

Goldman-Rakic, P.S. (1987) Circuitry of primate prefrontal cortex and regulation of behavior by representational memory. In F. Plum and V. Mountcastle (Eds.), *Handbook of Physiology. Vol. 5. The Nervous System,* American Physiological Society, Bethesda, MD, pp. 373–417.

Goldman-Rakic, P.S. and Porrino, L.J. (1985) The primate mediodorsal (MD) nucleus and its projection to the frontal lobe. *J. Comp. Neurol.,* 242: 535–560.

Goldstein, K. (1939) Clinical and theoretic aspects of lesions of the frontal lobes. *Arch. Neurol. Psychiat.,* 41: 865–867.

Gross, C.G., Bruce, C.J., Desimone, R., Fleming, J. and Gattass, R. (1981) Cortical visual areas of the temporal lobe: Three areas in the macaque. In C.N. Woolsey (Ed.), *Cortical Sensory Organization. Vol. 2. Multiple Visual Areas,* Humana Press, Clifton, NJ, pp. 187–216.

Hitzig, P. (1874) Untersuchungen über das Gehirn, Hirschwald, Berlin. Quoted in L. Bianchi (1922) *The Mechanism of the Brain and the Function of the Frontal Lobes,* William Wood, New York.

Hubel, D.H. and Wiesel, T.N. (1968) Receptive fields and functional architecture of monkey striate cortex. *J. Physiol.,* 195: 215–243.

Hubel, D.H. and Wiesel, T.N. (1972) Laminar and columnar distribution of geniculocortical fibers in the macaque monkey. *J. Comp. Neurol.,* 146: 421–450.

Hurst, E.M. (1959) Some cortical association systems related to auditory functions. *J. Comp. Neurol.,* 112: 103–119.

Hyvärinen, J. (1982) *The Parietal Cortex of Monkey and Man.* Springer, Berlin.

Jacobson, S. and Trojanowski, J.Q. (1977) Prefrontal granular cortex of the rhesus monkey. I. Intrahemispheric cortical afferents. *Brain Res.,* 132: 209–233.

Jones, E.G. and Burton, H. (1976) Areal differences in the laminar distribution of thalamic afferents in cortical fields of the insular, parietal and temporal regions of primates. *J.*

Comp. Neurol., 168: 197–248.

Jones, E.G. and Powell, T.P.S. (1969) Connexions of the somatic sensory cortex in the rhesus monkey. I. Ipsilateral cortical connexions. Brain, 92: 477–502.

Jones, E.G. and Powell, T.P.S. (1970a) Connexions of the somatic sensory cortex of the rhesus monkey. III. Thalamic connexions. Brain, 93: 37–56.

Jones, E.G. and Powell, T.P.S. (1970b) An anatomical study of converging sensory pathways within the cerebral cortex of the monkey. Brain, 93: 793–820.

Jones, E.G., Wise, S.P. and Coulter, J.D. (1979) Differential thalamic relationships of sensory-motor and parietal fields in monkeys. J. Comp. Neurol., 183: 833–882.

Kaas, J.H., Sur, M., Nelson, R.I. and Merzenich, M.M. (1981) The postcentral somatosensory cortex: Multiple representations of the body in primates. In C.N. Woolsey (Ed.), Cortical Sensory Organization. Vol. 1. Multiple Somatic Areas, Humana Press, Clifton, NJ, pp. 29–45.

Kennedy, T.T.K. (1955) An Electrophysiological Study of the Auditory Projection Areas of the Cortex in Monkey (Macaca mulatta), Thesis, University of Chicago, Chicago, IL.

Kievit, J. and Kuypers, H.G.J.M. (1977) Organization of the thalamocortical connexions to the frontal lobe in the rhesus monkey. Exp. Brain Res., 29: 299–322.

Künzle, H. and Akert, K. (1977) Efferent connections of cortical area 8 (frontal eye field) in Macaca fascicularis: A reinvestigation using the autoradiographic technique. J. Comp. Neurol., 173: 147–164.

Kuypers, H.G.J.M., Szwarcbart, M.K., Mishkin, M. and Rosvold, H.E. (1965) Occipitotemporal corticocortical connections in the rhesus monkey. Exp. Neurol., 11: 245–262.

Leinonen, L., Hyvärinen, J. and Sovijarvi, A.R.A. (1980) Functional properties of neurons in the temporo-parietal association cortex of awake monkey. Exp. Brain Res., 39: 203–215.

Lhermitte, F. (1986) Human autonomy and the frontal lobes. Part II. Patient behavior in complex social situations: The "environmental dependency" syndrome. Ann. Neurol., 19: 335–343.

Lhermitte, F., Pillon, B. and Serdaru, M. (1986) Human autonomy and the frontal lobes. Part I. Imitation and utilization behavior: A neuropsychological study of 75 patients. Ann. Neurol., 19: 326–334.

Luria, A.R. (1973a) The Working Brain. An Introduction to Neuropsychology, translated by B. Haigh. Basic Books, New York.

Luria, A.R. (1973b) The frontal lobes and the regulation of behavior. In K.H. Pribram and A.R. Luria (Eds.), Psychophysiology of the Frontal Lobes, Academic Press, New York, pp. 3–26.

Lynch, J.C. (1980) The functional organization of the posterior parietal association cortex. Behav. Brain Sci., 3: 485–499.

Merzenich, M.M. and Brugge, J.F. (1973) Representation of the cochlear partition on the superior temporal plane of the macaque monkey. Brain Res., 50: 275–296.

Merzenich, M.M., Sur, M., Nelson, R.J. and Kaas, J.H. (1981) Multiple cutaneous representations in areas 3b and I of the owl monkey. In C.N. Woolsey (Ed.), Cortical Sensory Organization. Vol. 1. Multiple Somatic Areas, Humana Press, Clifton, NJ, pp. 67–119.

Mesulam, M.-M. and Mufson, E.J. (1982) Insula of the Old World monkey. III. Efferent cortical output and comments on function. J. Comp. Neurol., 212: 38–52.

Mesulam, M.-M. and Pandya, D.N. (1973) The projections of the medial geniculate complex within the Sylvian fissure of the monkey. Brain Res., 60: 315–333.

Milner, B. (1982) Some cognitive effects of frontal lobe lesions in man. In D.E. Broadbent and L. Weiskrantz (Eds.), The Neuropsychology of Cognitive Function, The Royal Society, London, pp. 211–226.

Milner, B. and Petrides, M. (1984) Behavioural effects of frontal-lobe lesions in man. Trends Neurosci., 7: 403–407.

Mishkin, M. (1972) Cortical visual areas and their interactions. In A.G. Karczmar and J.C. Eccles (Eds.), Brain and Human Behavior, Springer, Berlin, pp. 187–208.

Mishkin, M., Ungerleider, L.G. and Macko, K.A. (1983) Object vision and spatial vision: Two cortical pathways. Trends Neurosci., 6: 414–417.

Morán, M.A., Mufson, E.J. and Mesulam, M.-M. (1987) Neural inputs into the temporopolar cortex of the rhesus monkey. J. Comp. Neurol., 256: 88–103.

Mountcastle, V.B., Lynch, J.C., Georgopoulos, A., Sakata, H. and Acuna, C. (1975) Posterior parietal association cortex of the monkey: Command functions for operations within extrapersonal space. J. Neurophysiol., 38: 871–909.

Mufson, E.J. and Mesulam, M.-M. (1982) Insula of the Old World monkey. II. Afferent cortical input and comments on the claustrum. J. Comp. Neurol., 212: 23–37.

Myers, R.E. (1967) Cerebral connectionism and brain function. In C.H. Millikan and F.L. Darley (Eds.), Brain Mechanisms Underlying Speech and Language, Grune and Stratton, New York, pp. 61–72.

Nauta, W.J.H. (1964) Some efferent connections of the prefrontal cortex in the monkey. In J.M. Warren and K. Akert (Eds.), The Frontal Granular Cortex and Behavior, McGraw-Hill, New York, pp. 397–409.

Nauta, W.J.H. (1971) The problem of the frontal lobe: A reinterpretation. J. Psychiat. Res., 8: 167–187.

Nauta, W.J.H. (1973) Connections of the frontal lobe with the limbic system. In L.V. Laitinen and K.E. Livingston (Eds.), Surgical Approaches in Psychiatry, University Park Press, Baltimore, MD, pp. 303–314.

Newsome, W.T., Maunsell, J.H.R. and Van Essen, D.C. (1986) Ventral posterior visual area of the macaque: Visual topography and areal boundaries. J. Comp. Neurol., 252: 139–153.

Olszewski, J. (1952) The Thalamus of the Macaca mulatta. S. Karger, Basel.

Pandya, D.N. (1987) General Discussion II. In *Motor Areas of the Cerebral Cortex, CIBA Foundation Symposium 132,* John Wiley, New York, pp. 165–170.

Pandya, D.N. and Barbas, H. (1985) Architecture and connections of the premotor areas in the rhesus monkey. *Behav. Brain Sci.,* 8: 595–597.

Pandya, D.N. and Barnes, C.L. (1987) Architecture and connections of the frontal lobe. In E. Perecman (Ed.), *The Frontal Lobes Revisited,* IRBN Press, New York, pp. 41–72.

Pandya, D.N. and Kuypers, H.G.J.M. (1969) Cortico-cortical connections in the rhesus monkey. *Brain Res.,* 13: 13–36.

Pandya, D.N. and Sanides, F. (1973) Architectonic parcellation of the temporal operculum in rhesus monkey, and its projection pattern. *Z. Anat. Entwicklungsgesch.,* 139: 127–161.

Pandya, D.N. and Seltzer, B. (1982) Intrinsic connections and architectonics of posterior parietal cortex in the rhesus monkey. *J. Comp. Neurol.,* 204: 196–210.

Pandya, D.N. and Vignolo, L.A. (1971) Intra- and interhemispheric projections of the precentral, premotor and arcuate areas in the rhesus monkey. *Brain Res.,* 26: 217–233.

Pandya, D.N. and Yeterian, E.H. (1985) Architecture and connections of cortical association areas. In A. Peters and E.G. Jones (Eds.), *Cerebral Cortex. Vol. 4. Association and Auditory Cortices,* Plenum, New York, pp. 3–61.

Pandya, D.N., Hallett, M. and Mukherjee, S.K. (1969) Intra- and interhemispheric connections of the neocortical auditory system in the rhesus monkey. *Brain Res.,* 14: 49–65.

Pandya, D.N., Dye, P. and Butters, N. (1971) Efferent cortico-cortical projections of the prefrontal cortex of the rhesus monkey. *Brain Res.,* 31: 35–46.

Pandya, D.N., Mufson, E.J. and McLaughlin, T.J. (1980) Some projections of the frontal operculum (gustatory area) in the rhesus monkey. *Anat. Rec.,* 196: 143A.

Pandya, D.N., Van Hoesen, G.W. and Mesulam, M.-M. (1981) Efferent connections of the cingulate gyrus in the rhesus monkey. *Exp. Brain Res.,* 42: 319–330.

Pandya, D.N., Rosene, D.L. and Galaburda, A.M. (1986) Thalamic connections of the superior temporal region in rhesus monkey. *Soc. Neurosci. Abstr.,* 12: 1368.

Pandya, D.N., Seltzer, B. and Barbas, H. (1988) Input-output organization of the primate cerebral cortex. In H.D. Steklis and J. Erwin (Eds.), *Comparative Primate Biology, Vol. 4. Neurosciences,* Alan R. Liss, New York, pp. 39–80.

Paul, R.L., Merzenich, M.M. and Goodman, H. (1975) Mechanoreceptor representation and topography of Brodmann's areas 3 and 1 of *Macaca mulatta.* In H.H. Kornhuber (Ed.), *The Somatosensory System,* Thieme, Stuttgart, pp. 262–269.

Petrides, M. and Pandya, D.N. (1984) Projections to the frontal cortex from the posterior parietal region in the rhesus monkey. *J. Comp. Neurol.,* 228: 105–116.

Powell, T.P.S. and Mountcastle, V.B. (1959) Some aspects of the functional organization of the postcentral gyrus of the monkey: A correlation of findings obtained in a single unit analysis with architecture. *Bull. Johns Hopkins Hosp.,* 105: 133–162.

Pribram, K.H. (1960) The intrinsic systems of the forebrain. In J. Field, H.W. Magoun and V.E. Hall (Eds.), *Handbook of Physiology. Vol. II. Neurophysiology,* American Physiological Society, Washington, DC, pp. 1323–1344.

Randolph, M. and Semmes, J. (1974) Behavioral consequences of selective subtotal ablations in the postcentral gyrus of *Macaca mulatta. Brain Res.,* 70: 55–70.

Reep, R. (1985) Relationship between prefrontal and limbic cortex: A comparative anatomical review. *Brain Behav. Evol.,* 25: 5–80.

Robinson, D.L. and Goldberg, M.E. (1978) Sensory and behavioral properties of neurons in posterior parietal cortex of the awake, trained monkey. *Fed. Proc.,* 37: 2258–2261.

Rosene, D.L. and Pandya, D.N. (1983) Architectonics and connections of the posterior parahippocampal gyrus in the rhesus monkey. *Soc. Neurosci. Abstr.,* 9: 222.

Rosenkilde, C.E. (1979) Functional heterogeneity of the prefrontal cortex in the monkey: A review. *Behav. Neural Biol.,* 25: 301–345.

Sakata, H. (1975) Somatic sensory responses of neurons in the parietal association area (area 5) in monkeys. In H.H. Kornhuber (Ed.), *The Somatosensory System,* Thieme, Stuttgart, pp. 250–261.

Sanides, F. (1969) Comparative architectonics of the neocortex of mammals and their evolutionary interpretation. *Ann. N.Y. Acad. Sci.,* 167: 404–423.

Sanides, F. (1972) Representation in the cerebral cortex and its areal lamination patterns. In G.F. Bourne (Ed.), *Structure and Function of Nervous Tissue, Vol. 5,* Academic Press, New York, pp. 329–453.

Selemon, L.D. and Goldman-Rakic, P.S. (1988) Common cortical and subcortical targets of the dorsolateral prefrontal and posterior parietal cortices in the rhesus monkey: Evidence for a distributed neural network subserving spatially guided behavior. *J. Neurosci.,* 8: 4049–4068.

Seltzer, B. and Pandya, D.N. (1976) Some cortical projections to the parahippocampal area in the rhesus monkey. *Exp. Neurol.,* 50: 146–160.

Seltzer, B. and Pandya, D.N. (1978) Afferent cortical connections and architectonics of the superior temporal sulcus and surrounding cortex in the rhesus monkey. *Brain Res.,* 149: 1–24.

Siwek, D. (1989) *The Topographic Organization of the Connections between the Mediodorsal Nucleus and the Prefrontal Cortex in the Rhesus Monkey.* Thesis, Boston University, Boston, MA.

Stuss, D.T. and Benson, D.F. (1986) *The Frontal Lobes,* Raven Press, New York.

Suzuki, H. (1985) Distribution and organization of visual and auditory neurons in the monkey prefrontal cortex. *Vision Res.,* 25: 465–469.

Suzuki, H. and Azuma, M. (1983) Topographic studies on visual neurons in the dorsolateral prefrontal cortex of the monkey. *Exp. Brain Res.*, 53: 47–58.

Teuber, H.L. (1972) Unity and diversity of frontal lobe functions. *Acta Neurobiol. Exp.*, 32: 615–656.

Trojanowski, J.Q. and Jacobson, S. (1975) A combined horseradish peroxidase-autoradiographic investigation of reciprocal connections between superior temporal gyrus and pulvinar in squirrel monkey. *Brain Res.*, 85: 347–353.

Ungerleider, L.G. and Mishkin, M. (1982) Two cortical visual systems. In: D.J. Ingle, M.A. Goodale and R.J.W. Mansfield (Eds.), *Advances in the Analysis of Visual Behavior,* MIT Press, Cambridge, MA, pp. 549–486.

Van Essen, D.C. and Maunsell, J.H.R. (1983) Hierarchical organization and functional streams in the visual cortex. *Trends Neurosci.*, 6: 370–375.

Van Hoesen, G.W. (1982) The parahippocampal gyrus. *Trends Neurosci.*, 5: 345–350.

Van Hoesen, G.W, Pandya, D.N. and Butters, N. (1972) Cortical afferents to the entorhinal cortex of the rhesus monkey. *Science,* 175: 1471–1473.

Vogt, C. and Vogt, O. (1919) Allgemeinere Ergebnisse unserer Hirnforschung. *J. Psychol. Neurol.*, 25: 279–461.

Von Bonin, G. and Bailey, P. (1947) *The Neocortex of Macaca mulatta,* University of Illinois Press, Urbana, IL.

Von Economo, C. and Horn, L. (1930) Über Windungsrelief, Masse und Rindenarchitektonik der Supratemporalfläche, ihre individuellen und ihr Seitenunterschiede. *Z. Gesamte Neurol. Psychiat.*, 130: 678–857.

Walker, A.E. (1938) *The Primate Thalamus,* University of Chicago Press, Chicago, IL.

Walker, A.E. (1940) A cytoarchitectural study of the prefrontal areas of the macaque monkey. *J. Comp. Neurol.*, 73: 59–86.

Walzl, E.M. and Woolsey, C.N. (1943) Cortical auditory areas of the monkey as determined by electrical excitation of nerve fibers in the osseous spiral lamina and by click stimulation. *Fed. Proc.*, 2: 52.

Weber, J.T. and Yin, T.C.T. (1984) Subcortical projections of the interior parietal cortex (area 7) in the stumptail monkey. *J. Comp. Neurol.*, 224: 206–230.

Wegener, J.G. (1969) Auditory discrimination behavior of brain damaged monkey. *J. Aud. Res.*, 4: 227–239.

Wegener, J.G. (1976) Auditory and visual discrimination following lesions of the anterior supratemporal plane in monkeys. *Neuropsychologia*, 14: 161–174.

Weiskrantz, L. and Mishkin, M. (1958) Effects of temporal and frontal cortical lesions on auditory discrimination in monkeys. *Brain,* 81: 406–414.

Weller, R.E. and Kaas, J.H. (1983) Retinotopic patterns of connections of area 17 with visual areas V-II and MT in macaque monkeys. *J. Comp. Neurol.*, 220: 253–279.

Whitlock, D.G. and Nauta, W.J.H. (1956) Subcortical projections from the temporal neocortex in *Macaca mulatta. J. Comp. Neurol.*, 106: 183–212.

Winfield, D.A., Gatter, K.C. and Powell, T.P.S. (1975) Certain connections of the visual cortex of the monkey shown by the use of horseradish peroxidase. *Brain Res.*, 92: 456–461.

Woolsey, C.N. (1958) Organization of somatic sensory and motor areas of the cerebral cortex. In H.F. Harlow and C.N. Woolsey (Eds.), *Biological and Biochemical Bases of Behavior,* University of Wisconsin Press, Madison, WI, pp. 63–81.

Woolsey, C.N. and Fairman, D. (1946) Contralateral, ipsilateral, and bilateral representation of cutaneous receptors in somatic areas I and II of the cerebral cortex of pig, sheep and other mammals. *Surgery,* 19: 684–702.

Yakovlev, P.I., Locke, S. and Angevine, J.B. (1966) The limbus of the cerebral hemisphere, limbic nuclei of the thalamus, and the cingulum bundle. In D.P. Purpura and M.D. Yahr (Eds.), *The Thalamus,* Columbia University Press, New York, pp. 77–91.

Yeterian, E.H. and Pandya, D.N. (1985) Corticothalamic connections of the posterior parietal cortex in the rhesus monkey. *J. Comp. Neurol.*, 237: 408–426.

Yeterian, E.H. and Pandya, D.N. (1988) Corticothalamic connections of paralimbic regions in the rhesus monkey. *J. Comp. Neurol.*, 269: 130–146.

Yukie, M. and Iwai, E. (1981) Direct projection from the dorsal lateral geniculate nucleus to the prestriate cortex in macaque monkeys. *J. Comp. Neurol.*, 201: 81–98.

Zeki, S.M. (1978) Functional specialization in the visual cortex of the rhesus monkey. *Nature,* 274: 423–428.

Discussion

R.W.H. Verwer: Could you, based on your hypothesis of evolutionary trends of cortical areas predict what new (future) cortical regions would look like with respect to their layering (not only beyond area 8 but also MII etc.)?

D.N. Pandya: It is difficult to predict what nature's plans are for further development of the brain. Nevertheless, we would presume for area 8 that the major features would be increasing granularization along with the acquisition of additional pyramidal neurons in layer III. With regard to MII, once again the pattern would be to acquire larger pyramidal neurons.

J.H. Kaas: How does your theory of the evolution of areas of cortex address the issue of the evolution of areas after the advent of primary visual (17) and somatic areas (3b), since these fields seem to exist in most mammals and therefore evolved early?

D.N. Pandya: Indeed, non-primate mammals, for example rats and rabbits, do have primary sensory and motor areas. From our perspective as well as the work of Sanides, it appears that primary areas, whether in primate or non-primate brains, hold a similar position within their respective trends of architectonic elaboration. However, these primary regions in non-primates do not seem to have precisely the same architectonic features, for example granularity, as corresponding areas in the primate brain.

CHAPTER 5

The anatomical relationship of the prefrontal cortex with the striatopallidal system, the thalamus and the amygdala: evidence for a parallel organization

Henk J. Groenewegen, Henk W. Berendse, Jan G. Wolters and Anthony H.M. Lohman

Department of Anatomy and Embryology, Vrije Universiteit, Van der Boechorststraat 7, 1081 BT Amsterdam, The Netherlands

Summary

Recent findings in primates indicate that the connections of the frontal lobe, the basal ganglia, and the thalamus are organized in a number of parallel, functionally segregated circuits. In the present account, we have focused on the organization of the connections between the prefrontal cortex, the basal ganglia and the mediodorsal thalamic nucleus in the rat. It is concluded that in this species, in analogy with the situation in primates, a number of parallel basal ganglia-thalamocortical circuits exist. Furthermore, data are presented indicating that the projections from particular parts of the amygdala and from individual nuclei of the midline and intralaminar thalamic complex to the prefrontal cortex and the striatum are in register with the arrangements in the parallel circuits. These findings emphasize that the functions of the different subregions of the prefrontal cortex cannot be considered separately but must be viewed as components of the integrative functions of the circuits in which they are involved.

Introduction

The prefrontal cortex (PFC) is generally defined as that part of the cortex that has reciprocal connections with the mediodorsal thalamic nucleus (MD), i.e. the mediodorsal projection cortex. The afferent and efferent connectivity of the PFC has been studied extensively. In addition to inputs from the MD (e.g. Leonard, 1969, 1972; Krettek and Price, 1977; Groenewegen, 1988), the PFC receives fibers from a wealth of other sources. These include the midline and intralaminar thalamic nuclei, the basal forebrain, the amygdala, the lateral hypothalamus, and several cell groups in the brain stem, among which the dopaminergic cell groups in the ventral mesencephalon are the most prominent (Divac et al., 1978; Gerfen and Clavier, 1979; Porrino and Goldman-Rakic, 1982; Thierry et al., 1983; Reep, 1984). The efferent projections from the PFC reciprocate most of the afferent projections and, in addition, are directed to the basal ganglia by way of the corticostriatal pathways.

On the basis of cytoarchitectonic and hodological criteria, the PFC can be parceled into several

H.J. Groenewegen, M.D., Ph. D., Department of Anatomy, Vrije Universiteit, van der Boechorststraat 7, 1081 BT Amsterdam, The Netherlands. Telephone: 31-20-5482709/5482723; Telefax: 31-20-6610751; Electronic mail: V49GANAT@HASARA11.BITNET.

areas that are thought to play different roles in prefrontal cortical functions (Eichenbaum et al., 1983; Kolb, 1984). A major criterium is the topographically organized relationship with the MD (Akert, 1964; Leonard, 1969, 1972; Krettek and Price, 1977; Goldman-Rakic and Porrino, 1985; Giguere and Goldman-Rakic, 1988; Groenewegen, 1988). In primates, the MD has three cytoarchitectonically different segments that maintain reciprocal connections with different parts of the PFC: the medial magnocellular segment is connected with the medial and orbitofrontal areas, the lateral, parvicellular segment is related to the dorsolateral PFC, and the paralamellar (multiform and densocellular) segment is associated with the frontal eye field (area 8). Although there are no clear cytoarchitectonic criteria to subdivide the MD of the rat, also in this species separate segments of MD are reciprocally connected with different subregions of the cortex of the frontal pole (Krettek and Price, 1977; Leonard, 1969, 1972; Groenewegen, 1988). These subregions exhibit structural differences and, since each of them has specific afferent and efferent relationships, it can be inferred that they also represent functionally different areas. This is also expressed by the fact that the various segments of the MD projecting to the PFC receive different sets of inputs (Price and Slotnick, 1983; Velayos and Reinoso-Suarez, 1982, 1985; Ilinsky et al., 1985; Russchen et al., 1987; Cornwall and Phillipson, 1988; Groenewegen, 1988).

An important step forward in our understanding of the functional role of the PFC was the recognition that through the MD the PFC receives information processed by the basal ganglia. This holds in particular for the ventral, limbic system-innervated parts of the striatum, i.e. the nucleus accumbens, the striatal elements of the olfactory tubercle, and the ventromedial part of the caudate-putamen. These structures project via the ventral pallidum and the medial part of the globus pallidus to the MD and subsequently to the PFC (Young et al., 1984; Groenewegen, 1988). As initially postulated by Heimer and Wilson (1975), it has now been demonstrated convincingly that parallel dorsal and ventral striatopallidofugal pathways exist which by way of the ventral anterior and the mediodorsal thalamic nuclei ultimately reach the premotor and the prefrontal cortical areas, respectively (Haber et al., 1985; Nauta, 1986; Zahm et al., 1987).

The concept of a parallel organization of the dorsal and ventral striatopallidothalamic systems, as briefly described above, is mainly based on experiments performed in the rat. The results of recent studies in primates indicate that also in this species a ventral striatopallidal system exists that through the MD leads to the PFC (Russchen et al., 1987; Haber et al., 1990). Moreover, it has been postulated that in primates additional parallel circuits can be recognized (Goldman-Rakic and Selemon, 1986; Alexander et al., 1986). Each circuit focuses on a specific, functionally distinct region of the frontal cortex and involves anatomically distinct sectors of the striatum, the pallidum and the thalamus. Thus, whereas virtually all cortical regions send projections to the basal ganglia, the output of the basal ganglia is exclusively directed, via a thalamic relay, towards the frontal cortex. The idea of a parallel organization of functionally segregated basal ganglia-thalamocortical circuits contrasts with the earlier belief that information from all parts of the cortical mantle converges via the basal ganglia on the (pre)motor cortex (DeLong and Georgopoulos, 1981; Goldman-Rakic and Selemon, 1986; Alexander et al., 1986).

A clear example of a "functional loop" is the "motor circuit", which focuses on the supplementary motor area and includes the putamen, the ventrolateral part of the internal segment of the globus pallidus, and the medial and oral parts of the ventrolateral thalamic nucleus (DeLong and Georgopoulos, 1981). In addition to the "motor circuit", Alexander et al. (1986) distinguish four other loops, including an "oculomotor circuit" and several "complex or associational circuits" that involve distinct regions of the PFC. As emphasized by Alexander et al. (1986), the "definition" of five principal basal ganglia-thalamocortical circuits does not exclude the possibility that on the basis of new anatomical and physiological data additional, similarly organized circuits may be defined. An example of this is the "limbic loop" which has recently been demonstrated and involves the medial magnocellular portion of the MD and the medial orbitofrontal cortex (Haber et al., 1990). Other components of this loop are the most ventral and medial part of the striatum and the medial part of the ventral pallidum (Russchen et al., 1987; Haber et al., 1990). Since there is at present relatively little detailed knowledge in primates of the topography in the striatopallidal and pallidothalamic connections, it is likely that the "composition" of some of the above-mentioned circuits will be further detailed or needs to be revised in the future. The "motor circuit" excepted, little is known about the functional implications of the parallel organization of the connections between the frontal cortex, the basal ganglia, and the thalamus. It has recently been shown with electrophysiological methods that within the "motor circuit" multiple levels of motor processing are conveyed concurrently, i.e. in parallel (Alexander et al., this volume). In view of the similarities in the anatomical organization of the different circuits, it might be assumed that the various motor, complex and limbic loops, although transferring different types of information, have comparable roles in different motor or behavioral processes. A further important conclusion is that the "functions" of the different components of the circuits, i.e. the frontal cortex, the striatum, the pallidum and the thalamic nuclei, cannot be considered separately, but that they are closely linked with each other (e.g. H.J.W. Nauta, 1979, 1986; Alexander et al., 1986; Goldman-Rakic and Selemon, 1986). Thus, each of the structural components of a circuit contributes a specific aspect to the integrative function of an individual circuit.

The aim of the present report is to review the available data on the relationships of the PFC with the basal ganglia and the thalamus in the rat. It will be concluded that in this species the forebrain structures are organized in a similar, parallel fashion as they are in primates. Furthermore, data will be presented indicating that individual nuclei of the intralaminar and midline thalamic complex and of the amygdala are associated with the basal ganglia-thalamocortical loops in a highly specific way.

Materials and methods

The material, on which the present observations are based, consists of a large collection of rat brains in which the anterograde tracer *Phaseolus vulgaris*-leucoagglutinin (PHA-L) was injected in different parts of the medial and lateral PFC, the midline and intralaminar thalamic nuclei, the ventral striatopallidal system, and the amygdala. The methods for the delivery of the tracer and the subsequent histological and immunohistochemical procedures have been described in extenso elsewhere (Gerfen and Sawchenko, 1984; Groenewegen et al., 1987; Groenewegen, 1988; Berendse and Groenewegen, 1990; Groenewegen and Wouterlood, 1990). In order to demonstrate the topography in the afferent connections of the ventral striatum of the rat, injections of the retrograde tracer choleratoxin subunit B (CTb) were placed in a separate group of animals. For the experimental details we refer to Ericson and Blomqvist (1988) and Berendse and Groenewegen (1990). In order to facilitate the comparison of the results of the different experiments, the distribution of the anterograde and retrograde labeling is illustrated in a number of standard transverse sections taken from a series of which the sections were alternately stained with Cresyl violet, with the Loyez fiber-staining method, and for acetylcholinesterase activity (cf. Groenewegen, 1988).

The PFC and the MD are subdivided according to the cytoarchitectonic and connectional criteria described previously (Van Eden and Uylings, 1985; Groenewegen, 1988). In brief, the PFC of the rat is composed of cortical subfields in the medial and the lateral aspects of the frontal lobe. These medial and lateral subfields are continuous with each other around the rostral tip of the hemisphere. The medial PFC is constituted by the ventrally located infralimbic and prelimbic areas, the dorsally and caudally adjacent ventral and dorsal anterior cingulate areas, and the medial precentral area located in the "shoulder region" of the hemisphere. The lateral PFC consists of the ventral and dorsal agranular insular areas. The medial, lateral, ventral, and ventrolateral orbital areas are also considered to belong to the PFC (Groenewegen, 1988). The MD in the rat can be parceled into medial, central, lateral and paralamellar segments, all four of which are present at mid-rostrocaudal levels of the nucleus only. Both rostrally and caudally, the MD is composed solely of the medial and lateral segments.

Results and discussion

The existence in the rat of several parallel cortico-striato-pallido-thalamic circuits involving the PFC

Corticostriatal projections

The results of several anterograde and retrograde tracing studies have shown that the PFC projects in a topographical manner to the striatum (Beckstead, 1979; Phillipson and Griffiths, 1985; McGeorge and Faull, 1989; Sesack et al., 1989). The dorsoventral axis in the medial PFC is maintained in these projections. For example, the anterior cingulate area projects to the dorsomedial corner of the caudate-putamen complex, whereas the ventral part of the prelimbic area sends fibers to the medial part of the ventral striatum (Sesak et al., 1989). There is also a medial-to-lateral topography: the medial PFC distributes its fibers medially in the striatum and the lateral PFC is connected with more lateral parts of this complex (Beckstead, 1979; Reep and Winans, 1982). The latter arrangement was confirmed in our own experiments with injections of the anterograde tracer PHA-L in the PFC (Fig. 1). An injection in the prelimbic area results in anterograde labeling of fibers and terminals mainly in the medial part of the ventral striatum, including the nucleus accumbens and the olfactory tubercle (Fig. 1A). By contrast, an injection in the dorsal agranular insular area gives rise to labeling in the lateral parts of both the caudate-putamen complex and the nucleus accumbens rostrally, whereas more caudally in the striatum the terminal field shifts to a more central position (Fig. 1B). This topography can also be appreciated from the results of injec-

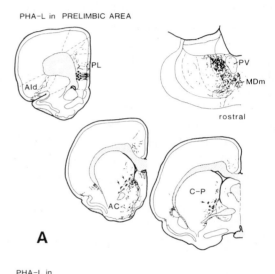

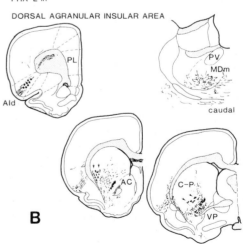

Fig. 1. Chartings of the anterograde labeled fibers and terminals in the striatum and the MD following an injection of the anterograde tracer PHA-L in the prelimbic area (A) and the dorsal agranular insular area (B). Arrows in the striatum indicate bundles of labeled passing fibers.

tions of the retrograde tracer choleratoxin subunit (CTb) in different parts of the ventral striatum (cf. also Phillipson and Griffiths, 1985). An injection in the medial part of the nucleus accumbens results in labeling of neurons virtually confined to the prelimbic area (Fig. 2). Following an injection which involves the lateral part of the nucleus accumbens and the ventrolateral part of the caudate-putamen complex, retrogradely labeled neurons are predominantly present in the dorsal and ventral agranular insular areas (Fig. 3). In this case, labeling, though sparser, is also found in the different subfields of the medial PFC. In agreement with other studies (e.g. primate: Selemon and Goldman-Rakic, 1985; rat: McGeorge and Faull, 1989; Sesack et al., 1989), the results of our anterograde tracing experiments indicate that the corticostriatal projections from a single cortical region in the PFC terminate throughout the rostrocaudal extent of the striatum.

On the basis of data in the literature (for references see above) and our own experiments

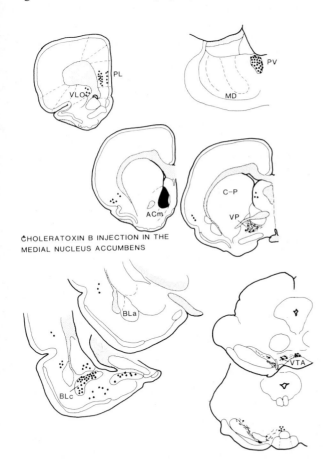

Fig. 2. Chartings of the distribution of retrogradely labeled neurons in the PFC, the basal forebrain, the amygdala, the midline thalamus, and the ventral mesencephalon (large dots) and of anterogradely labeled fibers in the ventral pallidum and the ventral mesencephalon (fiber-like representation) following an injection of the tracer CTb in the medial part of the nucleus accumbens.

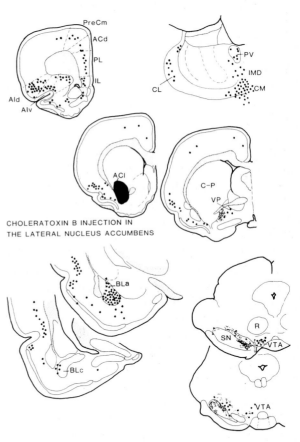

Fig. 3. Chartings of the distribution of retrogradely labeled neurons in the PFC, the basal forebrain, the amygdala, the midline and intralaminar thalamic nuclei, and the ventral mesencephalon (large dots) and of anterogradely labeled fibers in the ventral pallidum and the ventral mesencephalon (fiber-like representation) following an injection of the tracer CTb in the lateral part of the nucleus accumbens and the adjacent ventrolateral part of the caudate-putamen complex.

with large numbers of PHA-L injections in the PFC, Fig. 6A was composed, depicting the general topography of the corticostriatal projections orginating in the different subdivisions of the PFC. It must be realized that, although not indicated in this figure, there is a considerable overlap of terminal fields in the striatum of fibers from the individual PFC areas (cf. also Beckstead, 1979; Sesack et al., 1989). Also not accounted for in the schematic representation of Fig. 6A is the patchy distribution of corticostriatal terminations that has been described previously by several authors (e.g. Gerfen, 1984, 1989; Selemon and Goldman-Rakic, 1985; Sesack et al., 1989).

Striatopallidal projections

A topographical organization has been described for the connections from the striatum to the pallidum. In primates, the head of the caudate nucleus projects to the dorsolateral part of the globus pallidus, whereas the dorsal part of the putamen projects to the lateral part of the globus pallidus (Parent, 1986). Futhermore, the ventral portion of the putamen and the nucleus accumbens project in a medial-to-lateral topography to pallidal areas underneath the anterior commissure (Haber et al., 1990). In the rat, a comparable organization exists in the projections from the dorsal and the ventral striatum to the globus pallidus and the ventral pallidum (Nauta et al., 1978; Gerfen, 1985; Haber et al., 1985).

The results of a large number of experiments with small injections of PHA-L throughout both the dorsal and the ventral striatum confirm the general notion of strict dorsal-to-ventral, rostral-to-caudal and medial-to-lateral topographical arrangements in the striatopallidal connections. The medial-to-lateral topography in the ventral striatopallidal system can be appreciated from the two experiments with CTb injections in the ventral striatum, depicted in Figs. 2 and 3. Since this tracer is not only transported in retrograde but also in anterograde direction, in both cases a terminal field is found in the subcommissural ventral pallidum. Following the injection in the medial nucleus accumbens the fibers and terminals are located medially (Fig. 2), whereas after the lateral injection they are present more laterally in the ventral pallidum (Fig. 3). As shown by Heimer et al. (1987), the striatal parts of the olfactory tubercle project to the ventral parts of the ventral pallidum, maintaining a medial-to-lateral topography. In this part of the system the terminal fields have a considerable rostrocaudal extent. More dorsally, within the projections from the nucleus accumbens to the subcommissural part of the ventral pallidum, the topography in the projections is maintained along all 3 axes mentioned above (unpublished observations). The topographical organization in the striatopallidal projections is summarized in Fig. 6A by means of corresponding hatchings in striatal and pallidal areas.

Pallidothalamic projections

For a proper understanding of the organization of the pathways that lead from the striatum via the pallidum to the thalamus, it must be realized that the dorsal pallidal complex consists of an external and an internal segment. These are represented in the rat by the main body of the globus pallidus and the entopeduncular nucleus, respectively. The projections from the dorsal striatum to the entopeduncular nucleus are organized in a topographical way similar to those from the striatum to the main body of the globus pallidus, as described above (cat: Royce and Laine, 1984; rat: Fink-Jensen and Mikkelsen, 1989). The projection from the dorsal pallidal complex to the thalamus mainly originates in the entopeduncular nucleus and terminates in the ventral anterior nucleus (Haber et al., 1985; Nauta, 1986). A smaller contribution to this projection comes from the medial part of the main body of the globus pallidus (Haber et al., 1985; Groenewegen, 1988).

A distinction between external and internal segments within the ventral pallidum is difficult to make (cf. Haber and Nauta, 1983; Heimer et al., 1985; Nauta, 1986). The presence of a mixed plexus of substance P- and enkephalin-positive woolly fibers indicates that the ventral pallidum can be compared with both the internal and external segments of the dorsal pallidum. Joint characteristics are also clear from the outputs of the ventral pallidum which include, on the one hand, the subthalamic nucleus and the substantia nigra (cf. dorsal pallidum) and, on the other, the thalamus and the habenula (cf. the entopeduncular nucleus; Haber et al., 1985; Nauta, 1986; Groenewegen and Berendse, 1990). It has been hypothesized that a restricted region of the lateral hypothalamus, immediately medial to the entopeduncular nucleus, forms part of the internal segment of the ventral pallidal system (Groenewegen and Berendse, 1990; cf. also Heimer et al., 1985; Nauta et al., 1978; Groenewegen and Russchen, 1984). However, this lateral hypothalamic region does not project as abundantly to the thalamus as the subcommissural ventral pallidum, but instead directs its main output to the lateral habenula (Herkenham and Nauta, 1977; Groenewegen, 1988).

Projections from the ventral pallidum to the MD have already been hypothesized by Heimer and Wilson in 1975. Subsequent studies have demonstrated these projections and have further shown that they are topographically organized (Young et al., 1984; Haber et al., 1985; Zahm et al., 1987; Cornwall and Phillipson, 1988; Groenewegen, 1988). As an example of the restricted terminal fields in the MD of afferents from different parts of the pallidum, the distribution of fibers from the subcommissural ventral pallidum in the medial segment of the MD is shown in Fig. 4. The topographical organization of the pallidothalamic projections to the MD is illustrated in Fig. 5A. The subcommissural part of the ventral pallidum projects mainly to the medial segment of the MD, the fibers from the pallidal elements of the olfactory tubercle terminate in the central segment of the MD, and cells in the medial part of the globus pallidus emit fibers to the lateral segment of the MD. A peripheral "shell" of the MD, including its paralamellar segment, is innervated by non-dopaminergic fibers from the ventral tegmental area and the pars reticulata of the substantia nigra, but does not receive a significant pallidal input (Groenewegen, 1988). In Fig. 6A the topographical organization of the pallidal inputs to MD is shown by means of corresponding hatchings in pallidal areas and MD segments.

Reciprocal thalamocortical projections

The topography in the connections between the MD and the PFC has been described extensively in various tracing studies (Leonard, 1969, 1972; Krettek and Price, 1977; Beckstead, 1979; Cornwall and Phillipson, 1988; Groenewegen, 1988; Sesack et al., 1989). In general, a rather strict reciprocity

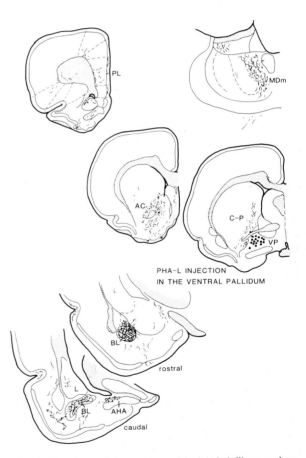

Fig. 4. Chartings of the anterogradely labeled fibers and terminals in the PFC, the ventral striatum, the MD, and the amygdala following an injection of the anterograde tracer PHA-L in the ventral pallidum.

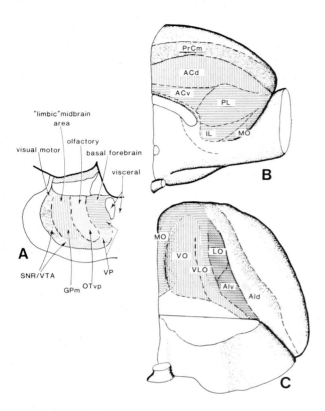

Fig. 5. Schematic representation of the topographical organization of the subcortical afferents of the different segments of the MD. In a transverse section through the MD (A), the afferents from pallidal or nigral origin are listed underneath. Above the section, the other inputs are indicated by single terms (for more details see Groenewegen, 1988). In medial (B) and ventral (C) views of the frontal lobe, the borders between the cytoarchitectonically different regions of the PFC are indicated with interrupted lines. The topography in the reciprocal relationships between the different segments of the MD (in A) and the various PFC subdivisions is specified by means of different hatchings.

agranular insular areas to the medial segment of the MD terminate in its rostral and caudal parts, respectively.

Identification of basal ganglia-thalamocortical circuits in the rat

The data discussed in the foregoing account provide the basis for the recognition of a number of parallel circuits in the rat. Four circuits associated with cytoarchitectonically different subregions of the PFC will be described and illustrated (Figs. 6B, C, 11, 12). As will be discussed later, these subregions of the PFC and, consequently, the circuits associated with them, may have different functions (see General discussion). It is possible, however, that the circuits now defined on the basis of the cytoarchitectonic differentiation within the PFC are not completely congruent with circuits that will be defined in the future based upon new functional data. Moreover, a further subdivision of the circuits may also be expected.

The dorsal anterior cingulate area is the main cortical component of a circuit which also includes the dorsomedial part of the caudate-putamen, the medial portion of the globus pallidus, and the lateral segment of the MD ("dorsal anterior cingulate circuit"; Fig. 6B). The ventrally adjacent prelimbic area is the cortical nodal point of a circuit which further involves the medial part of the nucleus accumbens, the medial portion of the subcommissural ventral pallidum, and the rostral part of the medial segment of the MD ("prelimbic circuit"; Fig. 11). In the lateral PFC, the ventral agranular insular area forms the cortical station of a circuit of which the other components are the striatal and pallidal elements of the olfactory tubercle, and the central segment of the MD ("ventral agranular insular circuit"; Fig. 6C). The fourth circuit is formed by the dorsal agranular insular area, the lateral part of the nucleus accumbens, the lateral part of the subcommissural ventral pallidum, and the caudal part of the medial segment of the MD ("dorsal agranular insular circuit"; Fig. 12).

between individual MD segments and particular PFC areas exists (Groenewegen, 1988). These reciprocal relationships between the MD and the PFC have been schematically depicted in Figs. 5A – C and 6A. From Fig. 1 it can be appreciated that within the medial segment of the MD a further differentiation exists, such that the projections from the prelimbic area and from the dorsal

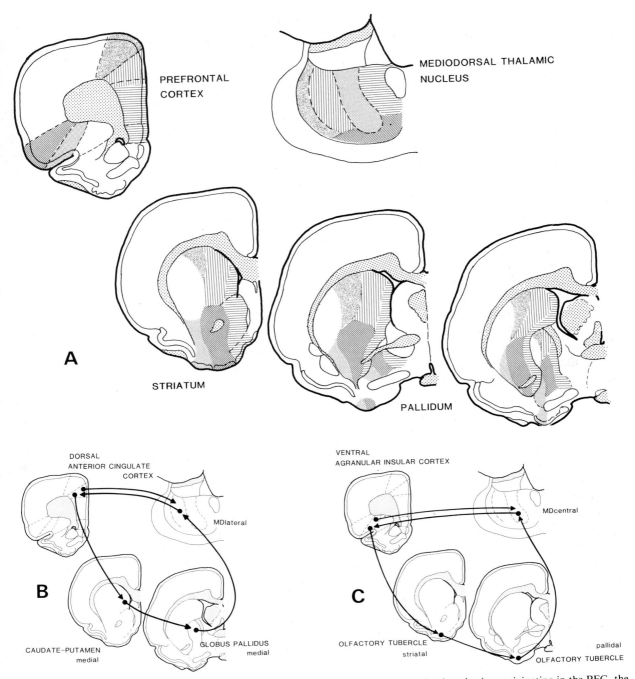

Fig. 6. (A) Schematic representation of the topographical organization in the corticostriatal projections originating in the PFC, the striatopallidal connections, the pallidothalamic pathways that terminate in the MD, and the reciprocal relationships between MD and PFC. The different areas, regions or segments in the cortex, striatum, pallidum, and thalamus that are connected with each other by way of the above-mentioned projections are indicated with corresponding hatchings. (B, C) Schematic representation of two parallel cortico-striato-pallido-thalamic circuits in the rat, i.e. the "dorsal anterior cingulate circuit" and the "ventral agranular insular circuit", which are "constructed" on the basis of the topography in the projections between the different components of the circuits (as shown in A). The circuit in B is focused on the dorsal anterior cingulate cortex, the circuit in C on the ventral agranular insular cortex. Other examples of such circuits are shown in Figs. 11 and 12.

The involvement of the midline/intralaminar thalamic nuclei and the amygdala in basal ganglia-thalamocortical circuits

It is obvious that, at all way stations of the above-described circuits, projections from other cortical and subcortical areas can contribute inputs to or may influence the information processed by these circuits. Furthermore, since the PFC, the striatum, and the pallidum all project to structures outside the circuits, it is clear that the basal ganglia-thalamocortical circuits cannot be considered as closed loops. Alexander et al. (1986) have pointed out that, at the level of the striatum, inputs from cortical areas other than the frontal cortex may converge with those from the frontal lobe. In addition, many areas in the parietal, occipital, and temporal lobes maintain corticocortical connections with the areas in the frontal lobe.

Subcortical areas that project to both the frontal cortex and the striatum are the midline and intralaminar thalamic nuclei, the mesencephalic dopaminergic cell groups, and various nuclei of the amygdaloid complex. Although the existence of such projections and their general organization have been known for quite some time, only the newly introduced, highly sensitive neuroanatomical tracers permit for a detailed description of their topographical relationships. The recent data discussed below reveal that the topography in the projections from the midline and intralaminar thalamic nuclei and from the basolateral amygdaloid nucleus to the PFC and the striatum is to a large extent in register with the organization of the basal ganglia-thalamocortical connections.

Midline and intralaminar thalamic nuclei

Projections from the midline and intralaminar thalamic nuclei to the cortex and the striatum have been described by many authors (e.g. Jones and Leavitt, 1974; Royce, 1978, 1983; Macchi et al., 1984; Beckstead, 1984; Jayaraman, 1985; Macchi and Bentivoglio, 1986; Herkenham, 1986; Royce et al., 1989; Berendse and Groenewegen, 1990). It has long been assumed that the cortical projections from these nuclei are widespread and, in this way, may have a diffuse influence on the cerebral cortex (Jones and Leavitt, 1974; Herkenham, 1986). However, the results of recent anterograde and retrograde tracing studies have demonstrated that the cortical projections of the midline/intralaminar complex are topographically organized and that, consequently, individual nuclei have rather restricted target areas in the cerebral cortex (Macchi and Bentivoglio, 1986; Royce et al., 1989; Berendse and Groenewegen, 1991). A topographical arrangement has also been described for the projections from the midline and intralaminar complex to the striatum (Veening et al., 1980; Groenewegen et al., 1980; Beckstead, 1984, Jayaraman, 1985; Phillipson and Griffiths, 1985; Berendse and Groenewegen, 1990). In a recent series of tracing experiments, we attempted to answer the question to what extent the topographical organization of the thalamostriatal and thalamocortical projections is related to the organization of the above-described basal ganglia-thalamocortical circuits.

The organizational principle in the relationships between, on the one hand, the thalamostriatal and thalamocortical projections and, on the other, the basal ganglia-thalamocortical circuits is illustrated by two series of experiments which involve the "prelimbic circuit" and the "dorsal agranular insular circuit", respectively. With respect to the "prelimbic circuit", an injection of CTb in the medial part of the nucleus accumbens (see above) results not only in the retrograde labeling of neurons in the prelimbic area, but also in the anterior part of the paraventricular nucleus of the thalamus (Fig. 2; cf. Berendse and Groenewegen, 1990). As can be expected on the basis of these retrograde tracing results, small injections of the anterograde tracer PHA-L in this part of the paraventricular nucleus result in heavy labeling of fibers and terminals in the medial part of the nucleus accumbens, except for a few patches of terminations in more lateral parts (Fig. 7; Berendse et al., 1988; Berendse and Groenewegen, 1990). In addition, in the frontal cortex a restricted, dense

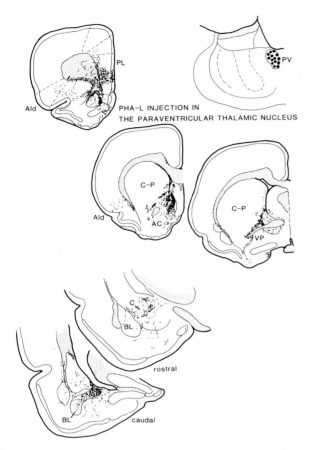

Fig. 7. Chartings of the anterogradely labeled fibers and terminals in the PFC, the ventral striatum, and the amygdala following an injection of the anterograde tracer PHA-L in the anterior part of the paraventricular thalamic nucleus.

terminal field is found in the ventral part of the prelimbic area, whereas terminations in other areas of the PFC are much sparser (Fig. 7). Interestingly, the areas in the PFC and the striatum that are reached by the fibers from the anterior part of the paraventricular thalamic nucleus are connected with each other by means of corticostriatal projections and furthermore form part of the "prelimbic circuit". In agreement with the results of other studies (Macchi et al., 1984; Price et al., 1987), additional projections from the paraventricular nucleus are found to numerous other forebrain areas. In the context of the present account, the most interesting are those to the amygdaloid complex. As illustrated in Fig. 7, paraventricular fibers terminate most prominently in the caudal part of the basolateral nucleus and further in the region of the central nucleus.

Concerning the "dorsal agranular insular circuit", it has been shown that the dorsal agranular area projects to the lateral part of the nucleus accumbens (Figs. 1B and 3). Following the injection of CTb in this part of the striatum, retrogradely labeled neurons are also present in the dorsal thalamus. In this case the labeling is mainly located in the central medial, intermediodorsal, paracentral, and central lateral nuclei (Fig. 3; cf. Berendse and Groenewegen, 1990). Injections of PHA-L that are restricted to the intermediodorsal thalamic nucleus result in anterograde labeling of fibers and terminals in the lateral part of the nucleus accumbens and the ventral part of the caudate-putamen at rostral levels and, more caudally, in the ventromedial part of the caudate-putamen (Fig. 8). In the PFC, terminations are predominantly present in the superficial layers of the dorsal agranular insular area (Fig. 8). Much sparser labeling is found in the medial PFC. Therefore, the areas in the PFC and the striatum that receive the strongest projections from the intermediodorsal thalamic nucleus are connected with each other by means of corticostriatal projections and are both way stations in the "dorsal agranular insular circuit". Projections from this midline thalamic region to the amygdala are directed to the central nucleus and the rostral part of the basolateral nucleus, whereas the caudal part of the latter nucleus is avoided (Fig. 8).

The conclusion from the present series of experiments is that individual nuclei of the intralaminar and midline thalamic system project to restricted parts of both the PFC and the striatum which, in turn, form way stations in a particular basal ganglia-thalamocortical circuit. In this way, the cortical and striatal components of a circuit can be influenced from a single thalamic locus. Whether the cortical and striatal projection fibers stem from the same or different neurons in the midline and intralaminar thalamic nuclei is still a

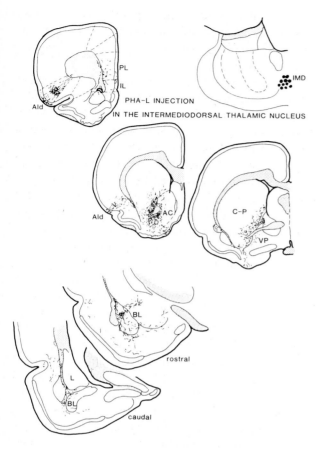

Fig. 8. Chartings of the anterogradely labeled fibers and terminals in the PFC, the ventral striatum, and the amygdala following an injection of the anterograde tracer PHA-L in the intermediodorsal thalamic nucleus.

matter of dispute (Royce, 1983; Macchi et al., 1984). However, since the neurons in the midline and intralaminar nuclei are in general closely spaced, the projections to the cortex and the striatum will probably both be influenced by the same inputs.

Amygdaloid nuclei

Projections from the amygdala to the striatum and the cortex have been described previously in different species, but only recently attention has been paid to the topographical organization that exists in these projection systems (for a review see Price et al., 1987; cf. also Ragsdale and Graybiel, 1988). In the rat, the amygdalostriatal fibers originate primarily in the basolateral amygdaloid nucleus, although also other amygdaloid nuclei contribute to these projections (Krettek and Price, 1978; Kelley et al., 1982; Russchen et al., 1985). The densest termination areas of the amygdalostriatal projections are the ventral striatum and the caudal part of the caudate-putamen complex, whereas the rest of the caudate-putamen complex receives a less dense amygdaloid innervation (Kelley et al., 1982; Russchen and Price, 1984; unpublished observations). The dorsolateral part of the anterior caudate-putamen, which is innervated by the sensorimotor cortex, receives no amygdaloid input (Kelley et al., 1982; cf. however Fass et al., 1984). The most detailed observations on the topographical organization of the amygdalostriatal projections were made by Russchen and Price (1984) who used the anterograde tracing of PHA-L. They showed, in agreement with the results of earlier studies (Krettek and Price, 1978; Kelley et al., 1982), that a rostral-to-caudal axis in the amygdala corresponds to a lateral-to-medial axis in the striatum. As illustrated in Figs. 2 and 3, the results of our own retrograde tracing experiments with injections of CTb in the ventral striatum confirm this topography. The injection in the medial part of the nucleus accumbens gives rise to abundant labeling of cells in the caudal part of the amygdala, primarily located in the basolateral nucleus (Fig. 2). Following the lateral injection in the nucleus accumbens virtually all neurons in the rostral part of the basolateral nucleus are retrogradely filled (Fig. 3). Further aspects of the projections from the basolateral nucleus to the striatum can be appreciated from experiments in which the anterograde tracer PHA-L was injected in different parts of the amygdala (Figs. 9 and 10). Following an injection in the caudal part of the basolateral nucleus, anterogradely labeled fibers terminate in medial parts of the caudate-putamen, the nucleus accumbens, and the olfactory tubercle (Fig. 9). An injection of PHA-L in the rostral part of the basolateral nucleus results in labeling in a more lateral zone of all 3 mentioned sectors of the striatum (Fig. 10).

The distribution of fibers from different parts of the basolateral amygdaloid nucleus to the PFC in our PHA-L experiments (Figs. 9 and 10) corresponds quite well with the distribution described by Krettek and Price (1978) using the anterograde autoradiographic tracing technique and by McDonald (1987) on the basis of retrograde tracing experiments. Thus, following the PHA-L injection in the caudal part of the basolateral nucleus labeled fibers and terminals are primarily found in the prelimbic area of the medial PFC (Fig. 9). Only a few terminals are present in the lateral PFC. By contrast, the injection of PHA-L in the rostral part of the basolateral nucleus results in anterograde labeling of fibers and terminals in the dorsal agranular insular area, whereas virtually no labeling is present in the medial PFC (Fig. 10).

It can now be concluded that the topographical organization in the amygdalostriatal and the amygdalocortical projections in the rat is such that the areas in the PFC and the striatum, that are reached by projections from a specific part of the basolateral nucleus, are also related to each other through corticostriatal connections. In this way, the caudal part of the basolateral nucleus is related to the "prelimbic circuit" and the rostral part of the basolateral nucleus to the "dorsal agranular in-

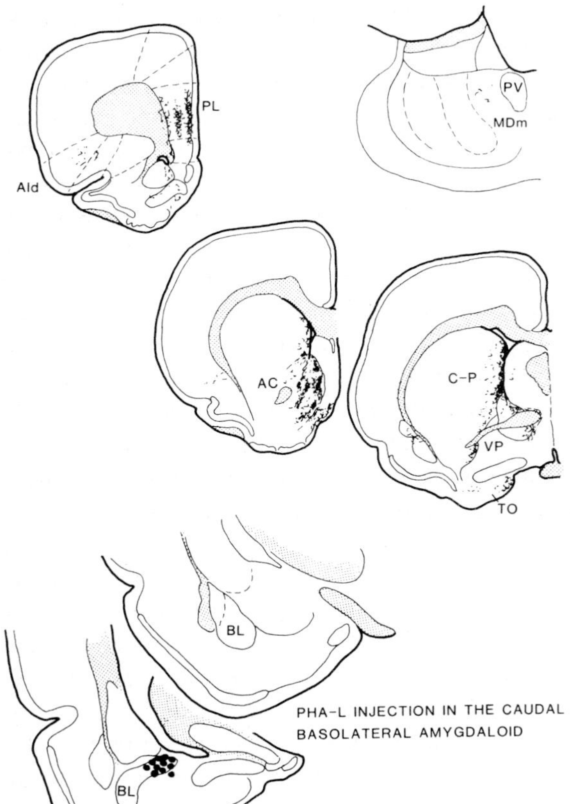

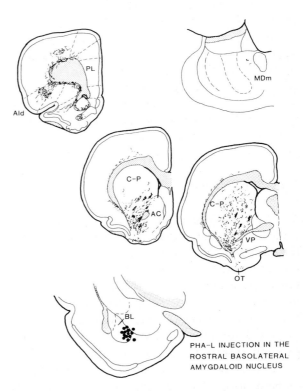

Fig. 9. Chartings of the anterogradely labeled fibers and terminals in the PFC, the ventral striatum, and the MD following an injection of the anterograde tracer PHA-L in the caudal part of the basolateral amygdaloid nucleus.

Fig. 10. Chartings of the anterogradely labeled fibers and terminals in the PFC, the striatum, and the MD following an injection of the anterograde tracer PHA-L in the rostral part of the basolateral amygdaloid nucleus.

sular circuit". In agreement with the results of other tracing studies (Krettek and Price, 1978; McDonald, 1987), following the present PHA-L injections in the basolateral amygdaloid nucleus labeled fibers and terminals are present in the medial segment of the MD (Figs. 9 and 10). These projections from the amygdala to the MD are much sparser than the amygdaloid projections to the striatum and the PFC. According to Krettek and Price (1978), the fibers from the caudal part of the amygdala terminate rostrally in the medial segment of the MD and those from the rostral part of the amygdala distribute more caudally and ventrally in the medial MD. Since in our PHA-L experiments the labeling in the MD is very sparse, a topography could not be established. The arrangement described by Krettek and Price (1978) fits in with the parallel organization of the basal ganglia-thalamocortical circuits. The fibers from the rostral part of the basolateral nucleus terminate in the caudal part of the medial segment of MD and the dorsal agranular insular area, both being part of the "dorsal agranular insular circuit". The rostral part of the medial segment of the MD and the prelimbic area are components of the "prelimbic circuit" and both receive amygdaloid projections from the caudal part of the basolateral nucleus.

General discussion

The main conclusion that can be drawn from the foregoing account is that the connections between the different components of the basal ganglia, the thalamus and the frontal cortex in the rat are organized in a series of parallel circuits. Therefore, the basic principle of a parallel organization of connections between these forebrain structures, as suggested for primates (Alexander et al., 1986), also holds for the rat. In the present study, we have focused our attention on the organization in the rat of basal ganglia-thalamocortical circuits that involve different parts of the PFC and on the association of the midline and intralaminar thalamic nuclei and the amygdala with these circuits. With respect to the first issue, the detailed knowledge of the topographical organization in the corticostriatal, striatopallidal, pallidothalamic and MD-PFC connections allows us to recognize at least 4 circuits that involve cytoarchitectonically distinct areas of the PFC. It seems likely that other areas in the PFC of the rat, such as the orbital and the medial precentral areas, are part of similar cortical-subcortical circuits. However, the precise topographical arrangements of the corticostriatal projections of these latter cortical areas and of the subsequent projections to the pallidum and the thalamus are as yet incompletely known.

As regards the association of the midline and intralaminar thalamic nuclei and the amygdala with the basal ganglia-thalamocortical circuits, it could be demonstrated that the projections of individual nuclei of the nonspecific thalamic complex and of particular parts of the amygdala to the prefrontal cortical and striatal way stations of the principal circuits are in register with the parallel arrangement of these circuits. For example, with respect to the "prelimbic circuit", the projections from the caudal part of the basolateral nucleus and those from the anterior part of the paraventricular thalamic nucleus overlap in both the prelimbic area and the medial part of the nucleus accumbens (Fig. 11). As mentioned above, the projection from the paraventricular nucleus to the caudal part of the basolateral amygdaloid nucleus is thus in register with the projections of this thalamic nucleus to the PFC and the striatum (Fig. 11). Comparable associations of the midline and intralaminar thalamus and the basolateral amygdaloid nucleus can be described for the "dorsal agranular insular circuit" (Fig. 12). The projections from the intermediodorsal thalamic nucleus and the rostral part of the basolateral nucleus overlap in both the dorsal agranular insular area and the lateral part of the nucleus accumbens. In addition, the intermediodorsal thalamic nucleus sends fibers to the rostral part of the basolateral nucleus, an arrangement which is therefore in concert with the projections of this thalamic nucleus to the PFC and the striatum (Fig. 12). For the "dorsal anterior cingulate" and the "ventral agranular insular" circuits, the available data are insufficient to conclude that similar arrangements with the midline and intralaminar thalamic and amygdaloid nuclei exist.

It may be noted here that additional projections of the structures involved in the principal basal ganglia-thalamocortical circuits are in a number of cases also in register with the parallel organization. In their description of cortical-subcortical circuits in primates, Alexander et al. (1986) included also the striatonigral projections to different parts of the pars reticulata of the substantia nigra and the nigrothalamic pathways. In the rat, the striatonigral pathways to the pars reticulata are topographically organized (Gerfen, 1985). However, since little is known about the topography in the nigrothalamic projections, it is difficult to indicate

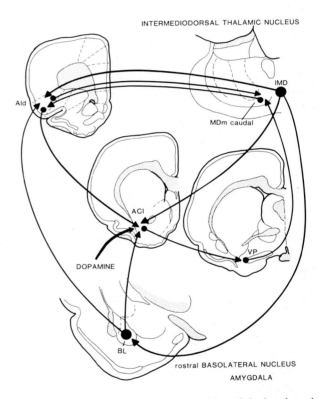

Fig. 11. Diagram to show the relationships of the basolateral amygdaloid nucleus and the midline and intralaminar thalamic nuclei with the "prelimbic circuit" (compare also Fig. 6). The projections from the paraventricular thalamic nucleus and those from the caudal part of the basolateral amygdaloid nucleus converge in the prelimbic area of the medial PFC and in the medial part of the ventral striatum. The latter two areas are connected with each other by means of corticostriatal projections. Note that the paraventricular nucleus projects, in addition, to the caudal part of the basolateral amygdaloid nucleus. Indicated by an arrow is the dopaminergic input to the ventral striatum.

Fig. 12. Diagram to show the relationships of the basolateral amygdaloid nucleus and the midline and intralaminar thalamic nuclei with the "dorsal agranular insular circuit" (compare also Fig. 6). The projections from the intermediodorsal thalamic nucleus and those from the rostral part of the basolateral amygdaloid nucleus converge in the dorsal agranular insular area of the lateral PFC and in the lateral part of the ventral striatum. The latter two areas are connected with each other by means of corticostriatal projections. Note that the intermediodorsal nucleus projects, in addition, to the rostral part of the basolateral amygdaloid nucleus. Indicated by an arrow is the dopaminergic input to the ventral striatum.

how these projections and the subsequent thalamocortical projections fit in with the parallel circuitry. Another example are the projections of the ventral pallidum to the nucleus accumbens, the medial PFC, and the basolateral amygdaloid nucleus, indicated in Fig. 4; they appear to fit rather well within the organization of the "prelimbic circuit" and the "dorsal agranular insular circuit". It must be realized that the PHA-L injection in this case covers both the medial and the lateral part of the ventral pallidum. Smaller injections give rise to more restricted terminal fields in the ventral striatum and the amygdala.

It has recently been emphasized that the corticocortical connections within one hemisphere are organized in parallel networks, that span various cortical centers in the different lobes of the hemisphere and subserve different functions (e.g., Selemon and Goldman-Rakic, 1988; Goldman-Rakic, 1988, and this volume; Cavada and Goldman-Rakic, 1989). This contrasts with, although it does not exclude, the earlier view of a stepwise pattern of convergence of corticocortical projections from primary sensory areas through sensory association areas to multimodal and limbic association areas and from prefrontal association, via premotor and supplementary motor to primary motor cortical areas (Pandya and Seltzer, 1982; Goldman-Rakic, 1988). The above-presented data clearly support the notion expressed by Goldman-Rakic (1988) of "a highly integrated but distributed cortical machinery whose resources are allocated to several basic parallel functional systems that bridge all major subdivisions of the cerebrum". As shown in the present study for cortical – subcortical relationships, not only are distinct parts of the PFC, the basal ganglia, and the specific thalamus organized in parallel loops (Alexander et al., 1986), but, in addition, parts of the limbic system and the nonspecific thalamus are specifically included in these circuits. Moreover, recent evidence suggests that a structure like the subthalamic nucleus which is traditionally associated with one functional system, i.e. the motor system, takes part in various basal ganglia circuits with different functions (Groenewegen and Berendse, 1990).

An important implication of the organization in parallel circuits of different subdivisions of the PFC and a number of subcortical structures is that we cannot consider the functions of the different subfields of the PFC as isolated features, but that we must regard them as components or aspects of the integrative functions of a circuit or a network. This can be illustrated for the "dorsal agranular insular circuit". As discussed by Scheel-Krüger and Willner (1990), different components of this circuit are involved in various aspects of "oral or perioral behavior". The dorsal agranular insular cortex is considered part of the visceral sensory cortex on the basis of its input from the parabrachial nucleus and its output to the nucleus of the solitary tract (Saper, 1982; Van der Kooy et al., 1982; Hurley-Gius and Neafsey, 1986; cf. also Neafsey, this volume). The area has been identified physiologically as part of the gustatory cortex (Kosar et al., 1986). The medial segment of the MD, projecting to the dorsal agranular insular area, receives inputs, among others, from autonomic centers such as the parabrachial and solitary nuclei in the brain stem (Groenewegen, 1988). Pharmacological manipulations of the cholinergic or of the dopaminergic transmission in the striatal projection area of the dorsal agranular insular cortex, i.e., the lateral part of the nucleus accumbens and the adjacent ventral part of the caudate-putamen complex, can elicit or suppress mouth and tongue movements (Kelley et al., 1989; Scheel-Krüger and Willner, 1990). The same applies to pharmacological manipulations of the anterior part of the basolateral amygdaloid nucleus (Scheel-Krüger, and Willner, 1990). Although it seems apparent that the different way stations of the "dorsal agranular insular circuit", and the part of the amygdala associated with this circuit, are involved in different aspects of "oral behavior", an integrative description of the behavior supported by this circuit and the specific functional roles of the constituents of the circuit has not been provided till date.

With respect to the prelimbic area, it has been suggested on the basis of neuroanatomical and physiological studies that this part of the medial PFC, together with the infralimbic area, represents the visceral motor cortex (Terreberry and Neafsey, 1983, 1987; Neafsey et al., 1987; Hurley-Gius and Neafsey, 1986; Neafsey, this volume). The role for the other constituents of the "prelimbic circuit", such as the medial parts of the nucleus accumbens and the ventral pallidum, in autonomic functions has received little attention in the literature till now. However, the involvement of the nucleus accumbens and the ventral pallidum in locomotor activity has been interpreted as an important component of feeding and drinking behavior (e.g., Mogenson et al., 1980, 1983) which is associated with visceral functions. In this context, it may be mentioned that, on the other hand, the prelimbic area has been associated with functions requiring spatial orientation (Kolb et al., 1982; Kolb, 1984). However, anatomical data suggest that these functions and the earlier mentioned visceral motor functions are probably invested in different parts of the prelimbic area and that the "prelimbic circuit" is not an entity but consists, at least, of 2 subcircuits. As shown by Hurley-Gius and Neafsey (1986), the neurons in the medial PFC projecting to the nucleus of the solitary tract are confined to the caudal half of the prelimbic and infralimbic areas. Furthermore, there is a clear distinction in the projections from the caudal and the rostral parts of the prelimbic area to the ventral striatum, the former projecting to the caudomedial nucleus accumbens and the latter to the rostral pole of this nucleus, including its lateral part (Sesack et al., 1989; own unpublished observations). It is important to note in the present context that the ventral pole of the subiculum, associated with visceral functions, projects most heavily to the caudomedial nucleus accumbens and the caudal part of the prelimbic area, whereas the dorsal pole of the subiculum, associated with sensory association functions, projects to the rostrolateral part of the nucleus accumbens and the rostral part of the prelimbic cortex (Groenewegen et al., 1987; Jay, Witter and Groenewegen, unpublished observations). Recent pharmacobehavioral data support the notion that the nucleus accumbens is not a functional entity. Administration of the peptide cholecystokinin (CCK8) in the caudomedial part of the nucleus accumbens induced hyperexploration, whereas this behavior could not be elicited from the rostral part of the nucleus (Daugé et al., 1989).

The principle of a parallel organization in the basal ganglia-thalamocortical circuits cortex is

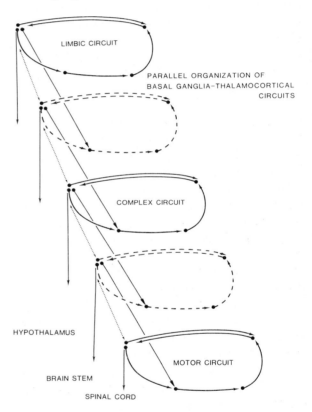

Fig. 13. Summary diagram of the principle of the parallel organization in the thalamocortical-basal ganglia circuits. The different way stations in the circuits are arranged in the same way as in Fig. 6B and C. The circuits indicated by interrupted lines may represent more than one circuit, since the precise number and identity of all possible circuits have not been established. Each of the circuits probably has its own output pathways to effector regions, such as the hypothalamus, the brain stem and the spinal cord. Concomitantly, information processed through the circuits may be integrated by way of corticocortical (stippled lines) and corticostriatal (solid lines) projections.

highly schematically depicted in Fig. 13. It must be emphasized here that these circuits leading from the cortex via the basal ganglia and the thalamus back to the cortex are exclusively directed towards the frontal cortex (i.e. the cortex rostral to the central sulcus in primates). Cortical areas caudal to the frontal cortex project in specific patterns towards both the frontal cortex and to the striatum, in this way feeding information into the circuits (Selemon and Goldman-Rakic, 1985; Alexander et al., 1986; Goldman-Rakic and Selemon, 1986). The way in which the caudal cortical projections to the frontal cortex and to the striatum are related to the organization in parallel circuits in the rat has not been determined (for primates cf. Selemon and Goldman-Rakic, 1985).

As indiated in Fig. 13, the circuits probably have their own outputs to the hypothalamus, the brain stem, and the spinal cord and may support different functions, ranging from highly complex behaviors, influenced by limbic structures, to pure motor acts. One of the main issues that needs to be studied in the future is how information processed by the various circuits is exchanged in order to produce integrated behavior. Integration between different circuits could take place through corticocortical connections, e.g. between the prefrontal, premotor, supplementary motor, and primary motor cortices (Fig. 13; for a brief discussion of this issue see Goldman-Rakic, 1988). Another possible way for the integration of information processed by different circuits is through corticostriatal projections (Fig. 13). In the rat, there is definitely overlap of the terminal fields in the striatum of the projections originating in cytoarchitectonically different PFC areas (Beckstead, 1979; Sesack et al., 1989; own unpublished observations). However, since the striatal projections from most cortical areas show a discontinuous, patchy way of termination, as an expression of the compartmental structure of the striatum, it is possible that the projections from different cortical areas converging in a certain striatal region interdigitate rather than overlap (cf. also Selemon and Goldman-Rakic, 1985). On the one hand, this could imply that the different cortical areas communicate with different striatal output channels. On the other hand, the integration between the different cortical inputs in a certain striatal region may take place through intercompartmental communication (Groenewegen et al., 1990). However, at present little is known about intrastriatal connections and, in particular, about the connections between the different compartments.

Finally, another possible way of integrating certain aspects of the different parallel circuits is through the dopaminergic system. As discussed elsewhere (Groenewegen et al., 1990), the topography in the striatonigral system is such that the ventral striatum and specific compartments of the dorsal striatum may affect the dopaminergic nigrostriatal input of extensive striatal regions (cf. also Nauta et al., 1978; Gerfen, 1985; Gerfen et al., 1987). Therefore, the organization of the reciprocal striatonigral and nigrostriatal pathways is not in register with the principal parallel circuits outlined above.

Acknowledgements

This work was supported by Medigon/NWO Program Grant No. 900-550-093.

Our thanks to Dr. Menno Witter for helpful discussions, Mrs. Joan Hage for secretarial assistance and Mr. Dirk de Jong for the photography.

Abbreviations

AC	nucleus accumbens
ACl	lateral part of the nucleus accumbens
ACm	medial part of the nucleus accumbens
ACd	dorsal anterior cingulate area
ACv	ventral anterior cingulate area
AHA	amygdalo-hippocampal area
AId	dorsal agranular insular area
AIv	ventral agranular insular area
BL	basolateral amygdaloid nucleus
BLa	anterior part of the BL
BLc	caudal part of the BL
C	central amygdaloid nucleus
CL	central lateral thalamic nucleus

CM	central medial thalamic nucleus
C-P	caudate-putamen complex
CTb	choleratoxin subunit B
GPm	medial part of the globus pallidus
IL	infralimbic area
IMD	intermediodorsal thalamic nucleus
L	lateral amygdaloid nucleus
LO	lateral orbital area
MD	mediodorsal thalamic nucleus
MDc	central segment of the MD
MDl	lateral segment of the MD
MDm	medial segment of the MD
MO	medial orbital area
OT	olfactory tubercle
OTvp	ventral pallidal elements of the olfactory tubercle
PFC	prefrontal cortex
PHA-L	*Phaseolus vulgaris*-leucoagglutinin
PL	prelimbic area
PreCm	medial precentral area
PV	paraventricular thalamic nucleus
R	red nucleus
SN	substantia nigra
SNR	substantia nigra, pars reticulata
VLO	ventrolateral orbital area
VO	ventral orbital area
VP	ventral pallidum
VTA	ventral tegmental area

References

Akert, K. (1964) Comparative anatomy of frontal cortex and thalamofrontal connections. In: J.M. Warren and K. Akert (Eds.), *The Frontal Granular Cortex and Behavior*, McGraw-Hill, New York, pp. 372–396.

Alexander, G.E., DeLong, M.R. and Strick, P.L. (1986) Parallel organization of functionally segregated circuits linking basal ganglia and cortex. *Annu. Rev. Neurosci.*, 9: 357–381.

Alexander, G.E., Crutcher, M.D. and DeLong, M.R. (1990) Basal ganglia-thalamocortical circuits: parallel substrates for motor, oculomotor, "prefrontal and "limbic" functions. *This Volume*, Ch. 6.

Beckstead, R.M. (1979) An autoradiographic examination of corticocortical and subcortical projections of the mediodorsal-projection (prefrontal) cortex in the rat. *J. Comp. Neurol.*, 184: 43–62.

Beckstead, R.M. (1984) The thalamostriatal projection in the cat. *J. Comp. Neurol.*, 223: 313–346.

Berendse, H.W. and Groenewegen, H.J. (1990) The organization of the thalamostriatal projections in the rat, with special emphasis on the ventral striatum. *J. Comp. Neurol.*, 299: 187–228.

Berendse, H.W. and Groenewegen, H.J. (1991) Restricted cortical terminal fields of the midline and intralaminar thalamic nuclei in the rat. In preparation.

Berendse, H.W., Voorn, P., te Kortschot, A. and Groenewegen, H.J. (1988) Nuclear origin of thalamic afferents of the ventral striatum determines their relation to patch/matrix configurations in enkephalin-immunoreactivity in the rat. *J. Chem. Neuroanat.*, 1: 3–10.

Cavada, C. and Goldman-Rakic, P. (1989) Posterior parietal cortex in rhesus monkey. II. Evidence for segregated corticocortical networks linking sensory and limbic areas with the frontal lobe. *J. Comp. Neurol.*, 287: 422–445.

Cornwall, J. and Phillipson, O.T. (1988) Afferent connections of the dorsal thalamus of the rat as shown by retrograde lectin transport. I. The mediodorsal nucleus. *Neuroscience*, 24: 1035–1049.

Daugé, V., Steimes, P., Derrien, M., Beau, N., Roques, B.P. and Féger, J. (1989) CCK8 effects on motivational and emotional states of rats involve CCKA receptors of the posteromedian part of the nucleus accumbens. *Pharmacol. Biochem. Behav.*, 34: 157–163.

DeLong, M.R. and Georgopoulos, A.P. (1981) Motor functions of the basal ganglia. In J.M. Brookhart, V.B. Mountcastle and V.B. Brooks (Eds.), *Handbook of Physiology, Section 1, The Nervous System, Vol. II, Part 2*, American Physiological Society, Bethesda, MD, pp. 1017–1061.

Divac, I., Kosmal, A., Björklund, A. and Lindvall, O. (1978) Subcortical projections to the prefrontal cortex in the rat as revealed by the horseradish peroxidase technique. *Neuroscience*, 3: 785–796.

Eichenbaum, H., Clegg, R.A. and Feeley, A. (1983) Reexamination of functional subdivisions of the rodent prefrontal cortex. *Exp. Neurol.*, 79: 434–451.

Ericson, H. and Blomqvist, A. (1988) Tracing of neuronal connections with cholera toxin subunit B: light and electron microscopic immunohistochemistry using monoclonal antibodies. *J. Neurosci. Meth.*, 24: 225–235.

Fass, B., Talbot, K. and Butcher, L.L. (1984) Evidence that efferents from the basolateral amygdala innervate the dorsolateral neostriatum in rats. *Neurosci. Lett.*, 44: 71–75.

Fink-Jensen, A. and Mikkelsen, J.D. (1989) The striatoentopeduncular pathway in the rat. A retrograde transport study with wheatgerm-agglutinin-horseradish peroxidase. *Brain Res.*, 476: 194–198.

Gerfen, C.R. (1984) The neostriatal mosaic: compartmentalization of corticostriatal input and striatonigral output systems. *Nature*, 311: 461–464.

Gerfen, C.R. (1985) The neostriatal mosaic. I. Compartmental organization of projections from the striatum to the substantia nigra in the rat. *J. Comp. Neurol.*, 236: 454–476.

Gerfen, C.R. (1989) The neostriatal mosaic: striatal patch-matrix organization is related to cortical lamination. *Science*, 246: 385–388.

Gerfen, C.R. and Clavier, R.M. (1979) Neural inputs to the prefrontal agranular insular cortex in the rat: horseradish peroxidase study. *Brain Res. Bull.,* 4: 347–353.

Gerfen, C.R. and Sawchenko, P.E. (1984) An anterograde neuroanatomical tracing method that shows the detailed morphology of neurons, their axons and terminals: immunohistochemical localization of an axonally transported plant lectin, *Phaseolus vulgaris*-leucoagglutinin (PHA-L). *Brain Res.,* 290: 219–238.

Gerfen, C.R., Herkenham, M. and Thibault, J. (1987) The neostriatal mosaic. II. Patch- and matrix-directed mesostriatal dopaminergic and non-dopaminergic systems. *J. Neurosci.,* 7: 3915–3934.

Giguere, M. and Goldman-Rakic, P.S. (1988) Mediodorsal nucleus: areal, laminar, and tangential distribution of afferents and efferents in the frontal lobe of rhesus monkeys. *J. Comp. Neurol.,* 277: 195–213.

Goldman-Rakic, P.S. (1988) Changing concepts of cortical connectivity: parallel distributed cortical networks. In P. Rakic and W. Singer (Eds.), *Neurobiology of Neocortex,* Wiley, Chichester, pp. 177–202.

Goldman-Rakic, P.S. (1990) This Volume, Ch. 16, p. 325.

Goldman-Rakic, P.S. and Porrino, L.J. (1985) The primate mediodorsal (MD) nucleus and its projection to the frontal lobe. *J. Comp. Neurol.,* 242: 535–560.

Goldman-Rakic, P.S. and Selemon, L.D. (1986) Topography of corticostriatal projections in nonhuman primates and implications for functional parcellation of the neostriatum. In E.G. Jones and A. Peters (Eds.), *Cerebral Cortex, Vol. 5,* Plenum, New York, pp. 447–466.

Groenewegen, H.J. (1988) Organization of the afferent connections of the mediodorsal thalamic nucleus in the rat, related to the mediodorsal-prefrontal topography. *Neuroscience,* 24: 379–431.

Groenewegen, H.J. and Berendse, H.W. (1990) Connections of the subthalamic nucleus with ventral striatopallidal parts of the basal ganglia in the rat. *J. Comp. Neurol.,* 294: 607–622.

Groenewegen, H.J. and Russchen, F.T. (1984) Organization of the efferent projections of the nucleus accumbens to pallidal, hypothalamic, and mesencephalic structures: A tracing and immunohistochemical study in the cat. *J. Comp. Neurol.,* 223: 347–367.

Groenewegen, H.J. and Wouterlood, F.G. (1990) Light and electron microscopic tracing of neuronal connections with *Phaseolus vulgaris*-leucoagglutinin (PHA-L), and combinations with other neuroanatomical techniques. In A. Björklund, T. Hökfelt, F.G. Wouterlood and A. VandenPol (Eds.), *Handbook of Chemical Neuroanatomy. Vol. 8. Analysis of Neuronal Microcircuits and Synaptic Interactions,* Elsevier, Amsterdam, pp. 47–124.

Groenewegen, H.J. Becker, N.E.H.M. and Lohman, A.H.M. (1980) Subcortical afferents of the nucleus accumbens septi in the cat, studied with retrograde axonal transport of horseradish peroxidase and bisbenzimid. *Neuroscience,* 5: 1903–1916.

Groenewegen, H.J., Vermeulen-Van der Zee, E., te Kortschot, A. and Witter M.P. (1987) Organization of the projections from the subiculum to the ventral striatum in the rat. A study using anterograde transport of *Phaseolus vulgaris*-leucoagglutinin. *Neuroscience,* 23: 103–120.

Groenewegen, H.J., Berendse, H.W., Meredith, G.E., Haber, S.N., Voorn, P., Wolters, J.G. and Lohman, A.H.M. (1990) Functional anatomy of the ventral, limbic system-innervated striatum. In P. Willner and J. Scheel-Krüger (Eds.), *The Mesolimbic Dopamine System: From Motivation to Action,* Wiley, Chichester, in press.

Haber, S.N. and Nauta, W.J.H. (1983) Ramifications of the globus pallidus in the rat as indicated by patterns of immunohistochemistry. *Neuroscience,* 9: 245–260.

Haber, S.N., Groenewegen, H.J., Grove, E.A. and Nauta, W.J.H. (1985) Efferent connections of the ventral pallidum: Evidence of a dual striatopallidofugal pathway. *J. Comp. Neurol.,* 235: 322–335.

Haber, S.N., Lynd, E., Klein, C. and Groenewegen, H.J. (1990) Topographic organization of the ventral striatal efferent projections in the rhesus monkey: an anterograde tracing study. *J. Comp. Neurol.,* 293: 282–298.

Heimer, L. and Wilson, R.D. (1975) The subcortical projections of the allocortex: similarities in the neural associations of the hippocampus, the piriform cortex, and the neocortex. In M. Santini (Ed.), *Golgi Centennial Symposium,* Raven Press, New York, pp. 177–193.

Heimer, L., Alheid, G.F. and Zaborsky, L. (1985) Basal ganglia. In G.W. Paxinos (Ed.), *The Rat Nervous System,* Academic Press, Australia, pp. 37–86.

Heimer, L., Zaborsky, L., Zahm, D.S. and Alheid, G.F. (1987) The ventral striatopallidothalamic projection. I. The striatopallidal link originating in the striatal parts of the olfactory tubercle. *J. Comp. Neurol.,* 255: 571–591.

Herkenham, M. (1986) New perspectives on the organization and evolution of nonspecific thalamocortical projections. In E.G. Jones and A. Peters (Eds.), *Cerebral Cortex, Vol. 5. Sensory-Motor Areas and Aspects of Cortical Connectivity,* Plenum Press, New York, pp. 403–445.

Herkenham, M. and Nauta, W.J.H. (1977) Afferent connections of the habenular nuclei in the rat. A horseradish peroxidase study, with a note on the fiber-of-passage problem. *J. Comp. Neurol.,* 173: 123–146.

Hurley-Gius, K.M. and Neafsey, E.J. (1986) The medial frontal cortex and gastric motility: microstimulation results and their possible significance for the overall pattern of organization of rat frontal and parietal cortex. *Brain Res.,* 365: 241–248.

Ilinsky, I.A., Jouandet, M.L. and Goldman-Rakic, P.S. (1985) Organization of the nigrothalamocortical system in the rhesus monkey. *J. Comp. Neurol.,* 236: 316–330.

Jayaraman, A. (1985) Organization of thalamic projections in the nucleus accumbens and the caudate nucleus in cats and its relation with hippocampal and other subcortical afferents. *J. Comp. Neurol.,* 231: 396–420.

Jones, E.G. and Leavitt, R.Y. (1974) Retrograde axonal transport and the demonstration of non-specific projections to the cerebral cortex and striatum from thalamic intralaminar nuclei in the rat, cat and monkey. *J. Comp. Neurol.*, 154: 349–378.

Kelley, A.E., Domesick, V.B. and Nauta, W.J.H. (1982) The amygdalostriatal projection in the rat — An anatomical study by anterograde and retrograde tracing methods. *Neuroscience*, 7: 615–630.

Kelley, A.E., Bakshi, V.P., Delfs, J.M. and Lang, C.G. (1989) Cholinergic stimulation of the ventrolateral striatum elicits mouth movements in rats: pharmacological and regional specificity. *Psychopharmacology*, 99: 542–549.

Kolb, B. (1984) Functions of the frontal cortex of the rat: A comparative review. *Brain Res. Rev.*, 8: 65–98.

Kolb, B., Pittman, K., Sutherland, R.J. and Whishaw I.Q. (1982) Dissociation of the contributions of the prefrontal cortex and dorsomedial thalamic nucleus to spatially guided behavior in the rat. *Behav. Brain Res.*, 6: 365–378.

Kosar, E., Grill, H.J. and Norgren, R. (1986) Gustatory cortex in the rat. I. Physiological properties and cytoarchitecture. *Brain Res.*, 379: 329–341.

Krettek, J.E. and Price, J.L. (1977) The cortical projections of the mediodorsal nucleus and adjacent thalamic nuclei in the rat. *J. Comp. Neurol.*, 171: 157–192.

Krettek, J.E. and Price, J.L. (1978) Projections from the amygdaloid complex to the cerebral cortex and thalamus in the rat and cat. *J. Comp. Neurol.*, 172: 687–722.

Leonard, C.M. (1969) The prefrontal cortex of the rat. I. Cortical projection of the mediodorsal nucleus. II. Efferent connections. *Brain Res.*, 12: 321–343.

Leonard, C.M. (1972) The connections of the dorsomedial nuclei. *Brain Behav. Evol.*, 6: 524–541.

Macchi, G. and Bentivoglio, M. (1986) The thalamic intralaminar nuclei and the cerebral cortex. In E.G. Jones and A. Peters (Eds.), *Cerebral Cortex. Vol. 5. Sensory-Motor Areas and Aspects of Cortical Connectivity*, Plenum Press, New York, pp. 355–401.

Macchi, G., Bentivoglio, M., Molinari, M. and Minciacchi, D. (1984) The thalamo-caudate versus thalamo-cortical projections as studied in the cat with fluorescent retrograde double labeling. *Exp. Brain Res.*, 54: 225–239.

McDonald, A.J. (1987) Organization of amygdaloid projections to the mediodorsal thalamus and prefrontal cortex: a fluorescence retrograde transport study in the rat. *J. Comp. Neurol.*, 262: 46–58.

McGeorge, A.J. and Faull, R.L.M. (1989) The organization of the projection from the cerebral cortex to the striatum in the rat. *Neuroscience*, 29: 503–537.

Mogenson, G.J., Jones, L.D. and Yim, C.Y. (1980) From motivation to action: Functional interface between the limbic system and the motor system. *Prog. Neurobiol.*, 14: 69–97.

Mogenson, G.J., Swanson, L.W. and Wu, M. (1983) Neural projections from nucleus accumbens to globus pallidus, substantia innominata, and lateral preoptic-lateral hypothalamic area: an anatomical and electrophysiological investigation in the rat. *J. Neurosci.*, 3: 189–202.

Nauta, H.J.W. (1979) A proposed conceptual reorganization of the basal ganglia and telencephalon. *Neuroscience*, 4: 1875–1881.

Nauta, H.J.W. (1986) A simplified perspective on the basal ganglia and their relation to the limbic system. In B.K. Doane and K.E. Livingston (Eds.), *The Limbic System: Functional Organization and Clinical Disorders*, Raven Press, New York, pp. 67–77.

Nauta, W.J.H. (1986) Circuitous connections linking cerebral cortex, limbic system, and corpus striatum. In B.K. Doane and K.E. Livingston (Eds.), *The Limbic System: Functional Organization and Clinical Disorders*, Raven Press, New York, pp. 43–54.

Nauta, W.J.H., Smith, G.P., Faull, R.L.M. and Domesick, V.B. (1978) Efferent connections and nigral afferents of the nucleus accumbens septi in the rat. *Neuroscience*, 3: 385–401.

Neafsey, E.J. (1990) Prefrontal cortical control of the autonomic nervous system: anatomical and physiological observations. *This Volume*, Ch. 7.

Neafsey, E.J., Hurley-Gius, K.M. and Arvanitis, D. (1987) The topographical organization of neurons in the rat medial frontal, insular and olfactory cortex projecting to the solitary nucleus, olfactory bulb, periaqueductal gray and superior colliculus. *Brain Res.*, 377: 261–270.

Pandya, D.N. and Seltzer, B. (1982) Association areas of the cerebral cortex. *Trends Neurosci.*, 5: 386–392.

Parent, A. (1986) *Comparative Neurobiology of the Basal Ganglia*, Wiley, Chichester.

Phillipson, O.T. and Griffiths, A.C. (1985) The topographic order of inputs to nucleus accumbens in the rat. *Neuroscience*, 16: 275–296.

Porrino, L.J. and Goldman-Rakic, P.S. (1982) Brainstem innervation of prefrontal and anterior cingulate cortex in the rhesus monkey revealed by retrograde transport of HRP. *J. Comp. Neurol.*, 205: 63–76.

Price, J.L. and Slotnick, B.M. (1983) Dual olfactory representation in the rat thalamus: an anatomical and electrophysiological study. *J. Comp. Neurol.*, 215: 63–77.

Price, J.L., Russchen, F.T. and Amaral, D.G. (1987) The limbic region. II. The amygdaloid complex. In A. Björklund, T. Hökfelt and L.W. Swanson, *Handbook of Chemical Neuroanatomy. Vol. 5. Integrated Systems of the CNS, Part I*, Elsevier, Amsterdam, pp. 279–388.

Ragsdale Jr., C.W. and Graybiel, A.M. (1988) Fibers from the basolateral nucleus of the amygdala selectively innervate striosomes in the caudate nucleus of the cat. *J. Comp. Neurol.*, 269: 506–522.

Reep, R.L. (1984) Relationship between prefrontal and limbic cortex: A comparative anatomical review. *Brain Behav. Evol.*, 25: 5–80.

Reep, R.L. and Winans, S.S. (1982) Efferent connections of dorsal and ventral agranular insular cortex in the hamster. *Mesocricetus auratus. Neuroscience,* 7: 2609–2635.

Royce, G.J. (1978) Cells of origin of subcortical afferents to the caudate nucleus: a horseradish peroxidase study in the cat. *Brain Res.,* 153: 465–475.

Royce, G.J. (1983) Single thalamic neurons which project to both the rostral cortex and caudate nucleus studied with the fluorescent double labeling method. *Exp. Neurol.,* 79: 773–784.

Royce, G.J. and Laine, E. (1984) Efferent connections of the caudate nucleus, including cortical projections of the striatum and other basal ganglia: An autoradiographic and horseradish peroxidase investigation in the cat. *J. Comp. Neurol.,* 226: 28–49.

Royce, G.J., Bromley, S., Gracco, C. and Beckstead, R.M. (1989) Thalamocortical connections of the rostral intralaminar nuclei: An autoradiographic analysis in the cat. *J. Comp. Neurol.,* 288: 555–582.

Russchen, F.T. and Price, D.L. (1984) Amygdalostriatal projections in the rat. Topographical organization and fiber morphology shown using the lectin PHA-L as an anterograde tracer. *Neurosci. Lett.,* 47: 15–22.

Russchen, F.T., Bakst, I., Amaral, D.G. and Price, J.L. (1985) The amygdalostriatal projections in the monkey. An anterograde tracing study. *Brain Res.,* 329: 241–257.

Russchen, F.T., Amaral, D.G. and Price, J.L. (1987) The afferent input to magnocellular division of the mediodorsal thalamic nucleus in the monkey, *Macaca fascicularis. J. Comp. Neurol.,* 256: 175–210.

Saper, C.B. (1982) Convergence of autonomic and limbic connections in the insular cortex of the rat. *J. Comp. Neurol.,* 210: 163–173.

Scheel-Krüger, J. and Willner, P. (1990) The mesolimbic system: principles of operation. In P. Willner and J. Scheel-Krüger (Eds.), *The Mesolimbic Dopamine System: From Motivation to Action,* Wiley, Chichester, in press.

Selemon, L.D. and Goldman-Rakic, P.S. (1985) Longitudinal topography and interdigitation of corticostriatal projections in the resus monkey. *J. Neurosci.,* 5: 776–794.

Selemon, L.D. and Goldman-Rakic, P.S. (1988) Common cortical and subcortical targets of the dorsolateral prefrontal and posterior parietal cortices in the rhesus monkey: Evidence for a distributed neural network subserving spatially guided behavior. *J. Neurosci.,* 8: 4049–4068.

Sesack, S.R., Deutch, A.Y., Roth, R.H. and Bunney, B.S. (1989) Topographical organization of the efferent projections of the medial prefrontal cortex in the rat: An anterograde tract-tracing study with *Phaseolus vulgaris*-leucoagglutinin. *J. Comp. Neurol.,* 290: 213–240.

Terreberry, R.R. and Neafsey, E.J. (1983) Rat medial frontal cortex: a visceral motor region with a direct projection to the solitary nucleus. *Brain Res.,* 278: 245–249.

Terreberry, R.R. and Neafsey, E.J. (1987) The rat medial frontal cortex projects directly to autonomic regions of the brain stem. *Brain Res. Bull.,* 19: 639–649.

Thierry, A.M., Chevalier, G., Ferron, A. and Glowinski, J. (1983) Diencephalic and mesencephalic efferents of the medial prefrontal cortex in the rat: Electrophysiological evidence for the existence of branched axons. *Exp. Brain Res.,* 50: 275–282.

Van der Kooy, D., McGinty, J.F., Koda, L.Y., Gerfen, C.R. and Bloom, F.E. (1982) Visceral cortex: a direct connection from prefrontal cortex to the solitary nucleus in rat. *Neurosci. Lett.,* 33: 123–127.

Van Eden, C.G. and Uylings, H.B.M. (1985) Cytoarchitectonic development of the prefrontal cortex in the rat. *J. Comp. Neurol.,* 241: 2253–267.

Veening, J.G., Cornelissen, F.M. and Lieven, P.A.J.M. (1980) The topical organization of the afferents to the caudatoputamen of the rat. A horseradish peroxidase study. *Neuroscience,* 5: 1253–1268.

Velayos, J.L. and Reinoso-Suarez, F. (1982) Topographic organization of the brainstem afferents to the mediodorsal thalamic nucleus. *J. Comp. Neurol.,* 206: 17–27.

Velayos, J.L. and Reinoso-Suarez, F. (1985) Prosencephalic afferents to the mediodorsal thalamic nucleus. *J. Comp. Neurol.,* 242: 161–181.

Young III, W.S., Alheid, G.F. and Heimer, L. (1984) The ventral pallidal projection to the mediodorsal thalamus: a study with fluorescent retrograde tracers and immunohistofluorescence. *J. Neurosci.,* 4: 1626–1638.

Zahm, D.S., Zaborszky, L., Alheid, G.F. and Heimer, L. (1987) The ventral striatopallidothalamic projection. II. The ventral pallidothalamic link. *J. Comp. Neurol.,* 255: 592–605.

Discussion

P. Goldman-Rakic: The topographic overlap of corticostriatal projections needs not to imply convergence. Double label studies in primates reveal that afferents from cortical areas interdigitate within the same portion of the neostriatum.

H.J. Groenewegen: I agree that topographic overlap does not necessarily indicate convergence on the same cell groups in the striatum, as has been shown by your own work (Selemon and Goldman-Rakic, 1985). Very little can be said about this specific point in rats since in this species the double label protocol of simultaneously labeling afferents from two cortical areas to a specific striatal region has not yet been applied.

E. Neafsey: What is the function of the lateral segment of MD?

H.J. Groenewegen: One of the strongest inputs to the lateral segment of MD arises in the dorsolateral tegmental nucleus of the caudal midbrain. In turn, this part of the dorsal tegmental region is afferented, on the one hand, through the lateral habenula by the internal segment of the globus pallidus (entopeduncular nucleus in the rat) and, on the other hand, by the limbic system through the medial habenula – interpeduncular nucleus (Herkenham and Nauta, 1977; Groenewegen et al., 1986). Therefore, in the dorsolateral tegmental nucleus the outputs of two important forebrain systems, i.e. the basal ganglia and the limbic system, seem to converge. The lateral segment of the MD appears to be an important way station in the pathways that lead from the dorsolateral tegmental nucleus back to the forebrain. Furthermore, the lateral segment further receives a significant input from the lateral hypothalamus (Groenewegen, 1988).

H.B.M. Uylings: The resemblance of your rat data with the primate data presented by Dr. Pandya in the dorsal (archicortical) – ventral (palaeocortical) trend connections with the MD subnuclei is striking (only the parvicellular portion of the MD is excepted). Does the so-called dorsal – ventral trend differentiation within the basal ganglia of the rat also correspond with data of the monkey?

H.J. Groenewegen: Dr. Pandya was mainly talking about the trends in the corticocortical connections in primates. I think that the "rules" underlying the corticocortical connections, as discussed by Dr. Pandya, have consequences also for the patterns of cortical – subcortical connections, including those with the MD and the basal ganglia. I think that the idea of "rules" underlying corticocortical and cortical – subcortical connections might help to understand why and may even predict which parts of the cortex and basal ganglia are connected with each other, also in the rat.

H.B.M. Uylings: According to Herkenham (1986), the MD is classified to belong to the group of thalamic nuclei which has specific projections mainly to the middle cortical layers. He also mentioned that those nuclei share the characteristic that they receive input from only a restricted number of subcortical sources. You show very convincingly that this does not fit in with the input of the MD. Is the MD an exception to this rule?

H.J. Groenewegen: Our data with respect to the laminae of termination of thalamocortical fibers from the MD are in agreement with the observations by Herkenham in that the strongest terminations are found in layer III of the PFC, although MD also has terminations in layer I. In contrast to other thalamic nuclei with a "layer III termination pattern", the afferents of MD originate in a wide array of subcortical structures, which is at variance with the ideas of Herkenham. However, also with respect to the termination pattern of the thalamocortical fibers from the midline and intralaminar nuclei, our data do not confirm those of Herkenham. According to Herkenham these projections are directed to the deep layers. The results of our experiments with restricted PHA-L injections in the intralaminar and midline nuclei suggest that the projections of these nuclei are very topographically organized and are directed to more superficial layers. In addition, sparse, smoothly appearing fibers are found in the deep layer VI in rather widespread cortical regions.

V.B. Domesick: I compliment your report on the numerous details of the circuitry of the medial infralimbic cortex and lateral "orbitofrontal" cortex. My question is whether the circuits of these two prefrontal areas involving the basolateral amygdala and the MD are separate.

H.J. Groenewegen: With respect to the projections to the lateral and the medial PFC, it is clear that the afferent projections from both the MD and the basolateral amygdaloid nucleus stem from largely separate cell groups in these two nuclei. The projections to the lateral PFC originate predominantly in the caudal part of the medial segment and in the central segment of MD and in the rostral part of the basolateral nucleus. The medial PFC is innervated by fibers from the rostral part of the medial segment of MD and the caudal portion of the basolateral nucleus. This indicates separation of the circuits through these different PFC areas. At the same time, the above-mentioned parts of the basolateral amygdaloid nuclei have additional cortical targest outside the PFC.

F. Clasca: In your presentation you have distinguished a caudal sector in the MD. What are the characteristic subcortical connections of this part of MD? Or is it only distinguishable based on the reciprocal connections with the dorsal agranular insular cortex?

H.J. Groenewegen: The subcortical connections of the lateral and dorsal part of the caudal sector of the MD are comparable with those of the lateral segment (dorsal tegmental region, substantia nigra, pretectal area) and the set of afferents to the medial and ventral part of the caudal sector is comparable with that of the medial segment (ventral pallidum, etc.). In the caudal part of the medial segment, few if any fibers terminate from the amygdala and olfactory tubercle. These afferents are prominent in the more rostral part of the medial segment and in the central segment.

D.A. Powell: (1) Does the MD project to the caudate nucleus? (2) Does the MD project to the thalamic midline nuclei (in particular the nucleus reuniens)?

H.J. Groenewegen: The answer to your first question is that we have seen anterograde labeling of fibers and terminals in the caudate-putamen complex following injections of WGA-HRP or PHA-L in the MD. However, such injections most probably involved cells of the adjacent midline and intralaminar thalamic nuclei which are known to project heavily to the striatum. Injections of retrograde tracers in the striatum reveal a small number of labeled cells within the boundaries of MD, but by far the most labeled neurons are present in the midline and intralaminar nuclei in these cases.

Concerning your second question, following injections of WGA-HRP or PHA-L in the MD, we have seen anterograde and retrograde labeling only in the reticular nucleus of the thalamus and we have not found clear evidence for further intrathalamic connections.

References

Groenewegen, H.J. (1988) Organization of the afferent connections of the mediodorsal thalamic nucleus in the rat, related to the mediodorsal-prefrontal topography. *Neuroscience,* 24: 379–431.

Groenewegen, H.J., Ahlenius, S., Haber, S.N., Kowall, N.W. and Nauta, W.J.H. (1986) Cytoarchitecture, fiber connections, and some histochemical aspects of the interpeduncular nucleus in the rat. *J. Comp. Neurol.,* 249: 65–102.

Herkenham, M. (1986) New perspectives on the organization and evolution of nonspecific thalamocortical projections. In E.G. Jones and A. Peters (Eds.), *Cerebral Cortex, Vol. 5. Sensory-motor Areas and Aspects of Cortical Connectivity.* Plenum Press, New York, pp. 403–445.

Herkenham, M. and Nauta, W.J.H. (1977) Afferent connections of the habenular nuclei in the rat. A horseradish peroxidase study, with a note on the fiber-of-passage problem. *J. Comp. Neurol.,* 173: 123–146.

Selemon, L.D. and Goldman-Rakic, P.S. (1985) Longitudinal topography and interdigitation of corticostriatal projections in the rhesus monkey. *J. Neurosci.,* 5: 776–794.

CHAPTER 6

Basal ganglia-thalamocortical circuits: Parallel substrates for motor, oculomotor, "prefrontal" and "limbic" functions

Garrett E. Alexander, Michael D. Crutcher and Mahlon R. DeLong

Department of Neurology, Johns Hopkins University School of Medicine, 600 North Wolfe Street, Baltimore MD, 21205, USA

Introduction

Although the basal ganglia had long been viewed primarily in terms of their role in motor control and movement disorders, it is now widely accepted that they contribute to a wide variety of behavioral functions, including skeletomotor, oculomotor, cognitive and even "limbic" processes. This functional diversity is suggested by several lines of evidence, including the striking and varied behavioral effects associated with experimental and disease-induced lesions of the basal ganglia, the extensive connections of these nuclei with the cerebral cortex, and the wide range of behavioral correlates documented by single cell recordings from basal ganglia neurons in experimental animals.

From a functional perspective, the basal ganglia should not be viewed in isolation. They receive topographic projections from all areas of the cerebral cortex, and in turn project their own influences back upon most areas of the frontal lobe via topographically organized pathways that pass through the thalamus. Earlier schemes suggested that projections from diverse cortical areas, including motor, sensory and "association" fields, converged within the basal ganglia and were then "funneled" back upon precentral motor areas (Evarts and Thach, 1969; Kemp and Powell, 1970; Allen and Tsukahara, 1974). Currently, however, the weight of evidence suggests a different type of organization: the basal ganglia, along with their connected cortical and thalamic areas, are viewed as components of a family of *"basal ganglia-thalamocortical" circuits* that are organized in a parallel manner and remain largely segregated from one another, both structurally and functionally (Alexander et al., 1986). Each circuit is thought to engage separate regions of the basal ganglia and thalamus, and the output of each appears to be centered on a different part of the frontal lobe: the *"motor"* circuit is focused on the precentral motor fields; the *"oculomotor"* circuit on the frontal eye fields; the *"prefrontal"* circuits on dorsolateral prefrontal and lateral orbitofrontal cortex; and the *"limbic"* circuit on anterior cingulate and medial orbitofrontal cortex (Fig. 1).

Since we last reviewed this topic in detail (Alexander et al., 1986), there has been considerable progress in clarifying the organization of the basal ganglia-thalamocortical circuits, and some of these newer data will be emphasized in the following discussion. Information pertaining to the "motor" and "oculomotor" circuits will also be emphasiz-

---

Correspondence: Dr. Garrett E. Alexander, Department of Neurology, Johns Hopkins University School of Medicine, 600 North Wolfe Street, Baltimore, MD 21205, USA.

ed, since at present these are the circuits whose functional characteristics are best understood. Nevertheless, much has been learned in recent years about the role of prefrontal cortical fields in certain complex behavioral processes, and coincident with these developments the anatomical connections linking prefrontal cortex and the basal ganglia have been defined with increasing precision. Although the behavioral/physiological significance of the prefrontal-basal ganglia connections has not been studied as extensively as that of the "motor" and "oculomotor" circuits, and the functions of the "limbic" circuit remain even more obscure, we suggest that future attempts to clarify the functional organization of the "prefrontal" and "limbic" circuits might be profitably based, at least in part, on current insights into the operations of their "motor" and "oculomotor" counterparts. As in our previous review, the following discussion will rely primarily upon data obtained in primates.

New data from a variety of disciplines have led to an expansion and refinement of the initial concept of parallel, segregated circuits. It now seems possible to define several different levels of functional differentiation within each of the basal ganglia-thalamocortical circuits. The evidence to date suggests that each circuit may contain a number of highly specialized "channels", and even "sub-channels", that permit parallel, multi-level processing of a vast number of variables to proceed concurrently. Within the "motor" circuit for example, a well-defined somatotopy is maintained throughout all stages of the circuit, thereby giving rise to clearly differentiated "leg", "arm" and "orofacial" channels. There is also evidence suggesting further subdivisions of the "motor" circuit in terms of (a) the types of behaviors subserved (e.g., sub-channels concerned with movement preparation vs. movement execution) and (b) the maintained segregation of influences from different cortical areas (e.g., separate sub-channels for each of the precentral motor fields). Moreover, included within each of the basal ganglia-thalamocortical circuits is a "direct" pathway that passes from the striatum directly to one of the basal ganglia output nuclei, and an "indirect" pathway which, though organized in parallel with the direct pathway, includes an intermediate relay through the external pallidum and subthalamic nucleus. The importance of the "indirect" pathway in overall circuit operations is now more fully appreciated. Together, these new developments suggest there may be multiple subsets of parallel pathways *within* each of the basal ganglia-thalamocortical circuits, thereby adding a new dimension to recent concepts of functional segregation and parallel processing within these structures.

"Motor" circuit (Fig. 2)

Clinical-pathological correlations provided some of the earliest evidence for a role of the basal ganglia in motor control. Subsequent anatomical and physiological studies have added further support. By the early 1970s, in fact, it was widely assumed that the entire output of the basal ganglia was directed to motor cortex via the ventrolateral

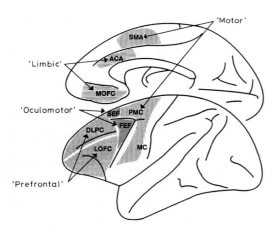

Fig. 1. Frontal lobe targets of basal ganglia output. This figure illustrates schematically the cortical areas which receive the output of the separate basal ganglia-thalamocortical circuits. ACA, anterior cingulate area; DLPC, dorsolateral prefrontal cortex; FEF, frontal eye field; LOFC, lateral orbitofrontal cortex; MC, primary motor cortex; MOFC, medial orbitofrontal cortex; PMC, premotor cortex; SEF, supplementary eye field; SMA, supplementary motor area.

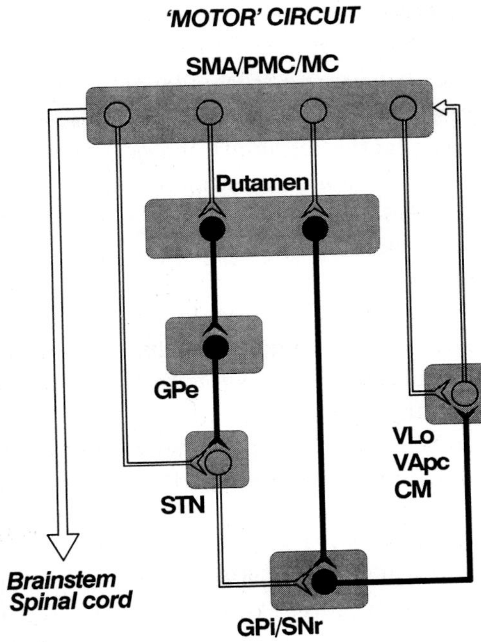

'MOTOR' CIRCUIT

Fig. 2. Simplified diagram of the "motor" circuit. The cortical areas shown projecting to the putamen include only the "closed loop" portion of the "motor" circuit. Additional "open loop" corticostriatal inputs to the "motor" circuit arise from the arcuate premotor area, primary somatosensory cortex and somatosensory association cortex. Inhibitory neurons are filled; excitatory are unfilled. CM, n. centrum medianum; GPe, external segment of the globus pallidus; GPi, internal segment of the globus pallidus; SNr, substantia nigra pars reticulata; STN, subthalamic nucleus; VApc, n.ventralis anterior, pars parvocellularis; VLo, n.ventralis lateralis, pars oralis.

thalamus, where basal ganglia and cerebellar influences were thought to converge (Evarts and Thach, 1969; Kemp and Powell, 1970; Allen and Tsukahara, 1974). Over the next decade, however, it became increasingly clear that the basal ganglia and cerebellar projections are directed to separate target zones within the thalamus (Asanuma et al., 1983), and that only part of the basal ganglia output, i.e., that of the "motor" circuit, is directed toward thalamic targets that project back upon precentral motor areas (DeLong and Georgopoulos, 1981).

In primates, the basal ganglia "motor" pathways are focused principally on the putamen and its connections. This part of the neostriatum receives topographic projections from primary motor cortex (MC) (Künzle, 1975; Jones et al., 1977) and from at least two premotor areas, including the arcuate premotor area (APA) (Künzle, 1978; Selemon and Goldman-Rakic, 1985) and the supplementary motor area (SMA) (Künzle, 1978; Jurgens 1984; Miyata and Sasaki 1984; Selemon and Goldman-Rakic, 1985). The putamen also receives topographic projections from primary somatosensory cortex (Brodmann's areas 3, 1 and 2) (Jones et al., 1977; Künzle, 1977) and from somatosensory association cortex (area 5) (Kemp and Powell, 1970). The terminal fields of these projections occupy the bulk of the putamen, apart from its most rostral and caudoventral extensions. While some of these projections may encroach slightly upon adjacent portions of the caudate nucleus, they are confined primarily to the putamen. These topographically organized projections result in a somatotopic organization which consists of a dorsolateral zone in which the leg is represented, a ventromedial orofacial region, and a territory in between in which there is representation of the arm (Künzle, 1975; Liles, 1975; Crutcher and DeLong 1984a; Alexander and DeLong, 1985b). Each of these representations extends along virtually the entire rostrocaudal axis of the putamen.

While the "arm" region of the putamen receives projections from the respective arm representations within the SMA, MC and APA (Künzle, 1975, 1977), a recent investigation using double anterograde labeling has shown that the terminal fields of these different projections, while contiguous, are essentially non-overlapping (Alexander et al., 1988). These findings raise the possibility, as yet untested, that such segregation may be maintained at subsequent stations in the pallidum and thalamus. If so, it would mean that there are separate SMA-, MC- and APA-specific sub-channels within each of the somatotopically defined (leg, arm, orofacial) channels of the "motor" circuit. The putamen projects topographically to the caudal and ventrolateral two-thirds of both the internal (GPi) and external (GPe) segments of the globus pallidus (Szabo, 1962; Cowan and Powell, 1966; W.J.H. Nauta and Mehler, 1966; Szabo, 1967; Johnson and Rosvold, 1971; DeVito et al., 1980; Parent et al., 1984b), and to a caudolateral territory within the substantia nigra, pars reticulata (SNr) (Szabo, 1962; W.J.H. Nauta and Mehler, 1966; Szabo, 1967; Parent et al., 1984a). Recent retrograde labeling studies in the primate have shown little evidence of collateralization of striatofugal projections to these different targets. Thus, striatal projections to GPe, GPi and SNr have been shown to arise from largely separate populations of striatofugal neurons (Beckstead and Cruz, 1986; Koliatsos et al., 1988). Moreover, those projecting to GPi and SNr have terminal fields that stain heavily for substance P, while the terminal fields of those projecting to GPe show heavy staining for enkephalin (Graybiel and Ragsdale, 1983).

The portion of GPi that receives input from the putamen sends projections to VLo (n. ventralis lateralis pars oralis) and to the lateral portion of VApc (n. ventralis anterior pars parvocellularis), as well as to CM (centromedian n.) (W.J.H. Nauta and Mehler, 1966; Kuo and Carpenter, 1973; R. Kim et al., 1976; DeVito and Anderson, 1982; François et al., 1988). Data concerning the nigrothalamic projections are less extensive, but it does appear that the lateral portions of the SNr, which receive inputs from the putamen (Szabo, 1967; Parent et al., 1984b), also project to the lateral portion of VAmc (n. ventralis anterior pars magnocellularis) (Carpenter and Peter, 1972; Carpenter et al., 1976; Ilinsky et al., 1985).

The "motor" circuit is closed by means of the thalamocortical projections from VLo and lateral VAmc to the SMA (Strick, 1976a; Kievit and Kuypers, 1977; Jürgens, 1984; Schell and Strick, 1984; Wiesendanger and Wiesendanger, 1985a), from lateral VApc (as well as VLo) to premotor cortex (PMC, exclusive of the APA, which appears to receive only cerebellar influences via thalamic area X) (Schell and Strick, 1984; Matelli et al., 1986, 1989), and from CM to MC (Kievit and Kuypers, 1977). In addition, VLo also projects to rostral MC (Strick, 1976a; Kievit and Kuypers, 1977; Wiesendanger and Wiesendanger, 1985a,b; Nambu et al., 1988; Matelli et al., 1989). Recently, it was shown that the continuity of a portion of the "motor" circuit could be demonstrated in individual animals by injecting combined anterograde and retrograde tracers into non-adjacent components of the circuit (viz., GPi and the SMA), which resulted in overlapping territories of anterograde and retrograde labeling within the intervening nuclei (viz., the putamen and VLo) (Hedreen et al., 1988).

It should be noted at this point that GPi projects to the pedunculopontine nucleus (PPN) as well as to the thalamus by means of collaterals (DeVito and Anderson, 1982; Harnois and Filion, 1982; Parent and De Bellefeuille, 1982). The functional role of this projection is unknown. A recent report in the rat (Spann and Grofova, 1989) indicates that the PPN projects to the medullary reticular formation, which in turn gives rise to descending projections to the spinal cord. This "descending" pathway may ultimately prove to play an important role in motor control and the pathophysiology of movement disorders, but its role is at present uncertain. It should be noted, however, that the PPN of the cat has been implicated in aspects of locomotion (Garcia-Rill, 1986). In addition to the non-thalamic efferents of the "motor" circuit, this circuit also receives input from the precentral motor fields by way of topographically organized projections to the STN (Hartmann-Von Monakow et al., 1978) and CM (Künzle, 1976). The "motor" circuit, then, has points of entry and exit other than the putamen and thalamus and has descending projections in addition to its prominent reentrant projection to the precentral motor fields.

While it is now generally accepted that the basal ganglia and cerebellar output nuclei project to different parts of the thalamus (Asanuma et al., 1983; Ilinsky et al., 1985; Jones, 1985; Ilinsky and Kultas-Ilinsky, 1987), the degree to which these two reentrant systems remain segregated at cortical levels is still a subject of some controversy.

Without embarking on an extended discussion of the largely technical issues involved, it now seems clear that there are some areas of the frontal lobe which receive both types of inputs from the thalamus, and others which receive thalamic inputs exclusively from nuclei that are primary targets of either basal ganglia or cerebellar projections (Strick, 1976a; Kievit and Kuypers, 1977; Schell and Strick, 1984; Wiesendanger and Wiesendanger, 1985a; Dum and Strick, 1989; Matelli et al., 1989). At present, the weight of evidence suggests that only the caudal half of the SMA (behind the genu of the arcuate sulcus) receives an exclusive projection from VLo, the thalamic target of pallidal projections (Schell and Strick, 1984; Wiesendanger and Wiesendanger, 1985a). VLo also projects to other precentral motor fields, however, including rostral MC, rostral SMA and parts of PMC exclusive of the APA, areas that also receive cerebellar influences relayed from VPLo and/or area X within the ventrolateral thalamus (Strick, 1976a; Kievit and Kuypers, 1977; Wiesendanger and Wiesendanger, 1985a; Matelli et al., 1989). But while these other cortical fields apparently receive both basal ganglia and cerebellar influences, it should be emphasized that they do not appear to receive input from any of the other (non-"motor") basal ganglia-thalamocortical circuits. Thus, the "motor" circuit receives a mixture of basal ganglia and cerebellar inputs (indeed, the APA, which sends substantial projections to the putamen, receives almost all of its thalamic input from a cerebellar recipient zone, area X (Schell and Strick, 1984; Matelli et al., 1989)), but appears to have little interaction with the "oculomotor", "prefrontal" or "limbic" circuits.

Additional evidence for the functional segregation of the "motor" circuit from the other basal ganglia-thalamocortical circuits includes the topographic specificity of projections fed back to the striatum from the thalamic and subthalamic components of this reentrant pathway. Thus, VLo and CM each send projections back to the putamen, but not to the caudate nucleus (Kalil, 1978; Parent, 1983, 1986). The dorsolateral two-thirds of the subthalamic nucleus (STN), which receives projections from MC, SMA and PMC (Hartmann-Von Monakow et al., 1978), also projects back to the putamen, rather than the caudate nucleus (Smith and Parent, 1986). This "motor" portion of the STN also projects to the "motor" regions of both pallidal segments (H.J.W. Nauta and Cole, 1978; DeLong et al., 1985).

The spatial convergence of the striatopallidal projection (H.J.W. Nauta and Mehler, 1966; Szabo, 1967; DeVito et al., 1980), and the fact that the large, disk-like dendritic arborizations of pallidal neurons are oriented orthogonally with respect to the incoming striatal projections (François et al., 1984), has prompted speculation, largely on anatomic grounds, that there must be a profound degree of functional integration within GPi (Percheron et al., 1984), which could result in a loss of specificity or relative "degradation" of certain types of functional information. However, single cell recording studies in primates have not shown the loss of specificity in the response properties of GPi neurons predicted by this proposal (Anderson and Horak, 1985; DeLong et al., 1985; Mitchell et al., 1987). In fact, physiological studies have revealed a pronounced degree of somatotopic organization and striking functional specificity at the level of individual neurons throughout the "motor" circuit. The somatotopic organization is evident at cortical (Murphy et al., 1978; Muakkassa and Strick, 1979; Strick and Preston, 1982a, b; Tanji and Kurata, 1982; Mitz and Wise, 1987), striatal (Künzle, 1975; Liles, 1975, 1979; Crutcher and DeLong, 1984a; Alexander and DeLong, 1985a, b; Liles and Updyke, 1985), pallidal (Szabo, 1967; DeLong et al., 1985), nigral (DeLong et al., 1983a), subthalamic (Hartmann-Von Monakow et al., 1978; DeLong et al., 1985) and thalamic (Kuo and Carpenter, 1973; Strick, 1976b; DeVito and Anderson, 1982; Hedreen et al., 1988) stages of the circuit (Fig. 3). At cortical, putaminal and pallidal stages neuronal responses to passive somatosensory examination and torque application have been found to be highly directional and invariably restricted to one segment of the contralateral body (leg, arm, face or trunk), and are usually confined to a single structure (often a single joint) within that segment. Most sensory responses of the subcortical components of the "motor" circuit are proprioceptive in nature, being evoked by active and/or passive movements of individual joints. Thus, while there must be some degree of functional integration within the "motor" circuit, such integration appears to be carried out along strictly somatotopic lines.

The maintained specificity of the neuronal response properties has also been demonstrated within this functional system by studies carried out at cortical, striatal and pallidal stages of the circuit, using motor tasks that effectively dissociated the direction of limb movement from the pattern of muscle activity (Crutcher and DeLong, 1984b; Mitchell et al., 1987; Alexander and Crutcher, 1990a,b; Crutcher and Alexander, 1990). In each case, the activity of substantial proportions of movement-related neurons was found to depend upon the direction of limb movement independent of the associated pattern of muscle activity. Within the SMA, MC, putamen, GPi and GPe, "directional" cells were found to comprise from 30 to 50% of the movement-related neurons, all of which showed sharply delineated somatopic features regardless of their level within the circuit (i.e., cortical, striatal, or pallidal). At the same time, however, significant proportions of "muscle-like" cells were also found in each of these areas.

Studies which have examined the timing of neuronal discharge at different levels of the "motor" circuit in relation to the onset of rapid, stimulus-triggered movements indicate that changes in neural activity tend to occur earlier at cortical than at subcortical stages of the circuit (Thach, 1978; Georgopoulos et al., 1982, 1989; Murphy et al., 1982; Tanji and Kurata, 1982; Crutcher and DeLong, 1984b; Anderson and Horak, 1985; Liles, 1985; Okano and Tanji, 1987; Crutcher and Alexander, 1990). These studies are therefore consistent with the concept that activity within the basal ganglia circuit is generally initiated at cortical levels, with feedback through the circuit occuring later. Obviously, such data do not rule out the possibility that under certain conditions (e.g., the execution of self-initiated rather than stimulus-triggered movements), neural activity changes might begin earlier within other parts of the circuit.

Recent findings indicate that the "motor" circuit may be involved not only in movement *execution*, but in the *preparation* for movement as well. Studies in primates have shown that the precentral motor fields, including PMC, SMA and MC, each contain neurons that show striking changes in discharge rate following presentation of an instructional stimulus that specifies the direction of an

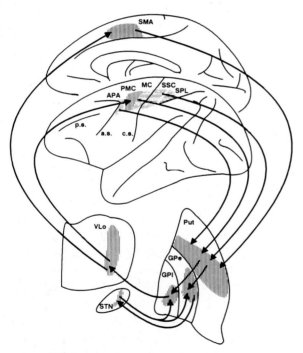

Fig. 3. Somatotopic organization of the "motor" circuit. The region of arm representation for each stage of the circuit is shaded. Somatotopy is maintained by virtue of topographically organized connections between each of these arm areas. The cortical areas which give rise to "open loop" inputs to the arm area of the putamen have horizontal hatching. All other arm areas have vertical hatching. a.s., arcuate sulcus; cs., central sulcus; p.s., principal sulcus; Put, putamen; SPL, superior parietal lobule; SSC, somatosensory cortex.

upcoming (stimulus-triggered) arm movement (Tanji and Evarts, 1976; Thach, 1978; Tanji et al., 1980; Kubota and Funahashi, 1982; Weinrich and Wise, 1982; Wise and Mauritz, 1983, 1985; Tanji, 1985; Tanji and Kurata, 1985; Lecas et al., 1986; Georgopoulos et al., 1989; Riehle and Requin, 1989; Alexander and Crutcher, 1990a). These directionally specific, instruction-dependent changes in activity are characteristically sustained until the occurrence of the movement-triggering stimulus, and appear to represent a neural correlate of one of the preparatory aspects of motor control referred to as "motor set" (Fig. 4). Similar directionally selective preparatory activity has also been documented within the putamen (Alexander, 1987; Alexander and Crutcher, 1990a), the target

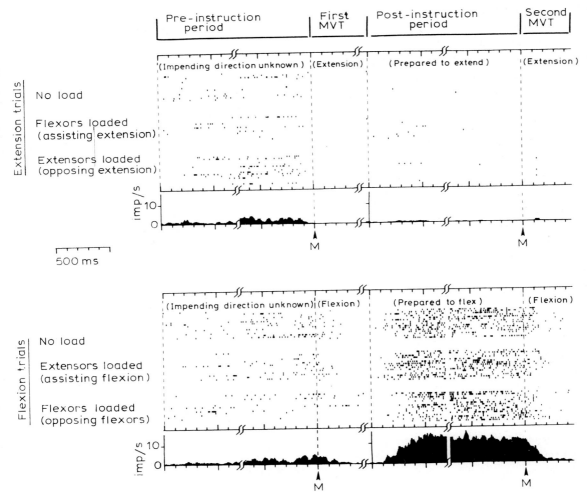

Fig. 4. Directionally selective, set-related discharge of a neuron located in the arm area of the putamen. This neuron showed a significant, sustained increase in activity prior to pre-planned elbow flexion movements. Each small tick indicates the occurrence of a single action potential, and each row represents the neuronal activity recorded during one trial. Histograms representing the average instantaneous discharge rates for all elbow extension trials ($N = 45$) and all flexion trials ($N = 45$) are located below the respective set of rasters. The split-plot rasters and histograms are aligned both on the onsets and terminations of the pre- and post-instruction periods. The terminations of these two periods corresponded to the onsets (M) of the first and second lateral movements, respectively, MVT, movement; MS, milliseconds; IMP/S, impulses/second. Reproduced with permission from Alexander (1987).

of projections from PMC, SMA and MC, suggesting that the basal ganglia-thalamocortical "motor" circuit may contribute to certain preparatory aspects of motor control, as well as to movement execution. The fact that individual neurons within these structures tend to exhibit either preparatory (set-related) or movement-related responses, rather than combinations of the two, suggests that preparatory and execution-related aspects of motor control may be mediated by separate functional channels within the "motor" circuit, analogous to the well-documented somatotopic channels (Alexander et al., 1986; Alexander, 1987; Alexander and Crutcher, 1990a). Furthermore, putamen neurons exhibiting preparatory activity for arm movements are found exclusively within the arm area of that nucleus (Alexander and Crutcher, 1990a).

There is also new evidence which indicates that during both the preparation and execution of limb movements multiple "levels" of motor processing may be carried out simultaneously, that is, *in parallel*, at different points within the "motor" circuit (Alexander and Crutcher, 1990a, b; Crutcher and Alexander, 1990). Thus, when monkeys were trained to perform a set of paradigms that dissociated several distinct functional "levels" of motor processing it was found that each of the 3 "motor" circuit structures studied (viz., SMA, MC and putamen) contained separate populations of neurons that discharged selectively in relation to either target-level variables (i.e., the location of the target in space), trajectory/kinematics-level variables (e.g., direction of limb movement, independent of muscle pattern of limb dynamics), or dynamics/muscle-level variables. Not only were the neural representations of these different levels of motor processing distributed across multiple structures within the circuit, but neuronal activity related to the various processing levels was found to proceed concurrently. The timing and specificity of these neuronal responses therefore suggest the possibility that within each of the somatotopic channels of the "motor" circuit (leg, arm, orofacial) there may be another level of organization comprising functionally specific sub-channels that encode selectively, but in parallel, information about such disparate motor behavioral variables as target location, limb kinematics and muscle pattern.

"Oculomotor" circuit (Fig. 5)

Clinical evidence strongly suggests a role of the basal ganglia in the control of eye movements. Characteristic oculomotor disturbances are seen in both Parkinson's disease (Flowers and Downing, 1978; White et al., 1983; Bronstein and Kennard, 1985; Hotson et al., 1986) and Huntington's disease (Leigh et al., 1983; Lasker et al., 1987, 1988). Some of these deficits appear to be attributable to dysfunction within oculomotor pathways that pass through the basal ganglia

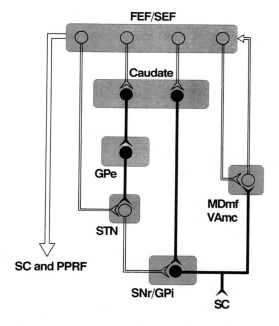

Fig. 5. Simplified diagram of the closed loop portion of the "oculomotor" circuit. Conventions are the same as in Fig. 2. MDmf, n. medialis dorsalis pars multiformis; PPRF, paramedian pontine reticular formation; SC, superior colliculus; VAmc, n. ventralis anterior pars magnocellularis.

(Hikosaka and Wurtz, 1985b; Alexander et al., 1986; Miller and DeLong, 1987; Albin et al., 1989).

The striatal portion of the basal ganglia-thalamocortical "oculomotor" circuit is centered on the body of the caudate nucleus, which receives projections from a number of interconnected cortical areas that are implicated in oculomotor control, including: the frontal eye fields (FEF, Brodmann's area 8), the supplementary eye fields (SEF) which lie dorsal to the superior limb of the arcuate sulcus, dorsolateral prefrontal cortex (areas 9 and 10), and posterior parietal cortex (area 7) (Künzle and Akert, 1977; Künzle, 1978; Yeterian and Van Hoesen, 1978; Selemon and Goldman-Rakic, 1985, 1988; Stanton et al., 1988a). Each of these cortical fields sends projections to the superior colliculus (SC) (Goldman and Nauta, 1976; Leichnetz et al., 1981; Fries, 1984; Huerta and Kaas, 1988; Stanton et al., 1988b), and each contains neurons that show selective activation in relation to specific aspects of oculomotor control (Lynch et al., 1977; Goldberg and Bushnell, 1981; Bruce and Goldberg, 1985; Schlag and Schlag-Rey, 1985; Andersen, 1989; Boch and Goldberg, 1989). In addition, eye movements can be evoked with microstimulation of either the FEF (Bruce et al., 1985) or the SEF (Schlag and Schlag-Rey, 1987).

The body of the caudate nucleus projects to the ventrolateral SNr and to a caudal and dorsomedial sector of GPi (Szabo, 1970; Parent et al., 1984b). The nigral component of the "oculomotor" circuit projects to a lateral territory in VAmc and MDmf (n. medialis dorsalis pars multiformis) (Carpenter et al., 1976; Ilinsky et al., 1985), and the GPi component projects to a lateral sector of VApc. Each of the thalamic areas sends projections back to the FEF and SEF (Akert, 1964; Kievit and Kuypers, 1977; Barbas and Mesulam, 1981; Huerta and Kaas, 1988). At least some of the nigrothalamic fibers also send collateral projections to the SC (Anderson and Yoshida, 1977; Beckstead et al., 1981; Parent et al., 1984a), suggesting that the nigrotectal projection provides an important output pathway for the "oculomotor" circuit.

As mentioned in the preceding section, the lateral SNr receives projections from both the putamen and the body of the caudate (Szabo, 1967; Parent et al., 1984b), and at a gross level there might appear to be some overlap between these putative motor and oculomotor projections. It is noteworthy, however, that the focus of the terminals originating in the putamen appears to lie somewhat dorsal to that of terminals arising from the body of the caudate (Szabo, 1970). That the two fiber systems are merely closely juxtaposed or interdigitated, rather than convergent, is further suggested by the fact that neurons in the lateral SNr have been found to discharge selectively, either in relation to eye movements or to orofacial movements, but not to both (DeLong et al., 1983a, b; Hikosaka and Wurtz, 1983a, b, c).

Single cell recording studies in primates have shown that FEF neurons may discharge in relation to visual fixation, saccadic eye movements, or passive visual stimuli (Mohler et al., 1973; Goldberg and Bushnell, 1981; Bruce and Goldberg, 1985). The visual receptive field properties of some FEF neurons have been shown to depend on the animal's behavioral set, with a cell's response to a visual stimulus showing enhancement when the stimulus serves as a target for a subsequent saccade (Goldberg and Bushnell, 1981). Similar types of activity have been demonstrated recently in single cell studies of the body of the caudate nucleus and the ventrolateral SNr (Hikosaka and Wurtz, 1983a, b, c; Hikkosaka et al., 1989a, b, c). Both of these structures have been shown to contain neurons that discharge selectively in relation to passive visual stimulation, to fixation of gaze, or to visually triggered or memory-contingent saccadic eye movements. Some neurons within the "oculomotor" circuit have also been found to have set-related properties (Fig. 6) that may be analogous to some of those seen within the "motor" circuit (see above) and "prefrontal" circuits (see below).

It should be noted that to date there have been no studies of neuronal activity within GPi using sophisticated oculomotor paradigms. Thus, the ex-

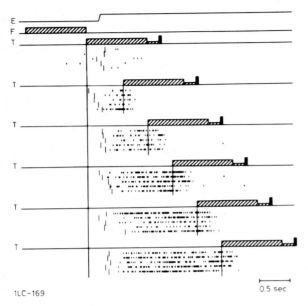

Fig. 6. Anticipatory activity of a neuron in the oculomotor region of the caudate. This neuron showed a sustained increase in activity prior to the appearance of an expected visual target. Each small tick represents a single action potential and each row a single trial. All trials are aligned on the offset of the fixation point (F). This time gap between the fixation point offset and the appearance of the target (T) was lengthened from top to bottom. Location of target was cued while the monkey was fixating (not shown). The large ticks represent the time of the saccade to the expected target location (E). Reproduced with permission from Hikosaka et al. (1989c).

istence of an "oculomotor" portion of GPi has yet to be confirmed physiologically. On the other hand, such confirmation is expected eventually, as it is quite clear from anatomical studies that the body of the caudate projects to the dorsomedial caudal GPi (Szabo, 1970; Parent et al., 1984b; Hedreen and DeLong, 1990).

Although the "oculomotor" circuit and the "motor" circuit have a number of structural and functional features in common, there are also some apparent differences between the two. For example, there is a reported paucity of excitatory (STN-projected, see below) responses within the oculomotor portion of the SNr (Hikosaka and Wurtz, 1982d), despite the fact that the STN is known to receive topographic projections from both the FEF and the SEF (Hartmann-Von Monakow et al., 1978; Huerta and Kaas, 1988) and to project to the lateral SNr (H.J.W. Nauta and Cole, 1978). In addition, the lateral SNr, which contributes primarily to the "oculomotor" circuit, projects heavily to the intermediate layers of the SC (Anderson and Yoshida, 1977; Beckstead et al., 1981; Parent et al., 1984a). This projection appears to have no exact equivalent in the connections of the "motor" circuit, unless one considers the projections from GPi/SNr to the PPN to be analogous.

The nigrocollicular pathway seems to play a significant role in the control of saccadic eye movements (Hikosaka and Wurtz, 1985a, b). But this is not the only route by which the "oculomotor" circuit could influence saccade-generating mechanisms. The thalamic portions of the circuit (MDmf and VAmc), by returning their projections to the FEF and the SEF, engage a set of cortical fields with their own descending connections to saccade-generating structures at midbrain and pontine levels (Goldman and Nauta, 1976; Fries, 1984; Leichnetz, 1986; Huerta and Kaas, 1988; Stanton et al. 1988b). Future studies will be needed to determine the relative importance of these parallel output pathways, and the degree to which they may operate in a coordinated manner to control different types of voluntary eye movements.

"Prefrontal" circuits (Fig. 7)

Considerable evidence from experimental studies in subhuman primates and clinical-pathological studies in man suggest a role of the basal ganglia in cognitive and other high-level processes (see. e.g., Cools et al., 1977; Brown and Marsden, 1987; DeLong et al 1989). These processes appear to depend upon connections between the frontal association areas and the caudate nucleus, connections that are now understood to contribute to the "prefrontal" basal ganglia-thalamocortical circuits. These may be conveniently divided into 2 parallel components, viz. a "dorsolateral prefrontal" circuit and a "lateral orbitofrontal" circuit.

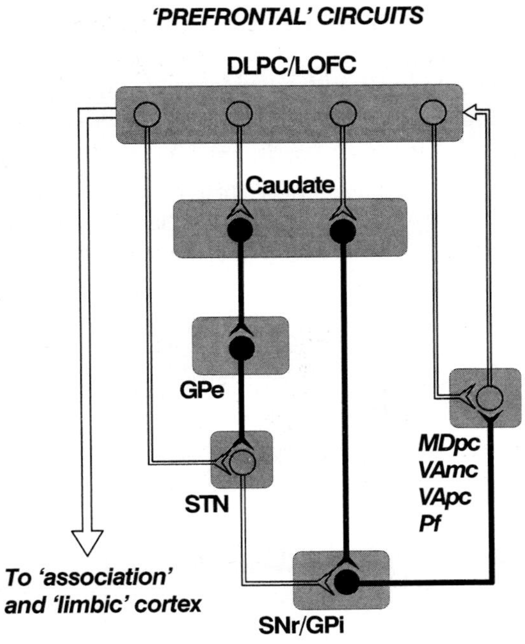

'PREFRONTAL' CIRCUITS

Fig. 7. Simplified diagram of the closed loop portions of the "prefrontal" circuits. Conventions are the same as in Fig. 2. MDpc, n. medialis dorsalis pars parvicellularis; Pf, n. parafascicularis.

At least down to the level of the pallidum, this division was well established anatomically by the early 1970s (Johnson and Rosvold, 1971). Lesioning studies had demonstrated differential effects on behavior associated with selective damage to dorsolateral vs. lateral orbitofrontal cortex (Goldman and Rosvold, 1970; Iversen and Mishkin, 1970; Butter and Snyder, 1972; Butters et al., 1973) and anatomical studies had shown that within the head of the caudate nucleus the two prefrontal fields had separate projection zones (Johnson, 1968), which in turn projected to different parts of the pallidum and nigra (Johnson and Rosvold, 1971).

"Dorsolateral prefrontal" circuit

The dorsolateral prefrontal cortex (DLPC), which includes tissue within and around the principal sulcus and on the dorsal prefrontal convexity (Brodmann's areas 9, 10; Walker's area 46) provides the closed loop portion of the corticostriate input to the "dorsolateral prefrontal" circuit. The projection from this cortical area terminates within the dorsolateral head of the caudate nucleus and throughout a continuous rostrocaudal expanse that extends to the tail of the caudate (Goldman and Nauta, 1977; Yeterian and Van Hoesen, 1978; Selemon and Goldman-Rakic, 1985, 1988). Portions of posterior parietal cortex (area 7) that are interconnected with the DLPC have also been shown to project to the dorsolateral head of the caudate nucleus (Yeterian and Van Hoesen, 1978). However, double-label anterograde transport studies have shown that although interconnected cortical areas may project to the same general areas in the neostriatum, the prefrontal and posterior parietal projections to the head of the caudate nucleus give rise to terminal fields that are largely interdigitating rather than overlapping (Selemon and Goldman-Rakic, 1985, 1988).

Rostral portions of the caudate nucleus project to the dorsomedial one-third of the globus pallidus and to rostral portions of the SNr (Szabo, 1962; Cowan and Powell, 1966; Johnson and Rosvold, 1971; Parent et al., 1984b; Hedreen and DeLong, 1990). Within each of these projections there is a mediolateral gradient, such that projections from the dorsolateral caudate are distributed to more lateral portions of the pallidum and nigra than are those from the ventromedial caudate. The dorsomedial one-third of GPi sends pallidothalamic projections to the medial part of VApc (Kuo and Carpenter, 1973; R. Kim et al., 1976). Rostrolateral portions of the SNr have been shown to project to MDpc (n. medialis dorsalis pars parvicellularis) (Ilinsky et al., 1985; Ilinsky and Kultas-Ilinsky, 1987). Each of these thalamic nuclei sends return projections to the DLPC, thus closing the circuit (Akert, 1964; Jacobson et al., 1978; Goldman-Rakic and Porrino, 1985).

Lesions of the DLPC result in characteristic deficits on tasks that require spatial memory (see Goldman-Rakic, 1987; Fuster, 1989, for reviews). There are indications, in fact, that this area of cortex may play a role in multiple modalities of short-term memory (Fuster, 1989). Single cell recording studies have documented substained changes in discharge rate in DLPC neurons that are selectively related to the retention of spatial (Fuster, 1973; Niki and Watanabe, 1976; Tanji and Kurata, 1979; Funahashi et al., 1989) or certain types of visual (Quintana et al, 1988; Yajeya et al., 1988) information. A recent study of DLPC neurons that seem to encode the remembered location of a spatial target has provided compelling evidence that this area of cortex may participate in the functional representation of external space (Funahashi et al., 1989) (Fig. 8 and see also Goldman-Rakic, this volume).

In light of the cognitive disturbances seen in

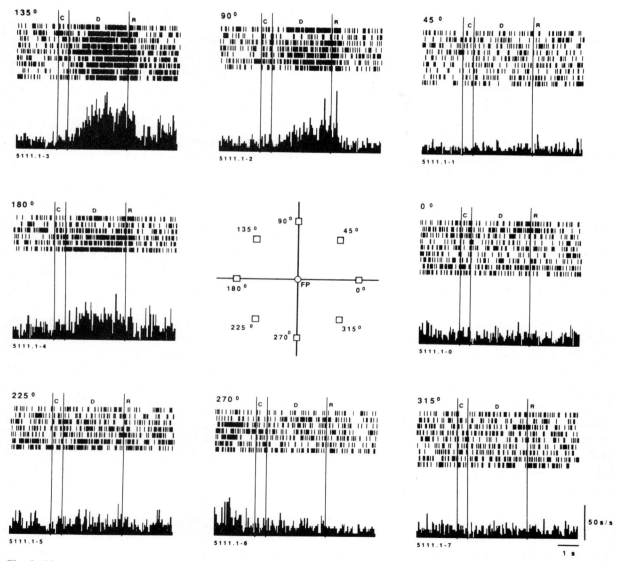

Fig. 8. Directional delay period activity of principal sulcus neuron during an oculomotor delayed-response task. This neuron exhibited directional delay period activity only when the cue had been presented in the upper quadrant of the contralateral visual field. C, time of presentation of the randomly selected visual cue (0.5 sec.); D. delay period (3 sec.); R, response period (0.5 sec.) Reproduced with permission from Funahashi et al. (1989).

humans with damage to the DLPC (see Goldman-Rakic, 1987; Fuster, 1989) it has long been assumed that this region plays an even larger role in human cognition than might be suggested by the findings obtained in experimental animals. Such a role is also consistent with evidence linking schizophrenia with dysfunction of the DLPC (Weinberger et al., 1986; Berman et al., 1988). Future research will doubtless reveal the extent to which the thought disorder of schizophrenia may be associated with specific disturbances within the "dorsolateral prefrontal" circuit.

The behavioral and cognitive expressions of damage to the prefrontal circuits that must occur in diseases of the basal ganglia is a subject of both considerable interest and uncertainty. Obviously in these disorders damage may extend beyond the basal ganglia, making associations less certain. In diseases directly involving the caudate, such as Huntington's disease, it has been found that cognitive impairments are highly correlated with caudate lesions or hypometabolism (Kuhl et al., 1982). In Parkinson's disease, it is striking that motor signs and symptoms typically develop early and without associated cognitive impairments. It is, thus, consistent, that striatal dopamine depletion is most marked in the putamen, i.e. the "motor" circuit (Nahmias et al., 1985; Hornykiewicz and Kish, 1987). The possibility that cognitive symptoms may develop in relation to dopamine depletion in the caudate, affecting the operation of the prefrontal circuits is supported by the report of a selective loss of medial SN neurons (which project to the caudate) in cases of PD with dementia. Another support for this possibility comes from the demonstration of largely frontal lobe-type deficits in parkinsonian subjects (Lees and Smith, 1983; Taylor et al., 1986; Brown and Marsden, 1987). Taylor et al. concluded from their findings that dysfunction of the "dorsolateral prefrontal" circuit was most likely responsible for the findings. It should not be forgotten, however, that loss of cortical DA could account as well for some of the behavioral and cognitive impairments.

"Lateral orbitofrontal" circuit

Lateral orbitofrontal cortex (LOFC, Brodmann's area 10; Walker's area 12) projects to a ventromedial sector of the caudate nucleus that extends from the head to the tail of that structure. This part of the caudate also receives input from the auditory and visual association areas of the superior and inferior temporal gyri, respectively (Yeterian and Van Hoesen, 1978; Van Hoesen et al., 1981; Selemon and Goldman-Rakic, 1985, 1988).

Ventromedial portions of the caudate nucleus project to a rostromedial portion of the SNr, and to a dorsomedial sector of GPi that lies just medial to the sector innervated by the dorsolateral caudate (Szabo, 1962; Johnson and Rosvold, 1971). The rostromedial SNr projects in turn to MDmc (n. medialis dorsalis pars magnocellularis), and to a medial portion of VAmc (Carpenter et al., 1976; Ilinsky et al., 1985; Ilinsky and Kultas-Ilinsky, 1987). These in turn project back to the LOFC, thus closing the circuit (Akert, 1964; Goldman-Rakic and Porrino, 1985; Ilinsky et al., 1985). The GPi component of the circuit appears to project to a medial portion of VApc, which also projects to the LOFC.

As with the dorsolateral "prefrontal" circuit, the functional characterization of the "lateral orbitofrontal" circuit has been carried out most extensively at the cortical level. Ablations of LOFC have been shown to result in perseverative interference with a monkey's capacity to make appropriate switches in behavioral set (Divac et al., 1967; Iversen and Mishkin, 1970; Butters et al., 1973; Mishkin and Manning 1978). There is evidence of similar impairments following lesions of the LOFC projection zone within the ventromedial head of the caudate nucleus (Johnson and Rosvold, 1971). It remains to be seen whether lesions placed selectively at other points in the circuit will produce the same disruptions of behavior.

Recent metabolic imaging studies of patients with obsessive/compulsive disorder have shown selective foci of cerebral hypermetabolism within orbitofrontal cortex and the caudate nucleus (Baxter et al., 1987, 1988; Swedo et al., 1989). These findings, in patients whose inability to control certain perseverative behaviors is reminiscent of some of the behavioral disturbances seen in experimental animals with LOFC lesions, suggest the possibility that obsessive/compulsive syndrome might be related to selective dysfunction within the "lateral orbitofrontal" circuit.

"Limbic" circuit (Fig. 9)

The basal ganglia-thalamocortical "limbic" circuit engages a number of cortical and subcortical structures that are traditionally considered to be "limbic" in nature, as well as certain portions of the basal ganglia whose "limbic" associations have been appreciated only relatively recently. Based upon important similarities between the connections and histochemical features of the neostriatum, and those of the nucleus accumbens and the medium-celled portion of the olfactory tubercle, the latter two structures are now referred to as the "ventral striatum" (Heimer and Wilson, 1975; Heimer et al., 1977; Heimer, 1978; H.J.W. Nauta, 1979). But because both subdivisions of the ventral striatum receive extensive projections from "limbic" structures, including the hippocampus, the amygdala, and entorhinal (Brodmann's area 28) and perirhinal cortices (area 35) (H.J.W. Nauta, 1961; Heimer and Wilson, 1975; Hemphill et al., 1981; Krayniak et al., 1981; Kelley and Domesick, 1981; Kelley et al., 1982), this portion of the striatum has also been designated the "limbic" striatum (W.J.H. Nauta and Domesick, 1977). The ventral striatum receives significant projections from the anterior cingulate area (ACA, area 24) and medial orbitofrontal cortex (MOFC, Walker's area 13), as well as from widespread sources within the temporal lobe (Powell and Leman, 1976; Van Hoesen et al., 1976, 1981; Yeterian and Van Hoesen, 1978; Baleydier and Mauguière, 1980; Hemphill et al., 1981; Selemon and Goldman-Rakic, 1985, 1988). The ventral striatum projects to the pre-commissural or ventral pallidum (Heimer, 1978; H.J.W. Nauta and Cole, 1978), which projects, in turn, to parts of MDmc (Kuo and Carpenter, 1973; Russchen et al., 1987). The "limbic" circuit is closed by thalamocortical projections from MDmc to the ACA and MOFC (Tobias, 1975; Vogt et al., 1979; Baleydier and Mauguière, 1980; Jurgens, 1983).

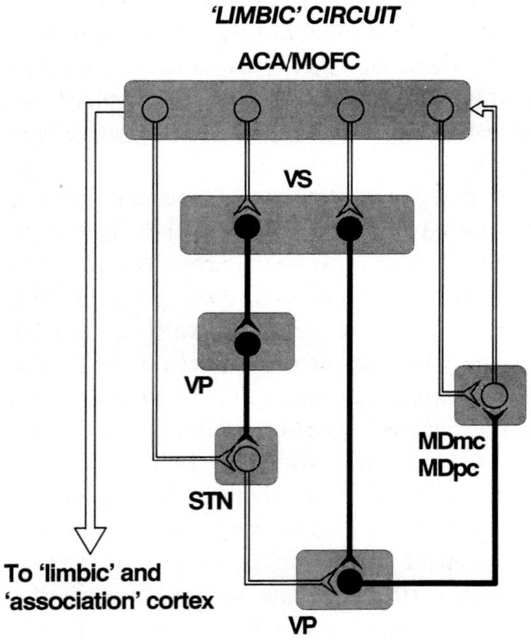

There are certain unique features of the "limbic" circuit that serve to distinguish its organization from that of the other basal ganglia-thalamocortical circuits. The ventral pallidum, for example, is not differentiated into an internal and external segment. Thus, it is not clear whether the "direct" and "indirect" pathways leading to the thalamus (see below) are well developed within this circuit, although it does appear that the ventral pallidum has reciprocal connections with the STN (H.J.W. Nauta and Cole, 1978; Haber et al., 1985). In addition, the "limbic" circuit receives

Fig. 9. Simplified diagram of the closed loop portions of the "limbic" circuit. Conventions are the same as in Fig. 2. MDmc. N. medialis dorsalis pars magnocellularis; VP, ventral pallidum; VS, ventral striatum.

substantial inputs from the amygdala at several points, including the ventral striatum (Krettek and Price, 1977; Kelley et al., 1982; Russchen et al., 1985), MDmc (W.J.H. Nauta 1962; Porrino et al., 1981; Aggleton and Mishkin, 1984), ACA and MOFC (Porrino et al., 1981). It is this last feature, in fact, that provides the rationale for our inclusion of corticostriatal projections from MOFC among the "limbic" inputs to the ventral striatum. There is a long tradition of distinguishing between "limbic" and "non-limbic" portions of prefrontal cortex based on the presence or absence of direct connections with the amygdala (W.J.H. Nauta, 1962; Porrino et al., 1981; Reep, 1984). The "limbic" circuit is also unique in having extensive efferent connections from the ventral pallidum to other subcortical structures that are generally considered as belonging to the "limbic" system, including the hypothalamus, the lateral habenula, and the ventromedial tegmental area (Haber et al., 1985).

The functions subserved by the "limbic" circuit are still largely unknown, although it seems likely that these pathways may play some role in emotional and/or motivational processes, given the large body of literature that has suggested such a role for various "limbic" structures (e.g., Livingston and Hornykiewicz, 1977; Doane and Livingston, 1986; LeDoux, 1987). Both cortical targets (ACA and MOFC) of the "limbic" circuit have been implicated in affective/motivational processes (Butter and Snyder, 1972; Butters et al., 1973) as well as selective attention. To date, however, there have been few attempts to analyze the functional aspects of this circuit using rigorous behavioral/neurophysiological techniques comparable to those that have been used in studies of the other circuits. This would seem to be a promising area for future research, in light of the relatively detailed information available regarding the structural features of the "limbic" circuit, and its organizational similarity to the other basal ganglia-thalamocortical circuits.

Despite the limited functional data available for the "limbic" circuit, it is tempting to speculate that this circuit might be involved in certain pathological conditions with significant behavioral manifestations. Thus, for example, it is frequently suggested that Tourette's syndrome, which is characterized by irrepressible motor and phonic tics, represents a disorder of the basal ganglia (Leckman et al., 1987; Trimble and Robertson, 1987). Although the evidence for this proposal is somewhat limited, the superficial resemblance between tics in Tourette's syndrome and dyskinesias (chorea, ballism, etc), in certain disorders of the "motor" circuit suggests the possibility that parallel mechanisms might be operating in both conditions. The fact that both tics and dyskinesia are improved by neuroleptics and made worse by dopamine agonists further suggests a common pathophysiologic basis for these disorders. Nevertheless, the relatively elaborate and stereotyped nature of tics, and the fact that with effort they can often be suppressed for brief periods, sets them apart from the more elemental and involuntary movements that are seen in the dyskinetic syndromes. This, and the fact that some tics involve gestures or vocalizations with strong emotional connotations (e.g., coprolalia), suggests that Tourette's syndrome might well be associated with dysfunction of "limbic" structures, rather than directly involving pathways or circuits that are strictly "motor" (Trimble and Robertson, 1987). If Tourette's syndrome is a disorder of basal ganglia origin, it could conceivably result from selective dysfunction of the "limbic" circuit. In that event, it might represent a condition directly analogous to the hyperkinetic disorders that arise from "motor" circuit dysfunction.

Operational features of basal ganglia-thalamocortical circuits

Much of our understanding of how the basal ganglia-thalamocortical circuits might function both normally and in certain pathologic states has come from studies of the "motor" and "oculomotor" circuits in primates, but important details regarding circuit operations have also emerged from pharmacologic and neurochemical studies in lower species. Each of the proposed basal ganglia-thalamocortical circuits share a number of features in common, several of which are illustrated in Fig. 10.

Within each circuit, specific cortical areas send excitatory glutamatergic projections to selected portions of the striatum (caudate nucleus, putamen or ventral striatum) (Spencer, 1976; Divac et al., 1977; J. Kim et al., 1977). By virtue of their high rates of spontaneous discharge (70–90 imp/sec) (DeLong, 1971), the basal ganglia output nuclei (GPi, SNr, ventral pallidum) exert a tonic, GABA-mediated, inhibitory effect on their target nuclei in the thalamus (Penney and Young, 1981;

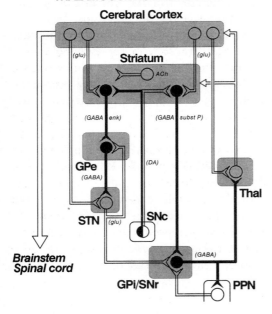

Fig. 10. Schematic diagram of the circuitry and neurotransmitters of the basal ganglia-thalamocortical circuitry. See text for discussion. (A few of the known connections have been omitted for the sake of clarity; e.g., striatal projections from the dorsal raphe, the STN and PPN.) Conventions are the same as in Fig. 2. ACh, acetylcholine; DA, dopamine; enk, enkephalin; GABA, gamma-aminobutyric acid; glu, glutamate; PPN, pedunculopontine nucleus; SNc, substantia nigra pars compacta; Thal, thalamus.

Chevalier et al., 1985; Deniau and Chevalier, 1985). This tonic inhibitory outflow from the basal ganglia appears to be differentially modulated by two opposing but parallel pathways that pass through each circuit. A *"direct pathway"* to GPi and SNr arises from striatal neurons that contain both GABA and substance P (Graybiel and Ragsdale, 1983). By means of this pathway, activation of striatal neurons, which are normally quiescent, tends to disinhibit the thalamus. An *"indirect pathway"* to GPi/SNr passes first to GPe via striatal neurons that contain both GABA and enkephalin (Graybiel and Ragsdale, 1983), then from GPe to the STN via a purely GABAergic pathway, and finally to GPi and SNr via an excitatory projection from the STN that appears to be glutamatergic. The high spontaneous discharge rate of most GPe neurons exerts a tonic inhibitory influence on the STN. Activation of the GABA-enkephalin striatal projection neurons tends to suppress the activity of GPe neurons and thereby disinhibit the STN, thus increasing the excitatory drive on GPi and SNr. This, in turn, results in increased inhibition of the thalamus. Thus, the two striatal efferent systems (the "direct" and "indirect" pathways) appear to have opposing effects upon the GPi and SNr and thus upon the thalamic targets of basal ganglia outflow. As yet, however, there is virtually no information available concerning the manner and degree to which these two systems may interact at the level of individual neurons within the basal ganglia output nuclei.

During the execution of specific limb or orofacial movements, movement-related neurons in GPi and SNr may show either phasic increases or phasic decreases in their high rates of spontaneous discharge (Georgopoulos et al., 1983; Anderson and Horak, 1985; Mitchell et al., 1987). There are mounting indications that phasic *decreases* in GPi/SNr discharge may play a crucial role in motor control by disinhibiting the ventrolateral thalamus and thereby gating or facilitating cortically initiated movements, and that phasic *increases* in GPi/SNr discharge may have the opposite effect (Klockgether et al., 1985).

Conditions associated with hypokinesia or akinesia, such as MPTP-induced parkinsonism, are characterized by tonic increases in the discharge rates of GPi and STN neurons and tonic decreases in the discharge of GPe neurons (Miller and DeLong, 1987; Filion et al., 1988). These findings are consistent with metabolic imaging studies that have provided indirect evidence of disinhibition of the STN, which presumably is secondary to excessive activation of striatofugal inhibitory (GABA/enkephalin) projections to GPe (Schwartzman and Alexander, 1985; Mitchell et al., 1986). In contrast, lesions of the STN, which result in dyskinesias (i.e., involuntary hyperkinesia), are characterized by tonic reductions in the discharge rates of GPi neurons (Hamada and DeLong, 1988). Metabolic imaging of monkeys with experimental dyskinesias induced by pharmacologic manipulations of the STN has shown reduced synaptic activity both in GPi/SNr and in the ventrolateral thalamus (Mitchell et al., 1985), as would be expected with reduced (inhibitory) input to the thalamus from GPi/SNr, and reduced (excitatory) input to the latter from the STN.

It should be noted that involuntary movements are not associated with all conditions that result in substantive reductions in inhibitory outflow to the ventrolateral thalamus from GPi/SNr. Lesions of GPi, for example, do not give rise to dyskinesias (Svennilson et al., 1960; Horak and Anderson, 1984), nor do lesions of the ventrolateral thalamus (Hassler et al., 1960). It would seem, therefore, that while akinesia/bradykinesia may represent a straightforward consequence of excessive output from GPi/SNr, dyskinesias or involuntary movements are likely to develop only when there is a selective reduction in the excitatory inputs to GPi/SNr from the "indirect" pathway, combined with relative preservation of the tonic and/or phasic inputs to these same structures from the "direct" pathway.

The role of dopamine within the basal ganglia appears to be complex, and many issues remain unresolved (see e.g., Kitai, 1981; Mercuri et al., 1985; Orr et al., 1986). There is evidence, however, that the nigrostriatal dopamine projections may exert contrasting effects on the "direct" and "indirect" striatal-GPi/SNr pathways. Dopaminergic inputs appear to have a net excitatory effect on striatal neurons that send GABA/substance P projections to GPi and SNr via the "direct" pathway, and a net inhibitory effect on those that send GABA/enkephalin projections to GPe via the "indirect" pathway (Hirata and Mogenson, 1984; Hirata et al., 1984; Bouras et al., 1986; Hong et al., 1985; Pan et al., 1985; Young et al., 1986). Thus, in effect, the overall influence of dopamine within the striatum may be to reinforce any cortically initiated activation of a particular basal ganglia-thalamocortical circuit by both facilitating conduction through that circuit's "direct" pathway (which has a net excitatory effect on the thalamus) and suppressing conduction through the "indirect" pathway (which has a net inhibitory effect on the thalamus).

These dual effects of dopamine at the level of the striatum may also help to explain some of the characteristic disorders of movement that are associated with abnormal levels of striatal dopamine. Thus, for example, dopamine deficiency (as in Parkinson's disease or MPTP-induced parkinsonism) should lead to excessive output from GPi/SNr (because of reduced conduction through the "direct" pathway, and increased conduction through the "indirect" pathway), thereby inhibiting their thalamic targets and resulting in hypokinesia. Conversely, dopamine excess or striatal hypersensitivity to dopamine (as in L-DOPA-induced dyskinesias) could lead to the opposite effect, namely diminished output from GPi/SNr, thereby disinhibiting those same thalamic targets and resulting in involuntary or irrepressible movements. In both cases, the predicted effects within the "motor" circuit are consistent with single cell recording and metabolic imaging data obtained in primate models of hypokinetic and hyperkinetic movement disorders (Mitchell et al., 1985, 1986; Schwartzman and Alexander, 1985; Miller and DeLong, 1987; Filion et al., 1988; Hamada and DeLong, 1988).

In monkeys performing saccade/fixation tasks, microinjections of GABA-related drugs into some of the output structures of the "oculomotor" circuit (viz., SNr and SC) have yielded oculomotor abnormalities that seem to be analogous to some of the skelotomotor disorders associated with dysfunction of the "motor" circuit. Thus, for example, just as involuntary movements involving the skelotomotor system appear to result from decreased GABAergic outflow from GPi to the ventrolateral thalamus, injections of muscimol, a GABA agonist, into the lateral SNr has been shown to result in irrepressible saccades (Hikosaka and Wurtz, 1985b). The resulting oculomotor effects appear to be mediated largely via the nigrocollicular pathway, with release of the SC from its tonic inhibitory inputs from the SNr. This is because injections of bicuculline, a GABA antagonist, into the SC also result in irrepressible saccades corresponding to the ocular motor fields of the regions injected (Hikosaka and Wurtz, 1985a). And, conversely, injections of muscimol into those same regions produce the opposite effect, viz., saccadic deficits within the same fields of movement (Hikosaka and Wurtz, 1985a).

The scheme depicted in Fig. 10 is, of course, greatly oversimplified. We have indicated a few of the feedback mechanisms associated with the basal ganglia-thalamocortical circuits, including the thalamostriatal projections and the reciprocal projections between the basal ganglia output nuclei and the PPN (W.J.H. Nauta and Mehler, 1966; Carpenter et al., 1976; DeVito et al., 1980; Parent and De Bellefeuille, 1982). Not shown, however, are a variety of structural details, such as the intrinsic feedback connections within each nucleus (Rafols and Fox, 1976; Wilson and Groves, 1980; Chang et al., 1982; Difiglia et al., 1982; Grofova et al., 1982; François et al., 1984; Percheron et al., 1984; Yelnick et al., 1984; Graveland and Difiglia, 1985), or the projections returned from the STN to the striatum (H.J.W. Nauta and Cole, 1978; Smith and Parent, 1986). We have also omitted discussion of several neurotransmitter systems (e.g., the cholinergic and serotonergic systems) that are believed to influence striatal operations.

These omissions do not imply a judgement that such features are inessential to an appreciation of the functional organization of basal ganglia-thalamocortical circuits. Rather, they merely reflect the recognition that our current understanding of these features is too limited to permit a meaningful discussion of their functional implications except in the most general sense of providing a mechanism for negative feedback and surround inhibition. It should also be noted that within the "limbic" circuit the demarcation of the "direct" (GABA/substance P) and "indirect" (GABA/enkephalin) pathways is not as clear as it is for the other circuits. While receiving projections from both the GABA/substance P and the GABA/enkephalin neurons of the ventral striatum, the ventral pallidum does not appear to be structurally differentiated in a manner comparable to the internal and external segments of the globus pallidus. Although the ventral pallidum does have reciprocal connections with the STN (H.J.W. Nauta and Cole, 1978; Haber et al., 1985), it is not yet known whether the STN-projecting neurons are selective recipients of GABA/enkephalin projections from the ventral striatum. However, this is a reasonable hypothesis given the high degree of structural/functional parallelism among the various basal ganglia-thalamocortical circuits.

Summary and conclusions

The central theme of the "segregated circuits" hypothesis is that structural convergence and functional integration occurs *within*, rather than between, each of the identified circuits. Admittedly, the anatomical evidence upon which this scheme is based remains incomplete. The hypothesis continues to be predicated largely on comparisons of anterograde and retrograde labeling studies carried out in different sets of animals. Only in the case of the "motor" circuit has evidence for the continuity of the loop been demonstrated directly in individual subjects; for the other circuits, such continuity is inferred from comparisons of data on different components of each circuit obtained in separate experiments. Because of the marked compression of pathways leading from cortex through basal ganglia to thalamus, comparisons of projection topography across experimental subjects may be hazardous.

Definitive tests of the hypothesis of maintained segregation await additional double- and multiple-label tract-tracing experiments wherein the continuity of one circuit, or the segregation of adjacent circuits, can be examined directly in individual subjects. It is worthy of note, however, that the few studies to date that have employed this methodology have generated results consistent with the segregated circuits hypothesis. Moreover, single cell recordings in behaving animals have shown striking preservation of functional specificity at the level of individual neurons throughout the "motor" and "oculomotor" circuits. It is difficult to imagine how such functional specificity could be maintained in the absence of strict topographic specificity within the sequential projections that comprise these two circuits.

This is not to say, however, that we expect the internal structure of functional channels (e.g., the "arm" channel within the "motor" circuit) to have cable-like, point-to-point topography. When the grain of analysis is sufficiently fine, anatomical studies have shown repeatedly that the terminal fields of internuclear projections (e.g., to striatum, pallidum, nigra, thalamus, etc.) often appear patchy and highly divergent, suggesting that neighboring groups of projection cells tend to influence interdigitating clusters of postsynaptic neurons. While more intricate and complex than simple point-to-point topography, however, this type arrangement should also be capable of maintaining functional specificity.

As discussed briefly above, it is not yet clear to what extent the inputs to the "motor" circuit from the different precentral motor fields (e.g., MC, SMA, APA) are integrated in their passage through the circuit. It now appears that at the level of the putamen such inputs remain segregated. Given the topographic nature of subsequent pathways linking putamen, pallidum and thalamus, it is distinctly possible that these may remain segregated throughout, with each sub-channel returning to its area of origin.

When viewed as a whole, the basal ganglia-thalamocortical circuits might be seen as having a unified role in modulating the operations of the entire frontal lobe and thereby influencing, by common mechanisms, such diverse "frontal lobe" processes as the maintenance and switching of behavioral sets and the planning and execution of limb and eye movements. If this view is correct, then future studies should confirm predictions that experimental and disease-induced perturbations of the "prefrontal" and "limbic" circuits, analogous to those described above for the "motor" and "oculomotor" circuits, will lead to circuit-specific behavioral alterations through common underlying mechanisms.

This scheme provides a useful substrate for understanding the varied signs and symptoms, i.e., motor, oculomotor, cognitive and "limbic", that result from diseases affecting the basal ganglia. It can easily be appreciated moreover, how lesions affecting different stations within a given circuit could result in a disruption of the same behavioral functions. It should be obvious, however, that the polarity of the resulting behavioral disturbance (e.g., hypo-vs. hyperkinetic disorders of movement) would depend on the specific structure and neurotransmitter system(s) affected, given what is now known about the operational features of these circuits. In addition, lesions that affect systems which appear to exert a modulatory influence (e.g., dopamine or serotonin) might produce more diffuse effects on several or all of the circuits.

Acknowledgements

This work was supported by grants NS-17678, NS-15417, and NS-23160 from the United States Public Healts Service.

Barbara A. Zuckerman provided expert assistance in the preparation of the manuscript.

Abbreviations

ACA	anterior cingulate area
ACh	acetylcholine
APA	arcuate premotor area
a.s.	arcuate sulcus
CM	n. centrum medianum
c.s.	central sulcus
DA	dopamine
DLPC	dorsolateral prefrontal cortex
enk	enkephalin
FEF	frontal eye field
GABA	gamma-aminobutyric acid
glu	glutamate
GPe	globus pallidus, external segment
GPi	globus pallidus, internal segment
LOFC	lateral orbitofrontal cortex
MC	primary motor cortex
MDmc	n. medialis dorsalis pars magnocellularis
MDmf	n. medialis dorsalis pars multiformis
MDpc	n. medialis dorsalis pars parvicellularis
MOFC	medial orbitofrontal cortex
MPTP	1-methyl-4-phenyl-1,2,3,6-tetrahydropyridine
Pf	n. parafascicularis
PMC	premotor cortex
PPN	pedunculopontine nucleus
PPRF	paramedian pontine reticular formation
p.s.	principal sulcus
Put	putamen
SC	superior colliculus
SEF	supplementary eye field
SMA	supplementary motor area
SNc	substantia nigra, pars compacta
SNr	substantia nigra, pars reticulata
SPL	superior parietal lobule
SSC	primary somatosensory cortex
STN	subthalamic nucleus
subst P	substance P
VA	n. ventralis anterior
VAmc	n. ventralis anterior, pars magnocellularis
VApc	n. ventralis anterior, pars parvicellularis
VLo	n. ventralis lateralis, pars oralis
VP	ventral pallidum
VPLo	n. ventralis posterior lateralis, pars oralis
VS	ventral striatum

References

Aggleton, J.P. and Mishkin, M. (1984) Projections of the amygdala to the thalamus in the cynomologus monkey. *J. Comp. Neurol.*, 222: 56–68.

Akert, K. (1964) Comparative anatomy of frontal cortex and thalamofrontal connections. In J.M. Warren and K. Akert (Eds.), *The Frontal Granular Cortex and Behavior*, McGraw-Hill, New York, pp. 372–396

Albin, R.L., Young, A.B. and Penney, J.B. (1989) The functional anatomy of basal ganglia disorders. *Trends Neurosci.*, 12: 366–375.

Alexander G.E. (1987) Selective neuronal discharge in monkey putamen reflects intended direction of planned limb movements *Exp. Brain Res.*, 67: 623–634.

Alexander, G.E. and Crutcher, M.D. (1990a) Preparation for movement: neural representations of intended direction in three motor areas of the monkey. *J. Neurophysiol.*, 64: 133–150.

Alexander, G.E. and Crutcher, M.D. (1990b) Neural representations of the target (goal) of visually guided arm movements in three motor areas of the monkey. *J. Neurophysiol.*, 64: 164–178.

Alexander, G.E. and DeLong, M.R. (1985a) Microstimulation of the primate neostriatum. I. Physiological properties of striatal microexcitable zones. *J. Neurophysiol.*, 53: 1401–1416.

Alexander, G.E. and DeLong, M.R. (1985b) Microstimulation of the primate neostriatum. II. Somatotopic organization of striatal microexcitable zones and their relation to neuronal response properties. *J. Neurophysiol.*, 53: 1417–1430.

Alexander, G.E., DeLong, M.R. and Strick, P.L. (1986) Parallel organization of functionally segregated circuits linking basal ganglia and cortex. *Annu. Rev. Neurosci.*, 9: 357–381.

Alexander, G.E., Koliatsos, V.E., Martin, L.J. Hedreen, J. and Hamada, I. (1988) Organization of primate basal ganglia "motor circuit". I. Motor cortex (MC) and supplementary motor area (SMA) project to complementary regions within matrix compartment of putamen. *Soc. Neurosci. Abstr.*, 14: 720.

Allen, G.I. and Tsukahara, N. (1974) Cerebrocerebellar communication systems. *Physiol. Rev.*, 54: 957–1006.

Andersen, R.A. (1989) Visual and eye movement functions of the posterior parietal cortex. *Annu. Rec. Neurosci.*, 12: 377–403.

Anderson, M.E. and Horak, F.B. (1985) Influence of the globus pallidus on arm movements in monkeys. III. Timing of movement-related information. *J. Neurophysiol.*, 54: 433–448.

Anderson, M. and Yoshida, M. (1977) Electrophysiological evidence for branching nigral projections to the thalamus and the superior colliculus. *Brain Res.*, 137: 361–364.

Asanuma, C., Thach, W.T. and Jones, E.G. (1983) Distribution of cerebellar terminations and their relation to other afferent terminations in the ventral lateral thalamic region of the monkey. *Brain Res. Rev.*, 5: 237–265.

Baleydier, C. and Mauguière, R. (1980) The duality of the cingulate gyrus in monkey, neuroanatomical study and functional hypothesis. *Brain*, 103: 525–554.

Barbas, H. and Mesulam, M.M. (1981) Organization of afferent input to subdivisions of area 8 in the rhesus monkey. *J. Comp. Neurol.*, 200: 407–431.

Baxter, L.R. Jr., Phelps, M.E., Mazziotta, J.C., Guze, B.H., Schwartz, J.M. and Selin, G.E. (1987) Local cerebral glucose metabolic rates in obsessive-compulsive disordeer. A comparison with rates in unipolar depression and in normal controls. (Published erratum appears in Arch. Gen. Psychiat. (1987) 44(9): 800.) *Arch. Gen. Psychiat.*, 44: 211–218.

Baxter, L.R. Jr., Schwartz, J.M., Mazziotta, J.C., Phelps, M.E., Pahl, J.J., Guze, B.H. and Fairbanks, L. (1988) Cerebral glucose metabolic rates in nondepressed patients with obsessive-compulsive disorder. *Am.J.Psychiat.*, 145: 1560–1563.

Beckstead, R.M. and Cruz, C.J. (1986) Striatal axons to the globus pallidus, entopeduncular nucleus and substantia nigra come mainly from separate cell populations in cat. *Neuroscience*, 19: 147–158.

Beckstead, R.M., Edwards, S.B. and Frankfurter, A. (1981) A comparison of the intranigral distribution of nigrotectal neurons labeled with horseradish peroxidase in the monkey, cat, and rat. *J. Neurosci.*, 1: 121–125.

Berman, K.F. and Weinberger, D.R. (1990) The prefrontal cortex in schizophrenia and other neuropsychiatric diseases: in vivo physiological correlates of cognitive deficits. *This Volume*, Ch. 26.

Berman, K.F., Illowsky, B.P. and Weinberger, D.R. (1988) Physiological dysfunction of dorsolateral prefrontal cortex en schizophrenia. IV. Further evidence of regional and behavioral specificity. *Arch. Gen. Psychiat.*, 45: 616–622.

Boch, R.A. and Goldberg, M.E. (1989) Participation of prefrontal neurons in the preparation of visually guided eye movements in the rhesus monkey. *J. Neurophysiol.*, 61: 1064–1084.

Bouras, C., Schulz, P., Constantinidis, J. and Tissot, R. (1986) Differential effects of acute and chronic administration of haloperidol on substance P and enkephalins in diverse rat brain areas. *Biol. Psychiat.*, 16: 169–174.

Bronstein, A.M. and Kennard, C. (1985) Predictive ocular motor control in Parkinson's disease. *Brain*, 108: 925–940.

Brown, R.G. and Marsden, C.D. (1987) Neuropsychology and cognitive function in Parkinson's disease: an overview. In: C.D. Marsden and S. Fahn (Eds.), *Movement Disorders 2*, Butterworth, London, pp. 99–123.

Bruce, C.J. and Goldberg, M.E. (1985) Primate frontal eye fields. I. single neurons discharging before saccades. *J. Neurophysiol.*, 53: 603–635.

Bruce, C.J. Goldberg, M.E. Bushnell, M.C. and Stanton, G.B. (1985) Primate frontal eye fields. II. Physiological and anatomical correlates of electrically evoked eye movements. *J. Neurophysiol.*, 54: 714–734.

Butter, C.M. and Snyder, D.R. (1972) Alterations in aversive and aggressive behaviors following orbital frontal lesions in rhesus monkeys. *Acta Neurobiol. Exp.*, 32: 525–565.

Butters, N., Butter, C., Rosen, J. and Stien, D. (1973) Behavioral effects of sequential and one-stage ablations of orbital prefrontal cortex in the monkey. *Exp. Neurol.*, 39: 204–214.

Carpenter, M.B. and Peter, P. (1972) Nigrostriatal and nigrothalamic fibers in the rhesus monkey. *J. Comp. Neurol.*, 144: 93–116.

Carpenter, M.B. Nakano, K. and Kim, R. (1976) Nigrothalamic projections in the monkey demonstrated by autoradiographic technics. *J. Comp. Neurol.* 165: 401–416.

Chang, H.T., Wilson, C.J. and Kitai, S.T. (1982) A Golgi study of rat neostriatal neurons: light microscopic analysis. *J. Comp. Neurol.*, 208: 107–126.

Chevalier, G., Vacher, S., Deniau, J.M. and Desban, M. (1985) Disinhibition as a basic process in the expression of striatal functions. I. The striato-nigral influence on tectospinal/tecto-diencephalic neurons. *Brain Res.*, 334: 215–226.

Cools, A.R., Lohman, A.H.M. and Van den Bercken, J.H.L. (1977) *Psychobiology of the Striatum*, North-Holland, Amsterdam.

Cowan, W.M. and Powell, T.P.S. (1966) Strio-pallidal projection in the monkey. *J. Neurol. Neurosurg. Psychiat.* 29: 426–439.

Crutcher, M.D. and Alexander, G.E. (1990) Movement-related neuronal activity coding direction or muscle pattern in three motor areas of the monkey. *J. Neurophysiol.*, 64: 151–163.

Crutcher, M.D. and DeLong, M.R. (1984a) Single cell studies of the primate putamen. I. Functional organization. *Exp. Brain Res.*, 53: 233–243.

Crutcher, M.D. and DeLong, M.R. (1984b) Single cell studies of the primate putamen. II. Relations to direction of movement and pattern of muscular activity. *Exp. Brain Res.*, 53: 244–258.

DeLong, M.R. (1971 Activity of pallidal neurons during movement. *J. Neurophysiol.*, 34: 414–427.

DeLong, M.R. and Georgopoulos, A.P. (1981) Motor functions of the basal ganglia. In: J.M. Brookhart, V.B. Mountcastle, V.B. Brooks and S.R. Geiger (Eds.), *Handbook of Physiology. The Nervous System. Motor Control. Sect. 1, Vol. II, Pt. 2,* American Physiological Society, Bethesda, MD, pp. 1017–1061.

DeLong, M.R., Crutcher, M.D. and Georgopoulos, A.P. (1983a) Relations between movement and single cell discharge in the substantia nigra of the bahaving monkey. *J. Neurosci.*, 3: 1599–1606.

DeLong, M.R., Georgopoulos, A.P. and Crutcher, M.D. (1983b) Cortico-basal ganglia relations and coding of motor preformance. In: J. Massion, J. Paillard, W. Schultz and M. Wiesendanger (Eds.), *Neural Coding of Motor Performance*, Springer, Heidelberg, pp. 30–40.

DeLong, M.R., Crutcher, M.D. and Georgopoulos, A.P. (1985) Primate globus pallidus and subthalamic nucleus: functional organization. *J. Neurphysiol.*, 53: 530–543.

DeLong, M.R., Alexander, G.E., Miller, W.C. and Crutcher,

M.D. (1989) Anatomical and functional aspects of basal ganglia-thalamocortical circuits. In: W. Winslow (Ed.), *Studies in Neuroscience*, Manchester University Press, Manchester.

Deniau, J.M. and Chevalier, G. (1985) Disinhibition as a basic process in the expression of striatal functions. II. The striatonigral influence on thalamocortical cells of the ventromedial thalamic nucleus. *Brain Res.*, 334: 227 – 233.

DeVito, J.L. and Anderson, M.E. (1982) An autoradiographic study of efferent connections of the globus pallidus in. *Exp. Brain Res.*, 46: 107 – 117.

DeVito, J.L., Anderson, M.E. and Walsh, K.E. (1980) A horseradish peroxidase study of afferent connections of the globus pallidus in: *Macaca mulatta*. *Exp. Brain Res.*, 38: 65 – 73.

Difiglia, M., Pasik, P. and Pasik, T. (1982) A Golgi and ultrastructural study of the monkey globus pallidus. *J. Comp. Neurol.,* 212: 53 – 75.

Divac, I., Rosvold, H.E. and Szwarcbart, M.K. (1967) Behavioral effects of selective ablation of the caudate nucleus. *J. Comp. Physiol. Psychol.*, 63: 184 – 190.

Divac, I., Fonnum, F. and Storm-Mathisen, J. (1977) High affinity uptake of glutamate in terminals of corticostriatal axons. *Nature*, 266: 377 – 378.

Doane, B.K. and Livingston, K.E. (1986) *The Limbic System: Functional Orgnization and Clinical Disorders*, Raven Press, New York.

Dum, R.P. and Strick, P.L. (1989) Premotor areas: nodal points for parallel efferent systems involved in the central control of movement. In D.R. Humphrey and H.J. Freund (Eds.), *Freedom to Move: Dissolving Boundaries in Motor Control. Dahlem Konferenzen*, John Wiley, Chischester, in press.

Evarts, E.V. and Thach, W.T. (1969) Motor mechanism of the CNS: Cerebrocerebellar interrelations. *Annu. Rev. Physiol.* 31: 451 – 498.

Filion, M., Tremblay, L. and Bedard, P.J. (1988) Abnormal influences of passive limb movements on the activity of globus pallidus neurons in parkinsonian monkeys. *Brain Res.*, 444: 165 – 176.

Flowers, K.A. and Downing, A.C. (1978) Predictive control of eye movements in Parkinson disease. *Ann. Neurol.*, 4: 63 – 66.

François, C., Percheron, G., Yelnick, J, and Heyner, S. (1984) A Golgi analysis of the primate globus pallidus. I. Inconstant processes of large neurons, other neuronal types, and afferent axons. *J. Comp. Neurol.*, 227: 182 – 199.

François, C., Percheron, G., Yelnick, J. and Tande, D. (1988) A topographic study of the course of nigral axons and of the distribution of pallidal axonal endings in the centre medianparafascicular complex of macaques. *Brain Res.*, 473: 181 – 186.

Fries, W. (1984) Cortical projections to the superior colliculus in the macaque monkey: A retrograde study using horseradish peroxidase. *J. Comp. Neurol.*, 230: 55 – 76.

Funahashi, S., Bruce, C.J. and Goldman-Rakic, P.S. (1989) Mnemonic coding of visual space in the monkey's dorsolateral prefrontal cortex. *J. Neurophysiol.*, 61: 331 – 349.

Fuster, J. (1973) Unit activity in prefrontal cortex during delayed-response performance: neuronal correlates of transient memory. *J. Neurphysiol.*, 36: 61 – 78.

Fuster, J.M. (1989) *The Prefrontal Cortex: Anatomy, Physiology and Neurophysiology of the Frontal Lobe*, Raven Press, New York, 276 pp.

Garcia-Rill, E. (1986) The basal ganglia and the locomotor regions. *Brain Res. Rev.*, 11: 47 – 63.

Georgopoulos, A.P., Kalaska, J.F., Caminiti, R. and Massey, J.T. (1982) On the relations between the direction of twodimensional arm movements and cell discharge in primate motor cortex. *J. Neurosci.*, 2: 1527 – 1537.

Georgopoulos, A.P., DeLong, M.R. and Crutcher, M.D. (1983) Relations between parameters of step-tracking movements and single cell discharge in the globus pallidus and subthalamic nucleus of the behaving monkey. *J. Neurosci.*, 3: 1586 – 1598.

Georgopoulos, A.P., Crutcher, M.D. and Schwartz, A.B. (1989) Cognitive spatial motor processes. III. Motor cortical prediction of movement direction during an instructed delay period. *Exp. Brain Res.*, 75: 183 – 194.

Goldberg, M.E. and Bushnell, M.C. (1981) Behavioral enhancement of visual responses in monkey cerebral cortex. II. Modulation in frontal eye fields specifically related to saccades. *J. Neurophysiol.*, 46: 773 – 787.

Goldman, P.S. and Nauta, W.J.H. (1976) Autoradiographic demonstration of a projection from prefrontal association cortex to the superior colliculus in the rhesus monkey. *Brain Res.*, 116: 145 – 149.

Goldman, P.S. and Nauta, W.J.H. (1977) An intricately patterned prefronto-caudate projection in the rhesus monkey. *J. Comp. Neurol.*, 171: 369 – 386.

Goldman, P.S. and Rosvold, H.E. (1970) Localization of function within the dorsolateral prefrontal cortex of the rhesus monkey. *Exp. Neurol.*, 27: 291 – 304.

Goldman-Rakic, P.S. (1987) Circuitry of primate prefrontal cortex and regulation of behavior by representational memory. In V.B. Mountcastle, F. Plum and S.R. Geiger (Eds.), *Handbook of Physiology. The Nervous System. Higher Functions of the Brain. Sect.1, Vol. V, Pt. 1,* American Physiological Society, Bethesda, MD, pp. 373 – 417.

Goldman-Rakic, P.S. (1990) Cellular and circuit basis of working memory in prefrontal cortex of non-human primates, this volume, Ch. 16.

Goldman-Rakic, P.S. and Porrino, L.J. (1985) The primate mediodorsal (MD) nucleus and its projection to the frontal lobe. *J. Comp. Neurol.*, 242: 535 – 560.

Graveland, G.A. and Difiglia, M. (1985) The frequency and distribution of medium-sized neurons with idented nuclei in the primate and rodent neostriatum. *Brain Res.*, 327: 307–311.

Graybiel, A.M. and Ragsdale, C.W. Jr. (1983) Biochemical anatomy of the striatum. In: P.C. Emsons (Ed.) *Chemical Neuroanatomy*, Raven Press, New York, pp. 427–504.

Grofova, I., Deniau, J.M. and Kitai, S.T. (1982) Morphology of the substantia nigra pars reticulata projection neurons intracellularly labelled with HRP. *J. Comp. Neurol.*, 208: 352–368.

Haber, S.N., Groenewegen, H.J., Grove, E.A. and Nauta, W.J.H. (1985) Efferent connections of the ventral pallidum: evidence of a dual striato pallidofugal pathway. *J. Comp. Neurol.*, 235: 322–335.

Hamada, I. and DeLong, M.R. (1988) Lesions of the primate subthalamic nucleus (STN) reduce tonic and phasic neural activity in globus pallidus. *Soc. Neurosci. Abstr.*, 14: 719.

Harnois, C. and Filion, M. (1982) Pallidofugal projections to thalamus and midbrain: a quantitative antidromic activation study in monkeys and cats. *Exp. Brain Res.*, 47: 277–285.

Hartmann-Von Monakow, K., Akert, K. and Künzle, H. (1978) Projections of the precentral motor cortex and other cortical areas of the frontal lobe to the subthalamic nucleus in the monkey. *Exp. Brain Res.*, 33: 395–403.

Hassler, R., Reichert, T., Mundinger, F., Umbach, W. and Gandleberger, J.A. (1960) Physiological observations in stereotaxic operations in extrapyramidal motor disturbances. *Brain*, 83: 337–350.

Hedreen, J.C. and DeLong, M.R. (1990) Organization of striatopallidal, striatonigral and nigrostriatal projections in the macaque. *J. Neurosci.*, in press.

Hedreen, J., Martin, L.J., Koliatsos, V.E., Hamada, I., Alexander, G.E. and DeLong, M.R. (1988) Organization of primate basal ganglia "motor circuit". IV. Ventrolateral thalamus links internal pallidum (GPi) and supplementary motor area (SMA). *Soc. Neurosci. Abstr.*, 14: 721.

Heimer, L. (1978) The olfactory cortex and the ventral striatum. In K.E. Livingstong and O. Hornykiewicz (Eds.), *Limbic Mechanism*, Plenum Press, New York, pp. 95–187.

Heimer, L. and Wilson, R.D. (1975) The subcortical projections of the allocortex. Similarities in the neural associations of the hippocampus, the piriform cortex, and the neocortex. In M. Santini (Ed.), *Golgi Centennial Symposium: Perspectives in Neurology*, Raven Press, New York, pp 177–193.

Heimer, L., Van Hoesen, G.W. and Rosene, D.L. (1977) The olfactory pathways and the anterior perforated substance in the primate brain. *Int. J. Neurol.*, 12: 42–52.

Hemphill, M., Holm, G., Crutcher, M., DeLong, M.R. and Hedreen, J. (1981) Afferent connections of the nucleus accumbens in the monkey. In: R. Chronister and J. DeFrance (Eds.) *Neurobiology of the Nucleus Accumbens*, Haer Inst. Press, Brunswick, ME, pp. 75–81,

Hikosaka, O. and Wurtz, R.H. (1983a) Visual and oculomotor functions of monkey substantia nigra pars reticulata. I. Relation of visual and auditory responses to saccades. *J. Neurophysiol.*, 49: 1230–1253.

Hikosaka, O. and Wurtz, R.H. (1983b) Visual and oculomotor functions of monkey substantia nigra pars reticulata. II. Visual responses related to fixation of gaze. *J. Neurophysiol.*, 49: 1254–1267.

Hikosaka, O. and Wurtz, R.H. (1983c) Visual and oculomotor functions of monkey substantia nigra pars reticulata. III. Memory-contingent visual and saccade responses. *J. Neurophysiol.*, 49: 1268–1284.

Hikosaka, O. and Wurtz, R.H. (1983d) Visual and oculomotor functions of monkey substantia nigra pars reticulata. IV. Relation of substantia nigra to superior colliculus. *J. Neurophysiol.*, 49: 1285–1301.

Hikosaka, O. and Wurtz, R.H. (1985a) Modification of saccadic eye movements by GABA-related substances. I. Effect of muscimol and bicuculline in monkey superior colliculus. *J. Neurophysiol.*, 53: 266–291.

Hikosaka, O. and Wurtz, R.H. (1985b) Modification of saccadic eye movements by GABA-related substances. II. Effects of muscimol in monkey substantia nigra pars reticulata. *J. Neurophysiol.*, 53: 292–308.

Hikosaka, O., Sakamoto, M. and Usui, S. (1989a) Functional properties of monkey caudate neurons. I. Activities related to saccadic eye movements. *J. Neurophysiol.*, 61: 780–798.

Hikosaka, O., Sakamoto, M. and Usui, S. (1989b) Functional properties of monkey caudate neurons. II. Visual and auditory responses. J. Neurophysiol., 61: 799–813.

Hikosaka, O., Sakamoto, M. and Usui, S. (1989c) Functional properties of monkey caudate neurons. III. Activities related to expectation of target and reward. *J. Neurophysiol.*, 61: 814–832.

Hirata, K. and Mogenson, G.J. (1984) Inhibitory response of pallidal neurons to cortical stimulation and the influence of conditioning stimulation of the substantia nigra. *Brai Res.*, 321: 9–19.

Hirata, K. Yim, C.Y. and Mogenson, G.J. (1984) Excitatory input from sensory motor cortex to neostriatum and its modification by conditioning stimulation of the substantia nigra. *Brain Res.*, 321: 1–8.

Hong, J.S., Yoshikawa, K., Kanamatsu, T. and Sabol, S.L. (1985) Modulation of striatal enkephalinergic neurons by antipsychotic drugs. *Fed. Proc.*, 44(9): 2535–2540.

Horak, F.B. and Anderson, M.E. (1984) Influence of globus pallidus on arm movements in monkeys. I. Effects of kainic-induced lesions. *J. Neurophysiol.* 52: 290–304.

Hornykiewicz, O. and Kish, S.J. (1987) Biochemical pathophysiology of Parkinson's disease. *Adv. Neurol.*, 45: 19–34.

Hotson, J.R., Langston, E.B. and Langston, J.W. (1986) Saccade responses to dopamine in human MPTP-treated parkinsonism; *Ann. Neurol.*, 20: 456–463.

Huerta, M.F. and Kaas, J.H. (1988) Connections of the physiologically defined supplementary eye field. *Soc. Neurosci. Abstr.*, 14: 159–159.

Ilinsky, I. and Kultas-Ilinsky, K. (1987) Sagittal cytoarchitectonic maps of the *Macaca mulatta* thalamus with a revised nomenclature of the motor-related nuclei validated by observations on their connectivity. *J. Comp. Neurol.* 262: 331–364.

Ilinsky, I., Jouandet, M.L. and Goldman-Rakic, P.S. (1985) Organization of the nigrothalamocortical system of the rhesus monkey. *J. Comp. Neurol.*, 236: 315–330.

Iversen, S.D. and Mishkin, M. (1970) Perseverative interference in monkeys following selective lesions of the inferior prefrontal convexity. *Exp. Brain Res.*, 11: 376–386.

Jacobson, S., Butters, M. and Tovsky, N.J. (1978) Afferent and efferent subcortical projections of behaviorally defined sectors of prefrontal granular cortex. *Brain Res.*, 159: 279–296.

Johnson, T.N. (1968) Projections from behaviorally-defined sectors of the prefrontal cortex to the basal ganglia, septum, and diencephalon of the monkey.*Exp. Neurol.*, 21: 20–34.

Johnson, T.N. and Rosvold, H.E. (1971) Topographic projections on the globus pallidus and the substantia nigra selectively placed lesions in the precommissural caudate nucleus and putamen in the monkey. *Exp. Neurol.*, 33: 584–596.

Jones, E.G. (1985) *The Thalamus*, Plenum Press, New York.

Jones, E.G., Coulter, J.D., Burton, H. and Porter, R, (1977) Cells of origin and terminal distribution of corticostriatal fibers arising in the sensory-motor cortex of monkeys. *J. Comp. Neurol.*, 173: 53–80.

Jürgens, U. (1983) Afferent fibers to the cingular vocalization region in the squirrel monkey. *Exp. Neurol.*, 80: 395–409.

Jürgens, U. (1984) The efferent and afferent connections of the supplementary motor area. *Brain Res.*, 300: 63–81.

Kalil, K. (1978) Patch-like termination of thalamic fibers in the putamen of the rhesus monkey: an autoradiographic study. *Brain Res.*, 140: 333-339.

Kelley, A.E. and Domesick, V.B. (1982) The distribution of the projection from the hippocamal formation to the nucleus accumbens in the rat: An anterograde-and retrograde-horseradish peroxidase study. *Neuroscience*, 7: 2321–2335.

Kelley, A.E., Domesick, V.B. and Nauta, W.J.H. (1982) The amygdalostriatal projection in the rat – an anatomical study by anterograde and retrograde tracing methods. *Neuroscience*, 7: 615–630.

Kemp, J.M. and Powell, T.P.S. (1970) The cortico-striate projection in the monkey. *Brain*, 93: 525–546.

Kievit, J. and Kuypers, H.G.J.M. (1977) Organization of the thalamo-cortical connexions to the frontal lobe in the rhesus monkey. *Exp. Brain Res.*, 29: 299–322.

Kim, J., Hassler, R., Hang, P. and Paik, K. (1977) Effect of frontal cortex ablation on striatal glutamic acid level in rat. *Brain Res.*, 132: 370–374.

Kim, R., Nakano, K., Jayaraman, A. and Carpenter, M.B. (1976) Projections of the globus pallidus and adjacent structures: an autoradiographic study in the monkey *J. Comp. Neurol.*, 169: 263–290.

Kitai, S.T. (1981) Electrophysiology of the corpus striatum and brain stem integrating systems. In J.M. Brookhart, V.B. Mountcastle, V.B. Brooks and S.R. Geiger (Eds.), *Handbook of Physiology: The Nervous System. Motor Control. Sect. 1, Vol. II, Pt. 2*, American Physiological Society, Bethesda, MD, pp. 997–1015.

Klockgether, T., Schwarz, M., Turski, L. and Sontag, K.H. (1985) Rigidity and catalepsy after injections of muscimol into the ventromedial thalamic nucleus: an electromyographic study in the rat. *Exp. Brain Res.*, 58: 559–569.

Koliatsos, V.E., Martin, L.J., Hedreen, J., Alexander, G.E., Hamada, I., Price, D.L. and DeLong, M.R. (1988) Organization of primate basal ganglia "motor circuit". II. Putamenal projections to internal (GPi) and external (GPe) globus pallidus originate in distinct neuronal projections within the matrix compartment. *Soc. Neurosci. Abstr.*, 14: 720.

Krayniak, P.F., Meibach, R.C. and Siegel, A. (1981) A projection from the entorhinal cortex to the nucleus accumbens in the rat. *Brain Res.*, 209: 427–431.

Krettek, J.E. and Price, J.L. (1977) Projections from the amygdaloid complex to the cerebral cortex and thalamus in the rat and cat. *J. Comp. Neurol.*, 172: 687–722.

Kubota, K. and Funahashi, S. (1982) Direction-specific activities of dorsolateral prefrontal and motor cortex pyramidal tract neurons during visual tracking. *J. Neurophysiol.*, 47: 362–376.

Kuhl, D.E., Phelps, M.E., Markham, C.H., Metter, E.J., Riege, W.H. and Winter, J. (1982) Cerebral metabolism and atrophy in Huntington's disease determined by 18FDG and computed topographic scan. *Ann. Neurol.*, 12: 425–434.

Künzle, H. (1975) Bilateral projections from precentral motor cortex to the putamen and other parts of the basal ganglia. An autoradiographic study in *Macaca fascicularis*. *Brain Res.*, 88: 195–209.

Künzle, H. (1976) Thalamic projections from the precentral motor cortex in *Macaca fascicularis*. *Brain Res.*, 105: 253–267.

Künzle, H. (1977) Projections from the primary somatosensory cortex to basal ganglia and thalamus in the monkey. *Exp. Brain Res.*, 30: 481–492.

Künzle, H. (1978) An autoradiographic analysis of the efferent connections from premotor and adjacent prefrontal regions (areas 6 and 9) in *Macaca fascicularis*. *Brain Behav. Evol.*, 15: 185–234.

Künzle, H. and Akert, K. (1977) Efferent connections of cortical area 8 (frontal eye field) *Macaca fascicularis*. A reinvestigation using the autoradiographic technique. *J. Comp. Neurol.*, 173: 147–163.

Kuo, J.S. and Carpenter, M.B. (1973) Organization of pallidothalamic projections in the rhesus monkey. *J. Comp. Neurol.*, 151: 201–236.

Lasker, A.G., Zee, D.S., Hain, T.C., Folstein, S.E. and Singer, H.S. (1987) Saccades in Huntington's disease: initiation defects and distratibility. *Neurology*, 37: 364–370.

Lasker, A.G., Zee, D.S., Hain, T.C., Folstein, S.E. and Singer, H.S. (1988) Saccades in Huntington's disease: slowing and dysmetria. *Neurology*, 38: 427–431.

Lecas, J.-C., Requin, J., Anger, C. and Vitton, N. (1986) Changes in neuronal activity of the monkey precentral cortex during preparation for movement. *J. Neurophysiol.*, 56: 1680–1702

Leckman, J.F., Walkup, J.T., Riddle, M.A., Towbin, K.E. and Cohen, D.J. (1987) Tic disorders. In: H.Y. Meltzer (Ed.) *Psychopharmacology: The Third Generation of Progress*, Raven Press, New York, pp. 1239–1246.

LeDoux, J.E. (1987) Emotion. In: V.B. Mountcastle, F. Plum and S.R. Geiger (Eds.), *Handbook of Physiology. The Nervous System. Higher Functions of the Brain. Sect. 1, Vol. V, Pt. 1*, American Physiological Society, Bethesda, MD, pp. 419–459.

Lees, J. and Smith, E. (1983) Cognitive deficits in the early stages of Parkinson's disease. *Brain*, 106: 257–270.

Leichnetz, G.R. (1986) Afferent and efferent connections of the dorsolateral precentral gyrus (area 4, hand/arm region) in the macaque monkey, with comparisons to area 8. *J. Comp. Neurol.*, 254: 460–492.

Leichnetz, G.R., Spencer, R.F., Hardy, S.G.P. and Astruc, J. (1981) The prefrontal corticotectal projection in the monkey: An anterograde and retrograde horseradish peroxidase study. *Neuroscience*, 6: 1023–1041.

Leigh, R.J., Newman, S.A., Folstein, S.E., Lasker, A.G. and Jensen, B.A. (1983) Abnormal ocular motor control in Huntington's disease. *Neurology*, 33: 1268–1275.

Liles, S.L. (1975) Cortico-striatal evoked potentials in the monkey. *(Macaca mulatta). Electroenceph. Clin. Neurophysiol.*, 38: 121–129.

Liles, S.L. (1979) Topographic organization of neurons related to arm movement in the putamen. In: T.N. Chase, N.S. Wexler and A. Barbeau (Eds.), *Advances in Neurology, Vol. 23*, Raven Press, New York, pp. 155–162.

Liles, S.L. (1985) Activity of neurons in putamen during active and passive movements of wrist. *J. Neurophysiol.*, 53: 217–236.

Liles, S.L. and Updyke, B. (1985) Projection of the digit and wrist area of precentral gyrus to the putamen: relation between topography and physiological properties of neurons in the putamen. *Brain Res.*, 339: 245–255.

Livingston, K.E. and Hornykiewicz, O. (1977) *Limbic Mechanisms: The Continuing Evolution of the Limbic System Concept*, Plenum Press, New York.

Lynch, J.C., Mountcastle, V.B., Talbot, W.H. and Yin, T.C.T. (1977) Parietal lobe mechanisms for directed visual attention. *J. Neurophysiol.* 40: 362–389.

Matelli, M. Camarda, R., Glickstein, M. and Rizzolatti, G. (1986) Afferent and efferent projections of the inferior area 6 in the macaque monkey. *J. Comp. Neurol.*, 251: 281–298.

Matelli, M., Luppino, G., Fogassi, L. and Rizzolatti, G. (1989) Thalamic input to inferior area 6 and area 4 in the macaque monkey. *J. Comp. Neurol.*, 280: 468–488.

Mercuri, N., Bernardi, G., Calabresi, P., Cotugno, A., Levi, G. and Stanzione, P. (1985) Dopamine decreases cell excitability in rat striatal neurons by pre- and postsynaptic mechanisms. *Brain Res.*, 358: 110–121.

Miller, W.C. and DeLong, M.R. (1987) Altered tonic activity of neurons in the globus pallidus and subthalamic nucleus in the primate MPTP model of parkinsonism. In: M.B. Carpenter and A. Jayaraman (Eds.), *The Basal Ganglia II*, Plenum Press, New York, pp. 415–427.

Mishkin, M. and Manning, F.J. (1978) Non-spatial memory after selective prefrontal lesions in monkeys. *Brain. Res.*, 143: 313–323.

Mitchell, I.J., Sambrook, M.A. and Crossman, A.R. (1985) Subcortical changes in the regional uptake of 2-deoxyglucose in the brain of the monkey during experimental choreiform dyskinesia elicited by injection of a gamma-aminobutyric acid antagonist into the subthalamic nucleus. *Brain*, 108: 405–422.

Mitchell, I.J., Cross, A.J., Sambrook, M.A. and Crossman, A.R. (1986) Neural mechanisms mediating 1-methyl-4-phenyl-1,2,3,6-tetrahydro-pyridine-induced parkinsonism in the monkey: relative contributions of the striatopallidal and striatonigral pathways as suggested bu 2-deoxyglucose uptake. *Neurosci. Lett.*, 63: 61–65.

Mitchell. S.J., Richardson, R.T., Baker, F.H. and DeLong, M.R. (1987) The primate globus pallidus: Neuronal activity to direction of movement. *Exp. Brain Res.*, 68: 491–505.

Mitz, A.R. and Wise, S.P. (1987) The somatotopic organization of the supplementary motor area: Intracortical microstimulation mapping.*J. Neurosci.*, 7: 1010–1021.

Miyata, M. and Sasaki, K. (1984) Horseradish peroxidase studies on thalamic and striatal connections of the mesial part of area 6 in the monkey. *Neurosci. Lett.*, 49: 127–133.

Mohler, C.W., Goldberg, M.E. and Wurtz, R.H. (1973) Visual receptive fields of frontal eye field neurons. *Brain Res.*, 61: 385–389.

Muakkassa, K.F. and Strick, P.L. (1979) Frontal lobe inputs to primate motor cortex: evidence for four somatotopically organized "premotor" areas. *Brain Res.*, 177: 176–182.

Murphy, J.T., Kwan, H.C., Mackay, W.A. and Wong, Y.C. (1978) Spatial organization of precentral cortex in awake primates. III. Input-output coupling. *J. Neurophysiol.*, 41: 1132–1139.

Murphy, J.T., Kwan, H.C., Mackay, W.A. and Wong, Y.C. (1982) Activity of primate precentral neurons during voluntary movements triggered by visual signals. *Brain Res.*, 236: 429–449.

Nahmias, C., Garnett, E.S., Firnau. G. and Lang, A. (1985) Striatal dopamine distribution in Parkinsonian patients during life. *J. Neurol. Sci.*, 69: 223–230.

Nambu, A., Yoshida, S. and Jinnai, K. (1988) Projection on the motor cortex of thalamic neuron with pallidal input in the monkey. *Exp. Brain Res.*, 71 658–662.

Nauta, H.J.W. (1979) A proposed conceptual reorganization of the basal ganglia and telencephalon. *Neuroscience*, 4: 1875 – 1881.

Nauta, H.J.W. and Cole, M. (1978) Efferent projections of the subthalamic nucleus: an autoradiographic study in monkey and cat. *J. Comp. Neurol.*, 180: 1 – 16.

Nauta, W.J.H. (1961) Fibre degeneration following lesions of the amygdaloid complex in the monkey. *J. Anat.*, 95: 515 – 531.

Nauta, W.J.H. (1962) Neural associations of the amygdaloid complex in the monkey. *Brain*, 85: 505 – 520.

Nauta, W.J.H. and Domesick, V.B. (1977) Cross-roads of limbic and striatal circuitry: hypothalamo-nigral connections. In: K.E. Livingston and O. Hoznykiewicz (Eds.), *Limbic Mechanisms: The Continuing Evolution of the Limbic System Concept*, Plenum Press, New York.

Nauta, W.J.H. and Mehler, W.R. (1966) Projections of the lentiform nucleus in the monkey. *Brain Res.*, 1: 3 – 42.

Niki, H. and Watanabe, M. (1976) Prefrontal unit activity and delayed response: relation to cue location versus direction of response. *Brain Res.*, 105: 79 – 88.

Okano, K. and Tanji, J. (1987) Neuronal activities in the primate motor fields of the agranular frontal cortex preceding visually triggered and self-paced movement. *Exp. Brain Res.*, 66: 155 – 166.

Orr, W.B., Gardiner, T.W., Stricker, E.M., Zigmond, M.J. and Berger, T.W. (1986) Short-term effects of dopamine-depleting brain lesions on spontaneous activity of striatal neurons: relation to local dopamine concentration and behavior. *Brain Res.*, 376: 20 – 28.

Pan, H.S., Penney, J.B. and Young, A.B. (1985) Gamma-aminobutyric acid and benzodiazepine receptor changes induced by unilateral 6-hydroxydopamine lesions of the medial forebrain bundle. *J. Neurochem.*, 45: 1396 – 1404.

Pandya, D.N. and Yeterian, E.H. (1990) Prefrontal cortex in relation to other cortical areas in rhesus monkey: architecture and connections. *This Volume*, Ch. 4.

Parent, A. (1983) The subcortical afferents to caudate nucleus and putamen in primate a flourescence retrograde double labeling study. *Neuroscience*, 10: 1137 – 1150.

Parent, A. (1986) *Comparative Neurobiology of the Basal Ganglia. Wiley Series in Neurobiology*, John Wiley, New York.

Parent, A and De Bellefeuille, L. (1982) Organization of efferent projections from the internal segment of globus pallidus in primate as revealed by fluorescence retrograde labeling method. *Brain Res.*, 245: 201 – 213.

Parent, A., Smith, Y., and De Bellefeuille, L. (1984a) The output organization the pallidum and substantia nigra in primate as revealed by a retrograde double-labeling method. In: J.S. McKenzie, R.E. Kemm and L.N. Wilcock (Eds.), *The Basal Ganglia, Structure and Function*, Plenum, New York, pp. 147 – 160.

Parent, A., Bouchard, C. and Smith, Y (1984b) The striatopallidal and striatonigral projections: two dinstinct fiber systems in primate. *Brain Res.*, 303: 385 – 390.

Penney, J.B. Jr. and Young, A.B. (1981) GABA as the pallidothalamic neurotransmitter: implications for basal ganglia function. *Brain Res.*, 207: 195 – 199.

Percheron, G., Yelnick, J. and François, C. (1984) A Gogli analysis of the primate globus pallidus. III. Spatial organization of the striato-pallidal complex. *J. Comp. Neurol.*, 227: 214 – 227.

Porrino, L.J., Crane, A.M. and Goldman-Rakic, P.S. (1981) Direct and indirect pathways from the amygdala to the frontal lobe in rhesus monkeys. *J. Comp. Neurol.*, 198: 121 – 136.

Powell, E.W. and Leman, R.B. (1976) Connections of the nucleus accumbens. *Brain Res.*, 105: 389 – 403.

Quintana, J., Yajeya, J and Fuster, J. (1988) Prefrontal representation of stimulus attributes during delay tasks. I. Unit in cross-temporal integration of sensory and sensory-motor information. *Brain Res.*, 474: 211 – 221.

Rafols, J.A. and Fox, C.A. (1976) The neurons in the primate subthalamic nucleus: a Golgi and electron microscopic study. *J. Comp. Neurol.*, 168: 75 – 112.

Reep, R. (1984) Relationship between prefrontal and limbic cortex: a comparative anatomical review. *Brain Behav. Evol.*, 25: 5 – 80.

Riehle, A. and Requin, J. (1989) Monkey primary motor and premotor cortex: single-cell activity related to prior information about direction and extent of intended movement. *J. Neurophysiol.*, 61: 534 – 548.

Russchen, F.T., Bakst, I., Amaral, D.G. and Price, J.L. (1985) The amygdalostriatal projections in the monkey. An anterograde tracing study. *Brain Res.*, 329: 241 – 257.

Russchen, F.T., Amaral, D.G. and Price, J.L. (1987) The afferent input to the magnocellular division of the mediodorsal thalamic nucleus in the monkey. *Macaca fascicularis*. *J. Comp. Neurol.*, 256: 175 – 210.

Schell, G.R. and Strick, P.L. (1984) The origin of thalamic inputs to the arcuate premotor and supplementary motor areas. *J. Neurosci.*, 4: 539 – 560.

Schlag, J. and Schlag-Rey, M. (1985) Unit activity related to spontaneous saccades in frontal dorsomedial cortex of monkey. *Exp. Brain Res.*, 58: 208 – 211.

Schlag, J. and Schlag-Rey, M. (1987) Does microstimulation evoke fixed-vector saccades by generating their vector or by specifying their goal? *Exp. Brain Res.*, 68: 442 – 444.

Schwartzman, R.J. and Alexander, G.M. (1985) Changes in the local cerebral metabolic rate for glucose in the 1-methyl-4-phenyl-1,2,3,6-tetrahydropyridine (MPTP) primate model of Parkinson's disease. *Brain Res.*, 358: 137 – 143.

Selemon, L.D. and Goldman-Rakic, P.S. (1985) Longitudinal topography and interdigitation of cortico-striatal projections in the rhesus monkey. *J. Neurosci.*, 5: 776 – 794.

Selemon, L.D. and Goldman-Rakic, P.S. (1988) Common cortical and subcortical targets of the dorsolateral prefrontal and posterior parietal cortices in the rhesus monkey: evidence for a distributed neural network subserving spatially guided behavior. *J. Neurosci.*, 8(11): 4049-4068.

Smith, Y. and Parent, A. (1986) Differential connections of caudate nucleus and putamen in the squirrel monkey *(Saimiri sciureus)*. *Neuroscience*, 18: 347–371.

Spann, B.M. and Grofova, I. (1989) Origin of ascending and spinal pathways from the nucleus tegmenti pedunculopontinus in the rat. *J. Comp. Neurol.*, 283: 13–27.

Spencer, H.J. (1976) Antagonism of cortical excitation of striatal neurons by glutamic acid diethylester: evidence for glutamic acid as an excitatory transmitter in the rat striatum. *Brain Res.*, 102: 91–101.

Stanton, G.B., Bruce, C.J. and Goldberg, M.E. (1988a) Frontal eye field efferents in the macaque monkey. I. Subcortical pathways and topography of striatal and thalamic terminal fields. *J. Comp. Neurol.*, 271: 473–492.

Stanton, G.B., Bruce, C.J. and Goldberg, M.E. (1988b) Frontal eye field efferents in the macaque monkey. II Topography of terminal fields in midbrain and pons. *J. Comp. Neurol.*, 271: 493–506.

Strick, P.L. (1976a) Anatomical analysis of ventrolateral thalamic input to primate motor cortex. *J. Neurophysiol.*, 39: 1020–1031.

Strick, P.L. (1976b) Activity of ventrolateral thalamic neurons during arm movement. *J. Neurophysiol.*, 39: 1032–1044.

Strick, P.L. and Preston, J.B. (1982a) Two representations of the hand in area 4 of a primate. I. Motor output organization. *J. Neurophysiol.*, 48: 139–149.

Strick, P.L. and Preston, J.B. (1982b) Two representations of the hand in area 4 of a primate. II. Somatosensory input organization. *J. Neurophysiol.*, 48: 150–159.

Svennilson, E., Torvik, A., Lowe, R. and Leksell, L. (1960) Treatment of parkinsonism by sterotactic thermolesions in the pallidal region. A clinical evaluation of 81 cases. *Acta Psychiat. Neurol. Scand.*, 35: 358–377.

Swedo, S.E., Schapiro, M.B., Grady, C.L., Cheslow, D.L., Leonard, H.L., Kumar, A., Friedland, R., Rapoport, S.I. and Rapoport, J.L. (1989) Cerebral glucose metabolism in childhood-onset obsessive-compulsive disorder. *Arch. Gen Psychiat.*, 46: 518–523.

Szabo, J. (1962) Topical distribution of the striatal efferents in the monkey. *Exp. Neurol.*, 5: 21–36.

Szabo, J. (1967) The efferent projections of the putamen in the monkey. *Exp. Neurol.*, 19: 463–476.

Szabo, J. (1970) Projections from the body of the caudate nucleus in the rhesus monkey. *Exp. Neurol.*, 27: 1–15.

Tanji, J. (1985) Comparison of neuronal activities in the monkey supplementary and precentral motor areas. *Behav. Brain Res.*, 18: 137–142.

Tanji, J. and Evarts, E.V. (1976) Anticipatory activity of motor cortex neurons in relation to direction of an intended movement. *J. Neurophysiol.*, 39: 1061–1068.

Tanji, J. and Kurata, K. (1979) Neuronal activity in the cortical supplementary motor area related with distal and proximal forelimb movements. *Neurosci. Lett.*, 12: 201–206.

Tanji, J. and Kurata, K. (1982) Comparison of movement-related activity in two cortical motor areas of primates. *J. Neurophysiol.*, 48: 633–653.

Tanji, J. and Kurata, K. (1985) Contrasting neuronal activity in supplementary and precentral motor cortex of monkeys. I. Responses to instructions determining motor responses to forthcoming modalities. *J. Neurophysiol.*, 53: 129–141.

Tanji, J., Taniguchi, K. and Saga, T. (1980) Supplementary motor area: Neuronal response to motor instructions. *J. Neurophysiol.*, 43: 60–68.

Taylor, A.E., Saint-Cyr, J.A. and Lang, A.E. (1986) Frontal lobe dysfunction in Parkinson's disease: the cortical focus of neostriatal outflow. *Brain*, 109: 845–883.

Thach, W.T. (1978) Correlation of neural discharge with pattern and force of muscular activity, joint position, and direction of intended next movement in motor cortex and cerebellum. *J. Neurophysiol.*, 41: 654–676.

Tobias, T.J. (1975) Afferents to prefrontal cortex from the thalamic mediodorsal nucleus in the rhesus monkey. *Brain Res.*, 83: 191–212.

Trimble, M.R. and Robertson, M.M. (1987) The psychopathology of tics. In: C.D. Marsden and S. Fahn (Eds.), *Movement Disorders 2*, Butterworth, London, pp. 406–422.

Van Hoesen, G.W., Mesulam, M.M. and Haaxma, R. (1976) Temporal cortical projections to the olfactory tubercle in the rhesus monkey. *Brain Res.*, 109: 375–381.

Van Hoesen, G.W., Yeterian, E.H. and Lavizzo-Mourney, R. (1981) Widespread corticostriate projections from temporal cortex of the rhesus monkey. *J. Comp. Neurol.*, 199: 205–219.

Vogt, B.A., Rosene, D.L. and Pandya, D.N. (1979) Thalamic and cortical afferents differentiate anterior from posterior cingulate cortex in the monkey. *Science*, 204: 205–207.

Weinberger, D.R., Berman, K.F. and Zec, R.F. (1986) Physiologic dysfunction of dorsolateral prefrontal cortex in schizophrenia. I. Regional cerebral blood flow evidence. *Arch. Gen. Psychiat.*, 43: 114–124.

Weinrich, M. and Wise, S.P. (1982) The premotor cortex of the monkey. *J. Neurosci.*, 2: 1329–1345.

White, O.B., Saint-Cyr, J.A., Tomlinson, R.D. and Sharpe, J.A. (1983) Ocular motor deficits in Parkinson's disease. *Brain*, 106: 571–587.

Wiesendanger, R. and Wiesendanger, M. (1985a) The thalamic connections with medial area 6 (supplementary motor cortex) in the monkey. *(Macaca fascicularis)*. *Exp. Brain Res.*, 59: 91–104.

Wiesendanger, R. and Wiesendanger, M. (1985b) Cerebellocortical linkage in the monkey as revealed by transcellular labeling with the lectin wheat germ agglutinin conjugated to the marker horseradish peroxidase. *Exp. Brain Res.*, 59: 105–117.

Wilson, C.J. and Groves, P.M. (1980) Fine structure and

synaptic connections of the common spiny neuron of the rat neostriatum: a study employing intracellular injection of horseradish peroxidase. *J. Comp. Neurol.*, 194: 599–615.

Wise, S.P. and Mauritz, K.-H. (1983) Motor aspects of cue-related activity in premotor cortex of the rhesus monkey. *Brain Res.*, 260: 301–305.

Wise, S.P. and Mauritz, K.-H. (1985) Set-related neuronal activity in the premotor cortex of rhesus monkeys: effects of changes in motor set. *Proc. R. Soc. Lond. B*, 223: 331–354.

Yajeya, J., Quintana, J. and Fuster, J. (1988) Prefrontal representation of stimulus attributes during delay tasks. II. The role of behavioral significance. *Brain Res.*, 474: 222–230.

Yelnick, J., Percheron, G. and François, C. (1984) A Gogli analysis of primate globus pallidus. II. Quantitative morphology and spatial orientation of dendritic aborizations. *J. Comp. Neurol.*, 227: 200–213.

Yeterian, E.H. and Van Hoesen, G.W. (1978) Cortico-striate projections in the rhesus monkey: the organization of certain cortico-caudate connections. *Brain Res.*, 139: 43–63.

Young III, W.S., Bonner, T.I. and Brann, M.R. (1986) Mesencephalic dopamine neurons regulate the expression of neuropeptide mRNAs in the rat forebrain. *Proc. Natl. Acad. Sci. USA*, 83: 9827–9831.

Discussion

H.J. Groenewegen: As has been shown by Parent et al. (1989), we also found a projection from small parts of the subthalamic nucleus to the external segment of the globus pallidus (GPe) which terminates in (3–4) bands. These bands in GPe overlap the parts in GPe which belong to different parallel circuits. Does this overlap interfere with the parallelism of the parallel circuits described?

G.E. Alexander: To the extent that bands of terminal labeling in GPe are found to cross functional boundaries between separate parallel circuits (e.g., to label both "motor" and "prefrontal" portions of GPe) following injections of anterograde tracers into the subthalamic nucleus (STN), we would predict that the injection sites had themselves straddled such boundaries within the STN. However, in order for studies of STN-GPe projections to provide an unambiguous test of the "segregated circuits" hypothesis, it would first be necessary to show that a given STN injection site had actually been confined to a single functional domain within that nucleus. This could be done either by characterizing the physiological properties of local neurons or by demonstrating the exclusivity of collateral projections from the same STN site to circuit-specific structures outside GPe, e.g., to the "motor" portion of the striatum.

Reference

Parent, A., Hazrati, L.N. and Smith, Y. (1989) The subthalamic nucleus in primates – a neuronatomical immunohistochemical study. In A.R. Crossman and M.A. Sambrook (Eds.), *Neural Mechanisms in Disorders of Movements*. John Libbey, London, pp. 29–35.

CHAPTER 7

Prefrontal cortical control of the autonomic nervous system: Anatomical and physiological observations

Edward J. Neafsey

Department of Anatomy, Loyola University Stritch School of Medicine, Maywood, IL 60153, USA

Introduction

"The highest nervous processes are potentially the whole organism" (J.H. Jackson, 1931).

What does this enigmatic statement mean? Since by "highest nervous processes" Hughlings Jackson was referring to the cerebral cortex, the statement can be paraphrased as "the cerebral cortex represents the whole organism". The identity between the cortex and the organism affirmed by Jackson implies not only that the entire organism is represented in the cerebral cortex, but also that there is no region of cortex, not even those concerned with the "highest processes", that is not also a representation of some part of the organism. This last statement summarizes the revolution now underway in our understanding of cerebral cortical organization. The traditional concept of strictly sequential processing of information from "primary" sensory cortical areas through a long series of "association" areas to its final destination in the "motor" cortex (Jones and Powell, 1970; Towe, 1973; Mesulam et al., 1977; Diamond,

1979; Swanson, 1983) is being challenged, or at least supplemented, with the concept of simultaneous, parallel processing of information in and through the many recently described multiple cortical representations of visual, auditory, or somatic inputs (e.g., Van Essen and Maunsell, 1982; Kaas, 1987; Livingstone and Hubel, 1988; Zeki and Shipp, 1988; Rodman et al., 1989; also see Ruch (1966) for an early description of this concept). As the cerebral cortex has filled up with these multiple sensory representations, the purely "association" areas of occipital, temporal, and parietal cortex have disappeared (Diamond, 1979; Kaas, 1987). Thus, in these regions, as well as in the precentral motor areas, Jackson would say the cerebral cortex "is" the eye, the ear, or the skin and muscle. The key to understanding the cerebral cortex, then, appears to be the body.

Only one major region of "association" cortex not identified with the body remains, the prefrontal cortex. However, its days as a purely "association" cortex also appear to be numbered because, as will be reviewed below, many recent as well as older studies have shown that most portions of the prefrontal cortex represent either the viscera or eye-head movement. The focus of this article is on prefrontal autonomic control of the viscera, postulating that the prefrontal cortex, at least in part, "is" or represents the heart, stomach, lungs, liver, kidney, etc. via its control over the autonomic nervous system or its processing of

Correspondence: Edward J. Neafsey, Ph.D., Department of Anatomy, Loyola University Stritch School of Medicine, 2160 S. First Avenue, Maywood, IL 60153, USA.

Dedication: To the memory of Percival Bailey, a pioneer in studies of frontal cortical control of the autonomic nervous system.

visceral afferent information. Some comments on prefrontal eye-head control will also be made.

History

Birth

The idea that the prefrontal cortex controls the autonomic nervous system is old (see reviews by Fulton, 1949; Kaada, 1960; Hoff et al., 1963), a fact which may be surprising considering the emphasis for the last 25 years on the role of this cortex in cognitive functions such as delayed response, temporal organization of behavior, and visuospatial memory (see Warren and Akert, 1964; and reviews by Fuster, 1980, 1981; Luria, 1980; Kolb, 1984; Goldman-Rakic, 1987). As early as 1869, J.H. Jackson (1931), on the basis of his observations of visceral physiological responses during epileptic seizures, stated that the "heart, arteries, and viscera, as well as the large muscles of the body, are represented in the units [gyri] of the cerebrum" (see Tharp (1972) for a recent description of visceral responses during a frontal seizure in man). Although he did not explicitly localize this visceral representation to the prefrontal cortex, Jackson did specify that it was associated with the parts of the cerebral cortex concerned with the highest functions, a view similar to that of Sechenov (1963) in Russia. By the turn of the century, several laboratories had reported that electrical stimulation of prefrontal cortical regions such as the gyrus proreus, orbital cortex, and cingulate gyrus elicited cardiovascular, gastric, and respiratory responses in monkeys, dogs, cats, and rabbits (Spencer, 1894; Winkler, 1899; Howell and Austin, 1900; Schafer, 1900). (Since anatomical studies have clearly demonstrated that the mediodorsal nucleus in primates, rats, and cats projects to both the anterior cingulate gyrus (Leonard, 1969; Divac and Kosmal, 1978; Niimi et al., 1978; Baleydier and Mauguière, 1980; Goldman-Rakic and Porrino, 1985; Vogt et al., 1987; Groenewegen, 1988) and the insular cortex (Leonard, 1969; Krettek and Price, 1977; Mufson and Mesulam, 1984; Van Eden and Uylings, 1985; Groenewegen, 1988), these cortical areas can also be considered as part of the prefrontal cortex.) According to Fulton (1939), these early studies linking the cortex with autonomic control were extended and amply confirmed by the work of Von Bechterew (1908 – 1911) and others, but "were largely discarded at the time of the Great War and lost sight of" (also see Kennard, 1944).

In 1925 Von Economo and Koskinas restated the connection between the frontal cortex and autonomic nervous system, hypothesizing that their cytoarchitectonic region FL, the most medial (subcallosal) portion of the band of agranular frontal cortex encircling the hemisphere in man, might have a visceromotor function. At this time, the results of the early electrical stimulation studies linking the frontal cortex with autonomic control were questioned because of concern over possible current spread to subcortical structures and because of the widespread epileptiform activity induced by some types of stimulation (Bard, 1929). However, subsequent, more carefully controlled studies confirmed the association between the frontal cortex and autonomic nervous system in primates, cats, and dogs (e.g., Hoff and Green, 1936; Bailey and Bremer, 1938; Smith, 1938; Bailey and Sweet, 1940).

Golden age

Prefrontal autonomic control was conclusively demonstrated in the classical experiments in the primate of W.K. Smith (1945), Ward (1948), Kaada (1951; also Kaada et al., 1949), and Wall and Davis (1951), who found that electrical stimulation of the anterior cingulate, posterior orbital, and insular regions of the prefrontal cortex elicited striking inhibition of respiration, both rises and falls in blood pressure and heart rate, and pupillary dilation and constriction. These findings were confirmed in man by electrical stimulation studies (Livingstone et al., 1948b; Pool and Ransohoff, 1949), as well as by the clear evidence for increased baroreflex sensitivity (Rinkel et al.,

1947), diminished bladder capacity (Rinkel et al., 1950), and other autonomic changes after prefrontal lobotomy (Greenblatt et al., 1950). Interestingly, prefrontal lobotomy also was said to provide relief from intractable, "visceral" pain (Falconer, 1948; Freeman and Watts, 1948; Fulton, 1949). This observation is consistent with recent findings that the spinothalamic tract, which carries visceral pain afferents in cats and primates (Cervero, 1985), terminates in the thalamic nucleus submedius in primates, cats, and rats (Craig and Burton, 1981; Craig et al., 1982; Price and Slotnick, 1983). The nucleus submedius in cats (Craig et al., 1982) and rats (Price and Slotnick, 1983) in turn has a strong projection to the orbital cortex. (There is some uncertainty about whether the primate has a nucleus submedius (Jones, 1985); in addition, the exact homology between rodent and primate orbital cortical regions is unclear (see Uylings and Van Eden, this volume).) The orbital cortex also respond to electrical stimulation of the vagus nerve in cats (Bailey and Bremer, 1938; Dell and Olson, 1951; also see later studies by Korn and Massion, 1964; Siegfried, 1965; Aubert and Legros, 1970). Finally, another visceral sensory representation was found in the insular cortex of man (Penfield and Rasmussen, 1950) and monkey (Hoffman and Rasmussen, 1953).

Thus, by 1951 a clear link was perceived between the anterior cingulate, orbital, and insular regions of the prefrontal cortex and virtually the entire spectrum of physiological systems under autonomic control (also see description of wide variety of internal organs which show classical conditioning by Bykov (1957)). In fact, the 20 years following 1945 were, in a sense, the "Golden Age" for this concept, an era which climaxed in several major, comprehensive reviews devoted to the topic of cortical autonomic control (Delgado, 1960; Kaada, 1960; Hoff et al., 1963).

Decline

Surprisingly, after the publication of these reviews, the importance of the prefrontal cortex in autonomic control gradually became less widely appreciated and, with a few exceptions, generated little research activity. What happened? Why was clear and well-established experimental evidence linking large portions of the frontal lobe to autonomic control forgotten and neglected? There are probably several factors, the most important of which may have been the increasing focus of research on the role of the prefrontal cortex in cognitive function, particularly on the deficits in delayed response performance seen after lesions of this region in animals (Jacobsen, 1935). For example, most of the papers presented at the two symposia on "Prefrontal Granular Cortex and Behavior" (Warren and Akert, 1964; Konorski et al., 1972) dealt in some way with its contribution to cognitive function.

A second factor contributing to neglect of prefrontal cortex relations with the autonomic nervous system was the increasing popularity of the concept of the limbic system, first described in 1952 by Paul MacLean as an extension of the Papez's (1937) concept of the central mechanism of emotion. The cingulate and orbitofrontal cortices were part of this system and at the outset, as components of what MacLean called the "visceral brain", their role in influencing the autonomic system was clearly acknowledged. However, as time passed, most elements of the limbic system were considered more as brain substrates for *experiencing*" emotion (Grossman, 1967; Nauta, 1971; Swanson, 1983; see recent review by LeDoux, 1987) rather than for *expressing* autonomic and endocrine responses associated with emotion. This latter function was almost exclusively associated with the hypothalamus (Ruch et al., 1966; Grossman, 1967), despite Wall and Davis" (1951) assertion that "the hypothalamus may *not* be the only head ganglion of the autonomic nervous system".

As a result of these two developments, the notion that the prefrontal cortex was involved in cognition and emotional experience became dominant, and most research in the field focused on investigation of this fascinating hypothesis. Not all investigators, however, agreed that this emphasis was correct. Nauta (1971), after reviewing the many cognitive studies published in *The Frontal Granular Cortex and Behavior* (Warren and Akert, 1964), called for future studies to focus on "the visceral and endocrine concomitants" of frontal lesions – not on the cognitive effects. In addition, there was an increasing awareness of the strong olfactory inputs to both the central portion of the mediodorsal nucleus in primates (Benjamin and Jackson, 1974), rabbits (J.C. Jackson and Benjamin, 1974) and rats (Price and Slotnick, 1983; Takagi, 1986) and to the orbital and insular subdivisions of the prefrontal cortex in primates (Takagi, 1986) and rats (Price and Slotnick, 1983). Lesions of this olfactory orbital and insular prefrontal cortex in rats produced behavioral deficits in olfactory discrimination (Eichenbaum et al., 1980) and in male sexual behavior (Sapolsky and Eichenbaum, 1980). Consistent with these findings, neurons in the orbital-insular cortex of the rabbit were found to respond to "biologically significant odors, such as feces, urine, and dry food" (Onoda et al., 1984). In the monkey orbitofrontal neurons also respond to odors (Takagi, 1986), as well as to the sight and taste of food (Thorpe et al., 1983); such lesions also produced visual recognition impairments (Bachevalier and Mishkin, 1986).

Rebirth

The neglect of prefrontal-autonomic relations lasted until the early 1980s, when renewed interest in the role of the prefrontal cortex in autonomic control was sparked by neuroanatomical evidence of *direct* projections from both the medial and lateral divisions of the rat prefrontal cortex to the vagal solitary nucleus in the medulla (Saper, 1982; Shipley, 1982; Terreberry and Neafsey, 1983, 1987; Van der Kooy et al., 1982, 1984; Neafsey et al., 1986b).

The rodent medial prefrontal projection to the solitary nucleus originated mainly from the ventral, infralimbic cortex (subcallosal cortex of the primate and cat (Room et al., 1985; Willett et al., 1986; Vogt et al., 1987)), with a smaller component arising in the more dorsal, prelimbic cortex, as is

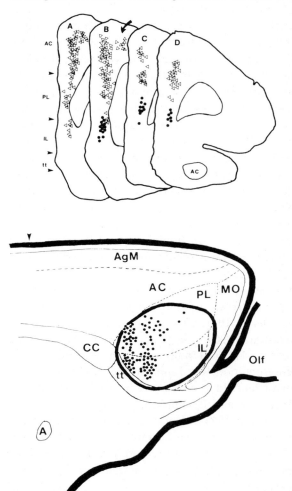

Fig. 1. Upper panel depicts a series of coronal sections of rat frontal cortex on which neurons retrogradely labeled from the solitary nucleus (solid circles, Fast Blue tracer) and superior colliculus (open triangles, Rhodamine-tagged microspheres) are indicated. Lower panel depicts a sagittal view of rat brain, with frontal pole at right. Solid dots indicate neurons retrogradely labeled with Fast Blue from solitary nucleus. Heavy solid line surrounds region where microstimulation altered gastric motility. Abbreviations: A or AC (upper panel) = anterior commissure; AC = anterior cingulate; AgM = agranular medial; CC = corpus callosum; IL = infralimbic; MO = medial orbital; Olf = olfactory bulb; PL = prelimbic; tt = taenia tecta. (Upper panel taken from Neafsey et al., 1986; lower panel taken from Hurley-Gius and Neafsey, 1986.)

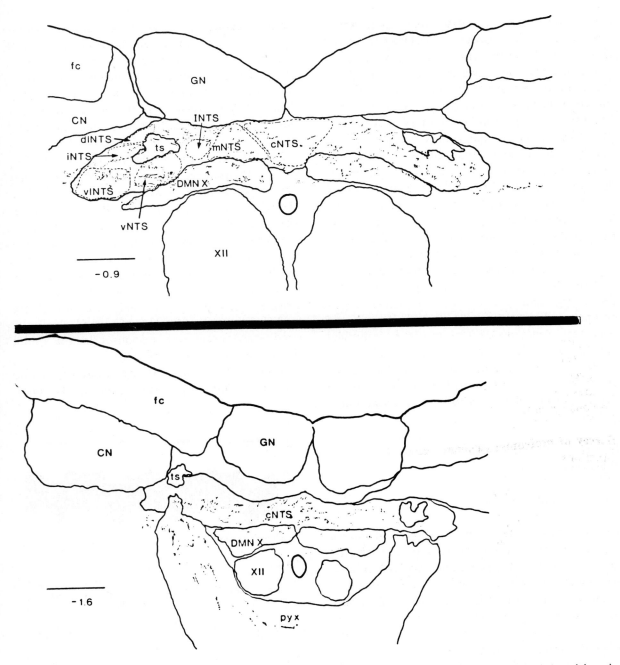

Fig. 2. Anterograde labeling pattern (fine lines and dots) in solitary nucleus (0.9 caudal to obex in upper panel, 1.6 caudal to obex in lower panel) after injection of wheat germ agglutinin conjugated with horseradish peroxidase into infralimbic region of medial frontal cortex. Abbreviations: CN = cuneate nucleus; cNTS = commissural nucleus; dINTS = dorsolateral nucleus; DMN X = dorsal motor nucleus of vagus; fc = fasciculus cuneatus; GN = gracilis nucleus; iNTS = interstitial nucleus; INTS = intermediate nucleus; mNTS = medial nucleus; pyx = pyramidal decussation; ts = tractus solitarius; vNTS = ventral nucleus; vlNTS = ventrolateral nucleus; XII = hypoglossal nucleus. Scale bars = 200 μm. (Figure taken from Terreberry and Neafsey, 1987.)

illustrated in Fig. 1. The descending axons course through the medial aspect of the internal capsule and cerebral peduncle, travel in the pyramidal tract in the medulla, and reach the solitary nucleus by ascending as part of the pyramidal decussation in the lower medulla (Terreberry and Neafsey, 1987). The fibers ramify within the solitary nucleus (Fig. 2), appearing to terminate in a number of subnuclei, including those that receive baroreceptor, gastric, and pulmonary stretch receptor afferents (Terreberry and Neafsey, 1987). The quantitatively larger projection from the rodent lateral prefrontal area to the solitary nucleus arises in the rostral dysgranular and posterior granular subdivisions of the insular cortex (Saper, 1982) and has a similar course and termination pattern to that just described for the medial prefrontal projection. My laboratory and others have undertaken experiments to determine the physiological and behavioral functions of these medial (infralimbic) and lateral (insular) regions of prefrontal cortex. These results are presented below in the context of an overall survey of prefrontal relations to the various organ systems controlled by the autonomic nervous system.

Survey of prefrontal autonomic control functions

Cardiovascular system

Anand and Dua (1956) and Delgado (1960) and his coworkers were among the first to confirm the alterations in blood pressure and heart rate following cingulate, orbital, and insular electrical stimulation in unanesthetized, awake, chronically prepared monkeys and cats.

Medial prefrontal cortex

Lofving's (1961) definitive study of blood pressure, heart rate, and regional blood flow responses to electrical stimulation of the medial prefrontal anterior cingulate cortex in the anesthetized cat identified a dorsal depressor region and a more ventral (subcallosal) pressor region. The depressor effects were due to both inhibition of tonic sympathetic outflow to the heart and blood vessels and activation of the vagus. These effects appeared to be mediated by way of cingulate projections to the anterior hypothalamus. The pressor effects were due to activation of sympathetic outflow to the heart and blood vessels.

Stimulation of the medial prefrontal cortex in the anesthetized rat has elicited both pressor and depressor responses, as well as heart rate decreases (Terreberry and Neafsey, 1984; Burns and Wyss, 1985; Hardy and Holmes, 1988). Fig. 3 illustrates examples of such responses from experiments performed in my laboratory. Cooling this same cortex in the rat reverses hypertension produced by mineralocorticoid-salt treatment or by aortic constriction (Szilagyi et al., 1987). Lesions of the medial frontal cortex also attenuated classically conditioned heart rate responses in the rabbit (Buchanan and Powell, 1982a, b).

In other studies done in my laboratory, blood pressure and heart rate responses during the condi-

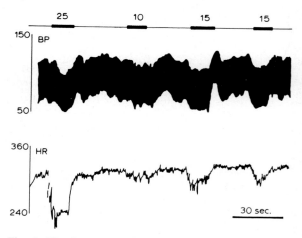

Fig. 3. Blood pressure (BP, mm Hg) and heart rate (HR, beats/min) responses from an anesthetized (ketamine HCl, 100 mg/kg, i.p.) and artificially ventilated rat during bilateral medial frontal cortical stimulation. Thick portions of top trace indicate stimulation periods, which are labeled with current intensity (μA) delivered from each electrode as a 50 Hz train of 0.5 msec pulses. Unpublished data from experiments by Dr. Robert Terreberry.

tioned emotional response paradigm were studied in both control rats and rats with bilateral suction or chemical lesions of the medial prefrontal cortex (Fig. 4). While such lesions did not affect the blood pressure increases to a tone that had been previously paired with footshock, they did change the heart response from a tachycardia to a bradycardia (Frysztak and Neafsey, 1987). Pharmacological blockade with methyl-atropine or atenolol (a β-blocker) revealed that this effect was primarily due to a reduction of sympathetic outflow to the heart in the lesioned animals. Another finding in lesioned animals was that the gain of the cardiac component of the baroreflex at rest was chronically reduced by a factor of 2, from -5.4 beats/min/mm Hg to about -2.5 beats/min/mm Hg, confirming the results of Verberne et al. (1987). This preexisting reduction in cardiac baroreflex gain also prevented the normal reductions in gain in this reflex that are seen during the conditioned emotional response.

In further work done in my laboratory, bilateral lesions of the medial prefrontal cortex of the rat also abolished the heart rate and blood pressure responses evoked by electrical or chemical stimulation of the ventral hippocampus (Ruit and Neafsey, 1988), perhaps by eliminating the infralimbic cortical target of the massive projection from the CA1 region of the ventral hippocampus (Swanson, 1981; Ferino et al., 1987).

Insular cortex

Both increases and decreases in blood pressure and heart rate have been evoked by electrical or chemical stimulation of the insular cortex of the rat or rabbit (Powell et al., 1985; Ruggiero et al., 1987; Hardy and Holmes, 1988). In addition, cells in rat insular cortex respond to baroreflex afferent stimulation (Cechetto and Saper, 1987). Insular cortex lesions, in contrast to medial prefrontal lesions, do not affect conditioned cardiac responses in rabbits (Powell et al., 1985).

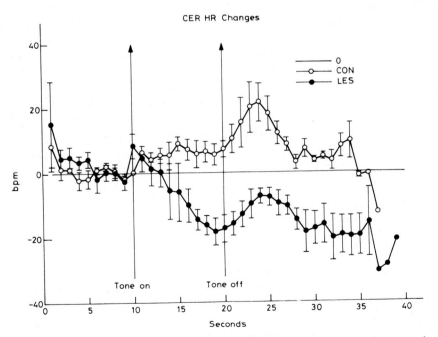

Fig. 4. Mean changes in heart rate from baseline during conditioned emotional response in sham-operated, control rats (CON, open circles, $n = 4$) and rats with bilateral, medial frontal cortex lesions that induced the infralimbic cortex (LES, solid circles, $n = 4$). (Figure taken from Frysztak and Neafsey, 1987.)

Orbital cortex

Electrical stimulation of the orbital cortex of awake, chronically prepared macaque monkeys evoked a 20–30% "parasympathetic" decrease in blood pressure, heart rate, cardiac output, and total systemic resistance, along with a 15% increase in stroke volume (Hall et al., 1977). At some points, opposite, "sympathetic" responses were evoked. This same study also found that the hearts of stimulated monkeys had focal myocardial lesions, including myocytolysis.

Gastro-intestinal system

Medial prefrontal cortex

Lesions or stimulation of the subcallosal anterior cingulate cortex or insular-orbital cortex inhibited or enhanced gastric and intestinal motility and tone in monkeys (Bailey and Sweet, 1940), dogs (Babkin and Kite, 1950; Babkin and Speakman, 1950), and cats (Strom and Uvnas, 1950; Eliasson, 1952; Hesser and Perret, 1960; Rostad, 1973). In our laboratory we have replicated these findings on gastric motility in the rat (Fig. 5), demonstrating that the effective region of medial frontal cortex corresponds to the prelimbic and infralimbic source of direct descending projections to the solitary nucleus (Hurley-Gius and Neafsey, 1986). Lesions of this medial cortex also markedly reduced the incidence of gastric ulcers induced by restraint stress in the rat (Henke, 1982).

Insular cortex

Neurons responsive to gastric distension have been found in the rostral lateral granular insular cortex of the rat (Cechetto and Saper, 1987), adjacent to the taste representation in this cortex (Yamamoto et al., 1980; Kosar et al., 1986). This area in rodents may be a region of convergence of olfactory and gustatory inputs (Shipley and Geinisman, 1984) and appears necessary for the establishment of illness-mediated taste-aversion learning or taste-potentiated odor-aversion learning (Lasiter et al., 1985a, b). Lesions here in rats also disrupt feeding, producing a transient aphagia (Kolb et al., 1977; Grijalva et al., 1985). In rats microstimulation of this same cortex also elicits movements of the intrinsic muscles of the tongue (Neafsey et al., 1986a), and lesions here impair tongue usage (Castro, 1975; Whishaw and Kolb, 1989).

Orbital cortex

Orbital and anterior cingulate electrical stimulation has been reported to cause increased salivation in primates (Kaada, 1951; Showers and Crosby, 1958).

Thermoregulatory system

Electrical stimulation of the orbitofrontal cortex increased the temperature of the contralateral ex-

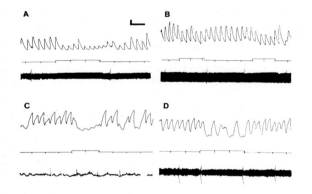

Fig. 5. Examples of different gastric responses evoked by bilateral medial frontal cortex stimulation in 4 different experiments. Upper trace is record of gastric motility, middle trace shows time marks every 30 sec and stimulation periods (upward deflection) and lower trace depicts respiratory activity in A, B, D and heart rate in C (upward deflection denotes slowing). Gastric responses were seen independently of respiratory and heart changes, although at some points gastric and respiratory or heart rate changes were evoked together. The vertical scale bar in A denotes 1 mm Hg for all motility records and 10 beats/min for heart rate record, and the horizontal bar denotes 30 sec. In all cases, stimulation parameters were 0.5 msec pulses of 50 μA at 10 Hz for 60–90 sec. (A) Stimulation (90 sec) produces an amplitude reduction. (B) Two periods of stimulation (75, 90 sec) produce a slight tone depression. (C) Stimulation (60 sec) produces a combined reduction in tone and amplitude. (D) Stimulation (90 sec) produces a reduction in tone and frequency with an increase in amplitude. (Figure taken from Hurley-Gius and Neafsey, 1986.)

tremities in primates and dogs (Delgado and Livingston, 1948; Livingston et al., 1948a), while decreases in skin temperature were reported following lobotomy in man (Fairman et al., 1950). Following a unilateral orbitofrontal lesion in the monkey, contralateral extremity skin temperature rose faster than ipsilateral extremity skin temperature in response to the heat challenge of being placed in an incubator at 32°C, indicating an enhanced vasodilation response (Livingston et al., 1948a). In contrast, lesions of the primate frontal premotor cortex produced the opposite response, i.e., contralateral extremity skin temperature increased less rapidly than ipsilateral extremity skin temperature in response to a heat challenge (Kennard, 1935). More recently, frontal pole lesions in rats have been found to produce hyperthermia (Blass, 1969). Finally, anterior cingulate stimulation also has elicited piloerection in primates (W.K. Smith, 1945; Ward, 1948; Kaada, 1951; Showers and Crosby, 1958).

Reproductive system

Anterior cingulate and orbital stimulation enhanced uterine contractions in rabbits (Setekleiv, 1964) and milk ejection in cats (Beyer et al., 1961), effects thought to be mediated by increased pituitary release of oxytocin (Beyer et al., 1961). In awake male squirrel monkeys, subcallosal anterior cingulate stimulation elicited penile erection at low thresholds (Dua and MacLean, 1964).

Adrenal, renal, and immune systems

Orbital cortex stimulation in the cat has produced a decrease in circulating levels of adrenaline (Von Euler and Folkow, 1958). In addition, stimulation of the orbital cortex in the monkey produced an eosinopenia (-76%) that appeared to be mediated via adrenal corticosteroid release stimulated by pituitary ACTH (Porter, 1954; also see Hall and Marr, 1975; in the rat, Dunn and Miller, 1987). Chronic stimulation of frontal lobe "pressor sites" in cats and dogs was reported to produce erythrocytosis, hyperplastic bone marrow, and an increase in blood volume (Kell et al., 1960), perhaps in response to enhanced secretion of erythropoietin by the kidneys due to the dramatically reduced renal blood flow produced by orbital cortex stimulation in cats (Cort, 1953). Orbital stimulation also stopped urine production in cats (Cort, 1953), and both orbital and cingulate stimulation could excite or inhibit urinary bladder activity in cats (Strom and Uvnas, 1950; Gjone and Setekleiv, 1963). Finally, prefrontal lesions in monkeys also produced massive hypertrophy of the thymus and lymph nodes (Messimy, 1939).

Discussion

Emotion and the prefrontal cortex

Visceral control

Since visceral function and emotion are closely associated (e.g., LeDoux, 1987), the relationship of the prefrontal cortex to emotion will be briefly reviewed. In an important early experiment, Jacobsen et al. (1935) described how bilateral frontal lobectomy eliminated a chimpanzee's emotional outbursts when she was unable to perform the delayed response task correctly (see this volume, Ch. 27, and Valenstein (1986) for a detailed analysis of this important experiment which is thought by some to have led Egas Moniz to develop the prefrontal lobotomy). In addition, the importance of the cingulate cortex in specific behaviors such as maternal care in rats (Stamm, 1954, 1955; Slotnick, 1967) and separation calls and play in the primate (MacLean and Newman, 1988) has also been documented. Evidence for prefrontal involvement in "reward" systems in the brain has been provided by demonstration of electrical (Routtenberg, 1971) or chemical (Goeders et al., 1986) self-stimulation in the prefrontal cortex in the rat. A role for the prefrontal cortex in emotional stress has been suggested by evidence for "specific" increases in dopamine turnover in the rat prefrontal cortex during stress (Thierry et al., 1976). Finally, a recent series of experiments has described increased "timidity" in rats following medial prefrontal lesions (Holson, 1986a, b; Holson and Walker, 1986).

However, in spite of this extensive range of studies, the importance of the prefrontal and cingulate cortices in emotion has recently been minimized (O.A. Smith and DeVito, 1984). Furthermore, the validity of even attempting to "localize" a psychological concept such as emotion to a specific brain region or even to the limbic system has been questioned (VanderWolf et al., 1988), as has the basic value of such a broad and all-encompassing concept as the limbic system (Brodal, 1981; Swanson, 1983). In addition, the recent "facial efference" concept of emotion (Adelmann and Zajonc, 1989), a modified version of William James's (1894) theory that emotion *is* our feeling of the bodily responses that follow directly upon perception of a stimulus, also has challenged the traditional view that limbic system structures, including the prefrontal cortex, comprise a set of brain regions whose activity is *directly experienced* as emotion (LeDoux, 1987).

Vocalization

My laboratory has recently reported findings of possible relevance to prefrontal function in emotion. In our experiments (Frystak et al., 1988), we made the observation that normal rats respond to a tone previously paired with footshock not only with a blood pressure and heart rate increase but also with a prolonged (2–4 min) period of ultrasonic vocalizations (ca. 25 kHz) that coincides with a period of immobility or "freezing behavior" (LeDoux et al., 1984). (Ultrasonic vocalizations also occur in more natural stress situations, such as when an "intruder" rat vocalizes after being placed in the cage of a "resident" rat (Fokkema et al., 1986; Thomas et al., 1983).) We have recently found that medial prefrontal lesions abolish these conditioned vocalizations, which presumably were an expression of some state of anxiety or fear in the rat. Similar to our findings in the rat, lesions of the subcallosal anterior cingulate region of prefrontal cortex in monkeys abolished "distress calls" (Sutton et al., 1974; MacLean and Newman, 1988). Thus, the medial prefrontal cortex is not concerned solely with autonomic regulation, but also contributes to somatic behaviors.

What about cognition?

Cognitive deficits observed after prefrontal lesions are in no small way related to deficits in the control of eye movements. Humans with unilateral prefrontal lesions are unable to suppress saccadic eye movements to stimuli in the visual field contralateral to the lesion (Guitton et al., 1985), and this distractibility seriously interferes with attention. These findings recall Malmo's (1942) interference hypothesis of the cause of the delayed response deficits seen in prefrontal monkeys. This hypothesis stated that the major difference between the normal and prefrontal monkeys was "one of degree of susceptibility to the interfering effects of extraneous stimuli during the delay interval". This notion was based on Malmo's discovery that he could restore delayed response performance in monkeys with frontal lesions by simply turning out the lights during the delay period to eliminate interfering visual distractions. Many subsequent studies (see Table I) have supported this explanation, and in 1973 Pribram affirmed that this hypothesis was "as it has been for three decades, the most viable and useful in explaining the effects of resection of the dorsolateral frontal cortex of primates" (also see Pribram et al., 1964, for more complete review of interference hypothesis). The increased interference of extraneous, distracting stimuli in prefrontal animals was thought to be due to the lack of prefrontal inhibitory control over the eye and head movements involved in the orienting response (distractibility, clearly demonstrated by Guitton et al., 1985) or over the most recently rewarded previous response movement, leading to perseverative errors. Thus, a strong case can be made for the loss of prefrontal eye movement control being a significant factor in the apparent cognitive or memory related deficit seen following prefrontal lesions (also cf. Boch and Goldberg, 1989).

TABLE I

List of some important findings consistent with Malmo's (1942) "interference hypothesis" of prefrontal function, including observations on prefrontal control of eye and head movements that are elements of the orienting response

1. Restoration of delayed response performance in prefrontal animals is achieved by turning off lights during delay (Malmo, 1942), by delivering an additional food reward at stimulus presentation before delay begins (Finan, 1942), by lightly sedating them with Nembutal (Wade, 1947), by making them hungrier or colder (Pribram, 1950), or by running sham release trials with no reward to extinguish responses to release after the delay (Konorski and Lawicka, 1964).
2. Perseverative errors after prefrontal damage are seen in humans on a variety of tasks (Luria and Homskaya, 1964) and on the Wisconsin Card Sort Test (Milner, 1964), in monkeys on delayed response (Mishkin, 1964), and in rats on other tasks (e.g., Sinnamon and Charman, 1988).
3. Profound hyperactivity is described in prefrontal monkeys (Ruch and Shenkin, 1943); hyperactive orienting reflexes are found in prefrontal monkeys (Mettler, 1944) and dogs (Konorski and Lawicka, 1964); enhanced startle responses are found in rats (Hammond, 1974).
4. Humans with frontal lobe lesions display reduced eye movements during visual scanning (Teuber, 1964; Tyler, 1969; Luria, 1980).
5. Lack of act inhibition and drive inhibition in prefrontal animals is attributed to their hyperreactivity and augmentation of various emotional states (Brutkowski, 1965).
6. A wide area of prefrontal cortex is found to project directly to the superior colliculus, a pathway that could control orienting responses (Domesick, 1969; Leonard, 1969; Goldman and Nauta, 1976; Leichnetz et al., 1981; Reep et al., 1987), consistent with reports that electrical stimulation of prefrontal cortex, particularly rostral cingulate regions, elicits orienting head and eye movements (e.g., Kaada, 1960; Hess, 1969; Sinnamon and Galer, 1984).
7. Prefrontal monkeys, in contrast to normals, use the last cue delivered during a delay to guide their response choice (Bartus and Levere, 1977).
8. Interference suppression is included as a key element of the temporal organization of behavior theory of prefrontal function (Fuster, 1980, 1981; Kolb, 1984), in Shallice's (1982) prefrontal "supervisory attentional system" theory, and in Stamm's (1987) solution to the riddle of the monkey's delayed response deficit.
9. Unilateral lesions produce contralateral visual neglect, particularly obvious when competing, bilateral stimuli are present (Crowne et al., 1981; Crowne and Pathria, 1982; Vargo et al., 1987).
10. Unilateral prefrontal lesions in man impair suppression of saccades to the contralateral side (Guitton et al., 1985; Butter, 1989).
11. A second, supplementary eye field is found in primate frontal cortex (Schlag and Schlag-Rey, 1987), further increasing the area of frontal lobe devoted to eye movements beyond the classical frontal eye field (e.g., van der Steen et al., 1986).
12. Inhibition of incorrect or inappropriate responses is included as an important element in the visuospatial representational memory theory of prefrontal function (Goldman-Rakic, 1987).
13. Because neurons in the principal sulcus region of monkey prefrontal cortex "show such a full range of activity in a series of oculomotor tasks, our results raise the cautionary note that some of the activity reported during various complicated behavioral tasks . . . may in fact be related only to the oculomotor correlates of these tasks" (Boch and Goldberg, 1989). This finding is consistent with Stamm and Rosen's (1969) finding that disturbance of principal sulcus activity by electrical stimulation interferes with delayed response performance only if the stimulation is delivered during the early phase of the delay when the monkey is still orienting his head and eyes to the cue.

Conclusion

What is the significance of prefrontal cortical control of the autonomic nervous system? Most generally, it is simply another instance of Jacksonian re-representation of a brain function at the cortical level. More specifically, lesion experiments have shown that the *medial*, anterior cingulate prefrontal region is necessary for the expression of the normal, sympathetically mediated increase in heart rate during stress. It also is necessary for the production of gastric ulcers that develop in rats under restraint stress. In addition, this cortex is needed to maintain the normal, resting level of the gain of the cardiac component of the baroreflex response. The *lateral*, insular prefrontal cortex contains a visceral sensory representation, including both taste and general visceral afferent inputs from a variety of organs, including the carotid sinus and stomach. This lateral prefrontal cortex is necessary for learning taste discriminations and taste aversions, but does not appear to be involved in expressing cardiovascular responses to stress. The *orbital* prefrontal cortex also influences the autonomic nervous system and receives olfactory, gustatory, visual, and pain inputs.

Nauta (1971) has suggested an additional, fascinating function for prefrontal autonomic control in man: it is required for the visceral, interoceptive, "gut feelings" that we experience in response to both real and imagined events. On the basis of reduced or absent galvanic skin responses (GSR) in frontal patients (Luria and Homskaya, 1964; also Kimble et al., 1965), Nauta postulated that loss of this cortex would produce an "interoceptive agnosia" due to the absence of any autonomic, "affective" responses to the various alternatives the individual may be considering, thereby contributing to the loss of foresight and behavioral anticipation in frontal lesion patients. Damasio and his coworkers (Anderson et al., 1988; Damasio and Tranel, 1988; Tranel et al., 1988) have recently documented just such a deficit in individuals with bilateral orbito-mesial frontal lesions: these subjects lack autonomic, skin conductance responses to "emotional" stimuli (Nauta's "interoceptive agnosia") and also behave inappropriately in social situations (Damasio's "acquired sociopathy").

These visceral regions of prefrontal cortex are also likely to be brain areas whose activity contributes to psychosomatic disease, perhaps justifying the Soviet term of "corticovisceral medicine" for this field (MacLean, 1949, 1958; Wolman, 1988). On the other hand, it is also possible that the activity of this same cortex may be modified by the process of biofeedback, thereby contributing to the health of the individual (Miller, 1983; Wolman, 1988).

Lastly, Fig. 6 illustrates a diagram of a scheme for the overall organization of the frontal and parietal cortices of the rat. Note the border between the agranular somatic motor and granular somatic sensory cortices on the convexity of the hemisphere; this corresponds to the rat's "central sulcus". Immediately lateral to the somatic sensory cortex lies the prefrontal insular visceral sensory cortex. Medial to the somatic motor cortex lie the two functional subdivisions of the medial prefrontal cortex: the more dorsal, agranular medial and anterior cingulate frontal eye field region; and the ventral prelimbic and infralimbic visceral motor cortex. Thus, symmetrical, "visceral" representations in prefrontal cortex anchor the medial and lateral edges of the neocortex. At the time I first sketched this diagram several years ago (Hurley-Gius and Neafsey, 1986), I was

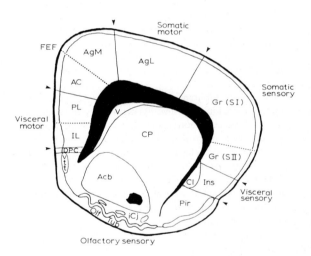

Fig. 6. A schematic representation of a coronal section taken through rat frontal and parietal cortex, with the major functional cortical areas (between arrowheads on cortical surface) labeled from medial to lateral as visceral motor, frontal eye fields, somatic motor, somatic sensory, and visceral sensory; olfactory sensory region lies most ventrally. Blackened areas represent subcortical white matter. Acb, accumbens; AC, anterior cingulate; AgL, lateral agranular cortex; AgM, medial agranular cortex; Cl, claustrum; CP, caudate-putamen; DPC, dorsal peduncular cortex; FEF, frontal eye fields; Gr(SI), primary sensory granular cortex; Gr(SII), secondary sensory granular cortex; ICj, islets of Calleja; IL, infralimbic cortex; Ins, insular cortex; Olf Tub, olfactory tubercle; Pir, pyriform cortex; PL, prelimbic cortex; V, ventricle; vtt, ventral taenia tecta. (Figure taken from Hurley-Gius and Neafsey, 1986.)

confident it provided a good "working" hypothesis for studies of the rat frontal cortex; now I think it may also be a useful guide to understanding the primate frontal lobe, including that of man. As Jackson's phrase discussed at the beginning of this paper implied, the key to understanding the prefrontal cortex is the body.

References

Adelmann, P.K. and Zajonc, R.B. (1989) Facial efference and the experience of emotion. *Annu. Rev. Psychol.,* 40: 249 – 280.

Anand, B.K. and Dua, S. (1956) Circulatory and respiratory changes induced by electrical stimulation of limbic system (visceral brain). *J. Neurophysiol.,* 19: 393 – 400.

Anderson, S.W., Damasio, H., Tranel, D. and Damasio, A.R. (1988) Neuropsychological correlates of bilateral frontal lobe lesions in humans. *Neurosci. Abstr.,* 14: 1288.

Aubert, M. and Legros, J. (1970) Topographie des projections de la sensibilité viscérale sur l'écorce cerebrale du chat. I. Etude des projections corticales du vague cervical chez le chat anesthésié au nembutal. *Arch. Ital. Biol.,* 108: 423 – 446.

Babkin, B.P. and Kite, W.C., Jr. (1950) Gastric motor effect of acute removal of cingulate gyrus and section of brain stem. *J. Neurophysiol.,* 13: 335 – 342.

Babkin, B.P. and Speakman, T.J. (1950) Cortical inhibition of gastric motility. *J. Neurophysiol.,* 13: 55 – 63.

Bachevalier, J. and Mishkin, M. (1986) Visual recognition impairment follows ventromedial but not dorsolateral prefrontal lesions in monkeys. *Behav. Brain Res.,* 20: 249 – 261.

Bailey, P. and Bremer, F. (1938) A sensory cortical representation of the vagus nerve. *J. Neurophysiol.,* 1: 405 – 412.

Bailey, P. and Sweet, W.H. (1940) Effects on respiration, blood pressure and gastric motility of stimulation of orbital surface of frontal lobe. *J. Neurophysiol.,* 3: 276 – 281.

Baleydier, C. and Maugière, F. (1980) The duality of the cingulate gyrus in monkey. Neuroanatomical study and functional hypothesis. *Brain,* 103: 525 – 554.

Bard, P. (1929) The central representation of the sympathetic system. *Arch. Neurol. Psychiat.,* 22: 230 – 241.

Bartus, R.T. and Levere, T.E. (1977) Frontal decortication in rhesus monkeys: a test of the interference hypothesis. *Brain Res.,* 119: 233 – 248.

Benjamin, R.M and Jackson, J.C. (1974) Unit discharges in the mediodorsal nucleus of the squirrel monkey evoked by electrical stimulation of the olfactory bulb. *Brain Res.,* 75: 181 – 191.

Beyer, F.C., Anguiano, L. and Mena, J. (1961) Oxytocin release in response to stimulation of cingulate gyrus. *Am. J. Physiol.,* 200: 625 – 627.

Blass, E.M. (1969) Thermoregulatory adjustments in rats after removal of the frontal poles of the brain. *J. Comp. Physiol. Psychol.,* 69: 83 – 90.

Boch, R.A. and Goldberg, M.E. (1989) Participation of prefrontal neurons in preparation of visually guided eye movements in the rhesus monkey. *J. Neurophysiol.,* 61: 1064 – 1084.

Brodal, A. (1981) *Neurological Anatomy in Relation to Clinical Medicine,* Oxford University Press, New York, pp. 689 – 690.

Brutowski, S. (1965) Functions of prefrontal cortex in animals. *Physiol. Rev.,* 45: 721 – 746.

Buchanan, S.L. and Powell, D.A. (1982a) Cingulate damage attenuates conditioned bradycardia. *Neurosci. Lett.,* 29: 261 – 268.

Buchanan, S.L. and Powell, D.A. (1982b) Cingulate cortex: Its role in Pavlovian conditioning. *J. Comp. Physiol. Psychol.,* 96: 755 – 774.

Burns, S.M. and Wyss, J.M. (1985) The involvement of the anterior cingulate cortex in blood pressure control. *Brain Res.,* 340: 71 – 77.

Butter, C.M., Rapcsak, S., Watson, R.T. and Heilman, K.M. (1989) Changes in sensory inattention, directional motor neglect and "release" of the fixation reflex following a unilateral frontal lesion: a case report. *Neuropsychologia,* 26: 533 – 545.

Bykov, K.M. (1957) *The Cerebral Cortex and the Internal Organs,* Transl. by W.H. Gantt, Chemical Publishing Co., New York.

Castro, A.J. (1975) Tongue usage as a measure of cerebral cortical localization in the rat. *Exp. Neurol.,* 47: 343 – 375.

Cechetto, D.F. and Saper, C.B. (1987) Evidence for a viscerotopic sensory representation in the cortex and thalamus in the rat. *J. Comp. Neurol.,* 262: 27 – 45.

Cervero, F. (1985) Visceral nociception: peripheral and central aspects of visceral nociceptive systems. *Phil. Trans. R. Soc. Lond. B,* 308: 325 – 337.

Cort, J.H. (1953) Effect of nervous stimulation on the arteriovenous oxygen and carbon dioxide differences across the kidney. *Nature,* 171: 784 – 785.

Craig, A.D. and Burton, H. (1981) Spinal and medullary lamina I projections to nucleus submedius in medial thalamus: a possible pain center. *J. Neurophysiol.,* 45: 443 – 466.

Craig, A.D., Weigand, S.J. and Price, J.L. (1982) The thalamo-cortical projection of the nucleus submedius in the cat. *J. Comp. Neurol.,* 206: 28 – 48.

Crowne, D.P. and Pathria, M.N. (1982) Some attentional effects of unilateral frontal lesions in the rat. *Behav. Brain Res.,* 6: 25 – 39.

Crowne, D.P., Yeo, C.H., and Russell, I.S. (1981) The effects of unilateral frontal eye field lesions in the monkey: visual guidance and avoidance behavior. *Behav. Brain Res.,* 2: 165 – 187.

Damasio, A.R. and Tranel, D. (1988) Domain-specific amnesia for social knowledge. *Neurosci. Abstr.,* 14: 1289.

Delgado, J.M.R. (1960) Circulatory effects of cortical stimulation. *Physiol. Rev.,* 40, Suppl. 4: 146–171.

Delgado, J.M.R. and Livingston, R.B. (1948) Some respiratory, vascular and thermal responses to stimulation of orbital surface of frontal lobe. *J. Neurophysiol.,* 11: 39–55.

Dell, P. and Olson, R. (1951) Projections thalamiques, corticales et cérebelleuses des afférences viscérales vagales. *C.R. Soc. Biol.,* 145: 1084–1088.

Diamond, I.T. (1979) The subdivisions of neocortex: A proposal to revise the traditional view of sensory, motor, and association areas. In J.M. Sprague and A.N. Epstein (Eds.), *Progress in Psychobiology and Physiological Psychology, Vol. 8,* Academic Press, New York, pp. 1–43.

Divac, I. and Kosmal, A. (1978) Subcortical projections to the prefrontal cortex in the rat as revealed by the horseradish peroxidase technique. *Neuroscience,* 3: 785–796.

Domesick, V.B. (1969) Projections from the cingulate cortex in the rat. *Brain Res.,* 12: 296–320.

Dua, S. and MacLean, P.D. (1964) Localization for penile erection in medial frontal lobe. *Am. J. Physiol.,* 207: 1425–1434.

Dunn, J. and Miller, D. (1987) Plasma corticosterone responses to electrical stimulation of the medial frontal cortex. *Neurosci. Abstr.,* 13: 416.

Eichenbaum, H., Shedlack, K.J. and Eckmann, K.W. (1980) Thalamocortical mechanisms in odor-guided behavior. I. Effects of lesions of the mediodorsal thalamic nucleus and frontal cortex on olfactory discrimination in the rat. *Brain Behav. Evol.,* 17: 255–275.

Eliasson, S. (1952) Cerebral influence on gastric motility in the cat. *Acta Physiol. Scand.,* 26, Suppl. 95: 1–69.

Fairman, D., Livingston, K. and Poppen, J.L. (1950) Skin temperature changes after unilateral and bilateral prefrontal lobotomy. In M. Greenblatt, R. Arnot and H.C. Solomon (Eds.), *Studies in Lobotomy,* Grune and Stratton, New York, pp. 380–385.

Falconer, M.A. (1948) Relief of intractable pain of organic origin by frontal lobotomy. *Res. Publ. Assoc. Res. Nerv. Ment. Dis.,* 27: 706–714.

Ferino, F., Thierry, A.M. and Glowinski, J. (1987) Anatomical and electrophysiological evidence for a direct projection from Ammon's horn to the medial prefrontal cortex in the rat. *Exp. Brain Res.,* 65: 421–426.

Finan, J.L. (1942) Delayed response with a predelay reinforcement in monkeys after removal of the frontal lobes. *Am. J. Psychol.,* 55: 202–214.

Fokkema, D.S., Koolhaas, J.M., van der Meulen, J. and Schoemaker, R. (1986) Social stress induced pressure breathing and consequent blood pressure oscillation. *Life Sci.,* 38: 569–575.

Freeman, W. and Watts, J.W. (1948) Pain mechanisms and the frontal lobes: a study of prefrontal lobotomy for intractable pain. *Ann. Intern. Med.,* 48: 747–754.

Frysztak, R.J. and Neafsey, E.J. (1987) Effects of rat medial frontal cortex lesions on conditioned emotional responses. *Neurosci. Abstr.,* 13: 1551.

Frysztak, R.J., Plopa, L. and Neafsey, E.J. (1988) Ultrasonic vocalizations are part of the rat's conditioned emotional response. *Neurosci. Abstr.,* 14: 448.

Fulton, J.F. (1939) Levels of autonomic function with particular reference to the cerebral cortex. In *The Interrelationship of Mind and Body,* Ch. XII. Res. Proc. Assoc. Res. Nerv. Ment. Dis., XIX: 219–236.

Fulton, J.F. (1949) *Physiology of the Nervous System,* 3rd Edn. Oxford University Press, New York, Ch. XXII. Cerebral cortex: the orbitofrontal and cingulate regions, pp. 447–467.

Fuster, J.M. (1980) *The Prefrontal Cortex,* Raven Press, New York.

Fuster, J.M. (1981) Prefrontal cortex in motor control. In Vernon Brooks (Ed.), *Handbook of Physiology. Section I. The Nervous System. Vol. II. Motor Control, Part 2,* American Physiological Society, Bethesda, MD, pp. 1149–1178.

Gjone, R. and Setekleiv, J. (1963) Excitatory and inhibitory bladder responses to stimulation of the cerebral cortex in the cat. *Acta Physiol. Scand.,* 59: 337–348.

Goeders, N.E., Dworkin, S.I. and Smith, J.E. (1986) Neuropharmacological assessment of cocaine self-administration into the medial prefrontal cortex. *Pharmacol. Biochem. Behav.,* 24: 1429–1440.

Goldman, P.S. and Nauta, W.J.H. (1976) Autoradiographic demonstration of a projection from prefrontal association cortex to the superior colliculus in the rhesus monkey. *Brain Res.,* 116: 145–149.

Goldman-Rakic, P.S. (1987) Circuitry of primate prefrontal cortex and regulation of behavior by representational memory. In F. Plum and V. Mountcastle (Eds.), *Handbook of Physiology. Section I. The Nervous System. Vol. V. Higher Functions of the Brain, Part 1,* American Physiological Society, Bethesda, MD, pp. 373–417.

Goldman-Rakic, P.S. and Porrino, L.J. (1985) The primate mediodorsal nucleus and its projection to the frontal lobe. *J. Comp. Neurol.,* 242: 535–560.

Greenblatt, M., Arnot, R. and Solomon, H.C. (1950) *Studies in Lobotomy,* Grune and Stratton, New York, pp. 466–467.

Grijalva, C.V., Kiefer, S.W., Gunion, M.W., Cooper, P.H. and Novin, D. (1985) Ingestive responses to homeostatic challenges in rats with ablations of the anterolateral neocortex. *Behav. Neurosci.,* 99: 162–174.

Groenewegen, H.J. (1988) Organization of the afferent connections of the mediodorsal thalamic nucleus in the rat, related to the mediodorsal-prefrontal topography. *Neuroscience,* 24: 379–431.

Grossman, S.P. (1967) *A Textbook of Physiological Psychology,* John Wiley, New York.

Guitton, D., Buchtel, H.A. and Douglas, R.M. (1985) Frontal

lobe lesions in man cause difficulties in suppressing reflexive glances and in generating goal-directed saccades. *Exp. Brain Res.*, 58: 455–472.

Hall, R.E. and Marr, H.B. (1975) Influence of electrical stimulation of posterior orbital cortex upon plasma cortisol levels in unanesthetized sub-human primate. *Brain Res.*, 93: 367–371.

Hall, R.E., Livingston, R.B. and Bloor, C.M. (1977) Orbital cortical influences on cardiovascular dynamics and myocardial structure in conscious monkeys. *J. Neurosurg.*, 46: 638–647.

Hammond, G.R. (1974) Frontal cortical lesions and prestimulus inhibition of the rat's acoustic startle reaction. *Physiol. Psychol.*, 2: 151–156.

Hardy, S.G.P. and Holmes, D.E. (1988) Prefrontal stimulus-produced hypotension in rat. *Exp. Brain Res.*, 73: 249–255.

Henke, P.G. (1982) The telencephalic limbic system and gastric pathology. *Neurosci. Biobehav. Rev.*, 6: 381–390.

Hess, W.R. (1969) *Hypothalamus and Thalamus. Experimental Documentation,* Georg Thieme Verlag, Stuttgart.

Hesser, F.H. and Perret, G.E. (1960) Studies on gastric motility in the cat. II. Cerebral and infracerebral influence in control, vagectomy and cervical cord preparations. *Gastroenterology*, 38: 231–246.

Hoff, E.C. and Green, H.D. (1936) Cardiovascular reactions induced by electrical stimulation of the cerebral cortex. *Am. J. Physiol.*, 117: 411–422.

Hoff, E.C., Kell, J.F., Jr. and Carroll, M.N., Jr. (1963) Effects of cortical stimulation and lesions on cardiovascular function. *Physiol. Rev.*, 43: 68–114.

Hoffman, B.L. and Rasmussen, T. (1953) Stimulation studies of insular cortex of *Macaca mulatta. J. Neurophysiol.*, 16: 343–360.

Holson, R.R. (1986a) Mesial prefrontal cortical lesions and timidity in rats. I. Reactivity to aversive stimuli. *Physiol. Behav.*, 37: 221–230.

Holson, R.R. (1986b) Mesial prefrontal cortical lesions and timidity in rats. III. Behavior in a semi-natural environment. *Physiol. Behav.*, 37: 239–247.

Holson, R.R. and Walker, C. (1986) Mesial prefrontal cortical lesions and timidity in rats. II. Reactivity to novel stimuli. *Physiol. Behav.*, 37: 231–238.

Howell, W.H. and Austin, M.F. (1900) The effect of stimulating various portions of the cortex cerebri, caudate nucleus and dura mater upon blood pressure. *Am. J. Physiol.*, 3: xxii–xxiii.

Hurley-Gius, K.M. and Neafsey, E.J. (1986) The medial frontal cortex and gastric motility: microstimulation results and their possible significance for the overall pattern of organization of rat frontal and parietal cortex. *Brain Res.*, 365: 241–248.

Jackson, J.C. and Benjamin, R.M. (1974) Unit discharges in the mediodorsal nucleus of the rabbit evoked by electrical stimulation of the olfactory bulb. *Brain Res.*, 75: 193–201.

Jackson, J.H. (1931) Clinical and physiological researches on the nervous system. I. On the anatomical and physiological localization of movements in the brain. In *Selected Writings of John Hughlings Jackson,* Hodder and Stoughton, London.

Jacobsen, C.F. (1935) Functions of the frontal association area in primates. *Arch. Neurol. Psychiat.*, 33: 558–569.

Jacobsen, C.F., Wolfe, J.B. and Jackson, T.A. (1935) An experimental analysis of the functions of the frontal association areas in primates. *J. Nerv. Ment. Dis.*, 82: 1–14.

James, W. (1984) What is emotion? *Mind,* 9: 188–205.

Jones, E.G. (1985) *The Thalamus,* Plenum Press, New York.

Jones, E.G. and Powell, T.P.S. (1970) An experimental study of converging sensory pathways within the cerebral cortex of the monkey. *Brain,* 93: 793–820.

Kaada, B.R. (1951) Somato-motor, autonomic and electrocorticographic responses to electrical stimulation of "rhinencephalic" and other structures in primates, cat and dog. *Acta Physiol. Scand.*, 24, Suppl. 83: 285 pp.

Kaada, B.R. (1960) Cingulate, posterior orbital, anterior insular and temporal pole cortex. In H.W. Magoun (Ed.), *Handbook of Physiology. Section 1. Neurophysiology, Vol. II,* American Physiological Society, Washington, DC, pp. 1345–1372.

Kaada, B.R., Pribram, K.H. and Epstein, J.A. (1949) Respiratory and vascular responses in monkeys from temporal pole, insula, orbital surface and cingulate gyrus. *J. Neurophysiol.*, 12: 347–356.

Kaas, J.H. (1987) The organization of neocortex in mammals: implications for theories of brain function. *Annu. Rev. Psychol.*, 38: 129–151.

Kell, J.F., Jr., Hoff, E.C. and Hennigar, G.R. (1960) Erythrocytosis or symptomatic polycythemia following chronic cerebral stimulation through indwelling electrodes. *J. Neurosurg.*, 17: 1028–1038.

Kennard, M.A. (1935) Vasomotor disturbances resulting from cortical lesions. *Arch. Neurol. Psychiat.*, 33: 537–545.

Kennard, M.A. (1944) Autonomic functions. In Paul Bucy (Ed.), *The Precentral Motor Cortex,* University of Illinois Press, Urbana, IL, pp. 293–306.

Kimble, D.P., Bagshaw, M.H. and Pribram, K.H. (1965) The GSR of monkeys during orienting and habituation after selective partial ablations of the cingulate and frontal cortex. *Neuropsychologia,* 3: 121–128.

Kolb, B. (1984) Functions of the frontal cortex of the rat: a comparative review. *Brain Res. Rev.*, 8: 65–98.

Kolb, B., Whishaw, I.Q. and Schallert, T. (1977) Aphagia, behavior sequencing, and body weight set point following orbital frontal lesions in rats. *Physiol. Behav.*, 19: 93–103.

Konorski, J. and Lawicka, W. (1964) Analysis of errors by prefrontal animals on the delayed-response test. In J.M. Warren and K. Akert (Eds.), *The Frontal Granular Cortex and Behavior,* McGraw-Hill, New York, pp. 271–294.

Konorski, J., Teuber, H.L. and Zernicki, B. (Eds.) (1972) *The Frontal Granular Cortex and Behavior. Acta Neurobiol.*

Exp., 32 (2).
Korn, H. and Massion, J. (1964) Origine et topographie des projections vagales sur le cortex antérieur chez le chat. *C.R. Acad. Sci. Paris*, 259: 4373 – 4375.
Kosar, E., Grill, H.J. and Norgren, R. (1986) Gustatory cortex in the rat. I. Physiological properties and cytoarchitecture. *Brain Res.*, 379: 329 – 341.
Krettek, J.E. and Price, J.L. (1977) The cortical projections of the mediodorsal nucleus and adjacent thalamic nuclei in the rat. *J. Comp. Neurol.*, 171: 157 – 192.
Lasiter, P.S., Deems, D.A. and Garcia, J. (1985a) Involvement of the anterior insular gustatory neocortex in taste-potentiated odor aversion learning. *Physiol. Behav.*, 34: 71 – 77.
Lasiter, P.S., Deems, D.A., Oetting, R.L. and Garcia, J. (1985b) Taste discriminations in rats lacking anterior insular gustatory neocortex. *Physiol. Behav.*, 35: 277 – 285.
LeDoux, J.E. (1987) Emotion. In F. Plum and V. Mountcastle (Eds.), *Handbook of Physiology. Section I. The Nervous System. Vol. V. Higher Functions of the Brain, Part 1*, American Physiological Society, Bethesda, MD, pp. 419 – 459.
LeDoux, J.E., Sakaguchi, A. and Reis, D.J. (1984) Subcortical efferent projections of the medial geniculate nucleus mediate emotional responses conditioned to acoustic stimuli. *J. Neurosci.*, 4: 683 – 698.
Leichnetz, G.R., Spencer, R.F., Hardy, S.G.P. and Astruc, J. (1981) The prefrontal corticotectal projection in the monkey; an anterograde and retrograde horseradish peroxidase study. *Neuroscience*, 6: 1023 – 1041.
Leonard, C.M. (1969) The prefrontal cortex of the rat. I. Cortical projection of the mediodorsal nucleus. II. Efferent connections. *Brain Res.*, 12: 321 – 343.
Livingston, R.B., Fulton, J.F., Delgado, J.M.R., Sachs, E., Brendler, S.J. and Davis, G.D. (1948a) Stimulation and regional ablation of orbital surface of frontal lobe. In *The Frontal Lobes. Res. Publ. Assoc. Res. Nerv. Ment. Dis.*, 8: 405 – 420.
Livingston, R.B., Chapman, W.P. and Livingston, K.E. (1948b) Stimulation of orbital surface of man prior to frontal lobotomy. In *The Frontal Lobes. Res. Publ. Assoc. Res. Nerv. Ment. Dis.*, 8: 421 – 432.
Livingstone, M. and Hubel, D. (1988) Segregation of form, color, movement, and depth: anatomy, physiology, and perception. *Science*, 240: 740 – 749.
Lofving, B. (1961) Cardiovascular adjustments induced from the rostral cingulate gyrus. *Acta Physiol. Scand.*, 53, Suppl. 184: 1 – 82.
Luria, A.R. (1980) *Higher Cortical Functions in Man*, Basic Books, New York.
Luria, A.R. and Homskaya, E.D. (1964) Disturbances in the regulative role of speech with frontal lobe lesions. In J.M. Warren and K. Akert (Eds.), *The Frontal Granular Cortex and Behavior*, McGraw-Hill, New York, pp. 353 – 371.

MacLean, P.D. (1949) Psychosomatic disease and the "visceral brain". Recent developments bearing on the Papez theory of emotion. *Psychosom. Med.*, 11: 338 – 354.
MacLean, P.D. (1952) Some psychiatric implications of physiological studies on frontotemporal portion of limbic system (visceral brain). *Electroenceph. Clin. Neurophysiol.*, 4: 407 – 418.
MacLean, P.D. (1958) Contrasting functions of limbic and neocortical systems of the brain and their relevance to psychophysiological aspects of medicine. *Am. J. Med.*, 25: 611 – 626.
MacLean, P.D. and Newman, J.D. (1988) Role of midline frontolimbic cortex in production of the isolation call of squirrel monkeys. *Brain Res.*, 450: 111 – 123.
Malmo, R.B. (1942) Interference factors in delayed response in monkeys after removal of frontal lobes. *J. Neurophysiol.*, 5: 295 – 308.
Messimy, R. (1939) Les effects, chez le singe, de l'ablation des lobes préfrontaux. *Rev. Neurol.*, 71: 1 – 37.
Mesulam, M.-M., Van Hoesen, G., Pandya, D.N. and Geschwind, N. (1977) Limbic and sensory connections of inferior parietal lobule (area PG) in the rhesus monkey: A study with a new method for horseradish peroxidase histochemistry. *Brain Res.*, 136: 393 – 414.
Mettler, F.A. (1944) Physiological effect of bilateral simultaneous frontal lesions in the primate. *J. Comp. Neurol.*, 81: 105 – 136.
Miller, N.E. (1983) Biofeedback and visceral learning. *Annu. Rev. Psychol.*, 29: 373 – 404.
Milner, B.R. (1964) Some effects of frontal lobectomy in man. In J.M. Warren and K. Akert (Eds.), *The Frontal Granular Cortex and Behavior*, McGraw-Hill, New York, pp. 313 – 334.
Mishkin, M. (1964) Perseveration of central sets after frontal lesions in monkeys. In J.M. Warren and K. Akert (Eds.), *The Frontal Granular Cortex and Behavior*, McGraw-Hill, New York, pp. 219 – 241.
Mufson, E.J. and Mesulam, M.M. (1984) Thalamic connections of the insula in the rhesus monkey and comments on the paralimbic connectivity of the medial pulvinar nucleus. *J. Comp. Neurol.*, 227: 109 – 120.
Nauta, W.J.H. (1971) The problem of the frontal lobe: a reinterpretation. *J. Psychiat. Res.*, 8: 167 – 187.
Neafsey, E.J., Bold, E.L., Haas, G., Hurley-Gius, K.M., Quirk, G., Sievert, C.F. and Terreberry, R.R. (1986a) The organization of the rat motor cortex: a microstimulation mapping study. *Brain Res. Rev.*, 11: 77 – 96.
Neafsey, E.J., Hurley-Gius, K.M. and Arvanitis, D. (1986b) The topographical organization of neurons in the rat medial frontal, insular, and olfactory cortex projecting to the solitary nucleus, olfactory bulb, periaqueductal gray and superior colliculus. *Brain Res.*, 377: 261 – 270.
Niimi, K., Niimi, M. and Okada, Y. (1978) Thalamic afferents to the limbic cortex in the cat studied with the method of

retrograde axonal transport of horseradish peroxidase. *Brain Res.,* 145: 225–238.
Onoda, N., Imamura, K., Obata, E. and Iino, M. (1984) Response selectivity of neocortical neurons to specific odors in the rabbit. *J. Neurophysiol.,* 52: 638–652.
Papez, J.W. (1937) A proposed mechanism of emotion. *Arch. Neurol. Psychiat.,* 38: 725–743.
Penfield, W. and Rasmussen, T.B. (1950) *The Cerebral Cortex of Man,* Macmillan, New York.
Pool, J.L. and Ransohoff, J. (1949) Autonomic effects on stimulating rostral portion of cingulate gyri in man. *J. Neurophysiol.,* 12: 385–392.
Porter, R.W. (1954) The central nervous system and stress-induced eosinopenia. *Recent Progr. Hormone Res.,* 10: 1–27.
Powell, D.A., Buchanan, S. and Hernandez, L. (1985) Electrical stimulation of insular cortex elicits cardiac inhibition but insular lesions do not abolish conditioned bradycardia in rabbits. *Behav. Brain Res.,* 17: 125–144.
Pribram, K.H. (1950) Some physical and pharmacological factors affecting delayed response performance of baboons following frontal lobotomy. *J. Neurophysiol.,* 13: 373–382.
Pribram, K.H. (1973) The primate frontal cortex — executive of the brain. In K.H. Pribram and A.R. Luria (Eds.), *Psychophysiology of the Frontal Lobes,* Academic Press, New York, pp. 293–314.
Pribram, K.H., Ahumada, A., Hartog, J. and Ross, L. (1964) A progress report on the neurological processes disturbed by frontal lesions in primates. In J.M. Warren and K. Akert (Eds.), *The Frontal Granular Cortex and Behavior,* McGraw-Hill, New York, pp. 28–55.
Price, J.L. and Slotnick, B.M. (1983) Dual olfactory representation in the rat thalamus: an anatomical and electrophysiological study. *J. Comp. Neurol.,* 215: 63–77.
Reep, R.L., Corwin, J.V., Hashimoto, A. and Watson, R.T. (1987) Efferent connections of the rostral portion of medial agranular cortex in rats. *Brain Res. Bull.,* 19: 203–221.
Rinkel, M., Greenblatt, M., Coon, G. and Solomon, H. (1947) Relation of the frontal lobe to the autonomic nervous system in man. *Arch. Neurol. Psychiat.,* 58: 570–581.
Rinkel, M., Solomon, H.C., Rosen, D. and Levine, J. (1950) Lobotomy and urinary bladder. In M. Greenblatt, R. Arnot and H.C. Solomon (Eds.), *Studies in Lobotomy,* Grune and Stratton, New York, pp. 370–379.
Rodman, H.R., Gross, C.G. and Albright, T.D. (1989) Afferent basis of visual response properties in area MT of the macaque. I. Effects of striate cortex removal. *J. Neurosci.,* 9: 2033–2050.
Room, P., Russchen, F.T., Groenewegen, H.J. and Lohman, A.H.M. (1985) Efferent connections of the prelimbic (area 32) and the infralimbic (area 25) cortices: an anterograde tracing study in the cat. *J. Comp. Neurol.,* 242: 40–55.
Rostad, H. (1973) Colonic motility in the cat. V. Influence of telencephalic stimulation and the peripheral pathways mediating the effects. *Acta Physiol. Scand.,* 89: 169–181.
Routtenberg, A. (1971) Forebrain pathways of reward in Rattus norvegicus. *J. Comp. Physiol. Psychol.,* 75: 269–276.
Ruch, T.C. (1966) The homotypical cortex — The "association areas". In T.C. Ruch, H.D. Patton, J.W. Woodbury and A.L. Towe (Eds.), *Neurophysiology,* W.B. Saunders, Philadelphia, PA, pp. 465–479.
Ruch, T.C. and Shenkin, H.A. (1943) The relation of area 13 on orbital surface of frontal lobes to hyperactivity and hyperphagia in monkeys. *J. Neurophysiol.,* 6: 349–360.
Ruch, T.C., Patton, H.D., Woodbury, J.W. and Towe, A.L. (1966) *Neurophysiology,* W.B. Saunders, Philadelphia, PA.
Ruggiero, D.A., Mraovitch, S., Granata, A.R., Anwar, M. and Reis, D.J. (1987) Role of insular cortex in cardiovascular function. *J. Comp. Neurol.,* 257: 189–207.
Ruit, K.G. and Neafsey, E.J. (1988) Cardiovascular and respiratory responses to electrical and chemical stimulation of the hippocampus in anesthetized and awake rats. *Brain Res.,* 457: 310–321.
Saper, C.B. (1982) Convergence of autonomic and limbic connections in the insular cortex of the rat. *J. Comp. Neurol.,* 210: 163–173.
Sapolsky, R.M. and Eichenbaum, H. (1980) Thalamocortical mechanisms in odor-guided behavior. II. Effects of lesions of the mediodorsal thalamic nucleus and frontal cortex on odor preferences and sexual behavior in the hamster. *Brain Behav. Evol.,* 17: 276–290.
Schafer, E.C. (1900) *Text-book of Physiology,* Y.J. Pentland, London.
Schlag, J. and Schlag-Rey, M. (1987) Evidence for a supplementary eye field. *J. Neurophysiol.,* 57: 179–200.
Sechenov, I. (1863) Reflex action of the brain. *Med. Vetnik. St. Petersb.,* 3: 461–493.
Setekleiv, J. (1964) Uterine motility of the estrogenized rabbit. V. Responses to brain stimulation. *Acta Physiol. Scand.,* 62: 313–322.
Shallice, T. (1982) Specific impairments of planning. *Phil. Trans. R. Soc. Lond. B,* 298: 199–209.
Shipley, M.T. (1982) Insular cortex projection to the nucleus of the solitary tract and brainstem visceromotor regions in the mouse. *Brain Res. Bull.,* 8: 138–148.
Shipley, M.T. and Geinisman, Y. (1984) Anatomical evidence for convergence of olfactory, gustatory, and visceral afferent pathways in mouse cerebral cortex. *Brain Res. Bull.,* 12: 221–226.
Showers, M.J.C. and Crosby, E.C. (1958) Somatic and visceral response from the cingulate gyrus. *Neurology,* 8: 561–565.
Siegfried, J. (1965) Topographie de projections corticales du nerf vague chez le chat. *Helv. Physiol. Acta,* 19: 269–278.
Sinnamon, H.M. and Charman, C.S. (1988) Unilateral and bilateral lesions of the anteromedial cortex increase perseverative head movements of the rat. *Behav. Brain Res.,* 27: 145–160.
Sinnamon, H.M. and Galer, B.S. (1984) Head movements

elicited by electrical stimulation of the anteromedial cortex of the rat. *Physiol. Behav.*, 33: 185–190.

Slotnick, B.M. (1967) Disturbances of material behavior in the rat following lesions of the cingulate cortex. *Behavior*, 29: 204–236.

Smith, O.A. and DeVito, J.L. (1984) Central neural integration for the control of autonomic responses associated with emotion. *Annu. Rev. Neurosci.*, 7: 43–65.

Smith, W.K. (1938) The representation of respiratory movements in the cerebral cortex. *J. Neurophysiol.*, 1: 55–68.

Smith, W.K. (1945) The functional significance of the rostral cingular cortex as revealed by its responses to electrical stimulation. *J. Neurophysiol.*, 8: 241–255.

Spencer, W.G. (1894) The effect produced upon respiration by faradic excitation of the cerebrum in the monkey, dog, cat and rabbit. *Phil. Trans.*, 185b: 609–657.

Stamm, J.S. (1954) Control of hoarding activity in rats by the median cerebral cortex. *J. Comp. Physiol. Psychol.*, 47: 21–27.

Stamm, J.S. (1955) The function of the median cerebral cortex in maternal behavior of rats. *J. Comp. Physiol. Psychol.*, 48: 347–356.

Stamm, J.S. (1987) The riddle of the monkey's delayed-response deficit has been solved. In E. Perecman (Ed.), *The Frontal Lobes Revisited,* IRBN Press, New York, pp. 73–89.

Stamm, J.S. and Rosen, S.C. (1969) Electrical stimulation and steady potential shifts in prefrontal cortex during delayed response performance by monkeys. *Acta Biol. Exp.*, 29: 385–399.

Strom, G. and Uvnas, B. (1950) Motor responses of gastrointestinal tract and bladder to topical stimulation of the frontal lobe, basal ganglia and hypothalamus in the cat. *Acta Physiol. Scand.*, 21: 90–104.

Sutton, D., Larson, C. and Lindeman, R.C. (1974) Neocortical and limbic lesion effects on primate phonation. *Brain Res.*, 71: 61–75.

Swanson, L.W. (1981) A direct projection from Ammon's horn to prefrontal cortex in the rat. *Brain Res.*, 217: 150–154.

Swanson, L.W. (1983) The hippocampus and the concept of the limbic system. In W. Seifert (Ed.), *Neurobiology of the Hippocampus,* Academic Press, New York, pp. 3–19.

Szilagyi, J.E., Taylor, A.A. and Skinner, J.E. (1987) Cryoblockade of the ventromedial frontal cortex reverses hypertension in the rat. *Hypertension*, 9: 576–581.

Takagi, S.F. (1986) Studies on the olfactory nervous system of the old world monkey. *Progr. Neurobiol.*, 27: 195–250.

Terreberry, R.R. and Neafsey, E.J. (1983) Rat medial frontal cortex: a visceral motor region with a direct projection to the solitary nucleus. *Brain Res.*, 278: 245–249.

Terreberry, R.R. and Neafsey, E.J. (1984) The effects of medial prefrontal cortex stimulation on heart rate in the awake rat. *Neurosci. Abstr.*, 10: 614.

Terreberry, R.R. and Neafsey, E.J. (1987) The rat medial frontal cortex projects directly to autonomic regions of the brainstem. *Brain Res. Bull.*, 19: 639–649.

Teuber, H.-L. (1964) The riddle of frontal lobe function in man. In J.M. Warren and K. Akert (Eds.), *The Frontal Granular Cortex and Behavior,* McGraw-Hill, New York, pp. 410–477.

Tharp, B.R. (1972) Orbital frontal seizures. An unique electroencephalographic and clinical syndrome. *Epilepsia*, 13: 627–642.

Thierry, A.M., Tassin, J.P., Blanc, G. and Glowinski, J. (1976) Selective activation of the mesocortical DA system by stress. *Nature*, 263: 242–244.

Thomas, D.A., Takahashi, L.K. and Barfield, R.J. (1983) Analysis of ultrasonic vocalizations emitted by intruders during aggressive encounters among rats (*Rattus norvegicus*). *J. Comp. Psychol.*, 97: 201–206.

Thorpe, S.J., Rolls, E.T. and Maddison, S. (1983) The orbitofrontal cortex: neuronal activity in the behaving monkey. *Exp. Brain Res.*, 49: 93–115.

Towe, A.L. (1973) Motor cortex and pyramidal system. In J.D. Maser (Ed.), *Efferent Organization and the Integration of Behavior,* Academic Press, New York, pp. 67–97.

Tranel, D., Damasio, A.R. and Damasio, H. (1988) Impaired autonomic responses to emotional and social stimuli in patients with bilateral orbital damage and acquired sociopathy. *Neurosci. Abstr.*, 14: 1288.

Tyler, H.R. (1969) Disorders of visual scanning with frontal lobe lesions. In S. Locke (Ed.), *Modern Neurology; Papers in Tribute to Derek Denny-Brown,* Little Brown, Boston, MA, pp. 381–393.

Valenstein, E.S. (1986) *Great and Desperate Cures: The Rise and Decline of Psychosurgery and Other Radical Treatments for Mental Illness,* Basic Books, New York.

Van der Kooy, D., McGinty, J.F., Koda, L.Y., Gerfen, C.R. and Bloom, F.E. (1982) Visceral cortex: direct connection from prefrontal cortex to the solitary nucleus in the rat. *Neurosci. Lett.*, 33: 123–127.

Van der Kooy, D., Koda, L.Y., McGinty, J.F., Gerfen, C.R. and Bloom, F.E. (1984) The organization of projections from the cortex, amygdala, and hypothalamus to the nucleus of the solitary tract in the rat. *J. Comp. Neurol.*, 224: 1–24.

van der Steen, J., Russel, I.S. and James, G.O. (1986) Effects of unilateral frontal eye-field lesions on eye-head coordination in monkey. *J. Neurophysiol.*, 55: 696–714.

Vanderwolf, C.H., Kelly, M.E., Kraemer, P. and Streather, A. (1988) Are emotion and motivation localized in the limbic system and nucleus accumbens? *Behav. Brain Res.*, 27: 45–58.

Van Eden, C.G. and Uylings, H.B.M. (1985) Cytoarchitectonic development of the prefrontal cortex in the rat. *J. Comp. Neurol.*, 241: 253–267.

Van Essen, D.C. and Maunsell, J.H.R. (1982) Hierarchical organization and functional streams in the visual cortex.

Trends Neurosci., 6: 370–375.

Vargo, J.M., Corwin, J.V., King, V. and Reep, R.L. (1987) Asymmetries in neglect and pattern of recovery following left vs. right medial precentral prefrontal lesions in rats. *Neurosci. Abstr.*, 13: 45.

Verberne, A.J.M., Lewis, S.J., Worland, P.J., Beart, P.M., Jarrott, B., Christie, M.J. and Louis, W.J. (1987) Medial prefrontal cortical lesions modulate baroreflex sensitivity in the rat. *Brain Res.*, 426: 243–249.

Vogt, B.A., Pandya, D.N. and Rosene, D.L. (1987) Cingulate cortex of the rhesus monkey. I. Cytoarchitecture and thalamic afferents. *J. Comp. Neurol.*, 262: 256–270.

Von Bechterew, W. (1908–1911) *Die Funktionen der Nervenzentra*, 3 Vols., Gustav Fisher, Jena.

Von Economo, C. and Koskinas, G. (1925) *Die Cytoarchitektonik der Hirnrinde des erwachsenen Menschen,* Verlag von Julius Springer, Wien and Berlin.

Von Euler, U.S. and Folkow, B. (1958) The effect of stimulation of autonomic areas in the cerebral cortex upon adrenaline and noradrenaline secretion from the adrenal gland in the cat. *Acta Physiol. Scand.*, 42: 313–320.

Wade, M. (1947) The effect of sedatives upon delayed responses in monkeys following removal of the prefrontal lobes. *J. Neurophysiol.*, 10: 57–61.

Wall, P.D. and Davis, G.D. (1951) Three cerebral cortical systems affecting autonomic function. *J. Neurophysiol.*, 14: 507–517.

Ward, A.A., Jr. (1948) The cingular gyrus: area 24. *J. Neurophysiol.*, 11: 13–23.

Warren, J.M. and Akert, K. (Eds.) (1964) *The Frontal Granular Cortex and Behavior,* McGraw-Hill, New York.

Whishaw, I.Q. and Kolb, B. (1989) Tongue protrusion mediated by spared anterior ventrolateral neocortex in neonatally decorticate rats: behavioral support for the neurogenetic hypothesis. *Behav. Brain Res.*, 32: 101–113.

Willett, C.J., Gwyn, D.G., Rutherford, J.G. and Leslie, R.A. (1986) Cortical projections to the nucleus of the tractus solitarius: an HRP study in the cat. *Brain Res. Bull.*, 16: 497–505.

Winkler, C. (1899) Attention and respiration. *Proc. Acad. Sci. Amst.*, 1: 121–138.

Wolman, B.B. (1988) *Psychosomatic Disorders,* Plenum, New York.

Yamamoto, T., Matsuo, R. and Kawamura, Y. (1980) Localization of cortical gustatory area in rats and its role in taste discrimination. *J. Neurophysiol.*, 44: 440–455.

Zeki, S. and Shipp, S. (1988) The functional logic of cortical connections. *Nature*, 335: 311–317.

Discussion

A. Kalsbeek: In your foot shock – or conditioned emotional response trials, did you try to lesion or manipulate the DAergic input to the PFC?

E.J. Neafsey: We have not done those experiments, but we also would like to know the effect of 6-OHDA lesions of the medial frontal cortex on these cardiovascular responses.

A.Y. Deutch: Our data indicate that prelimbic efferents avoid classical autonomic areas but infralimbic innervate these areas. Can the effects of medial cortical lesions occur through pathways indirect (suprarhinal cortex, lateral hypothalamus) but not direct?

E.J. Neafsey: My initial hope was that the direct pathway from the infralimbic cortex would be shown to be responsible for autonomic effects from this cortex. However, a recent study by Hardy and Holmes (1988) found that local lidocaine anesthesia along the course of this direct projection below the level of the midbrain did not block hypotension produced by MFC stimulation. In contrast, local anesthesia at the level of the hypothalamus did block these effects, suggesting that such indirect pathways may be important.

D.A. Powell: Were bradycardia and depressor responses ever elicited from freely moving rats?

E.J. Neafsey: In my lab, Dr. Robert Terreberry did elicit bradycardia from freely moving rats; we did not look at blood pressure in these experiments. (In one or two animals we also saw a tachycardia – the direction of the response seemed related to the pre-stimulus heart rate level.)

C. Cavada: It showed from your slides that the region of the solitary nucleus receives prefrontal projection from the most caudal part of the ventro-medial, prefrontal cortex. Later on, you have correlated the autonomic effects of hippocampal stimulation with the presence of hippocampo-prefrontal projections. Do you know whether the prefrontal site projecting to the solitary nucleus is a specific target of hippocampal projections?

E.J. Neafsey: I am glad you asked that question, and the answer is yes. In experiments in my laboratory recently completed by Dr. Ken Ruit, he found that at the light microscopic level anterogradely labeled hippocampal projections to the infralimbic cortex overlap the retrogradely labeled population of layer V cells which project to the solitary nucleus. At the electron microscopic level, he found many hippocampal terminals making synaptic contacts in layer V of the MFC (in one or two instances there appeared to be a direct synaptic contact between an anterogradely labeled hippocampal terminal and a retrogradely labeled soma of an infralimbic neuron projecting to the solitary nucleus). Using single unit recording techniques, he demonstrated that infralimbic neurons antidromically activated from the solitary nucleus could be orthodromically activated by hippocampal stimulation.

C.G. van Eden: Injections with retrograde tracers in spinal cord (HRD and FB) label cells in the medial PFC in exactly the same area as you showed the NTS projecting cells? Is this a second pathway for eliciting autonomic responses?

E.J. Neafsey: I do not know if it is the same cells or a different population. A few years ago at the Neuroscience meeting in the United States, Karen Hurley, David Cechetto and Cliff Saper also reported that infralimbic neurons in this same region projected to the thoracic cord, appearing to terminate in the intermediolateral cell column. No double labeling study has yet been carried out to determine if these are the same cells that project to the NTS or not.

References

Hardy, S.G.P. and Holmes, D.E. (1988) Prefrontal stimulus-produced hypotension in rat. *Exp. Brain Res.*, 73: 249 – 255.

Hurley-Gius, K.M., Cechetto, D.F. and Saper, C.B. (1986) Spinal connections of the infralimbic autonomic cortex. *Neurosci. Abstr.*, 12: 538.

SECTION II

Development and Plasticity in Prefrontal Cortex

CHAPTER 8

The development of the rat prefrontal cortex

Its size and development of connections with thalamus, spinal cord and other cortical areas

C.G. van Eden, J.M. Kros* and H.B.M. Uylings

Netherlands Institute for Brain Research, Amsterdam, The Netherlands

Introduction

The development of the brain has always attracted much interest, not only because of the intriguing processes which lead to a persons mental development, but also because of the intellectual challenge of understanding the biological mechanisms behind the development of such a complex and crucial organ as the brain. Many of these questions concern the generation, migration and differentiation of the constituent elements (the neurons and glial cells), and how these cells come to establish the "appropriate" contacts. Another kind of question concerns the functional development. When does a certain structure begin to operate, and when does its functional maturation stop (e.g. Goldman, 1971, 1972; Diamond and Goldman-Rakic, 1989). The following two chapters deal with the development of the cortical layers, the development of cortical neurons and ingrowth of afferent fibers in the

---

* Present address: Department of Pathology, Erasmus University, Dr. Molewaterplein 40, 3015 GD Rotterdam, The Netherlands.

Correspondence: Dr. C.G. van Eden, Netherlands Institute for Brain Research, Meibergdreef 33, 1105 AZ Amsterdam, The Netherlands.

human fetus (Kostović, and Mrzljak et al., this volume). The scope of the present paper is to give a description of some of these processes in the rat. On the basis of this description we will then focus upon the question of whether or not the prefrontal cortex has a later or prolonged development in comparison with other cortical areas. For a general review of the pre- and postnatal development of the rat cerebral cortex, the reader is referred to Uylings et al. (1990).

Development of the prefrontal cortex

The idea that the prefrontal cortical areas develop later than other cortical areas goes back to the work of the German neuroscientist Paul Flechsig. In 1898, Flechsig (see Flechsig, 1901) published his studies on the development of myelinization of the human cerebral cortex, which demonstrated that regional differences in the developmental myelinization fall into 3 groups. First of all, those cortical areas which attain an adult-like myelinization pattern already prior to birth all belong to the sensoric/motoric areas (areas 1–10 in Fig. 1). The second group, containing areas 11–30, also starts to develop before birth but continues myelinization into the first months of postnatal life. This group comprises, among others, the premotor areas and higher order sensory areas. The third group is formed by areas which do not show any signs of myelinization until 1 month after birth. All of the areas in this group belong to the so-called "association" areas. Fig. 1 summarizes these results and demonstrates that the prefrontal areas are among the latest of all to develop with only the superior temporal area being later. Furthermore, Yakovlev and Lecours (1967) claim that myelinization of the prefrontal areas has not stabilized until the fourth decade of human life.

Another indication for a late and/or prolonged development of the prefrontal cortex comes from more recent studies investigating the age-dependent effects of prefrontal lesions. Lesion studies have demonstrated that early lesions of the prefrontal cortex have little effect on those functions which in adulthood depend upon the integrity of the prefrontal cortex. In rats and monkeys the ability to functionally recover from prefrontal ablations persists until relatively late into postnatal life. In the monkey, for example, these investigations suggest that orbital prefrontal cortex may become funtionally mature by 1 year of age, whereas the dorsolateral cortex does not approach functional maturity until well into the second year of life (Goldman, 1971, 1972). In the rat, Kolb and Nonneman (1976) demonstrated that orbital prefrontal cortex lesions do not produce signs of adipsia or aphagia when inflicted before 60 days of age. Later studies showed that also the medial prefrontal cortex has a prolonged potential for recovery (Kolb and Nonneman, 1978); Nonneman and Corwin, 1981; Kolb, 1984).

This prolonged ability to recover functionally has been taken as being indicative of a relatively late maturation of these cortical areas, since the brain is considered to retain its plasticity as long as its connections have not yet become stabilized. In the immature brain, therefore, the function of the lesioned prefrontal cortex could be taken over by other parts of the (prefrontal) system (see also Kolb, this volume). Lesions performed at a later age have more serious effects because the prefrontal system would have lost its flexibility in the course of maturation (e.g. Goldman, 1972; Goldman-Rakic and Galkin, 1978; Finger and Amli, 1985).

All these examples, the old myelinization study as well as more recent neonatal lesion studies, point to a late or prolonged development of the prefrontal areas in rodents as well as in primates. This chapter will describe the structural development of these prefrontal areas in the rat in comparison with other cortical areas.

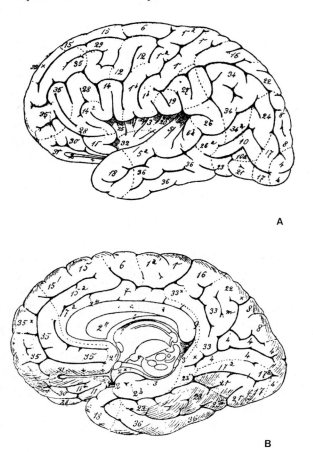

Fig. 1. Cortical areas on the lateral (a) and medial (b) aspects on the cerebral hemisphere. Numbering in sequence of myelogenesis, after Flechsig (1901).

Cytoarchitectonic development

The postnatal development of the rodent prefrontal cortex is characterized by the transformation of the cortical plate into the laminated adult cortex (see Fig. 2) (Van Eden and Uylings, 1985a). At birth, the cortical plate contains most of the neurons that will contitute the adult cortex but, until postnatal day (P) 2, some of the last generated neurons are still migrating towards the surface of the cortical plate (Raedler and Sievers, 1975; Uylings et al., 1990). On the first postnatal day (P1) several developing layers can be distinguished in the cerebral wall (Fig. 2). The outermost layer (layer I) is the former marginal zone, while most of the cerebral wall is formed by the cortical plate. Within the latter a certain degree of lamination can already be discerned. The most superficial part is the youngest. This dense part of the cortical plate contains immature neuronal elements of the future layers II – V. In the more mature, and less cell dense, lower part of the cortical plate, the developing layer VI and part of V can be seen. Underneath the cortical plate the future white matter can be found with in it the remnant neurons of the subplate zone as we know from human and cat studies (Kostović and Molliver, 1974; Kostović and Rakic, 1980; Valverde and Facal-Valverde, 1987; Chun et al., 1987; for rat see also Uylings et al., 1990).

By the end of the first postnatal week (P6 – P7), the upper dense part of the cortical plate is still present but is greatly reduced in width. Underneath the dense part of the cortical plate the CP has developed layers VI and V and, in the dorsolateral cortex, the granular layer IV can be recognized. In contrast to the human cortical development where a granular layer is initially formed even in areas that in adulthood are agranular, in the agranular rat prefrontal cortex no layer IV is ever formed during development. By P7 all layers II – VIb can recognized in the frontal pole of the cerebral hemisphere, although layers II and III are still very immature and tightly packed. Furthermore, by day 6, development, of the frontal cortex has progressed so far that using slightly adapted criteria the prefrontal areas can be distinguished (Van Eden and Uylings, 1985a).

At P10 the cortical plate has completely disappeared and layer II and III can clearly be recognized as two distinct layers. From day 10 through day 18 the cytoarchitectonic features of the cortical

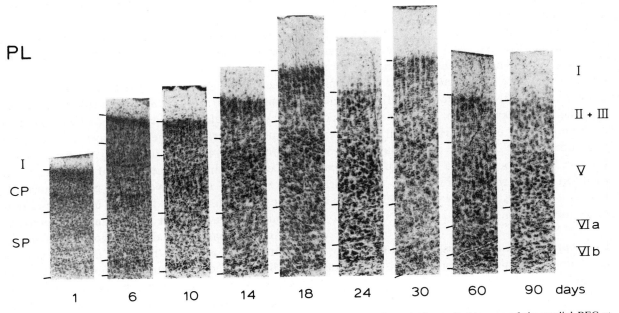

Fig. 2. Cyoarchitectonic development of the prefrontal cortex. Cross-sections through the prelimbic area of the medial PFC at various stages of postnatal development CP = cortical plate; SP = subplate layer; Roman numbers indicate cortical layers.

layers mature into the adult agranular pattern. Around day 18 the cortical laminae are most clearly delineated. Although between day 18 and 30 the laminar pattern faints somewhat, there is little change in the overall cytoarchitectonic pattern during this period. However, important changes do occur during this period (see "volumetry" below). Later, from day 30 onwards until day 90 the acquired characteristics blur further and the individual variation in the cytoarchitecture increases. These developmental changes in the appearance of the laminae are caused by a further dispersal of the cells, presumably as a result of differentiation and growth of the dendrites of the neurons present.

If we compare this account of the development of the prefrontal cortex with the descriptions of cytoarchitectonic development in other regions of the cortex, we must conclude that there is much similarity in the rate of development of cortical lamination in different cortical areas (e.g. Eayrs and Goodhead, 1959; Raedler and Sievers, 1975; Rice and van der Loos, 1977). These studies described the postnatal cytoarchitectonic development of the visual and somatosensory cortex in the rat. According to their observations the CP has differentiated into layers II and III at P7 and all cortical layers have attained their adult features by P14. Further changes, after P14, are only quantitatively dependent on a further decrease in cell density through continued increase of the neuropil. Leuba et al. (1976) and Heumann et al. (1976) give a similar rate of development for the maturation of a whole range of cortical areas. In all these areas (viz. areas 2, 3, 4, 10 an 17 and 18) layers II – VI become formed between P5 and P10. It is clear, therefore, that the differentiation of cortical lamination occurs not later in the PFC than in other (neo)cortical areas.

Development of connections between the mediodorsal thalamic nucleus and the prefrontal cortex

The development of what is considered the most important afferent system for the prefrontal cortex (Leonard, 1969; Krettek and Price, 1977), the mediodorsal nucleus of the thalamus (MDT) was studied in the postnatal period by means of an anterograde tracer (Van Eden, 1986). The development of other important afferent prefrontal fiber systems, such as the dopaminergic projection from the ventral tegmentum and the basolateral nucleus of the amygdala will be presented elsewhere (Kalsbeek et al., this volume; Verwer and van Vulpen, 1989). From other studies it is known that the specific thalamic fibers from, e.g., the lateral geniculate nucleus reach the visual cortex around ED18 (Lund and Mustari, 1977). They remain in the subplate layer (see also Kostović, this volume) until the day of birth (ED22), at which point they enter the cortical plate. After entering the cortex these fibers stay within the basal layers of the cortical plate until postnatal day 3. Only on day 4, when layer IV neurons have evolved from the dense part of the cortical plate, do they finally make contact with their adult target cells (Lund and Mustari, 1977; Wise and Jones, 1978). The ingrowth of the mediodorsal fibers into the postnatal prefrontal cortex differs only in details from this scheme (see Fig. 3). In contrast to the thalamocortical fibers in the granular sensory cortices, not all of the mediodorsal fibers in the prefrontal area wait in either the subplate layer or the basal cortical laminae until day 3. The majority of the MDT fibers waits but already by the day of birth a substantial number of the MDT fibers have penetrated the superficial, dense part of the cortical plate and some are observed in the future layer 1. During the following 3 – 4 days an increasing number of fibers accumulates in the upper cortical plate but it will take until P7 before layer II can be distinguished. The ingrowth of fibers from the mediodorsal thalamic nucleus, therefore, is not later in the prefrontal cortex than the thalamocortical fibers in the primary sensory areas. They are, in fact, relatively earlier in penetrating the immature cortical plate (dense part) and could thereby exert a potentially greater influence than do other thalamocortical projections on the development of the cortex. By P7, when the cortical plate is no longer distinguished and layers II and III have become visible as separate layers, the distribution of MDT fibers is essentially the same as in the adult. These fibers terminate densely in layers I, III and upper part of layer V of the medial and orbital prefrontal cortex. This scheme of ingrowth of thalamic fibers into the prefrontal area is comparable with the ingrowth into the visual or the somatosensory cortex, where

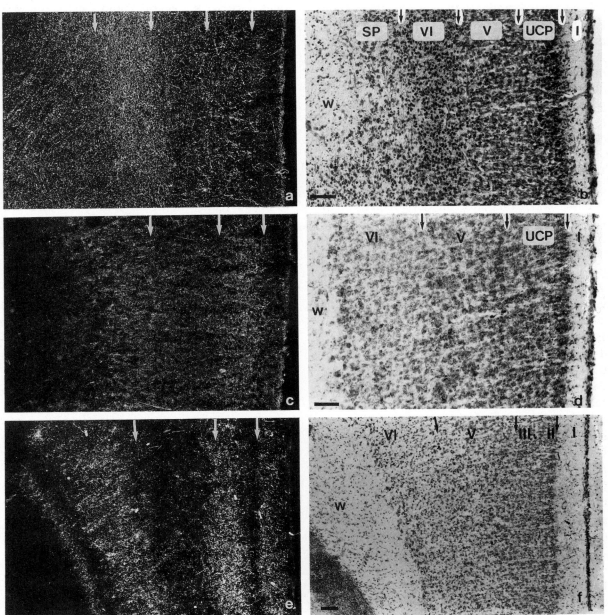

Fig. 3. Ingrowth of mediodorsal thalamic fibers into the medial prefrontal cortex. The thalamic fibers are labeled by iontophoretic WGA-HRP injection in the mediodorsal nucleus at P0 (a, b), P4 (c, d) and P7 (e, f); Dark-field illumination (a, c, e) and counterstained section (b, d, f). SP = subplate layer; UCP = upper part of the cortical plate; w = white matter; Roman numbers indicate cortical layers.

these fibers attain an adult-like termination pattern between P6 and P7 (Lund and Mustari, 1977; Wise and Jones, 1978).

The reciprocal projection from the prefrontal cortex to the mediodorsal thalamus is largely formed during the second week of postnatal life. The first retrogradely labeled cells are observed after injections of WGA-HRP in the mediodorsal thalamus on P4. Between P4 and P10 the number of pyramidal cells in layer VI increases and, by P10, retrogradely labeled cells begin to be observed in the hemisphere opposite to the injection site (Van Eden, 1986).

Volumetric development

Another way to monitor the development of the PFC is to measure its increase in size during the first postnatal months. Volumetry has advantages over measurement of cortical thickness because it is a parameter which is rather insensitive to minor deviations of the sectioning plane. Furthermore, it takes into account any developmentally occuring changes in the shape of the object (Uylings et al., 1984). The volumetric analysis of prefrontal cortex development reveals a different growth pattern than for cytoarchitectonic development. Since we were able to distinguish between prefrontal subareas from day 6, the areas could reproducibly be delineated (Van Eden and Uylings, 1985a), and, therefore, the volumes could be determined. The for shrinkage corrected, volumetric growth of the prefrontal areas (Fig. 4) shows a distinctly different pattern than that of the brain as a whole (Van Eden and Uylings, 1985b). All prefrontal areas irrespective of whether they are located in the medial or the lateral parts, show a period of transient overgrowth. At first, during the first weeks of postnatal life, the volume increases. For the medial PFC this increase lasts until P24, whereas in the orbital PFC growth continues until at least P30. After this period the volume is higher than in the adult rat. In the medial PFC the volume overshoot is approximately 30% higher than the adult value, whereas in the orbital PFC this is approximately 80%.

Comparing the individual subareas within the medial and orbital PFC, two kinds of growth patterns can be distinguished. The shoulder area of the hemisphere (i.e., the PFC subareas which are innervated by the lateral subnucleus of the MD) shows a different volumetric growth pattern than does the rest of the prefrontal areas. These PFC subareas, AC_d and PrC_m, differ from the PL and the orbital PFC subareas with respect to both, the maximum volume, and the rate of volumetric decline thereafter. In the PFC subareas of the "shoulder region", this decline is relatively fast and the adult volume is attained already at P30, whereas the other PFC subareas continue to decline until at least day 90.

Since day 90 is usually regarded as the beginning of the adult stage in the rat, a developmental process which continues until this age seems to point to a rather protracted maturation of at least some of the prefrontal areas. There are no volumetric data available on the development of other neocortical areas in the rat, although a transient volumetric overgrowth has been demonstrated to also occur in different brain regions in *Tupaia belangeri* (viz., striatum, visual cortex and subcor-

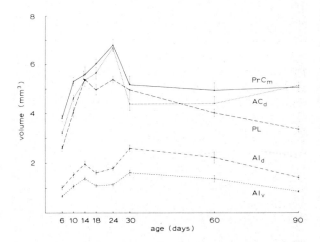

Fig. 4. Volumetric development of the medial and orbital prefrontal cortex. Upper 3 lines: growth curves of medial cortical areas, viz. prelimbic area (PL), dorsal anterior cingular area (AC_d) and medial precentral area (PrC_m). Lower lines: growth curves of the orbital prefrontal areas, viz. the ventral and dorsal agranular insular areas (AI_v and AI_d, respectively).

tical nuclei of the visual system (Zilles, 1978)) and area striata of the marmoset monkey (Fritschy and Garey, 1986). Therefore, a direct comparison with other cortical areas cannot be made in the rat. At the moment it is not clear which elements of the cortex cause the transient volumetric overgrowth. Counting of cells in the shoulder region of the PFC showed no significant changes in the neuron number during the period of overgrowth (Zijlstra er al., 1988). The volume of the cortex is mainly determined by the dendrites and the somata of the constituent neurons. In the rat visual cortex these elements also demonstrate a transient overgrowth. The neuronal somata reach their maximal volume between postnatal days 18 and 22 (Werner et al., 1981), and also the dendrites of pyramidal and non-pyramidal cells have their maximal extension around P13 (Parnavelas and Uylings, 1980; Uylings et al., 1990). Compared to these parameters the volumetric overgrowth in the prefrontal cortex seems to indicate a relatively late and prolonged maturation.

Since the transient volumetric overgrowth probably coincides with a transient extension of dendritic trees and neuronal somata, and not to a reduction in cell number, the possibility exists that reduction of volume is, somehow, related to interneuronal connectivity, e.g. a reduction or elimination of axonal branches or terminals. This possibility is further corroborated by reduction of plasticity of the mPFC after P25, i.e. during the period of volumetric decline (Kolb and Nonneman, 1978; Nonneman and Corwin, 1981).

Development and regression in the cortical afferents and spinal efferent systems

Regression and degeneration are processes which are frequently reported to occur during brain development. Programmed cell death, and the elimination of collateral axons and synapses have been observed in many parts of the developing brain and seem to represent a fundamental principle. A restriction in the projection pattern of neurons has been reported to occur during normal development in various areas of the CNS in a variety of spieces (Dehay and Kennedy, 1984; Jeffery et al., 1984; Innocenti, 1981, 1986; Ivy et al., 1984; Killackey and Chalupa, 1986; Olivarria and Sluyters, 1985; Provis et al., 1985). For example, developmental restriction has been demonstrated in retinal projections (Jeffrey et al., 1984; Provis et al., 1985), and in ipsilateral (Dehay and Kennedy, 1984), as well as commissural, corticocortical projections of the visual, parietal, temporal and somatosensory cortices (Innocenti, 1981, 1986; Ivy et al., 1984; Killackey and Chalupa, 1986; Olivarria and Sluyters, 1985). In corticocortical projections the developmental restriction probably serves to refine the initial distribution of connections between neurons. The development of corticocortical projections has been studied in different species: rat (Crandal and Caviness, 1984; Ivy et al., 1984), cat (Innocenti, 1981) and monkey (Swartz and Goldman-Rakic, 1982). In the rat the commissural corticocortical projections cross the corpus callosum soon after the first bridge between the hemispheres, the so-called "glial sling", is formed (Hankin and Silver, 1986). The corpus callosum is generated between embryonic days (E) 17 and 20 (Crandal and Caviness, 1984; Floeter and Jones, 1985; Uylings et al., 1990), and on E18 the cortical afferents can be found in the subplate layer underneath the cortical plate. In rat (Ivy et al., 1984), cat (Innocenti, 1981, 1986), and monkey (Killackey and Chalupa, 1986) it was shown that these initial pioneer fibers are more in number and have a more widespread distribution in the contralateral hemisphere than is found in the adult animal. Subsequently, during the early postnatal period (2nd postnatal week), these projections from the visual and somatosensory cortex are refined by elimination of axon collaterals as could be demonstrated by using fluorescent retrograde tracers as either long or short term markers. However, within the prefrontal corticocortical projection systems of the monkey no reduction could be demonstrated during development (Schwartz and Goldman-Rakic, 1982).

We have compared the neonatal and adult

distribution of cortical cells projecting to the medial prefrontal cortex in the rat, either through the corpus callosum or ipsilaterally. Fig. 5 shows cells retrogradely labeled after an injection with the fluorescent tracer Fast Blue made on P5. The injection site included the entire medial prefrontal cortex including the subareas PL, AC_d and P_rC_m (Fr2). Comparison of the distribution of the retrogradely labeled cells in the contra- and ipsilateral hemispheres with cells labeled by an injection of Fast Blue labeling the same mPFC subareas in the adult rat (Fig. 6) shows no obvious restrictions of the neonatal pattern, either in the commissural or in the ipsilateral projections to the prefrontal cortex. Therefore, it does not seem likely that elimination of transient collaterals plays an important role in development of prefrontal corticocortical connections in the rat.

On the other hand, a developmental reduction of fibers has also been described to occur in another cortical projection system: the corticospinal tract (CST) (Bates and Killackey, 1984; O'Leary et al., 1981; O'Leary and Stanfield, 1985, 1986; Stanfield and O'Leary, 1985; Stanfield et al., 1982). O'Leary and Stanfield showed that the layer V cells of the occipital rat cortex could be retrogradely labeled after second postnatal day injections with fluorescent tracers into the pyramidal decussation of the lower medulla (O'Leary et al.,

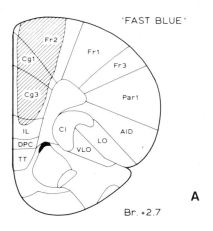

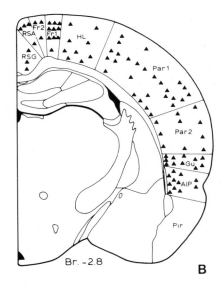

Fig. 5. Ipsilateral distribution of retrogradely labeled cortical cells after injection of 0.5 μl of Fast Blue into the medial PFC of a neonatal rat pup (P5). Injection site is comparable with Fig. 6.

Fig. 6. Ipsilateral distribution of retrogradely labeled cortical cells after a 1 μl injection of Fast Blue into the medial PFC. (a) Injection site; (b) distribution of FB-labeled cells at an anterior level comparable to that of Fig. 5. One symbol stands for 5 – 10 labeled cells.

1981). In adulthood this area of the cortex does not contribute to the CST. The fibers which constitute the CST in the adult rat originate in the rostral two thirds of the cortex. A major part of the CST fibers from the occipital cortex are retracted during the second postnatal week (Stanfield and O'Leary, 1985). As in the commissural systems of certain cortical areas, the restriction in CST projections from the occipital cortex is caused by selective collateral elimination rather than by cell death (O'Leary et al., 1981; Stanfield et al., 1982). By using retrogradely transported fluorescent dyes as either short term or long term markers, occipital neurons, which had a transiently extended pyramidal tract axon, appeared to have concurrent projections to either the superior colliculus or the pons (O'Leary and Stanfield, 1985). After retraction of the pyramidal tract collaterals, the occipital neurons maintain their connections with one of these subcortical nuclei.

In the adult rat, injections of anterograde tracers into prefrontal areas fail to identify fibers projecting toward the spinal cord (Beckstead, 1979). Recent reports suggest that more rostral areas of the cortex may also contain neurons which project to the CST during a limited period of postnatal development (Schleyer and Jones, 1988). By injecting the retrograde tracer Fast Blue into the cervical spinal cord, between C5 and C7, in newly born and young adult animals we have investigated a possible reduction of CST fibers from the prefrontal areas.

Since Fast Blue is taken up by damaged as well as undamaged axons passing through the injection site, cortical neurons that extend axons to the level C5 – C7 or further caudally are labeled by such injections which include the corticospinal tract (examples of injections are shown in Fig. 7). Animals in which the tracer spread through the central canal to more rostral levels were not used for analysis. Injections in the cervical spinal cord of neonatal rats result in a continuous band of cortical neurons in the frontal pole of the hemisphere (Fig. 8). Labeled cells are medium-sized and large pyramidal neurons. In the ventro-medial and the

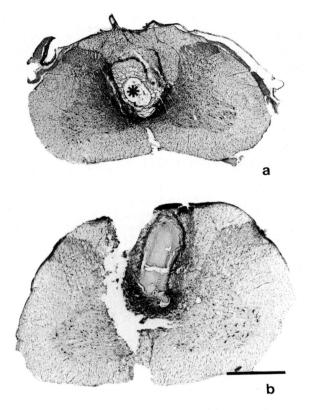

Fig. 7. Injection sites in the cervical spinal cord of (a) neonatal (P5) rat pup and (b) adult rat. Retrogradely labeled cells in the PFC are shown in Figs. 8 and 9, respectively.

medio-orbital parts of the frontal pole (areas: IL, MO, VO, VLO,) where layer V is thinner than in the dorsolateral and dorsomedial parts of the cortex, the labeled cells appear to be smaller and fewer in number. The medial half of VLO contains very few labeled cells. Further caudally, the band of labeled cells extends from the fundus of the rhinal sulcus to the ventral border of the medial cortex (Fig. 8c). At this level, however, the band is interrupted: no labeled cells are observed in the insular cortex immediately dorsal to the rhinal sulcus, i.e. the ventral subdivision of the agranular insular cortex (AI_v). The posterior agranular insular area (AI_p), which lies caudal of AI_v, contains many labeled cells. At this level, the band of labeled neurons is again uninterrupted extending from the rhinal sulcus to the cingulate cortical areas at the medial aspect of the hemisphere. This pattern of labeling continues in caudal direction, until at least the level of the anterior commissure.

In adult rats similar injections label cells in a much more restricted area than CST injections in the neonatal rat. In the adult rat, the orbital PFC areas in the frontal pole are completely devoid of

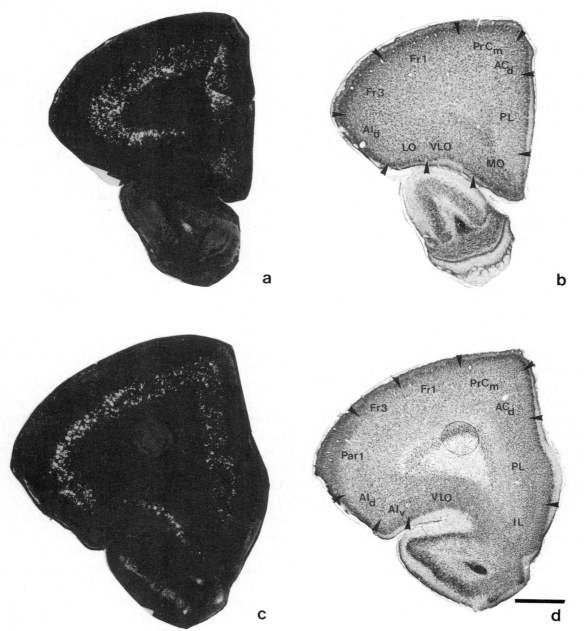

Fig. 8. Transverse sections through the frontal pole of the left hemisphere after neonatal injection with Fast Blue in the cervical spinal cord (see Fig. 7A). (a) FB-labeled cells in layer V at level A 7.5 mm (Sherwood and Timiras, 1970), (b) Same section as (a) cresyl-violet counterstained. (c) FB-labeled cells in at level A 7.0 mm. (d) Same section as (c) cresyl-violet counterstained.

labeled cells after CST injections (Fig. 9). No labeled cells can be observed in MO, VO, VLO, LO, AI_v, or AI_d. In dorsolateral parts of the frontal cortex, containing the motor cortices (Fr1, Fr3), and in the medial prefrontal areas (PrC_m, AC_d), a great number of labeled cells is still found within layer V. Few cells are located in the prelimbic (PL) and infralimbic (IL) areas on the medial wall of the

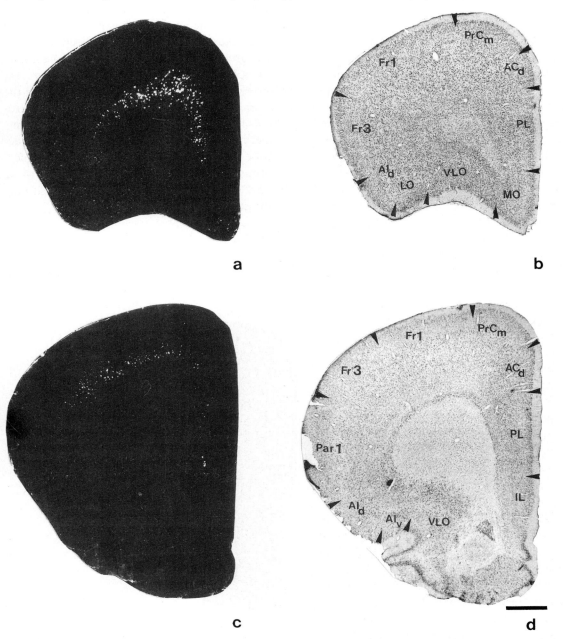

Fig. 9. Transverse sections through the frontal pole of the left hemisphere after injection with Fast Blue in the cervical spinal cord of an adult rat (see Fig. 7b). (a) FB-labeled cells in layer V at level comparable to Fig. 8a. (b) Same section as (a) cresyl-violet counterstained. (c) FB-labeled cells in at level comparable to Fig. 8c. (d) Same section as (c) cresyl-violet counterstained.

cortex, especially in the most rostral parts. Few labeled cells are also observed in the frontal parts of the parietal area (Par1). Another group of labeled cells is situated in AI_p lining the shallow parts of the rhinal sulcus.

Comparison of the labeling pattern after neonatal and adult corticospinal tract (CST) injections shows that several areas that have a corticospinal (CS) projection in neonatal stages fail to be labeled in adulthood. In the frontal cortex, this phenomenon is particularly evident in the orbital areas, where, after neonatal CST injections, numerous labeled layer V cells are found in all these orbital cortical areas, whereas the adult labeling pattern in the frontal cortex is similar to that previously reported (Hicks and D'Amato, 1977; Leong, 1983; Wise and Jones, 1977; Miller, 1987; Schleyer and Jones, 1988). Similar to these reports we failed to observe labeled cells in the following orbital areas: MO, LO, VO, VLO, AI_v or AI_d. The orbital prefrontal area AI_v is an exception in this respect since neither adult nor neonatal CST injections result in labeled CST neurons in this area. Thus, neurons in this area may never sustain a CST projection or, alternatively, they may grow a CS axon which fails to reach the level of the caudal cervical spinal cord. The present study can not be conclusive on this point.

The cells which can no longer be labeled in the adult rat either disappear in the course of development, or survive but do not longer project toward the spinal cord. In an attempt to make a distinction betwee these two mechanisms, we made use of the property that retrogradely transported FB remains visible in the cell soma for extended periods without leakage (e.g. O'Leary et al., 1981). Six pups received CST injections on day 5 and were allowed to survive until 35 days of age. This time interval was chosen because it is well beyond the period in which the transient CS neurons retract their spinal projections (Schleyer and Jones, 1988; Stanfield and O'Leary, 1985; Joosten and Van Eden, 1989). This experiment demonstrates that the cells which are labeled by CST injections during the first postnatal week are not eliminated by cell death (Fig. 10), and, therefore, are likely to retract their spinal collaterals similar to the developing neurons in the occipital cortex. Recent investigations (Joosten and Van Eden, 1989) using anterograde tracers support this observation by indicating that the number of axons in the spinal cord capable of being labeled by injections in the medial prefrontal cortex (AC_d, PL, IL) initially increases between day 3 and day 7. At day 14, however, very few, if any axons running down the CST can be labeled. Therefore, a number of projections from PL and IL also seem to be transient. Furthermore, the extension and subsequent retraction of axons from this part of the prefrontal cortex seems to follow a similar time course as those from the occipital cortex which are retracted during the second postnatal week.

Conclusions

These data on the structural development of the prefrontal cortex in the rat show that the prefron-

Fig. 10. Transverse section through the frontal pole of a 35-day-old rat, which received an FB-injection in the spinal cord at P5, showing a neonatal distribution of retrogradely labeled cells (see Fig. 7a).

tal association cortex develops according to a similar time schedule as do other (neo)- cortical areas such as the visual and parietal cortex. The development of the cortical layers, ingrowth of thalamic and dopaminergic fibers follow a scheme of development which is comparable to that of other cortical areas. Only the volumetric development seems to point to a delayed maturation of the prefrontal areas especially the orbital PFC. The reduction of volume which is observed after the maxima volume is attained at P24 and P30 for the medial and orbital PFC, respectively, is accompanied by a reduced potential for functional plasticity. However, this late volumetric decline and reduced functional plasticity are probably not correlated with an elimination of exuberant cortical afferents or corticospinal efferents. In the callosal and ipsilateral cortical afferent systems towards the prefrontal cortex, no such elimination seems to be present, whereas in the prefrontal CST system, where a considerable pruning of transient axons was demonstrated, this elimination of collateral projections takes place in the second postnatal week. This corresponds to the period, in which the CST collaterals are pruned from the occipital cortex, where the reduction of volume and plasticity occurs much earlier.

References

Bates, C.A. and Killackey, H.P. (1984) The emergence of a discretely distributed pattern of corticospinal projection neurons. *Dev. Brain Res.*, 13: 265–273.

Beckstead, R.M. (1979) An autoradiographic examination of corticocortical- and subcortical projections of the mediodorsal projection (prefrontal) cortex in the rat. *J. Comp. Neurol.*, 184: 43–62.

Chun, J.J.M., Nakamura, M.J. and Shatz, C.J. (1987) Transient cells of the developing mammalian telencephalon are peptide-immunoreactive neurons. *Nature*, 325: 617–620.

Crandal, J.E. and Caviness, V.S. (1984) Axon strata of the cerebral wall in embryonic mice. *Dev. Brain Res.*, 14: 185–195.

Dehay, C. and Kennedy, H. (1984) Transient projections from the fronto-parietal and temporal cortex to areas 17, 18 and 19 in the kitten. *Exp. Brain Res.*, 57: 208–212.

Diamond, A. and Goldman-Rakic, P.S. (1989) Comparison of human infants and rhesus monkeys on Piaget's AB task: Evidence for dependence of dorsolateral prefrontal cortex. *Exp. Brain Res.*, 74: 261–268.

Eayrs, J.T. and Goodhead, B. (1959) Postnatal development of the cerebral cortex in the rat. *J. Anat.*, 93: 385–402.

Finger, S. and Amli, C.R. (1985) Brain damage and neuroplasticity: Mechanisms of recovery or development *Brain Res. Rev.*, 10: 177–186.

Flechsig, P. (1901) Developmental (myelogenetic) localisation of the cerebral cortex in the human subject. *Lancet*, 2: 1027–1029.

Floeter, M.K. and Jones, E.G. (1985) The morphology and phased outgrowth of callosal axons in the fetal rat. *Dev. Brain Res.*, 22: 7–18.

Fritschy, J.M. and Garey, L.J. (1986) Quantitative changes in morphological parameters in the developing visual cortex of the marmoset monkey. *Dev. Brain Res.*, 29: 173–188.

Goldman, P.S. (1971) Functional development of the prefrontal cortex in early life and the problem of neuronal plasticity. *Exp. Neurol.*, 32: 366–387.

Goldman, P.S. (1972) Developmental determinants of cortical plasticity. *Acta Neurobiol. Exp.*, 32: 495–511.

Goldman-Rakic, P.S. and Galkin, Th.W. (1978) Prenatal removal of frontal association cortex in the fetal rhesus monkey: Anatomical and functional consequences in postnatal life. *Brain Res.*, 152: 451–485.

Hankin, M.H. and Silver, J. (1986) Mechanisms in axonal guidance. The problem of intersecting fiber systems. In: L.W. Browder (Ed.), *Developmental Biology, Vol. 2*, Plenum Press, New York, pp. 565–604.

Heumann, D., Leuba, G. And Rabinowicz, Th (1976) Postnatal development of the mouse cerebral neocortex. II. Quantitative cytoarchitectonics of visual and auditory areas. *J. Hirnforsch.*, 18: 483–500.

Hicks, S.P. and D'Amato, C.J. (1977) Locating corticospinal neurons by retrograde axonal transport of horseradish peroxidase. *Exp. Neurol.*, 56: 410–420.

Innocenti, G.M. (1981) Growth and reshaping of axons in the establishment of visual callosal connections. *Science*, 212: 824–827.

Innocenti, G.M. (1986) General organization of callosal connections in the cerebral cortex. In: E.G. Jones and A. Peters (Eds.), *Cerebral Cortex, Vol. 5*, Plenum, New York, pp. 291–353.

Ivy, G.O., Gould III, H.J. and Killackey, H.P. (1984) Variability in the distribution of callosal projection neurons in the adult rat parietal cortex. *Brain Res.*, 306: 53–61.

Jeffery, G., Arzymanow, B.J. and Lieberman, A.R. (1984) Does the early exuberant retinal projection to the superior colliculus in the neonatal rat develop synaptic connections? *Dev. Brain Res.*, 14: 135–138.

Joosten, E.A.J. and Van Eden, C.G. (1989) An anterograde tracer study on the development of corticospinal projections from the medial prefrontal cortex in the rat. *Dev. Brain Res.*,

45: 313–319.

Kalsbeek, A., De Bruin, J.P.C., Feenstra, M.G.P. and Uylings, H.B.M. (1990) Age-dependent effects of lesioning the mesocortical dopamine system upon prefrontal cortex morphometry and PFC-related behaviors. *This Volume*, Ch. 12.

Killackey, H.P. and Chalupa, L.M. (1986) Ontogenetic change in the distibution of callosal projecting neurons in the postcentral gyrus of the fetal rhesus monkey. *J. Comp. Neurol.*, 244: 331–348.

Kolb, B. (1984) Functions of the frontal cortex of the rat: A comparative review. *Brain Res. Rev.*, 8: 65–98.

Kolb, B. and Gibb, R. (1990) Anatomical correlates of behavioural change after neonatal prefrontal lesions in rat. *This Volume*, Ch. 11.

Kolb, B. and Nonneman, A.J. (1976) Functional development of prefrontal cortex in rats continues into adolescence. *Science*, 193: 335–336.

Kolb, B. and Nonneman, A.J. (1978) Sparing of function in rats with early prefrontal cortex lesions. *Brain Res.*, 151: 135–148.

Kostović, I (1990) Structural and histochemical reorganization of the human prefrontal cortex during perinatal and postnatal life. *This Volume*, Ch. 10.

Kostović, I. and Molliver, M.E. (1974) A new interpretation of the laminar development of the cerebral cortex: Synaptogenesis in different layers of the neopallium in the human fetus. *Anat. Rec.*, 178: 395.

Kostović, I and Rakic, P. (1980) Cytology and time of origin of interstitial neurons in the white matter in infant and adult human and monkey telencephalon. *J. Neurocytol.*, 9: 219–242.

Krettek, J.E. and Price, J.L. (1977) The cortical projections of the mediodorsal nucleus and adjacent thalamic nuclei in the rat. *J. Comp. Neurol.*, 171: 157–192.

Leonard, C.M. (1969) The prefrontal cortex of the rat. I. Cortical projection of the mediodorsal nucleus. II. Efferent connections. *Brain Res.*, 12: 321–343.

Leong, S.K. (1983) Localizing the corticospinal neurons in neonatal, developing and mature albino rat. *Brain Res.*, 265: 1–9.

Leuba, G., Heumann, D. and Rabinowicz, Th. (1976) Postnatal development of the mouse cerebral neocortex. II. Quantitative cytoarchitectonics of some motor and sensory areas. *J. Hirnforsch.*, 18: 461–481.

Lund, R.D. and Mustari, M.J. (1977) Development of the geniculocortical pathways in rats. *J. Comp. Neurol.*, 173: 289–306.

Miller, M.W. (1987) The origin of corticospinal projection neurons in rat. *Exp. Brain Res.*, 67: 339–351.

Mrzljak, L., Uylings, H.B.M., Van Eden, C.G. and Judáš, M. (1990) Neuronal development in human prefrontal cortex in prenatal and postnatal stages. *This Volume*, Ch. 9.

Nonneman, A.J. and Corwin, J.V. (1981) Differential effects of prefrontal cortex ablation in neonatal, juvenile, and young adult rats. *J. Comp. Physiol. Psychol.*, 95: 588–602.

O'Leary, D.D.M. and Stanfield, B.B. (1985) Occipital cortical neurons with transient pyramidal tract axons extend and maintain collaterals to subcortical but not intracortical targets. *Brain Res.*, 336: 326–333.

O'Leary, D.D.M. and Stanfield, B.B. (1986) A transient pyramidal tract projection from the visual cortex in the hamster and its removal by selective collateral elimination. *Dev. Brain Res.*, 27: 87–99.

O'Leary, D.D.M., Stanfield, B.B. and Cowan, W.M. (1981) Evidence that the early postnatal restriction of the callosal projection is due to the elimination of axonal collaterals rather than to the death of neurons. *Dev. Brain Res.*, 1: 607–617.

Olivarria, J. and Van Sluyters, R.C. (1985) Organization and postnatal development of callosal connections in the visual cortex of the rat. *J. Comp. Neurol.*, 239: 1–26.

Parnavelas, J.G. and Uylings, H.B.M. (1980) The growth of nonpyramidal neurons in the visual cortex of the rat: a morphometric study. *Brain Res.*, 193: 373–382.

Provis, J.M., Van Driel, D., Billson, F.A. and Russell, P. (1985) Human fetal optic nerve: Overproduction and elimination of retinal axons during development. *J. Comp. Neurol.*, 238: 92–100.

Raedler, A. and Sievers, J. (1975) The development of the visual system of the albino rat. *Adv. Anat. Embryol. Cell Biol.*, 50 (Fasc.3): 5–88.

Rice, F.L. and Van der Loos, H. (1977) Development of the barrels and barrelfield in the somatosensory cortex of the mouse. *J. Comp. Neurol.*, 171: 545–560,

Schleyer, D.J. and Jones, E.H.G. (1988) Topographic sequence of outgrowth of corticospinal axons in the rat: a study using retrograde axonal labeling with Fast Blue. *Dev. Brain Res.*, 38: 89–101.

Schwartz, M.L. and Goldman-Rakic, P.S. (1982) Single cortical neurons have axon collaterals to ipsilateral and contralateral cortex in fetal and adult primates. *Nature*, 299: 154–156.

Sherwood, N.M. and Timiras, P.S. (1970) *A Stereotaxic Atlas of the Developing Rat Brain*, University of California Press, Berkeley, CA.

Stanfield, B.B. and O'Leary, D.D.M. (1985) The transient corticospinal projection from the occipital cortex during the postnatal development of the rat. *J. Comp. Neurol.*, 238: 236–248.

Stanfield, B.B., O'Leary, D.D.M. and Fricks, C. (1982) Selective collateral elimination in early postnatal development restricts cortical distribution of rat pyramidal tract neurons. *Nature*, 298: 371–373.

Uylings, H.B.M., Van Eden, C.G. and Hofman, M.A.H. (1984) Morphometry of size/volume variables and comparison of their bivariate relations in the nervous system under different conditions. *J. Neurosci. Meth.*, 18: 19–37.

Uylings, H.B.M., Van Eden, C.G., Parnavelas, J.G. and

Kalsbeek, A. (1990) The pre- and postnatal development of rat cerebral cortex. In: B. Kolb and R. Tees (Eds.), *The Cerebral Cortex of the Rat*, MIT Press, Cambridge.

Valverde, F. and Facal-Valverde, M.V. (1987) Transitory population of cells in the temporal cortex of kittens. *Dev. Brain Res.*, 32: 283 – 288.

Van Eden, C.G. (1986) Development of connections between the mediodorsal nucleus of the thalamus and the prefrontal cortex in the rat. *J. Comp. Neurol.*, 244: 349 – 359.

Van Eden, C.G. and Uylings, H.B.M. (1985a) Cytoarchitectonic development of the prefrontal cortex in the rat. *J. Comp. Neurol.*, 241: 253 – 267.

Van Eden, C.G. and Uylings, H.B.M. (1985b) Postnatal volumetric development of the prefrontal cortex in the rat. *J. Comp. Neurol.*, 241: 268 – 275.

Verwer, R.W.H. and van Vulpen, E.H.S. (1989) Development of prefrontal cortex innervation from the amygdala in the rat. *Proc. 16th Summerschool Brain Res.*, N.I.B.R., Amsterdam, p. 84.

Wahlsten, D. (1981) Prenatal schedule of appearance of mouse brain commissures. *Dev. Brain Res.*, 1: 461 – 473.

Werner, L., Himmel, W. and Lohman, W. (1981) Postnatale Volumenentwicklung der Neurosomata in dorsalen Anteil des Corpus geniculatum laterale (Cgld) und in der Area striata der albino Ratte. In: K. Hecht, W. Ruediger, K. Seidel and M. Poppei (Eds.) *Zentralnervensystem, Entwickelung, Stoerungen und Motivation*, Deutches Verlag, Berlin, pp. 34 – 89.

Wise, S.P. and Jones, E.G. (1977) Cells of origin and terminal distribution of descending projections of the rat somatic sensory cortex. *J. Comp. Neurol.*, 175: 129 – 158.

Wise, S.P. and Jones, E.G. (1978) Developmental studies of thalamocortical and commissural connections in the rat somatic sensory cortex. *J. Comp. Neurol.*, 178: 187 – 208.

Yakovlev, P.I. and Lecours, A.-R. (1967) The myelogenetic cycles of regional maturation of the brain. In: A. Minkowski (Ed.), *Development of the Brain in Early Life*, Blackwell, Oxford, pp. 3 – 70.

Zilles, K.J. (1978) Ontogenesis of the visual system. *Adv. Anat. Embryol. Cell Biol.*, 54 (Fasc.3): 5 – 138.

Zijlstra, C., Verwer, R.W.H. and Van Eden, C.G. (1986) Comparison of numerical density of neurons before, during and after the volume overshoot of the medial prefrontal cortex. Research report Neth. Inst. Brain Res. 86 – 1. (In Dutch).

Discussion

P. Goldman-Rakic: Do you think that different principles govern development of callosal and corticospinal projections, or between prefrontal callosal as opposed to visual callosal neurons?

C.G. van Eden: From the work that I presented here today, I have no indications that callosal PFC fibers project to other areas in the neonatal period than in adulthood. This does not exclude a developmental reduction of callosal fibers *within* these areas. So far, attempts to use fluorescent tracers for a double labeling experiment separated in time in order to investigate such a developmental reduction in callosal PFC projections have failed because the form of the PFC changes markedly during neonatal development. This makes a second, overlapping injection with the second tracer almost impossible. Therefore, I do not know whether the development of callosal and CS projections are governed by a different principle. But when the mechanisms in the rat are comparable with those in the monkey and cat, it might be that a pruning of collateral axons occurs in the CS fiber systems and in the visual callosal system but not in the PFC callosal system (e.g., O'Leary and Stanfield, 1985; Innocenti, 1986; Schwartz and Goldman-Rakic, 1984).

R.W.H. Verwer: I would like to comment on the idea to make a first injection and later subsequent injection in exactly the same place. The first injection will damage the cortex and it is afterwards not comparable with a normal cortex at the same age.

C.G. van Eden: That is true; nevertheless it has been proved that it is possible to double label many cells in this way, indicating that necrosis is not preventing uptake of the second tracer in a large part of the initial injection site. I agree that a number of cells that fail to transport the second tracer may act like this because of this phenomenon. Therefore, a restriction of the projection area should also be demonstrated by other (anterograde) tracer methods.

References

Innocenti, G.M. (1986) General organization of callosal connections in the cerebral cortex. In: E.G. Jones and A. Peters (Eds.), *Cerebral Cortex, Vol.5*, New York, pp. 291 – 353.

O'Leary, D.D.M. and Stanfield, B.B. (1985) Occipital cortical neurons with transient pyramidal tract axons extend and maintain collaterals to subcortical but not intracortical targets. *Brain Res.*, 336: 326 – 333.

Schwartz, M.L. and Goldman-Rakic, P.S. (1984) Callosal and intrahemispheric connectivity of the prefrontal association cortex in rhesus monkey: relation between intraparietal and principal sulcal cortex. *J. Comp. Neurol.*, 226: 403 – 420.

CHAPTER 9

Neuronal development in human prefrontal cortex in prenatal and postnatal stages

Ladislav Mrzljak[1,2], Harry B.M. Uylings[2], Corbert G. Van Eden[2] and Miloš Judáš[1]

[1] *Section of Neuroanatomy, Department of Anatomy, Medical Faculty, University of Zagreb, Šalata 11, pp. 286, 41001 Zagreb, Yugoslavia, and* [2] *Netherlands Institute for Brain Research, Meibergdreef 33, 1105 AZ Amsterdam, The Netherlands*

Introduction

The mammalian cerebral cortex is organized in a complex way. Important histogenetic processes that lead to its formation are the proliferation, migration and differentiation of neurons and glial cells, growth of afferent and efferent fibers, synaptogenesis, and, finally, the elimination of certain cells and axonal collaterals (Sidman and Rakic, 1973; Rakic, 1982, 1988a, b, c; Cowan et al., 1984). Although several of these processes have already been described in classical studies of neuroanatomy (Vignal, 1888; Kölliker, 1896; Cajal 1899a, b, 1900, 1911, 1960; His, 1904; Lorente de Nó, 1933), the application of modern neuroanatomical methods to experimental primates has enabled us to obtain important new data about spatial and temporal parameters of the cellular basis of cortical histogenesis (for a review see Rakic, 1988b).

The neurons destined for the primate cerebral cortex originate prenatally in the germinal zones of the fetal telencephalic wall, i.e. the ventricular and subventricular zones (Sidman and Rakic, 1973; Rakic, 1982). According to the "radial unit"

model, the ventricular zone consists of proliferative units, and each of these units produces genealogically related neurons by mitosis (Smart and McSherry, 1982; Rakic, 1988a, b, c). After the completion of mitosis, the neurons derived from a common germinal cell migrate one after another along a common radial glia bundle to form so-called ontogenetic columns in the cortical plate (Rakic, 1971, 1972, 1988a, b, c). The size of any cortical area in the primate cortex depends on the number of ontogenetic columns, and, hence, on the number of proliferative units composing these columns (Rakic, 1988a, b, c). Therefore the cortical area that takes up the highest number of proliferative units for its formation is the human frontal association cortex, i.e. the prefrontal cortex (PFC), which is the largest area in the human cerebral cortex (Goldman-Rakic, 1982; Uylings and Van Eden, this volume). The PFC plays a crucial role in cognitive and polysensory functions, guidance of voluntary behavior by representational memory, and regulation of socially appropriate action (Goldman-Rakic et al., 1983; Goldman-Rakic, 1987a, this volume; Fuster, 1989, this volume). For correlation with its functional development it is useful to study the major structural progressive growth of dendrites, axons and dendritic spines, and the regressive elimination of so-called exuberant dendritic segments and spines during the differentiation of PFC neurons. Fur-

Correspondence: Harry B.M. Uylings, Netherlands Institute for Brain Research, Meibergdreef 33, 1105 AZ Amsterdam, The Netherlands.

thermore, knowledge of the non-pathologic development of the PFC may serve as a normative reference in the analysis of the developmental neuropathology of the cerebral cortex, especially in the syndromes in which cognitive functions are impaired.

Despite its unpredictability and limitations (for review see Braak and Braak, 1985), Golgi staining has been used in the majority of the developmental studies of neuronal differentiation in the human cerebral cortex at the light microscopic level (Vignal, 1888; Kölliker, 1896; Cajal, 1899a, b, 1900; Marin-Padilla, 1969, 1970a, b, 1972, 1988; Purpura, 1975a, b; Krmpotić-Nemanić et al., 1979; Paldino and Purpura, 1979; Takashima et al., 1980; Marin-Padilla and Marin-Padilla, 1982; Becker et al., 1984; Michel and Garey, 1984; Seress and Mrzljak, 1987; Kostović et al., 1989b). These studies primarily encompass motor, somatosensory, auditory and limbic cortical areas. Dendritic differentiation in the human frontal association cortex has received little attention. Only a few neuronal types have been studied, resulting from the analysis of a limited number of specimens (Poliakov, 1949, 1966, 1979; Conel, 1939 – 1967; Schadé and Van Groenigen, 1961; Schadé et al., 1964). Therefore, we have undertaken qualitative and quantitative studies in order to describe dendritic and axonal differentiation of PFC neurons, using a large number of specimens impregnated with various Golgi techniques. The general aim of this paper is to describe the differentiation of PFC neurons from very early fetal development (third month of gestation) until the adult stages in connection with the development of PFC lamination. Our data on the dendritic and axonal development of PFC neurons will be correlated with data on the development of afferent systems (Kostović and Goldman-Rakic, 1983; Kostović, this volume), on synaptogenesis (Huttenlocher, 1979; Huttenlocher et al., 1982), and on the histochemical maturation of PFC neurons (Kostović et al., 1988; Kostović, this volume). Correlations will also be made with data on dendritic differentiation of neurons in other areas of the cerebral cortex and on the metabolic and functional development of the PFC (Chugani and Phelps, 1986; Diamond and Door, 1989; Diamond and Goldman-Rakic, 1989). Our attention will be focused on the development of 3 neuronal populations: (1) the pyramidal neurons of layer III, which are the major source of callosal and association projections (Schwartz and Goldman-Rakic, 1984); (2) the nonpyramidal neurons, which, according to morphological criteria (Rakic, 1975), can generally be identified as local circuitry neurons; and (3) the transient neurons. These transient neurons comprise: (a) Cajal – Retzius cells in the marginal zone (fetal layer I) and (b) neurons in the subplate zone. This latter zone is a transiently present fetal lamina first described by Kostović and Molliver (1974). During development it serves as a "waiting" compartment for thalamocortical, basal forebrain, callosal and association fibers (Rakic, 1977; Goldman-Rakic, 1981, 1982; Kostović and Goldman-Rakic, 1983; Kostović and Rakic, 1984, 1990; Kostović, 1986, this volume). Later on, its neurons are either transformed into interstitial white matter and layer VI neurons or eliminated by cell death (Kostović and Rakic, 1980, 1990; Luskin and Shatz, 1985; Kostović et al., 1989b). In addition, we will discuss the presence, location and types of Alz-50-positive neurons during perinatal stages. It has been suggested that Alz-50 immunoreactivity could be a marker for degenerating neurons in the subplate zone and white matter (Wolozin et al., 1988; Al-Ghoul and Miller, 1989). However, see also the report by Hamre et al. (1989).

Material and methods

The material for this analysis included 41 Golgi-stained specimens, ranging from the very early stage of fetal development until the adult stages (see Table I). This material is part of the neuroembryological collection of the Section of Neuroanatomy, School of Medicine, Zagreb, which was collected with the approval of the Ethical Committee of the School of Medicine. Three different modifications of the Golgi staining were applied:

TABLE I

Age			Sex	Type of Golgi impregnation		
w.g.	months	years		Stensaas	Golgi – rapid	Golgi – Cox
Fetuses (N = 7)						
10.5			–	+		
13.5			–	+		
17			M	+		
21			M	+		
22			M	+		
22			F	+		
24			M	+		
Premature infants (N = 7)						
26			M	+	+	
27			F	+	+	
29			M	+	+	
32			F		+	
34			M		+	
36			M		+	
40 (Newborn)			M		+	
Infants (N = 7)						
	1		M		+	
	2		F		+	
	2.5		M		+	+
	3		F		+	
	5		F		+	
	7.5		M		+	
	12		M		+	+
Children (N = 8)						
	15		F		+	
	16		M		+	
		5	F		+	+
		5.5	M			+
		6	M		+	
		9	M		+	+
		10	M		+	
		11	M			+
Puberty period (N = 6)						
		14	M			+
		16	M			+
		16	M		+	
		17	M			+
		19	M		+	
		19	M			+
Young adults (N = 3)						
		27	M			+
		29	M			+
		30	M		+	
Adults (N = 3)						
		35	M			+
		46	M			+
		64	F		+	+

w.g. = weeks of gestation; M = male; F = female.

Golgi – Stensaas impregnation (Stensaas, 1967), the classical chrome – osmium Golgi – rapid impregnation according to Cajal (Cajal and De Castro, 1933), and Golgi – Cox impregnation according to Van der Loos (1959). The thickness of the Golgi sections analyzed was 150 μm in Golgi – Stensaas and Golgi – rapid materials and 175 μm in Golgi – Cox sections. Adjacent 30 – 50-μm-thick Golgi sections were counterstained with the Nissl method, or Nissl-stained sections of corresponding stages were taken from the neuroembryological collection at the Section of Neuroanatomy, Zagreb, for cytoarchitectonic landmarks.

The area of the PFC examined in specimens from the period of 10.5 weeks of gestation (w.g.) until 27 w.g. had to be determined by its relative position because of the immature cytoarchitectonic features of the cortical anlage and lack of sulci and gyri on the lateral telencephalic surface. Coronal slabs for Golgi impregnation were taken caudal to the polar cap from the dorsolateral and midlateral part of the frontal lobe (Fig. 1). After 26 – 27 w.g., gyri are formed in the frontal lobe in the preterm infant (Chi et al., 1977). From that period on, prepolar parts of the superior and middle frontal gyri were analyzed (Fig. 1). In the newborn, postnatal and adult brain, these parts of the medial and superior frontal gyri belong to the Brodmann areas 9 and 10 (Brodmann, 1909).

In addition to qualitative Golgi analysis, we also used the three-dimensional, semi-automatic dendrite measuring system developed at the Netherlands Institute for Brain Research (Overdijk et al., 1978; Uylings et al., 1986) for a quantitative assessment of the prenatal and perinatal development in the PFC. A total of 353 pyramidal neurons and 157 subplate neurons from the period of 10.5 w.g. until the second postnatal month were analyzed. The measurements were performed directly in 150-μm-thick Golgi – Stensaas and Golgi – rapid sections, using oil immersion objectives (Uylings et al., 1987; Mrzljak et al., 1990). In this report, only data on the total dendritic length of the pyramidal and subplate neurons will be included.

In the immunohistochemical part of the study, the monoclonal antibody Alz-50 (IgM, kindly donated by Abbott Laboratories Chicago, IL was used and the immunoperoxidase procedure was conducted on 50-μm-thick free-floating vibratome sections according to the protocol of Wolozin et al. (1988). Alz-50 immunoreactivity was analyzed in 2 fetuses (22 and 25 w.g.), 4 preterm infants (from 27 to 34 w.g.), 2 full-term newborns, 5 infant brains (2, 3, 5, 6, 7 and 15 months old), 3 brains

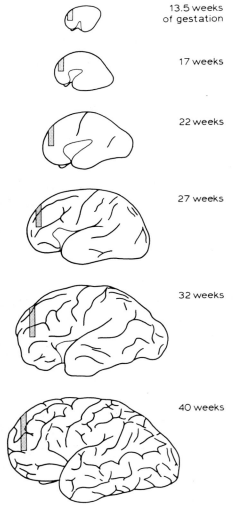

Fig. 1. The lateral surface of the human telencephalon during the prenatal stage. The hatched area indicates the parts of the prefrontal cortex studied. (From Mrzljak et al., 1988.)

from children (3.5 and 6 years old) and 2 adult brains. The material was obtained from the Brain Bank at the Netherlands Institute for Brain Research in Amsterdam, from the Department of Neuropathology (Dr. J.M. Kros) of the Erasmus University in Rotterdam and from the Section of Neuroanatomy in Zagreb. It was fixed for different periods of time in 4% paraformaldehyde in 0.1 M phosphate buffered saline, pH 7.4.

Results

The prenatal, perinatal and postnatal dendritic and axonal (neuronal) development in the human PFC will be traced through 6 different periods, on the basis of our data on changes in cortical histogenetic events (for details see Mrzljak et al., 1988; Kostović, 1990a, this volume).

Period 1: 10 – 25 w.g. Onset of dendritic differentiation of pyramidal neurons in the cortical plate (CP). Formation of the subplate zone and maturation of neuronal types in the subplate.

Period 2: 26 – 36 w.g. Late fetal or preterm infant period. Appearance of a 6-layered pattern in the cortical plate, peak in the development of the subplate zone (SP) followed by a reduction in its thickness. Period of rapid dendritic differentiation of pyramidal and nonpyramidal neurons in the cortical plate.

Period 3: First postnatal year. Neonatal period and infancy. Persistence and gradual transformation of the subplate zone. Intensive dendritic and spine growth on pyramidal neurons. Differentiation of nonpyramidal neurons (interneurons) in the supragranular layers.

Period 4: Second postnatal year. Early childhood period. Appearance of magnopyramidal neurons in layer III of the PFC. Persistence of fetal types of neurons (Cajal – Retzius cells and large numbers of interstitial neurons in the white matter).

Period 5: Period of childhood and adolescence. Gradual and prolonged elaboration of adult-like Golgi architecture.

Period 6: Period of adult morphology, i.e. late adolescence and adulthood.

Period 1: 10 – 25 w.g. Fetal period

The cortical plate (CP) is formed in different regions of the pallium between 7 and 9 w.g. (Kostović, 1990a, this volume). Between 10 and 12 w.g. the frontal pallial anlage consists of the marginal zone (MZ), the CP, the primordial subplate (SPp), the intermediate zone (IZ), the subventricular zone (SV) and the ventricular zone (VZ), as can be seen on Nissl-counter-stained sections (Fig. 2). At that time the neurons in the CP are immature, showing simple bipolar morphology with vertically oriented ascending ("leading") and descending ("trailing") processes. Ascending processes (dendrites) bifurcate in the MZ, while descending ones are either simply non-branched or

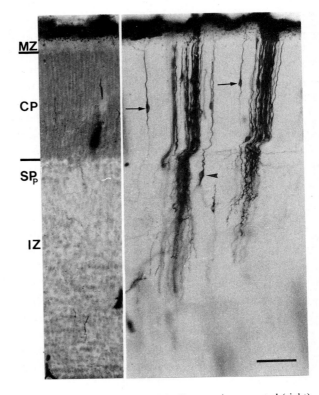

Fig. 2. Photograph of the Golgi – Stensaas impregnated (right) and adjacent Nissl-stained coronal section (left) of the prefrontal cortex in the earliest stage studied (10.5-week-old human fetus). The arrows point to the bipolar immature neurons in the cortical plate (CP). The arrowhead indicates a larger, more differentiated neuron in the primordium of the subplate zone (SPp). Bar scale is 50 µm. (Adapted from Mrzljak et al., 1988.)

form more complex root-like arborizations in the SPp and IZ (Fig. 2). In contrast to the simple morphology of the CP neurons, which have the potential to develop into any type of pyramidal or non-pyramidal neuron, the neurons in the MZ and below the CP have more differentiated dendrites. Bipolar horizontally oriented neurons with relatively large cell bodies in the superficial part of the MZ are early forms of Cajal – Retzius neurons. The fibrillar zone below the CP contains polymorphous neurons (typical of the subplate, see below) and is considered to be the primordium of the subplate zone (Mrzljak et al., 1988; Kostović, 1990a; Kostović and Rakic, 1990).

Between 13.5 and 15 w.g. a true subplate zone is formed in the prefrontal pallium which consists of an upper cell-dense and a lower more fibrillar part (Mrzljak et al., 1988). Numerous polymorphous, randomly oriented neurons are present in the subplate (Fig. 7, see also Fig. 3 in Mrzljak et al., 1988). In addition to polymorphous neurons, bipolar vertically oriented migrating neurons and horizontal bipolar or unipolar neurons are frequently observed throughout the subplate and IZ (Fig. 7). Although the majority of the CP neurons have simple bipolar morphology, the first incipient form of pyramidal neurons with very short basal and oblique dendrites is found in the CP between 13.5 and 15 w.g.

Between 17 and 25 w.g., the subplate is the thickest of all fetal cortical zones (MZ, CP and subplate zone; ratio between the CP and subplate zone thickness is 1:4 between 19 and 23 w.g.; Fig. 4A). Pyramidal neurons in the lower half of the CP are less immature during this fetal period than before (Fig. 3). In contrast to these large pyramidal neurons with clear basal and apical dendrites, pyramidal neurons in the more superficial part of the CP have short, barely recognizable basal and oblique dendrites (Fig. 3). Besides pyramidal neurons, the CP contains numerous transitional neuronal forms resembling pyramidal or non-pyramidal neurons (bipolar and immature multipolar neurons). None of the neuronal types of the local circuitry (interneurons) can yet be identified definitively on the basis of morphological criteria.

Apart from the enormous increase of its width, the major event within the subplate zone is the differentiation of several distinct neuronal types (Fig. 4A – E). In addition to polymorphous neurons (Fig. 4C), fusiform-like neurons (Fig. 4D, E), inverted pyramidal neurons, pyramidal neurons with usual shape and orientation, and multipolar-like neurons (Fig. 4B) are impregnated in the subplate. Quantitative morphometric analysis shows that until 26/27 w.g. the dendrites of subplate neurons are significantly longer than those basal dendritic

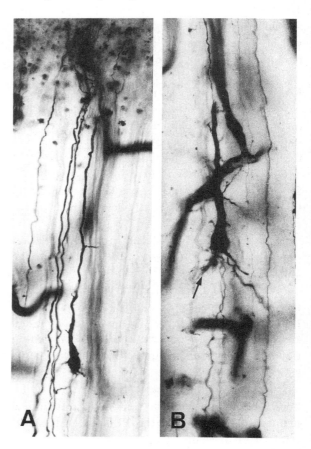

Fig. 3. Differentiation of pyramidal neurons in the cortical plate (CP) during midgestation (22-week-old human fetus; Golgi – Stensaas impregnation). (A) Immature pyramidal neuron in the superficial part of the CP. (B) More differentiated large pyramidal neurons from the middle third of the CP. The arrow points to a growing dendrite ending in a growth cone-like structure. (From Mrzljak et al., 1988.)

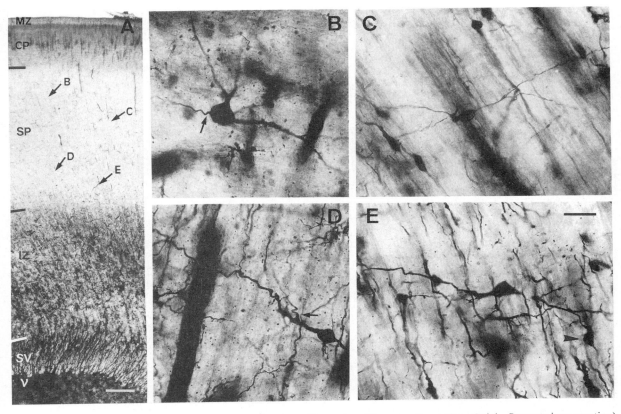

Fig. 4. Differentiation of prefrontal subplate neurons during midgestation (22 weeks of gestation; Golgi–Stensaas impregnation). (A) Nissl-counterstained Golgi–Stensaas impregnation. Arrows with letters point out to the positions of subplate neurons shown in B–E. Note that the ratio of the CP:SP thickness is approximately 1:4. (B) Multipolar subplate neuron. Arrow points to horizontally running axon. (C) Polymorphous subplate neurons. (D–E) Two fusiform-like neurons. Arrow in D points to axon. Arrowhead in E points to a polymorphous subplate neuron with short dendrites. Bar scales are 0.5 mm for A, 20 μm for B–E. (From Mrzljak et al., 1988.)

trees of pyramidal neurons in the CP, with the exception of inverted pyramidal subplate neurons (Fig. 8A, B). The subplate zone also contains migrating neurons, numerous immature neurons with small cell bodies and horizontal or oblique dendrites, radial glia fibers, and immature astrocytes.

Period 2: 26–36 w.g. Period of preterm infant

In this period the major cytoarchitectonic event in the PFC is the appearance of Brodmann's 6 basic cortical layers (1909) within the cortical plate. This process occurs as early as the beginning of the preterm infant period, between 26 and 29 w.g. Although the 6 basic layers have morphological characteristics of future cortical layers, they are called "fetal", because they still display many immature features, such as migrating and immature neurons. An important observation is that the internal granular layer (layer IV) cannot be used as a cytoarchitectonic criterion for delineation of the PFC in the preterm infant brain, or even in newborn and infant brains. During these periods of development layer IV is present throughout the frontal lobe and cerebral neocortex, even in areas that are agranular in adulthood, such as the premotor and motor cortices (Brodmann, 1909;

Kostović, 1990a, this volume).

In the preterm infant, the subplate zone reaches its maximum around 28–30 w.g. (the ratio between CP and subplate is 1:6; Kostović, this volume). However, between 32 and 36 w.g., the thickness of the subplate zone is reduced at the bottom of the cortical sulci. In the Golgi-impregnated sections, fetal layer I (former marginal zone) contains several types of Cajal–Retzius neurons which in this period and the next (newborn; see section on Period 3) reach their state of complete dendritic and axonal

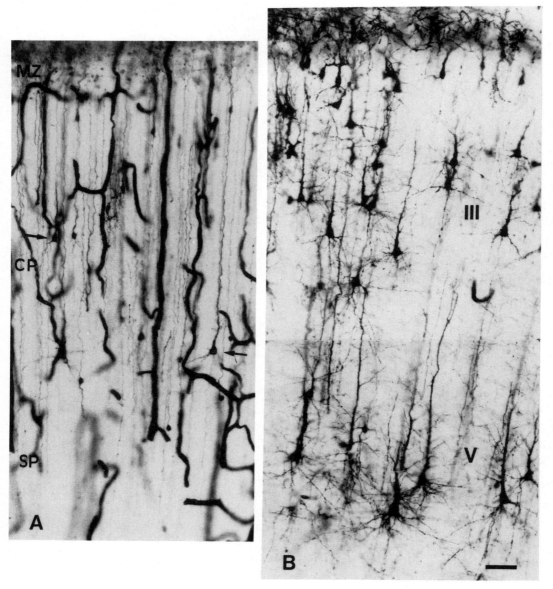

Fig. 5. (A) Pyramidal neurons in the cortical plate (CP) of a 22-week-old fetus (Golgi–Stensaas impregnation). (B) Pyramidal neurons in the fetal layers III and V of a 27-week-old preterm infant. Note significant dendritic differentiation in relation with fetal stages. Golgi–rapid impregnation. Bar scale is 20 μm.

development (see also Marin-Padilla, 1988). In the preterm infant, the Golgi architecture makes it possible to resolve cortical fetal layers on the basis of position and size of impregnated pyramidal neurons (Figs. 5 and 7). During this period the major observations in the Golgi picture are rapid dendritic and axonal differentiation of pyramidal neurons in the deep part of fetal layers III and V, and differentiation of nonpyramidal neurons. Pyramidal neurons in the deep part of layers III

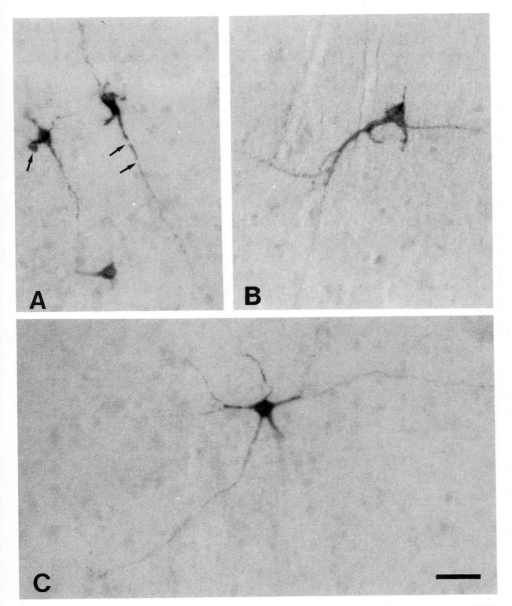

Fig. 6. Alz-50-immunoreactive neurons in the subplate zone of a 34-week-old preterm infant. (A) Two inverted pyramidal-like neurons which exhibit either swollen or interrupted dendrites (arrows). (B – C) Two immunoreactive neurons showing Golgi-like appearance: polymorphous neuron (B) and multipolar nonpyramidal neuron with stellate-like distribution of processes (C). Bar scale is 20 μm.

and V have grown much longer apical and basal dendritic trees with secondary and tertiary branches than was the case in the preceding period of development (Fig. 5). These qualitative observations are in agreement with quantitative data that show a rapid increase in the total length of basal dendritic trees of both layer III and layer V pyramidal neurons from 26/27 w.g. onwards (Fig. 8A). In contrast to these neurons, pyramidal neurons in the superficial half of layer III still display less developed basal and apical dendritic trees (Fig. 5). Between 26 and 29 w.g. the first dendritic spines are observed on apical and basal dendrites of fetal pyramidal neurons in layer V and, occasionally, on pyramidal neurons in the deep part of fetal layer III. In addition to pyramidal neurons, another type of projection neurons can now be discerned in infragranular layers, viz. fusiform neurons. Around 26 – 27 w.g., a rather developed type of nonpyramidal neuron can be identified in the CP in the supragranular layers and in layer IV. These interneurons are "double bouquet" neurons with characteristic columnar and axonal dendritic shape (Fig. 13; see also Fig. 12 in Mrzljak et al., 1988). The first immature basket interneurons are not observed in layer V until after 30 w.g. They are then characterized by a stellate distribution of dendrites and still sparsely developed axonal plexuses, without the "pericellular baskets".

All types of subplate neurons observed in the preceding developmental period are frequently impregnated throughout the subplate zone (Fig. 7). Measurements of the dendritic trees of subplate neurons demonstrate their continuous enlargement throughout this period (Fig. 8B). (Spine protrusions and growth features such as varicosities and growthcone-like terminal tips are observed on subplate neurons during this period.) About half the population of subplate neurons have axons ascending towards the cortical plate, while for the other half they descend towards subcortical structures. These observations suggest that among the multiple types of subplate neurons there are basically two populations with different functions: (a) projection neurons, and (b) interneurons for the cortical plate and subplate zone itself (see also Mrzljak et al., 1988; Shatz et al., 1988). At the end of the preterm infant period (36 w.g.) another interneuron type appears in the subplate zone, viz. a

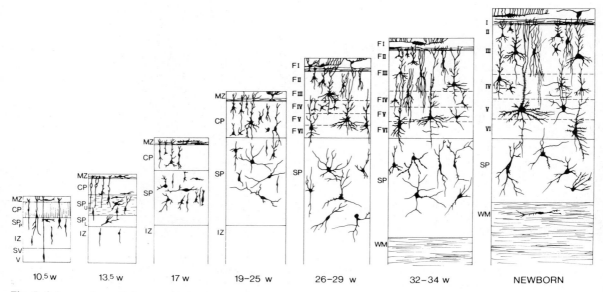

Fig. 7. Scheme of prenatal neuronal development in the prefrontal cortex. (From Mrzljak et al., 1988)

small type of neuron with strictly local axonal arborization, which probably corresponds to the "neurogliaform cell" of Cajal (Cajal, 1899a, b, 1900; Fairén et al., 1984; see also Fig. 16B and below, the section on Period 3).

Although immunoreactive neurons labeled with the monoclonal antibody Alz-50 are occasionally present in the subplate as early as 25 w.g., a large increase in their number is observed in the preterm infant period around 34 w.g. All types of subplate

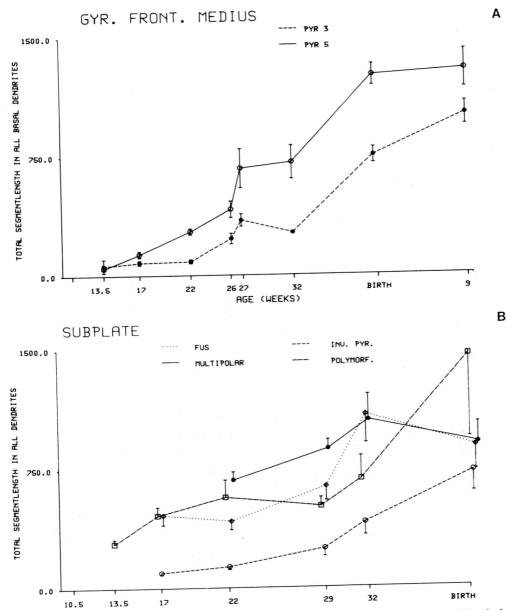

Fig. 8. Total dendritic length (mean with SEM in micra) of basal dendrites per pyramidal neuron (A) and of various types of subplate neurons (B) during prenatal and perinatal development. Pyr 3 and pyr 5 are pyramidal neurons from fetal layer III and V, respectively. FUS: fusiform neuron; multipolar: multipolar nonpyramidal neuron; inv. pyr.: inverted pyramidal cell; polymorf: polymorphous neuron.

neurons are stained with Alz-50, i.e. pyramidal, multipolar nonpyramidal, fusiform and polymorphous neurons (Fig. 6). The Alz-50 immunoreactivity is present in the axonal and dendritic domains and the majority of neurons have a Golgi-like appearance, but some Alz-50 immunoreactive (Alz50-IR) neurons have swollen and patchily stained dendrites (Fig. 6A). In the preterm infant, Alz50-IR neurons are present exclusively in the subplate zone. Only very rarely are these neurons found in the CP. In addition, the monoclonal antibody Alz-50 stains much fewer subplate neurons than the Golgi–Nissl staining, so that only a minority of the subplate neurons are visible in Alz-50 preparations. The highest frequencies of Alz50-IR neurons have been observed from 34 w.g. until birth. From birth to 6 years of age this frequency decreases until it virtually reaches the zero level.

Period 3: First postnatal year. Neonatal period and infancy

From birth onward the cortical layers (I – VI) cannot be regarded as "fetal" layers, since the

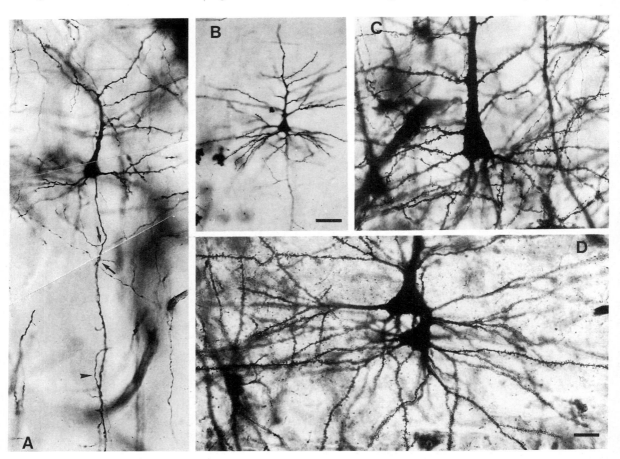

Fig. 9. Pyramidal neurons in layers III and V of the prefrontal cortex. Golgi – rapid impregnation. (A – C) 1-month-old infant. (A) Pyramidal neuron from the superficial part of layer III. Arrows point to the origin of recurrent axonal collaterals. The arrowhead points to a fiber of unknown origin contacting the pyramidal neuron axon. (B) Pyramidal neuron from the lower half of layer III. Note the well-differentiated basal and apical dendritic tree. (C) Layer V pyramidal neuron. Numerous dendritic spines are present on apical and basal dendrites. The advanced differentiation of this neuron is comparable with the layer V pyramidal neuron in a 15-month-old child (D). Bar scales are 50 μm for B and 20 μm for A, C, D.

neurons are positioned in their lamina of destination. In comparison with the preceding periods of development, the first postnatal year shows a large increase in size of dendritic trees and in spine growth on pyramidal neurons. This contributes to the further establishment of the 6-layered cortex by an increase in the amount of neuropil and a decrease in cell-packing density (see also Kostović, this volume). Furthermore, several types of interneurons can be distinguished in the supragranular layers that were not observed in the preterm infant. Another important observation in the PFC is the gradual resolution of the subplate zone during the first 6 postnatal months (Kostović et al., 1989a; Kostović, this volume). However, the part of this zone in the gyri of the PFC, in contrast with that in other cortical areas (i.e. visual and sensorimotor cortex), continues to contain neurons in the newborn infant (Mrzljak et al., 1988; Kostović et al., 1989a; Kostović, this volume).

Cajal – Retzius neurons in layer I reach their state of complete dendritic and axonal development in newborn and preterm infant. All types of Cajal – Retzius neurons described by Marin-Padilla and Marin-Padilla (1982) in the human motor cortex are also found frequently in layer I of the PFC. During the first 3 postnatal months, impregnation of various horizontal types of Cajal – Retzius neurons (Fig. 18A – C) shows that the pyriform, triangular type, with its characteristic descending process, prevails (Fig. 18D, E). After 3 postnatal months a reduced number of processes is observed in the horizontal type of Cajal – Retzius neurons, especially in those that emanate from horizontal dendrites towards the pial surface.

The progressive differentiation of pyramidal neurons during the infant period encompasses the ones in the supragranular as well as those in the infragranular layer. Even small pyramidal neurons in layer II and the superficial part of layer III exhibit well-developed basal and apical dendritic trees with growth of spines and axonal collaterals emanating from the main axonal arbor (Fig. 9A). Pyramidal neurons in layer V (Fig. 9C) still have larger cell bodies and dendritic trees than

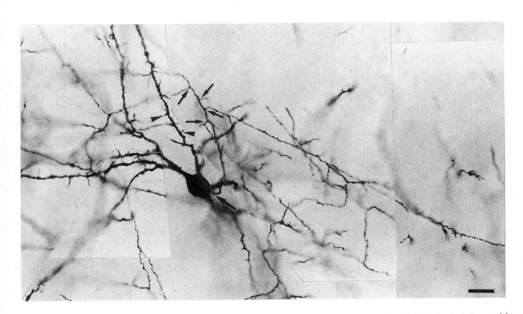

Fig. 10. Large multipolar basket neuron from layer V of the PFC of a 1-month-old infant. Golgi – rapid preparation. Large arrow points to the axonal origin and small arrows to the origin of horizontally running axonal collaterals. Arrowheads indicate dendritic spines. Bar scale is 20 μm.

pyramidal neurons in the lower half of layer III (Fig. 9B). The basal and apical dendrites of these layer V pyramidal neurons are covered with a large number of spines. Their cell bodies have only a few spines. During the first 6 postnatal months, pyramidal neurons in layer III, too, have spines on terminal tufts of the apical, oblique and basal dendrites, but fewer than the number found on layer V pyramidal dendrites. The spines found on pyramidal neurons are largely immature, i.e. long, hair-like spines prevail.

In parallel with the differentiation of the pyramidal neurons, an intensive development of interneurons occurs in both the supragranular and the infragranular layers. The *supragranular* layers now contain not only double bouquet neurons, but also basket neurons, multipolar and bitufted nonpyramidal neurons with extended axonal arborization, as well as chandelier and neurogliaform neurons. During the first postnatal year the axonal arbors of interneurons are unmyelated, allowing visible impregnation of large parts of them and of most of the fine terminal endings. The interneurons are covered with a significant number of long and short dendritic spines. Double bouquet neurons observed in layer III usually have a bitufted distribution of dendrites, while their axons have numerous ascending and descending collaterals, which run towards the superficial zone of the cortex and deep into the infragranular layers, respectively, forming column-shaped axonal arbors (Fig. 12). Axonal collaterals have characteristic short-stalked boutons (Fig. 12) that represent sites of synaptic contacts (Somogyi and Cowey, 1981). Such mature double bouquet neurons can be recognized as early as the first 3 months of postnatal life. In the same period, basket neuron-like multipolar nonpyramidal neurons, with ascending axons and widely distributed axonal trees, are impregnated in layer III next to the characteristic basket neurons (Fig. 11B). The appearance of their axonal arbors, with horizontal collaterals ending in terminal networks similar to those in basket neurons, confirms the obervations of Fairén et al. (1984) that there is no marked distinction between these neurons and basket neurons. The first "chandelier" neurons

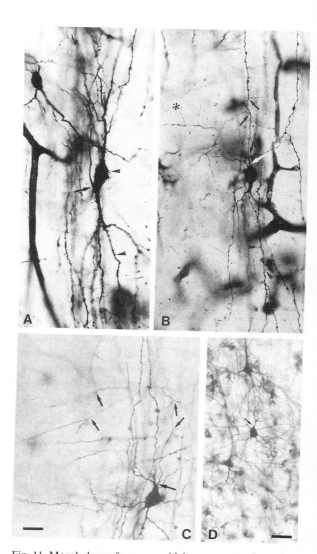

Fig. 11. Morphology of nonpyramidal neurons at various stages of development. (A) Bitufted nonpyramidal neuron at the interface of layer VI and subplate zone. The arrow points to the origin of ascending axon. Arrowheads point to spine-like protrusions on the cell body and dendrites. 1-month-old infant. Golgi – rapid impregnation. (B) Multipolar layer III nonpyramidal neuron with ascending axon (white arrow) and horizontal collaterals (arrows) from which further collaterals arise ending with terminal brackets (asterisk). Arrowhead points to the spines on the dendrite. 1-month-old infant. Golgi – rapid impregnation. (C) Multipolar layer III nonpyramidal (basket) neuron with ascending axon (large arrow points to its origin) and horizontal axonal collaterals (arrows). 5-year-old child. Golgi – Cox impregnation. (D) Multipolar neuron from layer III with stellate distribution of dendrites and only initial ascending part of axon impregnated (arrow). 19-year-old man. Golgi – rapid impregnation. Bar scales are 20 μm for A – C and 50 μm for D.

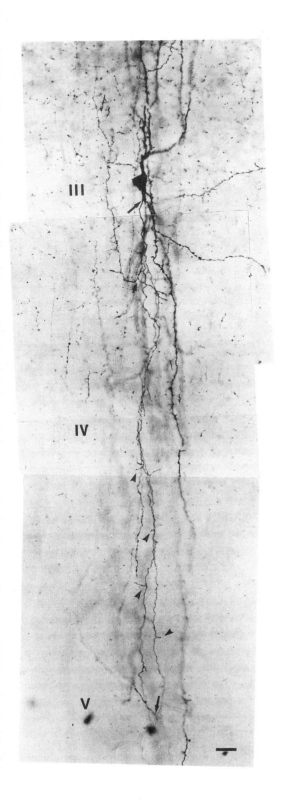

can be observed in our material in a 3-month-old child (Fig. 14A). Their cell bodies are oval, with a multipolar or bitufted arrangement of dendrites. The elongated axonal trees consist of complex rows, in each of which 50–70 vertical terminal sprouts were impregnated. These terminal sprouts have an immature appearance, with enlarged boutons and filopodia.

In *infragranular* layers the growth of dendrites and axons of interneurons in this period is significant in comparison with the preterm infant stage. Already during the first postnatal months, layer V basket neurons have long horizontal axonal collaterals (up to 700 μm), which give rise to further branches, ending in basket-like terminals (Fig. 10). An additional type of neuron can frequently be found in layer VI, at the border of the subplate zone and deep in the subplate zone (Fig. 11A). These nonpyramidal neurons have fusiform or oval cell bodies and a vertical bitufted or bipolar arrangement of very long dendrites (up to 800 μm in the subplate). The axon always ascends from the upper or lower pole of the cell body, with its collaterals widely distributed in the infragranular layers.

Neurogliaform neurons are frequently found throughout all cortical layers (I–VI) and in the subplate zone, but occur mostly at the interface between layers I and II. These small interneurons, which in the preterm infant are only found in the subplate zone, are already present in the cortical layers during the first 3 postnatal months. Neurogliaform neurons have very dense axonal plexuses, which are mainly distributed in the domain of the branched dendritic tree and are

---

Fig. 12. Double bouquet neuron from layer III in 1-month-old infant. Golgi–rapid impregnation. Axon descends from the lower main dendrite (large arrow) giving ascending and descending collaterals. Upper smaller arrow points to the origin of the ascending collateral, and the lower arrow to another ascending collateral in layer V. Note the columnar shape of the axonal arbor which both runs in the more superficial parts of the cortex and descends into infragranular layers. Arrowhead points to the short stalks on the axon which end in terminal boutons. Bar is 20 μm. 1-month-old infant. Golgi–rapid impregnation. Roman numbers indicate cortical layers.

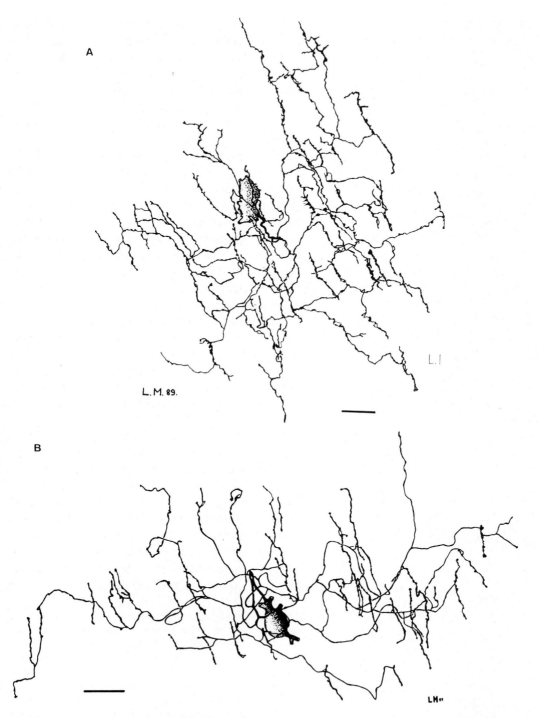

Fig. 13. Camera lucida drawing of cell bodies and axonal trees of 2 chandelier neurons based on Golgi – rapid impregnation. Dendrites are not shown. (A) Chandelier neuron in a 3-month-old child. Note the numerous vertical terminal endings, some of which consist of enlarged boutons and filopodia. The axonal tree has a more vertical extension. Arrow points to the axonal origin. (B) Chandelier neuron in a 6-year-old child. The axonal tree shows a more horizontal extension and is only partly impregnated due to the myelination. Bar scales are 30 μm for A and 40 μm for B.

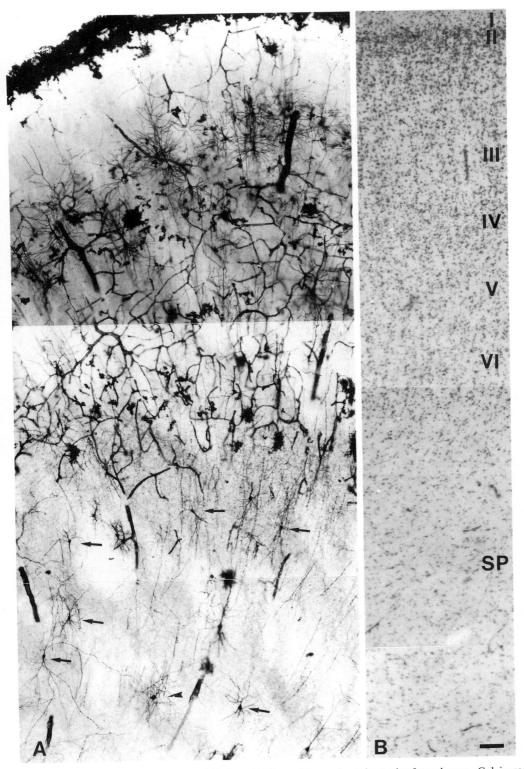

Fig. 14. Subplate neurons in a 1-month-old infant. (A) Part of the prefrontal superior frontal gyrus, Golgi – rapid stained. (B) Adjacent Nissl section. Arrows point to some of the numerous subplate neurons. Arrowhead points to the subplate neuron enlarged in Fig. 15B. Roman numbers in Nissl-stained section indicate layers. Bar scale is 100 μm.

therefore hard to trace (Fig. 15B).

During the first 6 postnatal months the subplate zone still persists below layer VI in the gyral zone (Fig. 16A, B). Most of it, however, is incorporated into the developing gyral white matter. Therefore, a part of the subplate neuron population will be located in the white matter and turn into interstitial neurons (Fig. 15C, D). The subplate and interstitial neurons in the gyral zone have a random orientation of dendrites and cell body, in which the vertical orientation prevails (Fig. 14A). However, in the sulcal region, but not at the bottom of the sulcus, the interstitial neurons below layer VI have a predominantly horizontal orientation (Fig. 15C, D). The subplate zone is characterized by a multiplicity of neuronal types (Figs. 14A and 15A, B). These include fusiform neurons, inverted pyramidal neurons (Fig. 15A), ordinarily oriented pyramidal neurons, neurogliaform neurons (Fig. 15B), multipolar and bitufted nonpyramidal neurons with ascending axons towards the cortex as well as bizarre polymorphous neurons with either short or extremely long curved dendrites. They are all frequently impregnated in the subplate zone, and all have spines. In the white matter the orientation of neuronal cell bodies and dendrites follows the direction of fiber bundles (Figs. 15C, D and 17A). Fusiform neurons with very long dendrites are the most frequently impregnated cell type in the white matter fiber bundles (Figs. 15C, D and 17A). A constant increase of spine number can be observed in these neurons between 1 and 7 months of age (see neuron in Fig. 17A). Besides fusiform neurons, slightly modified pyramidal neurons and multipolar nonpyramidal neurons can also be found in the white matter (see Fig. 15C). In contrast with the "healthy"-looking subplate and interstitial neurons, some of their neighboring neurons have disrupted dendrites. Fragments of cell bodies are missing, and they are in close contact with glial neurons (see Fig. 17B).

Alz50-IR neurons are frequently observed in the subplate zone and white matter between the newborn stage and the sixth postnatal month (Fig. 15E – G). Their occurrence is comparable with that of immunoreactive neurons in the late fetal period, around 34 w.g. The shape of Alz50-IR neurons in the subplate and white matter does not change from the preterm infant period either: various types of neurons are stained showing a Golgi-like image and a morphology identical to that of the Golgi-impregnated subplate and white matter neurons in the corresponding stages (Fig. 15E – G). Alz50-IR neurons are also found in the cortical layers II – VI during the first postnatal year. The frequency of Alz-50-positive neurons in the CP, however, is much lower than that in the subplate and white matter. After the sixth postnatal month Alz50-IR neurons are observed less frequently, both in the subplate and in the CP.

Period 4: Second postnatal year. Early childhood period

During the second postnatal year, significant changes occur in the Golgi architecture of the PFC. On average, the cell bodies of pyramidal neurons in the deep part of layer III become larger than those in layer V, giving the PFC its "magnopyramidal" appearance (see Period 6). Thus, from this stage on the deepest part of layer III, which predominantly contains these large pyramidal neurons, is regarded as sublayer IIIc (Fig. 16A). In addition to a further enlargement of the dendritic trees, a spatial reorientation of basal dendrites was observed, especially in pyramidal

---

Fig. 15. (A – C) Subplate and white matter neurons in 1-month-old infant. Golgi – rapid impregnation. (A) Inverted pyramidal subplate neuron. (B) Neurogliaform-like subplate neuron. Arrow points to axonal origin. (C) Three white matter neurons. Note long and "healthy"-looking dendrites oriented in direction of fiber bundles (2 lower neurons), and beading of dendrites (arrowhead) in the upper neuron. (D) White matter neurons in a 2.5-month-old infant. Golgi – rapid impregnation. (E – G) Alz-50-immunoreactive white matter neurons in a 3-month-old child. Note that their morphology is very similar to those shown in C and D. Bar scales are 50 μm for A – C, 100 μm for D and 20 μm for E – G.

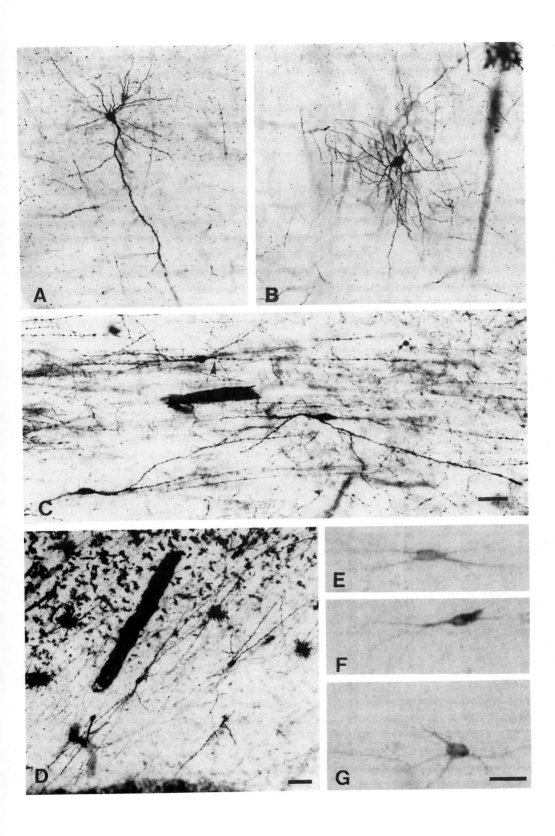

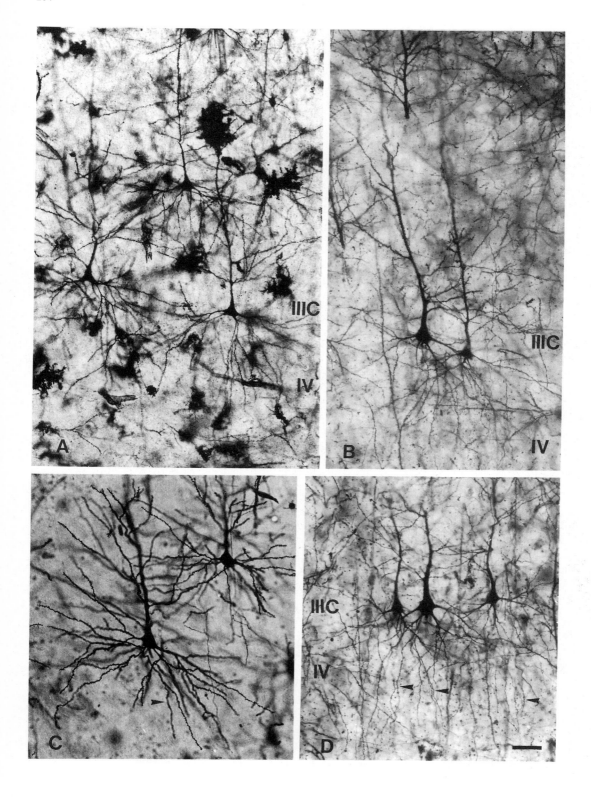

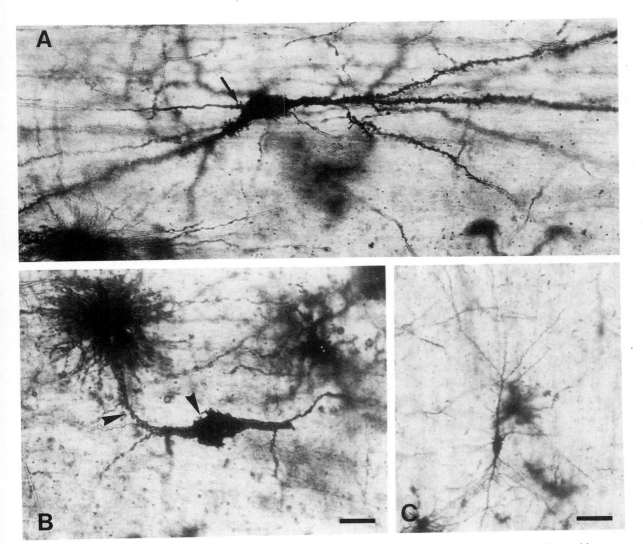

Fig. 17. Morphology of white matter neurons during postnatal development. Golgi – rapid impregnations. (A) Fusiform white matter neuron in a 7-month-old infant with orientation of primary dendrites in the direction of the white matter fiber bundles. Numerous dendritic and also somatic spines are visible. Arrows point to axonal origin. (B) Degeneratively looking neuron in the white matter below PFC of a 15-month-old child. Arrowheads point to the signs of degeneration of primary dendrites and cell body. Note that the neuron is in close contact with neighboring astrocytes. (C) Small white matter neuron 0.5 mm below layer VI. The orientation of the neuron is vertical to the pial surface. 19-year-old man. Bar scales are 20 μm for A and B and 50 μm for C.

neurons in layer V. This results in shape differences between layer III and layer V pyramidal neurons comparable to the ones observed in adult stages. Pyramidal neurons in layer IIIc distribute their basal dendritic trees in the deep part of layer III or deep into layer IV (Fig. 16A). Pyramidal neurons in layer V, by contrast, acquire their characteristic basal dendritic tree, with a horizon-

Fig. 16. Morphology of layer IIIc pyramidal neurons. (A) 15-month-old child. Golgi – rapid impregnation. (B) 19-year-old man. Golgi – rapid impregnation. (C) 5-year-old child. Arrowhead points to axon. Golgi – Cox impregnation. (D) 64-year-old woman. Golgi impregnation. In all figures the characteristic distribution of basal dendritic trees in layers IIIc and IV is visible. Arrowheads in Fig. 10D indicate the basal dendrites running into layer IV. Bar scale is 50 μm for A – D.

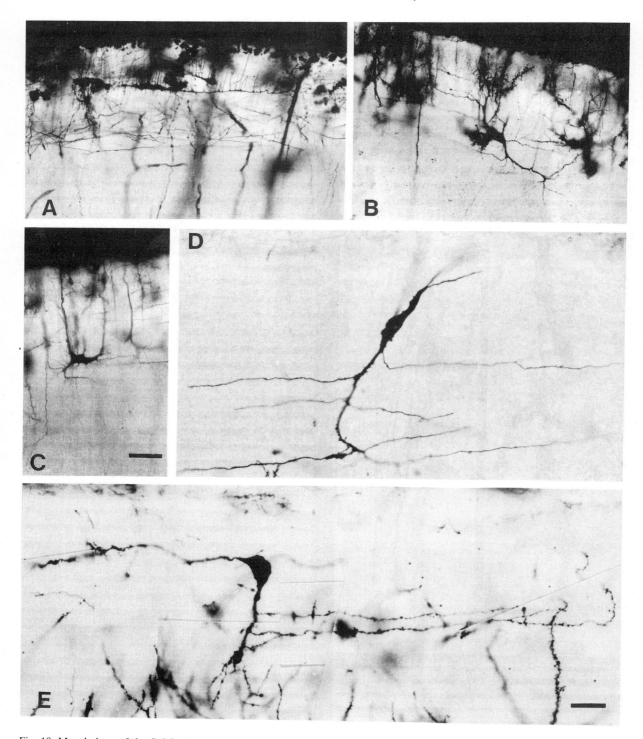

Fig. 18. Morphology of the Cajal–Retzius neurons during peri- and postnatal development. (A – C) Variety of Cajal–Retzius cell morphology in the newborn: classical horizontal bipolar type (A) and multipolar horizontal type (B, C). Golgi – rapid impregnation. (D) Bipolar vertically oriented Cajal–Retzius neuron from a 15-month-old child. Several very long horizontal processes derive from

tal extension of long, bundled dendrites, during early childhood (Fig. 9D). The number and density of the spines on the pyramidal neurons, especially in layer III, show a large increase in comparison with the previous period. No spines are observed on the pyramidal cell bodies after 1 year. More mature, short, "mushroom"-like spines predominate on pyramidal neurons in layers III and V. Contrary to the pyramidal neurons, nonpyramidal neurons show a distinct overall reduction in spine number. The remaining spines are now very sparse and short. The extent of impregnation of nonpyramidal axonal trees is less than in the previous stage.

In layer I, characteristic bipolar vertical (Fig. 18D) and horizontal types of Cajal–Retzius neurons are impregnated in a 1- and a 2-year-old child. The triangular types have the same morphology as in the previous period, having a descending process which gives rise to the horizontal axonal plexus (Fig. 18E). Horizontal forms of Cajal–Retzius neurons look somewhat different since they have lost most of their ascending and descending processes.

Within the white matter, all types of interstitial neurons, being derived from all types of subplate neurons, are frequently impregnated. Besides these neurons, "degenerative"-looking neurons are impregnated throughout the white matter (Fig. 17). Only a small number of Alz50-IR neurons are stained in the white matter during the second postnatal year.

Period 5: Period of childhood and adolescence

In the child brain, between postnatal years 2 and 10, the Golgi architecture gradually becomes adult-like. The presence of large pyramidal neurons in layer IIIc (Fig. 16C) is a dominant characteristic of the PFC in Golgi–Cox as well as Golgi–rapid stained material. The size of the basal and apical dendritic trees of layer IIIc pyramidal neurons is considerably larger at 5 years of age than in the second postnatal year (compare Fig. 10A and C). These dendrites are covered exclusively with adult-like short spines. In the puberty period, between 11 and 19 years of age, the morphology and size of the dendritic trees of layer III and layer V neurons cannot be qualitatively distinguished from the adult stages (Fig. 16B, D). Therefore, further morphometric analysis is necessary to see whether or not changes occur in the dendritic trees of these cortical neurons during the period of puberty.

Between 5 and 11 years of age, axonal plexuses of pyramidal and nonpyramidal neurons are only partly impregnated, due to myelination (Figs. 11 and 13B). The impregnated parts appear to be mature, just like the characteristic vertical axonal terminals of chandelier neurons, which are now composed of mature boutons. The latter are slender in comparison with the enlarged boutons and filopodia observed in the earlier period (Fig. 13B).

Surprisingly, we observed impregnated neurons with morphological characteristics of Cajal–Retzius neurons (see previous periods described above) in layer I of several child brains. The PFC of 5- and 6-year-old children contains neurons with a horizontal extension of dendrites, as well as types with a triangular or oval soma with a characteristic descending dendritic process, from which long horizontal axon-like processes arise (Fig. 18E). These neurons, which are considered to be primarily a fetal phenomenon, can be clearly distinguished from other layer I neurons, such as multipolar nonpyramidal neurons with an axon which arborizes only in the direct vicinity of the soma. On the basis of their resemblance with other fetal forms and their distinctly different morphology from other layer I neurons, we regard these neurons as late postnatal Cajal–Retzius neurons. Such horizontal and triangular or oval types of neurons are even found in a 19-year-old specimen.

During childhood, puberty and adult stages,

---

the descending dendrite. Golgi–Cox impregnation. (E) Layer I neuron in a 5-year-old child. The neuron has all morphological characteristics of a Cajal–Retzius cell. Note the similarity with the neuron in Fig. D. Golgi–Cox impregnation. Bar scales are 50 µm for A–C and 20 µm for D and E.

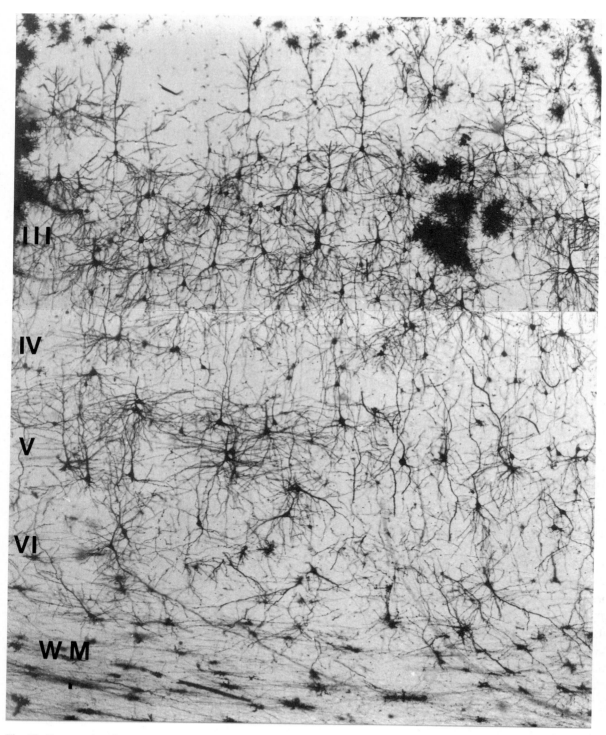

Fig. 19. Cross-section through the PFC of a 5-year-old child, Golgi – rapid impregnation. Note the difference in the orientation of the basal dendrites in pyramidal cells between layer III (horizontal and perpendicular) and layer V (primarily horizontal). Roman numbers indicate cortical layers. W.M. = white matter.

several types of interstitial neurons are observed in the white matter. These types, which have been described in detail by Kostović and Rakic (1980), comprise inverted pyramidal and normally oriented neurons, fusiform neurons (Fig. 17C), multipolar nonpyramidal neurons, and polymorphous neurons with bizarrely shaped dendrites of varying length. The frequency of impregnation of these neurons starts to decrease in childhood and continues to do so up to adult stages. In our material no type of interstitial neuron is preferentially impregnated. In the gyral zone, interstitial neurons immediately below layer VI have a more random orientation than the ones in the fiber bundles of sulci and gyri, which follow the direction of the myelinated fiber bundles. Our immunohistochemical analysis shows that Alz50-IR neurons are present only rarely in the white matter of 5- and 6-year-old children. In puberty and adult brain no Alz50-IR neurons are found.

Period 6: Late adolescence and adulthood

In conclusion, we will describe the major features of the Golgi architectonics of Brodmann's areas 9 and 10 in the adolescent and adult human brain. In the Golgi – rapid and Golgi – Cox impregnated material it is possible to delineate layers even without Nissl-counterstained sections, thanks to the limited neuronal impregnation of layer IV, and the distinct morphology and arrangement of the pyramidal neurons in layers III and V (Fig. 19).

The major feature of this part of the PFC is the abundance of large pyramidal neurons in the deepest part of layer III (sublayer IIIc, see Fig. 16B, D). According to Vogt and Vogt (1919) these pyramidal neurons are, on average, larger than layer V pyramidal neurons, which gives this type of cortex a "magnopyramidal" appearance. Large pyramidal neurons in layer IIIc have truely pyramidal or slightly oval cell bodies (length of the cell body base $23-32$ μm; height $32-38$ μm). The entire basal dendritic tree, which consists of $6-8$ primary dendrites, extends from sublayer IIIc into layer IV (Fig. 10B – D). Primary dendrites which emanate from the cell body in sublayer IIIc have no clear preferential direction, although no primary dendrites run into the direction of the superficial part of layer III. The secondary branches and primary dendrites that descend obliquely or perpendicularly towards layer IV show a maximal extension of $400-500$ μm (Fig. 16D). Terminal tufts of the apical dendrite of the pyramidal neurons with $10-13$ oblique dendrites arborize either in layer I or in layer II. In adult and adolescent brains the most frequently impregnated type of nonpyramidal neurons are the multipolar aspiny neurons with a stellate distribution of dendrites. Only the initial part of their axon, which usually ascends from the upper half of the cell body, is impregnated (Fig. 11D). In the lower part of layer III these nonpyramidal neurons reach the dimensions of their large neighboring pyramidal neurons, with cell body diameters of up to 30 μm. According to the criteria for classification of nonpyramidal neurons (Jones, 1975; De Felipe et al., 1986), these neurons belong to the type of basket interneurons.

Although it is possible to find pyramidal neurons in layer V with cell bodies as large as those in layer IIIc, such pyramidal neurons are impregnated less frequently. Furthermore, their apical and basal dendritic trees are less complex than those of layer III neurons. Basal dendrites of layer V pyramidal neurons ($5-6$ primary dendrites) often form horizontal bundles with a maximum lateral extension of $500-600$ μm (Fig. 19). The demarcation between layers V and VI is based on the presence of fusiform and modified pyramidal neurons. Very frequently these neurons send their dendrites into the white matter below layer VI.

Discussion

Development of Cajal – Retzius and subplate neurons

Cajal – Retzius and subplate neurons belong to the earliest differentiated neurons in the PFC and other mammalian cortical areas. Recent autoradiographic investigations in the cat show that

Cajal–Retzius and subplate neurons are generated at the same time, before any neuron which constitutes the layers of the adult cortex is generated (Luskin and Shatz, 1985).

The Cajal–Retzius neurons in the marginal zone (fetal layer I) develop simultaneously in the human primary visual cortex (Marin-Padilla, 1970a, 1988; Poliakov, 1979; Krmpotić-Nemonić et al., 1987) and in association areas (Poliakov et al., 1985; Mrzljak et al., 1988) as well as in other neo-, meso- and even archicortical areas (Mrzljak et al., unpublished observations). The transient nature of Cajal–Retzius neurons in the mammalian cortex has long been subject of discussion, since it is argued by several researchers that they do not disappear but only change their characteristic fetal forms, and continue to be present in their new forms in the adult layer I (Fox and Imman, 1966; Bradford et al., 1977; Parnavelas and Edmunds, 1983; Sas and Sanides, 1970; Cajal, 1899b, 1911; Conel, 1939–1967; but see also Derer and Derer, 1990). In the fetal human brain they exhibit various morphologies, and reach their developmental peak in the preterm infant and neonatal brain (Cajal, 1899a, b, 1900, 1911; Marin-Padilla, 1988; Mrzljak et al., 1988). According to Cajal (1899b, 1911), the typical fetal Cajal–Retzius neurons are no longer found in the human cortex after the second postnatal month. Conel (1939–1967) has described these neurons in the first 3 postnatal months and reported their degeneration by the end of the third month. We have frequently found Cajal–Retzius neurons throughout the first postnatal year and even in a 25-month-old child. Compared with typical fetal Cajal–Retzius neurons in the preterm infant and neonatal brain, the ones that are observed during the first postnatal year and in early childhood have undergone morphological changes, i.e. dendrites and axonal branches are reduced in number. During childhood and adolescence, layer I of the PFC was still found to contain neurons with a morphology similar to that of Cajal–Retzius neurons. This confirms Cajal's (1899b, 1911) original suggestion about their possible persistence in the mature cortex. Recent experimental evidence in the rodent brain shows that at least some Cajal–Retzius neurons remain in the adult cortex but that they have taken the form of nonpyramidal neurons (Parnavelas and Edmunds, 1983). In addition, Cajal–Retzius neurons are found in the adult primate paleocortex (Sanides and Sas, 1970). Although Cajal–Retzius neurons are part of the fetal cortical circuitry, their long axons may represent pioneer axon pathways for the establishment of later permanent pathways in layer I, a function that has also been suggested for subplate neurons (McConnell et al., 1989). The frequent persistence of Cajal–Retzius neurons in the PFC during the infant and childhood periods could suggest their prolonged involvement in the establishment of cortical circuitry. However, later regressive changes and transformation of their morphology may indicate a different and more limited role in the mature brain.

In the PFC the subplate zone is formed between 13.5 and 15 w.g., its neurons being derived from at least 2 fetal layers. Part of the population is derived from the primordium of the subplate zone before 13.5 w.g. The majority of the neurons, however, are probably derived from what is called the lower part of the cortical plate, between 13.5 and 15 w.g. These neurons will constitute the upper part of the subplate (SPu) (Mrzljak et al., 1988; Kostović and Rakic, 1990). The neuronal differentiation, dendritic outgrowth and ingrowth of afferent fibers into the subplate makes this fetal zone the most prominent part of the cortical anlage. The afferent fibers remain in the subplate for a certain period of time before they enter into the cortical plate (Rakic, 1977; Kostović and Goldman-Rakic, 1983; Kostović and Rakic, 1984, 1990). Following the ingrowth of afferent fibers into the subplate zone, between 17 and 25 w.g., several types of subplate neurons differentiate. These neurons make synaptic contacts with afferent fibers much earlier than the neurons in the cortical plate (Kostović and Molliver, 1974; Kostović and Rakic, 1990).

The major characteristics of the transient

subplate zone in the human PFC are differentiation of a large number of neuronal types in this zone and its postnatal transformation. In the visual and somatosensory cortices the subplate can no longer be distinguished at birth (Kostović and Rakic, 1990). By contrast, it is still visible in the newborn PFC and declines gradually during the first 6 postnatal months (Mrzljak et al., 1988; Kostović et al., 1989a; Kostović, this volume). Subplate neurons may play a role as interneurons for the developing cortical plate or the subplate zone itself, or they are projection neurons (Mrzljak et al., 1988; Shatz et al., 1988; McConnell et al., 1989). The axonal course and morphology of dendritic trees shows that many of the subplate neurons meet the criteria for local circuitry neurons (interneurons). True interneuronal types are found in the subplate, e.g. neurogliaform neurons, and multipolar and bitufted neurons with ascending axons and intrinsically distributed axonal plexuses. In addition, studies in experimental animals show that many of the subplate neurons synthesize GABA (Wahle et al., 1987; Van Eden et al., 1989; Lauder et al., 1987), known as the main transmitter substance of interneurons. Another role of subplate neurons, viz. that of subcortical and callosal projection neurons, is also well established from experiments in cats and monkeys (Chun et al., 1987; Schwartz and Goldman-Rakic, 1984, 1986; McConnell et al., 1989). Early subcortical projections of subplate neurons may play a pioneering role in establishing permanent subcortical projections (McConnell et al., 1989). One of our most important findings is the continuation of dendritic growth and differentiation of new neuronal types in the subplate in the preterm, newborn and early infant. This observation is interesting in the light of the prolonged functional role of subplate neurons in the developing PFC circuitry and the transient nature of these fetal neurons. Kostović and Rakic (1990) and Kostović (this volume) have suggested that the prolonged development and persistence of subplate neurons in the association cortices may be explained by the role of the subplate zone as a transient compartment of growing cortico-cortical connections during the formation of cerebral convolutions.

Eventually, subplate neurons are either incorporated into the white matter, where they persist as interstitial neurons (Kostović and Rakic, 1980; Chun and Shatz, 1989), or eliminated by cell death (Kostović and Rakic, 1980; Luskin and Shatz, 1985; Valverde and Facal-Valverde, 1987, 1988; Wahle and Meyer, 1987; Chun and Shatz, 1989). Giguere and Goldman-Rakic (1988) have found that in the adult monkey PFC many of these interstitial neurons have a functional role as projection neurons to the thalamus. The observation of "healthy"-looking spiny, spine-sparse and aspiny interstitial neurons without morphological signs of degeneration in the child, adolescent and adult brain supports the idea that in the human brain, too, part of the subplate neurons survive as interstitial neurons in the white matter (see also Kostović et al., 1989b). On the other hand, the presence of subplate and interstitial neurons that look "degenerated", and the constant decrease in frequency of impregnated interstitial neurons from the infant period onward may indicate a constant prolonged elimination of subplate neurons.

Alz-50 is a monoclonal antibody that recognizes a protein of 68 kDa in patients with Alzheimer disease and Down syndrome, but is not detectable in normal controls (Wolozin et al., 1986, 1988). Alz-50 immunoreactivity is also expressed in the normal human and rat developing cortex (Wolozin et al., 1988; Al-Ghoul and Miller, 1989; Hamre et al., 1989). The appearance of Alz-50 immunoreactivity in the PFC subplate and interstitial neurons is in agreement with recent suggestions that it can be either a "precursor" of their imminent degeneration (Wolozin et al., 1988; Al-Ghoul and Miller, 1989) or a sign of the transient expression of a specific protein during development, the function of which is as yet unknown (Wolozin et al., 1988; Hamre et al., 1989). The temporal and spatial appearance of Alz50-positive (Alz50-IR) neuronal types might suggest the former. In the period between 32 w.g. and the sixth postnatal month, when the gradual involution of the subplate occurs

(Kostović et al., 1989a; Kostović, this volume) and the subplate zone is a "high risk" region of cell death, the highest numbers of Alz-50-labeled neurons are found predominantly in the subplate and subjacent white matter (see also Wolozin et al., 1988). Later immunoreactive interstitial neurons in the late infant and childhood period could be markers for continued elimination of these transient neurons. Alz-50 antibody labels all kinds of subplate and interstitial neurons, so if Alz50-IR neurons really die during development, cell death would occur in all types of subplate neurons without predominance of any neuronal type during a particular period. On the other hand, immunoreactive subplate and interstitial neurons look "healthy", exhibiting Golgi-like morphology. In the developing rat cortex Hamre et al. (1989) have found transient expression of Alz-50 in a wide variety of cortical and subplate neurons, showing that Alz-50 recognizes an epitope that is altered during development. In contrast to rat brain (Al-Ghoul and Miller, 1989; Hamre et al., 1989), human brain shows only very rare labeling of Alz-50 antibody in cortical layers II – VI and far fewer labeled neurons in the subplate.

Development of pyramidal neurons

During development, pyramidal as well as non-pyramidal neurons follow the deep to superficial laminar gradient of dendritic maturation (Vignal, 1888; Kölliker, 1896; Cajal, 1911; Marin-Padilla, 1970a; Parnavelas et al., 1978), which has already been determined by their days of birth in the germinative zones of the cortical anlage (Rakic, 1972, 1975, 1982, 1988a, b). Neurons generated earlier take position deeper in the cortex and generally have more developed dendrites. The inside-out rule of differentiation, i.e. the rule that the dendritic and spine growth of layer V pyramidal neurons precedes the maturation of layer III neurons, is valid only until the first postnatal year. The prenatal and postnatal development of pyramidal neurons in the PFC is made up of several phases. Initially, between 13.5 and 15 w.g., only immature pyramidal neurons are found in the cortical anlage. Between 15 and 25 w.g., all basic dendritic features of pyramidal neurons are established. A dendritic growth spurt of pyramidal neurons in both the supragranular and infragranular layers occurs in the late fetal period of the preterm infant (around and after 26/27 w.g.), with the appearance of the first spines. Further dendritic outgrowth and spine formation continues throughout the first 2 postnatal years. In this period, the average size of pyramidal neurons in the deep part of layer III (sublayer IIIc) surpasses that of the layer V neurons, giving this part of the PFC its magnopyramidal characteristics. In Golgi architectonics this "magnopyramidality" and the further growth of the dendritic trees of pyramidal neurons become gradually more pronounced in the infant period, between postnatal years 2 and 10. The age at which the pyramidal neurons of layers III and V reach maximum extension and an adult-like size of dendritic fields cannot be determined by qualitative observations only. Our impression is that it might happen in the adolescent period, but a postnatal morphometric analysis is required to confirm this (Uylings et al., in prep.).

The major characteristic of pyramidal neuron maturation in Brodman's areas 9 and 10 of the human PFC is a prolonged postnatal elaboration of dendritic trees and spines, especially of layer III pyramidal neurons. Initially, during earlier fetal development (17 – 25 w.g.), the PFC lags behind the primary neocortical areas with regard to the differentiation of pyramidal neurons (for motor and visual cortices see Poliakov, 1949, 1979; Marin-Padilla, 1970a; Takashima et al., 1980). However, the late prenatal dendritic and spine growth spurt on pyramidal neurons, which is influenced by the ingrowth of afferent fibers in the late fetal period (26 – 34 w.g.), takes place simultaneously in the PFC and other neocortical areas (Mrzljak et al., 1988). The further intensive dendritic differentiation of pyramidal neurons in the PFC during the first 2 postnatal years also occurs in parallel to that in other neocortical areas, and a similar level of differentiation is obtained in this period (Poliakov, 1949, 1961, 1966; Conel, 1939–1967). Moreover, the appearance of the magnocellular character of the PFC during the second postnatal year occurs simultaneously in some other "association" neocortical areas (Poliakov, 1961). The establishment of the adult dendritic features of layer III pyramidal neurons, however, appears to be a more prolonged postnatal event in the PFC than in primary neocortical areas. In the visual cortex, the maximal dendritic length of layer III pyramidal neurons has been reported around postnatal year 2 (Becker et al., 1984). Schadé and Van Groenigen (1961) reported that layer III pyramidal neurons in the PFC (probably area 46; see Fig. 1 in their article) continues to grow after postnatal year 1. Our preliminary results on the development of spines on layer IIIc pyramidal neurons also indicate prolonged, fine remodeling of the pyramidal neurons (Mrzljak et al., in prep.). The maximal number of spines is reached during postnatal year 2 and a high spine number is retained during childhood and adolescence, followed by a decrease in adulthood. In addition, Kostović et al. (1988) and Kostović (this volume) have found that chemical maturation of layer III pyramidal neurons is a late and prolonged postnatal event. During this process, AChE-positive pyramidal neurons in layer III are only found after the first postnatal year, with the strongest activity in young adults.

Development of nonpyramidal neurons in cortical layers II – VI

In the present study we have largely limited our observations to the maturation of neuronal types whose morphology, chemical anatomy and neurophysiology is well explored (for review see Fairén et al., 1984; Jones 1975, 1988; Martin 1988), viz. double bouquet, chandelier, basket and neurogliaform neurons. Basket and chandelier cells are types of GABAergic neurons, first forming synapses with cell soma and proximal dendrites of pyramidal neurons, and then exclusively with initial segments of pyramidal neurons (Somogyi et al., 1982; Freund et al., 1989; De Felipe et al., 1985, 1986). Double bouquet and neurogliaform neurons accumulate GABA, display peptide immunoreactivity, and make synaptic contacts with more distant parts of dendrites of pyramidal and nonpyramidal neurons (Somogyi and Cowey, 1981, 1984; Somogyi et al., 1984; Kuljis and Rakic, 1989; Jones et al., 1988). Besides these interneurons we have only briefly described 2 other types, i.e. supragranular multipolar neurons and infragranular bitufted neurons. The former have an extended axonal field with "axonal arcades" (see also type 2 cells, Jones, 1975), and cannot be sharply distinguished from basket neurons (Fairén et al., 1984), while the latter have ascending neurons.

There are several characteristics of interneuronal development in the human PFC: (1) Differentiation of nonpyramidal neurons (interneurons) parallels differentiation of pyramidal neurons. (2) Various types of interneurons differentiate according to a temporal and spatial sequence. (3) During the development of interneurons, there is transitory spine growth on dendrites and soma, and remodeling of axonal plexuses.

Although the pyramidal cell types can be recognized at an earlier stage than the nonpyramidal ones, the later dendritic differentiation of nonpyramidal cells parallels that of pyramidal neurons (Cajal, 1899a, b, 1900; Poliakov, 1949, 1966, 1979; Marin-Padilla, 1970a; Takashima et al., 1980; Mrzljak et al., 1988; Parnavelas et al., 1978; Uylings and Parnavelas, 1981; Uylings et al., 1985, 1990; Miller 1986, 1988). The first interneurons which have acquired type-specific characteristics appear around 26/27 w.g. They are double bouquet neurons in supragranular layers, which contrasts the previous observations that large interneurons in infragranular layers mature earlier (Marin-Padilla, 1970a; Parnavelas et al., 1978; Meyer and Ferres-Torres, 1984; Miller, 1986). After double bouquet neurons, multipolar basket neurons develop in infragranular layers between 32 and 34 w.g. Basket neurons in the supragranular layers are still not differentiated by this time.

Mrzljak et al. (1988) found that during prenatal and perinatal development there is a caudorostral gradient in the differentiation of infragranular basket neurons in the frontal lobe, i.e. basket neurons in the motor cortex have more developed dendrites and axonal trees than those in the PFC. However, these differences no longer exist during the first 3 postnatal months, when an equal level of differentiation is reached for basket neurons in PFC and motor cortex (for the motor cortex see Marin-Padilla, 1969, 1970a, b, 1972). After the basket interneurons, the first forms of neurogliaform neurons are observed in both supragranular and infragranular layers in the neonatal period. In the subplate zone of the preterm infant these neurons are found before that period (see Discussion below). The early infant period is characterized by a further differentiation of interneurons, especially in the supragranular layers. In the first postnatal month this concerns layer III basket neurons and multipolar neurons with "arcade"-shaped extended axonal trees. The first chandelier neurons are found around the third postnatal month. In conclusion, all types of interneurons differentiate in the cortical layers during the first 6 months of postnatal life. The temporal and spatial appearance of various types of nonpyramidal neurons described above is quite similar to that in the visual cortex of the cat (Meyer and Ferres-Torres, 1984). The only exception is formed by the double bouquet neurons, which in humans first mature in the PFC. The intensive development of PFC interneurons in the supragranular layers can be related to the parallel maturation of commissural and association pyramidal neurons in layer III during infancy, since the connections and functional role of interneurons are closely related to those of pyramidal neurons (for review see Martin, 1988). During the infant period, overproduction of the cortico-cortical axons is expected (see Kostović, this volume) and newly differentiated layer III interneurons may be important for modulation of activity of layer III pyramidal neurons, which are a major source of these cortico-cortical connections. Chandelier neurons, which in the PFC mature in the infant period, are reported to be predominantly abundant in the association cortical areas. They are able to influence the activity of cortico-cortical pyramidal neurons in layer III through symmetric GABAergic synapses on initial segments of these neurons (Somogyi et al., 1982; Freund et al., 1989; Peters et al., 1982).

During development, the interneurons described above display transient morphological features: growth of somatic and dendritic spines that are not present in adulthood, and of complex, widespread axonal trees which have a simpler form in the mature cortex. Transient long and short spines are present on all types of interneurons from the preterm infant period (27 – 36 w.g.), reaching its maximum in infancy (first postnatal year). This observation confirms earlier findings about the presence of spines on interneurons during the development of the human cortex (Cajal 1899a, b, 1900; Marin-Padilla 1969, 1970a, b, 1972, 1987; De Carlos et al., 1987; Mrzljak et al., 1988) and other mammalian cortices (Lund et al., 1977; Parnavelas et al., 1978; Boothe et al., 1979; Mates and Lund, 1983; Meyer and Ferres-Torres, 1984; Miller, 1988). During development transitory spines are sites of synaptic contacts (Miller and Peters, 1981; Miller 1988). Later on, during childhood (after the first postnatal year) the spines are gradually eliminated. This can be influenced by reducing the number of exuberant afferent fibers, which are likely candidates for making synapses with transitory spines (see above and below). The axonal plexuses of interneurons, which are well stained during the late prenatal and the postnatal infant periods, can only be compared with those in child brains. In the adolescent and adult cortex, axons are largely resistant to Golgi impregnation because of myelination. Golgi staining leads us to suggest that the size and complexity of the axonal trees of basket, double bouquet and chandelier neurons in the PFC decline between the first and sixth postnatal years. Many axonal collaterals of the main axonal arbor are lost in that period. During the first postnatal year the characteristic candlestick-like axonal terminals of chandelier neurons in the human PFC, as well as in the human auditory and visual cortices (De Carlos et al., 1987; Marin-Padilla, 1987), exhibit immature enlargements and filopodia, and have a more complex structure than terminals of chandelier cells in a 6-year-old child. Such results concerning the reduction of axonal arbor elements of chandelier and other interneurons in the PFC during development confirm the observations in experimental animals (De Carlos et al., 1985; Somogyi et al., 1982; Meyer and Ferres-Torres, 1984).

Correlation of neuronal differentiation with development of afferent and efferent connections in the prefrontal cortex

There is general agreement among neuroscientists on the influence of developing afferent and efferent connections on dendrite and spine growth and on the expansion of the spatial orientation of dendritic trees (Cajal, 1911; Hamori, 1973; Rakic and Sidman, 1973; Morest, 1969; Pinto Lord and Caviness, 1979; Wise et al., 1979; Shatz and Rakic, 1981; for review see Rakic, 1975). Our recent results (Uylings et al., 1987, Mrzljak et al., 1988) show that prenatal and postnatal stages of intensive dendritic differentiation coincide with the development of afferent and efferent connections. The ingrowth of various types of afferent fibers into the primate cortical anlage follows a temporal and spatial sequence. The first afferent fibers approaching the cortical anlage are the monoaminergic ones (Olsen et al., 1973; Nobin and Björklund, 1973). They are followed by basal forebrain fibers and, immediately afterwards, by thalamic fibers from the mediodorsal nucleus (Kostović and Goldman-Rakic, 1983; Kostović, 1986, this volume). Later, during prenatal development, commissural and associative neurons arrive (Goldman-Rakic, 1981, 1982; Schwartz and Goldman-Rakic, 1983; Kostović and Rakic, 1990; Kostović, this volume). To date it is not known whether any type of these afferent systems has a specific influence on a particular process of dendritic differentiation (shaping of dendritic trees, growth of dendrites and spines) or whether various kinds of afferent systems participate equally in all processes of dendritic differentiation. Between 15 and 25 w.g., in parallel with the ingrowth of major afferent systems from the basal forebrain and the mediodorsal nucleus into the expanding subplate zone (Kostović, 1986, this volume; Kostović and Goldman-Rakic, 1983), a precocious maturation of subplate neurons takes place, i.e. dendritic trees of subplate neurons are longer and more differentiated than dendrites of the cortical plate neurons. This intensive growth and differentiation of several types of subplate neurons may be related to synaptic interaction with afferent fibers. The increased dendritic and cell body surfaces of subplate neurons represent numerous possible postsynaptic sites for basal forebrain and thalamic fibers before they enter the cortical plate. During the late fetal, newborn and postnatal infant periods, cortico-cortical (callosal and associative) fibers, which are major constituents of the late transforming subplate zone (Kostović and Rakic, 1990), may play an inductive role in further differentiating the persisting subplate neurons, i.e. appearance of new types of subplate neurons, elongation of dendrites, and spine growth.

The basic dendritic shape of pyramidal neurons develops between 13.5 and 25 w.g., before afferent fibers penetrate the cortical plate. Furthermore, in this period, immature pyramidal and even undifferentiated bipolar neurons in the cortical plate send their axons deep into the subplate zone and fetal white matter. These observations mean that differentiation of pyramidal neurons starts in the absence of large parts of their afferent input, and that outgrowth of axons precedes the dendritic differentiation of their neurons of origin. Such results are also well documented in monkey and rodent brains (Wise et al., 1979; Shatz and Rakic, 1981; Schwartz and Goldman-Rakic 1983, 1986). The development of basic morphological features of a neuron may be determined as early as the day it is generated in the ventricular zone before it reaches its position in the cortex (Rakic, 1988a, b, c; McConnell, 1989).

A correlation exists between growth and differentiation of the PFC pyramidal and nonpyramidal neurons and the development of thalamocortical connections. During the ingrowth of thalamocortical fibers into the cortical plate, between 26 and 34 w.g. (Kostović and Goldman-Rakic, 1983; Kostović, this volume), the dendritic length of pyramidal neurons is rapidly enlarged in both the supragranular and the infragranular layers (see also Uylings et al., 1987; Mrzljak et al., 1990). In this period, the first interneurons (double bouquet and basket neurons) are found, spines begin to grow predominantly on pyramidal neurons in infragranular layers, and the basic 6-layered cytoarchitectonic pattern is established in the PFC. This period of rapid dendritic growth and ingrowth of thalamocortical fibers into the cortical plate corresponds to the period of initial synaptogenesis in the cortical plate (Molliver et al., 1973; Kostović and Rakic, 1990). During the preterm infant period, developing thalamocortical fibers in the cortical plate are distributed in a more diffuse pattern (Kostović and Goldman-Rakic, 1983) than in the adult PFC (Giguere and Goldman-Rakic, 1989). Therefore, these fibers have the opportunity to make contacts, and have a stimulating effect on the differentiation of a wide range of different neuronal types through synaptic interaction with their dendrites.

It is well established in the adult mammalian cortex that thalamocortical fibers make contact with different morphological and functional types of neurons, such as spiny and smooth multipolar nonpyramidal neurons, but also with pyramidal neurons with cortico-cortical, callosal, corticothalamic, and corticostriatal axons, as well as pyramidal neurons projecting to other destinations (White, 1981; Hendry and Jones, 1983; for a review see Colonnier, 1981; Jones, 1988). Later on, in the newborn infant and early childhood periods, we expect the involvement of thalamocortical as well as commissurural and associative systems in the further differentiation and remodeling of dendritic trees of pyramidal and nonpyramidal neurons, such as growth of dendritic segments and spines, and the establishment of a spatial geometry of dendritic trees (epigenetic modulation).

The growth of commissural and associative pathways is expected to occur during the perinatal and infant period (Kostović, this volume), when we find a further differentiation and a growth spurt of pyramidal and nonpyramidal neurons in layer III. The interaction between the differentiation of layer III neurons and the callosal and associative afferents is in accordance with the layer III origin and terminal distribution of callosal and associative connections (Schwartz and Goldman-Rakic, 1984; for a review see Goldman-Rakic, 1987a). The establishment of these connections may influence the intensive growth of layer III pyramidal neurons during the late infant and early childhood periods. Similarly, Ramirez and Kalil (1985) have found in the rodent cortex that the most intensive growth of corticospinal neurons occurs during the establishment of their projections in the spinal cord. After the first postnatal year the large layer IIIc neurons display an adult-like spatial distribution of basal dendritic trees, while basal dendrites are distributed in the deep part of layers III and IV. In this way the dendrites tend to be targets for fibers from the mediodorsal thalamic nucleus, which in the primate PFC terminate in the pertinent layers (Giguere and Goldman-Rakic, 1988). Similar results, i.e. that neurons and their spatial dendritic extension are positioned together with thalamocortical fibers, are also observed in the monkey somatosensory (Hendry and Jones, 1983) and visual (Katz et al., 1989) cortices. Consequently, the thalamocortical system is assumed to be one of the factors for the neuronal shape (e.g. Ruiz-Marcos and Valverde, 1970; Borges and Berry, 1978).

The intensive spine growth on layer III pyramidal neurons occurs during the infant and early childhood period, in contrast to layer V pyramidal neurons. They develop a significant number of spines as early as the late fetal and neonatal period. These spines on pyramidal neurons and transitory spines on nonpyramidal neurons (interneurons) are possible targets of thalamocortical and cortico-cortical fibers, which are shown to synapse almost exclusively with spine protrusion in the adult cortex (Freund et al., 1989; for a review see Colonnier, 1981). The inductive role of afferent fibers on spine development and maintenance of spine morphology has been well established in experimental models (Hamori, 1973; Globus and Scheibel, 1966). Pyramidal neurons in layer III develop the highest number of spines in the early childhood period and retain a high frequency of spines through childhood and adolescence. Transitory spines on the interneuronal population, by contrast, gradually disappear during that period. Spines are probably lost as a consequence of the elimination of exuberant afferent fibers that make synaptic contacts with them during development (see for visual cortex Huttenlocher et al., 1982). The prolonged growth and maintenance of a high number of spines on layer III pyramidal neurons may be the result of prolonged development of cortico-cortical and other afferent fibers making synaptic contacts with spines of these neurons.

Correlation between neuronal and functional development of the human PFC

The results of this study show that all types of the PFC projection and local circuitry neurons are already developed in the early infant period (up to the sixth postnatal month), before the onset of cognitive functions. During that period, values of local glucose utilization (measure of neuronal functions) in the PFC are low (Chugani and Phelps, 1986). The onset of cognitive function occurs between the seventh and ninth postnatal months, which is the earliest possible period for testing functional competence of the infant PFC with the delayed response task and $A\overline{B}$ ("A" not "B") test (Diamond and Door, 1989; Diamond and Goldman-Rakic, 1989; for a review see Goldman-Rakic, 1987a). From that period on local glucose utilization is increased in the PFC (Chugani and Phelps, 1986) and infants are gradually improving on the delayed response task and the $A\overline{B}$ test (Diamond and Door, 1989; Goldman-Rakic, 1987a). Fine remodeling and differentiation of pyramidal neurons in layer III, which are assumed to be major cellular candidates for processing cognitive functions (Kostović, this volume), continue after the first postnatal year, and is concomitant with the development of cognitive functions. It seems that their full maturation is not completed until the puberty period, when adult-like functional competence of cognitive processing is revealed (Goldman-Rakic, 1987a). The prolonged structural maturation of layer III pyramidal neurons shows a clear correlation with the late development of their acetylcholinesterase innervation (Kostović et al., 1988; Kostović, this volume) and a possibly late establishment of an adult-like number of synapses in the PFC (Huttenlocher, 1979).

Acknowledgements

This work was supported by an IBRO/MacArthur Foundation Network Grant (Uylings, Kostović, Mrzljak and Van Eden) and by the Yugoslav – US Joint Board No. PN 855 and SIZ za Znanost SRH (Mrzljak and Judaš).

The antibody Alz-50 was kindly donated by Abbott Laboratories, Chicago, IL, USA. We thank Gerben van der Meulen for his photographic assistance, and Olga Pach and Aad Janssen for typing the manuscript and correcting the English.

References

Al-Ghoul, W.M. and Miller, M.W. (1989) Transient expression of Alz-50 immunoreactivity in developing rat neocortex: a marker for naturally occurring neuronal death? *Brain Res.*, 481: 361 – 367.

Becker, L.E., Armstrong, D.L., Chan, F. and Wood, M.M. (1984) Dendritic development in human occipital cortical neurons. *Dev. Brain Res.*, 13: 117 – 124.

Boothe, R.G., Greenough, W.T., Lund, J.S. and Wrege, K. (1979) A quantitative investigation of spine and dendritic development of neurons in the visual cortex (area 17) of

Macaca nemestrina monkeys. *J. Comp. Neurol.,* 185: 473–490.

Borges, S. and Berry, M. (1978) The effects of dark rearing on the development of the visual cortex of the rat. *J. Comp. Neurol.,* 180: 277–300.

Braak, H. and Braak, E. (1985) Golgi impregnation as a tool in neuropathology with particular reference to investigations of the human telencephalic cortex. *Prog. Neurobiol.,* 25: 93–139.

Bradford, R., Parnavelas, J.G. and Lieberman, A.R. (1977) Neurons in layer I of the developing occipital cortex of the rat. *J. Comp. Neurol.,* 176: 121–132.

Brodmann, K. (1909) *Vergleichende Lokalisationslehre der Grosshirnrinde,* Barth-Verlag, Leipzig, 324 pp.

Cajal, S.R. (1899a) Estudios sobre la corteza cerebral humane. Corteza visual. *Rev. Trim. Microgr.,* 4: 1–63.

Cajal, S.R. (1899b) Estudios sobre la corteza cerebral humane. Estructure de la corteza matriz del hombre y mamiferos. *Rev. Trim. Microgr.,* 4: 117–200.

Cajal, S.R. (1900) Estudios sobre la corteza cerebral humane. Estructure de la corteza acustica. *Rev. Trim. Microgr.,* 5: 129–183.

Cajal, S.R. (1911) *Histologie du Systeme Nerveux de l'Homme et des Vertébrés, Vol. 2,* transl. by S. Azoulay, Maloine, Paris, 993 pp.

Cajal, S.R. (1960) *Studies on Vertebrate Neurogenesis,* transl. by L. Guth (originally published in 1929), C.C. Thomas, Springfield, IL, 432 pp.

Cajal, S.R. and De Castro, F. (1933) *Técnica Micrográfica del Sistema Nervosio,* Tipografia Artistica, Madrid, 409 pp.

Chi, J.G., Dooling, E.C. and Gilles, F.H. (1977) Gyral development of the human brain. *Ann. Neurol.,* 1: 86–93.

Chugani, H.T. and Phelps M.E. (1986) Maturational changes in cerebral function in infants determined by ^{18}FDG position emission tomography. *Science,* 231: 840–843.

Chun, J.J.M. and Shatz, C.J. (1989) Interstitial cells of the adult neocortical white matter are the remnant of the early generated subplate neuron population. *J. Comp. Neurol.,* 282: 555–569.

Chun, J.J.M., Nakamura, M.I. and Shatz, C.J. (1987) Transient cells of the developing mammalian telencephalon are peptide-immunoreactive neurons. *Nature,* 325: 617–620.

Colonnier, M. (1981) The electron-microscopic analysis of the neuronal organization of the cerebral cortex. In: F.O. Schmitt, F.G. Worden, G. Adelman and S.G. Dennis (Eds.), *The organization of the Cerebral Cortes,* MIT Press, Cambridge, MA, pp. 199–235.

Conel, J.L. (1939–1967) *The Postnatal Development of the Human Cerebral Cortex, Vols. I–VIII,* Harvard University Press, Cambridge, MA.

Cowan, W.M., Fawcett, J.W., O'Leary, D.D.M. and Stanfield, B.B. (1984) Regressive events in neurogenesis. *Science,* 225: 1258–1265.

De Carlos, J.A., Mascarague, L.L. and Valverde, F. (1985) Development, morphology and topography of chandelier cells in the auditory cortex of the cat. *Dev. Brain Res.,* 22: 293–300.

De Carlos, J.A., Lopes-Mascanague, L.L., Cajal-Agueros, S.R. and Valverde, F. (1987) Chandelier cells in the auditory cortex of monkey and man: a Golgi study. *Exp. Brain Res.,* 66: 295–302.

De Felipe, J., Hendry, S.H.C., Jones, E.G. and Schmechel, D. (1985) Variability in the termination of GABAergic chandelier cell axons on initial segments of pyramidal cell axons in the monkey sensory-motor cortex. *J. Comp. Neurol.,* 231: 364–384.

De Felipe, J., Hendry, S.H.C. and Jones, E.G. (1986) A correlative electron microscopic study of basket cells and large GABAergic neurons in the monkey sensory-motor cortex. *Neuroscience,* 17: 991–1009.

Derer, P. and Derer, M. (1990) Cajal-Retzius cell ontogenesis and death in mouse brain visualized with horse radish peroxidase and electron microscopy. *Neuroscience,* 36: 839–856.

Diamond, A. and Door, A. (1989) The performance of human infants on a measure of frontal cortex function, the delayed response task. *Dev. Psychobiol.,* 22: 271–294.

Diamond, A. and Goldman-Rakic, P.S. (1989) Comparison of human infants and rhesus monkeys on Piaget's $A\overline{B}$ task: evidence for dependence on the dorsolateral prefrontal cortex. *Exp. Brain Res.,* 74: 24–41.

Fairén, À., De Felipe, J. and Redigor, J. (1984) Non-pyramidal neurons: a general account. In: A. Peters and E.G. Jones (Eds.), *Cellular Components of the Cerebral Cortex, The Cerebral Cortex, Vol. 1,* Plenum Press, New York, pp. 201–253.

Fox, M.W. and Inman, O. (1966) Persistence of Retzius–Cajal cells in the developing dog brain. *Brain Res.,* 3: 192–194.

Freund, T.F., Martin, K.A.C., Soltesz, I., Somogyi, P. and Whitteridge, D. (1989) Arborization pattern and postsynaptic targets of physiologically identified thalamocortical afferents in striate cortex of the macaque monkey. *J. Comp. Neurol.,* 289: 315–336.

Fuster, J.M. (1989) *The Prefrontal Cortex, Anatomy, Physiology and Neuropsychology of the Frontal Lobe,* 2nd Edn., Raven Press, New York, 255 pp.

Fuster, J.M. (1990) Behavioral electrophysiology of the prefrontal cortex of the primate. *This Volume,* Ch. 15.

Giguere, M. and Goldman-Rakic, P.S. (1988) Mediodorsal nucleus: areal, laminar and tangential distribution of afferents and efferents in the frontal lobe of rhesus monkeys. *J. Comp. Neurol.,* 277: 195–213.

Globus, A. and Scheibel, A.B. (1966) Loss of dendritic spine as an index of pre-synaptic terminal pattern. *Nature,* 212: 463–465.

Goldman-Rakic, P.S. (1981) Development and plasticity of primate frontal cortex. In: F.O. Schmitt, F.G. Worden, G. Adelman and S.G. Dennis (Eds.), *The Organization of the Cerebral Cortex,* MIT Press, Cambridge, MA, pp. 69–97.

Goldman-Rakic, P.S. (1982) Neuronal development and plasticity of association cortex in primates. *Neurosci. Res. Prog. Bull.*, 20: 520–532.

Goldman-Rakic, P.S. (1987a) Development of cortical circuitry and cognitive function. *Child Dev.*, 58: 601–622.

Goldman-Rakic, P.S. (1987b) Circuitry of primate prefrontal cortex and regulation of behavior by representational memory. In: F. Plum and V. Mountcastle (Eds.), *Handbook of Physiology. Vol. 5. The Nervous System*, Am. Physiol. Soc., Bethesda, MD, pp. 373–417.

Goldman-Rakic, P.S. (1990) Cellular and circuit basis of working memory in prefrontal cortex of nonhuman primates. *This Volume*, Ch. 16.

Goldman-Rakic, P.S., Isseroff, A., Schwartz, M.L. and Bugbee, N.M. (1983) Neurobiology of cognitive development. In: P. Mussen (Ed.), *Handbook of Child Psychology: Biology and Infancy Development*, Wiley, New York, pp. 281–344.

Hamori, J. (1973) The inductive role of presynaptic axons in the development of postsynaptic spines. *Brain Res.*, 62: 337–344.

Hamre, K.M., Hyman, B.T., Goodelt, C.R., West, J.R. and Van Hoesen, G.W. (1989) Alz-50 immunoreactivity in the neonatal rat: changes in development and co-distribution with MAP-2 immunoreactivity. *Neurosci. Lett.*, 98: 264–271.

Hendry, S.H.C. and Jones, E.G. (1983) Thalamic inputs to identified commissural neurons in the monkey somatic sensory cortex. *J. Neurocytol.*, 12: 299–316.

His, W. (1904) *Die Entwicklung des menschlichen Gehirns während der ersten Monate*, S. Hirzel, Leipzig.

Huttenlocher, P.R. (1979) Synaptic density in human frontal cortex. Developmental changes and effects of aging. *Brain Res.*, 163: 195–205.

Huttenlocher, P.R., De Courten, Ch., Garey, L.J. and Van der Loos, H. (1982) Synaptogenesis in human visual cortex – evidence for synapse elimination during normal development. *Neurosci. Lett.*, 33: 247–252.

Jones, E.G. (1975) Varieties and distribution of non-pyramidal cells in the somatic sensory cortex of the squirrel monkey. *J. Comp. Neurol.*, 160: 205–268.

Jones, E.G. (1988) What are the local circuits? In: P. Rakic and W. Singer (Eds.), *Neurobiology of Neocortex*, Wiley, New York, pp. 5–27.

Jones, E.G., Hendry, S.H.C. and De Felipe, J. (1988) GABA-peptide neurons of the primate cerebral cortex. A limited cell class. In: E.G. Jones and A. Peters (Eds.), *Further Aspects of Cortical Function Including Hippocampus, The Cerebral Cortex, Vol. 6*, Plenum Press, New York, pp. 237–266.

Katz, L.W., Gilbert, C.D. and Wiesel, T.N. (1989) Local circuits and ocular dominance columns in monkey striate cortex. *J. Neurosci.*, 9: 1389–1399.

Kölliker, A.V. (1896) *Nervensystem des Menschen und der Tiere. Handbuch der Gewebelehre des Menschen, Vol. 2*, 6th Edn., Engelmann, Leipzig.

Kostović, I. (1986) Prenatal development of nucleus basalis complex and related fibre system in man: a histochemical study. *Neuroscience*, 17: 1047–1077.

Kostović, I. (1990a) Entwicklung des Zentralnervensystems. In: K. Heinreichsen (Ed.), *Humane Embryologie*, Springer Verlag, Berlin, in press.

Kostović, I. (1990b) Structural and histochemical reorganization of the human prefrontal cortex during perinatal and postnatal life. *This Volume*, Ch. 10.

Kostović, I. and Goldman-Rakic, P.S. (1983) Transient cholinesterase staining in the mediodorsal nucleus of the thalamus and its connections in the developing human and monkey brain. *J. Comp. Neurol.*, 219: 431–447.

Kostović, I. and Molliver, M.E. (1974) A new interpretation of the laminar development of cerebral cortex: synaptogenesis in different layers of neopallium in the human fetus. *Anat. Rec.*, 178: 395.

Kostović, I. and Rakic, P. (1980) Cytology and time of origin of interstitial neurons in white matter in infant and adult human and monkey telencephalon. *J. Neurocytol.*, 9: 219–242.

Kostović, I. and Rakic, P. (1984) Development of prestriate visual projections in the monkey and human fetal cerebrum revealed by transient cholinesterase staining. *J. Neurosci.*, 4: 25–42.

Kostović, I. and Rakic, P. (1990) Developmental history of transient subplate zone in the visual and somatosensory cortex of the macaque monkey and human brain. *J. Comp. Neurol.*, 297: 441–470.

Kostović, I., Skavić, J. and Strinović, D. (1988) Acetylcholinesterase in the human frontal associative cortex during the period of cognitive development: early laminar shifts and late innervation of pyramidal neurons. *Neurosci. Lett.*, 90: 107–112.

Kostović, I., Lukinović, N., Judaš, M., Bogdanović, N., Mrzljak, L., Zecević, N. and Kubat, M. (1989a) Structural basis of the developmental plasticity in the human cerebral cortex: The role of the transient subplate zone. *Metab. Brain Dis.*, 4: 17–23.

Kostović, I., Seress, L., Mrzljak, L. and Judaš, M. (1989b) Early onset of synapse formation in the human hippocampus: A correlation with Nissl–Golgi architectonics in 15- and 16.5-week-old fetuses. *Neuroscience*, 30: 105–116.

Krmpotić-Nemanić, J., Kostović, I., Nemanić, Dj. and Kelović, Z. (1979) The laminar organization of the prospective auditory cortex in the human fetus. *Acta Otolaryngol.*, 87: 241–246.

Krmpotić-Nemanić, J., Kostović, I., Vidić, Z., Nemanić, D. and Kostović-Knezević, Lj. (1987) Development of Cajal–Retzius cells in the human auditory cortex. *Acta Otolaryngol.*, 103: 447–480.

Kuljis, R.O. and Rakic, P. (1989) Distribution of neuropeptide Y-containing perikarya and axons in various neocortical areas in the macaque monkey. *J. Comp. Neurol.,* 280: 383–392.

Lorente de Nó, R. (1933) Studies on the structure of the cerebral cortex. I. The area enthorhinalis. *J. Psychol. Neurol.,* 45: 381–438.

Lund, J.S., Boothe, R.G. and Lund, R.D. (1977) Development of neurons in the visual cortex (area 17) of the monkey (*Macaca nemestrina*): A Golgi study from fetal day 127 to postnatal maturity. *J. Comp. Neurol.,* 176: 149–188.

Luskin, M.B. and Shatz, C.J. (1985) Studies of the earliest generated cells of the cat's primary visual cortex: Cogeneration of the cells of the subplate and marginal zones. *J. Neurosci.,* 5: 1062–1075.

Marin-Padilla, M. (1969) Origin of the pericellular baskets of the pyramidal cells of the human motor cortex: A Golgi study. *Brain Res.,* 14: 633–646.

Marin-Padilla, M. (1970a) Prenatal and early postnatal ontogenesis of the human motor cortex: A Golgi study. I. The sequential development of the cortical layers. *Brain Res.,* 23: 167–183.

Marin-Padilla, M. (1970b) Prenatal and early postnatal ontogenesis of the human motor cortex: A Golgi study. II. The basket-pyramidal cell system. *Brain Res.,* 23: 185–191.

Marin-Padilla, M. (1972) Double origin of the pericellular baskets of the pyramidal cells of the human motor cortex: A Golgi study. *Brain Res.,* 38: 1–12.

Marin-Padilla, M. (1987) The chandelier cell of the human visual cortex: A Golgi study. *J. Comp. Neurol.,* 256: 61–70.

Marin-Padilla, M. (1988) Early ontogenesis of the human cerebral cortex. In: A. Peters and E.G. Jones (Eds.), *Development and Maturation of the Cerebral Cortex, The Cerebral Cortex, Vol. 7,* Plenum Press, New York, pp. 1–34.

Marin-Padilla, M. and Marin-Padilla, M.T. (1982) Origin, prenatal development and structural organization of layer I of the human cerebral (motor) cortex: A Golgi study. *Anat. Embryol.,* 164: 161–206.

Martin, K.A.C. (1988) From single cells to simple circuits in the cerebral cortex. *Quart. J. Exp. Physiol.,* 73: 637–702.

Mates, S.L. and Lund, J.S. (1983) Neuronal composition and development in lamina 4C of monkey striate cortex. *J. Comp. Neurol.,* 221: 60–90.

McConnell, S.K. (1989) The determination of neuronal fate in the cerebral cortex. *Trends Neurosci.,* 12: 342–349.

McConnell, S.K., Ghosh, A. and Shatz, C.J. (1989) Subplate neurons pioneer the first axon pathways from the cerebral cortex. *Science,* 245: 978–982.

Meyer, G. and Ferres-Torres, R. (1984) Postnatal maturation of non-pyramidal neurons in the visual cortex of the cat. *J. Comp. Neurol.,* 228: 226–244.

Michel, A.E. and Garey, L.J. (1984) The development of dendritic spines in the human visual cortex. *Hum. Neurobiol.,* 3: 223–227.

Miller, M.W. (1986) Maturation of rat visual cortex. III. Postnatal morphogenesis and synaptogenesis of local circuit neurons. *Dev. Brain Res.,* 25: 271–285.

Miller, M.W. (1988) Development of projection and local circuit neurons in neocortex. In: A. Peters and E.G. Jones (Eds.), *Development and Maturation of Cerebral Cortex, The Cerebral Cortex, Vol. 7,* Plenum Press, New York, pp. 133–175.

Miller, M.W. and Peters, A. (1981) Maturation of rat visual cortex. II. A Combined Golgi-electron microscopic study of pyramidal neurons. *J. Comp. Neurol.,* 203: 555–573.

Molliver, M.E., Kostović, I. and Van der Loos, H. (1973) The development of synapses in the human fetus. *Brain Res.,* 50: 403–407.

Morest, D.K. (1969) The differentiation of cerebral dendrites: a study of the post-migratory neuroblast in the medial nucleus of the trapezoid body. *Z. Anat. Entwickl.-Gesch.,* 128: 271–289.

Mrzljak, L., Uylings, H.B.M., Kostović, I. and Van Eden, C.G. (1988) Prenatal development of neurons in the human prefrontal cortex. I. A qualitative Golgi study. *J. Comp. Neurol.,* 271: 355–386.

Mrzljak, L., Uylings, H.B.M., Kostović, I. and Van Eden, C.G. (1990) Prenatal development of neurons in the human prefrontal cortex. II. A quantitative Golgi study. Submitted.

Nobin, A. and Björklund, A. (1973) Topography of the monoamine neuron systems in the human brain as revealed in fetuses. *Acta Physiol. Scand.,* 388: 1–30.

Olson, L., Boreus, L.O. and Seiger, A. (1973) Histochemical demonstration and mapping of 5-HT and catecholamine-containing neuron systems in human fetal brain. *Z. Anat. Enwickl.-Gesch.,* 139: 259–282.

Overdijk, J., Uylings, H.B.M., Kuypers, K. and Kamstra, A.W. (1978) An economical system for measuring cellular tree structures in three dimensions, with special emphasis on Golgi-impregnated neurons. *J. Microsc.,* 114: 271–284.

Paldino, A.M. and Purpura, D.P. (1979) Branching patterns of hippocampal neurons of human fetuses during dendritic differentiation. *Exp. Neurol.,* 64: 620–631.

Parnavelas, J.G. and Edmunds, S.M. (1983) Further evidence that Retzius–Cajal cells transform to nonpyramidal neurons in the developing rat visual cortex. *J. Neurocytol.,* 12: 863–871.

Parnavelas, J.G., Bradford, R., Mounty, E.J. and Lieberman, A.R. (1978) The development of non-pyramidal neurons in the visual cortex of the rat. *Anat. Embryol.,* 155: 1–14.

Peters, A., Proskauer, C.C. and Ribak, C.E. (1982) Chandelier cells in the rat visual cortex. *J. Comp. Neurol.,* 206: 397–416.

Pinto-Lord, M.C. and Caviness, V.S. (1979) Determinants of cell shape and orientation: A comparative Golgi analysis of cell-axon interrelationships in the developing neocortex of normal and reeler mice. *J. Comp. Neurol.,* 187: 49–70.

Poliakov, G.I. (1949) Structural organization of the human cerebral cortex during ontogenetic development. In: S.A. Sarkisov, I.N. Filimonov and N.S. Preobrazenskaya (Eds.), *Cytoarchitecture of the Cerebral Cortex in Man,* Medgiz,

Moscow, pp. 33–92. (In Russian.)
Poliakov, G.I. (1961) Some results of research into the development of the neuronal structure of the cortical ends of the analyzers in man. *J. Comp. Neurol.*, 117: 197–212.
Poliakov, G.I. (1966) Embryonal and postembryonal development of neurons of human cerebral cortex. In: R. Hassler and H. Stephan (Eds.), *Evolution of the Forebrain*, Thieme Verlag, Stuttgart, pp. 243–258.
Poliakov, G.I. (1979) *Entwicklung der Neuronen der menschlichen Grosshirnrinde*, VEB Georg Thieme, Leipzig, 320 pp.
Purpura, D.P. (1975a) Dendritic differentiation in human cerebral cortex: Normal and aberrant developmental patterns. In: G.W. Kreutzberg (Ed.), *Physiology and Pathology of Dendrites, Advances in Neurobiology, Vol. 12,* Raven Press, New York, pp. 91–116.
Purpura, D.P. (1975b) Normal and aberrant neuronal development in the cerebral cortex of human fetus and young infant. In: N.A. Buchwald and M.A.B. Brazier (Eds.), *Brain Mechanisms in Mental Retardation*, Academic Press, New York, pp. 141–169.
Rakic, P. (1971) Neuron-glia relationship during granule cell migration in developing cerebellar cortex. *J. Comp. Neurol.*, 141: 283–312.
Rakic, P. (1972) Mode of cell migration to the superficial layers of fetal monkey neocortex. *J. Comp. Neurol.*, 145: 61–84.
Rakic, P. (1975) Timing and major ontogenetic events in the visual cortex of the rhesus monkey. In: N.A. Buchwald and M.A.B. Brazier (Eds.), *Brain Mechanisms in Mental Retardation*, Academic Press, New York, pp. 3–40.
Rakic, P. (1977) Prenatal development of the visual system in the rhesus monkey. *Phil. Trans. R. Soc. Lond. (Biol.)*, 278: 245–260.
Rakic, P. (1982) Early developmental events: cell lineages, acquisition of neuronal positions and areal and laminar development. *Neurosci. Res. Prog. Bull.*, 20: 439–451.
Rakic, P. (1988a) Specification of cerebral cortical areas. *Science*, 241: 170–176.
Rakic, P. (1988b) Intrinsic and extrinsic determinants of neocortical parcellation: A radial unit model. In: P. Rakic and W. Singer (Eds.), *Neurobiology of Neocortex*, Wiley, New York, pp. 5–27.
Rakic, P. (1988c) Defects of neuronal migration and the pathogenesis of cortical malformations. In: G.J. Boer, M.G.P. Feenstra, M. Mirmiran, D.F. Swaab and F. van Haaren (Eds.), *Biochemical Basis of Functional Neuroteratology, Progress in Brain Research, Vol. 73,* Elsevier, Amsterdam, pp. 15–37.
Ramirez, L.F. and Kalil, K. (1985) Cortical stages for growth in the development of cortical neurons. *J. Comp. Neurol.*, 237: 506–518.
Ruiz-Marcos, A. and Valverde, F. (1970) Dynamic architecture of the visual cortex. *Brain Res.*, 19: 25–39.
Sanides, F. and Sas, E. (1970) Persistence of horizontal cells of the Cajal foetal type and of the subpial granular layer in parts of the mammalian paleocortex. *Z. Mikrosk.-Anat. Forsch.*, 82: 570–588.
Sas, E. and Sanides, F. (1970) A comparative Golgi study of Cajal foetal cells. *Z. Mikrosk.-Anat. Forsch.*, 82: 385–396.
Schadé, J.P. and Van Groenigen, W.B. (1961) Structural organization of the human cerebral cortex. *Acta Anat.*, 47: 74–111.
Schadé, J.P., Van Backer, H. and Colon, E. (1964) Quantitative analysis of neuronal parameters in the maturing cerebral cortex. In: D.P. Purpura and J.P. Schadé (Eds.), *Growth and Maturation of the Brain, Progress in Brain Research, Vol. 4,* Elsevier, Amsterdam, pp. 150–175.
Schwartz, M.L. and Goldman-Rakic, P.S. (1984) Callosal and intrahemispheric connectivity of the prefrontal association cortex in rhesus monkey: relation between intraparietal and principal sulcal cortex. *J. Comp. Neurol.*, 226: 403–420.
Schwartz, M.L. and Goldman-Rakic, P.S. (1986) Some callosal neurons of the fetal monkey frontal cortex have axons in the contralateral hemisphere prior to the completion of migration. *Soc. Neurosci. Abstr.*, 12: 1211.
Seress, L. and Mrzljak, L. (1987) Basal dendrites of granule cells are normal features of the fetal and adult dentate gyrus of both monkey and human hippocampal formations. *Brain Res.*, 405: 169–175.
Shatz, C.J. and Rakic, P. (1981) The genesis of afferent connections from the visual cortex of the fetal rhesus monkey. *J. Comp. Neurol.*, 196: 287–307.
Shatz, C.J., Chun, J.J.M. and Luskin, M.B. (1988) The role of the subplate in the development of the mammalian telencephalon. In: A. Peters and E.G. Jones (Eds.), *Development and Maturation of Cerebral Cortex, The Cerebral Cortex, Vol. 7,* Plenum Press, New York, pp. 35–58.
Sidman, R.L. and Rakic, P. (1973) Neuronal migration, with special reference to developing human brain: A review. *Brain Res.*, 62: 1–35.
Smart, I.H.M. and McSherry, G.M. (1982) Growth patterns in the lateral wall of the mouse telencephalon. II. Histological changes during and subsequent to the period of isocortical neuron production. *J. Anat.*, 134: 415–442.
Somogyi, P. and Cowey, A. (1981) Combined Golgi and electron microscopic study on the synapses formed by double bouquet cells in the visual cortex of the rat and monkey. *J. Comp. Neurol.*, 195: 547–566.
Somogyi, P. and Cowey, A. (1984) Double bouquet cells. In: A. Peters and E.G. Jones (Eds.), *Cellular Components of the Cerebral Cortex, The Cerebral Cortex, Vol. 1,* Plenum Press, New York, pp. 337–360.
Somogyi, P., Freund, T.F. and Cowey, A. (1982) The axo-axonic interneuron in the cerebral cortex of the rat, cat and monkey. *Neuroscience*, 11: 2577–2607.
Somogyi, P., Kisvarday, Z.F., Freund, T.F. and Cowey, A. (1984) Characterization by Golgi impregnation of neurons that accumulate ^{3}H-GABA in the visual cortex of the monkey. *Exp. Brain Res.*, 53: 295–303.
Stensaas, L.J. (1967) The development of hippocampal and dorsolateral pallial regions of the cerebral hemisphere in fetal

rabbits. I. 15 mm stage; spongioblast morphology. *J. Comp. Neurol.,* 129: 59–70.

Takashima, S., Chan, F., Becker, L.E. and Armstrong, D.L. (1980) Morphology of the developing visual cortex of the human infant. A quantitative and qualitative Golgi study. *J. Neuropathol. Exp. Neurol.,* 39: 487–501.

Uylings, H.B.M. and Parnavelas, J.G. (1981) Growth and plasticity of cortical dendrites. In: O. Feher and F. Joó (Eds.), *Cellular Analogues of Conditioning and Neuronal Plasticity, Adv. Physiol. Sci., Vol. 36,* Pergamon Press, London, pp. 57–64.

Uylings, H.B.M. and Van Eden, C.G. (1990) Qualitative and quantitative comparison of the prefrontal cortex in rat and in primates, including humans. *This Volume,* Ch. 3.

Uylings, H.B.M., Verwer, R.W.H., Van Pelt, J. and Parnavelas, J.G. (1985) Topological and metrical analysis of developing dendritic tree patterns of pyramidal and nonpyramidal neurons in the visual cortex in the rat. *Neurosci. Lett.,* Suppl., 22: 357.

Uylings, H.B.M., Ruiz-Marcos, A. and Van Pelt, J. (1986) The metric analysis of three-dimensional dendritic tree patterns: a methodological review. *J. Neurosci. Meth.,* 18: 127–151.

Uylings, H.B.M., Mrzljak, L., Van Eden, C.G. and Matthijssen, M.A.H. (1987) The quantitative development of neurons in the human prefrontal cortex during prenatal and perinatal stages. *Neuroscience,* 22: 393.

Uylings, H.B.M., Van Eden, C.G., Parnavelas, J.G. and Kalsbeek, A. (1990) The prenatal and postnatal development of rat cerebral cortex. In: B. Kolb and R.C. Tees (Eds.), *The Cerebral Cortex of the Rat,* MIT Press, Cambridge, MA, pp. 35–76.

Valverde, F. and Facal-Valverde, M.V. (1987) Transitory population of cells in the temperal cortex of kittens. *Dev. Brain Res.,* 32: 283–288.

Valverde, F. and Facal-Valverde, M.V. (1988) Postnatal development of interstitial (subplate) cells in the white matter of the temporal cortex of kittens: A correlative Golgi and electron microscopic study. *J. Comp. Neurol.,* 269: 168–192.

Van der Loos (1959) *Dendro-dendritic Junctions in the Cerebral Cortex,* Ph. D. Thesis, Stam, Haarlem, 99 pp. (In Dutch).

Van Eden, C.G., Mrzljak, L., Voorn, P. and Uylings, H.B.M. (1989) Prenatal development of GABA-ergic neurons in the neocortex of the rat. *J. Comp. Neurol.,* 289: 213–227.

Vignal, W. (1888) Recherches sur le développement des éléments des couches corticales du cerveau et du cervelet chez l'homme et les mommifères. *Arch. Physiol. Norm. Pathol.,* 2: 228–254.

Vogt, C. and Vogt, O. (1919) Allgemeine Ergebnisse unserer Hirnforschung. *J. Psychol. Neurol. (Lpz),* 25: 279–462.

Wahle, P. and Meyer, G. (1987) Morphology and quantitative changes of transient NPY-ir neuronal populations during early postnatal development of the cat visual cortex. *J. Comp. Neurol.,* 261: 165–192.

Wahle, P., Meyer, G., Wu, J.Y. and Albus, K. (1987) Morphology and axon terminal pattern of glutamate decarboxylase-immunoreactive cell types in the white matter of the cat occipital cortex during early postnatal development. *Dev. Brain Res.,* 36: 53–61.

White, E.L. (1981) Thalamocortical synaptic relations. In: F.O. Schmitt, F.G. Worden, G. Adelman and S.G. Dennis (Eds.), *The Organization of the Cerebral Cortex,* MIT Press, Cambridge, MA, pp. 153–161.

Wise, S.P., Fleshman, J.W. and Jones, E.G. (1979) Maturation of pyramidal cell form in relation to developing afferent and efferent connections of net somatic sensory cortex. *Neuroscience,* 4: 1275–1297.

Wolozin, B.L., Pruchnicki, A., Dickson, D.W. and Davies, P. (1986) A neuronal antigen in the brains of Alzheimer patients. *Science,* 23: 648–650.

Wolozin, B.L., Scientella, A. and Davies, P. (1988) Reexpression of a developmentally regulated antigen in Down syndrome of Alzheimer disease. *Proc. Natl. Acad. Sci. (U.S.A.),* 85: 6202–6206.

Discussion

J.M. Fuster: How reliable is the Golgi method for quantification of histological maturation?

L. Mrzljak: We are fully aware of the unpredictability of the Golgi method that stains 2–5% of the neurons. Our analysis was made on a large number of Golgi impregnated specimen. The material was obtained at as much as short postmortem delays and consistent impregnations were obtained. In addition, for any period of development we used the modification of the Golgi technique which was proved to be optimal for impregnation, i.e., Stensaas modification for fetal period, Golgi–rapid for the period of preterm infant and Golgi–Cox (together with Golgi–rapid) beyond the second postnatal year.

Only morphologically identified neurons that could be consistently followed throughout the entire development were used for measurements. All neuronal metric variables were expressed per neuron. In this way we consider that our analysis was made in the most optimal conditions for postmortem human material.

J.E. Pisetsky: Does the growth of blood vessels follow the growth and differentiation of cells?

L. Mrzljak: Development of the fetal cortical zones (cortical plate, subplate, etc.) is accompanied by the growth of a "network" of blood vessels which has a specific morphology in each layer. Transitory zones (like subplate) have their specific and temporary blood supply which differs from that in the cortical plate. The amount and branching frequency of blood vessels is higher in the subplate. (see also Marin-Padilla, 1988).

References

Marin-Padilla, M. (1988) Embryonic vascularization of the mammalian cerebral cortex. In: A. Peters and E.G. Jones (Eds.), *Development and Maturation of the Cerebral Cortex,* Vol. 7, Plenum Press, New York, pp. 479–509.

CHAPTER 10

Structural and histochemical reorganization of the human prefrontal cortex during perinatal and postnatal life

I. Kostović

Section of Neuroanatomy, Department of Anatomy, School of Medicine, University of Zagreb, 41000 Zagreb, Yugoslavia

Introduction

The development of the cerebral cortex proceeds through a series of progressive (proliferation, migration, differentiation) and "regressive" (naturally occurring neuronal death, retraction of axons etc.) histogenetic events. The extent of "regressive" events, their relationship with progressive histogenetic events and the continuity with the adult cortical organization are poorly understood. However, it is obvious that both progressive and "regressive" events lead to a more advanced level of cortical organization. Therefore, it seems appropriate to replace the term "regressive events" by the term "reorganization".

In the human brain, temporo-spatial relationships between progressive histogenetic events, reorganization and adult organization are quite complex. The most promising approach to this problem seems to be the study of transient patterns of cortical reorganization during the developmental period characterized by simultaneously occurring progressive histogenetic and reorganizational events.

Correspondence: Dr. Ivica Kostović, M.D., D.Sc., Professor of Anatomy, Head, Section of Neuroanatomy, Department of Anatomy, School of Medicine, University of Zagreb, 41000 Zagreb, Šalata 11, Yugoslavia.

We have evidence that during the development of the human cerebral cortex transient patterns of organization of cortical afferents, synapses and neurons are present from the 3rd intrauterine month to the 6th postnatal month (Molliver et al., 1973; Mrzljak et al., 1988; Kostović et al., 1989a). In addition, postnatal overgrowth (i.e. the overproduction of synapses and spines) begins to slow down only after the 2nd year of life (Huttenlocher, 1979; Huttenlocher and De Courten, 1987; Huttenlocher et al., 1982). All this indicates that transient patterns of cortical organization and reorganization may extend for the impressively long period of more than 3 years. Due to the prolonged duration and overlapping of both progressive and reorganizational events, it is very difficult to determine criteria for the "staging" of the cortical histogenesis in man. Results of our long-term study of all stages of the human cortical development (Molliver et al., 1973; Kostović and Molliver, 1974; Kostović and Rakic, 1980, 1984, 1990; Krmpotić-Nemanić et al., 1980, 1983; Kostović and Goldman-Rakic, 1983; Kostović and Štefulj-Fučić, 1985; Kostović, 1984, 1986, 1987, 1989; Judaš, 1987; Kostović et al., 1987, 1988, 1989a, b; Mrzljak et al., 1988, this volume, Ch. 9) led us to make a distinction between 4 broad phases in the cortical histogenesis (Table I):
(1) the early fetal phase of progressive

histogenetic events;
(2) late fetal phase, characterized by transient patterns of cortical organization;
(3) the early postnatal phase of reorganization and overproduction of cortical circuitry elements;
(4) the phase of late postnatal maturation.

In the frontal cortex the early phase begins during the seventh week of gestation. This phase shows a significant overlap with the phase of transient organization, which begins after 13 weeks of gestation. The most pronounced pattern of transient organization is present between 22 and 34 weeks of gestation. The early postnatal phase lasts for 2 years, while late postnatal maturation is completed between 23 and 26 years of age.

The major features of the first phase are intensive progressive histogenetic events: proliferation, migration, and neuronal differentiation (Poliakov, 1979; Rakic, 1982; Marin-Padilla, 1988; Mrzljak et al., 1988; Kostović, 1990; Kostović et al., 1989b; Mrzljak et al., this volume, Ch. 9).

In the late fetal phase, cortical cells, afferent axons and synapses form characteristic fetal patterns of cortical organization: transient layers and neuronal connections (Molliver et al., 1973; Kostović and Rakic, 1990; Kostović et al., 1989a) and display transient histochemical features (Kostović and Goldman-Rakic, 1983; Kostović and Rakic, 1984).

The early postnatal phase is characterized by intensive dendritic growth (Conell, 1939 – 1963; Poliakov, 1949, 1959, 1967, 1979; Schadé and Van Groenigen, 1961; Marin-Padilla, 1970a, b, 1988; Mrzljak et al., this volume, Ch. 9), overproduction of synapses (Huttenlocher, 1979; Huttenlocher et al., 1982; Huttenlocher and De Courten, 1987) and changes in transmitter-related properties (Johnston et al., 1985; Kostović et al., 1989a).

The final prolonged phase of cortical maturation (childhood and adolescence) is characterized by the very gradual chemical maturation of the associative pyramidal neurons of layer III (Kostović et al., 1988).

Very few data are available on the last 3 phases of the human cortical development, and yet these are the very periods that hold the key to the understanding of onset, timing and nature of the cognitive development in man.

This review is concerned primarily with perinatal and postnatal phases of cortical differentiation. I will focus my attention on two distinctive features of the histochemical maturation of the human frontal cortex:
(a) laminar shifts of histochemically identified thalamocortical and basal forebrain afferents, and
(b) the prolonged maturation of the "cholinergic" innervation of associative layer III pyramidal neurons.

The first of the following sections emphasizes the transient nature of laminar, columnar and areal organization of histochemically identified afferent systems in the late human fetus. The second deals with structural and histochemical changes that may explain the dramatic reorganization of the human prefrontal cortex during perinatal and early postnatal life. The third and final section concentrates on protracted maturational events occurring after the cortical reorganization. Special attention will be devoted to the possible significance of the late histochemical maturation of layer III associative neurons. The evidence is

TABLE I

The cortical histogenesis in man

Developmental phase	Major histogenetic feature
1. Early fetal	Progressive histogenetic events (proliferation, migration and differentiation)
2. Late fetal	Transient patterns of cortical organization
3. Early postnatal life and infancy	Reorganization & overproduction of cortical circuitry elements
4. Late postnatal (childhood puberty and adolescence)	Prolonged structural and chemical maturation

based on our analysis of the human prefrontal cortex obtained from postmortem human brain tissue (age range from 15 weeks of gestation to 26 years of age) and processed for Nissl staining, acetylcholinesterase (AChE) histochemistry and peptide immunocytochemistry. The methodological details have been described elsewhere (Kostović and Goldman-Rakic, 1983; Kostović, 1986).

Transient patterns of structural and histochemical organization of the cerebral cortex in the late human fetus and preterm infant

Cortical neurogenesis in man starts during the second intrauterine month (Sidman and Rakic, 1973; Rakic, 1978; Kostović, 1990). The first sign of laminar organization in the lateral telencephalon has been demonstrated as early as the 5th week of gestation (His, 1904; Kostović, 1990). After the formation of the cortical plate at the end of the 8th week of gestation (Kostović, 1990), the dynamic pattern of histogenetic events (proliferation, migration and neuronal differentiation) in the human cortex can be indirectly disclosed by the analysis of the sequential development of embryonic cellular zones. Basically, these transient embryonic zones have no direct counterpart in the mature brain (Rakic, 1982). The ingrowth of major cortical fiber systems induces the formation of a new transient zone, the so-called subplate zone (Kostović and Molliver, 1974; Kostović and Rakic, 1990), between 13 and 15 weeks of gestation.

At the peak of the subplate size (in the frontal cortex at 22 – 34 weeks of gestation), elements of the cortical circuitry (afferent axons, synapses and postsynaptic neurons) display laminar arrangement and transient chemical properties (Kostović and Štefulj-Fučić, 1985; Mrzljak et al., 1988; Shatz et al., 1988). The transient arrangement of the circuitry elements forms a spatial framework for cellular interaction between afferent fibers and migratory and postmigratory neurons, and may be important for the formation of radially oriented columns (Rakic, 1988b, c). On the other hand, transient arrangements during the late fetal period may play a crucial role in the development of transient functional activities of the late fetal cortex (see below). Therefore, the delineation of transient patterns of cortical organization seems to be of essential significance for the understanding of the cortical development in man.

Transient laminar organisation

Transient cytoarchitectonic pattern

The most prominent transient cortical compartment, situated below the cortical plate (Fig. 1A, B), is the subplate zone (Kostović and Molliver, 1974; Kostović and Rakic, 1980, 1984, 1990), which contains a "waiting" front of corticopetal axons (Rakic, 1976), large postmigratory polymorphous neurons and synaptic contacts (Kostović and Rakic, 1990; Mrzljak et al., 1988). The developmental peak of the subplate zone in the prefrontal cortex is between 22 and 34 weeks of gestation, when it is the thickest compartment of the cortical anlage (ratio between the width of the subplate zone and that of the cortical plate in the frontal cortex is 5:1).

Transient distribution and staining properties of histochemically identified afferents

During the period of the developmental peak of the subplate zone there is a transient spatiotemporal overlap of afferents originating sequentially from the basal forebrain, thalamus, and from the ipsi- and contralateral cerebral hemispheres (Kostović and Rakic, 1990). In the light of the functional significance of two major subcortical afferent systems, thalamocortical and basal forebrains, the possibility of their transient cellular interactions is particularly interesting. Such interactions may develop in two steps or phases, each of them corresponding to some transient physiological phenomena.

(1) Between 20 and 24 weeks of gestation (or even somewhat earlier) the intensive AChE staining of the superficial part of the subplate zone (Fig. 2A) results from the transient accumulation of thalamic and basal forebrain afferents

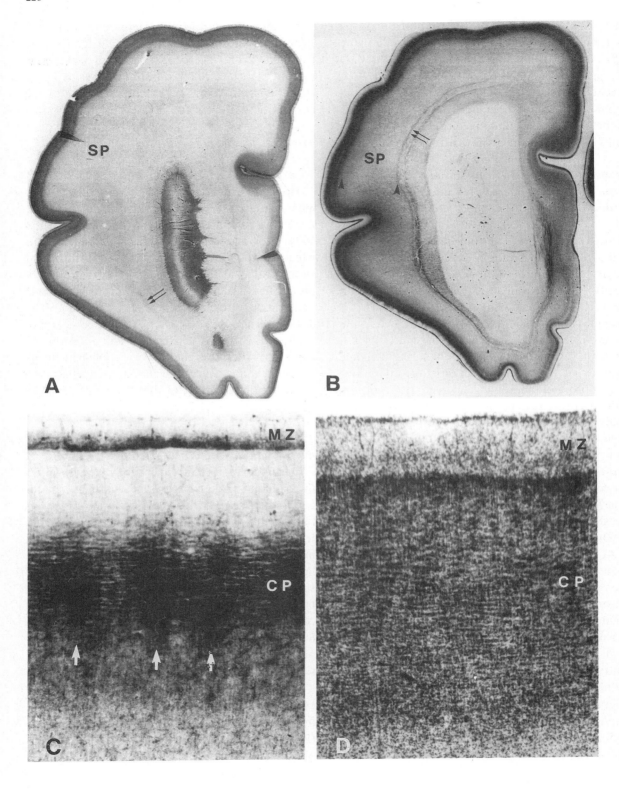

("double-dose" AChE staining) in this "waiting" compartment, just before the penetration of the cortical plate (Kostović and Goldman-Rakic, 1983; Kostović and Rakic, 1984; Kostović, 1986) This correlates well with the first, intermittent high-amplitude electro-encephalographic bursts in the cerebral hemisphere observed between 20 and 24 weeks of gestation (Dreyfus-Brisac, 1979).

(2) The second step of interaction may occur between 24 and 28 weeks of gestation, when thalamocortical fibers penetrate the cortical plate (Fig. 2B), and display intensive AChE reactivity and columnar (Fig. 2C) distribution (Kostović and Goldman-Rakic, 1983; Kostović et al., 1988). It should also be noted that the appearance of thalamocortical afferents in the prestriate visual (Kostović and Rakic, 1984), auditory (Krmpotić-Nemanić et al., 1980, 1983) and somatosensory (Kostović and Rakic, 1990) cortices correlates well with the development of synapses in the cortical plate after 23 weeks of gestation (Molliver et al., 1973). The establishment of thalamocortical connections seems to be the necessary prerequisite for the cortical analysis of sensory inputs, the perception of pain in the human fetus (Anand and Hickey, 1987) and the appearance of evoked potentials in preterm infants born before the 30th week of gestation (Vaughan, 1975). The basal forebrain afferent system may also participate in the establishment of transient behavioral patterns between 30 and 34 weeks of gestation (Kostović, 1986), characterized by the irregularity of sleep organization, with an earlier appearance of active REM sleep (Sterman and Hoppenbrouwers, 1971; Prechtl, 1974; Parmelee, 1975; Dreyfus-Brisac, 1979; Trevarthen, 1979; Wolff and Ferber, 1979; Leijon, 1982).

Transient vertical patterns

Between 20 and 24 weeks of gestation, the major feature of thalamocortical and basal forebrain afferents is the distribution within the subplate zone. After the 28th week of gestation (i.e. after their penetration into the cortical plate), the pattern of distribution of these afferents changes significantly: transient vertical patterns now become the major feature of the cortical organization (Figs. 1C, D and 2C).

Transient regional patterns

The presence of early areal borders between primary and secondary sensory areas clearly supports the notion of the pronounced regional specificity of transient cholinesterasic systems. In the preterm infant, the border between the prestriate and striate areas can be easily distinguished, due to strong AChE staining in the subplate zone and cortical plate of prestriate area 18 (Kostović and Rakic, 1984, 1990). It should be pointed out that this early areal pattern is just the reverse of the distribution in the postnatal brain, where stronger AChE staining is found in area 17. Furthermore, in the auditory cortex, transient AChE staining of subplate zone and cortical plate appears earlier within the primary auditory area than in the surrounding associative auditory cortex (Krmpotić-Nemanić et al., 1980, 1983). In the frontal cortex, transient regional differences are less discrete than in sensory areas (Fig. 2C). However, the prefrontal cortex displays stronger AChE staining than the premotor cortex, while in the precentral cortex AChE staining is restricted to the subplate and marginal zones.

Fig. 1. Transient cytoarchitectonic lamination in the prefrontal cortex of a 28-week-old human fetus (A: Nissl staining) compared with the transient laminar distribution of AChE-reactive staining (B: AChE histochemistry). Note that the thickness of the subplate zone (SP, between arrowheads) is 5 times that of the heavily stained cortical plate (CP). The subplate zone is defined as the layer between the cortical plate and the external capsule (double arrow). The AChE-reactive staining in layers IV and III of the cortical plate (CP) shows the transient "columnar" distribution (C, arrows). The adjacent Nissl section is shown in D.

228

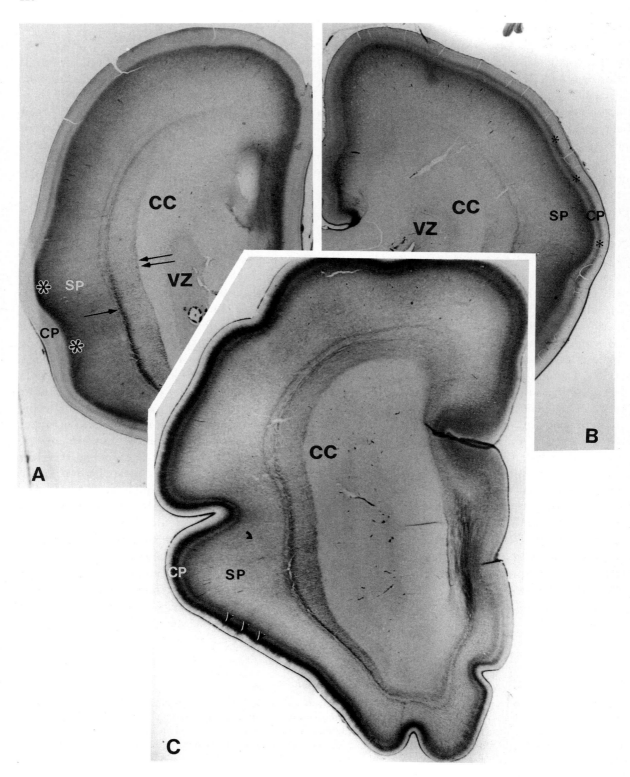

Since the appearance of regional differences in the chemoarchitectonics of the frontal, visual and auditory cortices correlates well with the AChE staining in the related thalamic nuclei and their projecting fibers, it seems reasonable to conclude that these thalamic afferents develop special chemical properties during the early differentiation of cortical areas. We believe that the transient arrangement of cortical afferents and cells and their subsequent reorganization are crucial events in the cortical histogenesis during perinatal life (see below).

Reorganization of the prefrontal cortex during the early postnatal period (neonatal period and infancy)

Cytoarchitectonic reorganization

The classical studies of the structural development of the human cortex during the early postnatal period and infancy (Filimonov, 1929; Kononova, 1940) were concerned with progressive differentiation of cytoarchitectonic parameters. There is general agreement that the relative amount of neuropil increases, while the cell-packing density decreases (Siwe, 1927; Zilles et al., 1986).

Little attention had been paid to the crucial observation of Korbinian Brodmann (1909) that in the late fetus (6–8 months of gestation) all prospective neocortical areas are characterized by the same basic pattern of cortical lamination (6-layered ontogenetic "Grundtypus"), with a well developed inner granular layer IV. According to Brodmann (1909), agranular or dysgranular cortical areas develop as a result of the process of the disappearance of granular layer IV. Since layer IV is a major recipient of thalamocortical afferents, the analysis of postnatal changes in its appearance and composition offers an opportunity to trace indirectly the elaboration of thalamocortical circuitry during the first postnatal year (Kostović et al., 1987). Our preliminary study has demonstrated that during infancy a well developed layer IV characterizes those frontal cortical areas which are dysgranular in the adult brain (Kostović et al., 1987). This confirms the notion that the reorganization of Brodmann's 6-layered ontogenetic "Grundtypus" includes the disappearance of granularity (Brodmann, 1909; Filimonov, 1929; Kostović et al., 1987). In some frontal areas, such as Broca's language region, this may be a very late event, and final laminar appearance is established as late as 4 years of age (Aldama, 1930; Judaš, 1987).

Obviously, we should explore more thoroughly the role of thalamocortical input in the differentiation of cortical areas. Recent experimental evidence in monkeys indicates that the thalamic input may be altered without leading to significant changes in the cytoarchitectonic organization of the innervated area (Rakic, 1988b). However, the diminished input from the appropriate thalamic nucleus during a critical period of development can influence the size of the cytoarchitectonic area (Rakic, 1988b, c).

Another major cytoarchitectonic event during the perinatal and postnatal development of the human frontal cortex is the transformation of the subplate zone. In the primary visual cortex the

---

Fig. 2. Laminar shifts and regional differences in AChE-reactive staining in the prefrontal cortex at different fetal stages. In a 22-week-old human fetus (A), AChE-reactive fibers originating in the basal forebrain (external capsule — arrow) and thalamus (internal capsule — double arrow) accumulate transiently in the superficial part (large asterisk) of the subplate zone (SP), displaying the earliest cortical regional differences. In the 24-week-old human fetus (B), AChE-reactive fibers gradually penetrate into the cortical plate (small asterisk) with a parallel decrease in staining of the subplate zone. At 28 weeks of gestation (C), strong AChE reactivity is evident in the cortical plate throughout the basolateral prefrontal cortex (arrowheads). Note 3 aspects of the AChE staining: (1) laminar preference with heavy staining of layer IV; (2) columnar pattern in the lateral and especially in the basal portion of the frontal pallium (arrowheads); and (3) regional differences related to the staining pattern of the cortical plate. CC = corpus callosum, VZ = ventricular zone.

subplate zone disappears during the last weeks of gestation (Kostović and Rakic, 1990), and it cannot be delineated as a cytoarchitectonic zone in the newborn somatosensory cortex, although numerous subplate neurons are present in the subjacent white matter (Kostović and Rakic, 1990). By contrast, the prefrontal cortex of the newborn infant has a well developed subplate (Mrzljak et al., 1988; Kostović et al., 1989a), which only gradually disappears during the first 6 postnatal months. The prolonged existence of the subplate zone in the prefrontal cortex is probably related to the prolonged growth of frontal cortico-cortical pathways through the subplate compartment and its role in postnatal shaping of tertiary gyri (Kostović and Rakic, 1990).

Histochemical reorganization and developmental shifts of cholinergic markers

The most prominent histochemical change in the thalamocortical system is the decrease in AChE reactivity in layer IV of the frontal cortex during the perinatal period (Figs. 3A, B, C and 4A, B). During this period the AChE reactivity in the mediodorsal nucleus of the thalamus gradually disappears (Kostović and Goldman-Rakic, 1983). The loss of AChE ractivity is an obvious sign of the change in chemical properties of thalamocortical axons. Whether this histochemical change plays a role in the functional reorganization of the thalamocortical input is not clear at present, owing to the uncertainties in the interpretation of the transient AChE reactivity (Robertson, 1987; Kristt, 1989). According to the hypothesis that AChE plays a developmental role (Kostović and Goldman-Rakic, 1983), postnatal disappearance of thalamic AChE simply indicates the end of active growth, of membrane surface recognition and of the differentiation in the afferent system. The interpretation of this decline in AChE reactivity is complicated by the fact that a dramatic reorganization of AChE was found in the marginal zone (future layer I), where the thalamic projection cannot be found in the adult cortex (Giguere and Goldman-Rakic, 1988). However, it is not very likely that AChE declines postnatally as a result of the reduction of synaptogenesis, because the early postnatal period is characterized by an overproduction of synapses (Huttenlocher, 1979; Huttenlocher et al., 1982; Huttenlocher and De Courten, 1987; Rakic et al., 1986). It is not to be excluded that changes in the AChE reactivity of basal forebrain afferents also contribute to the decrease in cortical AChE staining intensity. These cholinergic afferent fibers, however, maintain

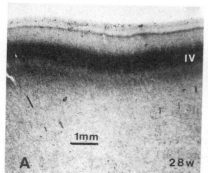
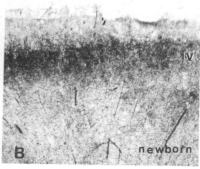
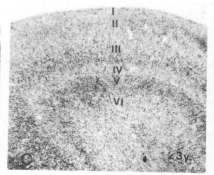

Fig. 3. The reorganization of the laminar distribution of the AChE reactivity in the human prefrontal cortex during perinatal and postnatal life. Roman numerals indicate cortical layers. Bar = 1 mm. Strong AChE reactivity in layer IV of the cortical plate of a 28-week-old premature infant (A) is followed by a significant decrease in layer IV reactivity in the newborn cortex (B). In a 23-year-ld adult (C) the AChE pattern is reversed: the granular layer IV (recipient of thalamocortical fibers) shows the weakest AChE reactivity.

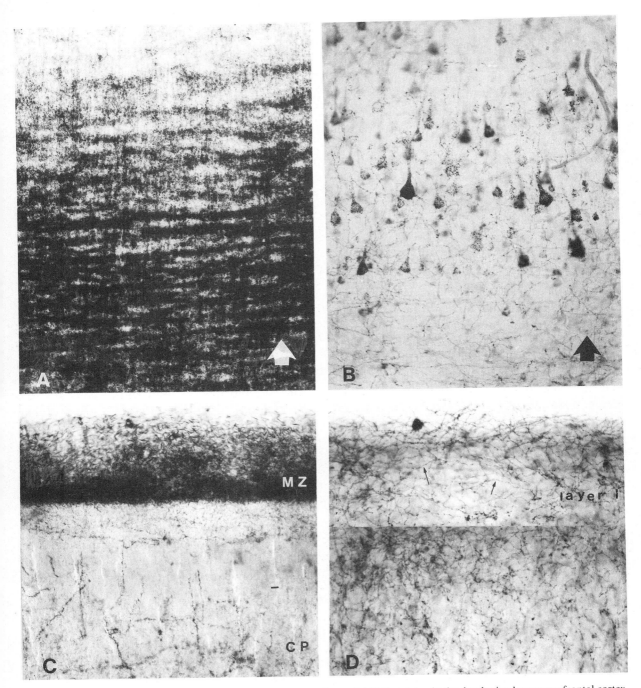

Fig. 4. Dramatic changes in laminar preferences and cellular correlates of AChE staining in the developing human prefrontal cortex. A shows a characteristic late fetal pattern with heavy staining in the neuropil of layer IV and the deep part of layer III (28 weeks of gestation). This neuropil staining pattern disappears postnatally (layer IV turns pale) and strongly reactive pyramidal perikarya develop within layer III during late postnatal life (23 years of age, B). C and D show the reorganization of the staining pattern in (fetal) layer I. In the cortex of a 28-week-old premature infant, the AChE-reactive band, composed of alternating "columnar" staining densities, occupies the middle portion of the marginal zone (MZ) (C). This dense band disappears during perinatal development. The molecular layer (layer I) of the young adult is characterized by a dense network of AChE fibers (D). Individual AChE-reactive fibers are readily encountered in the adult but not in the fetal cortex. Thick arrows indicate the direction to the pial surface. Thin arrows (D) indicate individual AChE-positive fibers in layer I.

their AChE reactivity throughout adulthood (Kostović et al., 1988; Mesulam and Geula, 1988).

In the late fetal cortex, most of the AChE found on cholinergic axons is probably loosely bound to the axonal membrane, and this may potentiate the histochemical reactivity of neuropil in innervated cortical layers. After the penetration of the cortical plate (late fetal phase), the histochemical differentiation of cholinergic fibers may gradually continue throughout the perinatal period. The major developmental trend may be the change in the ability of AChE to bind to an axonal membrane (Fig. 4C, D).

After birth, the AChE network, which is composed of well stained individual axons, gradually develops in a "deep to superficial" fashion (Kostović et al., 1988). These processes may actually represent a cellular aspect of the reorganization of this degrading enzyme on the cholinergic afferent axons. Developmental shifts and laminar reorganization of cortical AChE were also documented in the experimental studies in rodents and carnivores (Höhmann and Ebner, 1985; Bear et al., 1986; Prusky et al., 1988). It was also demonstrated that the changes in the AChE pattern are paralleled by a marked increase in nicotinic binding sites in rat (Prusky et al., 1988). The early postnatal reorganization of AChE-reactive basal forebrain afferents and cholinergic cortical receptors does not necessarily imply simultaneous changes in the synthetic cholinergic enzyme choline acetyltransferase (ChAT). However, recent immunocytochemical studies in experimental models have demonstrated a bimodal pattern of ChAT development in the cerebral cortex (Hendry et al., 1987; Huntley et al., 1987; Dori and Parnavelas, 1989), pointing suggestively to the transient expression of cholinergic markers in the developing neurons and their subsequent reorganization during development.

Redistribution of cortico-cortical systems

As was discussed in the preceding sections, current evidence supports the idea that changes in the distribution and chemical properties of the subcortico-cortical pathways play a significant role in the perinatal reorganization of the prefrontal cortex in man. Whether cortico-cortical pathways are involved in structural rearrangements of the human infant cerebral cortex is not known. Recent experimental studies in monkey have shown that commissural cortico-cortical fibers reside in the transient "subplate" zone between embryonic days E100 and E123 (Goldman-Rakic, 1982). Ipsilateral cortico-cortical fibers also "wait" in the subplate zone for several weeks before entering the cortical plate (Schwartz and Goldman-Rakic, 1990). In addition, some subplate neurons transiently send their axons through the corpus callosum (Schwartz and Goldman-Rakic, 1990). In the human cortex, "waiting" associative and commissural pathways are major constituents of the subplate zone after 28 weeks of gestation (Kostović and Rakic, 1990). Considering the fact that most thalamocortical axons enter the cortical plate before birth (Kostović and Goldman-Rakic, 1983; Kostović and Rakic, 1984), cortico-cortical fibers are the most likely constituents of the postnatal subplate zone. The prolonged persistence of the subplate zone in the frontal cortex may then be explained by the prolonged postnatal growth of the cortico-cortical pathways in man. Accordingly, a significant redistribution of cortico-cortical axons during subsequent development can be expected (Innocenti, 1982). Moreover, cortico-cortical axons are produced in excess during perinatal life. The final number is achieved by the process of competitive elimination during postnatal life (LaMantia and Rakic, 1984; Chalupa and Killackey, 1989).

The overproduction of circuitry elements and the reorganization during late infancy and early childhood

The phenomenon of overproduction of cortico-cortical axons is closely associated with the excessive synaptogenesis during the early postnatal period (Rakic et al., 1986; Goldman-Rakic, 1987).

In man, the overproduction of cortico-cortical axons is expected to occur during infancy. In the human visual cortex the maximum number of spines is reached at about 5 months of age (Michael and Garey, 1984). The peak of synaptogenesis begins around the 8th postnatal month and reaches a maximum at 2 years of age. Consequently, data on the maturation of circuitry elements in man indicate that the fine structural organization of the neocortex is changed during late infancy and early childhood. This evidence further supports the hypothesis of the cortical reorganization during development. With respect to the early, perinatal reorganization, these events, which begin during late infancy, may be called a "second" reorganization. However, there may be a significant overlap in time between these two phases of developmental reorganization. It is possible that the first one involves laminar, areal and vertical organization, with the parallel emergence of a fine (synaptic) reorganization, the latter having its developmental peak after the first year of life. The histogenetic events that lead to the "first" reorganization, i.e. formation of layers, growth of subcortical afferents, areal differentiation and the formation of columns, seem to be primarily under genetic control (Rakic, 1988a, b, c). The late reorganization seems to be open to interactions with the external world. Whether the human cortical reorganization involves a massive naturally occurring neuronal death, as is proposed for other mammalian species (Cowan et al., 1984), is not known at present. Naturally occurring neuronal death and related cellular events are classified as "regressive" in most of the current reviews on developmental biology (Cowan et al., 1984). However, the naturally occurring neuronal death and reorganizational histogenetic events described in the present study, rather than causing a regression of cortical organization, lead to the more advanced level of functional organization of the cerebral cortex. This is particularly obvious during human cortical development, in which almost 2 years are needed for the transformation of the transient cortical pattern of the preterm infant to the structurally adult-like pattern of cortical organization.

The "second" reorganization coincides with the emergence of cognitive functions in the human infant. It is very likely that the critical quantitative level of cortical circuitry is important for the emergence of cognitive function (Goldman-Rakic, 1987). However, the most significant period of cognitive development in man comes after the phase of overproduction of synapses. It is our working hypothesis that the maturation of cortical associative neurons is the major event in this last phase of cortical development in man. Evidence on the chemical maturation of associative elements during this important period in human cortical development will be presented in the following section.

Post-reorganization period: prolonged structural and chemical maturation of the cerebral cortex during childhood and adolescence

The period after the first postnatal year is characterized by the dramatic emergence of cognitive functions (Piaget, 1954; Trevarthen, 1979; Carey 1984). The maturational changes in cognitive functions (including language) can be observed up to the 14th postnatal year (Piaget, 1954). Very little is known about structural and chemical changes that may underlie the development of these complex cognitive functions. The classical Golgi (Vignal, 1888; Koelliker, 1896; Cajal, 1900, 1906; Poliakov, 1949, 1967, 1979; Marin-Padilla, 1970a, b; Purpura, 1975; Becker et al., 1984) and cytoarchitectonic (Von Economo and Koskinas, 1925; Siwe, 1927; Filimonov, 1929; Aldama, 1930; Kononova, 1940; Sarkisov, 1965) studies describe the progressive and gradual maturation of neurons and neuropil. From these studies it is difficult to determine the extent of structural change at the cellular level and the exact developmental timing of the prospective histogenetic events. Even if the appropriate methodology is available, it is difficult to study the maturation of transmitter-related systems due to the limited availability of well preserved human post-mortem tissue. A recent study on the histochemical maturation of the human frontal associative cortex focused attention on the association-commissural pyramidal neurons of layer III (Kostović et al., 1988). These neurons are the major source of associative and commissural projections (Jones, 1981; Schwartz and Goldman-Rakic, 1984; Caminiti et al., 1985; Innocenti, 1986) and are important neuronal elements underlying cognitive processing in the cerebral cortex (Goldman-Rakic, 1987). The initial histochemical analysis of the development of the layer III association neurons (Kostović, 1987; Kostović et al., 1988) demonstrated the late development of AChE reactivity in their cell bodies and surrounding fibrillar networks (Fig. 5A, B). The AChE reactivity of layer III pyramidal neurons begins to develop after the first postnatal year, increases gradually, and reaches its peak intensity in young adults (Kostović

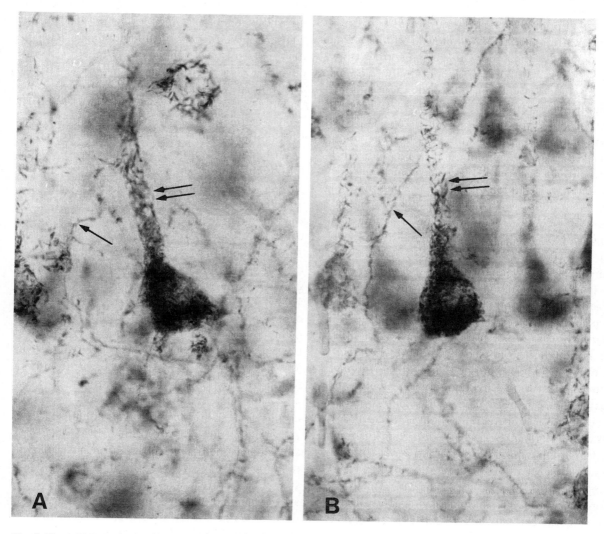

Fig. 5. The AChE-reactive pyramidal "associative" neurons of layer IIIc in the prefrontal cortex of a young man (23 years of age). At this stage the AChE-reactive product is present on the surface of cell bodies and proximal dendrites of pyramidal neurons (double arrow) as well as in approaching extrinsic fibers (arrows).

et al., 1988; Mesulam and Geula, 1988). These findings lead to the hypothesis that AChE-rich elements play a role in the innervation of cortical associative neurons during cognitive development in man. The AChE-rich innervation of layer III pyramidal neurons has so far been documented only in the human cortex, and this late developmental process seems to be a unique feature of the human, and possibly of the primate brain (Kostović et al., 1988). It is interesting that the AChE-reactive neurons are arranged in column-like clusters (Kostović et al., 1988). The late appearance of these clusters may be related to the late innervation of efferent columns of associative neurons. The prolonged maturation of AChE in the frontal cortex is in accordance with the late postnatal maturation of the cholinergic receptors in the human brain (Johnston et al., 1985).

There is surprisingly little information on the postnatal development of other transmitter systems in the human cortex. However, ontogenetic studies of the dopaminergic system in experimental primates indicate that the timetable for its development in the frontal cortex differs from that of other transmitters and that the dopamine concentration showed prolonged changes during postnatal development in experimental primates (MacBrown and Goldman, 1977; Goldman-Rakic et al., 1983). At present it is difficult to explore the functional significance of the developmental shifts and prolonged development of different transmitters and chemical systems during the late postnatal period. The fact that the late AChE innervation of layer III associative pyramidal neurons (Kostović et al., 1988) together with the late maturation of the cholinergic receptors (Johnston et al., 1985) develops after synaptic, structural and chemical reorganization is consistent with the idea that the cholinergic system plays a role in the stabilization and plasticity of synapses (Sillito, 1983). Although the process of synaptic reorganization may begin rather synchronously in different cortical regions (Rakic et al., 1986; Goldman-Rakic, 1987), the final maturation may be a very late event in frontal association areas (Kostović et al., 1988, 1989a).

Concluding remarks

Presented findings indicate that the frontal cortex of the late human fetus and preterm infant shows transient patterns of laminar, regional and vertical organization that are essentially different from those found in the cortex of infant, child or adult man. Furthermore, it appears that the establishment of normal cortical organization proceeds by means of a profound structural and transmitter-related reorganization and not exclusively by progressive elaboration of fetal patterns.

One of the major aspects of the transient organization described in the present review is the overlapping distribution of the cholinesterasic thalamocortical and cholinergic basal forebrain afferents in the "waiting" compartment of the transient subplate zone. Later on, strong, transient cholinesterase activity with a characteristic columnar distribution was also observed in the cortical plate. This histochemical reactivity originates in the mediodorsal nucleus of the thalamus and leads to the earliest areal parcelation of the prefrontal cortex. The transient arrangement of "waiting" thalamic and basal forebrain afferents within the subplate zone can enhance their potential interaction with early maturing peptidergic-GABAergic subplate neurons. More specifically, the interactions between the afferent fibers and subplate neurons appear to be at the synaptic level. We believe that the transient arrangement of cholinesterase-positive afferents from the thalamus and the basal forebrain, together with their transient postsynaptic elements in the subplate zone, cortical plate and marginal zone, forms the basic neural circuitry underlying transient electrical and behavioral phenomena observed in low-birth-weight infants. Several changes account for the tranformation of transient fetal patterns in the more mature cortical pattern during perinatal life and infancy. The first process, the dissolution of the subplate zone, begins during the last 2 months of gestation, when many fiber terminals relocate to the cortex and some subplate neurons die. The most dramatic histochemical change during perinatal life is the decrease in the AChE reactivity of granular layer IV, which runs parallel with the decrease in the staining intensity of the mediodorsal nucleus of the thalamus.

The histogenetic events in the frontal cortex during the first and second years of life are not yet fully explored. Current knowledge about other human cortical areas as well as experimental evidence in monkeys (Goldman-Rakic, 1987; Rakic, 1988b; Kostović and Rakic, 1990) seems to suggest that the terminal growth of cortico-cortical fibers and the overproduction of synapses and dendritic spines represent major histogenetic events during this period. The presence of an excessive quantity of cortical circuitry elements (callosal axons, synapses, dendritic spines) suggests that the development of the mature cortical organization depends on the relocation and/or retraction of cortico-cortical axons, on the elimination of excessive synapses and, eventually, on naturally occurring cell death. These events underlie the "second" reorganization of the infant cerebral cortex.

The major histochemical change after these reorganizational events is the appearance of the "cholinergic" innervation of layer III associative pyramidal neurons. This late developmental event, which seems to be characteristic of the human cortex, has its onset after the first postnatal year, proceeds gradually, and reaches its peak intensity in young adults. This late, prolonged phase of cortical histogenesis probably reflects a protracted development of the transmitter-related modulation of associative cortical circuitry and may be important for the neural basis of the normal cognitive development. These findings suggest that the neural substrate of cognitive development appears after structural and chemical reorganization of the infant association cortex.

Acknowledgements

Supported by Yugoslav – U.S. Joint Board Grant No. PN 855 and SIZ za znanost SRH.

It is a pleasure to thank Z. Cmuk, D. Budinšćak and B. Popović for their excellent technical assistance.

References

Aldama, J. (1930) Cytoarchitektonik der Grosshirnrinde eines 5-jährigen und eines 1-jährigen Kindes. *Z. Ges. Neurol. Psychiat.,* 130: 532 – 630.

Anand, K.J.S. and Hickey, P.R. (1987) Pain and its effects in the human neonate and fetus. *New Engl. J. Med.,* 317: 1321 – 1329.

Bear, M.F., Carnes, K.M. and Ebner, F.F. (1986) Postnatal changes in the distribution of acetylcholinesterase in kitten striate cortex. *J. Comp. Neurol.,* 237: 519 – 532.

Becker, L.E., Armstrong, D.L., Chan, F. and Wood, M.M. (1984) Dendritic development in human occipital cortical neurons. *Dev. Brain Res.,* 13: 117 – 124.

Brodmann, K. (1909) *Vergleichende Lokalisationslehre der Grosshirnrinde in ihren Prinzipien dargestellt auf Grund des Zellenbaues,* Barth, Leipzig, 324 pp.

Cajal, S.R. (1900) *Studien über die Hirnrinde des Menschen. 2. Heft. Die Bewegungsrinde,* Barth, Leipzig.

Cajal, S.R. (1906) *Studien über die Hirnrinde des Menschen. 5. Heft. Vergleichende Strukturbeschreibung und Histogenesis der Hirnrinde,* Barth, Leipzig.

Caminiti, R., Zeger, S., Johnson, P.B., Urbano, A. and Georgopoulos, P. (1985) Corticocortical efferent systems in the monkey; A quantitative spatial analysis of the tangential distribution of cells of origin. *J. Comp. Neurol.,* 241: 405 – 419.

Carey, S. (1984) Cognitive development. The descriptive problem. In M.S. Gazzaniga (Ed.), *Handbook of Cognitive Development,* Plenum Press, New York, pp. 37 – 66.

Chalupa, L.M. and Killackey, H.P. (1989) Process elimination underlies ontogenetic change in the distribution of callosal projection neurons in the postcentral gyrus of the fetal rhesus monkey. *Proc. Natl. Acad. Sci. (U.S.A),* 86: 1076 – 1079.

Conell, Le Roy (1939 – 1963) *The Postnatal Development of the Human Cerebral Cortex,* 6 Vols., Harvard University Press, Cambridge, MA.

Cowan, W.M., Fawcett, J.W., O'Leary, D.D.M. and Stanfield, B.B. (1984) Regressive events in neurogenesis. *Science,* 225: 1258 – 1265.

Dori, I. and Parnavelas, J.G. (1989) The cholinergic innervation of the rat cerebral cortex shows two distinct phases in development. *Exp. Brain Res.,* 76: 417 – 423.

Dreyfus-Brisac, C. (1979) Ontogenesis of brain bioelectrical activity and sleep organization in neonates and infants. In F. Falkner and J.M. Tanner (Eds.), *Human Growth. Vol. 3. Neurobiology and Nutrition,* Bailliere Tindall, London, pp. 157 – 182.

Filimonov, I.N. (1929) Zur embryonalen und postembryonalen Entwicklung der Grosshirnrinde des Menschen. *J. Psychol. Neurol.,* 39: 323 – 389.

Giguere, M. and Goldman-Rakic, P.S. (1988) Mediodorsal nucleus: Areal, laminar, and tangential distribution of afferents and efferents in the frontal lobe of rhesus monkeys. *J. Comp. Neurol.,* 277: 195 – 213.

Goldman-Rakic, P.S. (1982) Neuronal development and plasticity of association cortex in primates. *Neurosci. Res. Progr. Bull.*, 20: 520–532.

Goldman-Rakic, P.S. (1987) Development of cortical circuitry and cognitive function. *Child Dev.*, 58: 601–622.

Goldman-Rakic, P.S., Isseroff, A., Schwartz, M.L. and Bugbee, N.M. (1983) The neurobiology of cognitive development. In P. Mussen (Ed.), *Handbook of Child Psychology: Biology and Infancy Development*, Wiley, New York, pp. 281–344.

Hendry, S.H.C., Jones, E.G., Killackey, H.P. and Chalupa, L.M. (1987) Choline acetyltransferase-immunoreactive neurons in fetal monkey cerebral cortex. *Dev. Brain Res.*, 37: 313–317.

His, W. (1904) *Die Entwicklung des menschlichen Gehirns während der ersten Monate*, S. Hirzel, Leipzig.

Höhmann, C.F. and Ebner, F. (1985) Development of cholinergic markers in mouse forebrain. I. Choline acetyltransferase enzyme activity and acetylcholinesterase histochemistry. *Dev. Brain Res.*, 23: 225–241.

Huntley, S.H.C., Hendry, L.M., Jones, E.G., Chalupa, L.M. and Killackey, H.B. (1987) GABA, neuropeptide and ChAT expression in neurons of the fetal monkey cortex. *Soc. Neurosci. Abstr.*, 13: 76.

Huttenlocher, P.R. (1979) Synaptic density in human frontal cortex. Developmental changes and effects of aging. *Brain Res.*, 163: 195–205.

Huttenlocher, P.R. and De Courten, Ch. (1987) The development of synapses in striate cortex of man. *Hum. Neurobiol.*, 6: 1–9.

Huttenlocher, P.R., De Courten, Ch., Garey, L.J. and Van der Loos, H. (1982) Synaptogenesis in human visual cortex – evidence for synapse elimination during normal development. *Neurosci. Lett.*, 33: 247–252.

Innocenti, G.M. (1982) Development of interhemispheric cortical connections. *Neurosci. Res. Progr. Bull.*, 20: 532–540.

Innocenti, G.M. (1986) General organization of callosal connections in the cerebral cortex. In E.G. Jones and A. Peters (Eds.), *Cerebral Cortex, Vol. 5*, Plenum Press, New York, pp. 291–353.

Johnston, M.V., Silverstein, F.S., Reindel, F.O., Penney, J.B. Jr. and Young, A.B. (1985) Muscarinic cholinergic receptors in human infant forebrain: [^{3}H]quinuclidinyl benzilate binding in homogenates and quantitative autoradiography in sections. *Dev. Brain Res.*, 19: 195–203.

Jones, E.G. (1981) Anatomy of cerebral cortex: Columnar input-output organization. In F.O. Schmitt, F.G. Worden, G. Adelman and S.G. Dennis (Eds.), *The Organization of the Cerebral Cortex*, MIT Press, Cambridge, MA, pp. 199–235.

Judaš, M. (1987) *Perinatal Cytoarchitectonical Development of the Prospective "Motor Speech Area" in the Human Frontal Lobe*, M.Sc. thesis, University of Zagreb. (In Croatian.)

Koelliker, A. (1896) *Handbuch der Gewebelehre des Menschen. Vol. II. Nervensystem des Menschen und der Thiere*, 6 Edn., Engelmann, Leipzig.

Kononova, E.P. (1940) The development of the frontal region in postnatal period. *Contr. Mowcow Brain Inst.*, 5: 73–124. (In Russian.)

Kostović, I. (1984) Prenatal development of interstitial neurons in the "white" matter of the human telencephalon. *Soc. Neurosci. Abstr.*, 10: 47.

Kostović, I. (1986) Prenatal development of nucleus basalis complex and related fibre systems in man: a histochemical study. *Neuroscience*, 17: 1047–1077.

Kostović, I. (1987) Late postnatal development of the acetylcholinesterase-reactive innervation of the layer III pyramidal neurons in the human prefrontal cortex. *Neuroscience*, Suppl. 22: S228.

Kostović, I. (1990) Entwicklung des Zentralnervensystems. In K. Hinrichsen (Ed.), *Human Embryologie*, Springer, Berlin, in press.

Kostović, I. and Goldman-Rakic, P.S. (1983) Transient cholinesterase staining in the mediodorsal nucleus of the thalamus and its connections in the developing human and monkey brain. *J. Comp. Neurol.*, 219: 431–447.

Kostović, I. and Molliver, M.E. (1974) A new interpretation of the laminar development of cerebral cortex: synaptogenesis in different layers of neopallium in the human fetus. *Anat. Rec.*, 178: 395.

Kostović, I. and Rakic, P. (1980) Cytology and time of origin of interstitial neurons in white matter in infant and adult human and monkey telencephalon. *J. Neurocytol.*, 9: 219–242.

Kostović, I. and Rakic, P. (1984) Development of prestriate visual projections in the monkey and human fetal cerebrum revealed by transient cholinesterase staining. *J. Neurosci.*, 4: 25–42.

Kostović, I. and Rakic, P. (1990) Developmental history of transient subplate zone in the visual and somatosensory cortex of the macaque monkey and human brain. *J. Comp. Neurol.*, 297: 441–470.

Kostović, I. and Štefulj-Fučić, A. (1985) Distribution of somatostatin immunoreactive neurons in frontal neocortex and underlying "white" matter of the human fetus and preterm infant. *Soc. Neurosci. Abstr.*, 11: 352.

Kostović, I., Judaš, M. and Bogdanović, N. (1987) Postnatal maturation of human frontal cortex: new cytoarchitectonic criteria. *Soc. Neurosci. Abstr.*, 13: 1099.

Kostović, I., Škavić, J. and Strinović, D. (1988) Acetylcholinesterase in the human frontal associative cortex during the period of cognitive development: early laminar shifts and late innervation of pyramidal neurons. *Neurosci. Lett.*, 90: 107–112.

Kostović, I., Lukinović, N., Judaš, M., Bogdanović, N., Mrzljak, L., Zečević, N. and Kubat, M. (1989a) Structural basis of the developmental plasticity in the human cerebral cortex: The role of the transient subplate zone. *Metab. Brain*

Dis., 4: 17 – 23.
Kostović, I., Seress, L., Mrzljak, L. and Judaš, M. (1989b) Early onset of synapse formation in the human hippocampus: A correlation with Nissl – Golgi architectonics in 15- and 16.5-week-old fetuses. *Neuroscience,* 30: 105 – 116.
Kristt, D.A. (1989) Acetylcholinesterase in immature thalamic neurons: Relation to afferentation, development, regulation and cellular distribution. *Neuroscience,* 29: 27 – 43.
Krmpotić-Nemanić, J., Kostović, I., Kelović, Z. and Nemanić, D. (1980) Development of acetylcholinesterase (AChE) staining in human fetal auditory cortex. *Acta Otolaryngol.,* 89: 388 – 392.
Krmpotić-Nemanić, J., Kostović, I., Kelović, Z., Nemanić, D. and Mrzljak, L. (1983) Development of the human fetal auditory cortex: growth of afferent fibres. *Acta Anat.,* 116: 69 – 73.
LaMantia, A. and Rakic, P. (1984) The number, size, myelination and regional variation in the corpus callosum and anterior commissure of the developing rhesus monkey. *Soc. Neurosci. Abstr.,* 10: 1373.
Leijon, I. (1982) Assessment of behaviour on the Brazelton scale in healthy preterm infants from 32 conceptional weeks until full-term age. *Early Hum. Dev.,* 7: 109 – 118.
MacBrown, R. and Goldman, P.S. (1977) Catecholamines in neocortex of rhesus monkeys: regional distribution and ontogenetic development. *Brain Res.,* 124: 576 – 580.
Marin-Padilla, M. (1970a) Prenatal and early postnatal ontogenesis of the human motor cortex: a Golgi study. I. The sequential development of the cortical layers. *Brain Res.,* 23: 167 – 183.
Marin-Padilla, M. (1970b) Prenatal and early postnatal ontogenesis of the human motor cortex. II. The basket-pyramidal system. *Brain Res.,* 23: 185 – 191.
Marin-Padilla, M. (1988) Early ontogenesis of the human cerebral cortex. In A. Peters and E.G. Jones (Eds.), *The Cerebral Cortex. Vol. 7. Development and Maturation of the Cerebral Cortex,* Plenum Press, New York, pp. 1 – 34.
Mesulam, M.M. and Geula, C. (1988) Acetylcholinesterase-rich pyramidal neurons in the human neocortex and hippocampus: Absence at birth, development during the life span, and dissolution in Alzheimer's disease. *Ann. Neurol.,* 24: 765 – 773.
Michael, A.E. and Garey, L.J. (1984) The development of dendritic spines in the human visual cortex. *Hum. Neurobiol.,* 3: 223 – 227.
Molliver, M.E., Kostović, I. and Van der Loos, H. (1973) The development of synapses in the human fetus. *Brain Res.,* 50: 403 – 407.
Mrzljak, L., Uylings, H.B.M., Kostović, I. and Van Eden, C.G. (1988) Prenatal development of neurons in the human prefrontal cortex. I. A qualitative Golgi study. *J. Comp. Neurol.,* 271: 355 – 386.
Parmelee, A.H. (1975) Neurophysiological and behavioral organization of premature infants in the first months of life.

Biol. Psychiat., 10: 501 – 512.
Piaget, J. (1954) *The Construction of Reality in the Child,* Basic Books, New York.
Poliakov, G.I. (1949) Structural organization of the human cerebral cortex during ontogenetic development. In S.A. Sarkisov, I.N. Filimonov and N.S. Preobrazhenskaya (Eds.), *Cytoarchitectonics of the Cerebral Cortex in Man,* Medgiz, Moscow, pp. 33 – 92. (In Russian.)
Poliakov, G.I. (1959) Progressive differentiation of human cerebral cortical neurons during ontogenesis. In S.A. Sarkisov (Ed.), *Development of the Central Nervous System,* Medgiz, Moscow. (In Russian.)
Poliakov, G.I. (1967) Some results of research into the development of the neuronal structure of the cortical ends of the analysers in man. *J. Comp. Neurol.,* 117: 197 – 212.
Poliakov, G.I. (1979) *Entwicklung der Neuronen der menschlichen Grosshirnrinde,* VEB Georg Thieme, Leipzig.
Prechtl, H.F.R. (1974) The behavioural states of the newborn infant (a review). *Brain Res.,* 76: 185 – 212.
Prusky, G.T., Arbuckle, J.M. and Cynader, M.S. (1988) Transient concordant distributions of nicotinic receptors and acetylcholinesterase activity in infant rat visual cortex. *Dev. Brain Res.,* 39: 154 – 159.
Purpura, D.P. (1975) Normal and aberrant neuronal development in the cerebral cortex of human fetus and young infant. In N.A. Buchwald and M.A.B. Brazier (Eds.), *Brain Mechanisms in Mental Retardation,* Academic Press, New York, pp. 141 – 169.
Rakic, P. (1976) Prenatal genesis of connections subserving ocular dominance in the rhesus monkey. *Nature,* 261: 467 – 471.
Rakic, P. (1978) Neuronal migration and contact guidance in the primate telencephalon. *Postgrad. Med. J.,* 54: 25 – 40.
Rakic, P. (1982) Early developmental events: Cell lineages, acquisition of neuronal positions, and areal and laminar development. *Neurosci. Res. Progr. Bull.,* 20: 439 – 451.
Rakic, P. (1988a) Defects of neuronal migration and the pathogenesis of cortical malformations. In G.J. Boer, M.G.P. Feenstra, M. Mirmiran, D.F. Swaab and F. Van Haaren (Eds.), *Biochemical Basis of Functional Neuroteratology; Permanent Effects of Chemicals on the Developing Brain, Progr. Brain Res., Vol. 73,* Elsevier Science Publishers, Amsterdam, pp. 15 – 37.
Rakic, P. (1988b) Specification of cerebral cortical areas. *Science,* 241: 170 – 176.
Rakic, P. (1988c) Intrinsic and extrinsic determinants of neocortical parcellation: A radial unit model. In P. Rakic and W. Singer (Eds.), *Neurobiology of Neocortex,* Wiley, New York, pp. 5 – 27.
Rakic, P., Bourgeois, J.-P., Zečević, N., Eckenhoff, M.F. and Goldman-Rakic, P.S. (1986) Isochronic overproduction of synapses in diverse regions of the primate cerebral cortex. *Science,* 232: 232 – 235.
Robertson, R.T. (1987) A morphogenetic role for transiently

expressed acetylcholinesterase in developing thalamocortical systems. *Neurosci. Lett.*, 75: 259–264.
Sarkisov, S.A. (1965) *Development of the Child's Brain*, Meditsina/Medgiz, Moscow/Leningrad. (In Russian.)
Schadé, I.P. and van Groenigen, W.B. (1961) Structural organization of the human cerebral cortex. I. Maturation of the middle frontal gyrus. *Acta Anat.*, 47: 74–111.
Schwartz, M.L. and Goldman-Rakic, P.S. (1984) Callosal and intrahemispheric connectivity of the prefrontal association cortex in rhesus monkey: Relation between intraparietal and principal sulcal cortex. *J. Comp. Neurol.*, 226: 403–420.
Schwartz, M.L. and Goldman-Rakic, P.S. (1990) Early specification of callosal connections in the fetal monkey prefrontal cortex. Submitted.
Shatz, C.J., Chun, J.J.M. and Luskin, M.B. (1988) The role of the subplate in the development of the mammalian telencephalon. In A. Peters and E.G. Jones (Eds.), *The Cerebral Cortex. Vol. 7. Development and Maturation of the Cerebral Cortex,* Plenum Press, New York, pp. 35–58.
Sidman, R.L. and Rakic, P. (1973) Neuronal migration, with special reference to developing human brain: a review. *Brain Res.*, 62: 1–30.
Sillito, A.M. (1983) Plasticity in the visual cortex. *Nature*, 303: 477–478.
Siwe, S.A. (1927) Das Gehirn: Die mikroskopische Entwicklung des Grosshirns nach der Geburt. In K. Peter, G. Wetzel and F. Heindrich (Eds.), *Handbuch der Anatomie des Kindes*, Bargmann, Munich, pp. 609–632.
Sterman, M.B. and Hoppenbrouwers, T. (1971) The development of sleep-waking and rest-activity patterns from fetus to adult in man. In M.B. Sterman, D.J. McGinty and A.M. Adinolfi (Eds.), *Brain Development and Behavior*, Academic Press, New York, pp. 203–229.
Trevarthen, C. (1979) Neuroembryology and the development of perception. In F. Falkner and J.M. Tanner (Eds.), *Human Growth. Vol. 3. Neurobiology and Nutrition*, Bailliere Tindall, London, pp. 3–96.
Vaughan, H.G. Jr. (1975) Electrophysiologic analysis of regional cortical maturation. *Biol. Psychiat.*, 10: 513–526.
Von Economo, C. and Koskinas, G. (1925) *Cytoarchitektonik der Hirnrinde der erwachsenen Menschen.* Springer, Wien.
Vignal, M.W. (1888) Recherches sur le développement de la substance corticale du cerveau et du cervelet. *Arch. Physiol. Norm. Pathol. Ser. IV*, 2: 238–254, 311–338.
Wolff, P.H. and Ferber, R. (1979) The development of behavior in human infants, premature and newborn. *Annu. Rev. Neurosci.*, 2: 291–307.
Zilles, K., Werners, R., Büsching, U. and Schleicher, A. (1986) Ontogenesis of the laminar structure in areas 17 and 18 of the human visual cortex. A quantitative study. *Anat. Embryol.*, 174: 339–353.

Discussion

H.B.M. Uylings: In your last slide you mentioned that the structural development at 1 year might be related with cognitive development. Can you expand on the present knowledge of structural correlates with human (primate) cognitive development?

I. Kostović: The onset of cognitive functions in man begins during the overproduction of circuitry elements (postsynaptic spines, synapses, cortico-cortical pathways). At the moment it is not clear whether this overproduction phase of cortical development is directly related to the onset of cognitive functions. The late phase of structural and transmitter maturation of the cerebral cortex which occurs during intensive interaction with social environment (after the first year) involves primarily maturation of associative neuronal elements (cells, layers, areas). This phase of structural development correlates well with prolonged maturation of cognitive functions in man.

C. Cavada: (1) Given the prominent structural reorganization that takes place in the cortex in the perinatal period, which are your criteria to identify the various cortical areas at this stage?

(2) You have shown us heavy acetylcholinesterase (AChE) staining in the mediodorsal (MD) and pulvinar thalamic nuclei during the prenatal period. Does this staining fully disappear along development up to the adult age?

(3) Have you observed any temporal or topographic relationships between the differential increase in AChE staining in different regions of the frontal cortex (e.g. lateral and medial) and the corresponding decrease in the mediodorsal thalamic nucleus?

I. Kostović: (1) Due to the fact that the basic 6-layer pattern is present before birth it is possible to distinguish "primary" and secondary cortical areas during the perinatal period. However, within the prefrontal granular cortex areal differentiation is not complete and one has to consider both progressive areal differentiation and reorganization events.

(2) AChE staining remains in some neuronal populations of the mediodorsal and pulvinar thalamic nuclei. However, the overall staining (neuropil plus cells) decreases dramatically during the postnatal period.

(3) The laminar and vertical patterns of AChE staining are different in different regions of the prefrontal cortex. There is also a rostro-caudal gradient in staining intensity. Due to the very strong staining of the mediodorsal thalamus, fine regional differences are difficult to determine at the level of individual subdivisions of mediodorsal nucleus.

V.B. Domesick: Do you have evidence that the AChE-positive staining in the frontal cortex is a direct result of thalamocortical (MD) projection?

I. Kostović: In human material this evidence is indirect: there is close temporal correlation between staining of MD and frontal cortex. In addition, one can follow AChE-reactive fibers through the internal capsule to the cortex. In experimental monkey material a lesion of thalamic nuclei results in the decrease of cortical staining.

P.R. Lowenstein: What are the targets of asymmetric synapses in the transient subplate zone?

I. Kostović: The majority of synapses in the subplate zone were of asymmetrical type. The postsynaptic elements can be identified as small dendrites by ultrastructural criteria or as proximal dendrites by following them to the parent cell body.

P. Gaspar: I was very surprised and interested by you mentioning that layer IV (granular) was present throughout the cortex even in the prospective agranular areas. What were your criteria to identify layer IV? Nissl staining? Arrival of thalamocortical fibers?

I. Kostović: We delineate layer IV on the basis of cytoarchitectonic criteria. The prospective agranular premotor belt can be determined by its caudal border towards primary motor cortex where Betz cells were found after 32 weeks of gestation.

J.P.C. de Bruin: My question considers the "vigorous plasticity" and its temporal relation with the developmental processes (transient stages, etc.).

I. Kostović: "Vigorous plasticity" means significant structural reorganization of the cerebral mantle after a lesion (hemorrhagic or ischemic). The vigorous plasticity decreases after the disappearance of transient zones and establishment of connections.

CHAPTER 11

Anatomical correlates of behavioural change after neonatal prefrontal lesions in rats

Bryan Kolb and Robbin Gibb

University of Lethbridge, Lethbridge, Alberta, T1K 3M4, Canada

Introduction

At first look, the prefrontal cortex (PFC) appears almost hopelessly complicated. The cell typology, complexity of connectivity, and the intricacy of laminar organization have yet to be completely described, let alone understood, and the correlations with behaviour seem remote indeed. One way to approach such a complex system is to alter the organization of the cortex by perturbing it during development. The developing brain is vulnerable to a variety of insults that may alter the subsequent pattern of brain development and lead, in turn, to a variety of functional abnormalities. By correlating the changes in behaviour with those in structure it thus may be possible to open a window on the nature of neocortical organization. In addition, such studies may provide insight into the etiology of various developmental disorders, ranging from autism, schizophrenia, and mental retardation to hyperactivity and learning disabilities. The present paper focusses upon the behavioural and structural effects of damaging the prefrontal cortex of rodents at different developmental ages. In this paper we will begin by considering the effects of prefrontal cortical injury in adult rodents before considering the behavioural and anatomical effects of prefrontal injury at different ages. One of the overriding principles of our research has been to try to find anatomical correlates of behavioural changes *in the same animals*. We have gone down many false alleys but it appears that we are now in a position to offer some tentative hypotheses about the structural correlates of behavioural changes after perinatal PFC lesions in rodents.

The effects of prefrontal lesions in adult rats

It is possible to identify several cytoarchitectonically distinct regions of the rat cortex that receive afferents from the dorsal medial nucleus (MD) and thus have been described as prefrontal cortex (Fig. 1). These regions include several areas along the medial wall of the hemisphere (Zilles' Fr2, Cg1, Cg2, Cg3, IL), cortex along the ventral surface of the hemisphere, just above the olfactory bulb (Zilles' LO, VLO), and cortex just dorsal to the dorsal bank of the rhinal fissure (Zilles' AIV) (e.g., Divac et al., 1978). Using the lesion technique it is possible to functionally dissociate the medial regions, which we shall call medial frontal cortex (MF), from the ventral and lateral regions, which we shall call orbital frontal cortex (OF). Lesions in different studies have removed the entire frontal cortex (large frontal), the medial region, or the orbital region, allowing us to make some broad generalizations regarding the effects of MF, OF or

Correspondence: Bryan Kolb, Department of Psychology, University of Lethbridge, Lethbridge, Alberta T1K 3M4, Canada. Tel.: 403-329-2405; FAX: 403-329-2022.

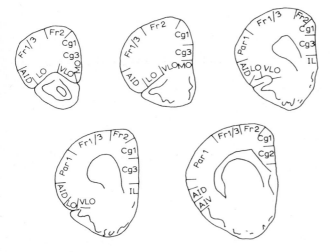

Fig. 1. Illustration of the different cytoarchitectonic fields in the frontal cortex of the rat. The anterior/posterior levels are +4.5, +3.7, +2.7, +2.2 and +1.7 mm anterior to bregma, respectively, with the most anterior plane being the top left. Nomenclature after Zilles (1985). Abbreviations: AId, dorsal agranular insular cortex; AIv, ventral agranular insular cortex; Cg1, cingulate cortex, area 1; Cg2, cingulate cortex, area 2; Cg 3, cingulate cortex, area 3; Fr1, frontal cortex, area 1; Fr2, frontal cortex, area 2; Fr3, frontal cortex, area 3; IL, infralimbic cortex; LO, lateral orbital cortex; MO, medial orbital cortex; Par 1, primary somatosensory cortex or parietal cortex, area 1; VLO, ventrolateral orbital cortex.

large lesions. Lesions of MF cortex produce chronic deficits in tests of spatial navigation, visual working memory, and habituation as well as deficits in food hoarding, maternal behaviour, nest building, play behaviour (in young animals), and reaching for, and handling, objects such as pieces of food (see Kolb, 1984, for a review). In contrast, rats with OF cortex removals have few impairments on tests of learning, although they do show impaired resistence to extinction, whereas they have significant deficits in feeding behaviour, which may be related to incidental damage to the somatosensory face area, a drop in body weight set point, increased activity, and altered social behaviour (see Kolb, 1984). Combined lesions of MF and OF cortex produce all of the above deficits but importantly, the deficits are significantly more severe than after either of the smaller lesions alone. Finally, we have made unilateral lesions of MF, OF, or both regions and found only mild chronic behavioural deficits.

These studies in adult rats have led us to devise a battery of behavioural tests with which to study the effects of frontal lesions at different ages. Our assumption in these studies has been that there are multiple possible behavioural outcomes after early lesions. Thus, (1) animals might appear relatively normal compared to animals with similar adult lesions, a result that we shall call *sparing of function;* (2) animals might be similar to adult operates; (3) animals might be worse off than adult operates; or (4) animals might have different behavioural symptoms than adult operates. In our studies we have tried to be mindful that it is only with a broad test battery that one will be able to distinguish these outcomes and make meaningful correlates between brain and behaviour in the developing brain. That is, single measures of behaviour are unlikely to provide meaningful correlations with structure, and our behavioural results consistently show that combinations of all 4 outcomes are common.

The behavioural effects of prefrontal lesions in neonatal rats

Since infant rats are small and naked it is possible to anaesthetize them with hypothermia, allowing the opportunity to surgically remove bits (or all) of the cortical mantle under a surgical microscope, and with little bleeding. Since the brain is so small and the skull sutures have a changing relationship to the underlying brain in infant rats (Kolb, 1987), it is difficult to make small precise lesions before about 5 days of age. Furthermore, since there is significant cell migration still occurring in the first week, especially in the medial wall of the hemisphere (Hicks and D'Amato, 1968; Uylings et al., 1990), it is difficult to ensure that lesions are removing tissue that is equivalent at different ages. We have attempted to control this problem by systematically varying lesion size to include large lesions, including both the MF and OF regions, and smaller lesions, largely restricted to the MF or OF areas. Further, we have compared the behaviour of these animals to those with complete decortications (e.g. Kolb and Whishaw, 1981a; Whishaw and Kolb, 1984), or hemidecortications (e.g. Kolb and Tomie, 1988) as well as more restricted lesions elsewhere in the cortex including motor cortex (Kolb and Holmes, 1983; Whishaw and Kolb, 1988), parietal cortex (Kolb et al., 1987), and temporal and visual cortex (e.g. Ladowski et al., 1988).

In our initial studies, we were impressed by the clear sparing of behaviours on various tests of learning such as delayed response, spatial reversals, and active avoidance (e.g. Kolb and Nonneman, 1978). This result was consistent with the Kennard principle, which suggested that the earlier the brain damage, the better the restitution of function. Nevertheless, at the same time we were gathering evidence that the Kennard principle did not always hold and that behaviour was sometimes either more disturbed by the earlier lesion or disturbed differently by the earlier lesion. As we pursued our studies it became apparent that a large number of factors interacted with the early brain damage and that there was no simple generalization regarding the effects of early cortical lesions to the PFC.

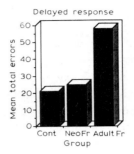
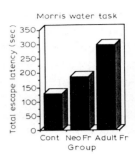
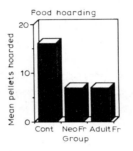
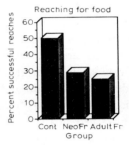

Fig. 2. Summary of the performance of rats with adult (90 days) or neonatal (3 days) medial frontal lesions on 2 tests of learned (delayed response, Morris water task) and 2 tests of species-typical behaviour (food hoarding, reaching). The neonatal operates show sparing of function on the learned but not on the species-typical behaviours. Abbreviations: Neo Fr, neonatal frontal lesion group; Adult, Fr, adult frontal lesion group; Cont, control group.

Behavioural measure

Although the distinction is arbitrary, it is possible to fractionate behaviour into 2 broad categories, which we shall call *learned* and *species-typical*. Most behavioural studies after early brain damage have focussed upon learned behaviours and there can be dramatic sparing on some tests of learning. For example, rats with MF lesions show striking sparing on tests such as delayed response and spatial reversals, tests which historically have been the mainstays of neuropsychological evaluation of the animal with a prefrontal lesion (Fig. 2). Sparing does not always occur on learning tests, however; it is only partial on the Morris water task and does not occur on the radial arm maze. In contrast to the sparing of learned behaviours, rats or hamsters with infant MF lesions show virtually no sparing on most tests of adult species-typical behaviour and even during development the play behaviour of juvenile rats is abnormal (Fig. 2). In sum, as a general rule it appears there is much better sparing of learned than of species-typical behaviours after neonatal lesions in rats. Perhaps this distinction reflects the possibility that learned behaviours can be solved by the brain in more than one way but that the neural organization of species-typical behaviours is relatively inflexible.

Age

There are dramatic changes in the rodent brain

over the first 2 weeks of life and one might predict major differences in the probability of sparing from lesions at different ages. When we began our studies, we anticipated that the earliest lesions were most likely to allow sparing but this was not the case. Rather, when we gave rats large bilateral lesions at different ages (e.g. postnatal days 1, 3, 5, 7, 10, 25) we found that whereas the animals with lesions in the first few days fared worse on test of learning and species-typical behaviours than animals with adult lesions, those animals with lesions around 10 days of age showed sparing of function on learning tests and on some species-typical tests (see Fig. 3). Similarly, when we studied animals with MF lesions (Table I) we found parallel results as there appears to be a window of time around 10 days of age when sparing is likely, whereas it is less likely if the damage is sustained in the first or third weeks of life (e.g. Kolb and Whishaw, 1981b; Kolb, 1987; Petrie and Kolb, unpublished). These results were not peculiar to prefrontal lesions as we have similar findings in animals with bilateral parietal lesions (Kolb et al., 1987).

Lesion size and location

We have considered several parameters of the lesion procedure by looking at MF and OF lesions, combined lesions, and unilateral lesions. Animals with MF lesions at any age over the first 10 days

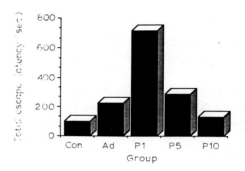

Fig. 3. Summary of the Morris water task performance of rats with large frontal lesions at different ages. There is sparing after lesions at 10 days of age but after lesions at 1 day of age the animals perform very poorly. (After Kolb, 1987.) Abbreviations: Con, control group; Ad, adult frontal lesion group; P1, 1 day frontal lesion group; P5, 5 day frontal lesion group; P10, 10 day frontal lesion group.

TABLE I

Summary of the behavioural effects of MF lesions at different ages

Lesion age (days)	Behavioural measure[a]				
	Body weight	Toe nail	Reaching	Beam	Morris task
3	X	X	X	X	X
6	X	X	XX	X	X
10	N	N	N	N	X
15	N	X	XX	N	X
30	X	N	X	N	X
90	X	X	XX	N	XX

[a] Abbreviations: N = equivalent to normal controls; X = significant impairment; XX = greater impairment than X.
Note: The behavioural measures include chronic body weight, toenail length, which measures the ability of animals to trim the nails, reaching, which measures the ability to reach for food, beam, which measures ability to traverse a narrow beam, and the Morris water task, which is a test of spatial navigation. The results show that animals with MF lesions at 10 days of age show remarkable sparing of function. All of the infant groups show sparing on the water task. The animals with lesions at 3 or 6 days show new deficits on beam traversing. (Data from Petrie and Kolb, unpublished.)

show sparing of learned behaviours but incomplete or absent sparing on many tests of species-typical behaviours (e.g. Kolb and Nonneman, 1978; Kolb and Whishaw, 1985a, b). Rats with OF lesions over the first 10 days also show sparing of learned behaviours and do show sparing on tests of tongue and mouth use and show normal body weight curves, in contrast to adult operates. In fact, it is not until the injury is incurred at about 60 days of age in adolescence that OF lesions produce a drop in body weight set point (Kolb and Nonneman, 1976).

In adult animals, lesions that include both MF and OF regions produce more severe deficits than more restricted lesions and this is also true in animals with injuries in infancy. Thus, animals with removal of both prefrontal fields in the first 5 days of life show severe behavioural deficits and virtually no sparing of function, even on tests of learning. Nevertheless, there is still sparing after lesions around 7 – 10 days of age, so that although the extent of injury reduces the likelihood of sparing, if the damage is at the right time, sparing is still possible even after large lesions (Fig. 4).

In contrast to the effect of large bilateral lesions in the first few days of life, unilateral lesions on the day of birth allow sparing of lateralized functions such as contralateral paw use (Hicks and D'Amato, 1970; Whishaw and Kolb, 1988). It has proven difficult to study the effects of unilateral lesions on most learned or species-typical behaviours, however, since the effects of even the largest frontal lesions are small and variable (Kolb et al., 1989a, b).

Finally, Whishaw and Kolb (1989) considered what would happen if only the OF cortex and the somatosensory face area remained in an otherwise complete neodecortication. The results were clear: there was no sparing of function but the animals still had normal tongue use and maintained a normal body weight. Thus, there was no evidence that functions of the medial PFC could be mediated by the remaining OF cortex in the absence of the rest of the neocortex.

Age at assessment

Since the behaviour of infants rats and hamsters is limited, the behavioural deficits observed later in life must emerge during development. Three studies have considered the development of these deficits. In the first we took advantage of the fact that hamsters are extremely efficient food hoarders and nest builders in adulthood and that the behaviours do not begin to emerge until shortly after weaning. Thus, hamsters were given MF or OF lesions at 4 days of age and their hoarding and nest building behaviours were studied from weaning (21 days of age) until adulthood (Kolb and Whishaw, 1985b). The results showed that although weanling hamsters with MF or OF lesions hoarded and built nests as well as normal littermates, the normal animals improved rapidly over the first few days whereas the operated animals did not (Fig. 5). This result is reminiscent of the data of Goldman (1974) who found that monkeys with neonatal PFC lesions also "grew into" their deficits on learning tests. Second, we took advantage of the finding that juvenile rats are unable to solve the Morris water task until about 22 – 25 days of age (Rudy et al., 1987), at which time they acquire it very rapidly. We gave rats large bilateral

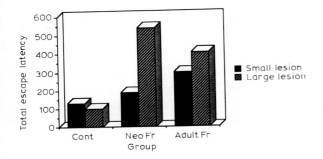

Fig. 4. Comparison of the effects of large and small frontal lesions in adulthood (90 days) or infancy (1 day) on the performance of the Morris water task. Larger lesions produce larger deficits but note that the difference is far larger in the neonatal operates who show sparing with small lesions but not with large lesions. Abbreviations: Cont, control groups; Neo Fr, neonatal frontal lesion groups; Adult Fr, adult frontal lesion groups.

frontal lesions at 4 days of age and then began water task training at 25 days of age. Surprisingly, right from the first trial block the operated animals differed significantly from their littermate controls (Fig. 5, lower panel). It therefore appears that rats do not develop into the deficits on at least this task. Third, we studied the play behaviour of rats with MF lesions on the day of birth and compared it to their littermate controls with whom they were housed until 60 days of age. The results were unequivocal: the animals with MF lesions initiated less play behaviour, engaged in less total play behaviour, were more active, and the males displayed more mounting behaviour than the littermate controls. Again, it appears that even before the neocortex is mature there is a marked behavioural effect of early frontal lesions. What is not clear, however, is whether these early behavioural effects are due directly to the absence of prefrontal tissue or might be due to changes in the rest of the cortex, which result from the prefrontal injury. Further, given that animals with lesions around 10 days of age show the best sparing, one wonders whether animals with lesions around 10 days of age might grow into their deficits.

Modulators: Environment, noradrenaline, and sex

In the course of doing our experiments we have begun to look at the role that various modulators might have upon behavioural outcome. We began by raising animals with large frontal lesions at different ages in either standard laboratory cages or in more complex environments that allowed greater activity and interaction with novel objects, much as Greenough and his colleagues have done (e.g. Greenough, 1976). Such treatments proved beneficial indeed as the animals raised in the enriched environments had greatly attenuated deficits. They showed environmental effects on every behaviour assessed, and even some behaviours for which the control animals showed no advantage from enrichment (e.g. tongue use) were markedly improved (Kolb and Elliot, 1987). Perhaps more importantly, animals with lesions at 5 days of age showed sparing after the enriched treatment and on many tests performed as well as animals with lesions around 10 days of age. This result is intriguing and needs to be pursued, especially with respect to structural correlates of both observations.

Our environmental studies led us to ask whether factors related to environment-induced neural plasticity might be related to these findings. In particular, we asked whether noradrenaline (NA) levels might influence behavioural or neural out-

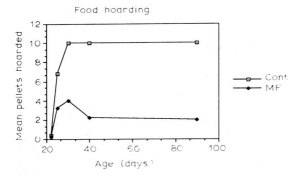
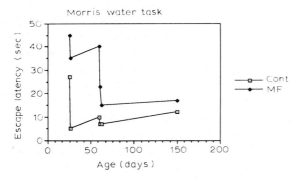

Fig. 5. Developmental studies of food hoarding in hamsters and water task performance in rats with neonatal frontal lesions at 4 days of age. The lesions in the hamsters are limited to the medial frontal regions whereas the lesions in the rats are larger and include damage to the adjacent motor cortex. The hamsters "grow into their deficits" on a species-typical behaviour whereas the rats show a large deficit as soon as they are able to solve the task. (After Kolb et al., 1989; Kolb and Whishaw, 1985a.)

come after early lesions since studies in the visual system had shown that NA might play a role in environmentally induced changes to ocular dominance columns in kittens (e.g. Kasamatsu et al., 1981). Total depletion of cortical NA in infant rats has yet to be shown to have a significant behavioural correlate but it seemed reasonable to suppose that it might play a role if the brain was to make a plastic response to early brain injury. This was the case as NA depletion prior to an MF lesion at 7 days of age completely blocked the normal sparing after this lesion (Fig. 6). Surprisingly, however, NA depletion did not have this effect if the lesion were at 4 or 10 days of age: rats with MF lesions at 4 days still show no sparing and rats with lesions at 10 days still showed sparing (Kolb and Sutherland, 1986; Sutherland et al., 1982). The fact that NA might influence the effect of infant PFC lesions may be especially important when one considers that many noradrenergic fibres pass through the frontal pole en route to the rest of the cortex (e.g. Morrison et al., 1981). Therefore, it is possible that early PFC lesions themselves lower NA and this in turn leads to greater behavioural impairments than would be expected. This appears unlikely, however, as we did NA assays on the posterior cortex of rats with frontal lesions at 4 days and found no correlation with behaviour of the same animals. There was a correlation with cortical thickness (Kolb et al., 1989a), however, which we shall return to below.

The central nervous system is modified in essentially irreversible ways by gonadal hormones secreted early in development (e.g. MacLusky and Naftolin, 1981). Although most is known about the subcortical effects of sex hormones, there are receptors in the MF cortex (e.g. MacLusky et al., 1979) and it has been shown in the mouse that there are aromatizing enzymes in the cortex that can convert testosterone into oestrogen, thus providing a mechanism for masculinization of the male cortex. We have at least preliminary evidence that sex may interact with early frontal lesions. We have found that early frontal lesions significantly reduce body weight in male but not in female rats, even though body weight reduction is present in both sexes following lesions in adulthood (e.g. Kolb, 1987). Further, it appears that medial frontal lesions at any age produce larger deficits in females than males in tests of spatial navigation (Petrie and Kolb, unpublished, 1989).

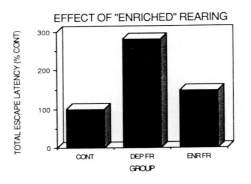

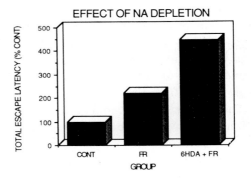

Fig. 6. Comparison of the effects of enriched rearing and noradrenaline (NA) depletion on sparing of function after frontal lesions at 7 days of age. The enriched rearing included 90 days of rearing (from weaning) in a large pen in which toy objects were changed twice weekly. The NA depletion was achieved by subcutaneous injections of 6-hydroxydopamine (6HDA) on the first 3 days of life. NA assays (HPLC) revealed virtually no dectectable cortical NA. Note that enrichment enhances performance whereas NA depletion worsens performance. The rats in the control group in the NA depletion study are also NA depleted in infancy, but have no frontal lesion. (After Kolb and Elliott, 1987; Sutherland et al., 1982.)

Summary

We can reach several conclusions regarding the behavioural effects of neonatal PFC lesions in rats and hamsters.

(1) Relative to the effects of PFC lesions in adulthood, early PFC lesions may lead to several behavioural outcomes ranging from sparing of function to a greater loss of function, the precise outcome varying with several factors. It is not possible to make a general statement that behaviour is more or less affected after neonatal PFC lesions without specifying the precise conditions in question.

(2) Conclusions regarding the behavioural effects of PFC lesions depend upon the behaviours assessed. In general, sparing is more likely on tests of learned than of species-typical behaviour, possibly because more than one strategy can be used on tests of learning. Nevertheless, sparing does not occur on all learning tests and it is unclear what the crucial differences might be between different learning tests such that sparing may or may not occur, even in the same animals.

(3) The behavioural effect of PFC lesions varies with age at damage: in general, it appears that sparing is most likely after lesions around 10 days of age and behaviour is most disturbed after lesions around the day of birth. Furthermore, by 25 days of age, sparing from PFC lesions is absent or greatly reduced for most behaviours after MF lesions, although this may not be true after OF lesions.

(4) As might be expected, the behavioural outcome varies with lesion size as larger lesions have greater behavioural effects but even large removals of both MF and OF cortex still allow sparing on some tests if the damage is around 10 days of age. This is puzzling since adults with MF lesions appear to have greater behavioural loss on some tests than animals with much larger lesions at 10 days of age.

(5) Several variables influence the behavioural outcome after early PFC lesions including environmental rearing conditions, noradrenaline, and sex. Furthermore, there is reason to believe that these factors may interact with age, behavioural measure and lesion size, which in turn may interact with one another. It is hardly surprising that it is proving difficult to make simple generalizations regarding the effect of neonatal PFC lesions!

The anatomical effects of prefrontal lesions in neonatal rats

It is our assumption that the different behavioural effects of early PFC lesions ought to be based upon consistent structural (or physiological) changes, and that variables that modulate the behavioural outcomes ought to modulate the structural changes as well. We began by looking at gross measures such as brain size and weight and found very large effects: the earlier the lesion the smaller and lighter the brain (Table II). Further analysis showed that these gross effects were correlated

TABLE II

Summary of chronic brain weight and cortical thickness measures after large frontal or MF lesions at different ages

Lesion age (days)	Brain weight	Cortical thickness
Large frontal		
1	68	77
2	NM	78
5	75	85
7	75	90
10	78	89
25	86	94
90	91	100
Medial frontal		
3	89	90
6	92	92
10	95	97
15	90	93
30	93	95
90	97	100

Data indicate percent littermate control values for male rats. NM = not measured.

with a shrinkage of the diencephalon as well as a thinning of the remaining cortex (e.g. Kolb et al., 1983; Kolb and Whishaw, 1981a, b). The diencephalic shrinkage was most likely due to retrograde cell loss after the cortical damage but it was unclear what had caused the cortical thinning, especially since the laminar structure appeared normal. Indeed, the cortical effects were all the more puzzling since it appeared to be related to age at surgery rather than behaviour. Recall that animals with lesions around 10 days of age showed the best behavioural outcome but their cortex was thinner than animals with lesions later in life (Table II). Since cortical thickness is a gross measure, which is made up many factors including, among others, cell numbers, myelin, and dendritic and axonal arbor, we decided to look at each of these components separately.

Cell number

We have estimated neurone numbers in two ways. First, using 1 μm thick sections through visual cortex, neurones were counted in a 500 μm column in Nissl-stained tissue from adult rats with bilateral frontal lesions on the day of birth. This analysis showed that there were only 83% as many neurones in the frontal as in the littermate control brains (Table III). Further, when cortical thickness was measured in the same tissue there was a significant difference in thickness as the tissue from operated animals was only 78% of control thickness, which approximates our data from thicker sections (Table II). Correlations between cell number and cortical thickness were revealing because for the frontal animals it was 0.88, suggesting that most of the variance in cortical thickness was due to cell number. In contrast, the correlation in control animals was 0.46, suggesting that more of the variance was accounted for by differences in neuropil.

Our second measure of cell number was done using Golgi–Cox embedded tissue cut at 120 μm. Neurones were counted in a 1 mm column in areas Par 1, TE2/3, Oc1, and Oc2L in tissue from adult rats with bilateral frontal lesions at 1 or 10 days of age. The results parallelled those from the Nissl-stained tissue as there were again about 80% as many cells in the visual cortex of the day 1 animals as in the littermate controls (Fig. 7). We take this as tentative evidence that the staining in the Golgi tissue was at least roughly equivalent in the operated and control brains. Importantly, the Golgi tissue showed 3 more intriguing results. First, there was an even larger difference in cell number in other cortical areas. In particular, there were only 60% as many cells as normal in primary somatosensory cortex (Par 1). Second, there was also a reduction in cell number in the tissue from animals with lesions at 10 days of age, but it was significantly less than in the earlier operates. It is tempting to speculate therefore that a principle reason that cortical thickness correlates with age at surgery is because cell loss correlates with age at surgery, although this needs to be pursued more systematically. Third, the number of cells varied across cortical areas as there were significantly

TABLE III

Summary of cell counts and cortical thickness

Group	Mean neurones	Neurones/mm	Mean thickness
Control	96.2 ± 2.7	85.7 ± 2.5	1.14 ± 0.4
1 day Fr	80.6 ± 4.5*	93.3 ± 4.2**	0.88 ± 0.4*

*Differs significantly from control group (2-tail t, $p < 0.01$).
**Differs significantly from control group (1-tail t, $p < 0.05$).

Note: Measurements were taken from a 240 μm column on 1 μm thick sections, which were drawn using a camera lucida tube. Mean neurones refers to the mean number of neurones counted in layers II–VI in each section. Only those neurones with an identifiable nucleus were counted. Neurones/mm refers to the mean number of neurones per mm of cortical tissue. This gives a relative density of the neurones. Thickness is the measurement (mm) of cortical thickness from the top of layer I to the bottom of layer VI. These data show that control animals have thicker cortex that has more neurones. If there is more neuropil in the control animals, it could be predicted that the cell packing density would appear to be larger in the operated brains, however, and this appears to be the case. Numbers represent means and standard errors.

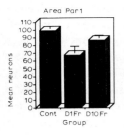
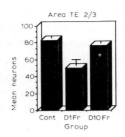
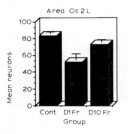
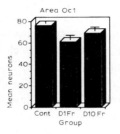

Fig. 7. Summary of neurone counts in 4 cortical areas in Golgi-stained tissue of adult rats (about 150 days) who sustained large frontal lesions at 1 or 10 days of age. The numbers represent the mean total number of neurones in layers I–VI in a 1 mm wide × 120 μm thick column. Neurones were counted if the cell body was present. The cortical regions are Zilles (1985) areas Par 1 (primary somatosensory cortex), TE 2/3 (posterior temporal cortex), Oc2L (visual cortex), and Oc1 (visual cortex). Abbreviations: CONT, control group; D1Fr, day 1 frontal lesion group; D10Fr, day 10 frontal lesion group.

more cells in parietal cortex than in posterior cortex in all groups. Although the latter result is consistent with the clear difference in cortical thickness between parietal and occipital-temporal cortex, it is not in accord with Rockel et al. (1980) who found that neurone number was constant across cortical areas. We cannot rule out the possibility, however, that the Golgi stain differed in its selectivity in different cortical areas.

When we analyzed the course of postoperative changes in cortical thickness we were struck by the fact that significant changes were visible 24 h after surgery and that the decrease in cortical thickness was related to age at surgery in the acute period as it had been chronically. Hence, the cortex was thinner the earlier the lesion, and it was unchanged in adults. This finding is especially provocative when one considers that there is virtually no neuropil in the newborn rat cortex so that a 20% drop in cortical thickness is unlikely to be due to a difference in neuropil at this age. Therefore, in view of our cell count data it is tempting to speculate that the early PFC lesions may lead to diffuse cell death throughout the cortex. This is unlikely to result merely from the trauma of surgery, however, because hemidecortication does not lead to a chronic reduction in neurone numbers in the remaining hemisphere even though there would be significant surgical trauma (e.g. Kolb and Tomie, 1988).

Dendritic arbor

The brains of rats with unilateral or bilateral removals of the MF, OF and anterior motor cortex at 1, 10 or 90 days of age were processed for Golgi–Cox staining. Examination of the dendritic arborization in layer II/III pyramidal cells and layer IV stellate cells in parietal cortex showed an unexpected correlation with behaviour: the early lesions produced a significant reduction in arborization whereas the 10 day lesions produced an increase in arborization and the adult lesions produced no change, relative to matched controls (Fig. 8). Furthermore, the dendritic changes were found only ipsilateral to the lesion in the unilateral cases. In order to evaluate the generality of the dendritic changes we have studied cells in a variety of cortical regions including Zilles' Par 1, Fr 1/3, TE 2/3, Oc1, and Oc2L. The greatest changes are in somatosensory and motor cortex (Par 1, Fr1/3) and there are virtually no changes in Oc1 and Oc2L. The latter results are intriguing in view of the loss of the significant PFC projection to the occipital cortex.

There are two important points to make regarding the results of our dendritic analyses. First, perhaps the most striking aspect of the dendritic data is that they correlate with the behavioural results: behavioural sparing is correlated with increased arborization and behavioural worsening is

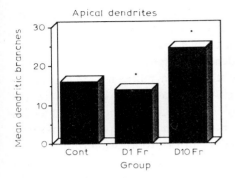

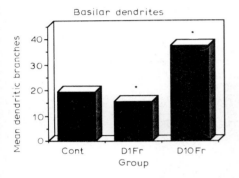

Fig. 8. Summary of the changes in layer II/III pyramidal neurones in area Par 1 of rats with neonatal frontal lesions. There is a significant drop in total branches in the day 1 operates and a significant increase in the day 10 operates. (After Kolb and Gibb, 1987.)

correlated with decreased arborization. Second, the changes in dendritic arborization make it easier to understand why sparing of function could be correlated with thinner than normal cortex: it appears that rats with PFC lesions at 10 days of age have increased cell death that is at least partly compensated for by increased dendritic arborization.

Cortical connections

It has been known for at least 20 years that unilateral cortical lesions alter descending cortical connections (e.g. Hicks and D'Amato, 1970, 1975). What is less clear, however, is how cortico-cortical connections might be changed by early lesions or how cortico-fugal connections might be changed by bilateral lesions. We have studied these questions by using retrogradely transported fluorescent dyes placed into the motor, somatosensory or occipital cortex of rats with large bilateral or unilateral frontal lesions that included not only the MF and OF fields but also extended posterior to include the anterior regions of motor cortex (Kolb and Van der Kooy, 1985).

Cortico-cortical connections

The principal cortico-cortical connections in the rat are summarized in Fig. 9. The PFC has extensive cortical connections, especially with the occipital cortex, and it seems likely that the loss of the PFC connections to the posterior cortex might alter the normal pattern of connectivity. It appears that large frontal lesions on the day of birth have little effect upon the pattern of cortico-cortical connections of either motor, somatosensory, or occipital cortex, although the connections appear to be heavier than normal, especially in animals with unilateral lesions. It may be that smaller lesions would have produced greater alterations in the connections although we found no alterations

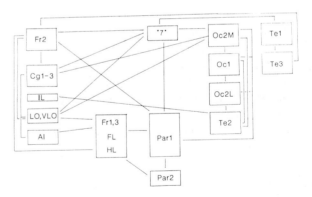

Fig. 9. Summary of the cortico-cortical connections in the rat. Data were collected by tracing injections of True Blue (0.05 – 0.1 μl) injected into the cortex. For simplicity a connection from Oc1 to Fr2 is ommitted. Abbreviations as in Fig. 1 and: FL, forelimb motor representation; HL, hindlimb motor representation; Par 2, second somatosensory area; "7", Krieg's area 7; Oc1, occipital cortex, area 1; Oc2L, lateral occipital cortex, area 2; Oc2M, medial occipital cortex, area 2; Te1, temporal cortex, area 1; Te2, temporal cortex, area 2; Te3, temporal cortex, area 3.

after bilateral posterior parietal lesions on the day of birth either.

Cortico-subcortical connections

There are clear changes in subcortical-cortical connections. First, injections of retrograde tracers into the somatosensory or visual cortex label thalamic nuclei that are not normally labelled. For example, several animals with somatosensory injections had labelled cells in the lateral geniculate nucleus. Injections into normal newborn rats produce similar labelling suggesting that these connections are usually present and later retract in normal brains. Perhaps the early lesions reduce competition and allow such cells to survive. Second, there are new connections from the amygdala to the visual and somatosensory cortex. Third, when the sensorimotor cortex adjacent to the lesion site is injected with retrograde tracers there are labelled cells in the substantia nigra and ventral tegmentum and we have verified these connections with injections of tritiated amino acids into the substantia nigra. Similar connections are sometimes seen in newborn rats as well, again suggesting that they normally die during development. Finally, when injections are made into the striatum of rats with unilateral lesions there is an increase in the extent of labelling in the intact PFC, much as has been reported for the rhesus monkey with a similar preparation (Goldman, 1978).

In summary, removal of the PFC in infant rats does significantly alter cortico-subcortical connections but there are surprisingly few changes in cortico-cortical connectivity.

Effect of NA depletion and environment

Both NA depletion and environmental stimulation modulate the effects of early PFC lesions so it is reasonable to expect them to also effect cortical structure. Indeed, both treatments affect brain weight, cortical thickness, and dendritic arborization. First, in animals with MF lesions around 7 days of age in whom sparing was blocked by NA depletion, there is a significant reduction in brain weight and cortical thickness (Kolb and Sutherland, 1986) and although our quantitative analysis is incomplete, there is a clear qualitative decrease in dendritic arborization too. Second, environmental stimulation markedly attenuates the behavioural effects of early PFC lesions, especially around 5 days of age, and this is correlated with a significant increase in brain weight, cortical thickness, and dendritic arborization (Fig. 10). Indeed, the increase in dendritic arborization was such that the lab-reared controls and enriched PFC animals did not differ.

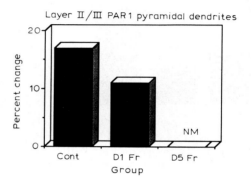

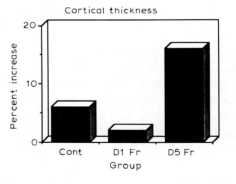

Fig. 10. A comparison of the dendritic arbor and cortical thickness in area Par 1 in rats given enrichment. Each bar represents (enriched environmental/lab reared) × 100. On the dendritic measure the D1Fr group shows significantly less increase than does the control group. On the cortical thickness measure, the D1Fr group shows significantly less increase than does the control or D5Fr group. The latter group shows significantly more increase than the control group. Abbreviations: CONT, control; D1Fr, day 1 frontal lesion; D5Fr, day 5 frontal lesion; NM, not measured.

In summary, it appears that treatments that modulate the behavioural effects of early PFC lesions also affect dendritic arborization in parallel ways: if behaviour is improved the arborization is increased; if behaviour is worsened, the arborization is decreased.

Summary

We can reach several conclusions regarding the effects of early PFC lesions in rats and hamsters.

(1) Early PFC lesions reduce brain size, weight, and cortical thickness. These effects are directly related to the age at which surgery is performed, the biggest reductions being observed in the youngest animals.

(2) Early PFC lesions reduce the number of neurones in the cortex, and again the biggest losses are seen in the youngest operates. Since the number of cells accounts for most of the variance in cortical thickness in rats with early PFC lesions it seems likely that the decreased cortical thickness is due primarily to neuronal death.

(3) Early PFC lesions have little effect upon cortico-cortical connectivity but do alter cortico-subcortical connections, especially with the amygdala and striatum.

(4) Early PFC lesions alter dendritic arborization in layer II/III and layer V pyramidal cells and layer IV stellate cells but the changes vary in magnitude and direction depending upon the age at surgery. PFC lesions on the day of birth lead to reduced arborization whereas PFC lesions at 10 days of age lead to increased arborization. These changes in arbor parallel changes in behaviour.

(5) Factors such as environmental stimulation and NA depletion alter brain weight, cortical thickness, and dendritic arborization. The changes in dendritic arborization parallel changes in behaviour.

Conclusions

We have approached the study of the prefrontal cortex by damaging the developing brain at different times during maturation and then studying the subsequent behavioural and anatomical consequences both in infancy and adulthood (Kolb and Whishaw, 1989). We believe that our behavioural and anatomical results point to a single, consistent correlate with behaviour after neonatal PFC lesions: the extent of dendritic arborization in the remaining cortex. We have also found various other changes in cortical structure after early PFC lesions (e.g. reduced brain weight, reduced cortical thickness, reduced cell numbers, anomalous subcortico-cortical connections) but these changes do not correlate in any meaningful way with factors that influence behaviour such as age of the animal at surgery or environmental conditions. It is likely that these changes in arborization reflect significant changes in cortical connectivity and perhaps in organization, but we have not yet found any evidence of what these changes might be.

Our behavioural studies are concordant with a growing body of evidence that suggests that the effect of early cortical injury is complex and poorly understood (e.g. Almli and Finger, 1984; Finger and Almli, 1984; Goldman-Rakic et al., 1983). Furthermore, it is consistent with a limited literature on the effects of perinatal PFC injury in the cortex of the rat and suggests that the rodent PFC may provide an excellent model for investigating cortical organization in mammals (e.g. Corwin et al., 1983; Kalsbeek et al., 1989a, b; Nonneman and Corwin, 1981; Nonneman et al., 1984). Furthermore, our results are consistent with, and add considerably to, the studies of the effects of early PFC lesions in primates by Goldman-Rakic and her colleagues (e.g. Goldman, 1974; Goldman and Galkin, 1978; Goldman et al., 1983) and in cats by Villablanca and his colleagues (e.g. Villablanca and Olmstead, 1981). Nevertheless, there are many questions that remain unanswered. In particular, we have little evidence regarding what the mechanisms of dendritic changes might be after early PFC lesions (for a speculative hypothesis, see Kolb and Whishaw, 1989). Furthermore, we have only begun to consider how the effects of other cortical lesions, such as posterior parietal lesions, might alter the development of the PFC, and thus alter behaviour (e.g. Kolb et al., 1987). Indeed, we do not even know how different lesion techniques such as the use of more selective neurotoxins, might compare with our suction ablations. We believe, however, that our paradigm, in which we make lesions in the rat at different ages and then study behavioural and anatomical changes in the same animals will prove to be a useful one to use in exploring the functions of the prefrontal cortex.

References

Almli, C.R. and Finger, S. (1984) *Early Brain Damage: Research Orientations and Clinical Observations,* Academic Press, New York, 368 pp.

Corwin, J.V., Leonard, C.M., Schoenfeld, T.A. and Crandall, J.E. (1983) Anatomical evidence for differential rates of maturation of the medial dorsal nucleus projections to prefrontal cortex in rats. *Dev. Brain Res.,* 8: 89–100.

Divac, I., Kosmal, A., Bjorklund, A. and Lindvall, O. (1978) Subcortical projections to the prefrontal cortex in the rat as revealed by the horseradish peroxidase technique. *Neuroscience,* 3: 785–796.

Finger, S. and Almli, C.R. (1984) *Early Brain Damage: Neurobiology and Behavior,* Academic Press, New York, 387 pp.

Goldman, P.S. (1974) An alternative to developmental plasticity: Heterology of CNS structures in infants and adults. In D.G. Stein, J.J. Rosen and N. Butters (Eds.), *Plasticity and Recovery of Function in the Central Nervous System,* Academic Press, New York, pp. 149–174.

Goldman, P.S. (1978) Neuronal plasticity in primate telencephalon: Anomalous crossed cortico-caudate projections induced by prenatal removal of frontal association cortex. *Science,* 202: 768–770.

Goldman, P.S. and Galkin, T.W. (1978) Prenatal removal of frontal association cortex in the fetal rhesus monkey: anatomical and functional consequences in postnatal life. *Brain Res.,* 152: 451–458.

Goldman-Rakic, P.S., Isseroff, A., Schwartz, M.L. and Bugbee, N.M. (1983) The neurobiology of cognitive development. In P. Mussen (Ed.), *Handbook of Child Psychology: Biology and Infancy Development,* Wiley, New York, pp. 311–344.

Greenough, W.T. (1976) Enduring brain effects of differential experience and training. In M.R. Rosenzweig and E.L. Bennett (Eds.), *Neural Mechanisms of Learning and Memory,* MIT Press, Cambridge, MA, pp. 255–278.

Hicks, S. and D'Amato, C.J. (1968) Cell migrations to the isocortex of the rat. *Anat. Rec.,* 160: 619–634.

Hicks, S.P. and D'Amato, C.J. (1970) Motor-sensory and visual behaviour after hemispherectomy in newborn and mature rats. *Exp. Neurol.,* 29: 416–438.

Hicks, S.P. and D'Amato, C.J. (1975) Motor-sensory cortex-corticospinal system and developing locomotion and placing in rats. *Am. J. Anat.,* 143: 1–42.

Kalsbeek, A., de Bruin, J.P.C., Feenstra, M.G.P., Matthijssen, M.A.H. and Uylings, H.B.M. (1989a) Neonatal thermal lesions of the mesolimbocortical dopaminergic projection decrease food-hoarding behavior. *Brain Res.,* 475: 80–90.

Kalsbeek, A., de Bruin, J.P.C., Matthijssen, M.A.H. and Uylings, H.B.M. (1989b) Ontogeny of open field activity in rats after neonatal lesioning of the mesocortical dopaminergic projection. *Behav. Brain Res.,* 32: 115–127.

Kasamatsu, I., Pettigrew, J.D. and Ary, M. (1981) Cortical recovery from effects of monocular deprivation: acceleration with norepinephrine and suppression with 6-hydroxydopamine. *J. Neurophysiol.,* 45: 254–266.

Kolb, B. (1984) Functions of frontal cortex of the rat: a comparative review. *Brain Res. Rev.,* 8: 65–98.

Kolb, B. (1987) Recovery from early cortical damage in rats. I. Differential behavioral and anatomical effects of frontal lesions at different ages of neural maturation. *Behav. Brain Res.,* 25: 205–220.

Kolb, B. and Elliott, W. (1987) Recovery from early cortical damage in rats. II. Effects of experience on anatomy and behavior following frontal lesions at 1 or 5 days of age. *Behav. Brain Res.,* 26: 47–56.

Kolb, B. and Gibb, R. (1987) Dendritic proliferation as a mechanism of recovery and sparing of function. *Soc. Neurosci. Abstr.,* 13: 1430.

Kolb, B. and Holmes, C. (1983) Neonatal motor cortex lesions in the rat: absence of sparing of motor behaviors and impaired spatial learning concurrent with abnormal cerebral morphogenesis. *Behav. Neurosci.,* 97: 697–709.

Kolb, B. and Nonneman, A.J. (1976) Functional development of the prefrontal cortex continues into adolescence. *Science,* 193: 335–336.

Kolb, B. and Nonneman, A.J. (1978) Sparing of function in rats with early prefrontal cortex lesions. *Brain Res.,* 157: 135–148.

Kolb, B. and Sutherland, R.J. (1986) A critical period for noradrenergic modulation of sparing from neocortical parietal cortex damage in the rat. *Soc. Neurosci. Abstr.,* 12: 322.

Kolb, B. and Tomie, J. (1988) Recovery from early cortical damage in rats. IV. Effects of hemidecortication at 1, 5, or 10 days of age on cerebral anatomy and behavior. *Behav. Brain Res.,* 28: 259–284.

Kolb, B. and Van der Kooy, D. (1985) Early cortical lesions alter cerebral morphogenesis and connectivity in the rat. *Soc. Neurosci. Abstr.,* 11: 989.

Kolb, B. and Whishaw, I.Q. (1981a) Decortication in rats in infancy or adulthood produced comparable functional losses on learned and species-typical behaviors. *J. Comp. Physiol. Psychol.*, 95: 468–483.

Kolb, B. and Whishaw, I.Q. (1981b) Neonatal frontal lesions in the rat: sparing of learned but not species-typical behavior in the presence of reduced brain weight and cortical thickness. *J. Comp. Physiol. Psychol.*, 95: 863–879.

Kolb, B. and Whishaw, I.Q. (1985a) Earlier is not always better: behavioral dysfunction and abnormal cerebral morphogenesis following neonatal cortical lesions in the rat. *Behav. Brain Res.*, 17: 25–43.

Kolb, B. and Whishaw, I.Q. (1985b) Neonatal frontal lesions in hamsters impair species-typical behaviors and reduce brain weight and neocortical thickness. *Behav. Neurosci.*, 99: 691–706.

Kolb, B. and Whishaw, I.Q. (1989) Plasticity in the neocortex: mechanisms underlying recovery from early brain damage. *Progr. Neurobiol.*, 32: 235–276.

Kolb, B., Sutherland, R.J. and Whishaw, I.Q. (1983) Abnormalities in cortical and subcortical morphology after neonatal neocortical lesions in rats. *Exp. Neurol.*, 79: 223–244.

Kolb, B., Holmes, C. and Whishaw, I.Q. (1987) Recovery from early cortical lesions in rats. III. Neonatal removal of posterior parietal cortex has greater behavioral and anatomical effects than similar removals in adulthood. *Behav. Brain Res.*, 26: 119–137.

Kolb, B., Day, J. Gibb, R. and Whishaw, I.Q. (1989a) Recovery from early cortical lesions in rats. VI. Cortical noradrenaline, cortical thickness, and development of spatial learning after frontal lesions or hemidecortications. *Psychobiology*, 17: 370–376.

Kolb, B., Zaborowski, J. and Whishaw, I.Q. (1989b) Recovery from early cortical damage in rats. V. Unilateral lesions have different behavioral and anatomical effects than bilateral lesions. *Psychobiology*, 17: 363–369.

Ladowsky, R., Gibb, R. and Kolb, B. (1988) The relationship of behavior and morphology following early lesions in rats. *Soc. Neurosci. Abstr.*, 14: 117.

MacLusky, N.J., Lieberberg, I. and McEwen, B.S. (1979) The development of estrogen receptor systems in the rat brain and pituitary: perinatal development. *Brain Res.*, 178: 129–142.

MacLusky, N.J. and Naftolin, F. (1981) Sexual differentiation of the central nervous system. *Science*, 211: 1249–1303.

Morrison, J.H., Molliver, M.E., Grzanna, R. and Coyle, J.T. (1981) The intracortical trajectory of the coeruleo-cortical projections in the rat: a tangentially organized cortical afferent. *Neuroscience*, 6: 139–158.

Nonneman, A.J. and Corwin, J.V. (1981) Differential effects of prefrontal cortex ablation in neonatal, juvenile, and young adult rats. *J. Comp. Physiol. Psychol.*, 95: 588–602.

Nonneman, A.J., Corwin, J.V., Sahley, C.L. and Vicedomini, J.P. (1984) Functional development of the prefrontal system. In S. Finger and C.R. Almli (Eds.), *Early Brain Damage, Vol. 2*, Academic Press, New York, pp. 139–153.

Rockel, A.J., Hiorns, R.W. and Powell, T.P.S. (1980) The basic uniformity in structure of the neocortex. *Brain*, 103: 221–244.

Rudy, J.W., Stadler-Morris, S. and Albert, P. (1987) Ontogeny of spatial navigation behaviors in the rat: Dissociation of "proximal" and "distal" cue-based behaviors. *Behav. Neurosci.*, 101: 62–73.

Sutherland, R.J., Kolb, B., Becker, J.B. and Whishaw, I.Q. (1982) Cortical noradrenaline depletion eliminates sparing of spatial learning after neonatal frontal cortex damage in the rat. *Neurosci. Lett.*, 32: 125–130.

Uylings, H.B.M., Van Eden, C.G., Parnavelas, J.G. and Kalsbeek, A. (1990) The pre- and postnatal development of rat cerebral cortex. In B. Kolb and R. Tees (Eds.), *The Cerebral Cortex of the Rat*, MIT Press, Cambridge, MA, in press.

Villablanca, J.R. and Olmstead, C.E. (1981) Conditions which may influence the effects of neonatal brain lesions. In P. Mittler (Ed.), *Frontiers of Knowledge in Mental Retardation*, University Park Press, Baltimore, MD, pp. 197–209.

Whishaw, I.Q. and Kolb, B. (1984) Behavioral and anatomical studies of rats with complete or partial decortication in infancy: functional sparing, crowding or loss, and cerebral growth or shrinkage. In S. Finger and C.R. Almli (Eds.), *Recovery from Brain Damage, Vol. 2*, Academic Press, New York, pp. 117–138.

Whishaw, I.Q. and Kolb, B. (1988) Sparing of skilled forelimb reaching and corticospinal projections after neonatal motor cortex removal or hemidecortication in the rat: support for the Kennard doctrine. *Brain Res.*, 451: 97–114.

Whishaw, I.Q. and Kolb, B. (1989) Tongue protrusion mediated by spared anterior ventrolateral neocortex in neonatally decorticate rats: behavioral support for the neurogenetic hypothesis. *Behav. Brain Res.*, 32: 101–113.

Zilles, K. (1985) *The Cortex of the Rat*. Springer, Berlin.

Discussion

A. Kalsbeek: Your data on the DAergic cortical innervation in the neonate fit nicely with our data on the development of the DAergic innervation, but do you think that the DAergic innervation in other cortical areas after an early PFC lesion is necessary for sparing of behavioral functions?

B. Kolb: No. In fact, the maximal changes in DAergic innervation appear to occur in animals with the earliest lesions, and these are the ones with the least sparing.

G.J. Boer: Referring to the apparent differences in the extent of branching of transplanted neurons, is there a difference between males and females in your early lesion/sparing effects?

B. Kolb: Yes. The data are very complex, however, as the sparing is task dependent. It will require more study to sort out the "rules" of the sex differences.

E.J. Neafsey: Increased dendrite arborization sounds like a "growth factor" effect. What do you think?

B. Kolb: The increased arborization could be a growth factor effect but I am betting that it is not as simple as a chemical such as nerve growth factor. I believe that it is also likely to be related to the changes in the afferents to the remaining cortical tissue. In particular, at P1 the neurons have not really begun to develop arbor, whereas at P10, they have really begun to sprout. The neurons are likely more susceptible to growth factors at P10.

H.B.M. Uylings: (1) Is the study of age dependent effects of bilateral frontal lesions complicated by the transmitter which pass through the frontal cortex after which they spread out over the entire cortex?

(2) Can you expand more on the variable of number of cells. Are these numbers absolute numbers or densities? And did you compare in every condition the number of neurons in Golgi with the number in Nissl sections?

B. Kolb: (1) Partly, yes. However, we have done amine assays in animals with early frontal lesions and there is certainly noradrenaline and serotonin present in the posterior cortex, and indeed, it is equivalent to the levels of adult operates. The question remains, however, just whether the reduction in these transmitters during development might affect development. One small bit of evidence for this is that there is an increase in noradrenaline in the cortex of neonatal hemidecorticates, and this is correlated with an increase in dendritic arbor in the cortex of these animals.

(2) The data on cells show that there are fewer total neurons and that the density is slightly higher in the neonatal operates. Thus, it appears that there is increased cell death, which leads to fewer neurons, and decreased dendritic arbor, which leads to a higher packing density in the neonatal operates. I should add, however, that the lesions in these studies were fairly large, as they included the medial frontal region and part of the motor cortex. It is possible that smaller lesions would have lesser effects but this remains to be seen.

V. Domesick: (1) Was there a difference in the size of the mediodorsal nucleus in the thalamus after P1 or P10 prefrontal lesions?

(2) Could this account for the great sparing of function of lesions at P10?

B. Kolb: (1) Yes, the MD and the entire thalamus were larger in those animals with P10 prefrontal lesions. But the MD seemed present in both cases, i.e., after both P1 and P10 lesions, and even after hemidecorticotion.

(2) No.

J.P.C. de Bruin: My question concerns the increased arborization in the P10 frontal lobe "lesioned" animals. Are the neurons with a "larger" dendritic arbor evenly distributed over the cortex, or are they found especially in areas closer to the lesion site?

B. Kolb: We so far have only measured neurons in the visual cortex (Oc1, Oc2L), temporal cortex (Te 2/3), and parietal cortex (Par 1). There appear to be very small changes in Oc1 and very large changes in Te 2/3, both of which are rather distal from the lesion site. The greatest changes, however, are found in the parietal cortex, which is close to the site. My guess is that as we measure more areas we will find the changes to be regionally specific, with a slight bias towards being larger near the lesion site.

A.Y. Deutch: Is there a "critical window" for the beneficial effects of "enriched environment" on behavioral performance at adult stage?

B. Kolb: Yes. It appears that the enrichment is most effective during development. In addition, the sensitivity of the cortex appears to be related to the age at which the lesion was incurred. The earlier the lesion in the first week of life, the less susceptible the cortex is to the environmental effects. I should note, however, that we do get environmental effects even in adulthood (i.e., about 120 days), so that it is really a "sensitive window" rather than a "critical one".

CHAPTER 12

Age-dependent effects of lesioning the mesocortical dopamine system upon prefrontal cortex morphometry and PFC-related behaviors

Andries Kalsbeek, Jan P.C. De Bruin, Matthijs G.P. Feenstra and Harry B.M. Uylings

Netherlands Institute for Brain Research, 1105 AZ Amsterdam, The Netherlands

Introduction

The anatomy of the cortical dopaminergic (DA) innervation appears to differ in at least one important respect from that of the other monoamines. Whereas noradrenaline (NA) and serotonin (5HT) projections are diffuse throughout the cortex, DA fibers innervate only selected cortical areas (Reader and Grondin, 1987; Reader et al., 1988). The abundant presence of other monoaminergic projections in the cortex (Fig. 1), and the fact that the content of DA in the cortex is only approximately 1% of that in the nearby striatum initially hampered the detection of a cortical DA innervation. Only recently have sensitive immunocytochemical techniques, using antibodies against DA itself (Geffard et al., 1984), enabled the description of the cortical DA innervation pattern to be completed (Van Eden et al., 1987; Plantjé et al., 1987). The results of these recent studies have revealed that DA fibers are to be found primarily in frontal, cingulate, pyriform and entorhinal cortices (Fig. 2).

It was not until 1973 that Thierry and coworkers presented the first evidence for a DA projection to the cortex, and a role for DA independent of NA. Interest for DA as an important factor in the functioning of the prefrontal cortex (PFC) was especially stimulated by the studies of Beckstead (1976), Berger et al. (1976) and Divac et al. (1978) who showed that in the rat the frontal field of the DA cortical projection coincided closely with the PFC, defined as the neocortical area which receives a projection from the mediodorsal nucleus of the thalamus (Rose and Woolsey, 1948). This overlap has been confirmed in other mammalian species, including primates (Björklund et al., 1978), indicating that the mesencephalic DA innervation is a conservative feature of the PFC in mammals.

The role of the DA projection to the PFC increasingly came into focus as some typical behavioral disturbances caused by lesions of the PFC could be duplicated by depletion of DA (see refs. in Stam et al., 1989). For instance, both local lesions of the PFC and lesions of the A10 cell group in the ventral tegmental area, where the mesocortical dopaminergic projection originates, caused a deficit in spatial delayed alternation (Kessler and Markowitsch, 1981; Nonneman and Corwin, 1981), increased aggressive behavior (Kolb and Nonneman, 1974; De Bruin et al., 1983;

Correspondence: Dr. A. Kalsbeek, Netherlands Institute for Brain Research, Meibergdreef 33, 1105 AZ Amsterdam, The Netherlands.

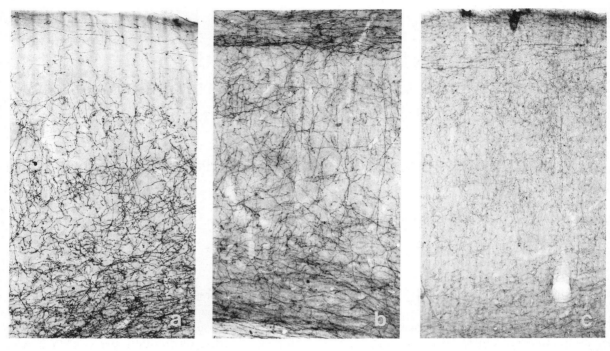

Fig. 1. The adult innervation pattern of dopamine (a), noradrenaline (b), and serotonin (c) in the prelimbic area (PL) of the prefrontal cortex of the rat (the pial surface is up). The catecholaminergic fibers (dopamine and noradrenaline) show a considerable overlap in their innervation pattern. The monoaminergic fibers and cell bodies in this and following figures were visualized with the immunocytochemical peroxidase-antiperoxidase procedure. All antibodies were kindly provided by Dr. R.M. Buijs.

Stam et al., 1989) and impaired food hoarding (Kolb, 1974; Nonneman and Corwin, 1981; Stam et al., 1989) in adult rats.

Experiments applying PFC lesions early in development have indicated that the effects of such treatment may be far less dramatic than a comparable injury sustained in adulthood. The mechanisms underlying this "sparing" of function are poorly understood. The correspondence between the behavioral effects of electrolytic lesions and selective DA depletion in PFC on the one hand, and the indications for a well established DA innervation in the frontal pole of the newborn rat (Schmidt et al., 1982), directed our interest to a possible developmental role of DA.

Only recently have the possible mechanisms underlying the developmental effects of DA and the other monoamines emerged to some degree. In principle, the transmission of developmental signals from the monoaminergic neurons may come about in either of two ways, viz. cell–cell contact or through "humoral" communication. Concerning the first point, it has been shown that postsynaptic receptor sites in the cerebral cortex develop prior to presynaptic specializations, and that they may induce synapse formation (Blue and Parnavelas, 1983). Although there are synaptic specializations between neurons in the rat cerebral cortex already before birth (Wolff, 1978; König and Marty, 1981), synapse formation and maturation is mainly a postnatal event in the rat. The first (temporary) synaptic contacts are formed prenatally in the subplate and marginal zone, i.e. the two cortical layers containing the first well developed cortical neurons, preceding the neurons which will eventually constitute the adult 6-layered cortex (for a review on the cortical development of the rat see: Uylings et al., 1990). A great deal of the synapsing

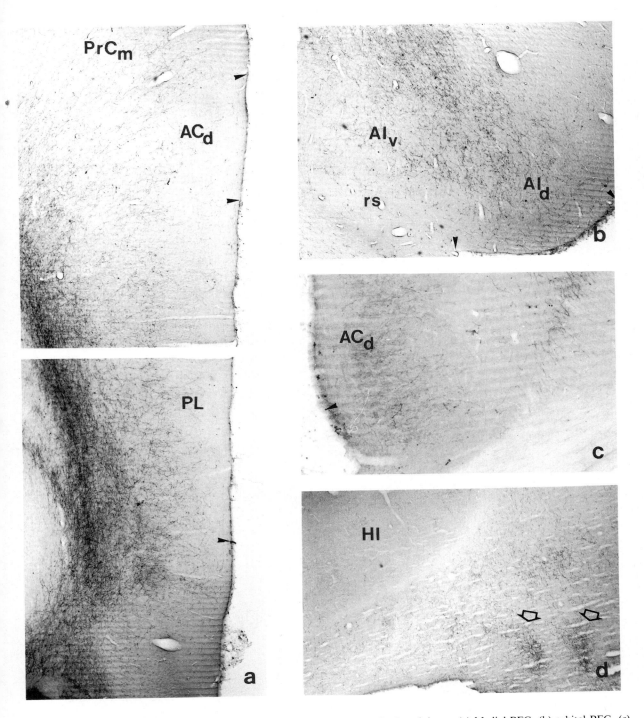

Fig. 2. The dopaminergic innervation pattern in a number of cortical areas in the adult rat. (a) Medial PFC, (b) orbital PFC, (c) supragenual PFC, (d) entorhinal cortex. Arrowheads indicate the borderlines between subareas of the PFC. Open arrows in (d) point to the typical cluster arrangement of dopaminergic fibers in the entorhinal cortex.

terminals possess monoamine containing granules (Coyle and Molliver, 1977; Zecevic and Molliver, 1978). Golgi studies have shown that the subplate neurons have large somas and extensive processes, and could therefore play an important role either as interneurons or as projection neurons during fetal development (Mrzljak et al., 1988). Lauder et al. (1982) have proposed that serotonergic fibers in the marginal zone could influence cell proliferation by contacting processes of cells in the ventricular zone. In the case of humoral transmission, the transmitter must be released from the developing system. The most suitable place for this seems to be the growth cone, and it has been shown in in vitro experiments that the growth cone is capable of spontaneous release of transmitter (Hume et al., 1983; Young and Poo, 1983). Immunocytochemical studies, furthermore, have shown that the leading growth cones are loaded with transmitter, e.g. dopamine (Fig. 3; Kalsbeek et al., 1988b) or serotonin (Aitken and Törk, 1988).

Both routes of signal transmission, however, require the presence of receptors on the recipient element. DA receptors have been reported to be already prenatally present in target areas of the DA innervation (Bruinink et al., 1983; Miller and Friedhof, 1988; Noisin and Thomas, 1988; Sales et al., 1989), but not before embryonic day 15. Between E15 and E18 the DA binding sites lack true phenotypic specificity, and do not show any stereoselectivity. Around E18, saturable binding kinetics can be demonstrated by Scatchard analysis (Miller and Friedhof, 1987). These findings, however, are derived mainly from striatal tissue, since the density of [^{3}H]spiroperidol binding sites in the PFC is quite low in comparison with the striatum, so that they are extremely difficult to measure. Furthermore, the accurate measure of these sites is complicated by the fact that [^{3}H]spiroperidol also binds to 5-HT$_2$ sites. As a result, few articles have been published on the DA receptors in this brain area and, to our knowledge, no detailed reports have yet been published on the development of DA receptors in the PFC. Two papers have reported cortical DA receptors appearing only postnatally (Murrin et al., 1985; Sales

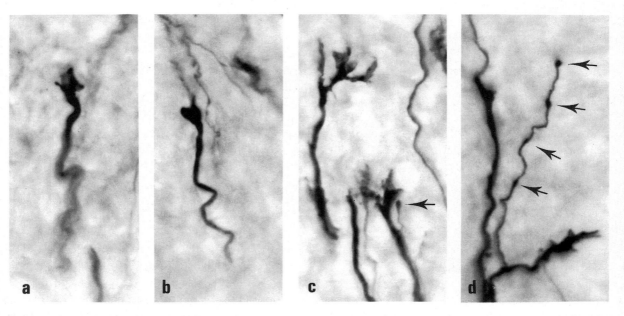

Fig. 3. In the growing axons DA is distributed all along the fiber, including the growth cone. Arrows point to the first appearing varicosities. (a, b) E18; (c) E20; (d) P0.

et al., 1989). Furthermore, the data of Levitt and coworkers on a limbic system associated membrane protein which is expressed in PFC cells as early as E15 give us a clue to how DA fibers may be able to recognize the appropriate cortical areas in which to form synaptic contacts (Levitt, 1984; Horton and Levitt, 1988). Combined with the results on intracellular cAMP accumulation (McMahon, 1974), growth cone motility and neurite outgrowth upon application of DA (McCobb et al., 1988; Lankford et al., 1987), these data may provide the beginning of an answer to the question of how DA exerts its influence on cortical development.

There have been quite a number of studies dealing with the behavioral consequences of an early depletion of DA, the great majority of which used 6-hydroxydopamine (6-OHDA) to lesion the developing DA projections. However, 6-OHDA primarily affects the projection areas of those DA neurons which originate in the substantia nigra, especially when applied intraventricularly or intracisternally as is usual in small animals. No data have previously been available, however, on the effect of an early depletion of DA from cortical areas. Concerning the morphological consequences of an early depletion of cortical DA no data are available either. However, there have been a number of studies examining the effects of an early depletion of the other catecholamine NA on cortical development, using 6-OHDA to destroy the cortical NA projection. A number of these studies suggests an impaired cortical development (Maeda et al., 1974; Onténiente et al., 1980; Felten et al., 1982; Segler-Stahl et al., 1982; König et al., 1985), although also negative results have been reported (Wendlandt et al., 1977; Ebersole et al., 1981; Lidov and Molliver, 1982). In addition, in vitro experiments support a trophic role for serotonin on the morpho-functional development of rat cerebral cortex (Chubakov et al., 1986).

The present chapter will give an overview of the behavioral and morphological effects of a depletion of the mesocortical DA system either in early life or in adulthood. In order to establish when, and where the DA containing fibers arrive in the PFC and at what age the adult innervation pattern is accomplished, this overview starts with a description of the pre- and postnatal development of the DA innervation in the PFC. Subsequently procedures are described to lesion these DA fibers as early and selectively as possible. After completion of the behavioral experiments different parameters of PFC development were measured and compared with sham operated littermates.

Development of the dopaminergic innervation of the prefrontal cortex

Almost since the demonstration that monoamines, among which DA, were transmitter candidates with a widespread distribution over the CNS, it has been recognized that these neurotransmitters could, in principle, play a role in the development of the nervous system. This idea was initiated by the pioneering work of Buznikow in the 1960s, when he demonstrated the presence of DA and the other monoamines in the sea urchin embryo already during cleavage and gastrulation (see the review of Buznikow, 1984). In vertebrates too, monoamines are present long before the onset of neurotransmission itself, and appear to be implicated in the control of embryogenesis, neural tube closure, palate formation and myoblast differentiation (Schlumpf et al., 1980; Lauder and Krebs, 1986; Lauder et al., 1988). The actual presence and synthesis of monoaminergic transmitters in neuronal systems start well before the neurogenesis of most of their target structures (Olson and Seiger, 1972).

For an extensive overview of the development of the different DAergic systems the reader is referred to the review of Kalsbeek et al. (1990). In short, the neurogenesis of DA cells in the ventral mesencephalon, which will eventually form the mesocortical projection, starts on embryonic days 12 – 15, with a peak on E14 – 15 (Altman and Bayer, 1981). This is somewhat later than the DA cells of the substantia nigra, which will form the mesostriatal (or nigrostriatal) projection (Hanaway et al., 1971; Lauder and Bloom, 1974). The transmitter itself is probably present from E13 on, as indicated by DA fluorescence (Olson and Seiger, 1972) and by immunocytochemistry (Voorn et al., 1988). Tyrosine hydroxylase, the rate limiting enzyme of the DA system, can be detected even earlier, viz. on E12.5 (Specht et al., 1981). Soon after the migrating cells have become fluorescent, they start sending out neuritic processes (Seiger and Olson, 1973; Tennyson et al., 1973). At E14 a massive bundle of DA fibers traverses the diencephalon and enters the telencephalon, where the forefront of this bundle reaches the ganglionic eminence, i.e. the prospective striatum (Voorn et al., 1988). On E16 the first DA-positive fibers reach the cortical anlage and enter the subplate of the frontal cortical pole (Fig. 4). In the presumptive medial prefrontal cortex the first DA containing fibers can be observed on E17 (Kalsbeek et al., 1988). The process of entering the cortical plate, containing the cells which will eventually form the adult 6-layered cortex, starts just before birth. Upon entering the cortical plate the thick, darkly stained DA containing fibers change their morphology: they become much thinner, while varicosities start to appear which could mean that functional contact sites start to develop around birth (Fig. 3). DA fibers invade exclusively that

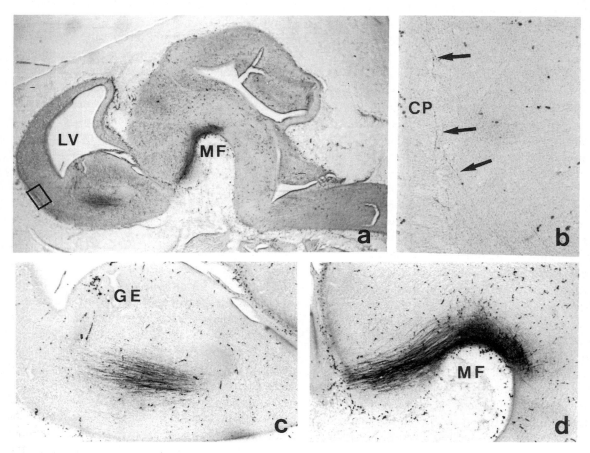

Fig. 4. Sagittal section through the brain of a 16-day-old embryo (a). The mesencephalic DA cell groups send projection fibers to the prosencephalon (d). Large bundles of DAergic fibers course through the diencephalon (d) and distribute in the striatum (c). Only a few fibers manage to enter the cortical plate at this time (b, boxed area in a). CP, cortical plate; GE, ganglionic eminence; LV, lateral ventricle, MF, mesencephalic flexure.

part of the developing cortex which contains already differentiated cells, while the upper cortical plate (containing only undifferentiated neurons) remains free of DA fiber ingrowth, a process which continues after birth. At postnatal day (PN) 10, when the cortical plate has disappeared, all cortical layers have received their appropriate DA innervation. The density of this innervation, however, increases until about 2 months of age.

The development of cortical DA innervation clearly deviates from that of the other monoaminergic systems. NA and 5HT fibers innervating the developing cortex show a bilaminar distribution forming a superficial and a deep band of immunoreactive fibers surrounding the cortical plate (Levitt and Moore, 1979; Wallace and Lauder, 1983; Verney et al., 1984). DA fibers occupy predominantly the subplate layer beneath the cortical plate and only rarely can a DA-positive fiber be observed in the superficial, marginal zone. Also the timetable of cortical ingrowth shows some marked differences. NA and 5HT fibers start to penetrate the cortical plate after their arrival, i.e. at E18 (Levitt and Moore, 1979; Wallace and Lauder, 1983), whereas DA containing fibers remain confined to the subplate for several days (E17 – E20) and only enter the cortical plate shortly before birth. The 5HT fibers, however, also show a delay in the innervation of the middle part of the cortical plate, containing the cells which will eventually differentiate in cortical layers II/III (Lidov and Molliver, 1982a; Aitken and Törk, 1988). The 5HT and DA innervation thus appears to follow the cortical maturation pattern, whereas NA containing fibers follow an innervation pattern which is relatively independent of the cortical maturation (Lidov and Molliver, 1982a; Verney et al., 1984; Kalsbeek et al., 1988b). These different schedules of development could underlie some of the complex interactions of the monoaminergic systems with cortical neurogenesis. So it seems that the DA fibers are in time to influence cortical development, since the first fibers arrive immediately after the cortical plate has been formed. However, the process of entering the cortical plate and forming synapses probably does not start until just before birth (Kalsbeek et al., 1989b; Uylings et al., 1990).

Perinatal techniques for lesioning the mesocortical dopaminergic projections

In order to study the possible developmental role of DA projections it was necessary to deplete the PFC from its DA input as early in development as possible. Since the first DA fibers already reach the subplate of the developing PFC at embryonic day 17, preferentially this depletion would need to be effected before birth. Experimentally one is faced immediately with the problem how to apply a lesion to perinatal animals which has both neuroanatomical specificity (i.e. restricted to the structure of interest), and pharmacological specificity (i.e. without interfering with other transmitter systems). The first experiments to lesion the DA input to the PFC were performed using 6-OHDA, which seemed the most specific and suitable method at that time. From the literature it was known that intraperitoneal application of 6-OHDA in early development preferentially depletes the NA system (Jonsson et al., 1974; Tassin et al., 1975). Similar to the subcutaneous application of 6-OHDA to neonatal animals (Schmidt and Bhatnagar, 1979), intracisternal or intraventricular administration, however, also lowers DA levels (Jonnson, 1983; Kostrzewa and Jacobowitz, 1974; Tassin et al., 1975). In a first attempt to lesion the DA input to the PFC 6-OHDA was delivered intraventricularly in utero on embryonic day 17 and on postnatal days 1 and 3. To further improve the selectivity of the lesion, the animals were pretreated with the NA uptake inhibitor desipramine. Approximately 50% of the injected rats showed obvious ventricular enlargements, presumably due to obstruction of the cerebrospinal fluid flow, secondary to damage caused by repeated injections. These rats were excluded from further comparison. The DA fiber density was estimated semiquantitatively in various brain regions of sham-treated and 6-OHDA-injected rats

33 days after birth. This treatment did not lead to an obvious decrease in DA fiber density in the PFC, in contrast to the caudate nucleus and the nucleus accumbens. It seems therefore that perinatal intraventricular application of 6-OHDA preferentially depletes the DA terminal fields in the striatum. The report of Peters et al. (1977) suggests similar results after intraventricular injections of 6-OHDA on postnatal days 1 and 2.

The next step, therefore, was to try and lesion the DA cells from which the DA containing fibers in the PFC originate. Through a new stereotaxic procedure (Hoorneman, 1985) it was possible to position the canula of the Hamilton syringe in the vicinity of the DA cell bodies in the ventral tegmental area of newborn rats in a reproducible way. Even when the 6-OHDA injection was placed in the center of the A10 cell group, the number of dopaminergic cells in the ventral tegmental area showed no substantial decrease, whereas the dopamine containing cells of the more laterally situated substantia nigra were almost completely lost on both sides (Fig. 5a). In the forebrain this resulted in a severe depletion of DA in the striatum (Fig. 5b), especially the dorsolateral parts, whereas the innervation density of the PFC was hardly affected (Fig. 5c). On the other hand, the cortical innervation of NA was almost completely depleted, confirming the high sensitivity of the cortical NA projection for 6-OHDA (Sievers et al., 1980). Although there have been reports of successful lesioning of the DA neurons of the A10 group with 6-OHDA in adult animals, in these cases, too, the substantia nigra seems to be more vulnerable to the destructive action of 6-OHDA (Voorn et al., 1987). Interestingly, the same hierarchy in sensitivities is found with MPTP (see further), but the reason for the differences has not yet been elucidated. One possible explanation is a difference in the uptake mechanism for DA. In addition, it has been shown that also the dendritic processes of the substantia nigra cells contain a DA uptake mechanism (Björklund and Lindvall, 1975; Silbergeld and Walters, 1979; Kelly et al., 1985), which could facilitate the accumulation of toxic

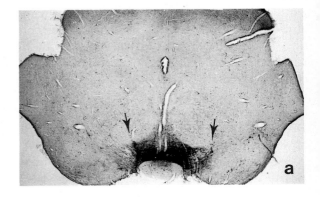

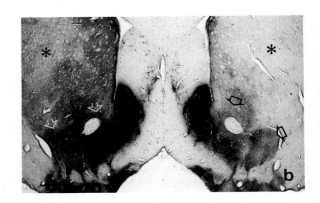

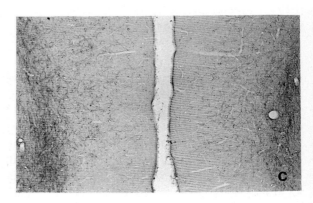

Fig. 5. Effects of neonatal application of 6-OHDA in the VTA at the first postnatal day. (a) Bilateral (arrows) infusion of 6-OHDA near the VTA causes an almost total destruction of more laterally situated substantia nigra cells, but leaves the dopaminergic cells in the VTA intact. (b) Depletion pattern in the striatum, affecting primarily the dorso-lateral part (asterisk) and leaving the nucleus accumbens relatively spared (open arrows). (c) Intact dopaminergic innervation in the PFC.

substances in the cell body. On the other hand, the presence of NA fibers in the VTA is implicated in the resistance of VTA-DA neurons against 6-OHDA (Hervè et al., 1986) and MPTP (D'Amato et al., 1986). But the colocalization with other transmitters may be involved too (Hökfelt et al., 1980; Seroogy et al., 1987), since it has been noted that the resistance of DA cells is most marked in areas in which cholecystokinin (CCK) and DA are coexistent.

In a later stage of our experiments we also tried the new catecholamine neurotoxin N-methyl-4-phenyl-2,3,5,6,-tetrahydropyridine (MPTP). Since intracerebral application of MPTP itself causes no permanent depletion in rats, we infused the active metabolite of MPTP, i.e. MPP^+, in the ventral mesencephalon (Feenstra et al., 1990). The depletion pattern was very much the same as the one observed following neonatal 6-OHDA infusions, viz. a virtual absence of DA fibers in the dorsolateral part of the striatum, and no evident reduction of neocortical DA fibers (Fig. 6). As can also be seen in Fig. 6, DA depletion reveals the patchy organization of the striatum more clearly. It is not clear whether this is only a consequence of the higher density of DA fibers inside patches, or if there is a difference in the vulnerability of DA neurons projecting in- and outside the patches. The same denervation pattern was observed by Wilson et al. (1987) after an MPTP infusion in

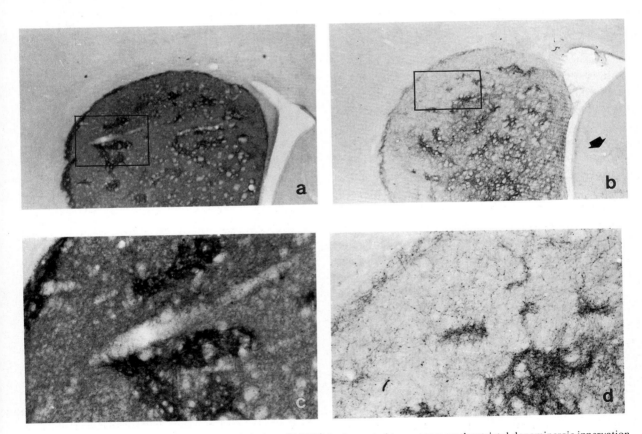

Fig. 6. Effects of neonatal (PN1) intracerebral infusion of MPP^+ in the ventral tegmentum on the striatal dopaminergic innervation 6 days after the infusion. Patches which are also visible in the control animal (a, c), seem relatively spared in the experimental animal (b, d). The dopaminergic innervation of the lateral septum (arrow in b) is still intact in the experimental animal. c and d are higher magnifications of the boxed areas in a and b.

adult dogs. Another major drawback of the intracerebral infusion of MPP$^+$ is that it causes a severe unspecific tissue damage, as had also been reported after infusions of MPP$^+$ in adult rats (Altar et al., 1986; Namura et al., 1987). As a consequence of the disappointing results in the experiments with catecholamine neurotoxins, we turned to the use of excitotoxic amino acid analogues. But infusion of either kainic or ibotenic acid into the VTA of neonatal animals did not produce a satisfactory depletion of cortical DA. These results are in accordance with the results of infusion experiments in the VTA of adult animals (Gebert et al., 1985; Schwarcz et al., 1979).

Because of the failure of the various chemical lesion techniques we next tried thermal lesions. In order to be able to lesion the DA neurons from which the mesocortical projection originates as selectively as possible, it was necessary to develop a method for the stereotaxical placement of bilateral electrodes in neonatal pups. The method as used in the present experiments is a modification of the procedure developed by Hoorneman (1985). By ensuring a stable position of the pup in the mould and a sufficiently deep anesthesia, the lesioned area could be restricted to the ventral tegmental area. The survival rate was almost 100%, no ventricular enlargements were observed, and the lesions had little or no effect on body weight throughout development. The effectiveness of the lesions was checked immunocytochemically, using antibodies against DA, NA and 5HT. Such thermal lesioning of the VTA introduced an extensive depletion of the DA innervation in the PFC (Fig. 7), whereas the massive DA input to the striatum was left intact. The most notable side-effect of this lesion technique was the concomitant elimination of the cortical 5HT innervation in the majority of the VTA lesioned animals and a partial depletion of DA in the ventral part of the striatum, i.e. the nucleus accumbens and the olfactory tubercle. The NA innervation of the forebrain was left intact (Kalsbeek et al., 1987, 1989a).

The timepoint of the thermal lesions (i.e. the first postnatal day) may seem to be fairly late, as the first DA fibers reach the subplate of the developing PFC already at E17, i.e. 4 days earlier (Kalsbeek et al., 1988b). However, the late invasion of the cortical plate by DA fibers and the morphological changes in these fibers as observed on the first postnatal days, make it likely that the ma-

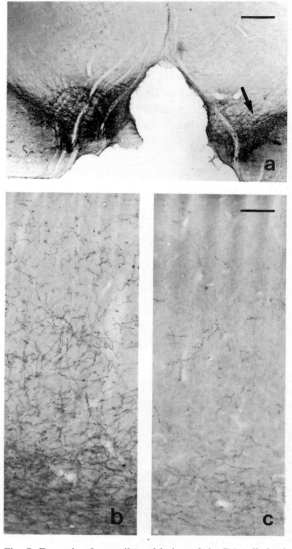

Fig. 7. Example of an unilateral lesion of the DA cells in the VTA, whereas the DA neurons in the neighboring substantia nigra (arrow) are spared. The resulting unilateral depletion of DA fibers in the medial PFC is visualized in Fig. 6c, the intact DA innervation in the unlesioned hemisphere is shown in Fig. 6b (pial surface is up in b and c).

jority of the DA fibers at PN1 are destroyed before synaptic contacts can be established in the PFC. Still, a developmental influence of DA released from the growth cones of fibers growing in the subplate cannot be excluded.

Sequelae of neonatal depletion of the dopaminergic innervation in the prefrontal cortex

To assess a possible early "inducer" effect of the mesocortical DA system on the cyto- and chemoarchitectonic development of the PFC, a number of experiments was performed comparing the biochemistry and morphology of sham-operated and neonatally VTA lesioned animals. Since the experimental procedure as described above inevitably damaged the cortical serotonergic innervation to a certain degree, a group of animals with only a depletion of cortical 5HT was included, or individual results were compared with the depletion rates of DA and 5HT. The former group consisted of animals bearing lesions confined to the medial part of the mesencephalon, just dorsal to the interpeduncular nucleus. Such lesions hardly affected the DAergic cells of the VTA, but did interrupt the fibers of the trans-tegmental 5HT pathway to the forebrain.

Biochemical consequences

In order to further characterize the neonatal lesions of the VTA, and to be able to compare the results with other studies, biochemical measurements of the monoamines and their metabolites in the PFC were performed, in addition to the earlier mentioned immunocytochemical characterization. Apart from the medial PFC these levels were also assayed in some other projection areas of the mesotelencephalic DA system, e.g. nucleus ac-

TABLE I

Influences of neonatal VTA lesions on the monoamine content in the prefrontal cortex, nucleus accumbens and the striatum in female rats

	DA		5-HT		NA	
	Mean	SEM	Mean	SEM	Mean	SEM
Medial prefrontal cortex						
Control	85.5	5.2	502.3	34.3	353.4	13.8
Sham	98.2	8.3	506.9	31.5	327.8	13.6
Exp.	17.2	2.4	122.0	18.9	329.4	25.0
Δ	83%***		76%***		0.5% ns	
Nucleus accumbens						
Control	5492.6	732.4	481.9	18.9	ND	
Sham	4882.3	446.1	475.8	22.1	ND	
Exp.	3041.1	309.8	178.5	19.4	ND	
Δ	38%***		63%***			
Caudate putamen						
Control	11567.8	648.2	435.0	46.4	ND	
Sham	12339.3	465.6	387.0	22.4	ND	
Exp.	7872.1	412.7	135.2	12.3	ND	
Δ	36%***		65%***			

Mean transmitter levels in ng/gr brain tissue.
Δ percentual difference as compared to the sham-group; ND not determined.
* $p < 0.05$, ** $p < 0.01$. *** $p < 0.001$ according to the Mann-Whitney-U test.

cumbens, olfactory tubercle and striatum (Table I). As was indicated before by the immunocytochemical experiments, the lesions were not always completely bilateral, and some part of the projection was generally spared. Likewise, the biochemical measurements showed a concomitant depletion of DA and 5HT in the nucleus accumbens and the olfactory tubercle. Not visible in the immunocytochemically stained sections was the 36% depletion of DA in the rostral striatum, detected by the biochemical measurements although the lesions were restricted to the VTA. Anatomical tracing studies have shown, however, that VTA cells also project to the ventro-medial part of the head of the striatal complex (Björklund and Lindvall, 1984; Gerfen et al., 1987).

The rise in metabolite/transmitter ratios, an index for transmitter utilization, showed an increased activity of the remaining DA and 5HT fibers, especially within the cortex (Kalsbeek et al., 1989b). A similar increase in the activity of surviving DAergic neurons has been described after adult lesions of the mesostriatal system in the rat (Westerink et al., 1978; Altar et al., 1987), and in the marmoset (Rose et al., 1989). These findings have been interpreted as reflecting adaptational mechanisms to compensate for the lesion-induced reduced innervation (Stachowiak et al., 1987). The same mechanisms may be responsible for the fact that no neurological impairments of Parkinson's disease become evident until 80% or more of the DA in the striatum is lost (Bernheimer et al., 1973; Hornykiewicz, 1973).

The next question which arose was whether an already hyperactive system would be able to react to another challenge. Different kinds of physical and emotional stress are able to produce fairly selective increases in the activity of the mesocortical DA system, as measured by indices of DA synthesis, metabolism and release (Bertolucci-D'Angio et al. and Deutch and Roth in this volume, but also see Antelman et al., 1988). In order to further investigate the "capacity" of the lesioned mesocortical DA system, neonatally VTA lesioned rats were exposed to 30 min footshock stress when they were 3 months of age (Feenstra et al., in prep.). In sham-lesioned animals the footshock stress caused a clear increase in the transmitter utilization of the DA projection to the mPFC, whereas the DAergic terminals in the nucleus accumbens show only a small, non-significant increase, and in the rostral striatum no change whatsoever occurs (Fig. 8). Comparable results and ratios were reported by Lavielle et al. (1978) in intact animals. In neonatally VTA lesioned animals without footshock the surviving DA fibers in the PFC increase their activity, whereas in the nucleus accumbens and striatum this increase is not marked, despite depletions down to 10% of control values. The DOPAC/DA ratios in the remaining mesocortical terminals were 2 – 3 times as high as those observed after 30 min of footshock stress in control animals. Fig. 8 also makes clear that the damaged mesocortical DA system in the neonatally VTA lesioned animals is unable to increase its activity any further in response to footshock stress, which could mean a disruption in neurotransmission. Similar results have been reported by others for the damaged mesostriatal DA system after neonatal application of 6-OHDA, by using α-methyl-p-tyrosine as a "challenge" factor (Rogers and Dunnett, 1989). This compensatory mechanism seems to operate also after adult application of 6-OHDA, as reported by others using either stress (Snyder et al., 1985), or amphetamine (Robinson and Whishaw, 1988) to stimulate DA activity. Previous findings in rats with large, 6-OHDA induced, DA depletions in the striatum showed that they behave normally in a neutral laboratory setting but show behavioral impairments when exposed to stressful stimuli (Snyder et al., 1985). Similar observations have been made during preclinical stages of Parkinson's disease (Schwab and Zieper, 1965). Owing to the concomitant reduction in the mesocortical and nigrostriatal DA systems in Parkinson's disease (Hornykiewicz, 1980; Scatton et al., 1983), and the higher vulnerability of the mesocortical DA system, it may be speculated that cognitive deficits will emerge before motor deficits (e.g. hypokinesia

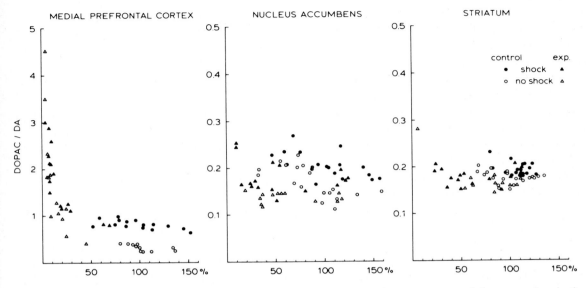

Fig. 8. Scatterplots showing the relation between DA concentrations (expressed as a percentage of the mean values in the control group) and the DOPAC/DA ratios in the medial PFC, nucleus accumbens and striatum of the different experimental groups.

and rigidity). Indeed, the results of Levin et al. (1988) suggest that Parkinson's disease of less than 2 years' duration is already associated with significant neuropsychological deterioration.

Morphological consequences

For the morphometric studies, neonatally lesioned and sham-operated animals were sacrificed between postnatal days 35 and 40, when the prefrontal cortex reaches its maximum volume (Van Eden and Uylings, 1985), and processed according to the specific histological procedures. Qualitative inspection of Nissl sections revealed no obvious differences between the two groups, although there was a tendency towards a less well-defined arrangement of cortical layers V and VI in DA depleted animals. Cortical thickness measurements and volumetric analysis can be used to gain insight into the more general alterations in brain morphology, and alterations of these variables have been found in a variety of experimental studies. Initially, the neonatal depletion of DA in the PFC seemed to cause a slight reduction in cortical thickness (Kalsbeek et al., 1987). In a second, more comprehensive, experiment, however, these results could not be replicated (Kalsbeek et al., 1989c). Also volumetric quantification of the different PFC subareas revealed no effect of the neonatal depletion of cortical DA. Previous studies measuring cortical thickness after neonatal depletion of the catecholaminergic (mainly noradrenergic) projections to the cortex, were not unequivocal either. Although König et al. (1985) reported an increase in cortical thickness, others have found no significant changes (Onténiente et al., 1980; Ebersole et al., 1981; Lidov and Molliver, 1982b), indicating that the effects, if they do exist at all, may be very small.

It might be, however, that changes take place in the arrangement of cortical layers which are not reflected in the total cortical thickness or volume. In order to examine whether a neonatal depletion of the DAergic input affects cortical cytoarchitecture, we compared the cortical distribution pattern of the GABA containing cell bodies and a number of neuropeptide containing neurons. These intrinsic neuropeptide containing cortical neurons all show a specific laminar distribution pattern (Fig. 9), which develops mainly postnatally. Recently, Naus and Bloom (1988) used the characteristic distribution pattern of somatostatin containing neurons to study the development of the "inverted" cortical cytoarchitecture in the mouse mutant reeler. In our study, neonatal depletion of cortical DA did not induce any obvious changes in the distribution pattern of the cortical transmitters which we examined (Kalsbeek and Uylings, 1989).

Since no changes could be detected in the overall architecture of the adult PFC, we subsequently focused on the primary target of the DA fibers in the PFC, layer V pyramidal cells. Already in 1974, Maeda et al. reported a disturbed outgrowth of the apical dendrites of pyramidal cells after neonatal depletion of cortical NA, but others could not replicate this finding (Wendlandt et al., 1977; Lidov and Molliver, 1982b). Only after a systematic and extensive analysis of dendritic morphology, were Felten et al. (1982) able to show a decreased branching frequency of pyramidal cells in the frontal and cingular cortex after neonatal depletion of cortical NA. In addition, it has been shown that ganglioside levels are reduced after neonatal depletion of cortical NA, indicating growth impairment of neuronal membrane structures (Segler-Stahl et al., 1982). In our experiments, the three-dimensional branching pattern of the large pyramidal cells in layer V of the medial PFC was measured (Fig. 10) using a semiautomatic dendrite measuring system developed at the Netherlands Institute for Brain Research (Overdijk et al., 1978; Uylings et al., 1986). The

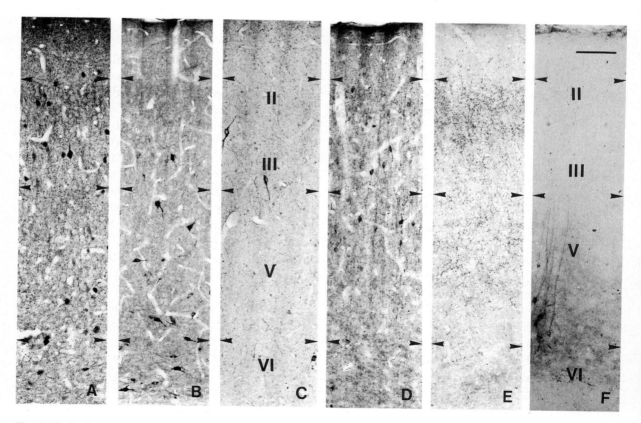

Fig. 9. The laminar distribution pattern in PL of a number of different cortical transmitters. Arrowheads indicate cortical layers. (a) GABA; (b) somatostatin; (c) neuropeptide Y; (d) vasoactive intestinal peptide; (e) substance P; (f) enkephalin. Bar = 0.2 mm.

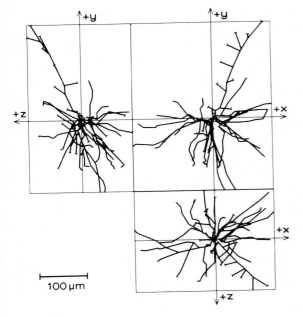

Fig. 10. Computer graphics (Overdijk et al., 1978) of a 3-dimensional basal dendritic field of a layer V pyramidal neuron in the medial prefrontal cortex of an adult rat, as measured in the present study. Displayed is the projection onto 3 orthogonal planes in which the y-axis is perpendicular to the pia.

most profound effect of neonatal DA depletion was upon the total number of dendritic segments of basal dendrites and the total length of basal dendrites per cell, showing a 26% and 30% reduction, respectively, as compared with the control group (Fig. 11). The decrease in total length of basal dendrites could be attributed to a decreased branching frequency since the length of the separate dendritic segments was hardly affected (Kalsbeek et al., 1989). The reduced branching frequency described here is compatible with the general idea that dendrite/axon interactions in early development provide an important stimulus to the growth and differentiation of pyramidal cell dendritic arbors (e.g. Berry et al., 1977; Pinto-Lord and Caviness, 1979; Wise et al., 1979). The maturation of dendritic trees is mainly a postnatal event, since the pyramidal cells of layer V have been generated just before birth and only have small dendrites at birth. The phase of rapid dendritic growth starts with the ingrowth of subcortical and callosal fibers, i.e. between PN6 and PN18 (Petit et al., 1988; Uylings et al., 1990). The question arises whether these changes are due solely to the reduction in the number of afferent fibers that reach the developing cortex, or if the particular transmitter that is depleted is also of importance. There is some evidence in favor of this second idea. In the first place, lesions predominantly depleting the 5HT innervation of the PFC in the present study had no effect on any of the dendritic parameters investigated (Fig. 11). Secondly, it has been shown that blocking DA activity without lesioning the fibers, i.e. with haloperidol, reserpine or α-methyl-p-tyrosine, retards the development of striatal target neurons (Tennyson et al., 1983; Iniguez et al., 1987).

Behavioral consequences

To test for possible functional consequences of the neonatally applied VTA lesions, tests were run on PFC mediated behaviors. Of course, behavioral changes can never be accounted for solely by the depletion of, in this case, the DAergic projection to the PFC. VTA lesions performed in the present studies, induce a concomitant partial depletion of DA in the rest of the mesolimbic system and the 5HTergic innervation in these areas. Different

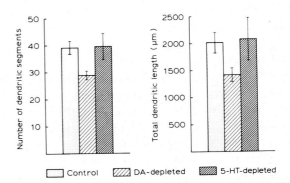

Fig. 11. Effect of cortical DA or 5HT depletion in early development on the number of segments in basal dendrites (left) and the total dendritic length of basal dendrites (right) of layer V pyramidal cells in the medial PFC. (Mean ± SEM are displayed.)

series of animals were tested for their performance in the open field, a food hoarding apparatus, and a spatial (delayed) alternation task, and for their sexual performance.

Open field

Tassin et al. (1980) showed very clearly in two strains of mice that differential activation of the mesocortical DA neurons may play a role in the differences observed in their open field behavior. Furthermore, in rats with bilateral electrolytic lesions of the VTA applied in adulthood, there is a strong correlation between the extent of the DA depletion in the PFC and their locomotor activity (Tassin et al., 1978). The neonatal VTA lesions as made in our studies caused a severe hypoactivity in the open field activity of young animals (PN25) as compared with sham-operated controls. On a second confrontation with the open field, 10 days later, these animals were still immobile most of the time. More restricted lesions of the medial part of the VTA, depleting predominantly 5HT in the forebrain, only caused a temporary hyperactivity, during the first exposure into the open field situation (Kalsbeek et al., 1989a). In adulthood (PN150), however, ambulation scores of neonatally VTA lesioned animals did no longer differ from that of their sham-operated littermates upon a first introduction in the open field (Fig. 12). It seems likely that biochemical compensatory mechanisms, as described above, play an important role in this restoration of function.

Food hoarding

Food hoarding is a highly organized, species-typical, behavior which is observed in various rodent species, and consists of the collection and transport of food pellets to the home cage in quantities greater than needed for immediate consumption (De Bruin, 1988). In adulthood both the dopaminergic activity of the mesolimbocortical neurons (Stinus et al., 1978; Oades, 1981; Kelley and Stinus, 1985; Herman et al., 1986; Blackburn et al., 1987; Choulli et al., 1987) and the integrity of the PFC (Kolb, 1974; Nonneman and Kolb, 1979; Nonneman and Corwin, 1981) are a prerequisite for a good performance of food hoarding behavior. Developmental studies have suggested that lesions of the PFC in early life might be far less effective in disturbing food hoarding behavior that are comparable lesions sustained in adulthood, e.g. "sparing" of cortical functions might occur (Nonneman and Corwin, 1981; Kolb and Whishaw, 1981, 1985). We therefore decided to eliminate the DAergic innervation of the PFC from birth on, and test the animals for their food hoarding behavior in adulthood (Kalsbeek et al., 1988a). The neonatal depletion of DA induced a clear reduction of food hoarding activity in experimental animals as compared with their sham-operated littermates (Fig. 13). After prolonged testing the difference disappeared, due to a decreased hoarding activity in the sham-operated animals. But the difference could be reinstated by reducing the time available for the hoarding of the pellets (week 5), or under ad libitum conditions (week 6). The hoarding scores showed a significant positive correlation only with DA levels in the mPFC, and not with NA or 5HT levels nor with any of the monoamine levels in the rest of the limbic forebrain (e.g. nucleus accumbens, septum and olfactory tubercle). These results make clear that depleting the PFC of its DAergic input is sufficient to disrupt food hoarding behavior, even when such depletion is performed shortly after birth.

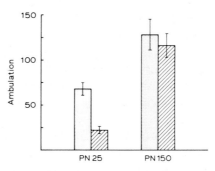

Fig. 12. Ambulatory activity (expressed as mean number of line crossings with SEM) of neonatally VTA lesioned and sham-operated animals when introduced for the first time in an open field situation.

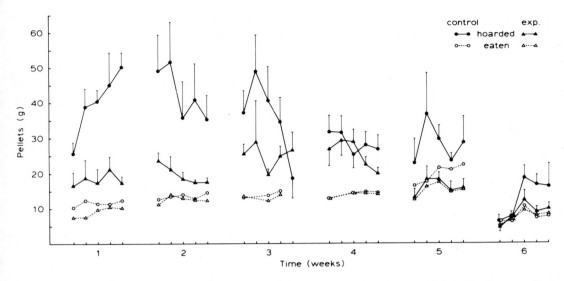

Fig. 13. Food hoarding of male Wistar rats (mean values ± SEM). The period during which food pellets could be hoarded was 60 min per session in weeks 1 and 4. The animals were able to eat during this period and during the ensuing 15 min period, when the hoarding alley was closed, but with previously hoarded food pellets present in the home cage. In week 5 these periods were 30 min and 2.5 h, respectively. During the last 3 days of week 6, when animals were no longer food-deprived, the hoarding alley was left open for the entire 3 h.

Spatial (delayed) alternation

Probably the best known feature of the PFC lesioned animal is its impaired performance in a delayed response task, especially those tasks which are spatially defined. As with food hoarding behavior, the performance in a spatial delayed alternation task can be disturbed in adult animals either by lesioning the PFC (Brito et al., 1982; Van Haaren et al., 1985; Doar et al., 1987) or by depleting its DAergic innervation (Simon et al., 1979, 1980; Kessler and Markowitsch, 1981; Taghzouti et al., 1985; Stam et al., 1989). Neonatal lesions of the PFC, in contrast, induced no deficits in a spatial delayed alternation task (Nonneman and Corwin, 1981). This result was recently replicated at our institute (De Brabander et al., in prep.). One developmental study, performing a DA depletion by way of an intraventricularly infusion of 6-OHDA at postnatal days 3 and 6, did report an impaired performance in a spatial delayed alternation task (Feeser and Raskin, 1987). To examine further the involvement of the mesocortical projection in the development of this typically PFC controlled behavior, we studied our neonatally VTA lesioned animals also in a T-maze for their performance in a delayed alternation task, when they were approximately 6 months old. After 1 week of adaptation and training, animals were tested for 2 weeks with a 0 sec intertrial interval (ITI). Then testing was continued for 1 more week with a 15 sec ITI. Each day consisted of 16 trials. After the experiment the lesion placement and the resulting depletion were checked using immunocytochemistry. Subsequent analysis of the behavioral data showed that bilaterally lesioned animals needed considerably more trials to reach criterion than did either sham-operated littermates or unilaterally lesioned animals. Although bilaterally VTA lesioned rats ultimately learn the task too, the introduction of a delay between subsequent choices again introduced a significant difference. Unilaterally lesioned animals did not differ significantly from sham-operated animals in either respect (Fig. 14).

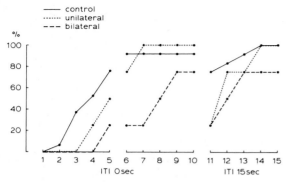

Fig. 14. Performance of neonatally VTA lesioned (unilateral and bilateral) and sham-operated animals in a spatial delayed alternation task. Results are expressed as the percentage of animals reaching the criterium (i.e. less than 3 mistakes) at different test days.

Therefore, a DA depletion of the PFC can only disturb the delayed alternation task effectively when the depletion is bilateral and (nearly) complete. Simon et al. (1979) reported on an unilaterally VTA lesioned rat in adulthood, which was the only one out of 10 lesioned rats to relearn the delayed alternation task. The neonatally induced deficits in a spatial delayed alternation task in our study seem to be less drastic and permanent than are the ones seen after adult lesions of the VTA (Simon et al., 1979). Combined with the sparing of such behavior which is seen after neonatal lesions of the PFC, this may indicate that these functions can be taken over by other parts of the cortex or maybe the caudate putamen (Nonneman and Kolb, 1979; Vicedomini et al., 1982, 1984; Kolb and Whishaw, 1989).

Sexual behavior

Treatment of pregnant or lactating rats with haloperidol, a DA receptor antagonist, has been shown to impair sexual behavior in the male offspring (Hull et al., 1984). In adult animals, manipulation of the DA system with systematically applied agonists or antagonists or with intraventricular application of 6-OHDA has been shown to affect sexual performance in both males and females. It is not clear, however, which DA system mediates these effects. Besides the hypothalamic DA containing cell groups, also the mesotelencephalic DA projection may be involved. Midbrain DA neurons have been implicated in the motivation and reward component of sexual behavior. Alderson and Baum (1981), and recently Mitchell and Stewart (1989) described a specific activation of the DA metabolism in the mesolimbic pathway of male rats after chronic exposure to testosterone or its neural metabolites. Adult lesioning of DA containing neurons in the VTA clearly affects sexual performance in both male (Barfield et al., 1975; Bracket et al., 1986) and female animals (Herndon, 1976; Sirinathinsinghji et al., 1986), although the effects are not unequivocal. To further examine the behavioral specificity of our neonatally applied VTA lesions, we tested the sexual behavior of male and female rats after neonatally applied VTA lesions (Matthijssen et al., 1990).

Experimental males all demonstrated normal mounting behavior, and had ejaculation positive tests. Moreover, there were no differences in any other measure of sexual performance between experimental and sham-operated animals (Fig. 15). In female animals too, neonatal VTA lesions induced no changes in sexual behavior. Experimental and sham-operated females showed equal rates of lordosis and proceptive behavior (Fig. 16). In addition, individual behavioral scores were compared with depletion rates of DA and 5HT. But no correlations were found between the amount of DA or 5HT depleted and the behaviors displayed. Therefore, neonatal VTA lesions fail to affect sexual performance in adulthood, contrary to VTA lesions applied in adult animals.

Concluding remarks

Neonatal depletion of DA and 5HT from the developing PFC, prior to synaptogenesis, has at best minor effects on the gross features of the cortical morphology, i.e. lamination, cortical thickness and volume. These results are in agreement with a number of transplantation studies which indicate a largely intrinsic determination of

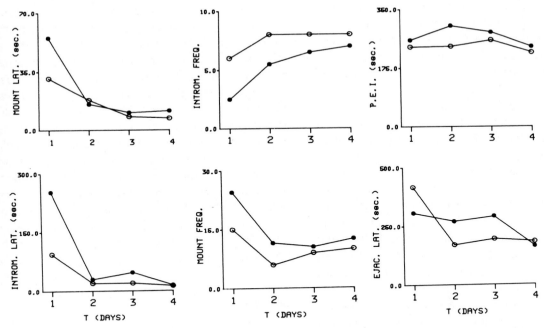

Fig. 15. The effects of neonatal VTA lesions on sexual behavior in adult male rats at successive days. Median values are given. PEI, post-ejaculatory interval. ○, experimental; ●, control.

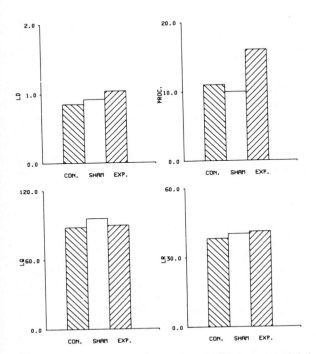

Fig. 16. Median values of female sexual behavior on the third test day for the 3 treatment groups. LQ, lordosis quotient (%); LD, lordosis duration (sec); LR, lordosis ratio (%); PROC, frequency of proceptive behavior.

the cortical cytoarchitecture (Chang et al., 1986; Fonseca et al., 1988; Mufson et al., 1987; Castro et al., 1988; Dunnett, this volume). Some features, however, seem to be dependent on specific aspects of the cortical area, e.g. ingrowing fiber systems and circulating humoral factors (Herman et al., 1988; Stanfield and O'Leary, 1985; Rakic, 1988). On the other hand, in a comparable experimental set-up, viz. depleting the cortical acetylcholine innervation by neonatal thermal lesions of the basal forebrain, Hohmann and co-workers (1988) showed that effects on cortical cytoarchitecture may be transient: they found the most severe effects on cortical morphogenesis in the first 2 postnatal weeks, but these differences were attenuated by the end of the first postnatal month, when the cortical acetylcholine innervation had reached control levels. Contrary to their results, we found no evidence of a restoration of the depleted transmitter systems, e.g. the mesocortical DA and 5HT innervation. It may be, however, that there are more subtle changes (such as a reduction in dendritic length) which are not reflected in the cortical thickness or volume, or else are obscured by compensatory changes in other systems. In any event, the dendritic measurements show that qualitative inspection is insufficient, it being necessary to perform a systematic quantitative analysis in order to be able to make any definitive statements about possible abnormalities.

The results of biochemical experiments show that the damaged DA system can compensate for lesion induced loss of DA containing terminals. They do this by increasing the DA metabolism of the remaining terminals. In an elegant study, using in vivo dialysis, Robinson and Whishaw (1988) have shown that the compensatory changes in the remaining terminals of the mesostriatal system are sufficient to normalize the extracellular (and presumably synaptic) concentrations of DA even after depletions of up to 95% of the DA terminals normally present in the striatum. Other studies, however, have not reported such a complete compensation of extracellular DA levels (Zetterström et al., 1986; Zhang et al., 1988; Brannan et al., 1989). Notwithstanding the possibility of an incomplete compensation, changes in the biochemistry of the damaged DA system, and the preferential DA depletion in the PFC but not the nucleus accumbens, probably account for the recovery of open field activity in adult animals, and the lack of changes in adult sexual behavior. Since, during normal development, there might be a condition of hypoinnervation (Stachowiak et al., 1987), these mechanisms can only become effective once the innervation has reached adult densities, i.e. after PN60 (Kalsbeek et al., 1988b). But the studies examining food hoarding behavior and the spatial delayed alternation task make clear that compensatory changes in DA metabolism are insufficient to restore all behavioral functions. It could be that an extra activation of the mesocortical neurons, necessary for the performance of the behaviors mentioned, is no longer possible in a damaged DA system which is already working at its highest capacity. This idea is supported by the footshock experiment (Fig. 8), showing that the remaining DA fibers are not able to further increase their activity in the face of a new challenge.

On the other hand, it may be that the reduction in DA and 5HT innervation during development causes such a disturbance in the "wiring" of the PFC that adequate performance of food hoarding and spatial delayed alternation tasks is no longer possible. Although our morphological studies found no gross alterations, there is a considerable impairment of dendritic arborization. The retarded growth of these layer V pyramidal cells, which form the origin of the major efferent projections of the PFC, indicate that such an idea is quite plausible. In fact, in their extensive review on cortical plasticity Kolb and Whishaw (1989) point to the extent of dendritic arborization in the remaining cortex as the most important correlate with sparing of function.

References

Alderson, C.A. and Baum, M.J. (1981) Differential effects of gonadal steroids on dopamine metabolism in mesolimbic and nigro-striatal pathways of male rat brain. *Brain Res.*, 218: 189–206.

Altar, C.A., Heikkila, R.E., Manzino, L. and Marien, M.R. (1986) 1-Methyl-4-phenylpyridine (MPP^+): regional dopamine neuron uptake, toxicity, and novel rotational behavior following dopamine receptor proliferation. *Eur. J. Pharmacol.*, 131: 199–209.

Altar, C.A., Marien, M.R. and Marshall, J.F. (1987) Time course of adaptations in dopamine biosynthesis, metabolism, and release following nigrostriatal lesions: implications for behavioral recovery from brain injury. *J. Neurochem.*, 48: 390–399.

Altman, J. and Bayer, S.A. (1981) Development of the brain stem in the rat. V. Thymidine-radiographic study of the time of origin of neurons in the midbrain tegmentum. *J. Comp. Neurol.*, 198: 677–716.

Aitken, A.R. and Törk, I. (1988) Early development of serotonin-containing neurons and pathways as seen in wholemount preparations of the fetal rat brain. *J. Comp. Neurol.*, 274: 32–47.

Antelman, S.M., Knopf, S., Caggiula, A.R., Kocan, D., Lysle, D.T. and Edwards, D.J. (1988) Stress and enhanced dopamine utilization in the frontal cortex: the myth and the reality. *Ann. N.Y. Acad. Sci.*, 537: 262–272.

Barfield, R.J., Wilson, C. and Mcdonald, P.G. (1975) Sexual behavior: extreme reduction of post ejaculatory refractory period by midbrain lesions in male rats. *Science*, 189: 147–149.

Beckstead, R.W. (1976) Convergent thalamic and mesencephalic projections to the anterior medial cortex in the rat. *J. Comp. Neurol.*, 166: 403–416.

Berger, B., Thierry, A.-M., Tassin, J.-P. and Moyne, M.A. (1976) Dopaminergic innervation of the rat prefrontal cortex: a fluorescence histochemical study. *Brain Res.*, 106: 133–145.

Bernheimer, H., Birkmayer, W., Hornykiewicz, O., Jellinger, K. and Seitelberger, F. (1973) Brain dopamine and the syndromes of Parkinson and Huntington. Clinical, morphological and neurochemical correlations. *J. Neurol. Sci.*, 20: 415–455.

Berry, M., Bradley, P. and Borges, S. (1977) Environmental and genetic determinants of connectivity in the central nervous system — an approach through dendritic field analysis. In M.A. Corner, R.E. Baker, N.E. Van De Poll, D.F. Swaab and H.B.M. Uylings (Eds.) *Maturation of the Nervous System, Prog. Brain Res., Vol. 48*, Elsevier, Amsterdam, pp. 133–149.

Bertolucci-D'Angio, M., Serrano, A., Driscoll, P. and Scatton, B. (1990) Involvement of mesocorticolimbic dopaminergic systems in emotional states. *This Volume*, Ch. 20.

Björklund, A. and Lindvall, O. (1975) Dopamine in dendrites of substantia nigra neurons: suggestions for a role in dendritic terminals. *Brain Res.*, 83: 531–537.

Björklund, A. and Lindvall, O. (1984) Dopamine-containing systems in the CNS. In A. Björklund and T. Hökfelt (Eds.): *Handbook of Chemical Neuroanatomy. Vol. 2. Classical Transmitters in the CNS, Part I*, Elsevier, Amsterdam, pp. 55–122.

Björklund, A., Divac, I. and Lindvall, O. (1978) Regional distribution of catecholamines in monkey cerebral cortex, evidence for a dopaminergic innervation of the primate prefrontal cortex. *Neurosci. Lett.*, 7: 115–119.

Blackburn, J.R., Phillips, A.G. and Fibiger, H.C. (1987) Dopamine and preparatory behavior. I. Effects of pimozide. *Behav. Neurosci.*, 101: 352–360.

Blue, M.E. and Parnavelas, J.G. (1983) The formation and maturation of synapses in the visual cortex of the rat. I. Qualitative analysis. *J. Neurocytol.*, 12: 599–616.

Brackett, N.J., Iuvone, P.M. and Edwards, D.A. (1986) Midbrain lesions, dopamine and male sexual behavior. *Behav. Brain Res.*, 20: 231–240.

Brannan, T., Knott, P., Kaufmann, H., Leung, L. and Yahr, M. (1989) Intracerebral dialysis monitoring of striatal dopamine release and metabolism in response to L-DOPA. *J. Neural Transm.*, 75: 149–157.

Brito, G.N.O., Thomas, G.H., Davis, B.J. and Gingold, S.I. (1982) Prelimbic cortex, mediodorsal thalamus, septum, and delayed alternation in rats. *Exp. Brain Res.*, 46: 52–58.

Bruinink, A., Lichtensteiner, W. and Schlumpf, M. (1983) Pre- and postnatal ontogeny and characterization of dopaminergic D2, serotonergic S2, and spirodecanone binding sites in rat forebrain. *J. Neurochem.*, 40: 1227–1236.

Buznikov, G.A. (1984) The action of neurotransmitters and related substances on early embryogenesis. *Pharmacol. Ther.*, 25: 23–59.

Castro, A.J., Tonder, N., Sunde, N.A. and Zimmer, J. (1988) Fetal neocortical transplants grafted to the cerebral cortex of newborn rats receive afferents from the basal forebrain, locus coeruleus and midline raphe. *Exp. Brain Res.*, 69: 613–622.

Chang, F.-L.F., Steedman, J.G. and Lund, R.D. (1986) The lamination and connectivity of embryonic cerebral cortex transplanted into newborn rat cortex. *J. Comp. Neurol.*, 244: 401–411.

Choulli, K., Herman, J.P., Rivet, J.M., Simon, H. and Le Moal, M. (1987) Spontaneous and graft-induced behavioral recovery after 6-hydroxydopamine lesion of the nucleus accumbens in the rat. *Brain Res.*, 407: 376–380.

Chubakov, A.R., Gromova, E.A., Konovalov, G.V., Sarkisova, E.F. and Chumasov, E.I. (1986) The effects of serotonin on the morphofunctional development of rat cerebral neocortex in tissue culture. *Brain Res.*, 369: 285–297.

Coyle, J.T. and Molliver, M.E. (1977) Major innervation of newborn rat cortex by monoaminergic neurons. *Science*, 196: 444–446.

D'Amato, R.J., Lipman, Z.P. and Snyder, S.H. (1986) Selectivity of the Parkinsonian neurotoxin MPTP: toxic metabolite MPP$^+$ binds to neuromelanin. *Science*, 231: 987–989.

De Brabander, H., De Bruin, J.P.C. and Van Eden, C.G. (1990) Comparison of neonatal and adult medial prefrontal cortex lesions on food hoarding and spatial delayed. Submitted.

De Bruin, J.P.C. (1988) Sex differences in food hoarding behaviour of Long Evans rats. *Behav. Process.*, 17: 191–198.

De Bruin, J.P.C., Van Oyen, H.G.M. and Van De Pol, N. (1983) Behavioural changes following lesions of the orbital prefrontal cortex in male rats. *Behav. Brain Res.*, 10: 209–232.

Deutch, A.Y. and Roth, R.H. (1990) The determination of stress-induced activation of the prefrontal cortical dopamine system. *This Volume*, Ch. 19.

Divac, I., Björklund, A., Lindvall, O. and Passingham, R.E. (1978) Converging projections from the mediodorsal thalamic nucleus and mesencephalic dopaminergic neurons to the neocortex in three species. *J. Comp. Neurol.*, 180: 59–72.

Doar, B., Finger, S. and Almli, C.R. (1987) Tactile-visual acquisition and reversal learning deficits in rats with prefrontal cortical lesions. *Exp. Brain Res.*, 66: 432–434.

Dunnett, S.B. (1990) Is it possible to repair the damaged prefrontal cortex by neural tissue transplantation? *This Volume*, Ch. 13.

Ebersole, P., Parnavelas, J.G. and Blue, M.E. (1981) Development of the visual cortex of rats treated with 6-hydroxydopamine in early life. *Anat. Embryol.*, 162: 489–492.

Feenstra, M.G.P., Voogel, A.J., Kalsbeek, A. and Rollema, H. (1990) Insensitivity of early postnatal rat brain dopaminergic systems to 1-methyl-4-phenylpyridinium ion (MPP$^+$). *Biog. Amines*, in press.

Feeser, H.R. and Raskin, L.A. (1987) Effects of neonatal dopamine depletion on spatial ability during ontogeny. *Behav. Neurosci.*, 101: 812–818.

Felten, D.L., Hallman, H. and Jonsson, G. (1982) Evidence for a neurotrophic role of noradrenaline neurons in the postnatal development of rat cerebral cortex. *J. Neurocytol.*, 11: 119–135.

Fonseca, M., Defelipe, J. and Fairen, A. (1988) Local connections in transplanted and normal cerebral cortex of rats. *Exp. Brain Res.*, 69: 387–398.

Gebert, I., Drescher, K. and Morgenstern, R. (1985) Kainic acid lesion of the ventral tegmental area: effect on mesolimbic-mesocortical dopamine terminals. *Biog. Amines*, 2: 163–169.

Geffard, M., Buijs, R.M., Sequela, P., Pool, C.W. and Le Moal, M. (1984) First demonstration of highly specific and sensitive antibodies against dopamine. *Brain Res.*, 294: 161–165.

Gerfen, C.R., Herkenham, M. and Thibault, J. (1987) The nigrostriatal mosaic. II. Patch- and matrix-directed mesostriatal dopaminergic and non-dopaminergic systems. *J. Neurosci.*, 7: 3915–3934.

Hanaway, J., Mcconnell, J.A. and Netsky, M.G. (1971) Histogenesis of the substantia nigra, ventral tegmental area of Tsai and interpeduncular nucleus: an autoradiographic study of the mesencephalon in the rat. *J. Comp. Neurol.*, 142: 59–74.

Herman, J.P., Choulli, K., Geffard, M., Nadaud, D., Taghzouti, K. and Le Moal, M. (1986) Reinnervation of the nucleus accumbens and frontal cortex of the rat by dopaminergic grafts and effects on hoarding behavior. *Brain Res.*, 372: 210–216.

Herman, J.P., Abrous, N., Vigny, A., Dulluc, J. and Lemoal, M. (1988) Distorted development of intracerebral grafts: long-term maintenance of tyrosine hydroxylase-containing neurons in grafts of cortical tissue. *Dev. Brain Res.*, 40: 81–88.

Herndon, J.G. (1976) Effects of midbrain lesions on female sexual behavior in the rat. *Physiol. Behav.*, 17: 143–148.

Hervé, D., Studler, J.-M., Blanc, G., Glowinski, J. and Tassin, J.-P. (1986) Partial protection by desmethylimipramine of the mesocortical dopamine neurones from the neurotoxic effect of 6-hydroxydopamine injected in ventral mesencephalic tegmentum. The role of the noradrenergic innervation. *Brain Res.*, 383: 47–53.

Hohmann, C.F., Brooks, A.R. and Coyle, J.T. (1988) Neonatal lesions of the basal forebrain cholinergic neurons result in abnormal cortical development. *Dev. Brain Res.*, 42: 253–264.

Hökfelt, T., Skirboll, L., Rehfeld, J.F., Goldstein, M., Markey, K. and Dann, O. (1980) A subpopulation of mesencephalic dopamine neurons projecting to limbic areas contains a cholecystokinin-like peptide: evidence from immunohistochemistry combined with retrograde tracing. *Neuroscience*, 5: 2093–2124.

Hoorneman, E.M.D. (1985) Stereotaxic operation in the neonatal rat; a novel and simple procedure. *J. Neurosci. Meth.*, 14: 109–116.

Hornykiewicz, O. (1973) Parkinsons's disease: from brain homogenate to treatment. *Fed. Proc.*, 32: 183–190.

Hornykiewicz, O. (1980) Biochemical abnormalities in some extrastriatal neuronal systems in Parkinson's disease. In U.K. Rinne, M. Klinger and G. Stamm (Eds.), *Parkinson's Disease – Current Progress, Problems and Management*, Elsevier, Amsterdam, pp. 109–119.

Horton, H.L. and Levitt, P. (1988) A unique membrane protein is expressed on early developing limbic system axons and cortical targets. *J. Neurosci.*, 8: 4653–4661.

Hull, E.M., Nishita, J.K., Bitran, D. and Dalterio, S. (1984) Perinatal dopamine-related drugs demasculinize rats. *Science*, 224: 1011–1013.

Hume, R.I., Role, L.W. and Fischbach, G.D. (1983) Acetylcholine release from growth cones detected with patches of acetylcholine receptor rich membranes. *Nature*, 305: 632–634.

Iniguez, C., Calle, F., Marshall, E. and Carreres, J. (1987) Morphological effects of chronic haloperidol administration on the postnatal development of the striatum. *Dev. Brain Res.*, 35: 27–34.

Jonsson, G. (1983) Chemical lesioning techniques: Monoamine neurotoxins. In A. Björklund and T. Hökfelt (Eds.), *Handbook of Chemical Neuroanatomy, Methods in Chemical Neuroanatomy, Vol. 1*, Elsevier/North-Holland, Amsterdam, pp. 463–507.

Jonsson, G., Pycock, Ch., Fuxe, K. and Sachs, C. (1974) Changes in the development of central noradrenaline neurones following neonatal administration of 6-hydroxydopamine. *J. Neurochem.*, 22: 419–426.

Kalsbeek, A. and Uylings, H.B.M. (1989) Immunocytochemical localization of the GABAergic and some peptidergic systems in the prefrontal cortex of the rat after neonatal depletion of cortical dopamine. *NIBR Res. Report* 1.

Kalsbeek, A., Buijs, R.M., Hofman, M.A., Matthijssen, M.A.H., Pool, C.W. and Uylings, H.B.M. (1987) Effects of neonatal thermal lesioning of the mesocortical dopaminergic projection on the development of the rat prefrontal cortex. *Dev. Brain Res.*, 32: 123–132.

Kalsbeek, A., De Bruin, J.P.C., Feenstra, M.G.P. and Uylings, H.B.M. (1988a) Neonatal thermal lesions of the mesolimbocortical dopaminergic projection decrease food-hoarding behaviour. *Brain Res.*, 475: 80–90.

Kalsbeek, A., Voorn, P., Buijs, R.M. and Uylings, H.B.M. (1988b) Development of the dopaminergic innervation in the prefrontal cortex of the rat. *J. Comp. Neurol.*, 269: 58 – 72.

Kalsbeek, A., De Bruin, J.P.C., Matthijssen, M.A.H. and Uylings, H.B.M. (1989a) Ontogeny of open field activity in rats after neonatal lesioning of the mesocortical dopaminergic projection. *Behav. Brain Res.*, 32: 115 – 127.

Kalsbeek, A., Feenstra, M.G.P., Van Galen, H. and Uylings, H.B.M. (1989b) Monoamine and metabolite levels in the prefrontal cortex and the mesolimbic forebrain following neonatal lesions of the ventral tegmental area. *Brain Res.*, 479: 339 – 343.

Kalsbeek, A., Matthijssen, M.A.H. and Uylings, H.B.M. (1989c) Morphometric analysis of prefrontal cortical development following neonatal lesioning of the dopaminergic mesocortical projection. *Exp. Brain Res.*, 78: 279 – 289.

Kalsbeek, A., Voorn, P. and Buijs, R.M. (1990) Development of dopamine-containing systems in the CNS. In A. Björklund, T. Hökfelt, and M. Tohyama (Eds.), *Handbook of Chemical Neuroanatomy. Ontogeny of Transmitters and Peptides in the Central Nervous System,* Elsevier, Amsterdam, in press.

Kelley, A.E. and Stinus, L. (1985) Disappearance of hoarding behavior after 6-hydroxydopamine lesions of the mesolimbic dopamine neurons and its reinstatement with L-Dopa. *Behav. Neurosci.*, 99: 531 – 545.

Kelly, E., Jenner, P. and Marsden, C.D. (1985) Evidence that [^{3}H]dopamine is taken up and released from non-dopaminergic nerve terminals in the rat substantia nigra in vitro. *J. Neurochem.*, 45: 137 – 144.

Kessler, J. and Markowitsch, H.J. (1981) Delayed alternation performance after kainic acid lesions of the thalamic mediodorsal nucleus and the ventral tegmental area in the rat. *Behav. Brain Res.*, 3: 125 – 130.

Kolb, B. (1974) Prefrontal lesions alter eating and hoarding behavior in rats. *Physiol. Behav.*, 12: 507 – 511.

Kolb, B. and Nonneman, A.J. (1974) Frontolimbic lesions and social behavior in the rat. *Physiol. Behav.*, 13: 637 – 643.

Kolb, B. and Whishaw, I.Q. (1981) Neonatal frontal lesions in the rat: sparing of learned but not species-typical behavior in the presence of reduced brain weight and cortical thickness. *J. Comp. Physiol. Psychol.*, 95: 863 – 879.

Kolb, B. and Whishaw, I.Q. (1985) Neonatal frontal lesions in hamsters impair species-typical behaviors and reduce brain weight and neocortical thickness. *Behav. Neurosci.*, 99: 691 – 706.

Kolb, B. and Whishaw, I.Q. (1989) Plasticity in the neocortex: mechanisms underlying recovery from early brain damage. *Progr. Neurobiol.*, 32: 235 – 276.

König, N. and Marty, R. (1981) Early neurogenesis and synaptogenesis in cerebral cortex. *Bibl. Anat.*, 19: 152 – 160.

König, N., Serrano, J.-J., Jonsson, G., Malaval, F. and Szafarczyk, A. (1985) Prenatal treatment with 6-hydroxydopa and DSP 4: biochemical, endocrinological and behavioural effects. *Int. J. Dev. Neurosci.*, 3: 501 – 509.

Kostrzewa, R.M. and Jacobowitz, D.M. (1974) Pharmacological actions of 6-hydroxydopamine. *Pharmacol. Rev.*, 26: 199 – 288.

Lankford, K., DeMello, G. and Klein, W.L. (1987) A transient embryonic dopamine receptor inhibits growth cone motility and neurite outgrowth in a subset of avian retina neurons. *Neurosci. Lett.*, 75: 169 – 174.

Lauder, J.M. and Bloom, F.E. (1974) Ontogeny of monoamine neurons in the locus coeruleus, raphe nuclei and substantia nigra of the rat. *J. Comp. Neurol.*, 115: 469 – 482.

Lauder, J.M. and Krebs, H. (1986) Do neurotransmitters, neurohumors, and hormones specify critical periods? In W.T. Greenough and J.M. Juraska (Eds.), *Developmental Neuropsychobiology,* Academic Press, New York, pp. 119 – 174.

Lauder, J.M., Wallace, J.A., Krebs, H., Petrusz and McCarthy, K. (1982) In vivo and in vitro development of serotonergic neurons. *Brain Res. Bull.*, 9: 505 – 625.

Lauder, J.M., Tamir, H. and Sadler, T.W. (1988) Serotonin and morphogenesis. I. Sites of serotonin uptake and binding protein immunoreactivity in the midgestation mouse embryo. *Development*, 102: 709 – 720.

Lavielle, S., Tassin, J.-P., Thierry, A.-M., Blanc, G., Herve, D., Barthelemy, C. and Glowinski, J. (1978) Blockade by benzodiazepines of the selective high increase in dopamine turnover induced by stress in mesocortical dopaminergic neurons of the rat. *Brain Res.*, 168: 585 – 594.

Levin, B.E., Llabre, M.M. and Weiner, W.J. (1988) Neuropsychological correlates of early Parkinson's disease: evidence for frontal lobe dysfunction. *Ann. N.Y. Acad. Sci.*, 537: 518 – 519.

Levitt, P. (1984) A monoclonal antibody to limbic system neurons. *Science*, 223: 299 – 301.

Levitt, P. and Moore, R.Y. (1979) Development of the noradrenergic innervation of neocortex. *Brain Res.*, 162: 242 – 259.

Lidov, H.G.W. and Molliver, M.E. (1982a) An immunocytochemical study of serotonin neuron development in the rat: Ascending pathways and terminal fields. *Brain Res. Bull.*, 8: 389 – 430.

Lidov, H.G.W. and Molliver, M.E. (1982b) The structure of cerebral cortex in the rat following prenatal administration of 6-hydroxydopamine. *Dev. Brain Res.*, 3: 81 – 108.

Maeda, T., Tohyama, M. and Shimizu, N. (1974) Modification of postnatal development of neocortex in rat brain with experimental deprivation of locus coeruleus. *Brain Res.*, 70: 515 – 520.

Matthijssen, M.A.H., Kalsbeek, A. and De Jonge, F.H. (1989) Sparing of male and female sexual behaviour after neonatal depletion of the dopaminergic mesocortical projection. *Neurosci. Res. Comm.*, 6: 95 – 103.

McCobb, D.P., Haydon, P.G. and Kater, S.B. (1988)

Dopamine and serotonin inhibition of neurite elongation of different identified neurons. *J. Neurosci. Res.,* 19: 19 – 26.

McMahon, D. (1974) Chemical messengers in development: A hypothesis. *Science,* 185: 1012 – 1021.

Miller, J.C. and Friedhoff, A.J. (1988) Prenatal neurotransmitter programming of postnatal receptor function. In G.J. Boer, M.G.P. Feenstra, M. Mirmiran, D.F. Swaab and F. Van Haaren (Eds.), *Biochemical Basis of Functional Neuroteratology, Progress in Brain Research, Vol. 73,* Elsevier, Amsterdam, pp. 509 – 522.

Mitchell, J.B. and Stewart, J. (1989) Effects of castration, steroid replacement, and sexual experience on mesolimbic dopamine and sexual behaviors in the male rat. *Brain Res.,* 491: 116 – 127.

Mrzljak, L., Uylings, H.B.M., Kostivic, I. and Van Eden, C.G. (1988) The postnatal development of neurons in the human prefrontal cortex. *J. Comp. Neurol.,* 271: 355 – 386.

Mufson, E.J., Labbe, R. and Stein, D.G. (1987) Morphologic features of embryonic neocortex grafts in adult rats following frontal cortical ablation. *Brain Res.,* 401: 162 – 167.

Murrin, L.C., Gibbens, D.L. and Ferrer, J.R. (1985) Ontogeny of dopamine, serotonin and spirodecanone receptors in rat forebrain – An autoradiographic study. *Dev. Brain Res.,* 23: 91 – 109.

Namura, I., Douillet, P., Sun, C.J., Pert, A.M., Cohen, R. and Chiueh, C.C. (1987) MPP$^+$ (1-methyl-4-phenyl-pyridine) is a neurotoxin to dopamine-, norepinephrine- and serotonin-containing neurons. *Eur. J. Pharmacol.,* 136: 31 – 37.

Naus, C.G. and Bloom, F.E. (1988) Immunohistochemical analysis of the development of somatostatin in the reeler neocortex. *Dev. Brain Res.,* 43: 61 – 68.

Noisin, E.L. and Thomas, W.E. (1988) Ontogeny of dopaminergic function in the rat midbrain tegmentum, corpus striatum and frontal cortex. *Dev. Brain Res.,* 41: 241 – 252.

Nonneman, A.J. and Corwin, J.V. (1981) Differential effects of prefrontal cortex ablation in neonatal, juvenile, and young rats. *J. Comp. Physiol. Psychol.,* 95: 588 – 602.

Nonneman, A.J. and Kolb, B. (1979) Functional recovery after serial ablation of prefrontal cortex in the rat. *Physiol. Behav.,* 22: 895 – 901.

Oades, R.D. (1981) Impairment of search behaviour in rats after haloperidol treatment, hippocampal or neocortical damage suggests a mesocorticolimbic role in cognition. *Biol. Psychol.,* 12: 77 – 85.

Olson, L. and Seiger, A. (1972) Early prenatal ontogeny of central monoamine neurons in the rat: fluorescence histochemical observations. *Z. Anat. Entwickl.-Gesch.,* 137: 301 – 316.

Onténiente, B., Nig, N., Sievers, S., Jenner, S., Klemm, H.P. and Marty, R. (1980) Structural and biochemical changes in rat cerebral cortex after neonatal 6-hydroxydopamine administration. *Anat. Embryol.,* 159: 245 – 255.

Overdijk, J., Uylings, H.B.M., Kuypers, K. and Kamstra, A.W. (1978) An economical semi-automatic system for measuring cellular tree structures in three dimensions, with special emphasis on Golgi-impregnated neurons. *J. Microsc.,* 114: 271 – 284.

Peters, D.A.V., Pappas, B.A., Taub, H. and Saari, M. (1977) Effect of intraventricular injections of 6-hydroxydopamine in neonatal rats on the catecholamine levels and tyrosine hydroxylase activity in brain regions at maturity. *Biochem. Pharmacol.,* 26: 2211 – 2215.

Petit, T.L., Leboutillier, J.C., Gregorio, A. and Libstug, H. (1988) The pattern of dendritic development in the cerebral cortex of the rat. *Dev. Brain Res.,* 41: 209 – 219.

Pinto-Lord, M.C. and Caviness, V.S., Jr. (1979) Determinants of cell shape and orientation: a comparative Golgi analysis of cell axon interrelationships in the developing neocortex of normal and reeler mice. *J. Comp. Neurol.,* 187: 49 – 70.

Plantjé, J.F., Steinbush, H.W.M., Schipper, J., Dijcks, F.A., Verheyden, P.F.H.M. and Stoof, J.C. (1987) D-2 dopamine-receptors regulate the release of [^{3}H]dopamine in rat cortical regions showing dopamine immunoreactive fibres. *Neuroscience,* 20: 157 – 168.

Rakic, P. (1988) Specification of cerebral cortical areas. *Science,* 241: 170 – 176.

Reader, T.A. and Grondin, L. (1987) Distribution of catecholamines, serotonin, and their major metabolites in the rat cingulate, piriform-enthorhinal, somatosensory, and visual cortex: a biochemical survey using high-performance liquid chromatography. *Neurochem. Res.,* 12: 1087 – 1097.

Reader, T.A., Ferron, A., Diop, L., Kolta, A. and Briere, R. (1988) The heterogeneity of the catecholamine innervation of cerebral cortex. Biochemical and electrophysiological studies. In M. Avoli, T.A. Reader, R.W. Dykes and P. Gloor (Eds.), *Neurotransmitters and Cortical Function,* Plenum, New York, pp. 333 – 355.

Robinson, T.E. and Whishaw, I.Q. (1988) Normalization of extracellular dopamine in striatum following recovery from a partial unilateral 6-OHDA lesion of the substantia nigra: a micro-dialysis study in freely moving rats. *Brain Res.,* 450: 209 – 224.

Rogers, D.C. and Dunnett, S.B. (1989) Hypersensitivity to α-methyl-*p*-tyrosine suggests that behavioural recovery of rats receiving neonatal 6-OHDA lesions is mediated by residual catecholamine neurones. *Neurosci. Lett.,* 102: 108 – 113.

Rose, J.E. and Woolsey, C.N. (1948) The orbitofrontal cortex and its connections with the mediodorsal nucleus in rabbit, sheep and cat. *Res. Publ. Assoc. Nerv. Ment. Dis.,* 27: 210 – 232.

Rose, S., Nomoto, M., Kelly, E., Kilpatrick, G., Jenner, P. and Marsden, C.D. (1989) Increased caudate dopamine turnover may contribute to the recovery of motor function in marmosets treated with the dopaminergic neurotoxin MPTP. *Neurosci. Lett.,* 101: 305 – 310.

Sales, N., Martres, M.P., Bouthenet, M.L. and Schwartz, J.C. (1989) Ontogeny of dopaminergic D-2 receptors in the rat nervous system: characterization and detailed autoradiographic mapping with [^{125}I]iodosulpride. *Neuroscience,* 28: 673–700.

Scatton, B., Javoy-Agid, F., Rouquier, L., Dubois, B. and Agid, Y. (1983) Reduction of cortical dopamine, noradrenaline, serotonin and their metabolites in Parkinsons's disease. *Brain Res.,* 275: 321–328.

Schlumpf, M., Lichtensteiner, W., Shoemaker, W.J. and Bloom, F.E. (1980) Fetal monoamine systems: early stages and cortical projections. In H. Parvez and S. Parvez (Eds.), *Biogenic Amines in Development,* Elsevier, Amsterdam, pp. 567–590.

Schmidt, R.H. and Bhatnagar, R.K. (1979) Assessment of the effects of neonatal subcutaneous 6-hydroxydopamine on noradrenergic and dopaminergic innervation of the cerebral cortex. *Brain Res.,* 166: 309–319.

Schmidt, R.H., Björklund, A., Lindvall, O. and Loren, I. (1982) Prefrontal cortex: dense dopaminergic input in the newborn rat. *Dev. Brain Res.,* 5: 222–228.

Schwab, R.S. and Zieper, I. (1965) Effects of mood, motivation, stress and alertness on the performance in Parkinson's disease. *Psychiat. Neurol.,* 150: 345–357.

Schwarcz, R., Hökfelt, T., Fuxe, K., Jonsson, G., Goldstein, M. and Terenius, L. (1979) Ibotenic acid-induced neuronal degeneration: A morphological and neurochemical study. *Exp. Brain Res.,* 37: 199–216.

Segler-Stahl, K., Rahmann, H. and Rosner, H. (1982) Effects of neonatal 6-hydroxydopamine treatment on cortical development in mice and rats as monitored by developmental changes of gangliosides. *Dev. Neurosci.,* 5: 554–566.

Seiger, A. and Olson, L. (1973) Late prenatal ontogeny of central monoamine neurons in the rat: Fluorescence histochemical observations. *Z. Anat. Entwickl.-Gesch.,* 140: 281–318.

Seroogy, K.B., Mehta, A. and Fallon, J.H. (1987) Neurotensin and cholecystokinin coexistence within neurons of the ventral mesencephalon: projections to forebrain. *Exp. Brain Res.,* 68: 277–289.

Sievers, J., Klemm, H.P., Jenner, S., Baumgarten, H.G. and Berry, M. (1980) Neuronal and extraneuronal effects of intracisternally administered 6-hydroxydopamine on the developing rat brain. *J. Neurochem.,* 34: 765–771.

Silbergeld, E.K. and Walters, J.R. (1979) Synaptosomal uptake and release of dopamine in substantia nigra: effects of aminobutyric acid and substance P. *Neurosci. Lett.,* 12: 119–126.

Simon, H., Scatton, B. and LeMoal, M. (1979) Definitive disruption of spatial delayed alternation in rats after lesions in the ventral mesencephalic tegmentum. *Neurosci. Lett.,* 15: 319–324.

Simon, H., Scatton, B. and LeMoal, M. (1980) Dopaminergic A10 neurones are involved in cognitive functions. *Nature,* 286: 150–151.

Sirinathinsinghji, D.J.S., Whittington, P.E. and Audsley, A.R. (1986) Regulation of mating behavior in the female rat by gonadotropin releasing hormone in the ventral tegmentum area: effects of selective destruction of the A10 dopamine neurons. *Brain Res.,* 374: 167–173.

Snyder, A.M., Stricker, E.M. and Zigmond, M.J. (1985) Stress-induced neurological impairments in an animal model of Parkinsonism. *Ann. Neurol.,* 18: 544–551.

Specht, L.A., Pickel, V.M., Joh, T.H. and Reis, D.J. (1981) Light-microscopic immunocytochemical localization of tyrosine hydroxylase in prenatal brain. I. Early ontogeny. *J. Comp. Neurol.,* 199: 233–253.

Stachowiak, M.K., Keller, R.W., Stricker, E.M. and Zigmond, M.J. (1987) Increased dopamine efflux from striatal slices during development and after nigrostriatal bundle damage. *J. Neurosci.,* 7: 1648–1654.

Stam, C.J., De Bruin, J.P.C., Van Haelst, A.M., Van Der Gugten and Kalsbeek, A. (1989) The influence of the mesocortical dopaminergic system on activity, food hoarding, aggression and spatial delayed alternation in male rats. *Behav. Neurosci.,* 103: 24–35.

Stanfield, B.B. and O'Leary, D.D.M. (1985) Fetal occipital cortical neurones transplanted to the rostral cortex can extend and maintain a pyramidal tract axon. *Nature,* 313: 135–137.

Stinus, L., Gaffori, O., Simon, H. and LeMoal, M. (1978) Disappearance of hoarding and disorganization of eating behavior after ventral mesencephalic tegmentum lesions in rats. *J. Comp. Physiol. Psychol.,* 92: 289–296.

Taghzouti, K., Louilot, A., Herman, J.P., LeMoal, M. and Simon, H. (1985) Alternation behavior, spatial discrimination, and reversal disturbances following 6-hydroxydopamine lesions in the nucleus accumbens of the rat. *Behav. Neural Biol.,* 44: 354–363.

Tassin, J.P., Velley, L., Stinus, L., Blanc, G., Glowinski, J. and Thierry, A.M. (1975) Development of cortical and nigro-neostriatal dopaminergic systems after destruction of central noradrenergic neurones in foetal or neonatal rats. *Brain Res.,* 83: 93–106.

Tassin, J.P., Stinus, L., Simon, H., Blanc, G., Thierry, A.M., Lemoal, M., Cardo, B. and Glowinski, J. (1978) Relationship between the locomotor hyperactivity induced by A10 lesions and the destruction of the fronto-cortical dopaminergic innervation in the rat. *Brain Res.,* 141: 267–281.

Tassin, J.-P., Herve, D., Blanc, G. and Glowinski, J. (1980) Differential effects of a two-minute open-field session on dopamine utilization in the frontal cortices of BALB/C and C57/6 mice. *Neurosci. Lett.,* 17: 67–71.

Tennyson, V.M., Mytilineou, C. and Barrett, R.E. (1973) Fluorescence and electron microscopic studies of the early development of the substantia nigra and area ventralis tegmenti in the fetal rabbit. *J. Comp. Neurol.,* 149: 233–258.

Tennyson, V.M., Gershon, P., Budinikas-Schoenebeck, M. and Rothman, T.P. (1983) Effects of extended periods of reserpine and α-methyl-p-tyrosine treatment on the development of the putamen in fetal rabbits. *Int. J. Dev. Neurosci.,* 1: 305–318.

Thierry, A.M., Blanc, G., Sobel, A., Stinus, L. and Glowinski, J. (1973) Dopaminergic terminals in the rat cortex. *Science,* 182: 499–501.

Uylings, H.B.M., Ruiz-Marcos, A. and Van Pelt, J. (1986) The metric analysis of three-dimensional dendritic tree patterns: a methodological review. *J. Neurosci. Meth.,* 18: 127–151.

Uylings, H.B.M., Van Eden, C.G., Parnavelas, J.G. and Kalsbeek, A. (1990) The pre- and postnatal development of rat cerebral cortex. In B. Kolb and R.C. Tees (Eds.), *The Cerebral Cortex of the Rat,* MIT Press, Cambridge, MA, pp. 35–76.

Van Eden, C.G. and Uylings, H.B.M. (1985) Postnatal volumetric development of the prefrontal cortex in the rat. *J. Comp. Neurol.,* 241: 268–274.

Van Eden, C.G., Hoorneman, E.M.D., Buijs, R.M., Matthijssen, A.A.H., Geffard, M. and Uylings, H.B.M. (1987) Immunocytochemical localization of dopamine in the prefrontal cortex of the rat at the light and electron microscopical level. *Neuroscience,* 22: 849–862.

Van Haaren, F., De Bruin, J.P.C., Heinsbroek, R.P.W. and Van De Poll, N.E. (1985) Delayed spatial response alternation: effect of delay-interval duration and lesions of the medial prefrontal cortex on response accuracy of male and female Wistar rats. *Behav. Brain Res.,* 18: 41–49.

Verney, C., Berger, B., Baulac, M., Helle, K.B. and Alvarez, C. (1984) Dopamine-β-hydroxylase-like immunoreactivity in the fetal cerebral cortex of the rat: noradrenergic ascending pathways and terminal fields. *Int. J. Dev. Neurosci.,* 2: 491–503.

Vicedomini, J.P., Corwin, J.V. and Nonneman, A.J. (1979) Role of residual anterior neocortex in recovery from neonatal prefrontal lesions in the rat. *Physiol. Behav.,* 28: 797–806.

Vicedomini, J.P., Isaac, W.L. and Nonneman, A.J. (1984) Role of the caudate nucleus in recovery from neonatal mediofrontal cortex lesions in the rat. *Dev. Psychobiol.,* 17: 51–65.

Voorn, P., Roest, G. and Groenewegen, H.J. (1987) Increase of enkephalin and decrease of substance P immunoreactivity in the dorsal and ventral striatum of the rat after midbrain 6-hydroxydopamine lesions. *Brain Res.,* 412: 391–396.

Voorn, P., Kalsbeek, A., Jorritsma-Byham, J. and Groenewegen, H.J. (1988) The pre- and postnatal development of the dopaminergic cell groups in the ventral mesencephalon and the dopaminergic innervation of the striatum of the rat. *Neuroscience,* 25: 857–877.

Wallace, J.A. and Lauder, J.M. (1983) Development of the serotonergic system in the rat embryo: An immunocytochemical study. *Brain Res. Bull.,* 10: 459–479.

Wendlandt, S., Crow, T.W. and Stirling, R.V. (1977) The involvement of the noradrenergic system arising from the locus coeruleus in the postnatal development of the cortex in rat brain. *Brain Res.,* 125: 1–9.

Westerink, B.H.C., Van Der Heyden, J.A.M. and Korf, J. (1978) Enhanced dopamine metabolism after small lesions in the midbrain of the rat. *Life Sci.,* 22: 749–756.

Wilson, J.S., Turner, B.H., Morrow, G.D. and Hartman, P.J. (1987) MPTP produces a mosaic-like pattern of terminal degeneration in the caudate nucleus of dog. *Brain Res.,* 423: 329–332.

Wise, S.P., Fleshman, J.W. and Jones, E.G. (1979) Maturation of pyramidal cell form in relation to developing afferent and efferent connections of rat somatic sensory cortex. *Neuroscience,* 4: 1275–1297.

Wolff, J.R. (1978) Ontogenetic aspects of cortical cytoarchitecture: lamination. In M.A.B. Brazier and H. Petsche (Eds.), *Architectonics of the Cerebral Cortex,* Raven Press, New York, pp. 159–173.

Young, S.H. and Poo, M. (1983) Spontaneous release of transmitter from growth cones of embryonic neurones. *Nature,* 305: 634–637.

Zecevic, N.R. and Molliver, M.E. (1978) The origin of the monoaminergic innervation of immature rat neocortex: an ultrastructural analysis following lesions. *Brain Res.,* 150: 387–397.

Zetterström, T., Herrera-Marschitz, M. and Ungerstedt, U. (1986) Simultaneous measurement of dopamine release and rotational behaviour in 6-hydroxydopamine denervated rats using intracerebral dialysis. *Brain Res.,* 376: 1–7.

Zhang, W.Q., Tilson, H.A., Nanry, K.P., Hudson, P.M., Hong, J.S. and Stachowiak, M.K. (1988) Increased dopamine release from striata of rats after unilateral nigrostriatal bundle damage. *Brain Res.,* 461: 335–342.

Discussion

A.Y. Deutch: (1) VTA electrolytic lesions destroyed VTA but not SN dopaminergic neurons, but 6-OHDA lesions involved SN but not VTA dopaminergic neurons. Could you explain this?

(2) Did you observe any increase in DA-immunoreactive axonal staining in the medial supralemniscal region following neonatal 6-OHDA lesions which resulted in loss of DA-immunoreactive perikarya in the SN, but not VTA DA-immunoreactive perikarya? I raise this question since evidence of increased staining of this region may indicate a morphological plasticity in the DA system in response to certain types of injury. The axons of mesencephalic DA neurons project from the midbrain to telencephalic targets through two pathways: the major system coursing through the VTA, and a very small projection running dorsal to the medial aspect of the medial lemniscus (supra-lemniscal pathway). In primates exposed to MPTP, there is a profound decrease in the number of DA neurons in the SN, and a marked increase in tyrosine hydroxylase-like immunoreactive fiber staining in the supralemniscal area.

A. Kalsbeek: (1) SN neurons seem to be more sensitive to 6-OHDA than the DA neurons of the VTA, therefore, infusion of 6-OHDA in the ventral mesencephalon preferentially depleted the SN and spared the VTA. In order to be able to lesion the DA neurons of the VTA we started to make thermal lesions. With the help of the stereotaxic procedure for neonates (as described in this chapter) it now was possible to lesion specifically the DA neurons of the VTA.

(2) Although lesions involving only the SN and not the VTA were not common, an increase in DA-immunoreactive axonal staining in the supralemniscal region was not obvious in these cases.

A.Y. Deutch: In your presentation you showed that neonatal thermal lesions of the VTA resulted in a decrease in PFC DA concentrations and a marked increase in metabolite:amine ratio in the PFC. Yesterday I presented data from our laboratory indicating that deafferentation of the DA innervation of the PFC resulted in a transsynaptic alteration in subcortical DA systems, such that the DA innervation of the nucleus accumbens (NAS) was rendered hyperresponsive to perturbation. Thus in animals ˃osed to mild footshock stress 2 weeks after 6-OHDA lesions ˃e PFC, an increase in NAS DA metabolism was observed OHDA-lesioned, but not sham-lesioned rats. You just nted data, however, illustrating that in your lesioned animals footshock did not evoke a further increase in metabolite:amine ratio in the NAS. What could be the reason for this difference?

A. Kalsbeek: There are a number of possible explanations for this difference. (1) In the first place there is a difference in lesion techniques; you applied 6-OHDA directly to the PFC of adult animals, whereas we depleted cortical DA by lesioning the VTA thermally in neonatal animals. Since both techniques affect other transmitter systems differently (for instance the cortical noradrenaline and serotonin systems), this could be the cause of a difference in the activation of subcortical DA systems. (2) Alternatively, as a result of our VTA lesions also part of the DA innervation in the NAS was depleted. This may have caused a compensatory hyperactivity in the surviving innervation preventing a further increase due to the footshock stress. The data as presented in Fig. 8, however, show no proof for this possibility. (3) Most probably, the difference lies in the fact that our footshock parameters were sufficient to activate the DA innervation of the NAS in control animals, in contrast to the current used in your study; if the NAS DA innervation was already activated to some degree, it is possible that it would not be hyperresponsive because the affinity of TH for its cofactor was already altered in the same manner as would result from cortical DA depletion.

CHAPTER 13

Is it possible to repair the damaged prefrontal cortex by neural tissue transplantation?

Stephen B. Dunnett

Department of Experimental Psychology, University of Cambridge, Cambridge CB2 3EB, UK

Introduction

The last 3 years have seen the first claims that transplanted neural tissues can provide a beneficial clinical treatment for patients with the neurodegenerative motor disorder of Parkinson's disease (Madrazo et al., 1987; Lindvall et al., 1987; Goetz et al., 1989). This has led to speculation about whether a similar strategy might be applicable for other neurological and psychiatric disorders. In the present context, this includes disorders in which the prefrontal cortex has been implicated, such as neuropsychological dysfunction that can result from traumatic or neurosurgical damage to the prefrontal cortex, and possibly also neurodegenerative diseases of ageing (in particular Alzheimer's disease), and conditions involving pathological disturbance of neurochemical function in the cortex (such as catecholamine disturbance in schizophrenia). The present chapter attemps to evaluate such speculation in the light of studies of grafting neural tissues in experimental animals.

Functional repair with neural grafts

The earliest attempts at transplantation of neural tissues in the mammalian brain used the neocortex as a key source of donor tissue and target site for the grafts. In 1890, Thompson took cortical tissue from adult cats and implanted the tissue pieces into the neocortex of adult dogs. He reported survival of the transplanted tissues for a period of several weeks in the host brain. However, with hindsight, it is likely that only glial and meningeal cells (rather than neurones) survived the transplantation process.

The first unequivocal cases of successful transplantation of mammalian brain tissue were reported by Elizabeth Dunn in 1917. She implanted wedges of cortical tissue taken from neonatal rat pups into cavities made in the neocortex of littermates. Although the success rate was less than 10%, her 4 successful cases all involved exposure of the lateral ventricles and apposition of the graft tissue to the choroid plexus, which has a rich vascular supply and so can provide nutrients for the isolated graft tissue. A second factor in her success was almost certainly the identification of the developing brain as a suitable source of donor tissue, with greater growth capacity and tolerance of anoxia than mature neurones. These two factors have subsequently been characterised as critical conditions for viability of neural grafts implanted in the central nervous system (Stenevi et al., 1976).

Once the conditions for reliable neural transplantation had finally become established in the early 1970s (Olson and Malmfors 1970; Das

---

Correspondence: Dr. S.B. Dunnett, Department of Experimental Psychology, University of Cambridge, Downing Street, Cambridge CB2 3EB, UK.

and Altman, 1971; Björklund and Stenevi, 1971; Lund and Hauschka, 1976), the techniques were initially used to investigate issues related to the mechanisms of development and the regenerative capacity of the central nervous system. The first demonstrations that such grafts can exert a functional influence over the behaviour of the host animal came nearly a decade later. Thus, grafts of embryonic dopamine cells to the rat striatum can reverse motor impairments induced by unilateral lesion of the intrinsic forebrain dopamine system (Björklund and Stenevi, 1979; Perlow et al., 1979). It remains the case that the most extensive studies of functional capacity of neural grafts have been conducted on model systems involving the diffusely projecting monoamine systems of the brainstem (Freed, 1983; Dunnett et al., 1985). However, neural grafts have subsequently been reported to provide a beneficial influence over host function in a variety of other model systems involving neuroendrocine, regulatory, sensory, motor, learning, memory and cognitive functions following genetic, foetal, toxic, traumatic, and neurodegenerative damage in the central nervous system (for reviews, see Sladek and Gash, 1984; Björklund and Stenevi, 1985; Azmitia and Björklund, 1987; Gash and Sladek, 1988; Dunnett and Richards, 1990).

Mechanisms of repair

Grafted tissues might influence the behavioural capacities of the host animal by a variety of mechanisms. Several schemes have been proposed related to the mechanisms of development and the regenerative capacity of the central nervous system (Freed et al., 1985; Gash et al., 1985; Dunnett and Björklund, 1987). The following alternatives are based on the review provided by Björklund et al. (1987):

(i) *Non specific or negative consequences of the implantation surgery*. These may include graft growth inducing space-occupying lesions, cyst and scar formation, changes in the blood – brain barrier, or induction of further degenerative changes in the host brain. Rosenstein and Brightman (1983) have proposed that the disturbance of the blood – brain barrier by neural grafts may provide a means to deliver drugs that do not normally penetrate the brain to discrete CNS targets. This issue is controversial, since other studies have suggested that neural grafts (at least when transplanted between animals of the same strain) reestablish an intact blood – brain barrier, which only becomes permeable when the grafts undergo rejection. As a second example of non-specific effects of transplantation, tumour-like growth of grafts may occasionally induce space-occupying lesions that can further disturb normal function (Ridley et al., 1988), although fortunately such observations are rare.

(ii) *Trophic actions on the host brain*. The acute secretion of trophic factors and migration of glial cells into the host brain may reduce lesion-induced cell death or promote functional reorganization and recovery within the host brain, independently of any sustained effect of the grafted neurones. For example, adrenal grafts have been reported to induce sprouting of host mesotelencephalic dopamine neurones, which may provide the mechanism by which these grafts reverse motor deficits in MPTP-treated mice (Bohn et al., 1987).

(iii) *Diffuse release of hormones or transmitters*. The grafted cells may provide a chronic secretion of deficient neuroactive chemicals such as hormones or neurotransmitters to the host brain. In effect, the graft acts like a biological mini-pump. Hypothalamic implantation of neuroendocrine secreting tissues may well work by such a mechanism. Similarly intraventricular placement of nigral grafts which do not establish connections with the host brain have been considered to exert their influence by diffuse release of deficient catecholamines into the host parenchyma (Freed et al., 1980).

(iv) *Diffuse reinnervation of the host brain by the grafts*. In this case the cells of the graft grow and extend neurites into the host brain and form synaptic connections with host targets. It is likely that the range of beneficial effects of intrastriatal dopamine grafts is dependent on the observation that their axons establish morphologically appropriate contacts with striatal medium spiny neurones (Freund et al., 1985; Clarke et al., 1988). However, although it is likely that such grafts receive some local influences from the host brain by receiving reciprocal axodendritic and axoaxonic contacts from the host striatum (Clarke et al., 1988; Sirinathsinghji and Dunnett, 1989), nigral grafts cannot fully reconstruct the damaged nigrostriatal circuitry when placed in an ectopic striatal location, and they do not reverse all symptoms associated whith forebrain dopamine depletion (Dunnett et al., 1989).

(v) *Reciprocal reformation of afferent and efferent connections between the graft and the host brain*. The ultimate goal in transplantation attempts to repair brain damage would be to achieve reconstruction of the damaged neural network in the brain, i.e. for the graft to develop a normal internal organisation and to develop a full set of appropriately organised afferent and efferent connections with the host brain such that it can fully restore the normal neuronal circuitry. This is probably unachievable, at least with present techniques. Perhaps the closest approach to such reconstruction has been with striatal grafts implanted in the lesioned neostriatum. Such grafts do develop both afferent and efferent connections with the host brain, and can ameliorate deficits in cognitive as well as motor tasks that are dependent on the integrity of transstriatal connectivity (Isacson et al., 1986; Björklund et al., 1987; Dunnett et al., 1988).

The variety of mechanisms by which grafts can provide their functional effects, which have been identified primarily in the study of subcortical model systems, provide a framework within which any structural repair and functional changes induced by grafts implanted into the prefrontal cortex may be evaluated.

Structural repair with cortical grafts

Anatomical reorganisation within cortical grafts

The earliest detailed descriptions of cortical grafts were provided by LeGros Clark (1940). He transplanted foetal rabbit cortex to 6-week-old hosts, using a cannula/plunger technique that is still widely used with considerable success (e.g. Smith and Ebner, 1986). The embryonic neurones within the grafts developed towards a mature cytologic phenotype and were seen (in several cases) to become organised into clusters or layers of cells of particular types in "a somewhat indistinct laminar pattern . . . characteristic of the normally developed cerebral cortex". Glees (1940) confirmed this capacity of embryonic cortical neuroblasts to continue differentiation when implanted onto the host cortical surface under the pia and observed one cortical graft to develop many features of the normal cortical organisation. For example, a wedge-shaped zone within this case was comprised of a narrow compact layer of pyramidal cells lying in immediate relation to a parallel row of small granular cells.

More recently, Golgi staining of neurones in the grafts has confirmed the occurrence of large pyramidal and both spiny and aspiny medium-sized neurones characteristic of the types seen in the intact neocortex (Jaeger and Lund, 1981; Floeter and Jones, 1984). However, a clear laminar organisation in cortical grafts is only rarely observed. A more common pattern is for grafted cortical cells to become organised into clusters (Floeter and Jones, 1984; Jaeger and Lund, 1980, 1981; Stein and Mufson, 1987). These clusters may be separated by bands of myelinated fibres in the depths of the graft and cell-poor zones reminiscent of a molecular layer, in particular at the graft host border, so that the neuronal clusters appear layered, but a clear lamination of cells within the grafts is generally denied. Although the majority of studies have employed solid grafts implanted into cortical cavities or as plugs into cortical or subcortical sites, a similar reaggregation of neurones into disorganised clusters is seen when the grafts are implanted as dissociated suspensions of embryonic cortical cells (Floeter and Jones, 1984; Sofroniew et al., 1986). To the extent that the cells within the grafts develop a laminar organisation, the layers do not show a particular alignment with the host neocortical laminae, unless the graft has been carefully implanted so as to maintain its normal orientation (Andres and Van der Loos, 1985; Chang et al., 1986).

Several recent studies have employed immunocytochemical staining to identify specific populations of neurones within cortical grafts implanted in the neocortex, including glutamic acid decarboxylase, choline acetyltransferase, NADPH-diaphorase, vasoactive intestinal polypeptide, neuropeptide Y, cholecystokinin, pancreatic polypeptide and somatostatin immunoreactive neurones (Ebner et al., 1984; Floeter and Jones, 1985; Gonzalez and Sharp, 1987; Sharp et al., 1987; Stein and Mufson, 1987), all of which are characteristic of normal cortex. Conversely, antibodies against other peptides that have not been found in the intact neocortex, including substance P, α-melanocyte or corticotropin-stimulating hormones, β-endorphin or arginine-vasopressin, do not label cells in the cortical grafts (Ebner et al., 1984).

Grafts of other neuronal populations implanted in the neocortex retain their characteristic cell populations. Thus, for example, Fine et al. (1985) implanted cholinergic-rich basal forebrain neurones into the neocortex of cholinergically depleted rats and confirmed good survival of choline acetyltransferase immunoreactive neurones. In this case somatostatin, enkephalin and neuropeptide Y immunoreactivity was observed within the basal forebrain grafts, but no staining for substance P, neurotensin or vasoactive intestinal polypeptide, the last of which is characteristic of the normal neocortex.

Thus a variety of morphological and immunohistochemical markers indicate that many different types of neurone which are included in the tissue graft retain predetermined expression of particular neurotransmitters, independently of whether those cell types are characteristically found in the normal cortex, whereas cortical tissues themselves develop the apparently full range of normal basic cell types.

Connectivity of cortical grafts

Fewer studies have investigated the formation of afferent and efferent connections between cortical grafts and the adult host brain. Most consistent are reports that the host brain can establish afferent inputs to cortical tissue grafts. Thus, cortical grafts have been seen to be reinnervated by NADPH-diaphorase fibres from the host cortex (Sharp et al., 1986) and by acetylcholinesterase-positive fibres, of presumed basal forebrain origin (Sofroniew et al., 1986; Dunnett et al., 1987; Gibbs and Cotman, 1987; Stein and Mufson, 1987). HRP labelling of the grafts has revealed retrograde labelling of cells in the contraleteral cortex and in subcortical thalamic, basal forebrain, locus coeruleus and raphe nuclei (Labbe et al., 1983, Gibbs et al., 1985; Dunnett et al., 1987). The thalamic inputs are generally sparse, but can be dramatically potentiated by basal forebrain lesions (Höhmann and Ebner, 1988).

By contrast to their clear attraction of afferent ingrowth, reciprocal efferent connections of cortical grafts implanted in the adult nervous system are less well established. However, sparse projections to the thalamus, amygdala and hippocampus in the adult brain have been suggested by both anterograde and retrograde HRP labelling (Gibbs et al., 1985; Dunnett et al., 1987; Gonzalez et al., 1988; Escobar et al., 1989). Conversely, extensive outgrowth from subcortical monoaminergic grafts implanted in the adult cortex has been well established by histochemical and immunohistochemical labelling, both for dopamine-rich (Dunnett et al., 1984; Herman et al., 1986) and cholinergic-rich graft tissues (Fine et al., 1985; Clarke and Dunnett, 1986; Dunnett et al., 1986).

Electrophysiological studies of cortical grafts

Electrophysiolocial techniques have been employed in several studies of the intrinsic organisation and functional connectivity of neuronal grafts in the neocortex. Attempts have been made to determine the degree to which instrinsic activity in the grafts is organised similar to that observed in the intact cortex. Moreover, as an adjunct to anatomical tracing techniques, physiological recording permits the determination of the degree to which patterned information can be relayed between grafts and the host brain.

In one remarkable series of studies, Bragin (1986; Bragin et al., 1987) made aspirative lesions of the barrelfield of adult rats' somatosensory cortex followed by transplantation of isotopic embryonic cortex to the lesion cavity. Microelectrodes were implanted into the grafts 2 – 3 months later to allow time for host – graft connections to become established. Background activity of cells in the graft consisted of low frequency randomly distributed discharges similar to those observed in the contralateral intact cortex. More dramatically, cells in the grafts responded by bursts of firing to vibrissae stimulation, indicating the convergence of inputs from peripheral receptors into the grafts. The main difference with intact neocortex was not the pattern of firing per se, but that grafted cells had larger receptive fields, compatible with the histological observations that grafted cells were not organised into detectable barrel-like zones. Confirmation of these observations has recently been obtained by Levin et al. (1987) using 2-deoxyglucose autoradiography to determine metabolic activity. These latter authors demonstrated that cortical grafts located in somatosensory cortex showed focal areas of increased 2-deoxyglucose uptake (by approx. 43%) in response to vibrissae stimulation. The specificity of this effect was demonstrated by the use of control grafts of non-cortical tissue in the somatosensory cortex or of cortical tissue in non-somatosensory sites, both of which failed to show any detectable response to stimulation.

In addition to the study of cortical tissue grafts, other tissues that normally innervate or are innervated by the neocortex have also been seen to reestablish functional connections with the host cortex following transplantation. For example, Hamasaki et al. (1987a, b) demonstrated reciprocal connections between lateral geniculate nucleus grafts and host occipital cortex by electrical stimulation and recording in slice preparations. Similarly, Harvey et al. (1982) found that when tectal tissue was implanted over the superior colliculus, electrical stimulation of the host occipital cortex orthodromically excited 25/214 single units within the grafts at latencies (< 15 msec, mean 7.3 msec) suggesting a monosynaptic input. In both of these studies the implants were made into neonatal hosts. However, even in adult hosts, physiologically effective sprouting of cortical neurones has been observed into striatal grafts (Rutherford et al., 1987).

The large majority of these anatomical and electrophysiological studies of various tissue implanted into the neocortex have employed motor, parietal, and occipital placements. Although it may be assumed that similar patterns of reorganisation may apply in prefrontal sites, this particular area of cortex has received scant attention, anatomically. Fortunately, there are a few more functional studies of cortical grafts, perhaps because of the rich psychological literature on the functions of this area in rats (for review, see Kolb, 1984).

Functional repair after prefrontal lesions

In view of the complex precision of the columnar organisation that has been thought to underlie the functional processing subserved by the neocortex (Mountcastle, 1979), one of the most dramatic suggestions of recent research has been the apparent capacity of cortical grafts to ameliorate some complex learning deficits resulting from prefrontal damage. Stein and colleagues first showed that rats' abilities to learn a delayed alternation task in a T maze, which are disrupted by aspirative lesions of the frontal cortex, could be substantially restored following transplantation of embryonic cortical tissue into the lesion cavity (Labbe et al., 1983; Stein et al., 1988). Similar effects following transplantation of embryonic cortical tissues have been reported on the recovery of visual brightness

(Stein et al., 1985) and pattern (Haun et al., 1985) discrimination following occipital cortex lesions, of taste aversion learning following gustatory neocortex lesions (Bermudez-Rattoni et al., 1987; Escobar et al., 1989), and of spatial maze learning following allocortical (hippocampal) lesions (Kimble et al., 1986).

Such dramatic recovery on complex learning tasks, when taken together with the observed formation of afferent and efferent connections between cortical grafts and the damaged host brain, make it tempting to suggest that the grafts influence recovery by means of a functional reconstruction of damaged cortical neural circuitries. However this conclusion is premature.

Temporal factors in recovery

In their first report, Labbe et al. (1983) implanted pieces of embryonic frontal cortex into an aspirative cavity of prefrontal cortex, then trained the rats on a T-maze alternation task for water reward. These animals were able to learn the task to criterion more rapidly than either rats with lesions alone or rats with control grafts of embryonic cerebellar tissue. In this study, 7 days separated the lesion and transplantation surgeries, and the behavioural training commenced 4 days later. Whereas the capacity of similar grafts to ameliorate cognitive learning has been replicated under similar specific conditions (Kesslak et al., 1986a; Dunnett et al., 1987; Kolb et al., 1988; Stein et al., 1988), both of these temporal factors have been found to be critical.

Firstly, the grafts are only effective in studies where behavioural testing commences within a few days of transplantation surgery and not when behavioural testing is delayed by 4 weeks or more. This dissociation was first observed by Stein et al. (1985) in a study involving occipital cortex lesions. Frontal cortex graft tissues implanted in the occipital cavity reduced the host animals' deficits in a brightness discrimination task 2 weeks after surgery, but did not influence their severe deficits in learning a pattern discrimination commencing 4 weeks later.

In this study by Stein et al. (1985), the passage of time is confounded with serial training effects and with task difficulty between the two blocks of testing. A more systematic comparison of this factor was therefore conducted by Kolb et al. (1988). They gave 4 groups of rats prefrontal lesions, two of which received foetal cortical transplants 2 weeks later. The animals were then trained in the Morris water maze commencing either immediately (early-test groups) or 4 weeks (late-test groups) after transplantation surgery. The lesions alone initially produced marked deficits in maze learning, which underwent some spontaneous recovery in the late-test groups. Whereas the early-test transplant rats showed a significant improvement over their lesion control group, they did not differ from the late-test transplant group which was significantly impaired with respect to the recovery that had taken place in the late-test controls. Thus, the benefit provided by the transplants is only apparent when assessed immediately after surgery, and at later times the grafts may actually add to the host animals' impairments.

In our own studies (Dunnett et al., 1987), we have employed the delayed alternation task, as in the original study by Labbe, Stein and colleagues. We replicated the improvement of delayed alternation learning in the rats with prefrontal lesions and grafts when training commenced 7 days after the lesion. However, as in the Kolb study, the rats with lesions alone underwent considerable spontaneous recovery, in our tests over 3 – 6 months, and the grafted animals were significantly impaired with respect to lesioned animals when tested long-term after lesion and graft surgery. This was true both for grafted animals that were naive to the test paradigm at the time of long-term test and for the rats which showed benefit on the short-term test when retested after a long-term delay.

The importance of these observations is twofold. Firstly, they suggest that recovery on these cognitive and learning tasks induced by grafting in the prefrontal cortex is only a transient phenomenon. Secondly, although cortical tissue grafts can establish afferent and efferent connections with the host brain, it is unlikely that the recovery is attributable to the reformation of such connections. Although conducted primarily on subcortical systems, all studies on the time course of axonal growth and synaptic connectivity between grafts and the host brain suggest that graft axons do not cross the graft–host border until 1–2 weeks after transplantation, and extensive connectivity does not approach asymptotic levels until 3–6 months later. Thus, the time course of graft–host connectivity matches the time of disappearance of benefit and the development of impairment, rather than the time when recovery is seen.

The reason that grafts in the prefrontal cortex induce long-term deficits over and above those seen in rats with prefrontal lesions alone is not clear, although several different mechanisms might be considered (Dunnett et al., 1987): (a) The grafts might inhibit spontaneous reorganisation and recovery processes in the host brain. (b) Growth and expansion in graft volume may compress or distort adjacent cortical and subcortical structures that are otherwise intact. (c) The neural activity provided by graft-derived ingrowth may contribute noise, rather than any useful patterned information, in the precisely organised circuits of host neocortex that remains intact. At present, there is little information that enables these alternatives to be resolved.

Secondly, recovery has only been observed when the graft surgery is conducted within 7–14 days of the lesions, and not with either shorter or longer intervals. Three studies have manipulated this factor specifically. First, Kesslak et al. (1986a) found that cortical tissue was only effective in ameliorating deficits in alternation learning following prefrontal damage when the implants were made 7 days after the lesions and not when implanted in the same surgical session. Secondly, in our own studies, we found recovery only when the implants were made 7 days after the lesion and not when made with a 28-day delay (Dunnett et al., 1987). Finally, Stein et al. (1988) considered 4 different delay intervals and found significant amelioration of the lesion deficits when the cortical implants were made 7 or 14 days after lesion surgery, but not when made with 30- or 60-day delays. These observations suggest that the grafts exert some trophic influence on the development of the lesion itself (Kesslak et al., 1986a; Stein et al., 1985, 1988; Dunnett et al., 1987).

Trophic factors in recovery

Cotman and colleagues have demonstrated that aspirative lesions of the cortex result in the secretion of both toxic and trophic factors that can influence the survival of cultured neurones in in vitro assays (Nieto-Sampedro et al., 1982; Manthorpe et al., 1983). The secretion of wound-derived trophic factors reaches a peak in the injured brain 7–10 days following the lesions, and it has been argued that timing of transplant surgery to coincide with this peak can markedly enhance graft viability (Manthorpe et al., 1983; Nieto-Sampedro et al., 1984).

It could therefore be the case that the reason why the lesion–transplantation delay is critical in the studies of prefrontal cortex is that this is necessary to achieve good graft viability. Indeed in the study by Stein et al. (1988), using embryonic donors of 19 days gestational age, the grafts manifested good survival only in the 7-day delay group, whereas survival was poor in the 14-, 30- and 60-day groups. However, survival alone does not appear to be the critical factor. For example, our own studies indicated better survival with 30-day delay than with 7-day delay if younger donor tissue (age 16–17 days) is used, although the former grafts had no functional benefit.

An alternative explanation is that the embryonic graft tissue itself secretes trophic factors that can influence both the acute toxic consequences and the course of recovery from the lesion. In support of this hypothesis, Kesslak et al. (1986b) found that implants of purified cultured astrocytes derived from embryonic brain or of adult tissue grafts predominantly comprised of glia were as effective as transplants of embryonic cortex in reducing deficits in delayed alternation learning. Perhaps the clearest evidence that cortical grafts are influencing the course of recovery in host systems is provided by a further study by Stein (1987). He trained 4 groups of rats with prefrontal lesions and cortical grafts in the Morris water maze task, under conditions in which the grafts provided a significant benefit. However, in one of the transplanted groups the grafts were removed just before training commenced. These animals performed as well as the group with grafts intact, and significantly better than the rats with lesions alone.

In conclusion, the limited number of studies that have so far investigated the functional capacity of neural grafts in the prefrontal cortex indicate that cortical tissue grafts can ameliorate some deficits induced by prefrontal lesions. However, the benefit is only provided in restricted circumstances, is not long-lasting, and does not apply to a wide range of tests. Moreover, it is likely that the acute benefit is not provided by any specific replacement of damaged neuronal circuitry, but rather by some trophic interaction with the dynamic development of the consequences of the primary lesion. Although the mechanisms by which this occurs are poorly understood, it is possible that they relate to the trophic support necessary to maintain connections in the developing and adult brain.

Trophic inhibition of retrograde degeneration

The potential for cortical grafts to provide trophic support for intrinsic neurones which have lost their normal targets was first studied extensively in the developing thalamus. Neonatal lesions of frontal or occipital cortex result in developmental atrophy of the corresponding afferent nuclei of the thalamus. Haun and Cunningham (1984, 1987) found that 5 days after transplanting embryonic neocortex into the cavity formed by a neonatal occipital lesion, atrophy in the corresponding dorsal lateral geniculate nucleus of the host was markedly attenuated. Non-cortical (cerebellar) control grafts were ineffective in preventing the lesion-induced atrophy. In these studies, using cell suspension grafts of the cortical cells, the protection was only temporary. However, Sharp and Gonzalez (1986) have achieved permanent prevention of retrograde thalamic atrophy using cortical implants into neonatal frontal cortical lesions. Although these two groups have employed different cortical lesion sites, Sharp and Gonzalez considered that the most likely reason for more lasting protection in their study was due to the use of solid tissue implants. These observations indicate that the trophic interactions necessary for survival of developing neurones and lost by removal of appropriate targets, can be reestablished by transplant-derived replacement of those targets. It may be necessary for developing host thalamic neurones to innervate the grafts to achieve such protection, although it cannot at present be excluded that the influence is entirely attributable to diffusable neurotrophic factors. Nor is it known if degenerative changes would be reinitiated if the graft were removed.

However, these observations relate primarily to a developmental context rather than protection against retrograde degeneration following axotomy of target removal in adulthood. This issue has been investigated in the magnocellular cholinergic cells of the nucleus basalis which atrophy in response to extensive loss of cortical targets, whether made by mechanical devascularisation or excitotoxic lesion (Sofroniew et al., 1983; Sofroniew and Pearson, 1985). Cortical cell suspensions implanted in the damaged cortex have the capacity to completely prevent this retrograde atrophy of the cholinergic neurones of the host nucleus basalis system, which sprouted to extensively reinnervate the cortical tissue grafts (Sofroniew et al., 1986). These observations support the notion that target-derived trophic factors are necessary for the maintenance of neural connections in the mature central nervous system as well for their normal development, and can be substituted by neural tissue grafts.

Summary and conclusions

The techniques are now well established for the viable transplantation of cortical and other neural tissues into the neonatal and adult cortex, at least in the laboratory rat. Under appropriate conditions such grafts survive well and can establish reciprocal connections with the host brain. On this basis, neural transplantation has become a powerful technique for the study of mechanisms involved in the development of the central nervous system and its capacity for regeneration after injury. Moreover, a variety of anatomical, electrophysiological and behavioural techniques suggest that grafted neural tissue may sustain functional interactions with the host brain. However, the extent and duration of recovery using present techniques is extremely limited. It remains undetermined whether such experimental observations may ever acquire therapeutic application.

Acknowledgements

The support of the Mental Health Foundation and the Medical Research Council is gratefully acknowledged.

References

Andres, F.L. and Van der Loos, H. (1985) Removal and reimplantation of the parietal cortex of the neonatal mouse: consequences for the barrelfield. *Dev. Brain Res.*, 20: 115 – 121.

Azmitia, E.C. and Björklund, A. (1987) *Cell and Tissue Transplantation into the Adult Brain. Ann. N.Y. Acad. Sci.*, Vol. 495.

Bermudez-Rattoni, F., Fernandez, J., Sanchez, M.A., Aguilar-Roblero, R. and Drucker-Colin, R. (1987) Fetal brain transplants induce recuperation of taste averions learning. *Brain Res.*, 416: 147 – 152.

Björklund, A. and Stenevi, U. (1971) Growth of central catecholamine neurones into smooth muscle grafts in the rat mesencephalon. *Brain Res.*, 31: 1 – 20.

Björklund, A. and Stenevi, U. (1979) Reconstruction of the nigrostriatal dopamine pathway by intracerebral nigral transplants. *Brain Res.*, 177: 555 – 560.

Björklund, A. and Stenevi, U. (1985) *Neural Grafting in the Mammalian CNS*. Elsevier, Amsterdam.

Björklund, A., Lindvall, O., Isacson, O., Brundin, P., Wictorin, K., Strecker, R.E., Clarke, D.J. and Dunnett, S.B. (1987) Mechanisms of action of intracerebral neural implants: studies on nigral and striatal grafts to the lesioned striatum. *Trends Neurosci.*, 10: 509 – 516.

Bohn, M.C., Cupit, L., Marciano, F. and Gash, D.M. (1987) Adrenal medulla grafts enhance recovery of striatal dopaminergic fibers. *Science*, 237: 913 – 916.

Bragin, A.G. (1986) Neuronal responses of embryonic rat somatosensory neocortex grafted into adult rat barrelfield. *Neurophysiologia*, 18: 833 – 836. (In Russian.)

Bragin, A.G., Bohne, A. and Pavlik, V.D. (1987) Electrophysiological indexes of the degree of grafted neural tissue integration with the host brain. *Neurophysiologia*, 19: 498 – 504. (In Russian.)

Chang, F.L., Steedman, J.G. and Lund, R.D. (1986) The lamination and connectivity of embryonic cerebral cortex transplanted into newborn rat cortex. *J. Comp. Neurol.*, 244: 401 – 411.

Clarke, D.J. and Dunnett, S.B. (1986) Ultrastructural organization of choline acetyltransferase-immunoreactive fibres innervating the neocortex from embryonic ventral forebrain grafts. *J. Comp. Neurol.*, 250: 192 – 205.

Clarke, D.J., Brundin, P., Strecker, R.E., Nilsson, O.G., Björklund, A. and Lindvall, O. (1988) Human foetal dopamine neurones grafted in a rat model of Parkinson's disease: ultrastructural evidence for synapse formation using tyrosine hydroxylase immunocytochemistry. *Exp. Brain Res.*, 73: 115 – 126.

Das, G.D. and Altman, J. (1971) Transplanted precursors of nerve cells: their fate in the cerebellums of young rats. *Science*, 173: 637 – 638.

Dunn, E.H. (1917) Primary and secondary findings in a series of attempts to transplant cerebral cortex in the albino rat. *J. Comp. Neurol.*, 27: 565 – 582.

Dunnett, S.B. and Björklund, A. (1987) Mechanisms of function of neural grafts in the adult mammalian brain. *J. Exp. Biol.*, 132: 265 – 289.

Dunnett, S.B. and Richards, S.-J. (1990) Neural Transplantation: From Molecular Basis to Clinical Applications. *Progr. Brain Res.*, Vol. 82, Elsevier, Amsterdam.

Dunnett, S.B., Bunch, S.T., Gage, F.H. and Björklund, A. (1984) Dopamine-rich transplants in rats with 6-OHDA lesions of the ventral tegmental area. I. Effects on spontaneous and drug-induced locomotor activity. *Behav. Brain Res.*, 13: 71 – 82.

Dunnett, S.B., Björklund, A., Gage, F.H. and Stenevi U. (1985) Transplantation of mesencephalic dopamine neurones to the striatum of adult rats. In A. Björklund and U. Stenevi (Eds.), *Neural Grafting in the Mammalian CNS*, Elsevier, Amsterdam, pp. 451 – 469.

Dunnett, S.B., Whishaw, I.Q., Bunch, S.T. and Fine, A. (1986) Acetylcholine-rich neuronal grafts in the forebrain of rats: effects of environmental enrichment, neonatal noradrenaline depletion, host transplantation site and regional source of embryonic donor cells on graft size and acetylcholinesterase-positive fibre outgrowth. *Brain Res.,* 378: 357–373.

Dunnett, S.B., Ryan, C.N., Levin, P.D., Reynolds, M. and Bunch, S.T. (1987) Functional consequences of embryonic neocortex transplanted to rats with prefrontal cortex lesions. *Behav. Neurosci.,* 101: 489–503.

Dunnett, S.B., Isacson, O., Sirinathsinghji, D.J.S., Clarke, D.J. and Björklund, A. (1988) Striatal grafts in rats with unilateral neostriatal lesions. III. Recovery from dopamine-dependent motor asymmetry and deficits in skilled paw reaching. *Neuroscience,* 24: 813–820.

Dunnett, S.B., Rogers, D.C. and Richards, S.J. (1989) Reconstruction of the nigrostriatal pathway after 6-OHDA lesions by combination of dopamine-rich nigral grafts and nigrostriatal "bridge" grafts. *Exp. Brain Res.,* 75: 523–535.

Ebner, F.F., Olschowska, J.A. and Jacobowitz, D.M. (1984) The development of peptide-containing neurons within neocortical transplants in adult mice. *Peptides,* 5: 103–113.

Escobar, M., Fernandez, J., Guevara-Aguilar, R. and Bermudez-Rattoni, F. (1989) Fetal brain grafts induce recovery of learning deficits and connectivity in rats with gustatory neocortex lesions. *Brain Res.,* 478: 368–374.

Fine, A., Dunnett, S.B., Björklund, A., Clarke D. and Iversen, S.D. (1985) Transplantation of embryonic ventral forebrain neurones to the neocortex of rats with lesions of nucleus basalis magnocellularis. I. Biochemical and anatomical observations. *Neuroscience,* 16: 769–786.

Floeter, M.K. and Jones, E.G. (1984) Connections made by transplants to the cerebral cortex of rat brains damaged in utero. *J. Neurosci.,* 4: 141–150.

Floeter, M.K. and Jones, E.G. (1985) Transplantation of fetal postmitotic neurons to rat cortex: survival, early pathway choices and long-term projections of outgrowing axons. *Dev. Brain Res.,* 22: 19–38.

Freed, W.J. (1983) Functional brain tissue transplantation: reversal of lesion-induced rotation by intraventricular substantia nigra and adrenal medulla grafts, with a note on intracranial retinal grafts. *Biol. Psychiat.,* 18: 1205–1267.

Freed, W.J., Perlow, M.J., Karoum, F., Seiger, Å., Olson, L., Boffer, B.J. and Wyatt, R.J. (1980) Restoration of dopaminergic function by grafting fetal rat substantia nigra to the caudate nucleus: long-term behavioral, biochemical, and histochemical studies. *Ann. Neurol.,* 8: 510–519.

Freed, W.J., de Medinaceli, L. and Wyatt, R.J. (1985) Promoting functional plasticity in the damaged nervous system. *Science,* 277: 1544–1552.

Freund, T.F., Bolam, J.P., Björklund, A., Stenevi, U., Dunnett, S.B., Powell, J.F. and Smith, A.D. (1985) Efferent synaptic connections of grafted dopaminergic neurones reinnervating the host neostriatum: a tyrosine hydroxylase immunocytochemical study. *J. Neurosci.* 5: 603–616.

Gash D.M. and Sladek, J.R. (1988) *Transplantation into the Mammalian CNS. Progress in Brain Research, Vol. 78,* Elsevier, Amsterdam.

Gash, D.M., Collier, T.J. and Sladek, J.R. (1985) Neural transplantation: a review of recent developments and potential applications to the aged brain. *Neurobiol. Aging,* 6: 131–150.

Gibbs, R.B. and Cotman, C.W. (1987) Factors affecting survival and outgrowth from transplants of entorhinal cortex. *Neuroscience,* 21: 699–706.

Gibbs, R.B., Harris, E.W. and Cotman, C.W. (1985) Replacement of damaged cortical projections by homotypic transplants of entorhinal cortex. *J. Comp. Neurol.,* 237: 47–65.

Glees, P. (1940) The differentiation of the brain and other tissues in an implanted portion of embryonic head. *J. Anat.,* 75: 239–247.

Goetz, C.G., Olanow, C.W., Koller, W.C., Penn, R.D., Cahill, D., Morantz, R., Stebbins, G., Tanner, C.M., Klawans, H.L., Shannon, K.M., Comella, C.L., Witt, T., Cox, C., Waxman, M. and Gauger, L. (1989) Multicenter study of autologous adrenal medullary transplantation to the corpus striatum in patients with advanced Parkinson's disease. *New Engl. J. Med.,* 320: 337–341.

Gonzalez, M.F. and Sharp, F.R. (1987) Fetal frontal cortex transplanted to injured motor/sensory cortex of adults rat. I. NADPH-diaphorase neurons. *J. Neurosci.,* 7: 2991–3001.

Gonzalez, M.F., Sharp, F.R. and Loken, J.E. (1988) Fetal frontal cortex transplanted to injured motor/sensory cortex of adult rats: reciprocal connections with host thalamus demonstrated with WGA-HRP. *Exp. Neurol.,* 99: 154–165.

Hamasaki, T., Hirakawa, K. and Toyama, K. (1987a) Electrophysiological and histological study of synaptic connections between lateral geniculate transplant and host visual cortex. *Appl. Neurophysiol.,* 50: 463–464.

Hamasaki, T., Komatsu, Y., Yamamoto, N., Nakajima, S., Hirakawa, K. and Toyama, K. (1987b) Electrophysiological study of synaptic connections between a transplanted lateral geniculate nucleus and the visual cortex of the host rat. *Brain Res.,* 422: 172–177.

Harvey, A.R., Golden, G.T. and Lund, R.D. (1982) Transplantation of tectal tissue in rats. III. Functional innervation of transplants by host afferents. *Exp. Brain Res.,* 47: 437–445.

Haun, F. and Cunningham, T.J. (1984) Cortical transplants reveal CNS trophic interactions in situ. *Dev. Brain Res.,* 15: 290–294.

Haun, F. and Cunningham, T.J. (1987) Specific neurotrophic interactions between cortical and subcortical visual structures in developing rat: in vivo studies. *J. Comp. Neurol.,* 256: 561–569.

Haun, F., Rothblat, L.A. and Cunningham, T.J. (1985) Visual cortex transplants in rats restore normal learning of a difficult visual pattern discrimination. *Invest. Ophthalmol. Vis. Sci.,* 26, Suppl. 3: 288.

Herman, J.P., Choulli, K., Geffard, M., Nadaud, D., Taghzouti, K. and LeMoal, M. (1986) Reinnervation of the nucleus accumbens and frontal cortex of the rat by dopaminergic grafts and effects on hoarding behavior. *Brain Res.*, 372: 210–216.

Höhmann, C.F. and Ebner, F.F. (1988) Basal forebrain lesions facilitate adult host fiber ingrowth into neocortical transplants. *Brain Res.*, 448: 53–66.

Isacson, O., Dunnett, S.B. and Björklund, A. (1986) Graft-induced behavioral recovery in an animal model of Huntington disease. *Proc. Natl. Acad Sci. (U.S.A.)*, 83: 2728–2732.

Jaeger, C.B. and Lund, R.D. (1980) Transplantation of embryonic occipital cortex to the tectal region of newborn rats: a light microscopic study of organization and connectivity of the transplants. *J. Comp. Neurol.*, 194: 571–597.

Jaeger, C.B. and Lund, R.D. (1981) Transplantation of embryonic occipital cortex to the tectal region of newborn rats: a Golgi study of mature and developing transplants. *J. Comp. Neurol.*, 194: 571–597.

Kesslak, J.P., Brown, L., Steichen, C. and Cotman, C.W. (1986a) Adult and embryonic frontal cortex transplants after frontal cortex ablation enhance recovery on a reinforced alternation task. *Exp. Neurol.*, 94: 615–626.

Kesslak, J.P., Nieto-Sampedro, M., Globus, J. and Cotman, C.W. (1986b) Transplants of purified astrocytes promote behavioral recovery after frontal cortex ablation. *Exp. Neurol.*, 92: 377–390.

Kimble, D.P., Bremiller, R. and Stickrod, G. (1986) Fetal brain implants improve maze performance in hippocampal-lesioned rats. *Brain Res.*, 363: 358–363.

Kolb, B.J. (1984) Functions of the frontal cortex of the rat. A comparative review. *Brain Res. Rev.*, 8: 65–98.

Kolb, B.J., Reynolds, B. and Fantie, B. (1988) Frontal cortex grafts have opposite effects at different postoperative recovery times. *Behav. Neural Biol.*, 5: 193–206.

Labbe, R., Firl, A., Mufson, E.J. and Stein, D.G. (1983) Fetal brain transplants: reduction of cognitive deficits in rats with frontal cortex lesions. *Science*, 221: 470–472.

LeGros Clark, W.E. (1940) Neuronal differentiation in implanted foetal cortical tissue. *J. Neurol. Psychiat.*, 3: 263–272.

Levin, B.E., Dunn-Meynell, A. and Sced, A.F. (1987) Functional integration of fetal cortical grafts into the afferent pathway of the rat somatosensory cortex (SmI). *Brain Res. Bull.*, 19: 723–734.

Lindvall, O., Dunnett, S.B., Brundin, P. and Björklund, A. (1987) Transplantation of catecholamine-producing cells to the basal ganglia in Parkinson's disease: experimental and clinical studies. In F.C. Rose (Ed.), *Parkinson's Disease: Clinical and Experimental Advances,* John Libbey, London, pp. 189–206.

Lund, R.D. and Hauschka, S.D. (1976) Transplanted neural tissue develops connections with host rat brain. *Science*, 193: 582–584.

Madrazo, I., Drucker-Colin, R., Diaz, V., Martinez-Mata, J., Torres, C. and Becerril, J.J. (1987) Open microsurgical autograft of adrenal medulla to the right caudate nucleus in two patients with intractable Parkinson's disease. *New Engl. J. Med.*, 316: 831–834.

Manthorpe, M., Nieto-Sampedro, M., Skaper, S.D., Lewis, E.R., Barbin, G., Longo, F.M., Cotman, C.W. and Varon, S. (1983) Neuronotrophic activity in brain wounds of the developing rat. Correlation with implant survival in the wound cavity. *Brain Res.*, 267: 47–56.

Mountcastle, V.B. (1979) An organizing principle for cerebral function: the unit module and the distributed system. In F.O. Schmitt and F.G. Worden (Eds.), *The Neurosciences Fourth Study Program,* MIT Press, Cambridge, MA, pp. 21–42.

Nieto-Sampedro, M., Lewis, E.R., Cotman, C.W., Manthorpe, M., Skaper, S.D., Barbin, G., Longo, F.M. and Varon, S. (1982) Brain injury causes a time-dependent increase in neuronotrophic activity at the lesion site. *Science*, 217: 860–862.

Nieto-Sampedro, M., Whittemore, S.R., Needels, D.L., Larson, J. and Cotman, C.W. (1984) The survival of brain transplants is enhanced by extracts from injured brain. *Proc. Natl. Acad. Sci. (U.S.A.)*, 81: 6250–6254.

Olson, L. and Malmfors, T. (1970) Growth characteristics of adrenergic nerves in the adult rat. Fluorescence histochemical and ^{3}H-noradrenaline uptake studies using tissue transplantation to the anterior chamber of the eye. *Acta Physiol. Scand. Suppl.*, 348: 1–112.

Perlow, M.J., Freed, W.J., Hoffer, B.J., Seiger, Å., Olson, L. and Wyatt, R.J. (1979) Brain grafts reduce motor abnormalities produced by destruction of nigrostriatal dopamine system. *Science*, 204: 643–647.

Ridley, R.M., Baker, H.F. and Fine, A. (1988) Transplantation of fetal tissues. *Br. Med. J.*, 296: 1469.

Rosenstein, J.M. and Brightman, M.W. (1983) Circumventing the blood-brain barrier with autonomic ganglion transplants. *Science*, 221: 879–881.

Rutherford, A., Garcia-Munoz, M., Dunnett, S.B. and Arbuthnott, G.W. (1987) Electrophysiological demonstration of host cortical inputs to striatal grafts. *Neurosci. Lett.*, 83: 275–281.

Sharp, F.R. and Gonzalez, M.F. (1986) Fetal cortical transplants ameliorate thalamic atrophy ipsilateral to neonatal frontal cortex lesions. *Neurosci. Lett.*, 71: 247–251.

Sharp, F.R., Gonzalez, M.F., Ferriero, D.M. and Sagar, S.M. (1986) Injured adult neocortical neurons sprout fibres into surviving fetal frontal cortex: evidence using NADPH-diaphorase staining. *Neurosci. Lett.*, 65: 204–208.

Sharp, F.R., Gonzalez, M.F. and Sagar, S.M. (1987) Fetal frontal cortex transplanted to injured motor/sensory cortex of adult rats. II. VIP-somatostatin-, and NPY-immunoreactive neurons. *J. Neurosci.*, 7: 3002–3015.

Sirinathsinghji, D.J.S. and Dunnett, S.B. (1989) Disappearance of the I-opiate receptor patches in the rat neostriatum following nigrostriatal dopamine lesions and their restoration after implantation of nigral dopamine grafts. *Brain Res.*, in press.

Sladek, J.R. and Gash, D.N. (1984) *Neural Transplants: Development and Function,* Plenum Press, New York.

Smith, L.M. and Ebner, F.F. (1986) The differentiation of non-neuronal elements in neocortical transplants. In G.D. Das and R.B. Wallace (Eds.), *Neural Transplantation and Regeneration,* Springer, New York, pp. 81 – 101.

Sofroniew, M.V. and Pearson, R.C.A. (1985) Degeneration of cholinergic neurons in the basal nucleus following kainic acid or N-methyl-o-aspartic acid application to the cerebral cortex in the rat. *Brain. Res.,* 339: 186 – 190.

Sofroniew, M.V., Pearson, R.C.A., Eckenstein, F., Cuello, A.C. and Powell, T.P.S. (1983) Retrograde changes in cholinergic neurons in the basal forebrain of the rat following cortical damage. *Brain Res.,* 289: 370 – 374.

Sofroniew, M.V., Isacson, O. and Björklund, A. (1986) Cortical grafts prevent atrophy of cholinergic basal nucleus neurons induced by excitotoxic cortical damage. *Brain Res.,* 378: 409 – 415.

Stanfield, B.B. and O'Leary, D.D.M. (1976) Fetal occipital cortical neurons transplanted to the rostral cortex can extend and maintain a pyramidal tract axon. *Nature,* 313: 135 – 137.

Stein, D.G. (1987) Transplant-induced functional recovery without specific neuronal connections. *Progr. Res. Am. Paralysis Assoc.,* 18: 4 – 5.

Stein, D.G. and Mufson, E.J. (1987) Morphological and behavioral characteristics of embryonic brain tissue transplants in adult, brain-damaged subjects. *Ann. NY Acad. Sci.,* 495: 444 – 463.

Stein, D.G., Labbe, R., Attella, M.J. and Rakowsky, H.A. (1985) Fetal brain tissue transplants reduce visual deficits in adult rats with bilateral lesions of the occipital cortex. *Behav. Neural Biol.,* 44: 266 – 277.

Stein, D.G., Palatucci, C., Kahn, D. and Labbe, R. (1988) Temporal factors influence recovery of function after embryonic brain tissue transplants in adult rats with frontal cortex lesions. *Behav. Neurosci.,* 102: 260 – 267.

Stenevi, U., Björklund, A. and Svendgaard, N.-Aa. (1976) Transplantation of central and peripheral monoamine neurons to the rat brain: techniques and conditions for survival. *Brain Res.,* 114: 1 – 20.

Thompson, W.G. (1890) Successful brain grafting. *N.Y. Med. J.,* 51: 701 – 702.

Discussion

C. Cavada: Which is your rationale to use E16 tissue for transplantation? I noticed that the previous speaker, Dr. Kolb, uses E17 grafts. May this difference be significant?

S.B. Dunnett: It can be expected that the slightly younger donor tissues have a somewhat greater viability and growth potential. However, both E16 and E17 are well within the acceptable range of donor ages for viable cortical tissue grafts.

G.J. Boer: If a short-term effect is seen upon grafting which may be due to trophic factors, the question is why this is not lasting? Have the trophic factors been applied once again in the later stages, to see if a short-term effect can be repeated? And if the enormous outgrowth of your grafts gives abnormalities in the host brain morphology, what about the functional results with grafts that are found to be smaller?

S.B. Dunnett: You must appreciate that present interpretations are based on extremely limited data. Thus, there have not been any systematic studies of the times at which growth factors may have an effect beyond the initial month after lesions, nor of situations of acute recovery, return of deficits and then reapplication of trophic factors. Similarly, there are insufficient data with which one may directly correlate graft growth with deficits over and above the extent observed in the rats with long-term lesion. I can only say that in the one study where we used the youngest donor tissues (Dunnett et al., 1987) and assessed the experimental animals both acutely after lesion and 6 months later, we found (a) extensive long-term graft growth in all animals, (b) detectable septo-hippocampal damage in all specimens where AChE staining was fortuitously collected that far back in the brain, and (c) long-term behavioural deficits in the grafted animals over and above those seen in the lesioned animal. It is therefore natural to suggest that the long-term deficits in the behavioural tests may have been attributable to the damage associated with the extent of long-term graft growth. However, these factors have not been directly manipulated in a systematic experimental study.

J.G. Parnavelas: What is the evidence that the axons which grow into the graft form connections? Has anyone done electron microscopy or recorded from the graft after stimulation?

S.B. Dunnett: There has not to my knowledge been either EM or electrophysiological studies on the prefrontal cortical model system I have described. However, there is substantial electrophysiological evidence that afferent projections can form functional synaptic contacts with neurones in grafted cortical tissues implanted in the somatosensory cortex of adult hosts (e.g., Bragin, 1986; Bragin et al., 1987).

H.B.M. Uylings: You mentioned that you took middorsal cortical tissue to graft. Was cingular cortex included in these grafts? The source of the cortical tissue might influence the effect of grafting.

S.B. Dunnett: The tissues were from the dorsal and lateral parts of embryonic cortex, and would topographically exclude the midline rim. I do not know if the site of neurogenesis of cingulate neurones has been established, in order to determine whether precursors of this population of cortical neurones would be excluded by our dissection. However, several studies indicate that it is the site of implantation rather than the particular regional source that determines the fate of cortical neurones (e.g. Stanfield and O'Leary, 1976).

References

Bragin, A.G. (1986) Neuronal responses of embryonic rat somatosensory neocortex grafted into adult rat barrelfield. *Neurophysiologia,* 18: 833–836. (In Russian.)

Bragin, A.G., Bohne, A. and Pavlik, V.D. (1987) Electrophysiological indexes of the degree of grafted neural tissue integration with the host brain. *Neurophysiologia,* 19: 498–504. (In Russian.)

Dunnett, S.B., Ryan, C.N., Levin, P.D., Reynolds, M. and Bunch, S.T. (1987) Functional consequences of embryonic neocortex transplanted to rats with prefrontal cortex lesions. *Behav. Neurosci.,* 101: 489–503.

Stanfield, B.B. and O'Leary, D.D.M. (1976) Fetal occipital cortical neurons transplanted to the rostral cortex can extend and maintain a pyramidal tract axon. *Nature,* 313: 135–137.

CHAPTER 14

Enhanced cortical maturation: Gangliosides in CNS plasticity

S.E. Karpiak and S.P. Mahadik

Division of Neuroscience, New York State Psychiatric Institute, and the Departments of Psychiatry, and Biochemistry and Molecular Biophysics, College of Physicians and Surgeons, Columbia University, New York, NY 10032, USA

Introduction

As part of the effort to define those processes which underlie the development and maturation of the mammalian central nervous system, neurobiologists have focused their attention on individual molecular processes associated with growth in neural plasma membranes. While the majority of research reports have concentrated on the function of plasma membrane proteins (e.g. adhesion molecules, ion channels, receptors), certain research strategies have explored the role of glycosphingolipids which comprise the characteristic lipid bilayer of plasma membranes (Goins et al., 1986). Gangliosides are one group of glycosphingolipids which have engendered particular research interest. These lipid molecules are found throughout all mammalian tissues, but, they are found in highest concentration in the central nervous system (Yamakawa, 1988; Ando, 1983) and are in even higher concentrations in synaptic plasma membranes (Skrivanek et al., 1982) The topographical configuration of gangliosides and their high CNS concentration make these molecules ideal candidates for critical processes during CNS development (cell recognition, migration, and adhesion) (Dreyfus et al., 1980).

Correspondence: Dr. Stephen Karpiak, Box 64, Department of Psychiatry, Division of Neuroscience, Columbia University, 722 W 168th Street, New York, NY 10032, USA.

Gangliosides are localized to the outer plasma membrane surface (Goins et al., 1986; Gregson et al., 1977; Hansson et al., 1977; Hungund and Mahadik, 1981) wherein the hydrophobic portion of the molecule is imbedded in the plasma membrane (constituting the lipid bilayer), and the hydrophilic portion of the molecule is exposed to the outer membrane surface. The hydrophilic portion of the ganglioside molecule is variable, being comprised of as many as 4 sugar moieties to which are attached sialic acid moieties. The number of sugar moieties, together with the number and position of the sialic acid molecules determine the type of ganglioside species (molecular specificity) (Yamakawa, 1988; Ando, 1983). Because of the high concentration and topographical localization of gangliosides in the central nervous system, these lipids have been hypothesized to mediate neuronal growth and repair processes (Rahmann, 1983; Ando, 1983; Cimino et al., 1987; Goins et al., 1986). Such hypotheses contend that ganglioside molecules may be involved in the action of trophic factors, the regulation of neuritogenesis, synaptogenesis, regeneration, cell – cell interaction, as well as synaptic transmission (DeFeudis et al., 1980; Ando, 1983; Domanska-Janik et al., 1986; Gardas and Nauman, 1981). These hypotheses are reinforced by observations that the concentration and topography of certain ganglioside species change markedly during brain development, aging, and in CNS pathologies.

Additional evidence to support a critical functional role for gangliosides in the CNS has been provided by the use of antibodies to gangliosides. In adult rats, intracortical injection of polyclonal or monoclonal antibodies to GM1 ganglioside induces recurrent epileptiform activity and seizure activity (Karpiak et al., 1976, 1981). Frieder and Rapport (1981, 1987) showed that antibodies to ganglioside GM1 enhanced the K^+ dependent depolarization and Ca-channel linked release of GABA from brain slices. There was no effect on the release of serotonin or norepinephrine. It has also been shown that polyclonal antibodies to GM1 can inhibit the consolidation phases of learning (Karpiak and Rapport, 1979), and inhibit morphine induced analgesia (Karpiak, 1982).

Ganglioside changes in CNS development and pathology

Several reports indicate that the levels of individual ganglioside species increase rapidly during the maturation of the mammalian brain (Suzuki, 1965b; Vanier et al., 1971; Dreyfus et al., 1984; Sonnino et al., 1981; Hilbig et al., 1984; DiCesare and Dain, 1972; Irwin et al., 1980; Yu et al., 1988). GD3 ganglioside levels increase during proliferation; GQ1b increases during cell migration and arborization; GD1a and GT1b increase during synaptogenesis; and GM1 and G4 increases parallel myelination. During CNS development the levels of the synthesizing enzymes for gangliosides, glycosyltransferases and sialyltransferases, are highest in postnatal development and lower during adult stages (DiCesare and Dain, 1972; Den et al., 1975). In addition, the topography of gangliosides, or surface exposure on the plasma membrane, changes during development. GM1 has been found to be differentially exposed on a variety of brain cells at various stages of development (Willinger and Schachner, 1980). The differential localization is also found in several rat brain nuclei (DeBaecque et al., 1976; Laev et al., 1978; Laev and Mahadik, 1989) These results emphasize that not only are changes in the net amount of individual ganglioside species important in establishing their function in the mediation of neuronal and glial cell surface interactions during development, but also, their surface topography and exposure must be considered.

In the rat forebrain development it has been shown that the level of gangliosides plateaus between the 4th and 6th postnatal week. In humans this plateau is preceded by an accelerated increase in ganglioside concentration which has been observed to begin at the 32nd week of gestation. The plateau is not reached until the 2nd postnatal month. Curiously, maximal ganglioside concentrations in the human cerebellum are not archieved until 1 – 2 years of age (Martinez and Ballabriga, 1978; Hess et al., 1976; Hilbig et al., 1984). The increases in ganglioside concentration are always associated with "growth spurts". During these increases, the ganglioside concentration of individual ganglioside species shows a shift from more complex ganglioside types (e.g. polysialogangliosides) to more simple ganglioside structures (e.g. mono- or disialogangliosides) (Ando, 1983; Yu et al., 1988). In summary, polysialogangliosides are highest during peak periods of neuronal growth (Yu et al., 1988; Dreyfus et al., 1980, 1984). The loss of polysialogangliosides in the mature CNS is consistent with the observed reduction in membrane surface charge as well as a decrease in synaptic density in brains of older animals. This process is thought to partially underlie the age dependent decline of cognitive capacity (Hilbig et al., 1984). There is no systematic study of the ganglioside metabolism in aging brain, but there is some evidence in rats, with increasing age, that levels of polysialogangliosides (GD1b and GT1b) decrease whereas levels of GM1 and GD1a slightly increase (Vanier et al., 1971; Hilbig et al., 1984). Analogous changes in ganglioside patterns have been observed in vitro. Using cultures of 8-day-old chick neurons, Dreyfuss et al. (1980) report higher levels of tri-and tetrasialogangliosides during cell growth, accompanied by lower levels of monosialogangliosides. They observed slight increases in the polysialogangliosides corresponding to a phase of cell division, followed by considerable increases during the phase of cell maturation.

A detailed analysis of growth cones from 18-day-old fetal rat brain showed increased levels of GD3 ganglioside and less of GD1a (Sbassching-Alger et al., 1988). In studies of developing rat cerebellar Purkinje cells in situ, Reynolds and Wilkin (1988) report that GD3 is expressed on the neuron's cell surface at postnatal day 7. This expression increases as the dendritic tree of the cell develops and enlarges.

Using several genetic mouse mutations which have selective losses of neuronal cell types or processes, Seyfried et al. (1979) report altered levels of some ganglioside species. For example, it was reported that GD1a levels were reduced in the weaver mutant which has considerable cerebellar granular cell loss. However, no ganglioside changes in the mouse mutant with Purkinje cell degeneration were detected. Bouvier and Seyfried (1989), conclude that the pathological changes in such mutant mice are a result of a partial deficiency in the activity of a specific sialyltransferase for GD3, GD1b, GT1b and GQ1b. The accumulation and deficiency of gangliosides in differentiating neural cells result in failed differentiation. These reports support the observations of Purpura and Walkey (1981) that the abnormal cortical cellular morphology (meganeurites) that is present in the ganglioside storage diseases, gangliosidosis (Suzuki, 1965a), supports the view that gangliosides are critical molecules in CNS cellular development.

In Tay–Sachs disease there is an accumulation of a single species of ganglioside (GM2). This disease is one of abnormal ganglioside metabolism (Purpura and Baker, 1977). Another ganglioside storage disease, GM1 gangliosidosis, is characterized by intracellular (neuronal) accumulation of ganglioside GM1 (Purpura and Baker, 1977). With the development of a feline model of human GM1 gangliosidosis the correlation between ganglioside storage and abnormal neuritic (meganeurites) development was established. This indirectly suggested a function for gangliosides in neuronal development. Since then, studies of the distribution of several ganglioside species in brains of patients with other central nervous diseases such as Jacob–Creutzfeldt disease (Suzuki and Chen, 1966), multiple sclerosis (Yu et al., 1974, 1982) and amyotrophic lateral sclerosis (Rapport et al., 1985a) have reported significant changes in ganglioside distribution. Of particular interest are the cortical ganglioside changes seen in Alzheimer's disease and dementia (Crino et al., 1989; Kamp et al., 1986). In these studies changes in ganglioside concentrations were seen in the prefrontal cortices. In this area, decreases in complex gangliosides (GD1b and GT1b) were observed, while there were no significant changes in GM1 or GD1a concentrations. In brains from amyotrophic lateral sclerosis (ALS) patients, changes in ganglioside patters and concentrations were seen in several cortical areas including the prefrontal cortex. In the frontal cortex from these ALS brains, levels of GD1b, GT1b and GQ1b were decreased and levels of GM2 and GD3 were increased (Rapport et al., 1985a).

Ganglioside antibody inhibits in vitro neuritogenesis

The ability of polyclonal antibody to brain ganglioside to alter the growth of neuronal cells and tissue in vitro has contributed to the view that these molecules are important elements in maturational processes. In these studies neurite outgrowth from neural cells and tissues is inhibited by antibody to ganglioside, particularly with antibody to GM1 ganglioside (Schwartz and Spirmman, 1982; Spirmman et al., 1982; Spoerri et al., 1988). Most recent experiments (Spoerri et al., 1988) report similar effects using several monoclonal antibodies to GM1 ganglioside. It was found that 5 monoclonal antibodies to GM1 ganglioside inhibited neurite outgrowth in vitro. However, each antibody had differential effects on neurite formation, length, number, and initiation. These differences in the type of inhibition produced by different types of monoclonal antibodies to GM1 suggest that the structural complexity of the ganglioside molecule and its surface exposure could contribute to the multiple complex processes underlying neural development.

While most studies of ganglioside involvement in CNS growth and development have focused on neurons, some studies have shown that gangliosides must also be important functional molecules in glial cells (Levine and Goldman, 1988). Facci et al. (1988) report that choleragenoid (the β-subunit of cholera toxin which binds with high specificity to GM1 ganglioside) induces a classic stellate morphology in astroglial cells accompanied by DNA inhibition and cell division.

Ganglioside antibody inhibits cortical CNS development

In a series of experiments our laboratories attempted to illustrate the critical function which gangliosides subsume in the central nervous system. These studies involved the injection into the CNS of polyclonal antibodies to brain ganglioside, followed by observations of any resulting functional, biochemical and morphological alterations (Rapport et al., 1979, 1980; Karpiak et al., 1976, 1981; Karpiak and Rapport, 1979). The control for these studies was the injection of antiserum to gangliosides from which antibodies to GM1 ganglioside were removed by absorption. Subsequent studies utilized affinity purified antibody to GM1, its fragments, and finally, monoclonal antibodies to GM1 ganglioside (Karpiak et al., 1976, 1981, 1982). In the adult mammalian system our labs reported that the injection of antibody to GM1 ganglioside could result in the induction of epileptiform activity, seizure activity and blockade of consolidation phases of learning processes. All other functions were not affected (Rapport et al., 1979, 1980, 1985; Karpiak and Rapport, 1979; Karpiak et al., 1976, 1978, 1981, 1982; Karpiak, 1982).

In an effort to better understand the action of antibody to ganglioside and therefore the function of endogenous ganglioside molecules in cortical CNS development, 5-day-old rats were injected intracisternally with 50 µl of either antiganglioside serum or an absorbed (with GM1) serum. At the age of 60–80 days these rats were tested on a complex operant behavior paradigm (DRL: differential reinforcements at low rates), and cortical samples were assessed for protein, DNA, RNA, galactocerebroside and sialic acid levels. Selected animals were stained for Golgi analyses to assess dendritic morphology (Kasarskis et al., 1981).

The rats exposed to the antibody to ganglioside showed normal weight gain development. There were no obvious neurological of behavioral abnormalities (e.g. normal food/water intake, activity levels, diurnal cycles, passive avoidance learning). However, when tested on a DRL operant learning behavior, it was found that as the complexity of the task was increased (increasing the DRL delay) those animals exposed to the antibody to ganglioside made significantly greater incorrect responses. Controls injected with absorbed antiserum (GM1) were not different from saline injected rats (Fig.1).

Biochemical analyses of the total cortical protein and DNA from animals exposed to the antiserum to ganglioside were not different from controls. However the cortical levels of RNA, galac-

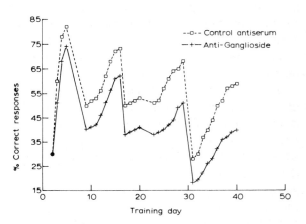

Fig. 1. Total AChE activity as measured in the cortices of rats treated (s.c.) with GM1 ganglioside (postnatal days 5–15) or injected with saline (control) ($p<0.01$).

tocerebroside and ganglioside sialic acid were reduced by 30%. The total ganglioside content of the entire brain from rats exposed to the antiserum was reduced by more than 10%. However the cortex was most affected since it showed a 30% ganglioside loss. Quantitative analyses of 11 ganglioside species showed no significant difference in the content distribution of these lipids. More detailed analyses of cortical laminar layers I, II and III indicated that levels of ganglioside sialic acid were reduced by 30% in those layers, and were less reduced in layers IV, V and VI. With galactocerebroside analyses, the greatest losses were found in layers IV, V and VI, with smaller losses seen in cortical layers I, II and III. In summary, the typical concentration gradients for ganglioside and galactocerebroside (indicative of myelin) across the cortical layers were largely abolished in animals injected with antiserum to ganglioside.

Routine histological examination of the cortex from rats exposed to antibody to GM1 revealed no significant changes or pathologies. Quantitative morphological analyses of dendritic spines on the oblique (secondary) dendrites of pyramidal cells (Golgi impregnation) showed there was a significant change in the number and type of spines. There was no change in the length of the spines. In rats injected with antiserum to ganglioside there was a 30% decrease in the number of spines, paralleled by a decrease in the number of this spines from 74% in controls and 28% in antiserum exposed rats. While controls only showed 22% stubby spines, the antiganglioside rats showed 68% of this spine type (Table I).

These studies showed that a single intracisternal injection antibody to brain ganglioside during the neonatal period of development could induce cortical and functional abnormalities detectable at adult age. While total cell and protein cortical contents were not affected, reductions in the levels of RNA, galactocerebroside, ganglioside sialic acid, accompanied by dendritic spine pathology, indicated that dendrogenesis was affected by the antibody administration. There was also hypomyelinization as indicated by decreased levels of galactocerebroside. Clearly the function of ganglioside in CNS development and maturation is critical for normal development.

Exogenous gangliosides increase neurite formation in vitro

Considerable interest was stimulated by the early observations (supra) that the addition of exogenous gangliosides into neural cell and tissue cultures resulted in increased neurite growth (Ferrari et al., 1983; Eberhardt, Maier and Singer, 1984; Leon et al., 1982, 1988). These observations became even more significant when it was reported that the stimulating effects of NGF on neurite outgrowth in vitro was enhanced (synergistically) by GM1 ganglioside. More precisely, NGF mediated neurite outgrowth was significantly enhanced by the addition of GM1 ganglioside. Moreover, GM1 and NGF synergistically, not additive, promote neurite outgrowth. The use of GM1 + NGF results in greater neurite outgrowth than the sum of effects of GM1 and NGF alone.

The mechanism(s) by which GM1 enhances NGF (or vice versa) remain unknown. It has been hypothesized that the ability of exogenous gangliosides to spontaneously insert into plasma membranes, and become functional, may be/underlie the mechanism by wich GM1 is effective in any of the paradigms (in vitro and in vivo) in which it has been used (O'Keefe and Cuatrecasas, 1977; Scheel et al., 1985).

TABLE I

Morphological analyses of secondary oblique cortical dendrites of pyramidal neurons in laminae III and V in rat isocortex

	Control	Anti-ganglioside sera
Spine length (μm)	1.28 + 0.06	1.13 + 0.04
Spine type (dist.)		
% Thin	74.8 + 3.5	27.8 + 4.8
% Stubby	22.5 + 3.6	67.6 + 4.7
% Mushroom	2.7 + 0.7	4.6 + 1.0

While many experiments were devised to explore the effectiveness of exogenous ganglioside treatment for both peripheral and central nervous system injury (Bianchi et al., 1986; Cuello et al., 1989; Sabel et al., 1985, 1988; Ramirez et al., 1987; Cahn et al., 1989; Karpiak et al., 1981, 1986, 1987; Mahadik and Karpiak, 1988; Karpiak and Mahadik, 1989), some have examined the effects of ganglioside administration in the normal mammalian system. Paralleling our first study which showed that the injection of antibody to ganglioside could alter CNS dendrogenesis, a study was done to investigate the effects of exogenous ganglioside on the developing rat CNS.

Exogenous ganglioside administration in rat neonates

Male rat neonates were injected on PN days 5 – 15 with daily subcutaneous injections of 2.5 mg of either total bovine brain ganglioside or one of the following ganglioside species (GM1, GD1a, GD1b, GT1b). On PN days 11, 12 and 14 the pups were tested and trained on a multidirectional avoidance paradigm. Performance was measured as the latency (sec) to escape from a gridded escape platform.

There were no differences in weight gain or eye opening (PN days 14 – 16) between ganglioside injected animals and controls. On the first day of training (PN 11) saline rats showed an average latency of 8 sec to escape. These escape latencies were reduced to 3 sec in pups treated with total ganglioside, GM1 or GD1b. On PN 13 the saline treated pups improved their latency score to 4.5 sec, and the total, GM1 and GD1b treated pups reduced their latencies to <2 sec. After skipping 1 day of testing, the saline pups showed no change in the escape latency as compared to PN day 12, but the total ganglioside, GM1 and GD1b treated pups reduced their latencies even further to 1 sec. Pups treated with GD1a and GT1b were not different at PN days 11, 12 and 14 from saline controls. The authors concluded that the neonates injected with ganglioside (total, GM1, GD1b) demonstrated improved acquisition and retention of the learning

TABLE II

Escape latencies on a multidirectional avoidance paradigm: Postnatal days 11, 12 and 14

Treatment	PN11	PN12	PN14
Saline	7.8 + 1.1	4.5 + 0.6	5.3 + 0.7
Total ganglioside	2.8 + 0.9	1.8 + 0.5	1.3 + 0.6
GM1	3.1 + 0.8	1.8 + 0.7	1.1 + 0.5
GD1b	3.7 + 0.7	1.9 + 0.5	1.2 + 0.4
GD1a	6.5 + 0.8	4.3 + 0.4	4.2 + 0.4
GT1b	6.4 + 0.8	4.4 + 0.6	4.9 + 0.5

task (Table II).

The improved learning performance of these ganglioside treated pups led to a study of the levels of acetylcholinesterase (AChE) (Fig. 2), and its two molecular forms (10s and 4s) in the cortices of these pups. It was thought that increased cholinergic activity might account for the increased learning performance of the pups. AChE was assayed in the cortices of pups at PN days 9, 14, 21 and 28. At all time points the levels of AChE as well as its two molecular forms 4s and 10s were

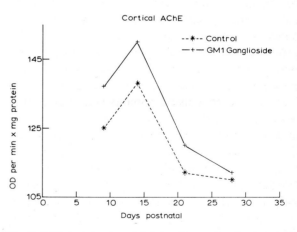

Fig. 2. DRL learning by adult rats after neonatal exposure to ganglioside antiserum or antiserum absorbed with GM1 ganglioside (control). Rates of learning were similar for both groups ($N = 6$/group), but, as the level (sec) of DRL increased (day 2 = 5 sec; day 9 = 7 sec; day 17 = 10 sec; day 30 = 15 sec) the anti-ganglioside group showed significant decreases in the percentage of correct responses (DRL = 10 sec $p<0.05$; DRL = 15 sec $p<0.01$).

significantly increased (10 – 13%) in pups treated with GM1 ganglioside. These increases were maximal at PN days 9 and 14. By PN day 28 the differences between controls and GM1 treated pups were smaller (though significant). In subsequent studies it was found that these differences were no longer evident by PN day 40.

In order to further clarify this effect of gangliosides on CNS maturation, RNA and DNA content analyses were done on the cortices of the GM1 treated pups at birth, 2 weeks and 4 weeks of age. No differences were found between controls of GM1 treated pups. It was concluded that there were no increases in cortical cell number.

In a parallel experiment the incorporation of [^{14}C]glucose carbon into neonatal cortical proteins (soluble and membrane) was done. At PN 10 days no differences were found between the saline and GM1 treated pups. However, at 15 days postnatally there was a 33% increase in incorporation into the membrane pool. No increases were seen in the soluble protein pool. This increased membrane protein turnover might reflect increased maturation (i.e. arborization, dendrogenesis and synaptogenesis). Preliminary studies had indicated that this increased incorporation was primarily into a 55 kDa protein (possibly tubulin).

While the results of these studies seem to have indicated that treatment of pups during the early postnatal period leads to enhanced CNS maturation, such a conclusion remains speculative. It is important to note that in these studies it was finally concluded that maturational processes were not increased, but accelerated. This conclusion is analogous to the observations made by investigators studying the facilitating effects of GM1 treatment on peripheral nerve regeneration after damage. In these studies, gangliosides clearly accelerated the rate of PNS regeneration, but the total net level of regenerating peripheral nerve innervation of muscle was no greater than that seen in controls. The rate of regrowth was enhanced, not the amount.

Ganglioside modulation of trophic factors: a unifying hypothesis

The effects of antibodies to ganglioside and exogenous ganglioside on the developing CNS, particularly cortex, are intriguing results which support the view that gangliosides play a critical role in neural development. The mechanism for this action remains undefined. One of the more appealing hypotheses contends that gangliosides are involved in the modulation of trophic factors.

Very recent developments in neurobiology have focused increased attention on the function of neuronotrophic factors in the CNS. The best know factor, NGF (nerve growth factor), has long been identified as being able to promote neuritogenesis in vitro, and most recently has been shown to markedly increase the survival of cholinergic neurons after axotomy of the medial septal dorsal/hippocampal pathway (Hefti, 1986). Consequently it has been hypothesized that the presence (or optimal levels) of these factors may be necessary for the maintenance of cell survival, and therefore, associated growth processes (i.e. maturation or regeneration). These observations are significant since GM1 ganglioside has been repeatedly shown to modulate (enhance) the effectiveness of NGF and other growth factors to promote neuritogenic processes in vitro and to promote cell survival in vivo after injury (Leon et al., 1984; Janigro et al., 1984; Di Patre et al., 1989; dal Toso et al., 1988; Cuello et al., 1989). Conceivably, exogenously administered monosialoganglioside's modulation of neuronotrophic factors in vivo may be the "link" by which CNS maturation and repair processes have been reported to be enhanced by the administration of these exogenous glycosphingolipids. It has been perplexing to see an enormous number of research reports substantiating the efficacy of exogenous gangliosides in promoting neuronal growth, plasticity and recovery in many different paradigms, with no evident unifying factor or "mechanism". The involvement of gangliosides in the modulation of neuronotrophic factors may be the "common denominator" by which both the normal developing CNS and the injured CNS are advantageously affected by exogenous ganglioside administration.

References

Ando, S. (1983) Gangliosides in the nervous system. *Neurochem. Int.*, 5: 507–537.

Bianchi, R., Janigro, D., Milan, F., Guidici, G. and Gorio, A. (1986) In vivo treatment with GM1 prevents the rapid decay of ATPase activities and mitochondrial damage in hippocampal slices. *Brain Res.*, 364: 400–404.

Bouvier, J.D. and Seyfried, T.N. (1989) Ganglioside composition of normal and mutant mouse embryos. *J. Neurochem.*, 52: 460–466.

Cahn, R., Borzeix, M.-G., Aldinio, C., Toffano, G. and Cahn, J. (1989) Influence of monosialoganglioside inner ester on neurologic recovery after global cerebral ischemia in monkeys. *Stroke*, 20: 652–656.

Cimino, M., Benfenati, F., Farabegoli, C., Cattabeni, F., Fuxe, K., Agnati, L.F. and Toffano, G. (1987) Differential effect of ganglioside GM1 on rat brain phosphoproteins: potentiation and inhibition of protein phosphorylation regulated by calcium/calmodulin and calcium/phospholipid-dependent protein kinase. *Acta Physiol. Scand.*, 130: 317–325.

Crino, P.B., Ullman, M.D., Vogt, B.A., Bird, E.D. and Volicer, L. (1989) Brain gangliosides in dementia of the Alzheimer type. *Arch. Neurol.*, 46: 398–401.

Cuello, A.C., Garofalo, L., Kenigsberg, R.L. and Maysinger, D. (1989) Gangliosides potentiate in vivo and in vitro effects of nerve growth factor on central cholinergic neurons. *Proc. Natl. Acad. Sci. (U.S.A.)*, 86: 2056–2060.

dal Toso, R., Giorgi, O., Soranzo, O., Kirschner, G. and Ferrari, G. (1988) Development and survival of neurons in dissociated fetal mesencephalic serum-free cell cultures. I. Effects of cell density and of an adult mammalian striatal-derived neuronotrophic factor (SDNF). *J. Neurosci.*, 8: 733–745.

DeBaecque, C., Johnson, A.B., Naiki, M., Schwarting, G. and Marcus, D.M. (1976) Ganglioside localization in cerebellar cortex: An immunoperoxidase study with antibody to GM1. *Brain Res.*, 114: 117–122.

DeFeudis, F.V., Yusufi, A.N.K., Ossola, L., Maitre, M., Wolff, P., Rebel, G. and Mandel, P. (1980) Antiserum to gangliosides inhibits [3H] GABA binding to a synaptosome-enriched fraction of rat cerebral cortex. *Gen. Pharmacol.*, 11: 251–254.

Den, H., Kaufman, B., McGuire, E.J. and Roseman, S. (1975) The sialic acids. XVIII. Subcellular distribution of seven glycosyltransferases in embryonic chicken brains. *J. Biol. Chem.*, 250: 739–746.

Di Patre, P.L., Casamenti, F., Cenni, A. and Pepeu, G. (1989) Interaction between nerve growth factor and GM_1 monosialoganglioside in preventing cortical choline acetyltransferase and high affinity choline uptake decrease after lesion of the nucleus basalis. *Brain Res.*, 480: 219–224.

DiCesare, J.L. and Dain, J.A. (1972) Localization, solubilization and properties of N-acetylgalactosaminyl ganglioside transferase and galactosyl ganglioside transferase in rat brain synaptic membrane. *J. Neurochem.*, 19: 403–410.

Domanska-Janik, K., Noremberg, K. and Lazarewicz, J. (1986) Gangliosides and synaptosomal calcium homeostasis. *Int. J. Tiss. Reac.*, 8: 373–382.

Dreyfus, H., Louis, J.C., Harth, S. and Mandel, P. (1980) Gangliosides in cultured neurons. *Neuroscience*, 5: 1647–1655.

Dreyfus, H., Ferret, B., Harth, S., Gorio, A., Durand, M., Freysz, L. and Massarelli, R. (1984) Metabolism and function of gangliosides in developing neurons. *J. Neurosci. Res.*, 12: 311–324.

Eberhardt Maier, C. and Singer, M. (1984) Gangliosides stimulate protein synthesis, growth, and number of regenerating limb buds. *J. Comp. Neurol.*, 230: 459–464.

Facci, L., Skaper, S.D., Favaron, M. and Leon, A. (1988) A role for gangliosides in astroglial cell differentiation in vitro. *J. Cell Biol.*, 106: 821–828.

Ferrari, G., Fabris, M. and Gorio, A. (1983) Gangliosides enhance neurite outgrowth in PC12 cells. *Dev. Brain Res.*, 8: 215–221.

Frieder, B. and Rapport, M.M. (1981) Enhancement of depolarization-induced release of gamma-aminobutyric acid from brain slices by antibodies to ganglioside. *J. Neurochem.*, 37: 634–639.

Frieder, B. and Rapport, M.M. (1987) The effects of antibodies on Ca^+ channel-linked release of gamma-aminobutyric acid in rat brain slices. *J. Neurochem.*, 48: 1048–1052.

Gardas, A. and Nauman, J. (1981) Presence of gangliosides in the structure of membrane receptor for thyrotropin. *Acta Endocr.*, 98: 549–555.

Goins, B., Masserini, M., Barisas, B.G. and Freire, E. (1986) Lateral diffusion of ganglioside GM1 in phospholipid bilayer membranes. *Biophys. J.*, 49: 849–856.

Gregson, N.A., Kennedy, M. and Liebowitz, S. (1977) Gangliosides as surface antigens on cells isolated from the rat cerebellar cortex. *Nature*, 266: 461–462.

Hansson, H.A., Holmgren, J. and Svennerholm, L. (1977) Ultrastructural localization of cell membrane GM1 ganglioside by cholera toxin. *Proc. Natl. Acad. Sci. (U.S.A.)*, 74: 3827–3786.

Hefti, F. (1986) Nerve growth factor promotes survival of septal cholinergic neurons after fimbrial transection. *J. Neurosci.*, 6: 2155-2162.

Hess, H.H., Bass, N.H., Thalheimer, C. and Devarakonda, R.

(1976) Gangliosides and the architecture of human frontal and rat somatosensory isocortex. *J. Neurochem.*, 26: 1115–1121.

Hilbig, R., Lauke, G. and Rahmann, H. (1984) Brain gangliosides during the life span (embryogenesis to senescence) of the rat. *Dev. Neurosci.*, 6: 260–270.

Hungund, B.L. and Mahadik, S.P. (1981) Topographic studies of gangliosides of intact synaptosomes from rat brain cortex. *Neurochem. Res.*, 6: 183–191.

Irwin, L.N., Michael, D.B. and Irwin, G.C. (1980) Ganglioside patterns of fetal rat and mouse brain. *J. Neurochem.*, 34: 1527–1530.

Janigro, D., Di Gregorio, F., Vyskocil, F. and Gorio, A. (1984) Gangliosides' dual mode of action: a working model *J. Neurosci. Res.*, 12: 499–509.

Kamp, P., Jager, W.D.H., Maathuis, J., Groot, P.D., Jong, J.D. and Bolhuis, P. (1986) Brain gangliosides in the presenile dementia of Pick. *J. Neurol. Neurosurg. Psychiat.*, 49: 881–885.

Karpiak, S.E. (1982) Antibodies to GM1 ganglioside inhibit morphine analgesia. *Pharmacol. Biochem. Behav.*, 16: 611–613.

Karpiak, S.E. and Mahadik, S.P. (1990) Ganglioside reduction of CNS ischemic injury. *CRC Crit., Rev. Neurobiol.*, 5: 221–237.

Karpiak, S.E. and Rapport, M.M. (1979) Inhibition of consolidation and retrieval stages of passive-avoidance learning by antibodies to gangliosides. *Behav. Neural Biol.*, 27: 146–156.

Karpiak, S.E., Graf, L. and Rapport, M.M. (1976) Antiserum to brain gangliosides produces recurrent epileptiform activity. *Science*, 194: 735–737.

Karpiak, S.E. Mahadik, S.P. and Rapport, M.M. (1978) Ganglioside receptors and induction of epileptiform activity: cholera toxin and choleragenoid (B subunits). *Exp. Neurol.*, 62: 256–259.

Karpiak, S.E., Mahadik, S.P., Graf, L. and Rapport, M.M. (1981) An immunological model of epilepsy seizures induced by antibodies to GM1 ganglioside. *Epilepsia*, 22: 189–196.

Karpiak, S.E., Huang, Y.L. and Rapport, M.M. (1982) Immunological model of epilepsy. *J. Neuroimmunol.*, 3: 15–21.

Karpiak, S.E., Li, Y.S. and Mahadik, S.P. (1986) Gangliosides reduce mortality due to global ischemia: membrane protection. *Clin. Neuropharmacol.*, 9: 338–340.

Karpiak, S.E., Li, Y.S. and Mahadik, S.P. (1987) Ganglioside treatment: reduction of CNS injury and facilitation of functional recovery. *Brain Injury*, 1: 161–170.

Kasarskis, E.J., Karpiak, S.E., Rapport, M.N., Yu, R.K. and Bass, N.H. (1981) Abnormal maturation of cerebral cortex and behavioral deficit in adult rats after neonatal administration of antibodies to ganglioside. *Brain Res.*, 227: 25–35.

Laev, H. and Mahadik, S.P. (1989) Topography of monosialoganglioside (GM1) in rat brain using monoclonal antibodies. *Neurosci. Lett.*, 102: 7–14.

Laev, H., Rapport, M.M., Mahadik, S.P. and Silverman, A.J. (1978) Immunohistological localization of ganglioside in rat cerebellum. *Brain Res.*, 157: 136–141.

Leon, A., Facci, L., Benvegnu, D. and Toffano, G. (1982) Morphological and biochemical effects of gangliosides in neuroblastoma cells. *Dev. Neurosci.*, 5: 108–114.

Leon, A., Benvegnu, D., dal Toso, R., Presti, D., Facci, L. and Giorgi, O. (1984) Dorsal root ganglia and nerve growth factor: a model for understanding the mechanism of GM1 effects on neuronal repair. *J. Neurosci. Res.*, 12: 277–287.

Leon, A., dal Toso, R., Presti, D., Benvegnu, D., Facci, L., Kirschner, G., Tettamanti, G. and Toffano, G. (1988) Development and survival of neurons in dissociated fetal mesencephalic serum-free cell cultures: modulatory effects of gangliosides. *J. Neurosci.*, 8: 746–753.

Levine, S.M. and Goldman, J.E. (1988) Ultrastructural characteristics of GD3 ganglioside-positive immature glia in rat forebrain white matter. *J. Comp. Neurol.*, 277: 456–464.

Mahadik, S.P. and Karpiak S.E. (1988) Gangliosides in treatment of neural injury and disease. *Drug Dev. Res.*, 15: 337–369.

Martinez, M. and Ballabriga, A. (1978) A chemical study on the development of the human forebrain and cerebellum during the brain "growth spurt" period. I. gangliosides and plasmalogens. *Brain Res.*, 159: 351–362.

O'Keefe, E. and Cuatrecasas, P. (1977) Persistence of exogenous, inserted ganglioside GM1 on the surface of cultured cells. *Life Sci.*, 21: 1649–1654.

Purpura, D.P. and Baker, H.J. (1977) Neurite induction in mature cortical neurones in feline GM1 ganglioside storage disease. *Nature*, 266: 553–554.

Purpura, D.P. and Walkey, S.U. (1981) Aberrant neurite and spine generation in mature neurons in the gangliodoses. In M.M. Rapport and A. Gorio (Eds.), *Gangliosides in Neurological and Neuromuscular Function, Development and Repair*, Raven Press, New York, pp. 1–16.

Rahmann, H. (1983) Functional implication of gangliosides in synaptic transmission. *Neurochem. Int.*, 5: 539–547.

Ramirez, J.J., Fass, B., Kilfoil, T., Henschel, B., Grones, W. and Karpiak, S.E. (1987) Ganglioside-induced enhancement of behavioral recovery after bilateral lesions of the entorhinal cortex. *Brain Res.*, 414: 85–90.

Rapport, M.M., Karpiak, S.E. and Mahadik, S.P. (1979) Biological activities of antibodies injected into the brain. *Fed. Proc.*, 38: 2391–2396.

Rapport, M.M., Karpiak, S.E. and Mahadik, S.P. (1980) Perturbation of CNS functions by antibodies to gangliosides. Speculations on biological roles of ganglioside receptors. *Adv. Exp. Med. Biol.*, 125: 335–338.

Rapport, M.M., Donnenfield, H., Brunner, W., Hungund, B.L. and Bartfeld, H. (1985a) Ganglioside patterns in amyotrophic lateral sclerosis brain regions. *Ann. Neurol.*,

18: 60–67.

Rapport, M.M., Mahadik, S.P., Vilim, F., Laev, H. and Karpiak, S.E. (1985b) Monoclonal antibodies to GM1 differ in their ability to induce convulsions. *J. Neurochem.,* 44: S56.

Reynolds, R. and Wilken, G.P. (1988) Expression of GD3 ganglioside by developing rat cerebellar Purkinje cells in situ. *J. Neurosci. Res.,* 20: 311–319.

Sabel, B.A., Dunbar, G.L., Fass, B. and Stein, D.G. (1985) Gangliosides, neuroplasticity, and behavioral recovery after brain damage. *Brain Plasticity. Learning, and Memory,* 481–493.

Sabel, B.A., Gottlieb, J. and Schneider, G.E. (1988) Exogenous GM_1 gangliosides protect against retrograde degeneration following posterior neocortex lesions in developing hamsters. *Brain Res.,* 459: 373–380.

Sbassching-Alger, M., Pfenninger, K.H. and Ledeen, R.W. (1988) Gangliosides and other lipids of the growth cone membrane. *J. Neurochem.,* 51: 212–220.

Scheel, G., Schwarzman, G., Hoffman-Bleihauer, P. and Sandhoff, K. (1985) The influence of ganglioside insertion into brain membranes on the rate of ganglioside degradation by membrane-bound sialidase. *Eur. J. Biochem.,* 153: 29–35.

Schwartz, M.A. and Spirmman, N. (1982) Sprouting from chicken embryo dorsal root ganglia induced by nerve growth factor is specifically inhibited by affinity-purified antiganglioside antibodies. *Neurobiology,* 79: 6080–6083.

Seyfried, T.N., Glasser, G.H. and Yu, R.K. (1979) Genetic variability for regional brain gangliosides in five strains of young mice. *Biochem. Genet.,* 17: 43–55.

Skrivanek, J.A., Ledeen, R.W., Margolis, R.U. and Margolis, R.K. (1982) Gangliosides associated with microsomal subfractions of brain: comparison with synaptic plasma membranes. *J. Neurobiol.,* 13: 95–106.

Sonnino, S., Ghidoni, R., Masserini, M., Aporti, F. and Tettamanti, G. (1981) Changes in the rabbit brain cytosolic and membrane-bound gangliosides during prenatal life. *J. Neurochem.,* 36: 227–233.

Spirmman, N., Sela, B.A. and Schwartz, M. (1982) Antiganglioside antibodies inhibit neuritic outgrowth from regenerating goldfish retinal explants. *J. Neurochem.,* 39: 874–877.

Spoerri, P.E., Rapport, M.M., Mahadik, S.P. and Roisen, F.J. (1988) Inhibition of conditioned media-mediated neuritogenesis of sensory ganglia by monoclonal antibodies to GM1 ganglioside. *Brain Res.,* 469: 71–77.

Suzuki, K. (1965a) Neuronal storage disease: a review. In H.M. Zimmerman (Ed.), *Progress in Neuropathology, Vol. 3,* Grune and Stratton, New York, pp. 173–202.

Suzuki, K. (1965b) The pattern of mammalian brain gangliosides. III. Regional and developmental differences. *J. Neurochem.,* 12: 969–979.

Suzuki, K. and Chen, G. (1966) Chemical studies on Jacob–Creutzfeldt disease. *J. Neuropathol. Exp. Neurol.,* 25: 396–408.

Vanier, M.T., Holm, M., Ohman, R. and Svennerholm, L. (1971) Developmental profiles of gangliosides in human and rat brain. *J. Neurochem.,* 18: 581–592.

Willinger, M. and Schachner, M. (1980) GM1 ganglioside as a marker for neuronal differentiation in mouse cerebellum. *Dev. Biol.,* 74: 101–117.

Yamakawa, T. (1988) Thus started ganglioside research. *Trends Biochem. Sci.,* 13: 452–454.

Yu, R.K., Ledeen, R.W. and Eng, L.F. (1974) Ganglioside abnormalities in multiple sclerosis. *J. Neurochem.,* 39: 464–477.

Yu, R.K., Macala, L.J., Taki, T., Weinfield, H.M. and Yu, F.S. (1988) Developmental changes in ganglioside composition and synthesis in embryonic rat brain. *J. Neurochem.,* 50: 1825–1829.

Discussion

C.G. van Eden: You showed that addition of an exogenous substance (gangliosides) to the developing brain accelerates development. Does this mean that normally there is not enough? Wat is the end of development: a better brain?

S.E. Karpiak: There are certainly "enough" gangliosides in the developing brain. Exogenous gangliosides do not seem to "substitute" for the endogenous molecules. Rather, the activity (membrane insertion and topographical exposure) of the exogenous gangliosides increases the responsiveness of the developing membranes to endogenous factors (e.g., trophic factors). While there is evidence that exogenous gangliosides accelerate development, there seems to be no net gain when the CNS reaches maturity. However, this has not been studied systematically. This has been the observation in PNS studies.

J.P.C. de Bruin: (1) You report (also when answering the question of Corbert van Eden) that gangliosides speed up the recovery of regeneration following peripheral nerve damage. Are the effects of gangliosides the same when CNS damage is concerned, i.e., *just* speeding up?

(2) Are gangliosides beneficial to recovery of function after *all* types of cortical damage? Specifically: motor or sensory cortex vs. associative cortex (learning task).

S.E. Karpiak: (1) The present view is that when gangliosides are used to treat CNS injury, their efficacy is derived from their ability to reduce the extent of injury during the acute phase of damage. By reducing the magnitude of injury, functional deficits are reduced, and the potential for recovery is increased – simply, less damage, more recovery.

(2) The efficacy of ganglioside therapy in animal models has been demonstrated in CNS damage caused by both electrical and chemical lesions (excitotoxins), trauma, ablations and ischemia. Some of these types of injuries have been studied in the cortex. I do not know of any studies which have differentially studied the motor/sensory/associative cortex. Presently, we are studying the effects of GM1 ganglioside on injury and recovery after ischemic damage to the parietal cortex in rats. In

part of these studies, we find a significant reduction in associated cognitive dysfunction (discrimination learning) in rats treated with GM1 ganglioside.

P.R. Lowenstein: (1) Do gangliosides penetrate the brain?

(2) Do antibodies administered to pregnant rats get into the fetal brain?

(3) Is there an increase in pain perception?

S.E. Karpiak: (1) Gangliosides do cross the intact blood–brain barrier in small but sufficient quantities. It is important to remember that in the injured CNS the blood barrier has been severely compromised, thus allowing greater access to the brain for systematically administered compounds.

(2) Yes, we have seen by using radiolabeled antibody to GM1 (affinity column purified), that antibody does enter the developing CNS in the pregnant rat. We could not find any antibody in the pregnant female's brain.

(3) This has not been systematically studied.

A.Y. Deutch: (1) Was asialoganglioside tried as control?

(2) Was GM1 administered continuously in the activity study?

S.E. Karpiak: (1) Available in sufficient quantities, asialoganglioside has never been shown to exert any effects in numerous test paradigms.

(2) GM1 was administered from postnatal days 5 through 15 (2.5 mg subcutaneously).

G.J. Boer: In relation to the question of transport, does GM pass the placental barrier? And if so, has this been investigated to see what (benificial) effects this can have?

S.E. Karpiak: To my knowledge no one has studied the transport of gangliosides across the placental barrier. I know of ongoing research in our labs which is investigating the ability of gangliosides to limit prenatal damage caused by a variety of insults. These studies have just begun.

E.J. Neafsey: (1) What is the mechanism of ganglioside effect on growth factors?

(2) What is the effective dose of GM1 for the adult rat?

S.E. Karpiak: (1) No one knows definitely the mechanisms by which gangliosides and growth factors interact (modulate). In our labs we have hypothesized that the increased exposure of ganglioside molecules on the plasma membrane surface of neuronal cell membranes may result in increased trophic activity. This increased ganglioside exposure may make the membrane more susceptible to interaction with endogenous trophic factors.

(2) The typical dose range is 10–30 mg/kg (i.m.). In our labs we find that 10 mg/kg is optimal for reducing injury after cortical focal ischemia.

H. Tanila: How can intravenously injected gangliosides reach the ischemic hemisphere used as a model?

S.E. Karpiak: In our model the middle cerebral artery is occluded. However, the hemisphere is still receiving blood supply from the anterior and posterior cerebral arteries (particularly in the peri-infarct zones). There is also blood supply from the contralateral hemisphere.

SECTION III

Functional Aspects of Prefrontal Cortex

CHAPTER 15

Behavioral electrophysiology of the prefrontal cortex of the primate

Joaquin M. Fuster

Department of Psychiatry and Brain Research Institute, School of Medicine, University of California at Los Angeles, Los Angeles, CA 90024, USA

Summary

The prefrontal cortex (PFC) is critical for temporal organization of behavior. It mediates cross-temporal sensorimotor contingencies, intergrating motor action (including speech) with recent sensory information. It performs this role through cooperation of 2 cognitive functions represented in its dorsolateral areas: short-term memory (STM) and preparatory set. Supporting data have been obtained from monkeys performing delay tasks, which epitomize the principle of cross-temporal contingency. In a given trial, the animal performs an act contingent on a sensory cue given a few seconds or minutes earlier. During the delay between cue and response, cells in dorsolateral PFC show sustained activation. Two cell categories can be identified in tasks in which cue and response are spatially separate. Cells of the first participate in STM: Their activation tends to diminish as the delay progresses; in some, the activation level depends on the particular cue received. Similar cells are found elsewhere in the cortex. Cells of the second category seem to take part in preparation of motor response: Their activation tends to increase in anticipation of it and may be attuned to the particular movement the cue calls for. This cell type is rare outside of the frontal cortex. The temporally integrative function of the PFC is probably based on local interactions between "memory" and "motor-set" cells, as well as on neural associations between PFC and posterior cortical areas.

Introduction

It is now well established that the prefrontal cortex plays an important role in the organization of behavior and, therefore, the order and timing of behavioral acts. There is mounting evidence that the essence of that role is what I have called the mediation of cross-temporal contingencies (Fuster, 1985), that is, the intergration of behavior in accord with sensory information that is temporally separate from the action itself. Accordingly, the prefrontal cortex would be that part of the neocortex that allows the organism to reconcile sensations and acts that are mutually contingent but temporally separate from each other. This view puts the prefrontal cortex at the top of neural stuctures involved in sensori-motor integration and in charge of bridging temporal gaps in the perception – action cycle. That is the cybernetic cycle of influences from sensory receptors to motor effectors, to the environment, and back to sensory receptors, that governs orderly behavior. I have postulated that the prefrontal cortex mediates

Correspondence: J.M. Fuster, M.D., Ph. D., UCLA Neuropsychiatric Institute, 750 Westwood Plaza, Los Angeles, CA 90024, USA. Tel. (213) 825 – 0247.

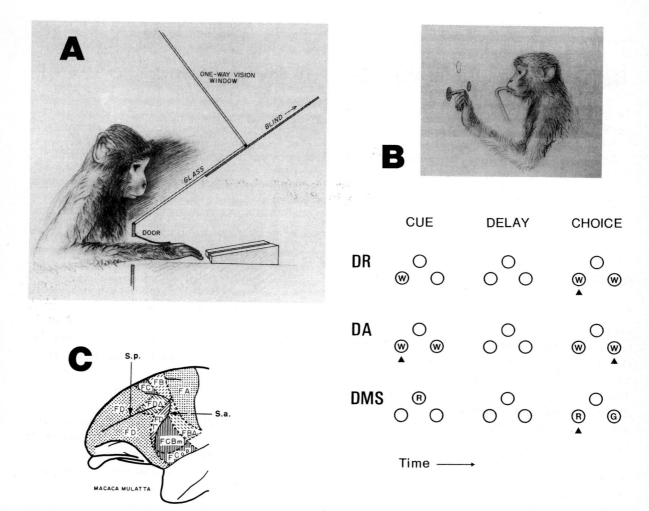

Fig. 1. Schematic diagram of delay tasks. (A) Typical direct-method delayed-response task. A trial begins with placement of a food morsel under one of the two identical objects in full view of the animal; the opaque screen (blind) is then lowered. After a delay of a few seconds or minutes, the screen is raised and the animal offered the choice of one object. If the choice is correct (object with concealed food), the animal retrieves the food as reward; an incorrect choice terminates the trial without reward. The position of the bait is changed at random from one trial to the next. (B) Diagram of test situation for indirect-method, automated, delay tasks. The subject faces a panel with 3 stimulus–response buttons. Below is the sequence of events for 3 different tasks (W, white light; G, green light; R, red light); triangles mark the site of correct response, which is rewarded with fruit juice. DR: indirect-method delayed response. The cue is the brief illumination of one of the lower buttons; after the delay, the 2 lower buttons are lit for choice; correct response is pressing the button lit before the delay. DA: delayed alternation. The subject must press 1 of the 2 lower buttons when they appear simultaneously lit between delays; the correct button alternates between right and left. DMS: delayed matching to sample. The sample (cue) is the brief colored illumination of the top button. After the delay, 2 colors appear simultaneously in the lower buttons; correct response is pressing the button with the sample color; the sample color and its position in the lower buttons are changed randomly from trial to trial. (C) Cytoarchitectonic map of the frontal cortex of the rhesus monkey according to Von Bonin and Bailey. The prefrontal cortex is labeled FD. Animals with prefrontal lesions show deficits in performance of the 4 tasks illustrated above. The magnitude of the deficit depends on the location and extent of the lesion.

those temporal gaps by critically supporting at least 2 cognitive functions that make that temporal bridging possible and, consequently, the temporal organization of behavior possible: (1) a temporally "retrospective" function of short-term memory for sensory information and (2) a temporally "prospective" function of preparatory motor set. Both these functions appear represented in the dorsolateral prefrontal cortex of the primate. In this chapter I briefly summarize some electrophysiological evidence for these propositions. In particular, my focus will be on single-unit activity from the prefrontal cortex of monkeys performing a class of behavioral tasks that epitomize cross-temporal integration: delay tasks.

Prefrontal cortex and the mediation of cross-temporal contingencies

The logic of delay tasks is the logic of cross-temporal contingencies. It can be summarized in 2 statements: "If now this, then later that; if earlier that, then now this". Fig. 1 illustrates examples of the most commonly used delay tasks. All these tasks require, on every trial, the performance of a discrete behavioral act in accord with a discrete item of sensory information that has been received in the recent past. All of them are impaired by lesions of the dorsolateral prefrontal cortex. Such impairments are a clear indication of the involvement of the dorsolateral prefrontal cortex in cross-temporal integration.

The first electrical indication of that prefrontal role was obtained by Walter and his colleagues in the human (Walter et al., 1964): a slow surface-negative potential which can be recorded from the frontal region in the interval of time, imposed by the investigator, between a stimulus and a motor act contingent on it. Those investigators called it the "contingent negative variation" (CNV). In the early 1970s, single-unit recordings from prefrontal areas in monkeys performing delay tasks provided the first indications at the neuronal level of the involvement of the prefrontal cortex in cross-temporal contingencies (Fuster, 1973). One of the most striking findings was that of neurons that fired continuously and at high levels during the period of delay that in those tasks is interposed between a sensory cue and the motor response contingent on it (Fig.2). Sustained inhibition was also encountered, although less commonly (Fig. 3). Both the sustained activation and the sustained inhibition of the delay period were shown to depend on the presence of a relationship of mutual contingency between the events that preceded and those that succeeded that period. For example, that sustained activation or inhibition disappeared in mock trials, when the animal was deprived of the information that determined the response at the end of the delay (Fig. 3).

Those early data clearly suggested the cross-temporal integrative role of prefrontal neurons. I attributed those kinds of activity to the involvement of those neurons in a form of transient or

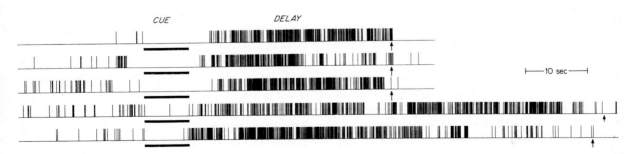

Fig. 2. Discharge of a prefrontal unit during 5 trials of a delayed-response task (direct method). A horizontal bar marks the cue period and an arrow the end of each trial. Note the activation of the cell during the delay (30 sec in the upper 3 trials, 60 sec in the lower 2).

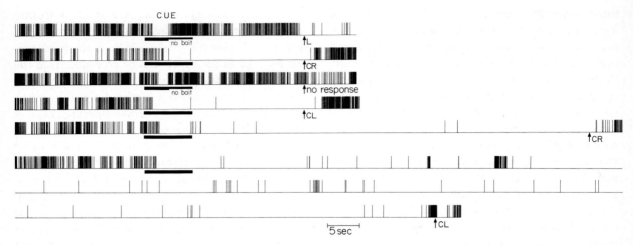

Fig. 3. Discharge of a prefrontal unit during 6 trials of a delayed-response task (direct method). On the first and third trials, the cue (bait under one object, as in Fig. 1A) is not given. In those 2 mock trials, with absence of cross-temporal contingency, the unit fails to show the normal inhibition during the delay that, in the last trial, is nearly 4 min. long. Abbreviations: C, correct response; R, right; L, left.

short-term memory. Such a memory involvement was subsequently made more evident by data from our lab and from others. Also later, it became clear that prefrontal cells do not engage only in short-term memory but also in a second temporally integrative function that I have postulated to be essential for the mediation of cross-temporal contingencies of behavior, that is, *preparatory motor set*. Both the memory role and the motor-set role of prefrontal neurons were essentially ascertained by the observation that in some units delay activity was coupled to the stimulus, while in others it was coupled to the motor response: stimulus coupling was observed for a considerable time after the stimulus, while response coupling was observed well in advance of the response. In the next 2 sections I briefly describe the evidence obtained in our laboratory and elsewhere for those two prefrontal processes of cross-temporal integration, that is, short-term or "working" memory and preparatory motor set. My work on both these aspects of prefrontal function has been carried out in collaboration with Gary Alexander, Richard Bauer, John Jervey, Javier Quintana, Carl Rosenkilde, and Javier Yajeya.

Prefrontal memory

By selectively cooling the dorsolateral convexity of the prefrontal cortex, we were able to demonstrate a reversible impairment of delay-task performance (Bauer and Fuster, 1976). That impairment had two outstanding features: (1) it was equally severe whether or not the cue to be remembered was spatially defined; (2) it increased in magnitude as a function of the delay (Fig. 4). In other words, the deficit became greater as the length of the period between cue and choice was increased. This second feature of the cryogenic prefrontal deficit was fully consistent with the assumption that the function impaired was one with a temporal decay. Of course, short-term memory is one such function.

Another characteristic of the prefrontal delay-task deficit is its supramodality. Not only does it occur whether or not the cue is spatially defined, but whether it is visual, auditory, or somesthetic. This supramodality of the deficit was first demonstrated in the human (Lewinsohn et al., 1972). Our recent cooling data from monkeys suggest that not only does the deficit apply to visual short-term memory but to haptic short-term memory as well. Furthermore, these data suggest that the deficit is not only supramodal but also cross-modal (i.e., it occurs in delay tasks requiring transfer across time from vision to touch, or vice versa).

During the delay (that is, during the retention period of delay tasks) many dorsolateral prefrontal units are coupled to the stimulus just presented. This coupling to the sensory stimulus after it has disappeared from the environment can be taken,

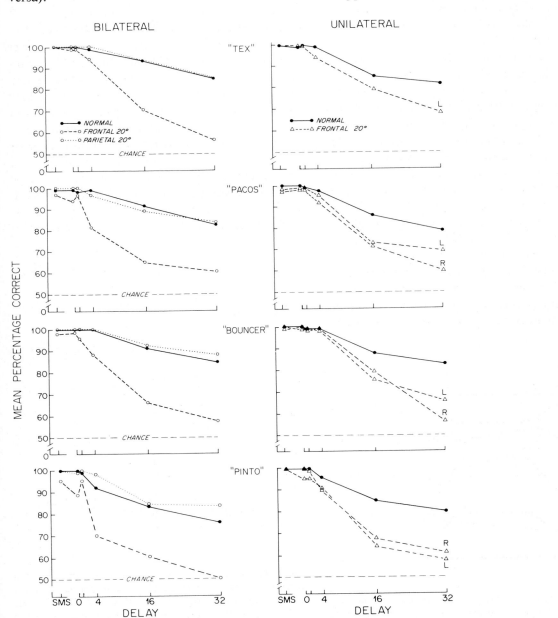

Fig. 4. Performance of 4 monkeys in delayed matching-to-sample (DMS: Fig. 2B) at normal cortical temperature and under cooling of dorsolateral prefrontal or posterior parietal cortex. SMS, simultaneous matching-to-sample; delay in seconds. Note increasing prefrontal deficit as a function of delay.

of course, as an indication of those units' involvement in retention of the stimulus. The coupling has 2 different forms or manifestations: (1) gradual descent of discharge as the delay progresses; and (2) differential stimulus-dependent firing of the unit in terms of the 2 or more alternative cues (e.g., directions, colors or patterns) conventionally utilized in delay tasks.

The gradual descent of delay activity after a peak at the time of the cue or immediately thereafter is, of course, reminiscent of mnemonic decay. It is exemplified by types C and D of my original classification (Fig. 5). However, in the absence of differential stimulus-dependent delay firing, the argument for cellular memory simply on the basis of a firing trend is only suggestive, not conclusive. At most, it suggests that that form of delay activity is related to the retention of the stimulus properties that are common to all the alternate cues in the task (like shape, brightness, etc.). Of course, there are very important components of the "engram", but they are not the only ones that determine the correctness of the choice. They are insufficient in themselves for the proper bridging of the cross-temporal contingency.

More convincing, in terms of their involvement in memory, is the differential stimulus-dependent discharge of prefrontal cells during the delay. Such units have been demonstrated in spatial delay tasks such as delayed response (Fig. 6) and delayed alternation, as well as in non-spatial delay tasks, such as delayed matching (Niki, 1974; Nike and Watanabe, 1976; Rosenkilde et al., 1981; Fuster et al., 1982; Kojima and Golman-Rakic, 1984; Quintanta et al., 1988; Funahashi et al., 1989). Even in such cells, however, the argument for memory may be inconclusive, expecially in spatial tasks. For in these tasks the direction of the correct response can be predicted by the animal either from the cue or from the direction of the previous response. As a result, unit activity related to memory cannot be easily distinguished from unit activity related to the motor response. This is especially true in units, such as the one in Fig. 6, in which both factors

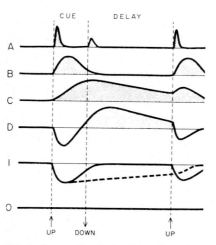

Fig. 5. Types of units in the prefrontal cortex of the monkey during delayed-response testing. The heavy line represents deviations of firing from inter-trial baseline. Arrows mark displacements of opaque screen between animal and test objects (Fig. 1A).

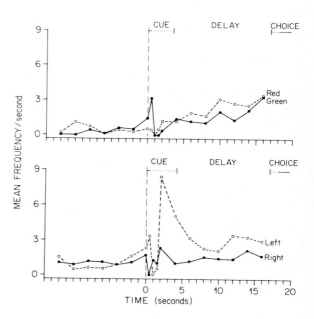

Fig. 6. Firing frequency of a prefrontal unit during testing in delayed response (indirect method) and delayed matching-to-sample with colors (Fig. 1B). (In these tests, the cue was terminated by the animal's pressing of a button; hence the variability of cue duration.) Note, during the delays, differential firing in delayed response and non differential but accelerating firing in red- and green-sample trials.

seem to play a role, in other words, units in which delay-activity is to some degree coupled to both the cue just presented and the impending response. In any event, the phenomenon of sensory coupling (i.e, memory-coupling) can be, and has been, demonstrated in tasks with spatial dissociation between cue and response.

It is of course debatable whether the sustained, cue-dependent, activation of some prefrontal neurons during the delay reflects the involvement of those neurons in the cell assemblies that Hebb postulated to be the essence of short-term memory. A reasonable case can be made for this, provided that a cell assembly is not construed simply as a cluster of neurons but as a network, as I have attempted to do elsewhere (Fuster, 1989); a network that extends well beyond the prefrontal cortex and includes other sectors of the association cortex. Which sector will be involved, in addition to the prefrontal cortex, will depend on the sensory character of the cue. If the cue is visual, as it is in visual delayed matching, the inferotemporal cortex will be involved (Fuster and Jervey, 1982; Fuster, in preparation), where neurons evidently take part in the retention of very specific visual information. If the cue is somesthetic, the posterior parietal cortex will be involved (Koch and Fuster, 1990), where some neurons seem to take part in retention of haptic information. In any case, those are cortical areas that interact functionally with the prefrontal cortex (Fuster et al., 1985, Quintana et al., 1990). The important point I wish to make here is that the prefrontal cortex is part of all neuronal networks that mediate the kind of cross-temporal contingencies that delay tasks contain; those other areas are also part of those networks inasmuch as the sensory information that needs to be transferred across time is, so to speak, within their specialty.

In conclusion, it seems indisputable that the neurons of the prefrontal cortex are very much engaged in a form or aspect of memory that can best be characterized as short-term and context-dependent memory. It is short term inasmuch as the action is to occur in the short term. It is context dependent in that it depends on the context of the action itself. Whatever the content of that memory, it is memory primarily (and perhaps even exclusively) in the service of behavioral action.

Prefrontal set

In addition to the retrospective, mnemonic, aspects of prefrontal function, its prospective, future-related, aspects have also been documented in the neuropsychological literature (Fuster, 1989). In the human, dorsolateral prefrontal lesions are associated with well-known difficulties in planning; they can be exposed by such formal tests as the Tower of London, which is a test of the ability to organize motor action in the short term (Shallice, 1982). It seems, therefore, that we are dealing with the failure of a function that is temporally symmetrical to the short-term memory function we have just discussed, though not by any means independent from it. Ingvar (1984) dubbed it "the memory of the future".

Just as there is electrophysiological evidence of the involvement of dorsolateral prefrontal neurons in short-term memory, there is evidence of their involvement in short-term preparation for movement or motor set. In delay tasks, such as delayed response, delayed matching and the like, motor-set cells behave in a manner directly opposite to that of the memory cells we have seen. Whereas memory cells seem to be looking backwards, to the cue, motor-set cells seem to be looking forward, to the impending motor response. Thus, we observe units that instead of gradually diminishing their firing between cue and response, accelerate it. In any event, units that show accelerating discharge during the delay are probably a special case of a general category of units that have been shown to do so while the monkey prepares for movement in a variety of tasks, including delay tasks (Kubota et al., 1974; Sakai, 1974; Niki and Watanabe, 1976; Fuster et al., 1982; Boch and Goldberg, 1989). A unit may accelerate its discharge regardless of which cue the monkey has been given or which response he is about to execute, provided that a motor act is in the making (Fig. 7). More direct

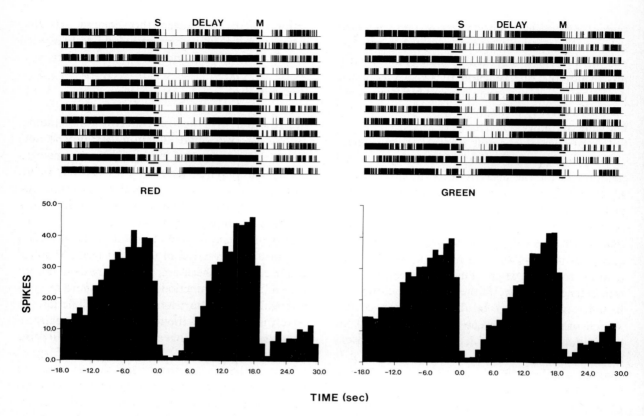

Fig. 7. Rasters and frequency histograms from a prefrontal unit during delayed matching-to-sample (red or green). Sample (S) and match (M) periods are marked by horizontal lines under the records. The unit accelerates its firing before the sample (which the animal is supposed to terminate by pressing a button) and during the 18-sec delays.

evidence of involvement in specific-response set can be found in units that show different rates of discharge, depending on the particular motor response that the monkey is preparing. This evidence can best be obtained by use of behavioral tasks in which the cue is spatially dissociated from the response.

Lately, with Quintana, we have been investigating the cellular phenomena associated with preparatory motor set in the dorsolateral prefrontal cortex. For that, we are using a number of delay tasks in which cue and response are spatially dissociated, for example, delayed matching to sample and delayed conditional discrimination. In one such task, the cue consists of a color always displayed in the same position of the field; the color determines the direction of the response (left or right) at the end of the delay. By use of reversible – cryogenic – lesions we are gathering further evidence that the integrative functions of the prefrontal cortex are important for performance of these tasks (Quintana and Fuster, 1988).

The most important microelectrode findings from these latest studies can be summarized as follows: Units that fire differentially at the time of the motor response (i.e., with a different firing frequency on right than on left response) commonly show selective response coupling well in advance of the response itself. In fact, response coupling may already appear at the time of the cue *if* that cue defines and determines the direction of the forthcoming response. In other words a prefrontal cell may reveal its tuning to a particular response direction already upon appearance of the sensory cue that symbolizes that direction; thus, the discharge of the cell, in accord with the color of the cue, "predicts" by several seconds the motor response of the animal. Furthermore, in the course of the delay, direction-selective cells show progressively greater discharge as their "preferred response" approaches, whether that response is to the right or to the left.

In conclusion, during the delay between cue and response, a substantial proportion of dorsolateral prefrontal units show motor coupling in anticipation of movement. Their discharge tends to increase as the time for expected movement approaches, whereas, as we have seen, the opposite is true for sensory-coupled cells, the discharges of which tend to diminish during the same period. It is reasonable to infer that sensory coupled cells are somehow involved in the retention of sensory information, while motor-coupled cells engage in preparation of the response that is consonant with that information according to the rules of the task.

Functional interactions

It appears from the foregoing that 2 general types of neurons in the dorsolateral prefrontal cortex participate in the 2 complementary processes of (1) holding sensory information, and (2) preparation for motor action in accordance with that information. It is not yet possible to determine the mechanisms by which the information is transferred across time and incorporated into the appropriate action. That transfer may at least in part occur locally within dorsolateral cortex. It should be noted that, during the delay, units with decelerating activity have been found there, in close proximity to units with accelerating activity (Fuster et al., 1982). Therefore, it is not inconceivable that information is relayed more or less directly from the former to the latter cells within small cortical confines, perhaps within functional modules or columns of the prefrontal cortex.

It seems also likely, however, that those transactions take place within and between widely distributed neuronal networks, of which the prefrontal cortex, along with other cortical areas, is a part. In order to understand this kind of interactions it is useful to look at the larger picture of cortical involvement in the perception – action cycle (Neisser, 1976; Arbib, 1981; Fuster, 1989). As mentioned above, this cycle is defined by the circular pattern of cybernetic influences running from the environment through sensory systems, through motor systems, and back to environment, that supports and regulates any orderly sequence of behavior. There is an extensive array of well-substantiated cortical and subcortical connections underlying that cycle of influences (Fig. 8).

Two general points need to be made here. The first is that the prefrontal cortex, together with polysensory association areas of posterior cortex with which it is well connected, constitutes the highest level of the sensory and motor hierarchies of cortical structures involved in the perception – action cycle. The prefrontal cortex, through its connections with those other cortical areas, closes at the top the ring of neural structures involved in sensorimotor integration. One implication of its supraordinate position, which I have attempted to spell out with physiological evidence, is that the prefrontal cortex (especially its dorsolateral portion) is in charge of closing, via those connections, the temporal gaps within the cycle. The dorsolateral prefrontal cortex is needed to integrate sensory input with later action, in other words, to close the cross-temporal contingencies within the cycle. In order to accomplish this, the prefrontal cortex probably interacts with other associative areas and with subcortical structures that are part of motor systems. Reciprocal interactions with these cortical and subcortical structures may, perhaps through some form of neural reverberation, support the short-term memory and motor-set functions of the prefrontal cortex that I postulate are needed for cross-temporal integration.

The second point that I want to make to close this discussion is that at least part of the output of prefrontal cortex that flows back upon sensory areas may well serve the role of corollary

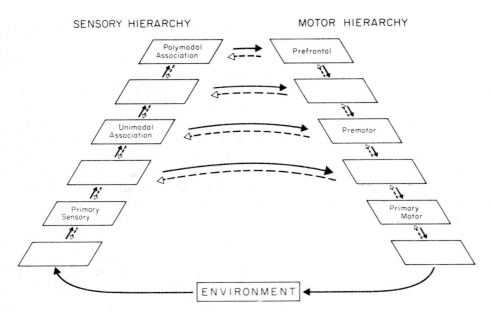

Fig. 8. Scheme of connectivity between cortical regions that are part of the sensory and motor hierarchies of information processing and are involved in the perception – action cycle. Blank boxes represent intermediate stages of cortical or subcortical processing within or between labeled regions. Note that connections are bidirectional. The basic scheme, as represented here (and especially well substantiated in visual and somesthetic systems), is consistent with a large body of anatomical and functional data (for details, see Fuster, 1989).

discharge. In this context, corollary discharge is another cybernetic function that serves cross-temporal integration and the perception – action cycle. As Teuber (1972) proposed it, the corollary discharge of the prefrontal cortex is some kind of output that this cortex sends to sensory structures concomitantly with action. It is supposed to prepare these structures for the anticipated effects of the action. Thus, whereas above we have dealt with motor set deriving from recent sensory information, here we are dealing with the converse, that is, sensory set deriving from recent action. In any event, the physiological evidence for corollary discharge is less compelling and the physiological mechanisms no less obscure than for motor set. At present, it seems appropriate to consider corollary discharge as another hypothetical mechanism, in addition to those of short-term memory and motor set, by which the prefrontal cortex ensures the integration of behavior in the temporal domain.

References

Arbib, M.A. (1981) Perceptual structures and distributed motor control. In: V.B. Brooks (Ed.), *Handbook of Physiology. Vol. II Nervous System,* Amer, Physiol. Soc. Bethesda, MD, pp. 1448 – 1480.

Bauer, R.H. and Fuster, J.M. (1976) Delayed-matching and delayed-response deficit from cooling dorsolateral prefrontal cortex in monkeys. *J. Comp. Physiol. Psychol.,* 90: 293 – 302.

Boch, R.A. and Goldberg, M.E. (1989) Participation of prefrontal neurons in the preparation of visually guided eye movements in the rhesus monkey. *J. Neurophysiol.,* 61: 1064 – 1084.

Funahashi, S., Bruce, C.J. and Goldman-Rakic, P.S. (1988) Memory fields: directional tuning of delay activity in the dorsolateral prefrontal cortex of rhesus monkey. *Soc. Neurosci. Abstr.,* 14: 860.

Fuster, J.M. (1973) Unit activity in prefrontal cortex during delayed-response performance: Neuronal correlates of transient memory. *J. Neurophysiol.,* 36: 61 – 78.

Fuster, J.M. (1985) The prefrontal cortex, mediator of cross-temporal contingencies. *Hum. Neurobiol.,* 4: 169 – 179.

Fuster, J.M. (1989) *The Prefrontal Cortex*, 2nd Edn., Raven Press, New York.

Fuster, J.M. and Jervey, J.P. (1982) Neuronal firing in the inferotemporal cortex of the monkey in a visual memory task. *J. Neurosci.*, 2: 361 – 375.

Fuster, J.M., Bauer, R.H. and Jervey, J.P. (1982) Cellular discharge in the dorsolateral prefrontal cortex of the monkey in cognitive tasks. *Exp. Neurol.*, 77: 679 – 694.

Fuster, J.M., Bauer, R.H. and Jervey, J.P. (1985) Functional interactions between inferotemporal and prefrontal cortex in a cognitive task. *Brain Res.*, 330: 299 – 307.

Ingvar, D.H. (1985) "Memory of the future": An essay on the temporal organization of conscious awareness. *Hum. Neurobiol.*, 4: 127 – 136.

Koch, K. and Fuster, J.M. (1990) Unit activity in monkey parietal cortex related to haptic perception and temporary memory. *Exp. Brain Res.*, in press.

Kojima, S. and Goldman-Rakic, P.S. (1984) Functional analysis of spatially discriminative neurons in prefrontal cortex of rhesus monkeys. *Brain Res.*, 291: 229 – 240.

Kubota, K., Iwamoto, T. and Suzuki, H. (1974) Visuokinetic activities of primate prefrontal neurons during delayed-response performance. *J. Neurophysiol.* 37: 1197 – 1212.

Lewinsohn, P., Zieler, R., Libet, J., Eyeberg, S. and Nielson, G. (1972) Short-term memory — a comparison between frontal and nonfrontal right- and left-hemisphere brain-damaged patients. *J. Comp. Physiol. Psychol.*, 81: 248 – 255.

Neisser, U. (1976) *Cognition and Reality: Principles and Implications of Cognitive Psychology*. Freeman, San Francisco, CA.

Niki, H. (1974) Differential activity of prefrontal units during right and left delayed response trials. *Brain Res.*, 70: 346 – 349.

Niki, H. and Watanabe, M. (1976) Prefrontal unit activity and delayed response: relation to cue location versus direction of response. *Brain Res.*, 105: 79 – 88.

Quintana, J. and Fuster, J.M. (1988) Effects of cooling parietal or prefrontal cortex on spatial and nonspatial visuo-motor tasks. *Soc. Neurosci. Abstr.*, 14: 160.

Quintana, J., Yajeya, J. and Fuster, J.M. (1988) Prefrontal representation of stimulus attributes during delay tasks. I. Unit activity in cross-temporal integration of sensory and sensory-motor information. *Brain Res.*, 474: 211 – 22.

Quintana, J., Fuster, J.M. and Yajeya, J. (1990) Effects of cooling parietal cortex on prefrontal units in delay tasks. *Brain Res.*, in press.

Rosenkilde, C.E., Bauer, R.H. and Fuster, J.M. (1981) Single cell activity in ventral prefrontal cortex of behaving monkeys. *Brain Res.*, 209: 375 – 394.

Sakai, M. (1974) Prefrontal unit activity during visually guided lever pressing reaction in the monkey. *Brain Res.*, 81: 297 – 309.

Shallice, T. (1982) Specific impairments of planning. *Phil. Trans. R. Soc. Lond. (Biol.)*, 298: 199 – 209.

Walter, W., Cooper, R., Aldridge, V., McCallum, W. and Winter, A (1964) Contingent negative variation: an electric sign of sensori-motor association and expectancy in the human brain. *Nature*, 203: 380 – 384.

Discussion

P. Goldman-Rakic: You seem to have associated "short-term memory" with retrospective function and "preparatory set" with prospective function, but is that so clear? I think a short-term memory as serving a prospective as well as a retrospective function.

J.M. Fuster: Nolo contendere — I agree. Short-term memory serves the prospective function of preparatory set. But the two functions, short-term memory and preparatory set, retain their individuality as two mutually complementary functions at the service of the supraordinate function of temporal integration. The two are temporally symmetrical: one is the memory of the recent past, the other the "memory" of the impending future. In more concrete terms, the latter is the set for the action that the former calls for.

H. Tanila: Cooling of inferotemporal cortex impaired the monkeys' performance in a DMS test. Did you find any effect of parietal cooling on the spatial DR test?

J.M. Fuster: By cooling the posterior parietal cortex we have observed a deficit in a visuo-spatial delay task with symbolic cueing of spatially separate responses, but not in the classical delayed-response (DR) task.

A. Kalsbeek: How do you think that the role of the prefrontal cortex as described by you is related to the specific activation of the mesocortical DAergic system?

J.M. Fuster: The activation of the mesocortical DA system is specific only in that it may be independent from that of the mesostriatal DA system and other neurotransmitter systems. However, the mesocortical system is non specific almost by definition. It originates in the reticular core of the brain stem and it innervates large areas of cortex, notably the prefrontal cortex. I envision DA activation as the precondition for the prefrontal cortex to perform its temporal integrative functions.

T. Paus: (1) What is the evidence of the interference control function by the prefrontal cortex? (2) In what way is it possible to dissociate this function from dorsolateral prefrontal functions?

J.M. Fuster: (1) The evidence lies mostly in ablation experiments: orbitofrontal lesions elicit deficits in performance of tasks that require the inhibition of internal or external interference. (2) Therefore, the interference control function appears primarily represented in the orbital prefrontal cortex, whereas the memory and set functions are primarily represented in the dorsolateral prefrontal cortex.

F. Mora: You mentioned a specific relationship between dorsolateral frontal cortex and aging. Is there a specific degeneration in that area that differs from that of other areas?

J.M. Fuster: I am not aware of any "specific" degeneration of the prefrontal cortex in aging. But the evidence is quite compelling, from many sources, that the prefrontal cortex is especially vulnerable to the aging process, at least as vulnerable as any other associative region of the neocortex.

J.E. Pisetsky: Do you think that in the execution of skillful actions with preparatory sets there is participation of long-term memory templates or elements?

J.M. Fuster: True, prefrontal short-term memory is in part the reactivation of old memories to be used for the preparatory motor set.

CHAPTER 16

Cellular and circuit basis of working memory in prefrontal cortex of nonhuman primates

Patricia S. Goldman-Rakic

Yale University School of Medicine, Section of Neuroanatomy, C303 SHM, New Haven, CT 06510, USA

Introduction

Working memory is the term applied by cognitive psychologists and theorists to the type of memory that is active and relevant only for a short period of time, usually on a scale of seconds (Baddeley, 1986). A common example of working memory is keeping in mind a newly read phone number until it is dialed and then immediately forgotten. This criterion — useful or relevant only transiently — distinguishes working memory from the very different process that has been called reference memory (Olton and Papas, 1979), semantic memory (Tulving, 1972), and procedural memory (Squire and Cohen, 1984), all of which have in common that their contents are, for all intents and purposes, stable over time (e.g., someone's name, the color of one's eyes, the shape of an apple, the manner of using a fork and knife). In contrast to working memory, all of these other forms of memory can be considered associative in the traditional sense; i.e., information is acquired by a fixed association between stimuli and responses and/or consequences.

I have argued that the capacity which cognitive psychologists call working memory is similar, if not identical, to the process which is measured by delayed-response tests in nonhuman primates (Goldman-Rakic, 1987). The ability to keep the location of an object "in mind" is not unlike the ability of humans to keep a phone number or a person's name in mind. Indeed, the classical delayed-response task was designed and introduced to comparative psychology by Walter Hunter to differentiate among animal species on intelligence — which he defined as the ability to respond to situations on the basis of stored information, rather than on the basis of "immediate" stimulation (Hunter, 1913).

A crucial feature of the delayed-response task is the need to update information on every trial. The correct response on trial *n* is not predictive for the correct response on trial *n + 1*. The lack of a predictive relationship between one trial and the next is unlike the procedure employed in associative learning or conditioning paradigms, in which the correct response or correct association (as in stimulus – stimulus learning) is the same on each subsequent trial. The underlying principle of delayed-response operates in other commonly used behavioral paradigms: spatial delayed alternation, object alternation, match-to-sample or nonmatch-to-sample tasks. All of these tasks rely on the common principle that the subject must keep "in mind" an item of information for only one trial, and update to a new memorandum on the next trial. Of course, an organism may need to store a

---

Correspondence: P.S. Goldman-Rakic, Ph. D., Yale University School of Medicine, Section of Neuroanatomy, C303 SHM, 333 Cedar Street, New Haven, CT 06510, USA. Tel.: (203) 785-4808; Fax: (203) 785-5263.

unique event or datum for future reference and, in this instance, additional mechanisms, including rehearsal, must operate. However, for purposes of discussion, the paradigm that I would like to focus on in this chapter is one in which erasure, or forgetting, of an event is as important as its registration. Baddeley's phrase, "scratch-pad memory" perfectly captures this feature of the working memory process (Baddely, 1973).

The objective of this chapter is to review the role of prefrontal cortex in working memory. There is a need to clarify this issue for purposes of constructive dialogue and discussion as well as to stimulate future research on the key mechanisms. Although there appears to be a growing consensus that the principal sulcal cortex in the monkey plays some role in memory processing (Passingham, 1985a, b; Goldman-Rakic, 1987, 1990; Fuster, 1989), there is far less agreement on the nature of this process and whether this is the cardinal function or one of many subserved by prefrontal cortex. Furthermore, in spite of good agreement about the transient nature of the memory trace in delayed-response tasks, terminology is still an issue (Passingham, 1985a, b; Goldman-Rakic, 1987, 1990; Fuster, 1989).

Terminology: what's in a name?

What is an appropriate designation for the process that is disturbed by dorsolateral prefrontal lesions in the form of impairment on delayed-response tasks? Historically, it was termed "immediate" memory to differentiate it from conventional memory (Jacobsen, 1936). It has also been simply called "spatial memory" (Mishkin, 1964). I have proposed that the term "working memory" is appropriate to describe this process. Representational memory is another very appropriate designation which has been used (Goldman-Rakic, 1987), but some objections to it can be raised because of its several meanings. For example, external stimuli in the environment often "represent" or signal other stimuli (e.g., the sound of a siren arouses expectation of a police car or a fire engine value) or other meanings that are indicated by the stimulus. Thus, memory processes based on external signals in non-delay tasks may also be considered representational. Further, the term is often used in the sense of central "representations" of the periphery.

The term "transient memory" is perfectly appropriate to designate the psychological process that is tapped by delayed-response tasks, but this term has the sense of a passive process that fades. Calling the process "short term memory" is also problematic because, traditionally, short-term memory has been considered an obligatory stage through which information must pass on its way to long-term memory (for review, see Roitblat, 1987; Baddeley, 1986). More recently, the term "cognitive" memory has been used by some to refer to representational memory. While working memory *is* a cognitive process, the name cognitive memory implies that all other types of memory, including episodic, semantic, and associative are *not* cognitive. The term which I advocate, "working memory", has the virtue of calling attention to the active, dynamic nature of the process, the similarity of the process in monkey and human, and it is easily distinguished from associative memory from which it can be dissociated in function, localization, and durability.

Fuster has proposed the terms "prospective" and "retrospective" memory and suggested that they may represent two distinct processes. These terms are attractive and appropriate. However, the term "working memory" serves both prospective *and* retrospective functions, depending on the point in time considered. For example, when the cue goes out of view and the animal has to keep its location in mind, we would call that prospective memory; the information is being held for a future event. However, at the end of the delay, when the animal selects a food well, the memory trace is considered retrospective since it reflects information presented in the past. Until these two terms can be shown to represent dissociable mechanisms (rather than serving several functions), I do not see the reason to postulate two distinct processes, nor do these terms particularly capture the transient and renewing nature of the process under review.

It may be argued that the process of working memory differs somewhat in humans and monkeys. A human can keep something in mind for a longer period of time by silently rehearsing it. However, rehearsal invokes repetitive associative processes, and information can be retained for longer periods by both laboratory animals and humans when rehearsal is allowed. It has been shown that monkeys can rehearse (Kojima et al., 1982) as can other prelinguistic organisms, chimps, children, the deaf, and others that do not have the benefit of language. Furthermore, individual differences may be expected among both humans and monkeys in the length of time that information can be held "in mind", as well as in the mental computations that can be performed on the internalized knowledge base. Species differences are large (Goldman-Rakic and Preuss, 1988) and may reflect the emergence of multiple parallel representational systems, each with independent mnemonic modules (Goldman-Rakic, 1987; 1988a, b; Rumelhardt and McClelland, 1988). However, the fundamental working memory process may be quite similar across mammalian species. There is no doubt that language adds another dimension to information processing, but the question to be resolved is whether linguistic analysis invokes a qualitatively new process rather than an additional, powerful channel of working memory. The concept of "working memory" has been developed in a linguistic framework (Baddeley, 1986).

Cellular mechanisms underlying working memory

Support for a memory interpretation of delayed-response deficits comes from many sources. I should like to call attention here to the recording studies in awake monkeys trained to perform delayed-response and other cognitive tasks (e.g., Fuster and Alexander, 1971; Kubota and Niki, 1971; Kojima and Goldman, 1982; Niki and Watanabe, 1986; Funahashi et al., 1989, 1990 a – c). As in other areas of the cerebral cortex, a variety of neuronal responses has been recorded from prefrontal neurons and there is every reason to believe that they all play some role in integrated delayed-response performance and in the fundamental process of working memory. For example, neuronal activities have been related to the cue, to the delay, to the response, or to some combination of these in manual delayed-response tasks (e.g., Fuster, 1989, for review). A similar variety of neuronal activation patterns has now been demonstrated in oculomotor tasks (Funahashi et al., 1989, 1990a, b; Boch and Goldberg, 1989; Barone and Joseph, 1989). The number of distinct types of neurons that participate in the working memory process is not fully known. Full characterization of prefrontal neuronal activity in delayed-response tasks requires assessing a neuron's pattern of activation under a variety of task permutations, and this is both time-consuming and technically difficult. Nevertheless, after 2 decades of research, we can draw several general conclusions: that all major classes of neuron (cue, delay, response) are found in the caudal portion of the dorsolateral prefrontal cortex; and all major classes of neuron can be identified in oculomotor as well as in manual delayed-response tasks. Thus, the functions of the prefrontal cortex pertain to several motor systems and are amenable to cellular and circuit analyses.

Delay-period activity: transient memory trace

It is very well established that neurons in the prefrontal cortex become activated during the delay period of a delayed-response trial; at issue is whether the neurons which express this activity are the cellular counterparts of a mnemonic event as suggested in seminal physiological studies of primate prefrontal cortex (Fuster and Alexander, 1971; Fuster, 1973; Kubota and Niki, 1971). It could be argued, for example, that neurons activated during the delay of a delayed-response trial are not necessarily representing specific information to be remembered but, rather, are engaged in some sort of general preparatory or motor set to respond. The act of preparing to respond could invoke postural mechanisms, both peripheral and central, and not necessarily implicate central registration, storage, or processing of stored information. However, recent evidence from studies in my laboratory are providing strong and convincing evidence for mnemonic processing in prefrontal cortex. Shintaro Funahashi, Charles Bruce, and I have been using an oculomotor delayed-response (ODR) paradigm to study prefrontal function (Funahashi et al., 1989, 1990a, b). The advantages of this paradigm over other methods of studying delayed-response performance are many. The animal is required to fixate a spot of light on a TV monitor and maintain fixation during the brief (0.5 sec) presentation of a stimulus followed by a delay period of variable length (Fig. 1). Visual stimuli can be presented in any part of the visual field, thus allowing complete control over the specific information that the animal has to remember on any given trial. The fixation spot is turned off only at the end of the delay period and its offset constitutes an instruction to the animal to break fixation and direct his gaze to where the target *had been* presented. Most important, because the animal is required to fixate during the delay period, he is discouraged from making anticipatory rehearsal responses to the cue location and, further, behavior during the delay is equated on every trial. Because the animal's behavior is rather strictly controlled in all phases of a trial, we believe that the animal can perform correctly only if it uses mnemonic processing.

Memory fields coded by prefrontal neurons

Although all the neurons in the caudal prefrontal cortex can be assumed to contribute to the functions of that cortex, the neurons that have particularly intrigued us are those that discharge vigorously when the stimulus goes out of view — in the delay period. As has been reported for manual delayed-response tasks (Kubota and Niki, 1971; Fuster, 1973), dorsolateral prefrontal neurons increase (or often decrease) their discharge rate during the delay period of a trial. The neuron displayed in the middle panel of Fig. 2 is an example: its activity rises sharply at the end of the stimulus, remains tonically active, and then ceases rather abruptly at the end of the delay, after the fixation spot disappears, and either just before, during, or after the response is initiated (see Funahashi et al., 1990b, for further discussion of saccade-related activity). Such neuronal activity must be as fundamental to the working memory process as orientation-tuned cellular activity of neurons in the visual cortex are to the perception of contour.

The results of our studies of neuronal activity during oculomotor delayed-response performance have advanced our understanding of delay-period activity by suggesting the concept of the "memory field". Although previous studies had shown that delay period activity of prefrontal neurons was directional (discharging more for left than for right trials or the reverse), such directional activity could be interpreted as a specialized code for left–right

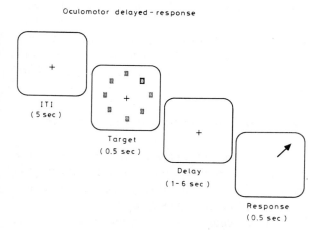

Fig. 1. Main phases of a trial in the oculomotor delayed-response (ODR) paradigm. As described in the text, the task is presented on a TV monitor. The monkey fixates the central spot and maintains fixation throughout the delay period until the fixation spot disappears, where upon it makes a response to the remembered target location. ITI, intertrial interval; arrow, correct direction for the response (arrow is for the reader, not the monkey).

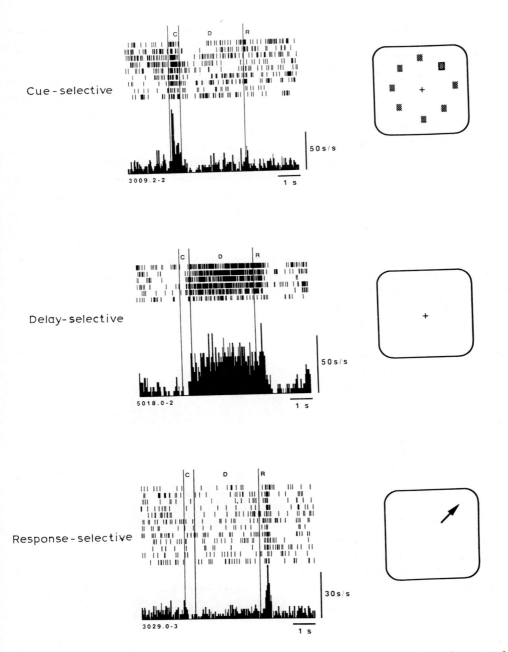

Fig. 2. Three distinct types of neuronal activation recorded from prefrontal neurons during performance of the oculomotor delayed-response task. Upper panel displays a neuron that exhibits a *phasic* response on appearance of the target; over 90% of such neurons are directionally selective; i.e., discharge selectively to a specific, usually contralateral, target location. The middle panel displays *tonic* activity during the delay period; in most prefrontal neurons displaying this activity, the activity is also directionally selective. The lower panel displays a neuron that is activated at the time of response; some oculomotor reponses occur before the response, but most occur after the response is initiated (Funahashi et al., 1990b). These responses are time-locked to the main events (cue, delay and response) of the ODR task.

position. And, indeed, this interpretation fits well the evidence that lesions of the principal sulcus produced deficits on left–right spatial delayed-response tasks. However, studies employing the oculomotor paradigms have shown clearly that prefrontal neurons can code the location of an object throughout the visual field; i.e., in all polar directions of space (Funahashi et al., 1989, 1990a). An important point is that the same neuron codes the same location repeatedly and different neurons code different locations. Thus, the memory field of a prefrontal neuron is quite analogous to the visual receptive fields of visual cortical neurons or the motor fields of motor system neurons.

Analysis of prefrontal function in the oculomotor paradigm is beginning to yield evidence of dynamic interactions among several neuronal classes. For example, we have obtained evidence that there are 2 types of delay-period activity. The first is a sustained pattern of tonic activity in the delay that commences only when the cue disappears and terminates abruptly after the response is initiated (the "D" cell) (Figs. 2 and 3c). This "D" profile can be contrasted with another, second type of activity pattern in which the cue elicits a phasic stimulus-locked response followed by a sustained tonic activity in the delay period (the "C + D" cell of Funahashi et al., 1990a). Further, the directional bias of the phasic "C" response and the cell's tonic "D" activity are highly correlated (Funahashi et al., 1990a). Finally, some prefrontal neurons respond phasically to the directional cues, but do not generate delay-period activity (the "C" cell) (Fig. 3). Our hypothesis is that each of these response patterns represents the activity of specific classes of prefrontal neurons with distinctive inputs and outputs, and that they are locally interconnected with one another. Thus, we could suppose that the C cell registers the direction of the cue at the time of presentation. Such information presumably originates outside of the prefrontal cortex and could be conveyed, for example, over parieto-prefrontal pathways (e.g., Cavada and Goldman-Rakic, 1989a, b; Leichnetz, 1980; Petrides and Pandya, 1985). Then, the C cell could input to the C + D cell, which computes a memory field (directional delay-period activity) from this "sensory" input. In a next step, the C + D cell may innervate the D cell which keeps the message "on line" and, possibly, is the efferent neuron in the circuit (Funahashi et al., 1990a). Of course, other scenarios and functional circuits are possible and all possibilities need to be analyzed by appropriate experiments in future studies. What we would underscore is that the caudal principal sulcal region contains local "memory" circuits or modules that may constitute the building blocks for numerous of the cognitive functions associated with prefrontal cortex (see Goldman-Rakic, 1987, for further discussion).

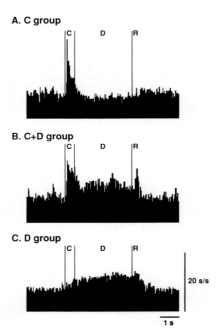

Fig. 3. Composite histograms summing over a large number of neurons recorded from the principal sulcus during the ODR task. Only trials for a neuron's preferred direction (largest response) are included. (A) Composite histogram of 27 neurons that responded to the cue. (B) Composite of 33 neurons that had both phasic cue-period activity and tonic delay-period activity. (C) Composite histogram of 78 neurons that exhibit only tonic delay-period activity. C = cue; D = delay; R = response periods. (From Funahashi et al., 1990a.)

Memory maps in prefrontal cortex

How are such neurons arranged in the prefrontal cortex? The answer to this question is still unclear. However, several lines of evidence indicate that there may be at least a crude topographic map of memory in prefrontal cortex. Circumscribed surgical lesions produce equally circumscribed memory loss for visuospatial targets. For example, we have evidence that a lesion in the posterior third of the principal sulcus, including the anterior bank of the arcuate sulcus, we have produced a deficit restricted to the upper contralateral quadrant of space; a more rostral lesion in the middle third of the principal sulcus produced a deficit in the lower contralateral quadrant (Fig. 4) (Funahashi et al., in preparation). Using the reversible lesion method of a pharmacological block, injections of bicuculline, a competitive antagonist of GABA, into the principal sulcus, we have produced a deficit restricted to one or two locations in space; a different placement of injection induced a deficit in a different location in the visual field (Sawaguchi and Goldman-Rakic, 1990) Furthermore, the deficits were most pronounced in the visual field contralateral to the hemisphere injected. Correspondingly, neurons which code the memory of left visual field targets are concentrated in the right hemisphere; those coding the right visual field are concentrated in the left hemisphere (Funahashi et al., 1989). Thus, the visuospatial memory system is lateralized and, very likely, the mechanisms for working memory are well integrated with the brain's mechanisms for spatial vision. Again, we presume that the relevant sensory information originates in the visual areas of the cortex and is transmitted via the parieto-prefrontal projections that provide a major input to the caudal principal sulcus.

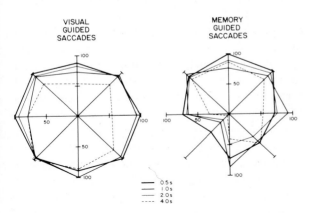

Fig. 4. Performance on oculomotor tasks after unilateral lesions of the middle third of the principal sulcus displayed on an octagonal graph. Percent correct performance is graphed separately for each radial position; the different lines represent different delays, as indicated. (Left panel) Performance during the ODR task when the cue remains on during the delay and is present at the end of the delay. The monkey performs without difficulty at all locations and delays. (Right panel) Performance on the ODR task in which the saccades are memory driven. The monkey exhibits a deficit that is mainly expressed in the lower quadrant of the visual field contralateral to the lesion. This deficit can be viewed as a "mnemonic scotoma" because the monkey has no difficulty in seeing or moving his eyes, as shown by his excellent performance when eye movements are visually guided.

Prefrontal function: working memory or motor set?

The activation of prefrontal neurons when a stimulus disappears from view, and the maintenance of that activation until a response is executed, is highly suggestive of a working memory process and, perhaps even underlies the conscious experience of that process (Goldman-Rakic, 1990). Nevertheless, it may be argued that the delay-period activity reflects simply the motor set of the animal to respond at the end of the delay, as has been described for neurons in premotor fields. However, prefrontal neurons exhibit increased or decreased discharge only for targets in memory fields. Thus, even though an animal is set to, and does, respond correctly at the end of the delay period on every trial (regardless of target location), any given neuron is attuned to only one or a few locations. A general motor set explanation cannot readily explain this result.

If prefrontal neurons are involved in mnemonic processing, one would except their activity to be sensitive to changes in the duration of the delay period. Indeed, it has been shown that the activity of prefrontal neurons that occurs during the delay period expands and contracts as the delay is lengthened or shortened (Kojima and Goldman, 1982). Again, this would be expected if the neuron is holding information "on line" that is to be retained until the end of the delay period.

Another argument for mnemonic coding comes from a recent analysis of prefrontal neuronal activity in an anti-saccade task (Funahashi and Goldman-Rakic, 1990). The task is identical to the oculomotor delay-response paradigm except that, at the end of the delay, the monkey is required to direct his gaze to the location *opposite* that in which the to-be-remembered cue was presented. Under this circumstance, the direction of the cue and that of the response are dissociated. We have compared the activity of 44 neurons that were recorded during both the conventional and anti-saccade versions of the oculomotor paradigm and have found that the great majority (66%) that responded to a specific direction in the conventional task maintained the same pattern of discharge, even when the response was to the opposite direction (Funahashi and Goldman-Rakic, 1990). Thus, this study indicates that delay-period activity is independent of the direction of the response in a large fraction of prefrontal neurons; so we can conclude that this activity codes a representation of sensory information and is not a motor code. However, a lower percentage of prefrontal cells *do* code for the direction of the impending motor response, and their firing patterns are influenced by the direction of the motor response, i.e., such cells increase their rate of discharge for saccades only in one or another direction. Such neurons also play a role in prefrontal function and emphasize the richness of neural coding in this cerebral area.

Finally, if prefrontal neurons were coding only a preparatory set to respond, one would expect them to be activated before incorrect as well as correct responses. However, neurons that have memory fields (i.e, have a "best direction" of discharge) either falter in activity during the delay or even completely fail to increase their rate preceding responses in the non-preferred incorrect direction (Funahashi et al., 1989). The fact that incremental firing precedes only correct responses indicates that the activity may be part of the internalized code needed to guide the correct response. Thus, it would serve a mnemonic function.

Working memory network

The principal sulcus in the prefrontal cortex is the anatomical focus for spatial delayed-response function, and knowledge of its connections with other structures is helping us to understand the circuit and cellular basis of working memory. It has become clear from our anatomical investigations that this prefrontal subdivision has reciprocal connections with more than a dozen distinct cortical association regions, with premotor centers, with the caudate nucleus, the superior colliculus, and brain stem centers (Fig. 5; for review see Goldman-Rakic, 1987). Each of these connections presumably contributes different subfunctions to the overall capacity to guide a response by the mental representation of a stimulus. The reciprocal connections between the prefrontal and parietal cortex carry information about the spatial aspects of the outside world. In recent studies, we have traced several pathways from distinct visual centers concerned with peripheral vision to the principal sulcus via relays in the posterior parietal cortex (Cavada and Goldman-Rakic, 1989a). Our anatomical analysis reveals that prefrontal neurons in and around the principal sulcus are but two synapses removed from the primary visual cortex. This network, additionally, contains several multisynaptic pathways between the prefrontal cortex and the hippocampal formation; we have speculated that these connections subserve a cooperative relationship between the hippocampus and the prefrontal cortex with regard to working memory (Goldman-Rakic, et al., 1984; Selemon and Goldman-Rakic, 1988). Our recent evidence of elevated metabolic activity in the dentate gyrus, and the several fields of Ammon's horn of monkeys performing working memory tasks, sup-

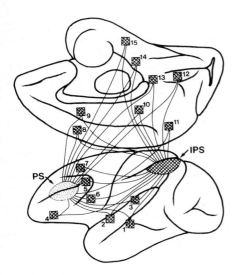

Fig. 5. Distributed circuit revealed by the double-anterograde tracing study described in Selemon and Goldman-Rakic (1988). One anterograde tracer, tritiated amino acids, was placed in the prefrontal cortex; another, WGA-HRP, was placed in the posterior parietal cortex (and the reverse). Alternate sections were processed for autoradiography and HRP histochemistry. Adjacent sections were charted and superimposed for determination of convergence, or lack thereof, in potential target areas. Area 46 and area 7 project to as many as 15 targets in common: (1) the depths of the middle third of the superior temporal sulcal cortex; (2) the insular cortex; (3) the fronto-parietal operculum in the dorsal bank of sylvian fissure; (4) the orbital prefrontal cortex; (5) the anterior arcuate cortex; (6 and 7) the ventral and dorsal subdivisions of premotor cortex; (8) the supplementary motor area; (9) the caudal half of the anterior cingulate cortex; (10) the entire posterior cingulate cortex; (11) the medial parietal cortex; (12) the medial prestriate cortex; (13) the caudomedial lobule; (14) the presubiculum; and (15) the parahippocampal gyrus. Thus, in our study, every cortical region that was innervated by the posterior parietal cortex also received a projection from the posterior prefrontal cortex. PS = principal sulcus; IPS = intraparietal sulcus. (From Selemon and Goldman-Rakic, 1988.)

port this line of thinking (Friedman and Goldman-Rakic, 1988). Also, increased neuronal activity during the delay period of delayed-response tasks has been recorded from neurons in Ammon's horn (Watanabe and Niki, 1985). Finally, working memory tasks like nonmatching-to-sample are just the type of task on which monkeys with lesions of the hippocampus are impaired. In the matching tasks, like in delayed-response tasks, information (the sample) is relevant only for one trial and each trial is independent from the last. It seems clear that the hippocampus and prefrontal cortex are functionally as well as anatomically related.

Connections with the caudate nucleus and superior colliculus, among other motor centers, are thought to play a role in transmitting response commands to the motor centers. The principal sulcus projects to premotor centers including the supplementary motor areas and, through these centers, has access to the primary motor cortex. Therefore, with respect to motor control also, the principal sulcus is only 2 synapses removed from primary motor neurons. Thus, we conclude that prefrontal cortex regulates behavior in collaboration with a large set of other cortical and subcortical structures which, all together, constitute the brain's machinery for spatial cognition (see Goldman-Rakic, 1987, for overview; also, Selemon and Goldman-Rakic, 1988).

Evolutionary significance of working memory

The significance of working memory for higher cortical function is not necessarily self-evident. Perhaps even the quality of its transient nature misleads us into thinking it is somehow less important than the more permanent archival nature of long-term memory. However, the brain's working memory function, i.e., the ability to bring to mind events in the absence of direct stimulation, may be its inherently most flexible mechanism and its evolutionarily most significant achievement. Thus, working memory confers the ability to guide behavior by representations of the outside world rather than by immediate stimulation, and thus to base behavior on ideas and thoughts.

The difference between guidance of behavior by symbols, concepts, or ideas and guidance by external stimuli cannot be overemphasized. At the most elementary level, our basic conceptual ability to understand that an object *out of view* nevertheless exists depends on the capacity to keep events in mind beyond the direct experience of those events.

For some organisms, including humans under certain conditions, "out of sight" is equivalent to "out of mind". Working memory has been invoked in all forms of cognitive processing, including language. Is has been pointed out that the failure to keep a word in mind after it has been uttered would lead to a grave restriction in the span over which contextual interactions can occur (Cohen and Servan-Schreiber, 1989). It is no less essential to the performance of mathematical operations (e.g., carrying over), to playing chess or bridge, to playing the piano without music, delivering a speech without reading (or by rote) and, finally, to fantasizing and planning ahead.

References

Baddeley, A.D. (1973) Working memory. *Phil. Trans. R. Soc. Lond. B. Biol. Sci.,* 302: 311 – 324.

Baddeley, A. (1986) *Working Memory,* Oxford University Press, London.

Barone, P. and Joseph, J.-P. (1989) Prefrontal cortex and spatial sequencing in macaque monkey. *Exp. Brain Res.,* 78: 447 – 464.

Boch, R. and Goldberg, M.E. (1989) Participation of prefrontal neurons in the preparation of visually guided eye movements in the rhesus monkey. *J. Neurophysiol.,* 61: 1064 – 1084.

Cavada, C. and Goldman-Rakic, P.S. (1989a) Posterior parietal cortex in rhesus monkey. I. Parcellation of areas based on distinctive limbic and sensory cortico-cortical connections. *J. Comp. Neurol.,* 286: 393 – 421.

Cavada, C. and Goldman-Rakic, P.S. (1989b) Posterior parietal cortex in rhesus monkey. II. Evidence for networks linking limbic and sensory areas to the frontal lobe. *J. Comp. Neurol.,* 286: 422 – 445.

Cohen, J.D. and Servan-Schreiber, D. (1989) A parallel distributed processing approach to behavior and biology in schizophrenia. *Techn. Rep. AIP-100,* Carnegie-Mellon University, Pittsburgh, PA.

Friedman, H.R. and Goldman-Rakic, P.S. (1988) Activation of the hippocampus by working memory: a 2-deoxyglucose study of behaving rhesus monkeys. *J. Neurosci.,* 8: 4693 – 4706.

Funahashi, S., Bruce, C.J. and Goldman-Rakic, P.S. (1989) Mnemonic coding of visual space in the monkey's dorsolateral prefrontal cortex. *J. Neurophysiol.,* 61: 1 – 19.

Funahashi, S., Bruce, C.J. and Goldman-Rakic, P.S. (1990a) Visuospatial coding of primate prefrontal neurons revealed by oculomotor paradigms. *J. Neurophysiol.,* 63 (4), 814 – 831.

Funahashi, S., Bruce, C.J. and Goldman-Rakic, P.S. (1990b) Pre- and postsaccadic activity in the dorsolateral prefrontal cortex of rhesus monkey. *J. Neurophysiol.,* in press.

Funahashi, S. and Goldman-Rakic, P.S. (1990) Delay-period activity of prefrontal neurons in delayed saccade and antisaccade tasks. *Soc. Neurosci. Abstr.,* Vol. 16, p. 1223.

Fuster, J.M. (1973) Unit activity in prefrontal cortex during delayed-response performance: neuronal correlates of transient memory. *J. Neurophysiol.,* 36: 61 – 78.

Fuster, J.M. (1989) *The Prefrontal Cortex,* 2nd Edn., Raven Press, New York, p. 255.

Fuster, J.M. and Alexander, G.E. (1971) Neuron activity related to short term memory. *Science,* 173: 652 – 654.

Goldman-Rakic, P.S. (1987) Circuitry of the prefrontal cortex and the regulation of behavior by representational knowledge. In: F. Plum and V. Mountcastle (Eds.), *Handbook of Physiology, Vol. 5,* American Physiological Society, Bethesda, MD. p. 373.

Goldman-Rakic, P.S. (1988a) Changing concepts of cortical connectivity: parallel distributed cortical networks. In: P. Rakic and W. Singer (Eds.), *Neurobiology of Neocortex,* Dahlem Konferenzen, John Wiley, New York, pp. 177 – 202.

Goldman-Rakic, P.S. (1988b) Topography of cognition: parallel distributed networks in primate association cortex. *Annu. Rev. Neurosci.,* 11: 137 – 156.

Goldman-Rakic, P.S. (1990) The prefrontal contribution to working memory and conscious experience. *Exp. Brain Res., Suppl.,* in press.

Goldman-Rakic, P.S. and Preuss, T. (1988) Wither comparative psychology? *Behav. Brain Sci.,* 10: 666.

Goldman-Rakic, P.S., Selemon, L.D. and Schwartz, M.L. (1984) Dual pathways connecting the dorsolateral prefrontal cortex with the hippocampal formation in the rhesus monkey. *Neuroscience,* 12: 719 – 743.

Hunter, W.S. (1913) The delayed reaction in animals and children. *Behav. Monogr.,* 2: 1 – 86.

Jacobsen, C.F. (1936) Studies of cerebral function in primates. *Comp. Psychol. Monogr.* 13: 1 – 68.

Kojima, S. and Goldman-Rakic, P.S. (1982) Delay-related activity of prefrontal neurons in rhesus monkeys with prefrontal lesions. *Brain Res.,* 248: 43 – 49.

Kojima, S., Kojima, M. and Goldman-Rakic, P.S. (1982) Operant behavioral analysis of delayed-response performance in rhesus monkeys with prefrontal neurons. *Brain Res.,* 248: 51 – 59.

Kubota, K. and Niki, H. (1971) Prefrontal cortical unit activity and delayed cortical unit activity and delayed alternation performance in monkeys. *J. Neurophysiol.,* 34: 337 – 347.

Leichnetz, G.R. (1980) An intrahemispheric columnar projection between two cortical multi-sensory convergence areas (inferior parietal lobule and prefrontal cortex): an anterograde study in macaque using HRP gel. *Neurosci. Lett.,* 18: 119 – 124.

Mishkin, M. (1964) Perseveration of central sets after frontal lesions in monkeys. In: J.M. Warren and K. Akert (Eds.), *The Frontal Granular Cortex and Behavior,* New York, McGraw Hill, New York, p. 219.

Niki, H, and Watanabe, M. (1986) Prefrontal unit activity and delayed response: relation to cue location versus direction of response. *Brain Res.,* 105: 78 – 88.

Olton, D.S. and Papas, B.C. (1979) Spatial memory and hippocampal function. *Neuropsychologia,* 17: 669 – 782.

Passingham, R.E. (1985a) Memory of monkeys *(Macaca mulatta)* with lesions in prefrontal cortex. *Behav. Neurosci.,* 99: 3 – 21.

Passingham, R.E. (1985b) Cortical mechanisms and cues for action. *Phil. Trans. R. Soc. Lond. B. Biol. Sci.,* 308: 101 – 111.

Petrides, M. and Pandya, D.N. (1985) Projections to the frontal cortex from the posterior parietal region in the rhesus monkey. *J. Comp. Neurol.,* 228: 105 – 116.

Roitblat, H.L. (1987) *Introduction to Cognition,* Freeman, New York, pp. 146 – 190.

Rumelhardt, D.E. and McClelland, J.L. (1988) PDP models and general issues in cognitive science in parallel distributed processing. In: J.L. McClelland, D.E. Rumelhardt and the PDP Research Group (Eds.), MIT Press, Cambridge, MA, pp. 110 – 146.

Sawaguchi, T. and Goldman-Rakic, P.S. (1990) Topographic representation of memory in the prefrontal cortex of monkeys revealed by local application of bicusculline. *Soc. Neurosci. Abstr.,* in press.

Selemon, L.D. and Goldman-Rakic, P.S. (1988) Common cortical and subcortical target areas of the dorsolateral prefrontal and posterior parietal cortices in the rhesus monkey: A double label study of distributed neural networks. *J. Neurosci.,* 8: 4049 – 4068.

Squire, L.R. and Cohen, N.J. (1984) Human memory and amnesia. In: G. Lynch, J.L. McGaugh and N.M. Weinberger (Eds.), *Neurobiology of Memory and Learning,* Guilford Press, New York, p. 3.

Tulving, E. (1972) *Episodic and Semantic Memory,* Academic Press, New York, pp. 381 – 403.

Watanabe, T. and Niki, H. (1985) Hippocampal unit activity and delayed response in the monkey. *Brain Res.,* 325: 241 – 254.

Discussion

M. Godschalk: In your Oculomotor Delayed Response paradigm, you consider the delay period-related excitation to be related to memory processes. To show that that is indeed the case, and to exclude a relation to preparation of the upcoming eye movement, it would be interesting to know what would happen if the cue were to remain on (till the fixation light dimmed). Do you have any data on that?

P. Goldman-Rakic: We have examined cellular activity during the delay when the cue remains on. In some cases, but not always, we are able to show that the cell's enhanced discharge occurs specifically when the cue is out of view. The strongest evidence for the memory hypothesis is our finding that errors are likely to occur whenever enhanced neuronal discharge is not sustained during the delay (see. Fig. 13, Funahashi et al., 1989). Further a given neuron exhibits enhanced activity in the delay only for a cue in a specific direction, and not for all cues and directions, even though the animal is prepared to respond on every trial, indicating a specific mnemonic process. Finally, prefrontal lesions produce performance deficits on the memory-guided version, but not the sensory-guided version of the task.

T. Paus: Did the "anti-saccade" task follow the normal ones? If so, did monkeys have trouble suppressing this over-learned behavior pattern ("simple habit")?

P. Goldman-Rakic: The anti-saccade task often followed the standard task, but once the animal had been trained there was no evidence of interference between the two tasks. The animal quickly learned to know which task he was performing.

H.B.M. Uylings: In trying to combine some data presented by Mortimer Mishkin and your data, especially those on the connections of the temporal lobe with the PFC via the parietal cortex, I wonder what will be the effects of parietal cortex lesions in comparison with those of lesions in the PFC?

P. Goldman-Rakic: It seems generally that posterior parietal lesions disrupt perceptual processes leading to problems like spatial neglect, whereas dorsolateral prefrontal lesions disrupt memory of spatial events. However, without appropriately refined methods of testing, it is not easy to differentiate the nature of the deficit expressed either in humans or experimental animals, and the deficits following prefrontal and parietal lesions have often been confused. This is an important area of research.

C.G. van Eden: Dr. Kolb has stated and Dr. Milner has shown that the PFC is not involved in spatial memory per se. On the other hand, Dr. Fuster and you yourself have shown that the area contains cells whose activity is related to either the left or the right stimulus. Is PFC implicated in L/R discrimination or is it in your opinion also involved in more complex spatial memory paradigmata?

P. Goldman-Rakic: Considerable evidence indicates that it is the memory for spatial stimili rather than their discrimination which presents difficulty to the animal with prefrontal lesions, i.e., they are not impaired on left-right discrimination problems. The common deficits expressed by frontal patients and monkeys with prefrontal lesions is their inability to keep events in mind for even a few or more seconds. For example, as Milner has shown, frontal lobe patients, like the lesioned monkeys, cannot remember what they did last, whether they selected one or another item on a page, or which stimulus they saw most recently. To have direct comparisons between patients and animals, we need to use comparable tests and compare comparable lesions. I have argued that different regions of prefrontal cortex perform the same working memory functions, but in different informational domains (Goldman-Rakic, 1987).

M.A. Corner: Does the available evidence justify speaking of prefrontal cortex (PFC) as "accessing (specific) information" for behavioral control, or could a simple inhibiting "hold" function for the PFC explain the available data?

P. Goldman-Rakic: What Funashi's recordings have shown most dramatically is that a large fraction of individual prefrontal neurons hold directionally specific information "on line" (e.g. 125 degree position). We have also observed cells that discharge in the delay about equally for all cue directions, so-called "omnidirectional" neurons, but most have what we have termed "memory fields", i.e. they carry information about a specific direction.

CHAPTER 17

Distributed neuroelectric patterns of human neocortex during simple cognitive tasks

Alan Gevins

EEG Systems Laboratory, San Francisco, CA 94107, USA

Introduction

Scalp-recorded neural potentials have been used widely for studying brain activity associated with higher mental functions in humans for over 50 years. With the development of better tools, more specific information about the spatial and temporal features of neurocognitive processes has been forthcoming. Such tools include neuroelectric recordings with many channels, advanced signal processing techniques, and correlation of neuroelectric measures with anatomical information from magnetic resonance (MR) images. These tools are described in the first part of this chapter; applications of the tools to studying human higher cognitive functions are presented in the second part.

Tools for measuring the split second components of neurocognitive processes

(A) Improving spatial sampling

(1) Electrode arrays with 124 channels

One of the requirements for extracting detailed information about cognitive processes from the scalp-recorded EEG is to have adequate spatial sampling. The 19 channels customarily employed in clinical recordings provide an interelectrode distance of about 6 cm. While this is sufficient for detecting signs of gross pathology, it is obviously insufficient for resolving functional differences within small cortical regions. To improve spatial resolution, we have been making 59-channel recordings for the past several years. This provides an interelectrode distance of about 3.5 cm on a typical adult head, which is still not good enough. To improve sampling, we recently developed a 124-channel recording system which provides an interelectrode distance of about 2.25 cm. Subjects wear a stretchable EEG recording cap, with electrodes placed on the cap according to an expanded version of the standard International 10-20 System (Gevins, 1988). Prior to each EEG recording session, the 3-D position of each electrode on the individual subject's head is measured precisely with a commercial 3-D digitizer (Fig. 1). To date (9/1989), 12 full-scale 124-channel recordings have been made from subjects receiving visual, auditory and somatic stimuli.

(2) Registration of scalp electrode positions with underlying anatomical structures

In order to visualize the brain areas underlying the scalp electrodes, a procedure is needed for aligning scalp electrode positions and underlying anatomical structures. This first requires producing an accurate anatomical representation of a subject's brain.

Correspondence: Dr. A. Gevins, EEG Systems Laboratory, 51 Federal St., San Francisco, CA 94107, USA. Tel.: (415) 957-1600.

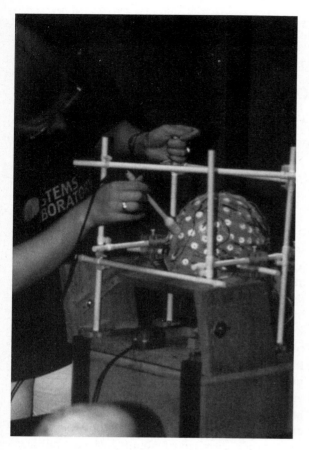

Fig. 1. Digitization of scalp electrode positions. The subject is resting in a head rest designed to minimize head movement. The technician touches the stylus (which contains electromagnetic field sensors for x, y and z axes) to each of the electrodes in turn. The 3-D coordinates of each electrode position are transmitted to the data collection and analysis computer.

(a) Distortion correction of magnetic resonance images. While magnetic resonance (MR) images are invaluable because of the anatomical differentiation they provide, they contain inherent distortion that, if not corrected, may cause quantitative measurements of position, length, area and volume to be erroneous. The distortion can exceed 10%, and commonly arises from calibration errors and inhomogeneities of the magnetic field gradients. We have been working on correcting this problem using both phantom calibration data and data recorded from human subjects wearing specially constructed helmets with spherical fiducial markers which are easily visualized on MR images. It is necessary to correct both image intensity and image position.

To test our methods, scans from a Diasonics MTS MR system were made on 3 subjects. The maximum total distortion measured on this machine was 8%. Variation between images with TR = 600 msec and TE = 20 msec, and images with TR = 2000 msec and TE = 35 and 70 msec were found to be less than 2%. We compared the location of the spherical fiducial markers in coronal, sagittal and axial images and various image transforms were then used to bring the measured points into alignment. We found that by computing a separate scale factor for each direction combined with a translation and rotation, 2 sets of images could be brought into reasonably close alignment. An example of a set of Diasonics MRIs before correction is shown in Fig. 2a. It is clear that the anatomical positions corresponding to the scalp and cranium do not line up. The coronal sections shown in blue are shifted to the left of their correct position. The sagittal sections shown in reddish brown are "stretched" by approximately 17% in the anterior/posterior direction compared to the horizontal or transaxial sections which are shown in dark green. The same images after correcting for distortion are shown in Fig. 2b.

(b) Alignment of EEG electrode positions with MR surface reconstructions. A linear transformation is calculated to superimpose electrode positions and the scalp-surface contours obtained from MR images. This transformation is initially determined by visually adjusting a graphical display of the electrodes and scalp contours. An optimal transformation is then calculated by a program that adjusts each parameter in the transformation until the average distance between all electrode positions and the closest scalp point is minimized (Fig. 3). The aligned surface model can also be superimposed onto composite MR images showing various orientations. Additionally, surfaces may

(3) Reduced blur distortion of brain potentials

(a) Laplacian derivation. Neuroelectric signals recorded at the scalp are principally distorted by transmission through the low-conductance skull. This distortion manifests as a spatial low-pass filtering which causes the potential distribution at the scalp to appear blurred or out of focus. There are a number of methods for reducing this distortion, among which the spatial Laplacian operator is perhaps the simplest and most effective. This method, which is often referred to as the *Laplacian derivation*, is derived by computing the second derivative in space of the potential field at each electrode. This converts the potential into a quantity proportional to the current entering and exiting the scalp at each electrode site, and eliminates the effect of the reference electrode used during recording. An approximation to the Laplacian derivation, introduced by Hjorth (1975, 1980) assumes that electrodes are equidistant and at right angles to each other. Although this approximation is fair-

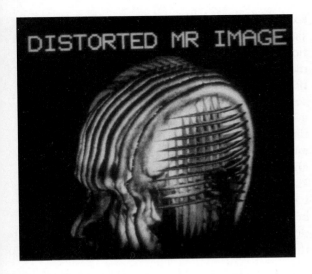

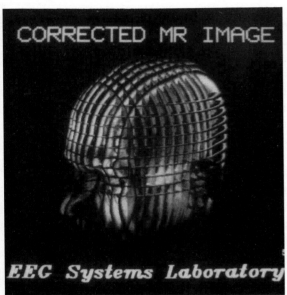

Fig. 2. (a) Composite of MR images in horizontal, sagittal, and coronal orientations as originally recorded. The location of the scalp is not consistent for different orientations. (b) A composite of the same images after transformation to correct for distortion. The scalp surface now appears at the same location in all orientations.

be constructed by manually tracing other structures on each MR image and then calculating the polygonal surface which fits these contours (Fig. 4).

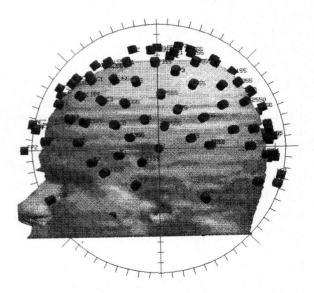

Fig. 3. Rough sagittal view of 124-electrode montage positioned on the scalp surface constructed from horizontal MR images. Also shown is the coordinate system with the origin located halfway between the T3 and T4 temporal electrodes. Some electrodes are not properly aligned with the scalp surface due to a mechanical problem which has subsequently been corrected.

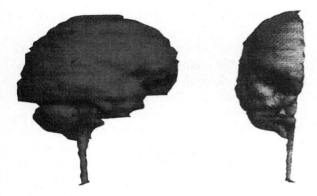

Fig. 4. Reconstruction of the right hemisphere of the brain and spinal cord viewed from the right and anteriorly.

ly good for some electrode positions such as midline central (Cz), it is less accurate for others such as midtemporal (T5). We have been using a more accurate estimate of the Laplacian that is based on projecting the measured electrode positions onto a two-dimensional surface. Although this produces a dramatic improvement in topographic detail, some problems remain because of the assumptions that surrounding electrodes used to estimate the Laplacian of an electrode are near that electrode and that the current gradient is uniform over the region encompassed by the surrounding electrodes. Furthermore, it is not possible to estimate the Laplacian at peripheral electrodes since the surrounding electrodes are incomplete.

(b) Spatial deconvolution using spherical head model. By modeling the tissues between brain and scalp as surfaces with different thicknesses and resistances, we have performed a deblurring operation that, in principle, makes the potential appear as if it were recorded just above the level of the brain surface (Doyle and Gevins, 1986), without assumptions about the actual (cortical or subcortical) source locations. The deblurring operation, however, requires detailed modeling of the tissues which, when the exact shape of the head is taken into account, is a great deal of work. The operation is even further complicated by the fact that a solution to estimating the local resistance of the skull precisely does not yet exist. With the conduction of potentials from a localized source spread over a considerable area of scalp, the summation of signals at any given scalp site may reflect many sources over much of the brain. In the context of a 4-shell spherical head model, we have estimated the amount of spread — the "point spread" — for a radial equivalent dipole source in the cortex to be about 2.5 cm. If the conductance of the skull is known, a deblurring operation using a model-based deconvolution can, in principle, achieve a better signal enhancement than a Laplacian derivation when the distance between electrodes is less than about the point spread distance of 2.5 cm.

(c) Finite element method. Another method of increasing spatial resolution for those cases for which the source generators can be modeled as current dipoles, and for which MR data is available, is the finite element method (FEM). The entire volume of the head, as found in MR images, is broken up into many small elements representing various tissues: the scalp, skull, and brain. By assigning each element a conductivity constant (obtained from textbook values) and for a known source, it is possible to calculate the potential at each vertex of all the finite elements using Maxwell's equations. Because the number of vertices is approximately 10,000, an efficient algorithm is necessary to make this practical on a small computer. Using a SUN Sparc-1 workstation rated at about 12 MIPS, the initial matrix decomposition based on a set of finite elements takes about 90 min, while the potential computation for each source takes 6 min. If a practical method can be developed for estimating local skull conductance, the FEM deblurring method has the capability of producing highly enhanced representations of the current distribution on the exposed surface of the cortex.

(4) Artifact detection

The usual practice in evoked potential studies of cognition is to automatically reject artifacted trials in which the voltage of the eye-movement measurement channels exceeds a fixed threshold (Barlow, 1986). While this procedure catches large contaminants, it entirely misses small ones. This can lead to a spurious result if there are small, but consistent saccades or microblinks approximately time-locked to stimulus presentation. Although we also use an on-line artifact detection procedure to automatically flag portions of trials and individual electrodes that have unusually high or low amplitude, all data are examined visually on a graphics terminal to confirm and improve the computer's detections as needed. In our studies with clinically healthy, young adult subjects, there is about 10% data attrition due to artifacts.

(5) Data set formation: Controlling for spurious sources of variance

After the data has been cleared of instrumental and subject-related artifacts, data sets are usually formed in pairs to test specific hypotheses. In forming these data sets, it is imperative that the major difference between two sets be related to the hypothesis being tested. It is, of course, standard practice to try to eliminate spurious differences by careful experimental design, but there is always the chance that some remaining factors differ between sets. These uncontrolled factors can include small residual eye-movement contaminants, arousal level, and response movement parameters (e.g., force or reaction time), all of which are known to affect neuroelectric signals.

To ascertain that the major source of variance is actually related to the hypothesis, the two sets of artifact-free trials are submitted, usually on a subject-by-subject basis, to an interactive program which displays the means, t-tests, and histogram distributions of up to 50 behavioral and physiological event variables. These include stimulus parameters, reaction time and movement magnitude and duration, error, EEG arousal index, eye-movement and muscle potential indices, and so on. The data sets are inspected for significant differences in variables which are not related to the hypothesis, and outliers are discarded. As an example, an unintentional difference between experimental conditions in response force may be present. In the data set with the larger response force, the associated movement-related potentials could overlap the P300 evoked potential peak causing a spurious between-condition difference in P300 amplitude. After careful balancing of the data sets for such movement parameters, valid assessments about P300 peak effect may be made. In balancing our data we are careful not to truncate the histogram distribution severely. The unrelated variables are reduced to a between-condition α significance of 0.2, or if this is not possible without seriously affecting the distribution, to just over 0.05. The net effect of these procedures is the certainty that when a neuroelectric difference between experimental conditions is found, the difference actually relates to the hypothesis under consideration.

(6) Neurocognitive pattern analysis

We have been using the term "neurocognitive pattern analysis" (NCP analysis) to refer to our procedures for extracting task-related spatiotemporal patterns from the unrelated background activity of the brain. In the first of 3 generations of NCP analysis, we measured background EEG spectral intensities while people performed complex tasks, such as arithmetic problems lasting up to 1 min (Fig. 5). These patterns had sufficient specificity to identify the type of task (Gevins et al., 1979a, b), but when the tasks were controlled for stimulus, response and performance-related factors, they had identical, spatially diffuse EEG spectral scalp distributions (Gevins et al., 1979a, c). This study suggested that complex tasks involving a variety of sensory, cognitive and motoric processes activate large widespread areas of cortex to a degree proportional to the subject's effort. It also strongly suggested that most studies of EEG correlates of cognitive activities, including those of hemispheric lateralization, may have confounded electrical activity related to limb and eye movements, stimulus properties, and task difficulty with those of mental activity per se (Gevins et al., 1979a).

In the second generation of NCP analysis, we measured cross-correlations between electrodes recorded during performance of simple visuo-

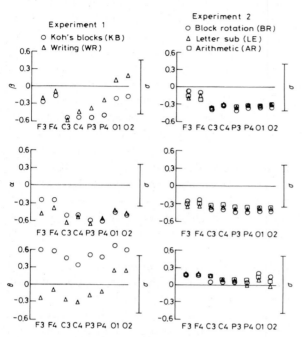

Fig. 5. Results of experiments designed to assess EEG correlates of higher cognitive functions. (Left) Tasks of Expt. 1 were 1 min long and involved limb movements and uncontrolled differences in stimulus characteristics and performance-related factors. (Right) Tasks of Expt. 2 were less than 15 sec long and required no motion of the limbs; stimulus characteristics and performance-related factors were also relatively controlled. The graphs display means over all subjects of standard scores of EEG spectral intensities (expressed as changes from visual fixation values for clarity of display) recorded during performance of 2 tasks in Expt. 1 and 3 tasks in Expt. 2. Upper, middle, and lower sets of graphs are for spectral intensities in the beta, alpha, and theta bands. The abscissas show scalp electrode placements: F3, left frontal; F4, right frontal; C3, left central; C4, right central; P3, left parietal; P4, right parietal; O1, left occipital; and O2, right occipital. Standard deviations, which differed but slightly between electrode placements, are indicated at the right. Although there are prominent EEG differences between the uncontrolled tasks of Expt. 1, EEG differences between the relatively controlled tasks of Expt. 2 were lacking. Each of the controlled tasks is, however, associated with a remarkably similar bilateral reduction in alpha and beta band spectral intensity over occipital, parietal and central regions. (From Gevins et al., 1979a.)

motor judgment tasks (Gevins et al., 1981). From this experiment, in which rapidly shifting focal patterns were extracted from two similar spatial tasks (Fig. 6), it was clear that a split-second temporal resolution is imperative for isolating the rapidly shifting neurocognitive processes associated with successive information processing stages.

In the third generation, we extended our methods to include event-related covariances (ERCs). The ERC approach is based on the hypothesis that when regions of the brain are functionally related, their event-related potential (ERP, another name for evoked potential) components are related in shape and in time (Gevins and Bressler, 1988). The idea is that the ERP waveform delineates the time course of event-related mass activity of a neural population, so that if two populations are functionally related, their ERPs should line up in time, perhaps with some delay. If so (and if the relationships are linear as they often appear to be), this could be measured by the lagged covariance between the ERPs, or portions of the ERPs, from different regions. This is the event-related covariance method.

The procedures that are followed for ERC analysis are described here. Procedures 1–4 have been discussed in detail above.

(1) A sufficient amount of data are recorded using as many electrodes as possible.
(2) Data with artifact contamination are removed.
(3) Pairs of conditions to be compared are selected, and trials with extreme values of behavioral variables are eliminated.
(4) The Laplacian operator is applied to the potential distribution of each non-peripheral scalp electrode location.
(5) Analysis intervals and digital filter characteristics are determined. The analysis intervals are usually either centered on an ERP peak, or are positioned just before or after a stimulus or response.
(6) Enhanced, filtered and decimated averaged Laplacian ERPs for each condition are com-

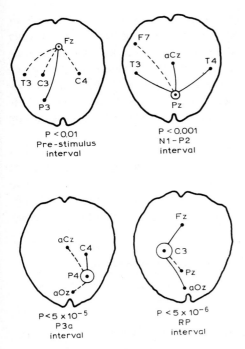

Fig. 6. Spatial brain potential differences between 2 split-second tasks requiring a spatial judgment are shown in a top down outline of the brain with the front at the top. A response was required in one task, while the other required withholding the response. The most significant differing areas, their significance level, and the most prominent correlations with other electrodes are shown. A solid line between 2 electrodes indicated that the correlations were higher in the response task, while a dotted line indicates higher no-response task correlations. The appearance of very localized cognitive activity can be created by examining differences between 2 similar split-second tasks. Note how the lateralization shifts from right to left in less than a tenth of a second. (From Gevins et al., 1985.)

characterize the ERC. The covariance analysis interval is the width of one period of the band-center frequency of each filter. Down-sampling factors are determined by the 20 dB rejection point, and the covariance function is computed up to a lag-time of one-half period of the high frequency for each band. For example, we often use a filter with 3 dB cutoffs at 4 and 7 Hz, and with 20 dB attenuation at 1.5 and 9.5 Hz. The filtered time series are decimated from 128 Hz to 21 Hz for each covariance calculation. Covariance is estimated over a 187 msec window, which corresponds to one period of a 5.5 Hz sinusoid. Each window is lagged by up to 8 lags at the original undecimated sampling rate, i.e., one-hundred-twenty-eighth of a second per lag.

(8) The significance of ERCs is determined by reference to an estimate of the standard deviation of the "noise" ERC. The noise ERC is computed by averaging random intervals in each single trial of the ensemble of trials. ERC analysis is then performed on a filtered and decimated version of the resulting "noise" averages, yielding a distribution of "noise" ERCs. The threshold for significance is reduced according to the dimensionality of the data with Duncan's correction procedure. The number of channels is used as a conservative estimate of the number of independent dimensions. The most significant ERCs in each interval are graphed.

puted. (In an optional procedure, used when the signal-to-noise ratio is very low, a statistical procedure is used to identify trials with measurable event-related signals, and averages are formed only from those trials — Gevins et al., 1986.)

(7) Multilag cross-covariance functions are computed between all pairwise channel combinations of these averaged ERPs in each selected analysis window. The magnitude of the maximum value of the cross-covariance function and its lag time are the features used to

(9) ANOVA and post-hoc t tests are used to compare ERC patterns between conditions. The similarity of appearance of two ERC graphs is measured with an estimate of the correlation between them. The estimate comes from a distribution-independent "bootstrap" Monte Carlo procedure (Efron, 1982), which also yields a confidence interval for the estimates.

(10) The between-subject variability of ERC patterns is tested by determining whether each pair of experimental conditions of a particular subject can be distinguished using discriminating equations generated on the other subjects.

(11) The within-subject reliability is assessed by attempting to discriminate the experimental conditions for each session using equations generated on that subject's other sessions.

The tests of both between- and within-subject variability and reliability are performed on sets of single trials. This quantifies the extent to which the condition-specific patterns from the ERC analysis of the average ERPs can be observed in each trial. Although this procedure could be done with any type of discriminant analysis, we have developed the use of distribution-independent, layered, artificial "neural network" pattern classification algorithms for this purpose (Gevins, 1980, 1984, 1987; Gevins and Morgan, 1986, 1988). We have shown that this method has better sensitivity than stepwise or full-model linear or quadratic discriminant analysis (Gevins, 1980). The pattern recognition approach has the advantage of testing how well a subject's individual trials conform to those of the group in discriminating two behavioral conditions of interest. In the same way, the trials of each session of a subject are tested by conformity to trials from the other sessions of that subject. Requiring trial-by-trial discriminability is a strict condition for deciding between-subject variability and within-subject reliability.

Each subject's classification yields a score which is the percent of trials that are correctly classified by the group discrimination equations. The score is assessed for significance by comparison to the binomial distribution (Gevins, 1980). A significant classification score for a subject indicates that the group equations are successful in discriminating the two conditions in his or her trials.

Within-subject (between-session) reliability is tested in a similar manner. The trial set (consisting of the two conditions) from each of a subject's sessions is tested with equations developed on the trial sets from his or her other sessions. The single-trial ERC values come from channel pairs that are significant in the ERC pattern formed from the average over all his or her sessions. Post-hoc comparisons are valuable in determining whether effects of learning and/or habituation are evident over sessions, by indicating which sessions are alike, and where transitions occur between sessions.

Figure 7 is a block diagram of the data collection and analysis process discussed above. For studies not requiring pattern recognition analysis, the event-related covariances are computed on averaged event-related potentials after bandpass filtering.

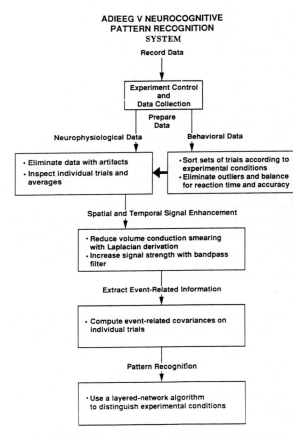

Fig. 7. ADIEEG-V system for pattern recognition of event-related brain signals. Separate subsystems perform on-line experimental control and data collection, data selection and evaluation, signal processing and pattern recognition. Current capacity is 128 channels.

In the next section, results of a study of bimanual visuomotor performance and a study of the effects of mental fatigue on human cognitive networks are presented. Preliminary results of a study of elementary language processes are also described.

Applying the tools

(A) Bimanual visuomotor task

One of the goals of the bimanual visuomotor experiment was to study prefrontal involvement while subjects prepared to perform a task accurately and used feedback about their accuracy to gauge their responses (Gevins et al., 1989a, b).

(1) Subjects and task

Seven healthy, right-handed, male adults participated in this study. A visual cue, slanted to the right or to the left, prompted the subject to prepare to make a response pressure with the right or left index finger. One second later, the cue was followed by a visual numeric stimulus (numbers 1 – 9) indicating that a pressure of 0.1 – 0.9 kg should be made with the index finger of the previously indicated hand. A 2 digit number, presented 1 sec after the peak of the response pressure, provided feedback that indicated the subject's exact pressure. On a random 20% of the trials, the stimulus number was slanted in the opposite direction to the cue; subjects were to withhold their responses on these "catch trials". The next trial followed 1 sec after disappearance of the feedback. Each subject performed several hundred trials, with rest breaks as needed.

(2) Recordings

Twenty-six channels of EEG data, as well as vertical and horizontal eye movements and flexor digitori muscle activity from both arms, were recorded. All single-trial EEG data were screened for eye movement, muscle potential, and other artifacts, and contaminated data were discarded.

(3) Analysis and results

Intervals used for ERC analysis were centered on major event-related potential peaks. ERCs were computed between each of the 120 pairwise combinations of the 16 non-peripheral channels. Intervals were set from 500 msec before the cue to 500 msec after the feedback.

We first calculated the mean error (deviation from the required finger pressure) over all trials from the recording session. Individual trials were then classified as accurate (trial error less than mean error) or inaccurate (error greater than mean error).

ERC patterns during a 375 msec interval centered 687 msec post cue (spanning the late contingent negative variation (CNV), regardless of subsequent accuracy, involved left prefrontal sites, as well as appropriately lateralized central and parietal sites (Fig. 8a). Inaccurate performance by the right hand was preceded by a very simple pattern, while inaccurate performance by the left hand was preceded by a complex, spatially-diffuse pattern (Fig. 8b). The relative lack of ERCs preceding inaccurate right-hand performance may simply reflect inattention on those trials, while the strong and complex patterns preceding inaccurate performance with the left hand may reflect effortful, but inappropriate, preparation by the right-handed subjects.

ERC patterns related to feedback about accurate and inaccurate performances were similar immediately after the onset of feedback, but began to differ in an interval centered at 281 msec that spanned the early P3 peak (P3E) (Fig. 9a, b). The ERC patterns for feedback to accurate performance by the two hands were very similar (bootstrap correlation = 0.91 ± 0.01), involving midline antero-central, central, antero-parietal, parietal and antero-occipital sites, left antero-parietal and antero-central sites, and right parietal, antero-parietal, antero-central, and frontal sites. These accurate patterns involved many long-delay (32 – 79 msec) ERCs. The waveforms of the frontal and antero-central sites consistently lagged

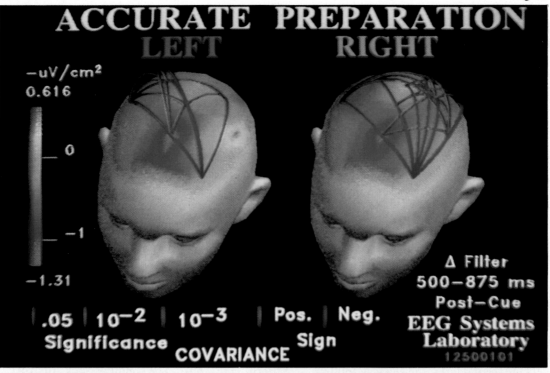

Fig. 8a

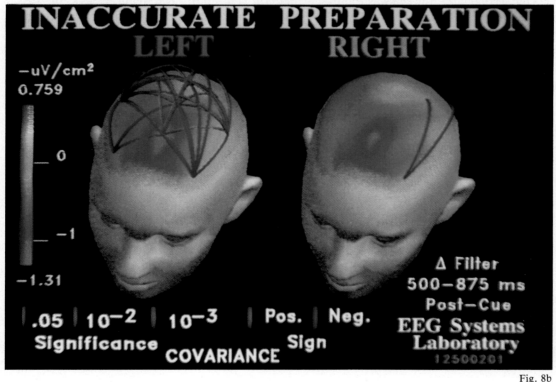

Fig. 8b

those of more posterior sites. For feedback to inaccurate performance, patterns for both hands were also very similar (bootstrap correlation = 0.90 ± 0.02), and involved most of the same sites as the accurate patterns, with the striking exception of the left and midline frontal sites. Again, frontal waveforms lagged those of the more posterior sites with which they covaried. There were even more long-delay ERCs than in the accurate patterns.

(4) Summary

The pre-stimulus ERC patterns seem to characterize a distributed preparatory neural set that is related to the accuracy of subsequent task performance. This network involves distinctive cognitive (frontal), integrative-motor (midline precentral) and lateralized somesthetic-motor (central and parietal) components. The involvement of the left-frontal site is consistent with Teuber's (1964) notions of corollary discharge, and with other experimental and clinical findings suggesting the synthesis and integration of functional networks in prefrontal cortical areas (Fuster, 1989; Goldman-Rakic, 1988a, b; Stuss and Benson, 1986). A midline anterocentral integrative motor component is consistent with known involvement of premotor and supplementary motor areas in initiating motor responses. The finding of an appropriately lateralized central and parietal component is consistent with evidence from primates and humans for neuronal firing in motor and somatosensory cortices prior to motor responses.

Since ERC feedback patterns of accurate or inaccurate performance (involving either hand) were more similar than those between accurate and inaccurate patterns for one hand, it may be inferred that the feedback patterns were related more to performance accuracy than to the hand used. The fact that ERC patterns following disconfirming feedback involved more frontal sites than did patterns following confirming feedback is consistent with the idea that greater resetting of performance-related neural systems is required following disconfirming feedback. Likewise, the front focus of these differences is consistent with the importance of the frontal lobes in the integration of sensory and motor activities (Fuster, 1989; Stuss and Benson, 1986).

(B) Effects of mental fatigue on functional brain topography

In this study, the effects of mental fatigue on preparation and memory, stimulus recognition and stimulus processing were studied (Gevins et al., 1990).

(1) Subjects and task

Five healthy, right-handed, male subjects performed a task that required that they remember 2 continuously changing numbers, in the presence of numeric distractors, and produce precise finger pressures. Each trial consisted of a warning symbol, followed by a single-digit visual stimulus to be remembered, followed by the subject's finger-pressure response to the stimulus number presented 2 trials ago, followed by a 2-digit feedback number indicating the accuracy of the response. For example, if the stimulus numbers in 5 successive trials were 8, 6, 1, 9, 4, the correct response would be a pressure of 0.8 kg when seeing the 1, 0.6 kg for the 9, and 0.1 kg for the 4. To increase

---

Fig. 8a. Preparatory event-related covariance (ERC) patterns (colored lines). Measurements are from an interval 500–875 msec after the cue for subsequently accurate right- and left-hand visuomotor task performance by 7 right-handed men. The thickness of a covariance line is proportional to its significance (from 0.05 to 0.005). A violet line indicates the covariance is positive, while a blue line is negative. ERCs involving left frontal and appropriately contralateral central and parietal electrode sites are prominent in patterns for subsequently accurate performance of both hands.

Fig. 8b. The magnitude and number of preparatory ERCs are greater preceding subsequently inaccurate left-hand performance than those preceding inaccurate right-hand performance by the right-handed subjects. Inaccurate left-hand preparatory ERCs are more widely distributed compared with the left-hand accurate pattern. For the right hand, fewer and weaker ERCs characterize subsequently inaccurate performance.

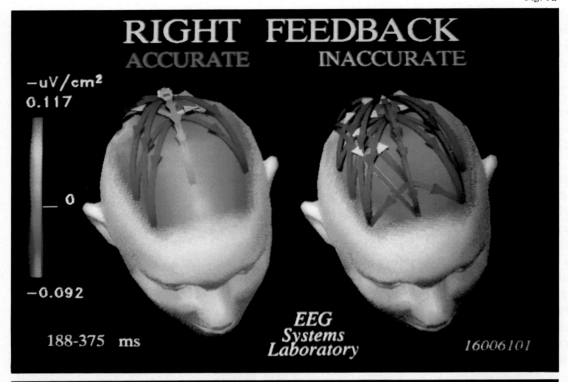
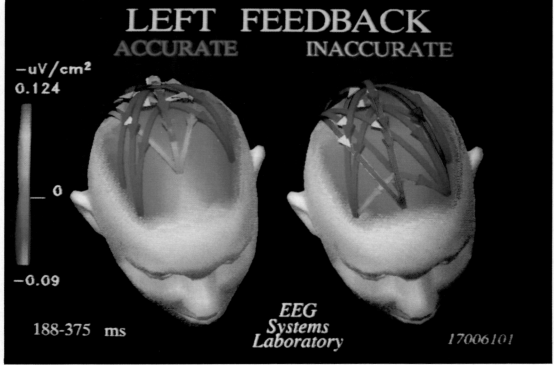

Fig. 9a

the task difficulty, subjects were required to withhold their response on a random 20% of the trials. These "no-response catch trials" were trials in which the current stimulus number was identical to the stimulus 2 trials ago. Subjects were given ample practice to stabilize accuracy and reaction time.

Subjects performed the task over a 10–14 h period. Sets of trials with equally accurate performance and response movement parameters were selected from 3 periods: an early period during the first 7 h (Alert), a middle period just prior to any decline in overall performance accuracy (Incipient Performance Impairment), and a late period after performance had significantly degraded.

(2) Recordings

EEGs were recorded with either 33 or 51 channels set in a nylon mesh cap. Vertical and horizontal eye movements were also recorded, as were the responding flexor digitori muscle potentials, electrocardiogram, and respiration. Three-axis magnetic resonance image scans were made of 3 of the 5 subjects.

(3) Analysis and results

Neuroelectric effects were observed during 2 fraction-of-second intervals when subjects: (1) prepared to receive a new stimulus number while holding the 2 previous stimulus numbers in working memory (CNV interval), and (2) when they withheld their response in the instance where the current stimulus number was the same as the 2-back stimulus number (P300 interval). Significant differences were seen between the early Alert period and the middle Incipient Performance Impairment (IPI) period ($p \ll 0.0001$). While the magnitude of the patterns was reduced during both preparatory and response inhibition intervals, the topographic distribution of the pattern was only affected during preparation (Fig. 10). The preparatory pattern shifted from one strongly focused on midline central and pre-central sites to one focused primarily on right-sided precentral and parietal sites. It appeared as though extended task-performance altered the "neural strategy" used to perform the same behavior. Thus, prolonged mental work differentially affected 2 successive split-second information processing intervals.

The extent to which each subject's patterns corresponded to the group's, and the extent to which it was possible to distinguish individual trials from the Alert and Incipient Performance Impairment periods of the session, were determined next using pattern recognition analysis. ERCs common to the group (see Fig. 10, left and middle) were considered as possible variables. For the preparatory interval, they consisted of ERCs computed over the 500 msec prestimulus epoch of each trial. For separate groups of 3 and 2 subjects whose resting EEG characteristics differed, 5 equations were formed on four-fifths of the trials, and tested on the remaining one-fifth. The average test set accuracy of Alert vs. IPI discrimination was then computed and tested for significance by reference to the binomial distribution. Discrimination accuracy was 62% ($p < 0.001$). Individualized equations were generated on the subject with the most usable data, still using the variables from the group pattern. Discrimination accuracy climbed to 81% ($p < 0.0001$).

(4) Summary

Striking changes occurred in the ERC patterns after subjects performed the difficult memory and fine-motor control task for an average of 7–9 h, but before performance deteriorated. Pattern strength was reduced in a fraction-of-a-second-

---

Fig. 9a, b. Most significant (top 2 S.D.) feedback ERC patterns elicited when subjects were given information about the accuracy of their finger pressure response by the right hand (a) and left hand (b). ERCs were derived from a 187-msec-wide interval, centered 281 msec after feedback onset, on theta-band-filtered, 7-subject-averaged evoked potential waveforms. Since ERC feedback patterns of accurate or inaccurate performance (involving either hand) were more similar than those between accurate and inaccurate patterns for one hand, the feedback patterns seem to be more related to accuracy than to the hand used to respond.

long response preparation interval over midline precentral areas and over the entire left hemisphere. By contrast, pattern strength in a succeeding response inhibition interval was reduced over all areas. The pattern changed least in an intervening interval associated with visual-stimulus processing. This suggests that, in addition to the well-known global reduction in neuroelectric signal strength, functional neural networks are selectively affected by sustained mental work in specific fraction-of-a-second task intervals. For practical application, these results demonstrate the possibility of detecting leading indicator neuroelectric patterns which precede degradation of performance due to sustained mental work.

(C) Neurocognitive analysis of elementary language processes

Preliminary results of a recent experiment demonstrate good spatial and temporal differentiation of basic linguistic functions using 59-channel EEG recordings.

(1) Subjects and task

Nine right-handed, healthy male subjects per-

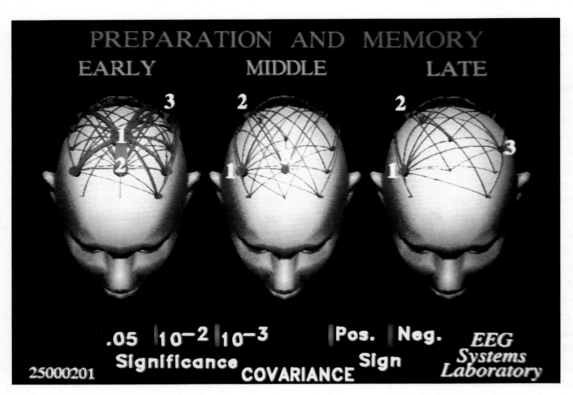

Fig. 10. Pattern recognition analysis using an artificial layered neural network distinguished ERC neuroelectric patterns recorded during Baseline (early), Incipient Performance Impairment (middle), and Impaired Performance (late) periods from 5 Air Force test pilots performing a difficult visuomotor-memory task over a 14 h period. Baseline data were obtained during the first 7 h; incipient performance impairment data during hours 7–10 preceded impaired performance; the impaired performance data were obtained during hours 10–14. The ERCs were measured during a 500 msec interval when the subjects were remembering 2 numbers and preparing for the next stimulus. ERCs greatly decline in magnitude from Baseline to Incipient Performance Impairment to Impaired Performance epochs. The patterns also changed, with the emphasis shifting from the (1) midline central, (2) midline precentral, and (3) left parietal sites to right hemisphere sites.

formed a language task in which they had to judge whether the second visually presented stimulus of a given condition (S2) formed a match with the first stimulus (S1). There were 4 conditions, fully randomized in the experiment. In the graphic *non-letter* condition, the stimuli were characters of Katakana, a script of Japanese with which none of the subjects were familiar. Subjects were required to judge whether or not the stimuli were identical. In the phonemic condition, subjects were required to judge if the pronounceable but neologistic word stimuli sounded alike. In the semantic condition, subjects were required to decide if the high frequency, open class monosyllabic words were opposite or not. Finally, in the grammatical condition, subjects judged if the verb of the S2 formed a meaningful and grammatically correct sentence with the S1 pronoun. Eighty-five percent of the trials were "match" trials and no response was required; subjects responded to the 15% mismatch trials with a button press using the left index finger.

(2) Recordings

EEGs referenced to the midline anterior-parietal (aPz) electrode were recorded from 59 scalp electrodes. The montage was an extended 10–20 system (Gevins, 1988), and included the frontal sites aF1 and aF2, Fz, F3–F8, and Fpz; the anterior central aCz, aC1–aC6; central Cz, C3 and C4; anterior temporal aT5 and aT6; temporal T3–T8; lower temporal lT1, lT2, lT5, lT6; ventral temporal vT5 and vT6; anterior parietal aP1–aP6; parietal Pz, P3–P6; anterior occipital aO1 and aO2; occipital Oz, O1 and O2; ventral occipital vO1 and vO2; and the inion (I) and both mastoids (M1 and M2). Vertical eye movements were recorded bipolarly from an electrode pair placed supra- and suborbitally; horizontal eye movements were recorded bipolarly between electrodes at the outer canthus of each eye. Other bipolar pairs were placed over flexor digitori muscles of left and right arms to record EMG, and at the submentalis to record subvocal movements of the larynx and mouth. The EEG was amplified 8333 times, band-pass filtered from 0.5 to 50 Hz and recorded at 128 samples/sec.

(3) Analysis and results

Analysis of both the ERP and ERC data is ongoing. Among the condition differences observed to date, syntactic and non-syntactic trials were clearly differentiated by ERP topography. After S1, the grammatic condition alone had a substantial negative peak at 442 msec at lateral frontal and anterior central sites. N442 was most robust at aC1, F5, and F3, where it was significantly larger than the semantic condition ($p < 0.05$). Its left-sided lateralization was not significant between F3 and F4, but did reach significance between F5 and F6 ($p < 0.05$), being almost absent at F6. After S2, the grammatic condition was again distinguished from the semantic condition by a positive peak at 279 msec at left frontal electrodes. At F3 (F5 in 2 subjects), the grammatic P279 was larger in amplitude than that in the semantic condition ($p < 0.05$).

(4) Summary

Stimuli in both conditions in this experiment were words with similar physical characteristics. The main difference between them post S1 was that the semantic (non-syntactic) condition used open class words (content words such as nouns and adjectives) and the grammatic (syntactic) condition used closed class words (function words), in this case pronouns. It is possible that the N442 observed for the syntactic condition is related to processing the closed class words, to the initiation of a "syntactic parser" (Garrett, 1982) or to both processes simultaneously. The location of the "syntactic" effect at left frontal sites (F3, F5 and aC3) after both S1 and S2 is consistent with clinical observations of syntactic deficits and difficulties in handling closed class words in aphasia patients whose lesions involve and extend deep to Broca's area (e.g., Gordon, 1985; Metter et al., 1983).

Conclusions

(A) Methodological

Improved neuroelectric recording and analysis technologies are providing new information about brain-behavior relationships. Sharper spatial resolution is provided by an increased number of electrodes and use of the Laplacian derivation. Information about common activity and its temporal relationships is provided by the event-related covariance measure, while neural network pattern recognition analysis provides a powerful method of detecting neurocognitive signals in sets of single trial data. Given the consistency between the neuroelectric results presented above and known clinical and experimental findings, the new findings suggesting functional interdependencies among prefrontal and other cortical areas during preparatory and feedback processes are especially interesting. Our current research is focused on refining and elaborating these findings.

(B) Models of neural information processing in cognitive electrophysiology

Because of the stimulus – response design inherent in most experimental designs, most models of cognitive functioning have a passive tone. The brain reacts to a given stimulus, and the stages leading to response are inferred from measures of reaction time, ERP peak latencies, and so on. However, we know from experience, observation, and inference that cognitive processes are highly interactive. Our environment is, in a sense, altered by our perception of it, since perception itself is a synthesis of sensation, current brain state, and past cognitive experience. This synthesis involves a continuously updated, dynamic internal representation of what we imagine our self and environment to be like at any given moment. Moreover, we use our effector and sensory systems to actively probe the environment for information relevant to the maintenance and updating of the self/world model (Fig. 11). Each perception, each action, is incorporated into the internal model, and new perceptions and actions are in turn influenced through the model's role in directing attentional and conceptual processes. It is challenging, but not impossible, to design experimental situations which emphasize this dynamic and interactive nature of cognition. Two areas that have been of particular interest to us are *preparatory* processes, which precede the stimulus and are directed by the internal model, and *feedback*, which governs the updating of the model after behaviors have been carried out. Although it is likely that the frontal cortex plays a pivotal role in both processes, it would be simplistic to consider the frontal lobes as a mere executor. Rather, it is likely that the entire brain is involved in a constellation of rapidly changing functional networks which provide the delicate balance between stimulus-locked behavior and

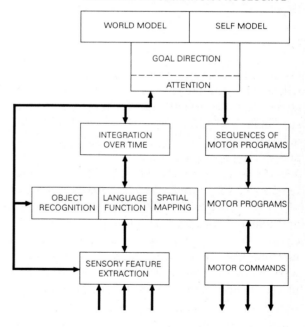

Fig. 11. Sketch of parallel, sequential, hierarchially organized information processing in functional networks of the human neocortex. The previous moment's internal model influences the current moment's goal direction and attention, which in turn influences other stages of processing.

purely imaginary ideation. The pre-stimulus and the feedback-associated "processing networks" observed in the above studies may be signs of such inter-related activity. With even further advances in brain imaging in neuropsychophysiology, we can hope to achieve increasingly detailed and direct measurements of the organization and inter-relationships of sensory and higher cognitive behaviors in health and in disease.

Acknowledgments

This work was supported by grants from the National Institute of Neurological Diseases, The National Institute of Mental Health, The Air Force Office of Scientific Research and The National Science Foundation.

The author gratefully acknowledges the efforts of his scientific collaborators, Drs. S. Bressler, P. Brickett, B. Cutillo, J. Illes, J. Le and B. Reutter for their contributions to the original research presented here.

References

Barlow, J.S. (1986) Artifact processing (rejection and minimization) in EEG data processing. In: F.H. Lopes da Silva, W. Storm van Leeuwen and A. Rémond (Eds.), *Clinical Applications of Computer Analysis of EEG and other Neurophysiological Signals: Handbook of Electroencephalography and Clinical Neurophysiology, Vol. 2,* Elsevier, Amsterdam, pp. 15–65.

Doyle, J.C. and Gevins, A.S. (1986) Spatial filters for event-related brain potentials, *EEGSL, Technical Report,* 86–001.

Efron, B. (1982) *The Jackknife, the Bootstrap, and Other Resampling Plans,* Society for Industrial and Applied Mathematics, Philadelphia, PA.

Fuster, J.M. (1989) *The Prefrontal Cortex: Anatomy, Physiology, and Neuropsychology of the Frontal Lobe,* Raven Press, New York.

Garrett, M.F. (1982) Production of speech: Observations from normal and pathological use. In: A.W. Ellis (Ed.), *Normality and Pathology in Cognitive Functions, Vol. 9,* Academic Press, New York.

Gevins, A.S. (1980) Application of pattern recognition to brain electrical potentials. *IEEE Trans. Pattern Anal. Mach. Intell.,* PAM-12, 383–404.

Gevins, A.S. (1984) Analysis of the electromagnetic signals of the human brain. Milestones, obstacles and goals. *IEEE Trans. Biomed. Eng.,* BME-31(12), 833–850.

Gevins, A.S. (1987) Statistical pattern recognition. In: A. Gevins and A. Rémond (Eds.), *Methods of Analysis of Brain Electrical and Magnetic Signals: Handbook of Electroencephalography and Clinical Neurophysiology, Vol. 1,* Elsevier, Amsterdam.

Gevins, A.S. (1988) Recent advances in neurocognitive pattern analysis. In: E. Basar (Ed.), *Dynamics of Sensory and Cognitive Processing of the Brain,* Springer, Heidelberg, pp. 88–102.

Gevins, A.S. (1989) Signs of model making by the human brain. In: E. Basar and T. Bullock (Eds.), *Dynamics of Sensory and Cognitive Processing by the brain,* Springer, Heidelberg.

Gevins, A.S. and Bressler, S.L. (1988) Functional topography of the human brain. In: G. Pfurtscheller (Ed.), *Functional Brain Imaging,* Hans Huber, Bern, pp. 99–116.

Gevins, A.S. and Morgan N.H. (1986) Classifier-directed signal processing in brain research. *IEEE Trans. Biomed. Eng.,* BME-33(12), 1054–1068.

Gevins, A.S. and Morgan, N.H. (1988) Applications of neural network (NN) signal processing in brain research. *IEEE ASSP Trans.,* 36(7), 1152–1161.

Gevins, A.S., Zeitlin, G.M., Doyle, J.C., Yingling, C.D., Schaffer, R.E., Callaway, E. and Yeager, C.L. (1979a) Electroencephalogram correlates of higher cortical functions. *Science,* 203, 665–668.

Gevins, A.S., Zeitlin, G.M., Yingling, C.D., Doyle, J.C., Dedon, M.F., Schaffer, R.E., Roumasset, J.T. and Yeager, C.L. (1979b) EEG patterns during "cognitive" tasks. I. Methodology and analysis of complex behaviors. *Electroenceph. Clin. Neurophysiol.,* 47, 693–703.

Gevins, A.S., Zeitlin, G.M., Doyle, J.C., Schaffer, R.E. and Callaway, E. (1979c) EEG patterns during "cognitive" tasks. II. Analysis of controlled tasks. *Electroenceph. Clin. Neurophysiol,* 47, 704–710.

Gevins, A.S., Doyle, J.C., Cutillo, B.A., Schaffer, R.F., Tannehill, R.L., Ghannam, J.H., Gilcrease, V.A. and Yeager, C.L. (1981) Electrical potentials in human brain during cognition: New method reveals dynamic patterns of correlation. *Science,* 213, 918–922.

Gevins, A.S., Schaffer, R.E., Doyle, J.C., Cutillo, B.A., Tannehill, R.L. and Bressler, S.L. (1983) Shadows of thought: Rapidly changing, asymmetric, brain potential patterns of a brief visuomotor task. *Science,* 220, 97–99.

Gevins, A.S., Doyle, J.C., Cutillo, B.A., Schaffer, R.E., Tannehill, R.S. and Bressler, S.L. (1985) Neurocognitive pattern analysis of a visuomotor task: Rapidly-shifting foci of evoked correlations between electrodes. *Psychophysiology,* 22, 32–43.

Gevins, A.S., Morgan, N.H., Bressler, S.L., Doyle, J.C. and Cutillo, B.A. (1986) Improved event-related potential estimation using statistical pattern classification. *Electroenceph. Clin. Neurophysiol.,* 64, 177–186.

Gevins, A.S., Morgan, N.H., Bressler, S.L., Cutillo, B.A.,

White, R.M., Illes, J., Greer, D.S., Doyle, J.C. and Zeitlin, G.M. (1987) Human neuroelectric patterns predict performance accuracy. *Science,* 235, 580–585.

Gevins, A.S., Cutillo, B.A., Bressler, S.L., Morgan, N.H., White, R.M., Illes, J. and Greer, D.S. (1989a) Event-related covariances during a bimanual visuomotor task. Part I. Methods and analysis of stimulus-and response-locked data. *Electroenceph. Clin. Neurophysiol.,* 74, 58–75.

Gevins, A.S., Cutillo, B.A., Bressler, S.L., Morgan, N.H., White, R.M., Illes, J. and Greer, D.S. (1989b) Event-related covariances during a bimanual visuomotor task. Part II. Preparation and feedback. *Electroenceph. Clin. Neurophysiol.,* 74, 147–160.

Gevins, A.S., Bressler, S.L., Cutillo, B.A., Illes, J., Miller, J.C., Stern, J. and Jex, H.R. (1990) Effects of prolonged mental work on functional brain topography. *Electroenceph. Clin. Neurophysiol.,* in press.

Goldman-Rakic, P.S. (1988a) Topography of cognition: Parallel distributed networks in primate association cortex. *Annu. Rev. Neurosci.,* 11, 137–156.

Goldman-Rakic, P.S. (1988b) Changing concepts of cortical connectivity: Parallel distributed cortical networks. In: P. Rakic and W. Singer (Eds.), *Neurobiology of Neocortex,* John Wiley, New York, pp. 177–202.

Gordon, B. and Caramazza, A. (1985) Closed- and open-class words: Failure to replicate differential frequency sensitivity. *Brain Language,* 15, 143–160.

Hjorth, B. (1975) An on-line transformation of EEG scalp potentials into orthogonal source derivations, *Electroenceph. Clin. Neurophysiol.,* 39, 526–530.

Hjorth, B. (1980) Source derivation simplifies topographical EEG interpretation. *Am. J. EEG Technol.,* 20, 121–132.

Metter, E.J., Riege, W.H., Hanson, W.R., Kuhl, D.E., Phelps, M.E., Squire, L.R., Wasterlain, C.G. and Benson, D.F. (1983) Comparison of metabolic rates, language and memory in subcortical aphasias. *Brain Language,* 19, 33–47.

Stuss, D. and Benson, D.F. (1986) *The Frontal Lobes,* Raven Press, New York.

Teuber, H.L. (1964) In: J. Warren and K. Akert (Eds.), *The Frontal Granular Cortex and Behavior,* McGraw-Hill, New York.

Discussion

C.G. van Eden: Anatomy shows extensive connections between PFC in both hemispheres. What can you tell about differential functioning of both PFCs? And what is the time relation between them in terms of activity?

A. Gevins: We have only studied a few tasks so far, so anything I say should be considered anecdotal. We have seen signs of left PFC involvement in right-handed subjects during a preparatory interval of a task requiring fine motor control by either hand. Signs of right PFC involvement were evident during intervals when subjects were given information about the accuracy of their response, irrespective of which hand was used. We have not yet systematically measured timing relations between left and right PFC.

G.E. Alexander: Have you been able to document the development of covariance between different regions during learning or skill acquisition?

A. Gevins: Not yet. In our first attempts at this, subjects adapted too quickly to changes in task requirements so that there was an insufficient number of trials to measure functional patterns during learning.

F.H. Lopes da Silva: My question concerns the problem of whether the signals recorded from different sites may be said to reflect sequential or simultaneous activation of different brain areas. I believe that the information that you have about "some delays" must be relevant to answer this question. However, in your lecture you emphasized the statistical values of covariance but said little about the estimates of delay. How do you interpret those estimations of delay? Are all covariances that you showed coupled to a delay difference from zero, or not!

A. Gevins: I did not discuss the meaning of the time delays since we have not yet had an opportunity to study them in detail. Indeed, many of them are non-zero, but I would hesitate to speculate on their meaning at this time.

R.W.H. Verwer: How deep into the brain can you penetrate with the probing? And is there a difference in the signals received from cortex parallel to the skull and perpendicularly oriented cortex? What would be the consequence of attributing a function to particular cortical areas (i.e., more or less convoluted areas)?

A. Gevins: We have optimized our experiments and analysis to record signals from superficial "association" cortex by, for example, using tasks that require a considerable degree of attention and computing the Laplacian derivation to reduce volume conduction smearing. These factors nothwithstanding, I am very cautious in attributing a function to a cortical area underlying an electrode. I use phrases such as "signs of cortical involvement" or "results were consistent with cortical involvement" to emphasize the fact that the three-dimensional origin of these complex scalp neuroelectric patterns is not known.

E.J. Neafsey: Do your data favor a parallel or sequential scheme for cortical processing?

A. Gevins: I favor both. The microstructure of neurocognitive processes has both parallel and sequential elements.

F. Reischies: High covariances would be expected between electrodes which are near each other. How do you correct for this?

A. Gevins: We correct for this in several ways. First, we reduce volume conduction effects with spatial filters. Second, we reduce background activity unrelated to the stimulus or response with signal averaging and digital filters. Then, we measure event-related covariance, that is, covariance between electrodes which is specifically related to a subject's task.

W. Wijker: How do you test for statistical significance? Applying 128 leads could easily lead to chance capitalization.

A. Gevins: We determine the significance of event-related covariances by comparing each measured value with a distribution of non-event-related covariances, that is, covariances between the same two electrodes computed from the same data without stimulus registration.

CHAPTER 18

Influence of the ascending monoaminergic systems on the activity of the rat prefrontal cortex

A.-M. Thierry, R. Godbout, J. Mantz and J. Glowinski

Inserm U. 114, Collège de France, Chaire de Neuropharmacologie, 75231 Paris Cedex 05, France

Introduction

The influence of the brain stem on the activity of the cerebral cortex was originally demonstrated by Moruzzi and Magoun (1949) who described the "ascending reticular activating system". This influence was believed to be exclusively indirect until noradrenergic (NA) and serotoninergic (5HT) neurons in the brainstem, NA and 5HT fibers in the cerebral cortex were visualized (Dahlström and Fuxe, 1964). Then, the use of new anterograde and retrograde tracing techniques and the improvements in histochemical fluorescence methods allowed to demonstrate the existence of direct reticulo-cortical NA and 5HT projections (Lindvall and Björklund, 1984; Fallon and Loughlin, 1987). It was also shown that dopaminergic (DA) terminals are present in the cerebral cortex (Thierry et al., 1973) and that the cortical DA innervation originates from the brainstem-mesencephalon (Lindvall and Björklund, 1984).

The rat medial prefrontal cortex (PFC) is one of the cortical regions innervated by the 3 aminergic systems, i.e., DA, NA, and 5HT (Fig. 1). The DA innervation of the PFC, which is particularly dense in deep layers (V and VI), mainly originates from the A10 group of DA cells located in the ventral tegmental area (VTA) and its adjacent regions (Lindvall and Björklund, 1984). Whereas DA neurons of the VTA also innervate limbic subcortical regions, distinct cells project specifically either to cortical or to subcortical structures (Thierry et al., 1984). In contrast to the restricted distribution of DA terminals in specific cortical areas, the NA and 5HT fibers distribute throughout the whole cerebral cortex (Fallon and Loughlin, 1987). Ascending NA fibers, which originate from the locus coeruleus (LC), collateralize extensively and innervate the antero-posterior axis of the cerebral cortex as well as subcortical structures. In the PFC, as in most other cortical areas, NA terminals are distributed in all cortical layers with a predilection for molecular layer I. The 5HT projections to the cerebral cortex arise from the dorsal (DRN) and median (MRN) raphe nuclei of the mesencephalon. Similarly to the LC neurons, individual cells of the DRN or MRN may project not only to cortical but also to subcortical structures. The 5HT innervation of the cerebral cortex is more dense than the NA innervation; 5HT fibers are distributed in all cortical layers and their density in the PFC is significantly greater in layer I (Audet et al., 1989).

Analysis of the influence of the aminergic ascending systems on cerebral functions has generated considerable interest since several psychoactive

Correspondence: Dr. A.-M. Thierry, INSERM U. 114, Collège de France, Chaire de Neuropharmacologie, 11, Place Marcelin Berthelot, 75231 Paris Cedex 05, France.

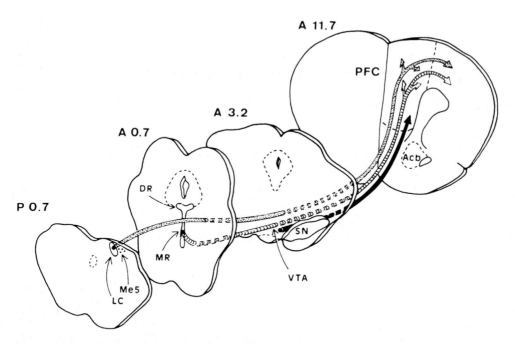

Fig. 1. Schematic representation, on coronal sections, of the monoaminergic pathways from the locus coeruleus (LC), dorsal and median raphe nuclei (DR, MR), and ventral tegmental area (VTA) to the prefrontal cortex (PFC). Me5, mesencephalic trigeminal nucleus; SN, substantia nigra; Acb, accumbens nucleus.

drugs (such as antidepressants, neuroleptics, amphetamine, etc.) are known to interfere with DA, NA or 5HT neurotransmission. However, the role of monoamines in the PFC has not yet been studied extensively. The PFC has a determining influence in the regulation of emotional states, in the control of motor activity and in cognitive processes such as representational memory (Kolb, 1984; Goldman-Rakic, 1987; Fuster, 1989). Experimental evidence suggests that DA and NA neuronal systems exert a major control in these functions. First, the particularly high reactivity of the mesocortical DA system to stressful situations (as compared to the mesolimbic and the nigrostriatal DA systems) has been underlined by various authors (Thierry et al., 1984). Second, lesion of the mesocortical DA system has been shown to induce an increase in locomotor activity in the rat; this effect is observed only if the NA innervation is preserved, suggesting that important interactions between DA and NA systems take place at the cortical level (Taghzouti et al., 1988). Third, the specific loss of DA in the PFC of rats or monkeys was associated with severe impairments in a cognitive test, the delayed alternation task (Brozoski et al., 1979; Simon et al., 1980). Moreover, it has been recently reported that the administration of an α_2-adrenergic agonist improved spatial delayed-response performance in aged rhesus monkeys (Arnsten and Goldman-Rakic, 1985). Finally, no direct study has analyzed the influence of 5HT in PFC functions. However, it is known that the 5HT ascending system may interfere with the mesocortical DA system, since the lesion of MRN induced a decrease of DA turnover in the PFC (Hervé et al., 1981). It thus appeared of importance to analyze the respective roles of the ascending aminergic systems in the control of the activity of PFC neurons. We will review recent electrophysiological studies performed in the rat in which we have compared the modulatory influence of the DA, 5HT and NA ascending systems on the spontaneous or evoked activity of PFC cells.

(I) Electrophysiological characteristics of DA, 5HT and NA neurotransmission in the PFC

The electrophysiological effects of iontophoretic application of DA, NA or 5HT on the activity of cortical cells are still controversial. However, inhibition of the spontaneous firing is the most common effect reported in the cerebral cortex (Foote et al., 1983; Phillis, 1984). For comparison, it was thus of interest to analyze the influence of the stimulation of the DA, 5HT and NA ascending systems on the spontaneous activity of PFC neurons in anesthetized rats.

(A) Influence of the DA system

The electrical stimulation of the VTA (at a frequency of 1 Hz) induced an inhibitory response in the majority (80%) of PFC cells recorded in layers III – VI. The mean duration and latency of these responses were 110 msec and 18 msec respectively (Fig. 2). Some of the inhibited cells could be identified as output PFC neurons by antidromic activation (Ferron et al., 1984; Peterson et al., 1987). Whether the inhibition is exerted directly or via intracortical interneurons in contact with efferent PFC cells, as suggested in a recent in vitro study, is not yet established (Penit-Soria et al., 1987).

Several data indicate that the responses induced by VTA stimulation are mediated by the activation of the mesocortical DA neurons (Ferron et al., 1984): (1) The latency of the inhibitory responses was compatible with the slow conduction velocity of mesocortical DA fibers. (2) The inhibitory responses were markedly reduced after pharmacological depletion of catecholamines with α-methyl-paratyrosine (α-MPT) treatment or following destruction of the ascending catecholaminergic systems by local microinjection of 6-hydroxydopamine (6-OHDA). After specific lesions of the NA system which spared DA neurons, the inhibitory responses of PFC neurons induced by VTA stimu-

lation persisted and their duration was even slightly longer. (3) The iontophoretic application of DA inhibited the spontaneous activity of PFC cells (Bunney and Aghajanian, 1976).

Biochemical studies have revealed the existence of 2 types of DA receptors in the central nervous system: the D_1, which is positively coupled to adenylate cyclase, and the D_2, which is negatively coupled or not linked to adenylate cyclase (Stoof and Kebabian, 1984). These 2 types of receptors are present in the PFC (Bockaert et al., 1977; Bouthenet et al., 1987). Experiments have been done in order to characterize the type of DA receptors involved in the DA induced electrophysiological responses in the PFC. The systemic administration of neuroleptics such as

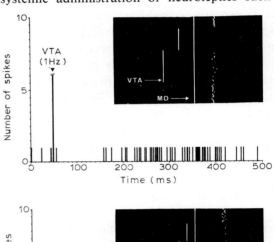

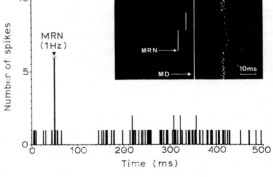

Fig. 2. Effect of VTA and MRN stimulation on spontaneous and evoked activity of PFC cells. Peristimulus time histograms showing the inhibitory responses in 2 neurons of the PFC to single pulse stimulation (1 Hz) of the VTA (upper pannel) and MRN (lower pannel). Insets: Raster dot-displays showing the excitatory responses induced by MD stimulation (5 Hz) on these cells and the inhibition of the MD evoked responses by previous stimulation of the VTA (upper inset) and MRN (lower inset).

sulpiride, spiroperidol, fluphenazine or cis-flupentixol markedly decreased the inhibitory responses of PFC neurons to VTA stimulation (Thierry et al., 1986; Peterson et al., 1987). On the other hand, other neuroleptics, levomepromazine and the palmitic ester of pipothiazine did not antagonize, and even slightly increased, the duration of the VTA induced inhibition. The inhibition of the spontaneous activity of cells in deep layers of the PFC induced by the iontophoretic application of DA was specifically blocked by sulpiride, a selective D_2 antagonist, but not by SCH23390, a selective D_1 antagonist (Sesack and Bunney, 1989). These data are in favor of the involvement of D_2 receptors in the inhibitory effect of DA on PFC cells. However, some data suggest that the PFC D_2 receptor is not identical to the D_2 subtype described in subcortical structures. Indeed, the iontophoretic application of a specific D_2 agonist, LY171555, with or without the presence of a D_1 agonist, failed to inhibit the firing of PFC cells (Sesack and Bunney, 1989). Furthermore, the systemic administration of the neuroleptic haloperidol blocked the inhibitory responses induced by VTA stimulation in the nucleus accumbens but not in the PFC (Thierry et al., 1986).

(B) Influence of the 5HT systems

Electrical stimulation of DRN and MRN has been shown to inhibit the spontaneous activity of cingulate, frontoparietal and sensorimotor cortical neurons (Olpe, 1981; Jones, 1982). Recently, we have found that PFC neurons mainly recorded in layers III – VI were also inhibited, and that a greater proportion of these neurons were affected by MRN (53%) than by DRN (35%) stimulation (Mantz et al., 1990). The mean durations and latencies of the inhibitory responses were 82 msec and 18 msec respectively after MRN stimulation and 75 msec and 18 msec after DRN stimulation (Fig. 2). The greater potency of MRN stimulation could be due to a preferential activation of MRN than DRN fibers when stimulated at a frequency of 1 Hz, since 2 distinct types of 5HT fibers arising from DRN and MRN respectively innervate the cerebral cortex (Kosofsky et al., 1987). A coactivation of efferent DRN fibers could also occur when the MRN is stimulated. Indeed, all neurons inhibited by DRN stimulation were also inhibited by MRN stimulation, but the reverse was not systematically observed. A convergence of the effect of MRN and VTA stimulation on a same PFC cell was often found. However, the effect of MRN stimulation appeared to be less potent, the duration of the inhibitory responses was of shorter duration (82 msec and 110 msec after MRN and VTA stimulation respectively) and a smaller population of PFC neurons was effected by MRN (53%) than by VTA (80%) stimulation.

The inhibition of the spontaneous firing of PFC neurons induced by raphe nuclei stimulation is very likely related to the activation of 5HT ascending systems. Indeed, microiontophoretic application of 5HT on PFC cells has been reported to decrease their firing rate (Lakovski and Aghajanian, 1985). The selective lesion of 5HT ascending fibers by local microinjection of 5,7-dihydroxytryptamine (5,7-DHT) markedly reduced the number of PFC neurons inhibited by DRN or MRN stimulation (Mantz et al., 1990). Using radioligand binding techniques 2 main subtypes of 5HT receptors ($5HT_1$ and $5HT_2$) have been described on the basis of their differential affinities for particular ligands (Peroutka, 1988). Both types of 5HT receptors are present in the PFC; however, of the different brain areas, the PFC is one of the structures which contains the highest density of $5HT_2$ receptors (Pazos and Palacios, 1985; Pazos et al., 1985). Acute systemic administration of specific $5HT_2$ receptor antagonists such as ketanserin or ritanserin blocked the inhibitory responses of PFC cells induced by MRN stimulation suggesting that these responses could be mediated through $5HT_2$ receptors (Mantz et al., 1990).

(C) Influence of NA ascending system

In contrast to the effects observed following electrical stimulation of the VTA and of the raphe nuclei, single-pulse stimulation (1 Hz) of LC did not induce reliable modifications in the spontaneous activity of PFC cells (Mantz et al., 1988). However, when a higher frequency of stimulation was used, a marked decrease in the firing rate of PFC neurons was observed. Trains of pulses at a frequency of 20 Hz applied for 10 sec in the LC produced a long-lasting post-stimulus inhibition (mean duration = 45 sec) of the spontaneous discharge in 57% of the PFC cells tested (Fig. 3). This effect was decreased markedly following depletion of cortical catecholamines by α-MPT pretreatment or selective destruction of the NA ascending pathways by local 6-OHDA injections, suggesting that these inhibitory responses are mediated by NA neurons (Mantz et al., 1988). Moreover, microiontophoretic application of NA has been shown to decrease the spontaneous activity of PFC cells (Bunney and Aghajanian, 1976). Inhibitory responses induced by local application of NA or stimulation of the LC in different cortical areas were shown to be antagonized by β-adrenergic blocking agents (Olpe et al., 1981). However, various types of electrophysiological responses to NA have been reported in the cortex (Foote et al., 1983). The respective contribution to these effects of the different adrenoreceptor subtypes linked to distinct transductional mechanisms has not yet been clearly delineated.

Some PFC neurons are sensitive to both NA and DA, but cells in layers II and III are more sensitive to NA than DA, whereas in layers V and VI the reverse is found (Bunney and Aghajanian, 1976). In agreement with this observation and with the fact that, in our study, some PFC neurons showed typical inhibitory responses to both LC and VTA stimulations, it can be concluded that there is a convergence of the effects of NA and DA ascending systems on PFC cells.

(II) Respective roles of the ascending DA, 5HT and NA ascending systems in the control of evoked responses in the PFC

Several authors have analyzed the effect of iontophoretic applications of DA, 5HT or NA on evoked responses in cortical cells (Foote et al., 1983; Phillis, 1984). DA or 5HT have been shown to reduce the excitatory responses induced by the

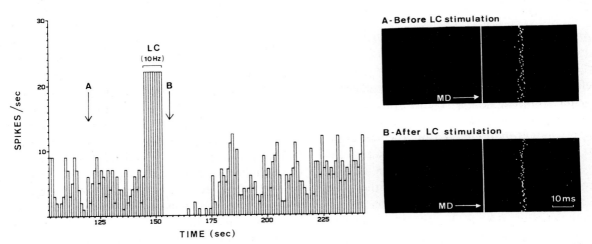

Fig. 3. Effect of LC stimulation on the spontaneous and evoked activity of a PFC cell. (Left panel) Time – frequency histogram showing the long lasting post-stimulus inhibition induced by LC stimulation (20 Hz, 10 sec) on the spontaneous activity of a PFC neuron. The period of stimulation is indicated by the horizontal bar and the concomitant peak corresponds to the stimulus artefact. (Right panel) Raster dot-displays showing the excitatory responses induced by MD stimulation (5 Hz) applied before (arrow A) or after (arrow B) LC stimulation. Note that when MD stimulation is applied during the post-stimulus inhibitory period induced by LC stimulation, the MD evoked responses are preserved.

iontophoretic application of glutamate or acetylcholine. In contrast, NA decreased the basal firing of the cells but did not block the effect of acetylcholine. In addition, the iontophoretic application of NA reduced the spontaneous firing to a greater extent than the increased activity evoked by acoustic stimulation in the auditory cortex of the awake monkey (Foote et al., 1975). More recently, a facilitating effect of NA on somatosensory responses was described in primary sensory areas of the rat neocortex (Waterhouse and Woodward, 1980). In this model, 5HT typically exerted a different effect, i.e. the iontophoretic application of 5HT on somatosensory cortical neurons preferentially suppressed the excitatory responses to tactile stimuli (Waterhouse et al., 1986).

In order to further understand the impact of DA, 5HT and NA neurons on their target cells in the PFC we have analyzed their influence on 2 types of evoked responses in the PFC of ketamine anesthetized rats: (1) the excitatory responses induced by electrical stimulation of the mediodorsal nucleus of the thalamus (MD), the main thalamic afferent to the PFC, and (2) the excitation produced by a noxious peripheral stimulus (intense tail pinch).

(A) Influence of the monoaminergic systems on MD evoked responses

The electrical stimulation of the MD at a frequency of 5 – 10 HZ elicited a single response (mean latency; 16 msec) in 80% of PFC cells. These excitatory responses very likely originate from the activation of MD neurons since they were markedly reduced after local microinjection of kainic acid in the MD (Ferron et al., 1984). A convergence on the same PFC cell of the effects of MD stimulation and of the activation of DA, 5HT and NA ascending systems was observed. Most of the cortical cells inhibited by VTA or MRN stimulation could be excited by the stimulation of the MD. When VTA or MRN stimulation were triggered respectively 3 – 45 msec or 5 – 35 msec prior to MD stimulation, the excitatory responses were blocked (Fig. 2) (Ferron et al., 1984; Mantz et al., 1990). In contrast to what was observed with the DA and 5HT systems, the activation of the NA ascending system did not affect the MD-evoked responses (Fig. 2). The excitatory responses induced by MD stimulation applied at different time intervals after LC stimulation (performed at 20 Hz for 10 sec) were always preserved even though the spontaneous firing of PFC cells was markedly decreased (Mantz et al., 1988).

(B) Influence of monoaminergic systems on noxious tail pinch evoked responses

The application of an intense tail pinch (applied for 10 sec) led to an activation of 25% of PFC cells. The usual pattern of the response was an increase in the firing rate occurring 1 – 5 sec after the onset of the pinch which lasted throughout the pinch application and in some cases for a longer period (2 – 20 sec). Units activated by noxious tail pinch also responded to noxious thermic stimulation (immersion of the tail in hot water). On the other hand, the activity of these cells was not affected by hair movements, light touch, slight persistent pressure or passive joint movements (Mantz et al., 1988).

VTA stimulation (10 Hz) inhibited completely the spontaneous activity of most of the PFC cells sensitive to tail pinch. When tail pinch was applied during VTA stimulation the excitatory response to the painful stimulus was also completely blocked (Mantz et al., 1988). Similarly, the increased firing evoked by tail pinch was markedly reduced when the stimulus was applied during MRN stimulation (10 Hz) (Mantz et al., 1990). In all cases, the evoked response to tail pinch reappeared after the VTA or MRN stimulation period. In contrast, when tail pinch was applied during the inhibitory period induced by LC stimulation (20 Hz during 10 sec) the excitatory response to the noxious stimulus was always preserved (Mantz et al., 1988).

(III) Conclusion

Ascending DA, 5HT and NA neurons which originate respectively from the VTA, the raphe nuclei and the locus coeruleus modulate markedly neuronal activity in the PFC. The main differences in the influence of DA and 5HT compared to NA afferents on the spontaneous activity or evoked responses of PFC cells were underlined in this brief review. Indeed, activation of the DA or 5HT systems induces a phasic inhibition of the spontaneous firing rate of PFC neurons and blocks the excitatory responses elicited by thalamic stimulation or by a noxious peripheral stimulus. In contrast, the activation of the NA system elicits a long lasting inhibition of the basal firing of cortical cells without blocking the evoked responses, thus enhancing the signal-to-noise ratio.

The influence of the 3 aminergic systems was observed on efferent PFC neurons. A characteristic of the output PFC neurons, in the rat, is the extensive collateralization of their axons (Ferino et al., 1987). By modulating the activity of these efferent PFC neurons, DA, 5HT and NA afferents indirectly control the activity of distinct subcortical structures such as the nucleus accumbens or the striatum. For example, it is known that DA denervation of the nucleus accumbens or the striatum leads to the development of D_1 receptor hypersensitivity in these structures. It has been shown that destruction of DA terminals in the PFC prevented this phenomenon (Hervé et al., 1989).

In conclusion, it appears that the DA, 5HT and NA ascending systems control neuronal activity in the PFC and modulate the transfer of information towards subcortical structures. By modifying the responsiveness of PFC neurons to afferent synaptic inputs, disregulation of the aminergic neurotransmissions may, therefore, induce disorders of motor activity, emotion or cognitive processes, functions in which the PFC plays a prominent role.

References

Arnsten, A.F.T. and Goldman-Rakic, P.S. (1985) Alpha-2-adrenergic mechanisms in prefrontal cortex associated with cognitive decline in aged nonhuman primates. *Science,* 230: 1273 – 1276.

Audet, M.A., Descarries, L. and Doucet, G. (1989) Quantified regional and laminar distribution of the serotonin innervation in the anterior half of adult rat cerebral cortex. *J. Chem. Neuroanat.,* 2: 29 – 44.

Bockaert, J., Tassin, J.P., Thierry, A.M., Glowinski, J. and Prémont, J. (1977) Characteristics of dopamine and beta-adrenergic sensitive adenylate cyclases in the cerebral cortex of the rat. Comparative effects of neuroleptics on frontal cortex and striatal dopamine sensitive adenylate cyclases. *Brain Res.,* 122: 71 – 86.

Bouthenet, M.L., Martres, M.P., Sales, N. and Schwartz J.C. (1987) A detailed mapping of dopamine D-2 receptors in rat central nervous system by autoradiography with (125 – I) iodosulpiride. *Neuroscience,* 20: 117 – 155.

Brozoski,T.J., Brown, B.M., Rosvold, H.E. and Goldman, P.S. (1979) Cognitive deficit caused by regional depletion of dopamine in prefrontal cortex of rhesus monkey. *Science,* 205: 929 – 932.

Bunney, B.S. and Aghajanian, G.K. (1976) Dopamine and norepinephrine innervated cells in the rat prefrontal cortex: pharmacological differentiation using microiontophoretic techniques. *Life Sci.,* 19: 1783 – 1792.

Dahlström, A. and Fuxe, K. (1964) Evidence for the existence of monoamine-containing neurons in the central nervous system. I. Demonstration of monoamines in the cell bodies of the brainstem neurones. *Acta Physiol. Scand.,* Suppl. 232: 1 – 55.

Fallon, J.H. and Loughlin, S.E. (1987) Monoamine innervation of cerebral cortex and a theory of the role of monoamines in cerebral cortex and basal ganglia. In: E.G. Jones and A. Peters (Eds.), *Cerebral Cortex, Vol. 6,* Plenum, New York, pp. 41 – 127.

Ferino, F., Thierry, A.M., Saffroy, M. and Glowinski J. (1987) Interhemispheric and subcortical collaterals of medial prefrontal cortical neurons in the rat. *Brain Res.,* 417: 257 – 266.

Ferron, A., Thierry, A.M., Le Douarin, C. and Glowinski, J. (1984) Inhibitory influence on the mesocortical dopaminergic system on spontaneous activity or excitatory response induced from the thalamic mediodorsal nucleus in the rat medial prefrontal cortex. *Brain Res.,* 302: 257 – 265.

Foote, S.L., Freedman, R. and Oliver, A.P. (1975) Effects of putative neurotransmitters on neuronal activity in monkey auditory cortex. *Brain Res.,* 86: 229 – 242.

Foote, S.L., Bloom, F.E. and Aston-Jones, G. (1983) Nucleus locus coeruleus: new evidence of anatomical and physiological specificity. *Physiol. Rev.,* 844 – 914.

Fuster, J.M. (1989) *The Prefrontal Cortex,* Raven Press, New York.

Goldman-Rakic, P.S. (1987) Circuitry of primate prefrontal cortex and regulation of behavior by representational memory. In: F. Blum (Ed.), *Handbook of Physiology. Vol. V. The Nervous System: Higher Function of the Brain,*

American Physiological Society, Bethesda, MD, pp. 373–417.

Hervé, D., Simon, H., Blanc, G., Le Moal, M., Glowinski, J. and Tassin, J.P. (1981) Opposite changes in dopamine utilization in the nucleus accumbens and the frontal cortex after electrolytic lesion of the median raphe in the rat. *Brain Res.*, 216: 422–428.

Hervé, D., Trovero, F., Blanc, G., Thierry, A.M., Glowinski, J. and Tassin, J.P. (1989) Non dopaminergic prefrontocortical efferent fibers modulate D1 receptor denervation supersensitivity in specific regions of the rat striatum. *J. Neurosci.*, 9: 3699–3708.

Jones, R.S.G. (1982) Responses of cortical neurones to stimulation of nucleus raphe medianus: a pharmacological analysis of the role of indoleamines. *Neuropharmacology*, 21: 511–520.

Kolb, B. (1984) Functions of the frontal cortex of the rat: a comparative review. *Brain Res. Rev.*, 8: 65–98.

Kosofsky, B.E. and Molliver, M.E. (1987) The serotoninergic innervation of cerebral cortex: different classes of axon terminals arise from dorsal and median raphe nuclei. *Synapse*, 1: 153–168.

Lakovski, J.M. and Aghajanian, G.K. (1985) Effects of ketanserin on neuronal responses to serotonin in prefrontal cortex, lateral geniculate and dorsal raphe nucleus. *Neuropharmacology*, 24: 265–273.

Lindvall, O. and Björklund, A. (1984) General organization of cortical monoamine systems. In: L. Descarries, T.R. Reader and H.H. Jasper (Eds.), *Monoamine Innervation of Cerebral Cortex*, Alan R. Liss, New York, pp. 9–40.

Mantz, J., Milla, C., Glowinski, J. and Thierry, A.M. (1988) Differential effects of ascending neurons containing dopamine and noradrenaline in the control of spontaneous activity and of evoked responses in the rat prefrontal cortex. *Neuroscience*, 27: 517–526.

Mantz, J., Godbout, R., Tassin, J.P., Glowinski, J. and Thierry, A.M. (1990) Inhibition of spontaneous and evoked unit activity in the rat medial prefrontal cortex by mesencephalic raphe nuclei. *Brain Res.*, 524: 22–30.

Moruzzi, G. and Magoun H.W. (1949) Brainstem reticular formation and activation of the cortex. *Electroenceph. Clin. Neurophysiol.*, 1: 455–473.

Olpe, H.R. (1981) The cortical projection of the dorsal raphe nucleus: some electrophysiological and pharmacological properties. *Brain Res.*, 216: 61–71.

Pazos, A. and Palacios, J.M. (1985) Quantitative autoradiographic mapping of serotonin receptors in the rat brain. I. Serotonin-1 receptors. *Brain Res.*, 346: 205–230.

Pazos, A., Cortés, R. and Palacios, J.M. (1985) Quantitative autoradiographic mapping of serotonin receptors in the rat brain. II. Serotonin-2 receptors. *Brain Res.*, 346: 231–249.

Penit-Soria, J., Audinat, E. and Crepel, F. (1987) Excitation of rat prefrontal cortical neurons by dopamine: an in vitro electrophysiological study. *Brain Res.*, 425: 263–274.

Peroutka, S. (1988) 5-Hydroxytryptamine receptor subtypes: molecular, biochemical and physiological characterization. *Trends Neurosci.*, 11: 496–500.

Peterson, S.L., St Mary, J.S. and Harding N.R. (1987) Cisflupentixol antagonism of the rat prefrontal cortex neuronal response to apomorphine and ventral tegmental area input. *Brain Res. Bull.*, 18: 723–729.

Phillis, J.W. (1984) Microiontophoretic studies of cortical biogenic amines. In: L. Descarries, T.R. Reader and H.H. Jasper (Eds.), *Monoamine Innervation of Cerebral Cortex*, Alan R Liss, New York, pp. 175–194.

Sesack, S.R. and Bunney, B.S. (1989) Pharmacological characterization of the receptor mediating electrophysiological responses to dopamine in the rat medial prefrontal cortex: a microiontophoretic study. *J. Pharmacol. Exp. Ther.*, 248: 1323–1333.

Simon, H., Scatton, B. and Le Moal, M. (1980) Dopaminergic A10 neurones are involved in cognitive functions. *Nature*, 286: 150–151.

Stoof, J.C. and Kebabian, J.W. (1984) Two dopamine receptors: biochemistry, physiology, and pharmacology. *Life Sci.*, 35: 2281–2296.

Taghzouti, K., Simon, H., Hervé, D., Blanc, G., Studler, J.M., Glowinski, J., Le Moal, M. and Tassin, J.P. (1988) Behavioural deficits induced by an electrolytic lesion of the rat ventral mesencephalic tegmentum are corrected by a superimposed lesion of the dorsal noradrenergic system. *Brain Res.*, 440: 172–176.

Thierry, A.M., Blanc, G., Sobel, A., Stinus, L. and Glowinski, J. (1973) Dopaminergic terminals in the rat cortex. *Science*, 182: 499–501.

Thierry, A.M., Tassin, J.P. and Glowinski, J. (1984) Biochemical and electrophysiological studies of the mesocortical dopamine system. In: L. Descarries, T.R. Reader and H.H. Jasper (Eds.), *Monoamine Innervation of Cerebral Cortex*, Alan R. Liss, New York, pp. 233–261.

Thierry, A.M., Le Douarin, C., Penit, J., Ferron, A. and Glowinski, J. (1986) Variation in the ability of neuroleptics to block the inhibitory influence of dopaminergic neurons on the activity of cells in the prefrontal cortex. *Brain Res. Bull.*, 16: 155–160.

Waterhouse, B.D. and Woodward D.J. (1980) Interaction of norepinephrine with cerebro-cortical activity evoked by stimulation of somato-sensory afferent pathways. *Exp. Neurol.*, 67: 11–34.

Waterhouse, B.D., Moises, H.C., and Woodward, D.J. (1986) Interaction of serotonin with somatosensory cortical neuronal responses to afferent synaptic inputs and putative neurotransmitters. *Brain Res. Bull.*, 17: 507–518.

Discussion

J.Scheel-Krüger: What is the possible contribution of locus coeruleus and MRN/DRN raphe innervation of VTA dopamine cells in VTA on the response of dopamine inhibitory effect on PFC cells? A direct effect on PFC or indirect via VTA?

A.M. Thierry: After lesioning of the VTA area, raphe stimulation and locus coeruleus still produce their distinct inhibitory effects in PFC, indicating the direct projection to PFC is very important. The technique of stimulation, however, may be too crude to provide detailed analyses of the delicate interactions between locus coeruleus or raphe and VTA cells.

M.A. Corner: Is the differential effect of NA vs. 5HT stimulation specific for the PFC, or do other cortical regions respond in a similar fashion?

A.M. Thierry: Results from other labs, mostly on visual and somatosensory cortex, indicate similar principles operating over the entire (neo) cortex.

C.G. van Eden: Are the inhibitory effects of monoamines (especially NA which needs a prolonged stimulation) on PFC cells produced by monosynaptic contacts?

A.M. Thierry: From our results it cannot be concluded that the effects are produced by monosynaptic contacts: the implication of cortical GABAergic interneurons is currently under investigation. Concerning the effect of LC stimulation, similar data have been obtained in other target structures of the LC, using this particular pattern of stimulation. This could be related to the characteristics of NA fibers: they show abundant collateralization, variability in the excitability of cell bodies and fibers, and inhibitory recurrent collaterals.

E.F. Neafsey: What about conduction velocity from PFC to subcortical targets?

A.M. Thierry:: The conduction velocity of efferent cortical fibers is very slow (0.5 – 1 m/sec), corresponding to a conduction time of 10 – 20 msec.

H.B.M. Uylings: Does the successive order of stimulation in the different brain regions (like VTA after MD or MD after VTA) influence the electric activity response in the PFC cells?

A.M. Thierry: In order to block the excitatory response induced by MD stimulation VTA stimulation has to be applied 3 – 45 msec before that of MD.

A.Y. Deutch: Do neurons contributing to corticofugal projections to different subcortical structures (e.g., mediodorsal thalamus, striatum) respond to different degrees of evoked inhibition?

A.M. Thierry: Most corticofugal projections collateralize to multiple subcortical targets, so at the present time it is not possible to determine if differences exist — present data would suggest no differences.

J.P.C. de Bruin: (1) How can 6-OHDA lesions of the VTA affect either mesocortical or mesolimbic DA systems in a specific way? (2) The recordings were all made in medial PFC. Can the data be generalized to orbital PFC?

A.M. Thierry: (1) Pretreatment with DMI partially protects DA neurons which project to the PFC but not those which project to nucleus accumbens. (2) We have not made recordings in the lateral PFC, but I would suggest that similar results would have been obtained.

C. Vidal: How do you explain the inhibition seen after stimulation of MD at low frequency stimulation?

A.M. Thierry: We do not have any explanation yet, because we need intracellular recordings for that. There may be two possibilities: one is a different sequence in EPSP – IPSP according to the frequency of stimulation; the other could be that different populations of cells are activated.

CHAPTER 19

The determinants of stress-induced activation of the prefrontal cortical dopamine system

Ariel Y. Deutch and Robert H. Roth

Departments of Psychiatry and Pharmacology, Yale University School of Medicine, New Haven, CT 06508, USA

Introduction

The dopamine (DA) innervation of the prefrontal cortex (PFC) differs from other mesotelencephalic DA terminal field regions (including other mesocortical areas) in that it responds in a quantitatively different manner to a number of pharmacological and environmental manipulations. For example, mild footshock stress results in the preferential metabolic activation of the DA innervation of the PFC (Thierry et al., 1976); more severe stressors result in the concurrent activation of other forebrain DA terminal fields (Antelman et al., 1988; Dunn, 1988; Roth et al., 1988a). Similarly, pharmacological challenge with phencyclidine results in DA release from all mesocortical and mesolimbic terminal fields, but increases activation of the PFC DA innervation to a significantly greater degree than is observed in other sites (Deutch et al., 1987a). Conversely, acute administration of neuroleptics such as haloperidol results in a significantly less robust activation of the PFC DA innervation than is seen in the striatum or mesolimbic areas. The quantitatively different response characteristics of the PFC DA system and the resultant pattern of changes across the mesotelencephalic DA terminal fields presumably reflect differences in the regulatory features of mesencephalic neurons which give rise to the DA innervations of these different forebrain regions.

The telencephalic projections of the midbrain DA neurons are topographically organized (Fallon, 1988). However, the topography is not absolute: different forebrain regions (e.g., the striatum) receive DA afferents from a number of different regions within the midbrain DA neuron zone. Since biochemical measurements in a given region such as the striatum therefore reflect a summed biochemical response of DA neurons originating in different parts of the midbrain, it appears likely that both features specific to the midbrain DA neurons (including autoregulatory features and afferents regulating neuronal activity) as well as features which are restricted to specific terminal fields (synaptic arrangements controlling DA release which is evoked by actions at the nerve terminal in the absence of impulse flow — so-called impulse-independent or presynaptic regulation of DA release) contribute to the observed patterns of response of the mesotelencephalic terminal fields to pharmacological and environmental manipulations.

A number of regulatory controls over DA neurons are known. These can be crudely divided into those mechanisms through which DA release is regulated by alterations in impulse flow along the DA neurons, as opposed to mechanisms which

---

Correspondence: Dr. Ariel Y. Deutch, Department of Psychiatry, Yale University School of Medicine, Connecticut Mental Health Center, 34 Park Street, New Haven, CT 06508, USA.

regulate DA neurons independent of impulse flow (Chesselet, 1984; Roth et al., 1987; Wolf et al., 1987). The regulatory controls involved in the normal impulse-dependent release of DA from the nerve terminal range from intrinsic regulatory features (such as somatodendritic autoreceptors which regulate nerve firing, or the release of co-localized transmitters) to extrinsic regulatory features (such as afferents which terminate on the somatodendritic regions of the mesencephalic DA neurons). In addition to these mechanisms, DA release from terminals can occur in the absence of impulse flow; such release is mediated by heteroceptors located on DA terminals which respond to neurotransmitters released from local circuit neurons or afferents impinging on the DA axon. The regulatory features which control DA synthesis and release are shown in Fig. 1.

Although the initial description of the characteristic response of the PFC DA system to stress was made over 15 years ago, we have only recently begun to understand the mechanisms which govern the response of DA neurons to stress. What are the regulatory features of DA neurons which contribute to the observed pattern of responses to a noxious environmental event, such as exposure to footshock, or induction of a fearful state? Moreover, what are the features of mesoprefrontal cortical DA neurons which render their response characteristics to a number of pharmacological and environmental challenges different from those of other mesotelencephalic DA neurons? Using stress-induced alterations in mesotelencephalic DA neurons as a model, we will explore some of the features which regulate the mesoprefrontal cortical DA neurons, and examine both the similarities and differences between DA neurons projecting to the PFC and those innervating other telencephalic sites.

Characterization of the PFC DA response to stress

Thierry and co-workers (1973a, b) initially described the DA innervation of the PFC, and in subsequent functional studies demonstrated that mild footshock stress resulted in the biochemical activation of the anteromedial PFC DA innervation (Thierry et al., 1976). They noted that stress exposure resulted in a striking increase in DA utilization in the PFC, and an increase of lesser magnitude in the nucleus accumbens septi (NAS); stress-induced alterations in DA utilization in the olfactory tubercle or striatum were not observed. A large number of reports which confirmed these initial observations soon followed, and indicated that exposure to mild stress (such as footshock stress which does not elicit overt escape behavior or vocalization, or alternatively conditioned fear) results in the preferential metabolic activation of the DA innervation of the PFC (Claustre et al., 1986; Deutch et al., 1985a; Herman et al., 1982; Krammercy et al., 1984; Reinhard et al., 1982). At the present time it seems most appropriate to state that stress results in the metabolic activation of the mesoprefrontal cortical DA innervation, since a variety of biochemical measures (such as changes in DA metabolite concentrations, increases in DA turnover, activity of the catecholamine biosynthetic enzyme tyrosine hydroxylase) indicative of an increase in transmitter-related metabolism are observed following stress. Moreover, recent in vivo

INTRINSIC	EXTRINSIC
Impulse-modulating autoreceptors*	Precursor availability*
Synthesis-modulating autoreceptors*	Afferent innervation*
Release-modulating autoreceptors	
Firing pattern	
End-product regulation of TH	

Fig. 1. Regulatory features of the mesotelencephalic dopamine neurons. The regulatory controls have been subdivided into those which are extrinsic as opposed to those intrinsic to the neuron. While a number of these regulatory mechanisms are present in all of the mesotelencephalic dopamine neurons (e.g., release-modulating autoreceptors), certain of the regulatory mechanisms are operative only in subsets of the midbrain dopamine neurons (e.g., synthesis- and impulse-modulating autoreceptors). * Regulatory controls which are operative only in subsets of mesotelencephalic dopamine neurons.

dialysis studies indicate that DA is indeed released in the PFC in animals subjected to acute stress (Abercrombie et al., 1989; Damsma et al., 1989; Sorg and Kalivas, 1989). While it is clear that DA release occurs in the PFC in response to stress, at the present time there is no direct evidence to indicate that mild stressors result in an increase in the firing rate of identified mesoprefrontal cortical DA neurons in the awake, freely moving animal. A single report has suggested that stress does increase the firing rate of some feline midbrain DA neurons (Trulson and Preussler, 1984); this report awaits confirmation.

Under certain circumstances DA and its deaminated metabolites appear to be released from noradrenergic neurons (Anden and Grabowska-Anden, 1983; Curet et al., 1986; Scatton et al., 1984). As such, it is conceivable that the stress-elicited increase in DA utilization that occurs following stress could result from activation of noradrenergic axons in the PFC. However, this does not appear to be the case, since lesions interrupting the noradrenergic innervation of the cortex do not attenuate the DA response to footshock stress (Claustre et al., 1986; Thierry et al., 1976). Moreover, other cortical regions which receive a rich noradrenergic innervation do not exhibit an increase in DA metabolism in response to mild stress. Finally, the stress-induced activation of tyrosine hydroxylase in the PFC appears in large part to be attributable to activation of the enzyme in dopaminergic and not noradrenergic terminals (Iuvone and Dunn, 1986).

Mild footshock and restraint stress increase indices of DA release and synthesis in the PFC (Thierry et al., 1976; Reinhard et al., 1982; Roth et al., 1988a), as do conditioned fear and food deprivation (Carlson et al., 1988; Clauste et al., 1986; Deutch et al., 1985a; Herman et al., 1982). Since a neutral auditory tone previously paired with footshock (conditioned fear) elicits an increase in PFC DA utilization, the characteristic PFC response probably cannot be attributable to response to a painful stimulus. However, the mag-

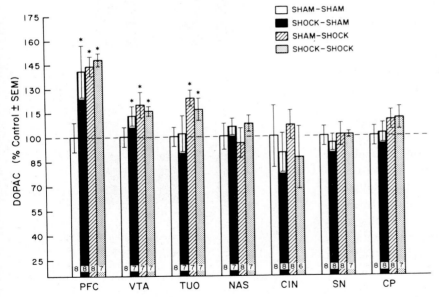

Fig. 2. The effects of mild footshock (0.2 mA) stress and conditioned fear on on mesotelencephalic DA function. Both the footshock stress and exposure to a neutral tone previously paired with the footshock stress result in the metabolic activation of the medial prefrontal cortex (PFC). Similarly, DA metabolism is augmented in the ventral tegmental area (VTA), source of the cortical DA innervation, by both conditioned fear and footshock stress. However, in other telencephalic DA terminal fields, and in the substantia nigra (SN), DA metabolism is not increased by conditioned fear. Abbreviations: TUO, olfactory tubercle; NAS, nucleus accumbens; CIN, cingulate cortex; CP, striatum. Reproduced from Deutch et al. (1985a), with the permission of Elsevier Science Publishers (Amsterdam). * $p \leq 0.05$.

nitude of the increase in DA turnover appears to be greater in animals subjected to footshock than in animals examined under the conditioned fear paradigm, and thus the presence of a painful stimulus may be perceived as a more severe stressor; an alternative explanation is simply that some degree of habituation occurs during the CS-UCS pairing trials. Significantly, the increased DA utilization in response to conditioned fear is observed only in the PFC (Claustre et al., 1986; Deutch et al., 1985a; Herman et al., 1982) and in the ventral tegmental area (VTA; see Fig. 2). The enhanced utilization in the VTA presumably reflects the biochemical activation of those A10 DA neurons projecting to the PFC (Deutch et al., 1985a; Ida and Roth, 1987; Herman et al., 1988). While only relatively mild stressors result in the selective activation of the PFC DA innervation, it should be noted that such stressors are nonetheless of greater intensity than required to elicit a corticosterone response (Antelman et al., 1988). These findings may indicate that the PFC DA response to stress shares with certain hormonal responses to stress (e.g., the prolactin response – see Kant et al., 1983) a graded involvement, possibly reflecting the necessity for different coping mechanisms.

The selective increase in DA utilization in the PFC observed after exposure to very mild stressors, such as conditioned fear, is not observed after exposure to stressors of either greater intensity (e.g., increased footshock currents – see Fig. 3) or after exposure of longer duration to relatively mild stressors (e.g., increased time in restraint – see Fig. 3). Thus, increases in either the intensity or duration of stress result in biochemically measurable changes in DA utilization in mesolimbic areas (such as the NAS); still more severe stress effects increased DA utilization in the striatum (Cabib et al., 1988; Dunn, 1988; Roth et al., 1988a, Speciale et al., 1986; see Fig. 3). Moreover, the pattern of response across the forebrain DA terminal fields appears to depend as well upon the precise strain of experimental animal. Animals selectively bred for rapidity of acquisition of an avoidance task exhibit markedly different DA response characteristics to stress (Cabib et al., 1988; D'Angio et al., 1987, 1988; Scatton et al., 1988; Tassin et al., 1980). The strain-dependent nature of the degree of stress-elicited activation of the mesoprefrontal cortical DA neurons thus parallels the strain dependency of behavioral and hormonal responses to stress (McCarty and Kopin, 1978; Shanks and Anisman, 1988). Indeed, even Sprague–Dawley rats supplied from different vendors exhibit different degrees of PFC DA activation after exposure to footshock stress of the same delivered intensity and duration (unpublished observations).

The functional significance of the PFC DA response to stress is unclear. The augmentation of PFC DA utilization may simply reflect a bio-

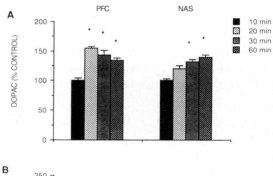

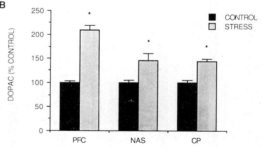

Fig. 3. The effects of increasing stress duration (A) or stress intensity (B) on telencephalic DA function. While the prefrontal cortex responds to 20 min of restraint stress, DA metabolism is not increased in the nucleus accumbens (NAS) until 30 min of restraint (A). DA metabolism can be augmented by increasing the intensity of stress as well as duration: 0.26 mA footshock stress (as compared to 0.2 mA, as shown in Fig. 1) results in the metabolic activation of the DA innervation of both the mesolimbic site (NAS) as well as the striatum (CP; B). These observations were made in collaboration with Dr. See-Ying Tam. * $p \leq 0.05$.

chemical response to stress similar to the adrenal corticosterone response. Certain recent data argue against such an interpretation. D'Angio et al. (1988) observed that Roman low avoidance ("high emotionality") rats do not respond to certain environmental stressors with an increase in prefrontal cortical concentrations of the DA metabolite 3,4-dihydroxyphenylacetic acid (DOPAC) as measured by in vivo voltammetry, while conversely Roman high avoidance (low emotionality) strain rats do respond. These data were interpreted to suggest that the metabolic activation of the PFC DA innervation may be associated with increased vigilance and attention of the animal in an attempt to cope with the stressor (Claustre et al., 1986; D'Angio et al., 1988; Scatton et al., 1988).

However, Hjorth et al. (1986, 1987a) demonstrated that systemic administration of the direct DA agonists apomorphine and 3-(3-hydroxyphenyl)-N-n-propylpiperidine (3-PPP), in a range of doses thought to selectively interact with autoreceptors and thus inhibit DA release, resulted in anxiolytic activity in the rat. Furthermore, the selective DA D_2 antagonist sulpiride has been reported to be anxiolytic in doses at which postsynaptic receptors are occupied (Costall et al., 1987; Kawano et al., 1975). These data thus suggest that inhibition of effective DA activity (either through inhibition of release or blockade of postsynaptic receptors) results in a reduction in anxiety. These data may therefore be interpreted to indicate that the function of DA activation in the PFC is to serve to *initiate* anxiety, or alternatively to recruit other neuronal systems (for example, serotonergic or noradrenergic systems, which do respond in a biochemically measurable manner to more severe stress than that required to elicit mesocortical DA activation). Both suggestions are consistent with the speculations of Scatton and coworkers, since anxiety results in heightened vigilance, and recruitment of other neuronal systems may participate in the acquisition of coping strategies aimed at ameliorating the anxiety.

Benzodiazepine modulation of the stress-induced activation of the PFC

The stress-induced metabolic activation of the prefrontal cortical DA innervation can be prevented by pretreatment with clinically effective anxiolytic agents, including benzodiazepine agonists. Thus, diazepam has been shown to prevent both the stress-induced increase in DA utilization and the stress-elicited increase in DA synthesis in the PFC and mesolimbic DA terminal fields (Claustre et al., 1986; D'Angio et al., 1987; Fadda et al., 1978; Lavielle et al., 1978; Reinhard et al., 1982; Roth et al., 1988a). Other benzodiazepine agonists, including non-sedating analogues, also block the stress-induced activation of mesocorticolimbic DA systems (Hirada et al., 1989; Roth et al., 1988a). The ability of benzodiazepine agonists to prevent DA activation of the PFC following exposure to stress is prevented by benzodiazepine antagonists such as Ro 15-1788, suggesting that benzodiazepine modulation of the DA stress response occurs through anxiolytic action at the benzodiazepine/$GABA_A$ receptor complex.

Benzodiazepine agonists enhance the probability that GABA will elicit chloride flux across the membrane; antagonists have no intrinsic action of their own. In contrast, a new group of drugs which interact with the benzodiazepine receptor and which block the effects of benzodiazepine agonists have intrinsic action at the receptor which is opposite to that exerted by agonists. These compounds have been called inverse agonists; among the β-carboline benzodiazepine inverse agonists is methyl-β-carboline-3-carboxyamide (FG-7142). Since FG-7142 has physiological effects opposite to benzodiazepine agonists, it follows that this agent and other β-carboline inverse agonists would possess anxiogenic, rather than anxiolytic, properties; such is indeed the case (Corda et al., 1983; Crawley et al., 1985; Dorow et al., 1983; File et al., 1985; Ninan et al., 1982).

Administration of benzodiazepine inverse agonists to rats results in the metabolic activation of the PFC DA innervation (Claustre et al., 1986; Giorgio et al., 1987; Ida and Roth, 1987; Roth et al., 1988a; Tam and Roth, 1985). The pattern of changes induced by administration of FG-7142 is strikingly similar to that elicited by mild footshock stress. Thus, the metabolic activation of forebrain DA systems is restricted to the PFC, and not observed in the mesolimbic areas (Tam and Roth, 1985). Indeed, high doses of FG-7142 appear to actually reduce DA metabolite concentrations in the striatum (Tam and Roth, 1985). FG-7142 administration also results in a small but significant increase in DA utilization and synthesis in the ventral tegmental area (Knorr et al., 1989; see Fig. 4). In contrast, DA utilization is reduced in the substantia nigra (see Fig. 4), and thus parallels the β-carboline-induced decrease in utilization in the striatum, projection field of the substantia nigra.

The metabolic activation of the prefrontal cortical DA innervation that occurs in response to FG-7142 and other benzodiazepine inverse agonists indicates that both environmental events (stress) and pharmacological challenges can activate the cortical DA system, and suggest that the preferential activation of PFC by both environmental and pharmacological challenges may be subserved by similar mechanisms. Moreover, clinically effective anxiolytic agents, in doses lower than those which elicit sedation, reduce both footshock-elicited and conditioned fear-induced activation of the PFC DA system, and prevent the FG-7142-elicited activation of these DA neurons. The benzodiazepine modulation of the stress response suggests that the DA stress response reflects, at least in part, a non-human condition equivalent to anxiety.

Impulse-dependent regulation of DA neurons and stress

There are a number of regulatory controls over DA neurons. It is not clear to what degree the quantitatively different response of the PFC dopaminergic innervation to stress, as well as the pattern of activation observed across the mesotelencephalic DA terminal fields, reflect differences between the regulatory controls over the various mesotelencephalic neurons. However, certain findings allow one to eliminate, as least tentatively, involvement of some regulatory controls over DA neurons as contributing to the observed pattern of responses of mesotelencephalic DA neurons. Conversely, other findings strongly suggest that certain regulatory controls, such as certain afferents impinging on the VTA DA neurons, may be specifically involved in the regionally selective enhancement of mesotelencephalic DA neurons by stress.

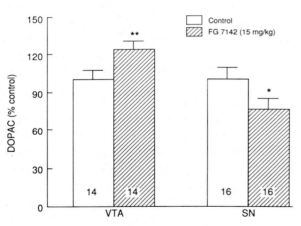

Fig. 4. The effects of the β-carboline benzodiazepine inverse agonist FG-7142 (15 mg/kg, i.p.) on the DA midbrain neurons in the ventral tegmental area (VTA) and substantia nigra (SN). The β-carboline, which increases DA metabolism in the PFC but not in other cortical or mesolimbic terminal fields, increases DA metabolism in the VTA. The effects of the inverse agonist are therefore similar to those observed following footshock stress. However, this dose of the β-carboline decreased DA metabolism in the substantia nigra (paralleling a decrease in metabolism in the striatum), and in this manner differs from footshock stress. These observations were made in collaboration with Dr. See-Ying Tam. * $p \leq 0.05$.

Impulse- and synthesis-modulating autoreceptors

DA D_2 autoreceptors are present on both the somatodendritic and axon terminal regions of certain DA neurons. Those located on the somatodendritic regions respond to DA by inhibiting firing (impulse-modulating autoreceptors) and also appear to regulate synthesis (synthesis-modulating); synthesis-modulating autoreceptors are also present on axon terminals.

Certain mesotelencephalic DA neurons appear to lack impulse- and synthesis-modulating autoreceptors. Originally demonstrated by Bannon and Roth using biochemical methods (Bannon et al., 1980), synthesis-modulating autoreceptors are not present, or present in very low density, on DA axons in the prefrontal and cingulate cortices (Bannon and Roth, 1983; Bannon et al., 1983b). Subsequently, electrophysiological techniques confirmed the absence of impulse-modulating somatodendritic autoreceptors on mesencephalic DA neurons projecting to the prefrontal and cingulate cortices (Chiodo et al., 1984). More recently, other electrophysiological experiments have revealed that a small population of identified mesoprefrontal cortical CA neurons do possess DA autoreceptors which modulate firing rate (Gariano et al., 1989; P.D. Shepard and German, 1984). This small group of DA neurons differs from the majority of the mesoprefrontal cortical DA neurons in that they are situated in the far lateral aspects of the VTA (adjacent to the medial terminal nucleus of the accessory optic tract, in the nucleus paranigralis (see Phillipson, 1979a)); in light of the location of these neurons within the VTA, it is possible that these neurons correspond to the very small population of midbrain DA neurons which collateralize to innervate a number of telencephalic sites, including the PFC (Loughlin and Fallon, 1984). The small number of mesoprefrontal DA neurons which possess impulse- and (presumably) synthesis-modulating autoreceptors probably provide so small a proportion of the PFC DA innervation as to preclude detection using biochemical methodologies.

Stress increases indices of both DA release (Abercrombie et al., 1989; Deutch et al., 1985a; Sorg and Kalivas, 1989; Thierry et al., 1976) and DA synthesis (Hirada et al., 1989; Reinhard et al., 1982) in the PFC; under conditions of low-intensity footshock such an effect is restricted to the PFC, and not observed in other mesocortical regions such as the cingulate cortex. Since there do not appear to be synthesis-modulating autoreceptors on dopaminergic afferents to either the prefrontal or cingulate cortices, it appears that the increase in PFC synthesis elicited by mild footshock stress cannot be attributable to the lack of synthesis-modulating autoreceptors. Similarly, the stress-evoked increase in DA release in the PFC is also probably not attributable to the lack of impulse-modulating autoreceptors on midbrain DA neurons, since these autoreceptors are lacking from both meso-prefrontal and meso-cingulate neurons.

Release-modulating autoreceptors

In contrast to the lack of impulse- and synthesis-modulating autoreceptors on most DA neurons innervating the prefrontal and cingulate cortices, release-modulating autoreceptors are present on all mesocortical DA neurons (Altar et al., 1987; Galloway et al., 1986; Plantjé et al., 1987; Wolf and Roth, 1987; Wolf et al., 1986). These D_2 DA receptors are present on DA axons innervating the pyriform, entorhinal, cingulate, and prefrontal cortices, and appear to functionally regulate release of DA from nerve terminals (Plantjé et al., 1987).

Since DA release in the PFC, but not the cingulate cortex, is augmented by exposure to mild stressors, yet in both sites release-modulating autoreceptors are present on DA axons, any functional regional differences attributable to release-modulating D_2 autoreceptors would not appear to contribute to the stress-induced activation of the PFC DA system. However, there are constraints upon such a conclusion. Recent electrophysiological data suggest that postsynaptic D_2 receptors in the PFC may be functionally different from D_2 receptors in other regions, such as the striatum. The inhibition of postsynaptic neurons in the PFC that occurs in response to stimulation of the VTA is not blocked by all neuroleptics (Thierry et al., 1986). Furthermore, the inhibition of postsynaptic PFC neurons that is induced by iontophoretic application of DA is not blocked in a stereospecific manner by the D_2 antagonist sulpiride, and the D_2 agonist LY 171555 does not inhibit PFC neurons (Sesack and Bunney, 1989). While it may be premature to extrapolate from postsynaptic receptors to presynaptic autoreceptors, the electrophysiological data at least suggest that D_2 postsynaptic receptors in the PFC may be a functionally distinct subtype of D_2 site from D_2 receptors in other brain regions. Biochemical differences also suggest that D_2 receptors may be regionally distinct, since there are regional differences in the apparent size of the D_2 receptor as labeled by an azidospiperone (Amlaiky and Caron, 1985, 1986). Such differences may be attributable to differences in post-translational modification (e.g., glycosylation) or alternatively may reflect different forms of the D_2 receptor coded for by mRNAs which result from alternative splicing (dal Toso et al., 1989; Sokoloff et al., 1990). Finally, apparent regional differences in D_2 receptors may reflect differences in the coupling of the receptor to intracellular processes (G proteins), since the various G proteins are regionally heterogeneous, even within the striatum (De Keyser et al., 1989). It should be mentioned that it is not completely clear to what degree the observed biochemical changes in the PFC, in contrast to other regions which appear relatively insensitive to stress (such as the striatum), reflect site-specific differences in activation or differences attributable to homeostatic mechanisms aimed at returning the function of a perturbed system to baseline conditions. For example, the increase in DA metabolite concentrations in the VTA following footshock stress and conditioned fear is thought to reflect release of DA from dendrites of the mesotelencephalic neurons (Deutch et al., 1985a). DA released in the VTA would interact with impulse-modulating autoreceptors to dampen activity in mesolimbic and certain mesocortical neurons. However, DA neurons innervating the PFC lack impulse-modulating autoreceptors and hence activity in this subset of neurons would not return as rapidly to control levels. Thus, the observed selective activation of the DA innervation of the PFC may in part reflect a decreased capacity of the mesoprefrontal cortical DA neurons to rapidly dampen activation through autoregulatory controls, as compared to DA neurons innervating other cortical and limbic sites.

Consistent with such an interpretation, in vivo voltammetric data indicate that the striatal DA innervation responds to stress with an increase release of DA, but that the increase in amine release rapidly abates (Ikeda and Nagatsu, 1985; Keller et al., 1983). These observations suggest that the inability to observe a stress-evoked increase in striatal dopamine release, when using biochemical indices such as metabolite accumulation, may be indicative of the strong transynaptic regulatory controls over certain DA cells (e.g., the nigrostriatal neurons), resulting in the rapid return to baseline of the firing rate of these DA neurons, or alternatively a presynaptic regulation of DA release. These observations also may suggest that the inability to ascribe the preferential activation of the PFC to a lack of impulse-modulating or synthesis-modulating autoreceptors may be somewhat misleading: the lack of these autoregulatory receptors on mesoprefrontal cortical DA neurons may nonetheless amplify the observed biochemical activation by rendering these neurons less readily capable of returning to normal function following an initial perturbation. However, a recent in vivo dialysis study (Abercrombie et al., 1989) indicates that while relatively severe tail shock stress evokes an increase in striatal DA release, the increase occurs in the dialysate sample collected *after* termination of the tail shock period. In contrast, the DA innervations of the PFC and the NAS respond to stress by augmenting DA release as measured in the two 15 min samples collected *during,* but not after, tail shock.

At present, it seems most reasonable to posit that the absence of synthesis- and impulse-modulating D_2 autoreceptors on mesoprefrontal cortical DA neurons does not contribute to the quantitatively different responses of the PFC DA system to stress. However, it is clear that additional studies will be required to evaluate to what degree differences in activation as opposed to response maintenance contribute to the stress-evoked biochemical augmentation of DA utilization in the PFC. Similarly, studies which have examined regional heterogeneities in the *pre*-synaptic D_2 autoreceptor are lacking; until these data are gathered it will not be possible to eliminate differences in release-modulating autoreceptors as a contributory regulatory feature to the stress-induced activation of the mesoprefrontal cortical DA system.

Precursor regulation of the PFC DA innervation

Catecholamine synthesis in various CNS sites has been reported to covary with local tyrosine concentration (Fernstrom, 1983). While all catecholamine neurons are ultimately dependent upon tyrosine as a precursor substance, the degree to which relatively small changes in precursor availability are rapidly reflected by changes in metabolism of dopaminergic neurons has only recently been determined. The relationship between local tyrosine concentration and catecholamine synthesis appears to be operative only under conditions in which tyrosine hydroxylase (TH) is in an activated state. One factor regulating the state of TH activation (phosphorylation) is the intracellular concentrations of end product (DA); neuronal stores of DA are in turn related to the firing rate of the neurons. The higher basal firing rate of mesoprefrontal DA neurons (Chiodo et al., 1984) and the greater frequency with which burst firing is observed in these neurons (Grace, 1988) presumably contribute to a greater proportion of TH being in the activated form in the PFC than in other mesotelencephalic regions (Iuvone and Dunn, 1986). Such a relative predominance of phosphorylated enzyme in the PFC may contribute to the enhanced responsiveness of the cortical DA innervation to precursor (tyrosine) availability (Roth et al., 1988b; Tam and Roth, 1985).

The DA innervation of the prefrontal and cingulate cortices, but not other mesotelencephalic sites, respond to administration of tyrosine by increasing synthesis (Roth et al., 1988b). Moreover, administration of the large neutral amino acid valine, which competes with tyrosine for uptake and thereby lowers central concentrations of tyrosine (Fernstrom, 1983), significantly decreases catecholamine synthesis in the PFC (Roth et al., 1988b). Animals rendered diabetic by administration of streptozotocin, in which central tyrosine concentrations decline, exhibit decreased in vivo tyrosine hydroxylation (an index of DA synthesis) in the PFC; the decrease in both endogenous tyrosine concentrations and in synthesis is reversed by treatment with insulin (Bradberry et al., 1989).

It has recently been demonstrated that the degree of activation of the PFC DA innervation elicited by administration of FG-7142, as reflected by DA metabolite concentrations, is increased in animals administered physiologically relevant doses of tyrosine (Roth et al., 1988b). This finding is consistent with the suggestion that the increased activity of the PFC DA innervation renders this system unusually dependent upon precursor availability. Thus, the increased responsiveness of the PFC DA system to stress is probably not attributable to differences in precursor availability — if anything, such an increased dependence upon precursor in the face of stress-induced heightened precursor demands might be expected to result in a lower degree of responsiveness of the mesoprefrontal cortical DA system.

Afferent control of mesoprefrontal cortical DA neurons

The DA neurons innervating the medial PFC of the rat are located in the ventral tegmental area (VTA). The VTA receives a large number of afferents from telencephalic, diencephalic, and hindbrain regions (Phillipson, 1979b). Moreover, afferents to the VTA do not innervate the region diffusely, but rather innervate specific nuclei within the ventral tegmental area (Phillipson, 1979a, b), although DA neurons are distributed throughout the region. For example, the nucleus interfascicularis, situated in the most medial aspect of the ventral mesencephalic tegmentum, receives afferents from the medial habenula and raphe, but does not appear to receive prominent projections from telencephalic DA rich areas (Phillipson, 1979b). Projections to the VTA from the nucleus accumbens septi (NAS) are relatively sparse (Conrad and Pfaff, 1976; Nauta et al., 1978), and GABAergic projections from the nucleus accumbens to the VTA are predominantly restricted to the anteromedial VTA (Walaas and Fonnum, 1980). The relatively distinct termination pattern of afferents to the VTA, coupled with the topographic organization of the A10 DA neurons onto the telencephalon (Fallon, 1988; Scheibner and Tork, 1987) suggests that specific afferents to the VTA may regulate subsets of midbrain DA neurons, such as those innervating the PFC.

Exposure to stress must be perceived by an animal in order for the DA neurons to respond to the stressor, and such information conveyed to the DA neurons. While diffuse hormonal signals may be of significance in the periphery, it seems likely that neural pathways carrying information concerning the stress (be this information exteroceptive or interoceptive) must reach the DA neurons in some manner in order for stress to effect alterations in DA utilization. A recent study by Herman et al. (1988) suggests that afferents to the dopamine neurons rather than intrinsic regulatory features are primarily involved in determining the responsiveness of the DA neurons to stress. Herman and associates transplanted fetal mesencephalic DA neurons to either the NAS or VTA several weeks after 6-hydroxydopamine lesions of the A10 DA neurons were made. Six months later, grafted animals and normal controls were subjected to footshock stress. Success of the grafts in replacing DA was determined using biochemical (DA concentrations) and behavioral (locomotor response to amphetamine) criteria. Previous studies had indicated that striatal transplants of fetal mesencephalic DA neurons can indeed exhibit appropriate functional responses to pharmacological challenges (such as augmenting DA release in response to haloperidol administration); such responses are presumably due to functionally coupled release-modulating autoreceptors (Meloni et al., 1988). Herman and associates (1988) demonstrated that footshock stress elicited an increase in DA utilization in both the NAS and VTA in normal subjects, but not in the grafted DA neurons placed in the NAS and VTA. These data suggest that the lack of afferent innervation establishing appropriate synaptic contact with the grafted DA neurons results in the lack of responsiveness of the graft to stress.

In contrast to the DA neurons of the substantia nigra (SN), the VTA DA neurons do not appear to be tightly regulated by a transynaptic feedback pathway. Indeed, as noted above, projections from the mesolimbic DA terminal fields to the VTA are relatively sparse; the majority of the projections to the A10 DA neurons which convey information concerning the functional status of neurons in the mesolimbic terminal fields appear to originate in the ventral pallidum (Groenewegen and Russchen, 1984; Mogenson et al., 1983; Phillipson, 1979b; Zahm, 1989). Similarly, projections from the PFC to the VTA (which predominantly originate in the caudal prelimbic and infralimbic cortices) are sparse (Beckstead, 1979; Sesack et al., 1989).

Two strategies are useful in determining if afferents to the VTA are regulating the PFC DA stress response. Lesions can be made in sites from which afferents to the VTA originate, and subsequently the response of the mesotelencephalic DA system to stress assessed. While such an approach has proven of considerable use in determining the central systems subserving the stress-induced activation of the hypothalamo-hypophyseal system, this approach has not been extensively used in studies of the mesotelencephalic DA system, since transynaptic alterations in the function of the DA neurons may ensue, complicating interpretation of the data. An alternative approach to assessing the involvement of afferent regulation in determining the DA response to stress is to use pharmacological methods to obtain data concerning specific transmitter systems which may impact on DA neurons and thereby regulate the DA stress response. This approach has been used extensively, and has yielded a large amount of valuable information.

Tachykinin peptides

Peptidergic systems which may regulate the midbrain DA neurons have been among the most intensively studied of the transmitter systems. In particular tachykinin and opiate peptides have been examined, and most recently attention has focused on corticotropin-releasing factor. The tachykinin peptides (which possess a conserved carboxy terminal sequence -Phe-X-Gly-Leu-Met-NH$_2$) include substance P (SP; X = Phe) and the aliphatic tachykinin substance K (SK; X = Val); these peptides arise from alternative splicing of the preprotachykinin gene. Substance P and substance K appear to be colocalized in the nervous system. In addition, another aliphatic tachykinin, neuromedin K, is present in the mammalian central nervous system (Maggio, 1988). The initial demonstration of a modulatory role of SP on mesotelencephalic DA neurons came from experiments demonstrating that SP augments DA-dependent behavior following local administration of the peptide into the DA cell body regions (A.E. Kelley and Iversen, 1978; A.E. Kelley et al., 1979a, b). Subsequently, Lisaprawski et al. (1981) reported that concentrations of the aromatic tachykinin SP were decreased in the VTA and interpeduncular nucleus, but not PFC or SN, following footshock stress, and suggested that neurons in the habenula which contain SP project to the VTA and govern the responsiveness of the mesoprefrontal cortical DA neurons to stress. Bannon et al. (1983a) consolidated the regulatory position of SP by demonstrating that immunoneutralization of VTA SP (accomplished by direct VTA injection of a SP monoclonal antibody) prevented the stress-induced activation of the PFC DA innervation, but did not affect basal DA turnover in the PFC. These data provided strong support for the role of tachykinin afferents in regulating the mesoprefrontal cortical DA stress response.

Subsequent examinations focused on the role of the colocalized aliphatic tachykinin SK. In contrast to SP, which upon direct administration to the VTA increases DA metabolism in the PFC (Deutch et al., 1985b; Elliot et al., 1986), SK infusion into the VTA does not appear to augment cortical DA metabolism, but acts prominently on the mesolimbic DA system (Deutch et al., 1985b; Kalivas et al., 1985). These data suggest that modulation of the cortical stress response by tachykinin peptides might be more specific than hitherto realized. This proved to be the case: after 20 min of mild footshock, SP concentrations in the VTA are decreased, suggesting that SP-containing afferents to the VTA release the peptide in response to the stress; in contrast, VTA concentrations of SK are unaltered by stress (Bannon et al., 1986; Deutch et al., 1985b). These data suggest that substance P participates in the stress-induced activation of the mesocortical DA system, but that the colocalized tachykinin substance K may not. Moreover, since levels of SK in the VTA do not decline after footshock stress, it appears that features of the afferent input to the VTA (the release processes governing the colocalized peptides) rather than characteristics of the post-synaptic DA neuron, such as the different tachykinin receptors (Kalivas et al., 1985), are the key regulatory processes involved.

The dissociation between VTA concentration of the aromatic and aliphatic tachykinins after mild footshock stress may also help explain the pattern of DA activation observed after mild footshock stress, i.e., an increase in DA utilization in the PFC but not in the mesolimbic or striatal sites. As noted above, the differential actions of the tachykinins in modulating activity in mesocortical and mesolimbic systems is consistent with specific afferents (such as SP- and SK-containing afferents) regulating different mesotelencephalic neurons. However, it would be expected that at some point SK as well as SP would be released in the VTA from afferents which contain both tachykinins; such release of the aliphatic tachykinin may occur upon exposure of the animal to more severe stressors, and thereby contribute to the activation of limbic DA systems.

The possible role of the tachykinins in the anxiogenic β-carboline-elicited activation of the mesoprefrontal cortical DA innervation has also been examined. In contrast to the footshock stress-induced decrease in VTA concentrations of SP, administration of FG-7142 does not result in decreased levels of either SP or SK in the VTA or SN (unpublished observations). These data suggest that the tachykinins are not involved in β-carboline-induced activation of the PFC DA system. Since the involvement of other peptidergic systems in the actions of the anxiogenic β-carbolines has not been examined, it is possible that the β-carbolines act directly through benzodiazepine receptors located on VTA DA neurons (see below).

Opioid peptides

Opioid peptides have been suggested to play a role in modulating the stress-induced activation of the PFC DA system. Footshock stress clearly results in an increase in blood levels of preopiomelanocortin (POMC)-derived peptides, including β-endorphin (Kant et al., 1983; Rossier et al., 1977). Interactions between dopamine and opioid peptides have been well documented, and include dopaminergic regulation of opioid peptides in mesocorticolimbic terminal fields (Deutch and Martin, 1983; Palmer and Hoffer, 1980; Sivam et al., 1987) and opioid regulation of dopamine systems (Wood, 1983). Initial electrophysiological reports indicated that morphine potently activates A10 DA neurons (Gysling and Wang, 1983; Matthews and German, 1984; Ostrowski et al., 1982). Subsequently, Kalivas and associates demonstrated that enkephalins activate mesocorticolimbic DA neurons through a μ-opiate receptor (Kalivas and Richardson-Carlson, 1986; Kalivas et al., 1983a; Latimer et al., 1987).

Miller et al. (1984) reported that pretreatment with the opioid antagonist naloxone blocked the stress-elicited increase in PFC DA metabolism, and suggested that exposure to stress resulted in the release of endogenous opiates which activated the mesoprefrontal cortical DA neurons. Subsequently, Bannon et al. (1986) demonstrated that dynorphin B concentrations decrease in the VTA following footshock stress, but failed to observe a change in VTA levels of the preproenkephalin A-derived peptides Leu-enkephalin and Met-enkephalin-Arg-Gly-Leu. However, Kalivas and Abhold (1987) reported that Met-enkephalin levels were selectively decreased in the medial (but not lateral) VTA after footshock stress. Moreover, they demonstrated that direct VTA infusion of the opioid antagonist naltrexone attenuated the stress-induced activation of the PFC DA system. The regional specificity of the alterations in Met-enkephalin in the VTA following footshock stress may suggest that the opioid peptide is acting upon specific subsets of DA neurons.

Corticotropin-releasing factor, neurotensin, and somatostatin

Other peptides present in the central nervous system have not been as thoroughly evaluated as the tachykinin and opioid peptides for their possible modulation of the mesoprefrontal cortical DA neurons. Corticotropin-releasing factor (CRF) is a critical link in the stress-induced activation of the hypothalamo-hypophyseal-adrenal axis. Immunoneutralization of CRF prevents the stress-induced release of adrenal hormones (Nakane et al., 1985). Similarly, the CRF antagonist α-helical CRF potently inhibits stress-induced behavioral events (Berridge and Dunn, 1987; Kalin et al., 1988; Britton et al., 1986). Intraventricular CRF administration has been reported to result in a stress-like enhancement of DA utilization in the cortex of mice (Dunn and Berridge, 1987); this observation has recently been confirmed in rats (Matsuzaki et al., 1989). However, intraventricular CRF administration enhances DA utilization in a number of mesotelencephalic terminal field, including the PFC, septum, and bed nucleus of the stria terminalis, but not in the NAS or striatum. This pattern of activation is distinct from that observed following most stressors (Deutch et al., 1985a; Dunn, 1988). Moreover, direct administration of CRF to the VTA does not increase DA utilization in the PFC (Kalivas et al., 1987), also suggesting that the increase in prefrontal cortical DA utilization observed after intraventricular CRF may differ from the stress-evoked activation of this DA system. Acute stress does not alter CRF concentrations in the VTA, but does tend to decrease concentrations of the peptide in the PFC (Chappel et al., 1986; Deutch et al., 1987b). These data may suggest that intraventricular administration of CRF activates the PFC DA innervation through local circuit interactions in the PFC (see below).

Somatostatin and neurotensin innervate and regulate midbrain DA neurons (Deutch et al., 1988; Kalivas et al., 1983b; Pinnock, 1985; Nemeroff and Cain, 1985; Woulfe and Beaudet, 1989). Accordingly, stress-induced alterations in mesotelencephalic somatostatin and neurotensin concentrations have been examined (Deutch et al., 1987b). No significant alterations in peptide concentrations in any mesotelencephalic area, including the VTA, were observed after 20 min of footshock stress, with one exception: neurotensin concentrations in the VTA were significantly elevated by 20 min of footshock stress. It is difficult to interpret such an increase in peptide concentration. Release of the peptide (and subsequent action on DA neurons) would typically be reflected in decreased concentrations of the peptide. Neurotensin perikarya in the VTA also contain DA (Hökfelt et al., 1984). Since DA metabolite concentrations were increased in the VTA, suggesting increased release of the amine, it appears unlikely that decreased release of neurotensin contributes to the increase in VTA peptide concentration after stress. Moreover, peptide synthesis from prohormones generally occurs on a slower temporal scale. Perhaps the most parsimonious explanation for an increase in VTA neurotensin concentrations is increased release of the peptide, coupled with degradation of the parent peptide to yield active neurotensin fragments; the major degradative peptide fragment of NT has been reported to augment DA release (Markstein and Emson, 1988).

Neurotensin differs from the other peptides examined in that it is colocalized with certain midbrain DA neurons, including some of those that project to the PFC (Hökfelt et al., 1984; Studler et al., 1988; Tassin et al., 1988). Thus, neurotensin may conceivably regulate those midbrain DA neurons in which it is present by acting at neurotensin autoreceptors after impulse-dependent release (Hervé et al., 1986; Quirion et al., 1985; Szigethy and Beaudet, 1989). The nature of such interactions is only beginning to be explored, and attention has primarily focused on these interactive events at the terminal field level (Bean et al., 1989a, b; Hervé et al., 1986). Future studies will be required to assess the contribution of this autoregulatory feature to regulation of the PFC DA stress response.

Excitatory amino acids

Classical transmitters as well as peptides have been implicated in modulation of the PFC DA stress response. Among the excitatory amino acids the most attention has focused on glutamate. The efferents of the PFC are thought to use an excitatory amino acid as a transmitter (Carter, 1982; Divac et al., 1977; Girault et al., 1986; Spencer, 1976). In particular, PFC efferents to the midbrain DA cell group areas have been suggested to contain an excitatory amino acid transmitter (Christie et al., 1985a; Kornhuber et al., 1984; Nieoullon and Dusticier, 1983). At present it is not clear whether the relevant transmitter is glutamate, aspartate, or N-acetylaspartyl-glutamate (NAAG) or combinations of these amino acids, or alternatively if different amino acids are localized to different corticofugal projections.

Because of the ubiquitous distribution of glutamate and aspartate, very few studies have focused on assessing stress-induced alterations in concentrations of these amino acids after stress. Acute severe stress has been reported to selectively increase glutamate and aspartate concentrations in cortical regions and the substantia nigra (Palkovits et al., 1986). Mount et al. (1989) recently reported that glutamate stimulates [^{3}H]DA release from dissociated cell cultures of the ventral midbrain in a specific and dose-dependent manner, and that the effects of glutamate were blocked by *cis*-2,3-

piperidine dicarboxylic acid.

Consistent with potent excitatory effects of glutamate on VTA neurons (which are independent of toxicity of the amino acid), Kalivas et al. (1989) found increased locomotor activity following direct VTA injection of glutamate or kainic acid, but not N-methyl-D-aspartate (NMDA) or quisqualic acid. Furthermore, Kalivas and co-workers demonstrated that glutamate infusion into the VTA increased DA release in both PFC and NAS, as well as the VTA, but that NMDA infusions resulted in a selective increase in DA utilization in the PFC; kainate increased DA metabolism in the NAS and VTA only. Finally, VTA infusion of 3-($\pm$)-2-carboxypiperazine-4-yl)propyl-1-phosphonic acid (CPP), a NMDA antagonist, completely prevented the stress-induced activation of the PFC DA system. These studies suggest that glutamate regulates the VTA DA neurons, and that a specific excitatory amino acid receptor (NMDA) system modulates the mesoprefrontal cortical DA system; in contrast, the kainate receptor appears to modulate mesolimbic (NAS) DA neurons. Future studies will be required to determine if specific receptors on the different mesotelencephalic projections are postsynaptic to afferents which preferentially utilize a specific excitatory amino acid. Such a mechanism may help to explain the pattern of changes in the mesotelencephalic DA terminal fields following stress, and specifically the change in that pattern upon exposure of animals to more severe stressors.

As noted above, the projections from the PFC to the midbrain DA cell group regions are thought to utilize an excitatory amino acid as a transmitter (Abarca and Bustos, 1985; Carter, 1982; Christie et al., 1985a; Kornhuber et al., 1984). The firing patterns of DA neurons in the midbrain have been well documented, and include both an irregular firing of these neurons at approximately 4–7 Hz and a pattern of discharge characterized by rapid bursts of firing at a significantly higher frequency (Freeman et al., 1985; Grace and Bunney, 1984; Grace, 1988). The "bursting" pattern of firing is thought to significantly enhance DA release from axons in the terminal field (Gonon, 1988). Mesoprefrontal cortical DA neurons exhibit the greatest degree of bursting of the midbrain DA neurons. The shift from the irregular slow firing pattern of DA neurons to the rapid bursting mode can be induced by stimulation of the PFC (Gariano and Groves, 1988); to date, stimulation of no other CNS site has been demonstrated to induce burst firing of midbrain DA neurons. Since the bursting mode of activity of DA neurons appears to be associated with increased transmitter release from terminals, it is possible that the preferential biochemical activation of the mesoprefrontal cortical DA innervation is associated with an initial activation of corticofugal neurons which utilize an excitatory amino acid transmitter; current studies are examining this possibility.

GABA

The benzodiazepine modulation of stress-induced activation of the PFC DA innervation suggests that γ-aminobutyric acid (GABA) systems are a key component of the DA stress response, since the benzodiazepine receptor is a part of a macromolecular complex including the GABA receptor and a chloride channel. There have been a number of reports suggesting the direct involvement of GABA mechanisms in stress, and a role for GABA as an anxiolytic (Sanger, 1985; R.A. Shepard, 1987). Thus, studies have documented changes in GABA receptor-mediated chloride flux following acute stress (Biggio et al., 1984; Concas et al., 1987; Kuriyama et al., 1984; Schwartz et al., 1987). The rather conflicting literature concerning the function of GABA in stress may partially reflect the multiple sites at which GABA may exert its actions.

GABA-containing neurons in the globus pallidus and ventral pallidum, as well as the NAS and striatum, project to the ventral mesencephalic DA cell body regions and provide information concerning the actions of DA at mesolimbic and striatal terminal fields (Araki et al., 1985; Fonnum et al., 1978; Ribak et al., 1980; Walaas and Fonnum, 1980). Electrophysiological data indicate that

GABA acts directly on midbrain DA neurons to potently inhibit firing, whereas there is an indirect action of GABA to stimulate DA neurons (Grace and Bunney, 1979; Pinnock, 1984; Lacey et al., 1988). In the substantia nigra, the direct action of GABA is believed to occur through $GABA_B$ receptors located on DA neurons, while the indirect excitatory effects are thought to occur through $GABA_A$ receptors located on zona reticulata neurons proximal to the DA cells (Grace and Bunney, 1984; Lacey et al., 1988; Pinnock, 1984).

Recent data indicate that microinjection of muscimol, a $GABA_A$ agonist, into the VTA increases locomotor activity, whereas administration of the $GABA_B$ agonist baclofen does not (Kalivas et al., 1990). These data are consistent with $GABA_A$ agonists acting presynaptically to inhibit GABA release and thus remove a GABAergic inhibition over DA neurons, the latter effect mediated by $GABA_B$ receptors. The localization of $GABA_B$ receptors to DA neurons suggests that the $GABA_B$ receptor may modulate the stress-induced activation of the PFC. This is indeed the case: Kalivas et al. (1990) infused the $GABA_B$ agonist baclofen into the VTA, and prevented the stress-induced activation of the mesoprefrontal cortical DA innervation.

While the ability of baclofen to prevent the stress-induced activation of the PFC DA system clearly suggests that a GABAergic mechanism is involved in the regulation of the mesoprefrontal cortical DA neurons, it should be noted that attempts to systemically (as opposed to locally) manipulate GABAergic function can yield data suggesting that GABAergic systems are not involved in the stress-induced enhancement of the PFC DA system (Claustre et al., 1986). The ubiquitous distribution of GABA and the myriad ways through which it can modulate dopaminergic systems, both via impulse-dependent and impulse-independent (presynaptic) mechanisms, suggest that systemic manipulations impact at multiple points along the dopamine neuron and thus may obscure certain specific effects.

GABAergic neurons are undoubtably involved in the ability of the anxiogenic β-carbolines to activate biochemically mesoprefrontal cortical DA neurons. However, there has been little direct examination of the specific function of GABAergic systems in modulating the β-carboline enhancement of cortical DA metabolism in vivo. The role of GABA in modulating the actions of the anxiogenic β-carbolines on cortical DA function has recently been examined in vitro, and is discussed below (see below: impulse-independent regulation of DA neurons).

Serotonin

A large number of studies have indicated that central serotonin (5-HT) systems interact with the mesotelencephalic DA system; such studies have examined serotonin regulation of DA release in striatal and mesolimbic areas and have described changes in forebrain DA function following lesions of the raphe nuclei. In particular, lesion studies have suggested that pontine raphe neurons regulate mesolimbic and mesocortical DA systems. Electrolytic lesions of the median raphe effect a time-dependent increase in DA utilization in the NAS, but reduce DA utilization in the PFC (Hervé et al., 1981). Lesions of the DR result in an increase in DA utilization in the NAS, but do not alter DA function in the PFC (Hervé et al., 1979).

Serotonergic neurons respond to stress, and the firing rate of 5-HT raphe neurons can be modified by stress and by clinically effective anxiolytic agents, including the benzodiazepines. These observations have led to the suggestion that 5-HT neurons may be a critical locus of action of anxiolytic drugs. Stress results in regionally heterogeneous changes in various biochemical parameters of 5-HT function (Boadle-Biber et al., 1989; Culman et al., 1980; De Souza and Van Loon, 1986; Lee et al., 1987; Paris et al., 1987); the specific regions affected and the exact direction of change depend on the type of stressor (Lee et al., 1987; Paris et al., 1987).

Direct manipulation of 5-HT systems by microinjections of serotonergic drugs and anxiolytic agents have led to the hypothesis that serotonergic neurons in the dorsal raphe are a site of action of the anxiolytic agents (Iversen, 1984; Thiébot et al., 1982; Thiébot and Soubrié, 1983). Electrophysiological data indicate that benzodiazepines potentiate GABA inhibition of dorsal raphe serotonergic neurons (Gallager, 1978); moreover, non-benzodiazepine anxiolytic agents also inhibit the firing of 5-HT neurons in the raphe (VanderMaelen et al., 1986). Consistent with the suggestion that dorsal raphe neurons are part of a central anxiolytic system is the observation that administration of low doses of the serotonin precursor 5-hydroxytryptophan (5-HTP) exerts anxiolytic effects, while higher doses exert pro-conflict effects in a lick-suppression test (Hjorth et al., 1987b). The emergence of pro-conflict effects after high doses of 5-HTP may reflect a slight augmentation of 5-HT levels and the interaction of these relatively small amounts with presynaptic autoreceptors, thus decreasing subsequent firing of raphe neurons, or alternatively may reflect differential involvement of different 5-HT receptor subtypes.

Norepinephrine, epinephrine, and acetylcholine

Surprisingly little attention has been devoted to examination of the involvement of norepinephrine (NE), epinephrine (Epi), or acetylcholine (ACh) in the stress-induced activation of the mesocortical DA system. While a large number of studies have examined the responses of NE and Epi to various stressors (Kvetnansky et al., 1977; Weiss et al., 1980; Anisman et al., 1981; Ida et al., 1982, 1985; Glavin et al., 1983; Sudo, 1983; Stanford et al., 1984; for reviews, see Stone, 1975, and Glavin, 1985), these studies have focused on the role of the noradrenergic system in stress, rather than any role noradrenergic systems may play in the regulation of the cortical DA systems after stress. As has been mentioned previously, the stress-induced activation of the prefrontal cortical DA system can occur in the apparent absence of effects on cortical NE systems (see above). However, it is somewhat paradoxical that clonidine, the α_2 agonist, is extremely potent in its ability to prevent the stress-induced activation of the PFC DA innervation. This may be explained by two different mechanisms. Clonidine has recently been shown to decrease burst firing in midbrain DA neurons without altering firing rate of the neurons (Grenhoff and Svensson, 1988). Burst firing appears to be associated with an increased release of transmitter relative to the normal firing pattern (Gonon, 1988); it is possible that clonidine may be effective in reducing the augmented DA release by preventing a stress-elicited shift in the firing pattern of DA neurons to a burst-firing mode. Moreover, clonidine may also act by altering GABA release in a regionally specific manner (Pittaluga and Raiteri, 1988) and therefore modulate impulse-independent regulation of DA release in the PFC (see below).

Changes in cholinergic systems induced by stress have received little attention. This is attributable in part to the difficulty in assessing alterations in cholinergic function using traditional measures of cholinergic function (such as changes in high affinity choline uptake). With the advent of in vivo dialysis methods which allow the determination of ACh in dialysates from brain, the lack of knowledge concerning stress-induced changes in central cholinergic systems may soon by corrected.

Impulse-independent regulation of DA neurons and stress

The afferent regulation of DA neurons at the somatodendritic level appears to be the major mechanism through which stress-induced activation of the mesoprefrontal cortical DA neurons occurs. Compelling evidence for a significant contribution of autoregulatory features to the observed pattern of stress-induced activation of the mesotelencephalic DA system, at the present time, is lacking. Some data suggest that impulse-independent (presynaptic) regulatory influences may be involved in stress-evoked changes in DA metabolism; the degree to which such presynaptic regulatory influences are involved in the environmental stress-induced activation of the prefrontal cortical DA system is unclear. However, such mechanisms do apparently contribute to the β-carboline-induced activation of the mesoprefrontal cortical DA innervation.

As noted above, pretreatment with anxiolytic benzodiazepine agonists prevents the stress-induced activation of the PFC DA innervation. A key question concerning the benzodiazepine modulation of mesotelencephalic DA neurons is at what site(s) within the CNS do the benzodiazepines exert their anxiolytic effects. Similarly, at what sites do FG-7142 and other anxiogenic benzodiazepine inverse agonists act to increase PFC DA metabolism?

Claustre et al. (1986) demonstrated that systemic administration of zolpidem, a non-benzodiazepine anxiolytic agent which interacts with central benzodiazepine receptors (Dennis et al., 1988; Depoortere et al., 1986), prevents both the footshock stress-induced biochemical activation of the PFC DA system, and the activation of the PFC DA innervation that occurs in response to administration of the anxiogenic β-carbolines methyl-β-carboline-3-carboxylate (β-CCM) and ethyl-β-carboline-3-carboxylate (β-CCE). Direct administration of zolpidem into the PFC failed to prevent the stress-induced increase in cortical DA utilization, as did direct infusion of the anxiolytic into the VTA. These data led to the speculation that regions other than the PFC and VTA represented the site of action of zolpidem and other anxiolytics.

However, in preliminary experiments we have been able to partially reverse footshock-induced activation of the PFC DA innervation by direct intra-VTA administration of chlordiazepoxide, a benzodiazepine anxiolytic (see Fig. 5). VTA injection of chlordiazepoxide significantly reduced the ability of footshock to augment cortical DA metabolism; however, cortical DA utilization in footshock-exposed animals was still significantly greater than in controls. These data suggest that an anxiolytic benzodiazepine may act at the VTA as well as at other sites to reduce stress-evoked activation of the PFC DA innervation. The reason for the discrepancy between these data and those of Claustre et al. (1986) is not clear, but may partially reflect the fact that we used a benzodiazepine agonist, whereas Claustre and colleagues injected a non-benzodiazepine which interacts with central benzodiazepine receptors but with a highly unusual pharmacological profile.

We have also recently examined the effects of direct VTA administration of the β-carboline inverse agonist DMCM on mesocorticolimbic DA function. Injection of DMCM into the VTA resulted in a dose-related increase in DA metabolism in the PFC, but not in the NAS. These results thus parallel the sequelae of systemic administration of the β-carbolines.

The reduction in stress-induced cortical DA metabolism in animals treated with VTA infusions of chlordiazepoxide suggests that benzodiazepine receptors are associated with midbrain DA neurons. Benzodiazepine receptors are present in high density in the PFC, and are also present in the

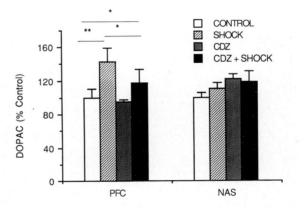

Fig. 5. The attenuation of the effects of mild footshock stress by direct administration of chlordiazepoxide into the ventral tegmental area (VTA). Injection of the anxiolytic benzodiazepine into the VTA significantly attenuated, but did not completely reverse, the stress-induced increase in PFC DOPAC. This is reflected by the finding that PFC DOPAC concentrations in footshock-stressed animals which received intra-VTA chlordiazepoxide were significantly lower than in stressed animals not pretreated with the benzodiazepine, but were significantly higher than PFC DOPAC levels in control animals. These observations were made in collaboration with William A. Clark. ** $p \leq 0.01$, * $p \leq 0.05$, for respective comparisons between groups (marked by the bars).

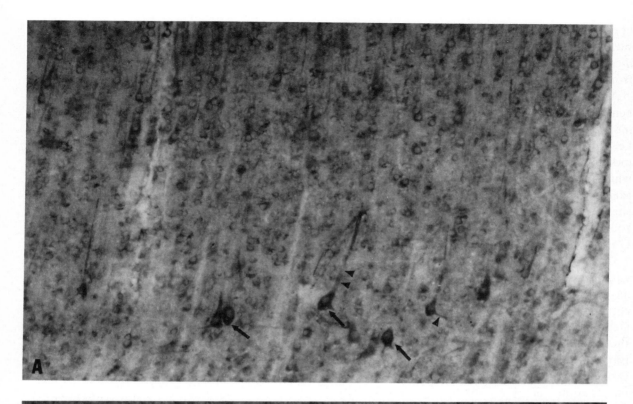

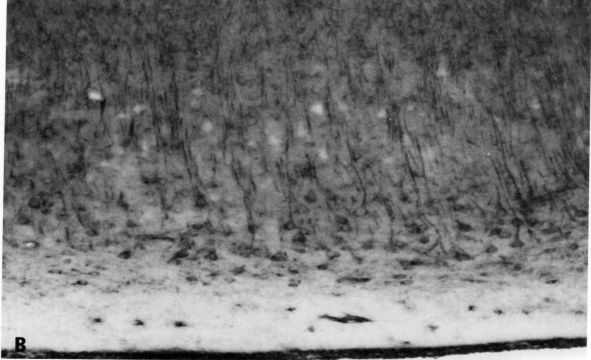

Fig. 6A, B

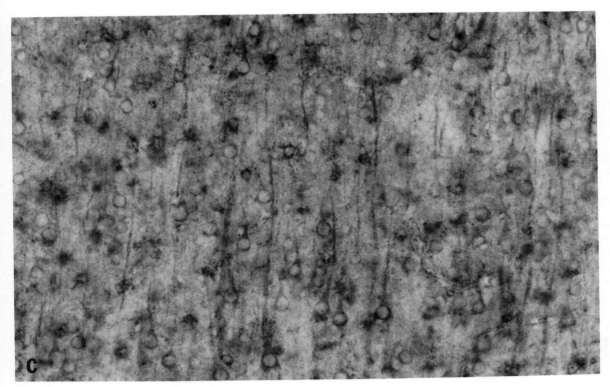

Fig. 6C

Fig. 6. Distribution of the benzodiazepine receptor (BZDr) as revealed by immunohistochemistry using a monoclonal antibody directed against an alpha subunit of the receptor. BZDr immunoreactivity (BZDr-ir) is found on cortical neurons (A–C), and also appears to be associated with certain midbrain DA neurons (D–F). The benzodiazepine receptor monoclonal antibody was kindly supplied by Dr. John Tallman. (A) Dense labeling of a few large pyramidal neurons in the deep layers of rhesus monkey frontal cortex (arrows). In these neurons the BZDr apears to be intraneuronal, in contrast to the surface labeling of most of the neurons in the more superficial layers. The BZDr-ir can be seen in both the apical dendrites (double arrowhead) and the basilar dendrites (single arrowhead). (B) BZDr labeling in the superficial layers of the entorhinal cortex of the rhesus monkey. The receptor appears to be localized to the surface of most neurons. The pial surface is at the bottom of the figure. (C) Labeling of BZDr on the surface of deep layer pyramidal neurons in the entorhinal cortex; in this figure the pial surface would be up. The labeling of the somata and apical dendrites can be clearly seen. In contrast to the frontal cortex, in which a few very densely stained pyramidal neurons can be seen in the deep layers, the labeling in the entorhinal cortex appears to be confined to the cell surface. (D) Distribution of the BZDr in the normal human VTA. The melanin-containing neurons (brown) serve as a marker for DA neurons. Some of these DA neurons can be seen to exhibit BZDr staining (arrows), whereas other neurons do not appear to be BZDr-immunoreactive. The number of DA neurons which also exhibit BZDr-ir is greater in the VTA and lateral substantia nigra (SN) than in the main body of the SN. (E) DA neurons of the SN in the same patient; the majority of the melanized neurons do not appear to be BZDr-ir. (F) Example of a double-labeled melanin-positive BZDr-ir neuron in the lateral VTA.

VTA (Young and Kuhar, 1980; Young et al., 1981). Marcel et al. (1987) did not observe a significant decrease in anterior cingulate cortical benzodiazepine binding sites after lesions disrupting the DA innervation of this region. However, we have recently used a monoclonal antibody directed against the benzodiazepine receptor in immunohistochemical studies and observed that some DA neurons in the human mesencephalon appear to stain heavily for the benzodiazepine receptor (see Fig. 6); electron microscopic examination will be required to determine if the receptor is being synthesized in the DA neurons, as opposed to the receptor being associated with af-

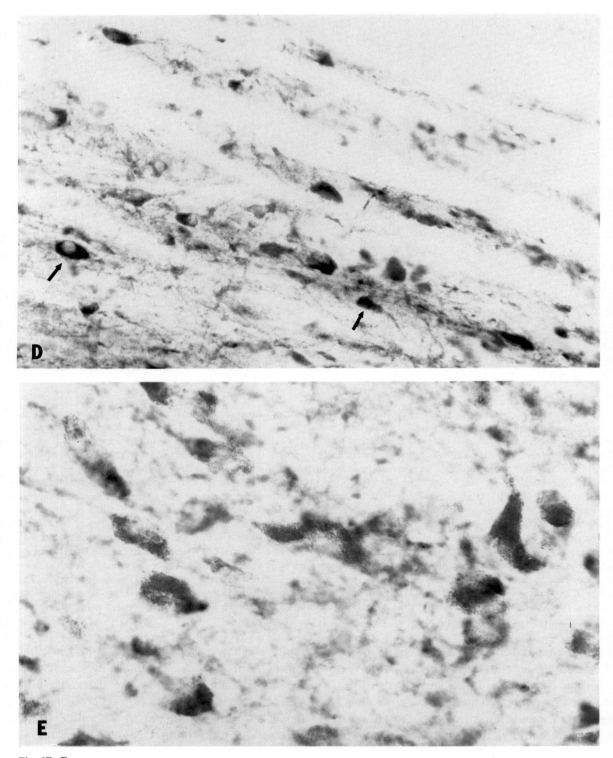

Fig. 6D, E

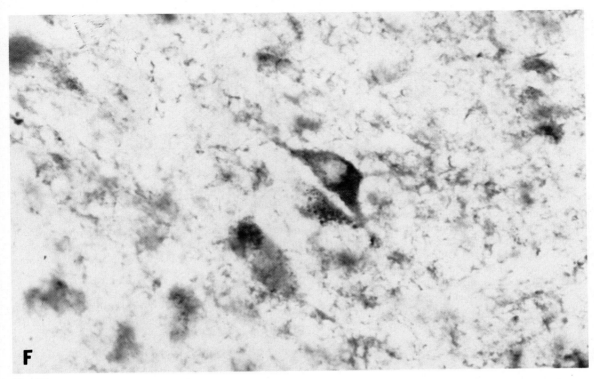

Fig. 6F

ferents. Biochemical and electrophysiological data suggest that the $GABA_A$ receptor complex-synthesizing neuron is proximal to the DA neurons. In addition to the presence of benzodiazepine receptor-like immunoreactivity in the midbrain, heavy staining can be seen in cortical regions (see Fig. 6).

The suggestion that benzodiazepines act at multiple sites, including the VTA, to reduce the stress-induced activation of the PFC DA system, is corroborated by recent data indicating the sites at which FG-7142 acts to alter DA synthesis. Systemic administration of FG-7142 augments in vivo tyrosine hydroxylation in the PFC and VTA, but not in other mesotelencephalic sites (Knorr et al., 1989) (Fig. 7); these effects parallel those observed in response to footshock stress. However, in vitro tyrosine hydroxylation is increased by the β-carboline in slices prepared from the PFC, but decreased in striatal slices; the effects of FG-7142 were reversed by co-incubation with the benzodiazepine antagonist Ro 15-1788 (see Fig. 8). Incubation of striatal and prefrontal cortical slices with GABA resulted in an increase and decrease, respectively, in tyrosine hydroxylation, effects opposite to those obtained with the β-carboline (see Fig. 9).

These data complement the finding that direct VTA injection of the benzodiazepine chlordiazepoxide significantly, but not completely, reduces the stress-induced augmentation of cortical DA metabolism, and suggest that the anxiolytic β-carbolines act at multiple sites to alter DA function. These data also suggest that the β-carbolines may act through both impulse-dependent and impulse-independent mechanisms to increase DA synthesis. Systemic administration of FG-7142 increased synthesis in the PFC and VTA; however, incubation of cortical slices with the β-carboline also increased tyrosine hydroxylation, suggesting that part of the effect on DA synthesis elicited by the benzodiazepine inverse agonist occurs through an impulse-independent, presynaptic effect on the DA terminals in the PFC.

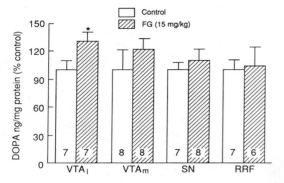

Fig. 7. The effects of systemic administration of the β-carboline benzodiazepine inverse agonist FG-7142 on in vivo tyrosine hydroxylation in the midbrain dopamine cell body regions. FG-7142 significantly increased DA synthesis in the lateral aspects of the ventral tegmental area (VTA_l); the change in the medial VTA (VTA_m) approached but did not achieve statistical significance. FG-7142 did not significantly alter in vivo tyrosine hydroxylation in either the substantia nigra (SN) or retrorubral field (RRF), site of the A8 DA neurons. These observations were made in collaboration with Dr. Amy M. Knorr. * $p \leq 0.05$.

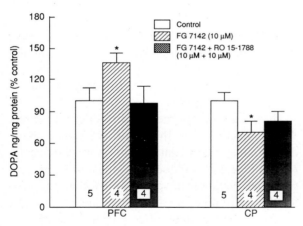

Fig. 8. The effects of the anxiolytic benzodiazepine inverse agonist FG-7142 on in vitro tyrosine hydroxylation in PFC and striatal (CP) slices. FG-7142 significantly increased in vitro tyrosine hydroxylation in the PFC, but in contrast effected a decrease in the striatum. The effects of the β-carboline inverse agonist were reversed by co-incubation with the benzodiazepine antagonist Ro 15-1788. The changes in in vitro tyrosine hydroxylation in the PFC and striatum correspond to the changes in metabolic activation of the PFC and striatum after systemic administration of FG-7142. Reproduced from Knorr et al. (1989) with permission of Elsevier Science Publishers. * $p \leq 0.05$.

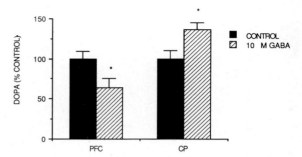

Fig. 9. The effects of GABA on in vitro tyrosine hydroxylation in the PFC and striatum (CP). Whereas FG-7142, which is an inverse agonist at the benzodiazepine receptor portion of the $GABA_A$/benzodiazepine receptor complex, increases tyrosine hydroxylation in the PFC but decreases tyrosine hydroxylation in the striatum, GABA results in opposite effects. These observations were made in collaboration with Dr. Amy M. Knorr. * $p \leq 0.05$.

The impulse-independent regulation of DA release in a regionally specific manner by GABA contributes to our understanding of a number of behavioral events, such as disruption of DA-dependent behavioral tasks after selective alteration in local GABA function (Brailowsky et al., 1989; Sawaguchi et al., 1988a, b). Moreover, the regionally specific effects of GABA in modulating DA metabolism may help to explain the anxiolytic actions of clonidine. As noted previously, clonidine administration has been reported to regularize the firing rate of substantia nigra DA neurons (Grenhoff and Svensson, 1988) as well as modulate stress-altered activity of locus coeruleus neurons (Quintin et al., 1986); the latter actions may partially underlie the ability of the α_2 adrenoceptor agonist to potently reverse both footshock-activation of the PFC DA innervation and the FG-7142-elicited behavioral activation (Crawley et al., 1985). In addition, a recent report indicates that clonidine induces GABA release in cortex in a regionally selective manner: enhanced release of the amino acid is observed in the frontal and parietal, but not temporal or occipital, cortices (Pittaluga and Raiteri, 1988). These data therefore suggest that clonidine may reverse the stress-induced activation of the PFC DA system by enhancing GABA release in the frontal cortex.

DA system interactions and the pattern of stress-induced changes

The pattern of stress-induced activation of the mesotelencephalic DA neurons is clearly dependent upon both stress intensity and duration of exposure to the stressor. Thus, sufficiently mild footshock or conditioned fear results in a selective increase in DA utilization in the PFC, and not in the mesolimbic or striatal regions. As stress intensity is increased (for example, by increasing the footshock current), additional DA terminal fields display evidence of increased DA utilization. While regionally specific local circuit interactions may account for the pattern of activation across the mesotelencephalic DA terminal fields, it is also possible that interactions between the different DA terminal fields may contribute to this distinct pattern.

Recent data suggest that DA release in a specific site may impact on DA function in a spatially segregated CNS area by altering the function of projections connecting the two regions, and thus transynaptically altering the function of distant DA axons (see Deutch et al., 1988, 1990a; Louilot et al., 1987, 1989; Simon and Le Moal, 1984). The best characterized interaction between mesotelencephalic DA systems is that between the PFC DA innervation and the striatal (including the ventral striatum, or nucleus accumbens) DA innervations.

The DA innervation of the PFC appears to hold corticofugal projection neurons, including those to the striatum, NAS, and VTA, under tonic inhibition (Ferron et al., 1984; Peterson et al., 1987; Sesack and Bunney, 1989; Thierry et al., 1979). The corticostriatal projection neurons appear to use glutamate or a related excitatory amino acid as a neurotransmitter (Carter, 1982; Christie et al., 1985b, 1987; Divac et al., 1977; Spencer, 1976), and these excitatory projections to the striatum converge with DA inputs as well as thalamic afferents (Kocsis et al., 1977). Glutamate release in the striatum is thought to enhance DA release through a presynaptic (impulse-independent) influence (Girault et al., 1986; Glowinski et al., 1988; Jhamandas and Marien, 1987). Thus, reduction of dopaminergic tone in the PFC, whether accomplished through interruption of the DA innervation or through pharmacological intervention should augment DA release in the striatum.

Carter and Pycock and associates reported that 6-hydroxydopamine lesions of the DA innervation of the PFC augment apparent DA release in the striatum and NAS (Carter and Pycock, 1978a, b, 1980; Pycock et al., 1980). Subsequently, certain aspects of this work have been replicated and expanded (Haratounian et al., 1988; Leccese and Lyness, 1987; Martin-Iversen et al., 1986). However, certain other attempts to replicate this work have not succeeded in obtaining corroborating data (Joyce et al., 1983; Oades et al., 1986; Rosin et al., 1990). We have not observed any increases in basal DA metabolism in response to DA deafferentation of the PFC, but have observed an enhanced responsiveness of the DA innervation of the NAS to pharmacological challenge after 6-hydroxydopamine PFC lesions (Rosin et al., 1990). Moreover, using in vivo voltammetric assessment, Louilot et al. (1989) have reported that manipulations of the PFC DA innervation modulate apparent DA release in the striatum. While it remains unclear to what degree the changes in subcortical function reflects changes in basal release, as opposed to changes in the responsiveness of striatal and mesolimbic DA innervations to perturbation, these data suggest that the frontal cortical DA innervation may transynaptically regulate subcortical DA function.

Louilot et al. (1989) noted that changes in DA function in the NAS were observed after administration of either DA agonists and antagonists into the PFC. Thus, not only did administration of antagonists increase apparent DA release in the NAS (as measured by voltammetric determinations of extracellular DOPAC), but agonist administration decreased the DOPAC signal. The stress-induced pattern of activation in the subcortical (limbic and striatal) DA terminal fields may be related to such interactions between DA terminal fields. If mild footshock stress results in a preferential and primary metabolic activation of the PFC DA innervation, than the inability to observe a similar increase in subcortical sites (for example, the NAS) may partially reflect a transynaptic inhibition of mesolimbic DA release. However, as the intensity or duration of stress increases, either greater direct actions on the midbrain DA neurons or recruitment of local circuit interactions may partially overcome the transynaptic inhibition of DA tone in the subcortical sites.

This scenario is clearly speculative. However, certain recent data are consistent with such an interpretation. We have recently noted that 6-hydroxydopamine lesions of the PFC alter subsequent response to mild footshock stress, such that a biochemically measurable increase in DA utilization in the NAS can be observed in the lesioned, but not vehicle-infused, subjects (Deutch et al., 1990; see Fig. 10). We have also recently observed changes in the pattern of stress-induced mesotelencephalic DA activation in animals treated during the perinatal period with diazepam, an anxiolytic benzodiazepine. Exposure of these animals to footshock stress at a current intensity sufficient to increase DA utilization in the PFC and the NAS, but not the striatum, resulted in a biochemically measurable increase in DA utilization in the striatum but markedly dampened the prefrontal cortical DA response (Deutch et al., 1989; see Fig. 11). These findings may reflect decreased cortical DA function allowing the expression of a heighten-

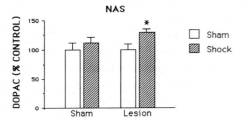

Fig. 10. The effects of dopamine depletion of the PFC on stress-induced changes in the metabolic activation of the nucleus accumbens (NAS) dopamine innervation. Animals sustaining a 6-hydroxydopamine lesion of the PFC DA innervation exhibit a metabolic activation of the NAS DA innervation. In contrast, vehicle-infused animals do not respond to the mild footshock stress with an increase in mesolimbic DA concentrations. These findings suggest that removal of the DA innervation of the PFC results in a transynaptic inhibitory influence over the NAS DA innervation. Modified from Deutch et al. (1990), with the permission of Elsevier Science Publishers. * $p \leq 0.05$.

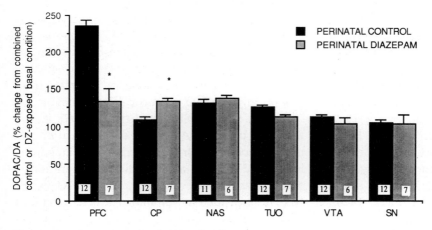

Fig. 11. The effects of perinatal benzodiazepine administration on the stress-induced activation of the mesotelencephalic dopamine innervation. Pregnant dams were implanted with subcutaneous diazepam pellets which released the benzodiazepine from day E8 of gestation through the first postnatal week. Animal which received perinatal diazepam and were subsequently exposed to footshock stress as adults exhibited a marked blunting of the PFC DA response, and in contrast exhibited a striatal metabolic activation not seen in combined (normal control and sham-pellet-implanted) controls. Modified from Deutch et al. (1989), with the permission of Elsevier Science Publishers. * $p \leq 0.05$.

ed subcortical DA response. It should be emphasized that a number of mechanisms may contribute to such changes in the pattern of mesotelencephalic activation after stress. However, interactions between the different DA terminal fields must be considered as contributing to the observed pattern.

Le Moal and colleagues have demonstrated interactions not only between cortical and subcortical DA systems, but in addition between separate subcortical DA innervations (Deminière et al., 1988; Louilot et al., 1985, 1989; Oades et al., 1986; Simon et al., 1988; see also P.H. Kelley and Moore, 1976). In addition, interconnections of the mesencephalic DA neurons at the somatodendritic level (Deutch et al., 1986, 1988) may permit functional interactions between the different DA neurons through an impulse-dependent mechanism. The interactions of the mesotelencephalic DA neurons suggest that future studies will have to attend to such somatodendritic, impulse-dependent regulation of DA release in interpretation of changes in DA systems, as well as regulation of DA release at the impulse-independent level by afferents impinging on the DA axons (Jones et al., 1988a, b; Nieoullon et al., 1985).

Conclusions

There are clearly a multiplicity of controls over mesoprefrontal cortical DA neurons; a number of these regulatory features modulate the response of the mesoprefrontal cortical DA neurons to stress. At the current time there are no compelling direct data which indicate that autoregulatory features govern the preferential metabolic activation of the PFC by acute stress. Rather, a number of distinct afferents to the DA neurons of the VTA appear to modulate the stress-induced activation of the PFC DA innervation. Moreover, impulse-independent phenomena may also contribute to the activation of the PFC DA system by stress, although at the present time definitive data are available only for the β-carboline-induced activation of the PFC DA axons. Finally, interactions between the different mesotelencephalic DA projections may contribute to the observed pattern of activation of the forebrain DA systems by stress.

It is unclear to what degree the distinctly different systems and mechanisms which regulate the stress-induced activation of the PFC DA innervation interact. For example, several peptides appear to be involved in the enhancement of PFC DA metabolism observed after acute mild footshock stress. While different afferents to the VTA appear to release substance P and enkephalin in response to acute stress, it is conceivable that different degrees of stress result in the two types of peptides being differentially released. Alternatively, different types of stressors (for example, those involving a painful stimulus, such as footshock, as opposed to stressors that do not have an immediate painful stimulus, such as conditioned fear) may evoke different transmitter release characteristics in the afferents to the VTA which contain tachykinin and opioid peptides.

In addition to differences in the degree or type of stress, the different afferent systems may exhibit different temporal patterns (time-course) of response to acute stress. Most of the investigations of different afferent systems in stress have focused on the response to relatively brief (for example, 20 min) periods of stress. However, most peptides do not exhibit a high turnover rate (although there are clearly exceptions; see Bean et al., 1989b), since either catabolism of prohormones or de novo synthesis of peptides is required for peptide replacement. Neurons using peptide transmitters may therefore be expected to respond to stress for a relatively brief period before ceasing to release the peptide. This situation stands in sharp contrast to classical transmitters. Thus, peptide and classical transmitter afferents which regulate DA function may be important for contributing to stress exposure of different durations. In addition, differences in autoregulatory features between different mesotelencephalic DA neurons may contribute to the observed pattern of changes in these DA neurons in response to stress.

Studies of the response of animals to relatively

long-term exposure (in the order of days or longer) to stress indicate that different hormonal response patterns develop, dependent upon the duration of the stress. Such differences may also be present in the central mesotelencephalic DA response to stress. For example, we have observed that animals exposed to footshock for 5 daily 20 min sessions do not respond to footshock (or the presentation of a neutral auditory stimulus previously paired with footshock) with an increase in PFC DA utilization on the fifth day; this may reflect the development of behavioral coping strategies aimed at minimizing the shock. In contrast, a number of investigators have noted behavioral and biochemical sensitization to various DA agonists, as well as to peptides which regulate the VTA DA neurons (Dougherty and Ellinwood, 1981; Kalivas, 1985; Kalivas and Taylor, 1985; LaHoste et al., 1988). Moreover, there is a cross-sensitization between pharmacological treatments which augment DA function and stress (Kalivas and Duffy, 1989; Kalivas et al., 1986; Robinson and Becker, 1982; Wilcox et al., 1986). In animals sensitized to stress by prior exposure to agents which cross-react with stress, multiple regulatory features (such as multiple afferent controls) may be required for the heightened response of the PFC DA innervation to stress (Kalivas et al., 1988). Moreover, under such conditions autoregulatory mechanisms may play a much greater role in determining the enhancement of PFC DA metabolism.

The stress-induced augmentation of the PFC DA innervation has been extensively investigated in the rat; these animals can be easily studied, are relatively inexpensive, and have distinctly organized brain systems which are amenable to experimental manipulation. However, the organization of the area designated the medial prefrontal cortex in the rat differs from that in primates. It has not been possible to arrive at a universally accepted definition of what areas can be considered homologous to the primate PFC in the rat. The frontal cortical projection field of the midbrain DA neurons overlaps the mediodorsal thalamic projection field (Beckstead, 1976; Divac et al., 1978); it has therefore been proposed that the prefrontal cortex in various species can be defined on the basis of receiving these convergent projections. However, given the disparity between cytoarchitectonic and hodological criteria, such a definition of the PFC has been resisted. However, it does appear that those cortical zones receiving overlapping projections from the mediodorsal thalamus and ventral mesencephalon are homologous, even if the regions so defined extend beyond the cytoarchitectonic boundaries of the primate prefrontal cortex.

In contrast to the primate cortex, in which a DA innervation is present in many cortical regions outside of the mediodorsal thalamic projection area, the DA innervation of the rodent cortex is relatively restricted. In the frontal cortical areas DA afferents are present in the anteromedial PFC and the suprarhinal cortex (Berger et al., 1976; Thierry et al., 1973a, b; Van Eden et al., 1987). The anteromedial PFC includes a number of distinct cytoarchitectonic fields, including the prelimbic cortex (area 32), dorsal anterior cingulate cortex (area 24b), and the medial precentral ("shoulder") cortex. DA fibers also invade the infralimbic cortex (area 25), a region for which unambiguous evidence of a mediodorsal thalamic input is lacking (Krettek and Price, 1977; Leonard, 1969; Rose and Woolsey, 1948). Even within some of these PFC regions further subdivisions are possible. For example, behavioral data suggest that the medial precentral (shoulder) cortex in the rodent can be considered homologous to the premotor cortex, supplementary motor cortex, and frontal eye field of the primate (Passingham et al., 1988). When stress-induced augmentation of the PFC has been examined, the entire medial PFC of the rodent, which may correspond to a number of distinct cortical regions in the primate, is generally dissected. However, it is possible that the stress-elicited increase in DA release and metabolism may be restricted to different divisions of the anteromedial PFC. Autoregulatory features on the DA axons invading these different PFC areas may differ, as has been suggested by the findings indicating distinct subsets of mesoprefrontal cortical DA neurons classified on the basis of impulse-regulating autoreceptors. Similarly, given regional differences in the organization of interneuronal populations as well as distinct regional heterogeneities in non-dopaminergic afferents, the presynaptic regulation of DA function through heteroceptors located on the DA axons may differ across the different areas comprising the PFC.

It will be of increasing importance to define the neuronal systems through which stress operates. Little attention has been devoted to a detailed analysis of the organization of the neural substrates responsible for the "alerting" of the DA systems. Moreover, interactions between certain mesotelencephalic DA terminal fields may dictate the pattern of DA response across the forebrain regions. The frontal cortical efferents, which are topographically organized both within and across the different regions which comprise the medial PFC (Sesack et al., 1989), project either directly or via single relays to a large number of sites which may be involved in stress, including regions of the brain thought to subserve autonomic function (Beckstead, 1979; Sesack et al., 1989; Terreberry and Neafsey, 1987). Stress results in autonomic arousal as well as the augmentation of PFC DA metabolism; however, the sympathetic arousal induced by stress can occur with very mild stressors which are of insufficient intensity to increase PFC DA metabolism. If there are separable DA innervations of the medial PFC, it is conceivable that the DA fibers in register with distinct PFC efferents (for example, those to the amygdala as opposed to those to the VTA) form distinct but interactive systems which are activated by stress — yet when assessing the concentration of DA and its metabolites across the entire medial PFC no activation is observed. Such an organization of separable PFC systems involved in stress would be similar to the parallel organization model of functionally segregated corticofugal systems proposed by Alexander et al. (1986; see this volume). Future studies will be required to more clearly define distinct PFC circuits which are regulated by DA and which are involved with different aspects of the stress response.

Acknowledgements

We gratefully acknowledge our colleagues William A. Clark, Amy M. Knorr, Diane L. Rosin, and See-Ying Tam, with whom many of these studies were conducted.

These studies were supported in part by NIMH grants MH-45124 (AYD) and MH-14092 (RHR), by grants from the Scottish Rite Schizophrenia Research Program and the American Parkinson Disease Association, by support of the Abraham Ribicoff Research Facilities of the Connecticut Mental Health Center, and by support from both the National Center for Post-traumatic Stress Disorder and the Schizophrenia Research Center, Veterans Administration Medical Center, West Haven, CT.

References

Abarca, J. and Bustos, G. (1985) Release of D-[^{3}H]aspartic acid from the rat substantia nigra: effect of veratridine-evoked depolarization and cortical ablation. *Neurochem. Int.*, 7: 229 – 236.

Abercrombie, E.D., Keefe, K.A., DiFrischia, D.S. and Zigmond, M.J. (1989) Differential effect of stress on in vivo dopamine release in striatum, nucleus accumbens, and medial frontal cortex. *J. Neurochem.*, 52: 1655 – 1658.

Alexander, G.E., DeLong, M.R. and Strick, P.L. (1986) Parallel organization of functionally segregated circuits linking basal ganglia and cortex. *Annu. Rev. Neurosci.*, 9: 357 – 381.

Altar, C.A., Boyar, W.C., Oei, E. and Wood, P.L. (1987) Dopamine autoreceptors modulate the in vivo release of dopamine in the frontal, cingulate, and entorhinal cortices. *J. Pharmacol. Exp. Ther.*, 242: 115 – 120.

Amlaiky, N. and Caron, M.G. (1985) Photoaffinity of the D2-dopamine receptor using a novel high affinity radioiodinated probe. *J. Biol. Chem.*, 260: 1983 – 1986.

Amlaiky, N. and Caron, M.G. (1986) Identification of the D2-dopamine receptor binding subunit in several mammalian tissues and species by photoaffinity labelling. *J. Neurochem.*, 47: 196 – 204.

Andén, N.-E. and Grabowska-Andén, M. (1983) Formation of deaminated metabolites of dopamine in noradrenergic neurons. *Naunyn-Schmiedeberg's Arch. Pharmacol.*, 324:

1–6.

Anisman, H., Ritch, M. and Sklar, L.S. (1981) Noradrenergic and dopaminergic interactions in escape behavior: analysis of uncontrollable stress effects. *Psychopharmacology*, 74: 263–268.

Antelman, S.M., Knopf, S., Caggiula, A.R., Kocan, D., Lysle, D.T. and Edwards, D.J. (1988) Stress and enhanced dopamine utilization in the frontal cortex: the myth and the reality. *Ann. N.Y. Acad. Sci.*, 537: 262–272.

Araki, M., McGeer, P.L. and McGeer, E.G. (1985) Striatonigral and pallidonigral pathways studied by a combination of retrograde horseradish peroxidase tracing and a pharmacohistochemical method for γ-aminobutyric acid transaminase. *Brain Res.*, 331: 17–24.

Bannon, M.J. and Roth, R.H. (1983) Pharmacology of mesocortical dopamine neurons. *Pharmacol. Rev.*, 35: 53–68.

Bannon, M.J., Michaud, R.L. and Roth, R.H. (1980) Mesocortical dopamine neurons. Lack of autoreceptors modulating dopamine synthesis. *Mol. Pharmacol.*, 19: 270–275.

Bannon, M.J., Elliot, P.J., Alpert, J.E., Goedert, M., Iversen, S.D. and Iversen, L.L. (1983a) Role of endogenous substance P in stress-induced activation of mesocortical dopaminergic neurons. *Nature*, 306: 791–792.

Bannon, M.J., Wolf, M.E. and Roth, R.H. (1983b) Pharmacology of DA neurons innervating the prefrontal, cingulate, and pyriform cortices. *Eur. J. Pharmacol.*, 92: 119–125.

Bannon, M.J., Deutch, A.Y., Tam, S.-Y., Zamir, N., Eskay, R.L., Lee, J.-M., Maggio, J.E. and Roth, R.H. (1986) Mild footshock stress dissociates substance P from substance K and dynorphin from Met- and Leu-enkephalin. *Brain Res.*, 381: 393–396.

Bean, A.J., Adrian, T.J., Modlin, I.M. and Roth, R.H. (1989a) Dopamine and neurotensin storage in colocalized and noncolocalized neuronal populations. *J. Pharmacol. Exp. Ther.*, 249: 681–687.

Bean, A.J., During, M.J., Deutch, A.Y. and Roth, R.H. (1989b) Effects of dopamine depletion on striatal neurotensin: Biochemical and immunohistochemical studies. *J. Neurosci.*, 9: 4430–4438.

Beckstead, R.M. (1976) Convergent thalamic and mesencephalic projections to the anterior medial cortex in the rat. *J. Comp. Neurol.*, 166: 403–416.

Beckstead, R.M. (1979) An autoradiographic examination of corticocortical and subcortical projections of the mediodorsal-projection (prefrontal) cortex in the rat. *J. Comp. Neurol.*, 184: 43–62.

Berger, B., Thierry, A.M., Tassin, J.P. and Moyne, M.A. (1976) Dopaminergic innervation of the rat prefrontal cortex: a fluorescence histochemical study. *Brain Res.*, 106: 133–145.

Berridge, C.W. and Dunn, A.J. (1987) A corticotrophin-releasing factor antagonist reverses the stress-induced changes of exploratory behavior in mice. *Hormones Behav.*, 21: 393–401.

Biggio, G., Concas, A., Serra, M., Salis, M., Corda, M.G., Nurchi, V., Crisponi, C. and Gessa, G.L. (1984) Stress and β-carbolines decrease the density of low affinity GABA binding sites; an effect reversed by diazepam. *Brain Res.*, 305: 13–18.

Boadle-Biber, M.C., Corley, K.C., Graves, L., Phan, T.-H. and Rosecrans, J. (1989) Increase in the activity of tryptophan hydroxylase from cortex and midbrain of male Fisher 344 rats in response to acute or repeated sound stress. *Brain Res.*, 482: 306–316.

Bradberry, C.M., Deutch, A.Y., Karasic, D.H. and Roth, R.H. (1989) Regionally-specific alterations in mesotelencephalic dopamine synthesis: association with precursor tyrosine. *J. Neur. Trans.*, 78: 221–229.

Brailowsky, S., Silva-Barrat, C., Ménini, C., Riche, D. and Naquet, R. (1989) Effects of localized, chronic GABA infusions into different cortical areas of the photosensitive baboon, *Papio papio*, *Electroenceph. Clin. Neurophysiol.*, 72: 147–156.

Britton, K.T., Lee, G., Vale, W., Rivier, J. and Koob, G.F. (1986) Corticotropin releasing factor (CRF) receptor antagonist blocks activating and "anxiogenic" actions of CRF in the rat. *Brain Res.*, 369: 303–306.

Cabib, S., Kempf, E., Schleef, C., Oliverio, A. and Puglisi-Allegra, S. (1988) Effects of immobilization stress on dopamine and its metabolites in different brain areas of the mouse: role of genotype and stress duration. *Brain Res.*, 441: 153–160.

Carlson, J.N., Glick, S.D., Hinds, P.A. and Baird, J.L. (1988) Food deprivation alters dopamine utilization in the rat prefrontal cortex and symmetrically alters amphetamine-induced rotational behavior. *Brain Res.*, 454: 373–377.

Carter, C.J. (1982) Topographical distribution of possible glutamatergic pathways from the frontal cortex to the striatum and substantia nigra in rats. *Neuropharmacology*, 21: 383–393.

Carter, C.J. and Pycock, C.J. (1978a) Studies on the role of catecholamines in the frontal cortex. *Br. J. Pharmacol.*, 62: 402P.

Carter, C.J. and Pycock, C.J. (1978b) Lesions of the frontal cortex of the rat: changes in neurotransmitter systems in subcortical regions. *Br. J. Pharmacol.*, 62: 430P, 1978.

Carter, C.J. and Pycock, C.J. (1980) Behavioral and biochemical effects of dopamine and noradrenaline depletion within the medial prefrontal cortex of the rat. *Brain Res.*, 192: 163–176.

Chappel, P.B., Smith, M.A., Kilts, C.D., Bissette, G., Ritchie, J., Anderson, C. and Nemeroff, C.B. (1986) Alterations in corticotropin-releasing factor-like immunoreactivity in discrete brain regions after acute and chronic stress. *J. Neurosci.*, 6: 2908–2914.

Chesselet, M.-F. (1984) Presynaptic regulation of

neurotransmitter release in the brain: facts and hypotheses. *Neuroscience*, 12: 347–375.

Chiodo, L.A., Bannon, M.J., Grace, A.A., Roth, R.H. and Bunney, B.S. (1984) Evidence for the absence of impulse-regulating somatodendritic and synthesis-modulating nerve terminal autoreceptors on subpopulations of mesocortical dopamine neurons. *Neuroscience*, 12: 1–16.

Christie, M.J., Bridge, S., James, L.B. and Beart, P.M. (1985a) Excitotoxic lesions suggest an aspartatergic projection from rat medial prefrontal cortex to ventral tegmental area. *Brain Res.*, 333: 169–172.

Christie, M.J., James, L.B. and Beart, P.M. (1985b) An excitant amino acid projection from the medial prefrontal cortex to the anterior part of the nucleus accumbens in the rat. *J. Neurochem.*, 45: 477–482.

Christie, M.J., Summers, R.J., Stephenson, J.A., Cook, C.J. and Beart, P.M. (1987) Excitatory amino acid projections to the nucleus accumbens septi in the rat: a retrograde transport study utilizing D[^{3}H]aspartate and [^{3}H]GABA. *Neuroscience*, 22: 425–429.

Claustre, Y., Rivy, J.P., Dennis, T. and Scatton, B. (1986) Pharmacological studies on stress-induced increase in frontal cortical dopamine metabolism in the rat. *J. Pharmacol. Exp. Ther.*, 238: 693–700.

Concas, A., Mele, S. and Biggio, G. (1987) Foot shock stress decreases chloride efflux from rat brain synaptoneurosomes. *Eur. J. Pharmacol.*, 135: 423–427.

Conrad, L.C.A. and Pfaff, D.W. (1976) Autoradiographic tracing of nucleus accumbens efferents in the rat. *Brain Res.*, 113: 589–596.

Corda, M.G., Blaker, W., Mendelson, W. and Guidotti, A. (1983) β-Carbolines enhance the shock-induced suppression of drinking in the rat. *Proc. Natl. Acad. Sci. (U.S.A.)*, 80: 2072–2078.

Costall, B., Hendrie, C.A., Kelly, M.E. and Naylor, R.J. (1987) Actions of sulpiride and tiapride in a simple model of anxiety in mice. *Neuropharmacology*, 26: 195–200, 1987.

Crawley, J.N., Ninan, P.T., Pickar, D., Chrousos, G.P., Linnoila, M., Skolnick, P. and Paul, S.M. (1985) Neuropharmacological antagonism of the β-carboline-induced "anxiety" response in rhesus monkeys. *J. Neurosci.*, 5: 477–485.

Culman, J., Kvetnansky, R., Torda, T. and Murgas, K. (1980) Serotonin concentration in individual hypothalamic nuclei of rats exposed to acute immobilization stress. *Neuroscience*, 5: 1503–1506.

Curet, O., Dennis, T. and Scatton, B. (1986) The formation of deaminated metabolites of dopamine in the locus coeruleus depends on noradrenergic neuronal activity. *Brain Res.*, 335: 297–301.

Dal Tose, R., Sommer, B., Ewert, M., Pritchett, D.B., Bach, A., Chivers, B.D., and Seeberg, P. (1989) The dopamine D_2 receptor: two molecular forms generated by alternative splicing. *EMBO J.*, 8: 4025–4034.

Damsma, G., Yoshida, M., Wenskstern, D., Nomikos, G.G. and Phillips, A.G. (1989) Dopamine transmission in the rat striatum, nucleus accumbens, and pre-frontal cortex is differently affected by feeding, tail pinch, and immobilization. *Soc. Neurosci. Abstr.*, 15: 557, 1989.

D'Angio, M., Serrano, A., Rivy, J.P. and Scatton, B. (1987) Tail pinch stress increases extracellular DOPAC levels (as measured by in vivo voltammetry) in the rat nucleus accumbens but not frontal cortex: antagonism by diazepam and zolpidem. *Brain Res.*, 409: 169–174.

D'Angio, M., Serrano, A., Driscoll, P. and Scatton, B. (1988) Stressful environmental stimuli increase extracellular DOPAC levels in the prefrontal cortex of hypoemotional (Roman high-avoidance) but not hyperemotional (Roman low-avoidance) rats. An in vivo voltammetric study. *Brain Res.*, 451: 237–247.

De Keyser, J., Walraevens, H., De Backer, J.-P., Ebinger, G. and Vauquelin, G. (1989) D_2 dopamine receptors in the human brain: heterogeneity based on differences in guanine nucleotide effect on agonist binding, and their presence on corticostriatal terminals. *Brain Res.*, 484: 36–42.

Deminière, J.M., Tagzhouti, K., Tassin, J.P., Le Moal, M. and Simon, H. (1988) Increased sensitivity to amphetamine and facilitation of amphetamine self-administration after 6-hydroxydopamine lesions of the amygdala. *Psychopharmacology*, 94: 232–236.

Dennis, T., Dubois, A., Benavides, J. and Scatton, B. (1988) Distribution of central ω_1 (benzodiazepine$_1$) and ω_2 (benzodiazepine$_2$) receptor subtypes in the monkey and human brain. An autoradiographic study with [^{3}H]flunitrazepam and the ω_1 selective ligand [^{3}H]zolpidem. *J. Pharmacol. Exp. Ther.*, 247: 309–322.

Depoortere, H., Zivkovic, B., Lloyd, K.G., Sanger, D., Perrault, G., Langer, S.Z. and Bartholini, G. (1986) Zolpidem, a novel nonbenzodiazepine hypnotic. I. Neuropharmacological and behavioral effects. *J. Pharmacol. Exp. Ther.*, 237: 649–658.

DeSouza, E.B. and Van Loon, G.R. (1986) Brain serotonin and catecholamine responses to repeated stress in rats. *Brain Res.*, 367: 77–86.

Deutch, A.Y. and Martin, R.J. (1983) Mesencephalic dopamine modulation of pituitary and central β-endorphin: relation to food intake regulation. *Life Sci.*, 33: 281–287.

Deutch, A.Y. and Roth, R.H. (1987) Calcitonin gene-related peptide in the ventral tegmental area: selective modulation of prefrontal cortical dopamine metabolism. *Neurosci. Lett.*, 74: 169–174.

Deutch, A.Y., Tam, S.-Y. and Roth, R.H. (1985a) Footshock and conditioned stress increase 3,4-dihydroxyphenylacetic acid (DOPAC) in the ventral tegmental area but not substantia nigra. *Brain Res.*, 333: 143–146.

Deutch, A.Y., Maggio, J.E., Bannon, M.J., Kalivas, P.W., Tam, S.-Y., Goldstein, M. and Roth, R.H. (1985b) Substance K and substance P differentially modulate mesolimbic and mesocortical systems. *Peptides*, Suppl. 2, 6:

113–122.

Deutch, A.Y., Kalivas, P.W., Goldstein, M. and Roth, R.H. (1986) Interconnections of the mesencephalic dopamine cell groups. *Soc. Neurosci. Abstr.*, 12: 875.

Deutch, A.Y., Tam, S.-Y., Freeman A.S., Bowers M.B. Jr. and Roth, R.H. (1987a) Mesolimbic and mesocortical activation induced by phencyclidine: contrasting pattern to striatal response. *Eur. J. Pharmacol.*, 134: 257–264.

Deutch, A.Y., Bean, A.J., Bissette, G., Nemeroff, C.B., Robbins, R.J. and Roth, R.H. (1987b) Stress-induced alterations in neurotensin, somatostatin, and corticotropin-releasing factor in mesotelencephalic dopamine system regions. *Brain Res.*, 417: 350–354.

Deutch, A.Y., Goldstein, M., Baldino, F. Jr. and Roth, R.H. (1988) Telencephalic projections of the A8 dopamine cell group. *Ann. N.Y. Acad. Sci.*, 537: 27–50.

Deutch, A.Y., Gruen, R.J. and Roth, R.H. (1989) The effects of perinatal diazepam exposure on stress-induced activation of the mesotelencephalic dopamine system. *Neuropsychopharmacology*, 2: 105–114.

Deutch, A.Y., Clark, W.A. and Roth, R.H. (1990) Prefrontal cortical dopamine depletion enhances the responsiveness of mesolimbic dopamine neurons to stress. *Brain Res.*, 521: 311–315.

Divac, I., Fonnum, F. and Storm-Mathisen, J. (1977) High affinity uptake of glutamate in terminals of corticostriatal axons. *Nature*, 266: 377–378.

Divac, I., Björklund, A., Lindvall, O. and Passingham, R.E. (1978) Converging projections from the mediodorsal thalamic nucleus and mesencephalic dopaminergic neurons to the neocortex in three species. *J. Comp. Neurol.*, 180: 59–72.

Dorow, R., Horowski, R., Paschelke, G., Amin, M. and Braestrup, C. (1983) Severe anxiety induced by FG 7142, a β-carboline ligand for benzodiazepine receptors. *Lancet*, 41: 98–99.

Dougherty, G.G. and Ellinwood, E.H., Jr. (1981) Chronic D-amphetamine in nucleus accumbens: lack of tolerance or reverse tolerance of locomotor activity. *Life Sci.*, 28: 2295–2298.

Dunn, A.J. (1988) Stress-related activation of cerebral dopaminergic systems. *Ann. N.Y. Acad. Sci.*, 537: 188–205.

Dunn, A.J. and Berridge, C.W. (1987) Corticotropin-releasing factor administration elicits a stress-like activation of cerebral catecholaminergic systems. *Pharmacol. Biochem. Behav.*, 27: 685–691.

Elliott, P.J., Alpert, J.E., Bannon, M.J. and Iversen, S.D. (1986) Selective activation of mesolimbic and mesocortical dopamine metabolism in rat brain by infusion of a stable substance P analogue into the ventral tegmental area. *Brain Res.*, 145–147.

Fadda, F., Argiolas, A., Melis, M.R., Tissari, A.H., Onali, P.L. and Gessa, G. (1978) Stress-induced increase in 3,4-dihydroxyphenylacetic acid (DOPAC) levels in the cerebral cortex and N. accumbens: reversal by diazepam. *Life Sci.*, 23: 2219–2224.

Fallon, J.H. (1988) Topographic organization of ascending dopaminergic projections. *Ann. N.Y. Acad. Sci.*, 537: 1–9.

Fernstrom, J.D. (1983) Role of precursor availability in control of monoamine biosynthesis in brain. *Physiol. Rev.*, 63: 484–545.

Ferron, A., Thierry, A.M., Le Douarin, C. and Glowinski, J. (1984) Inhibitory influence of the mesocortical dopaminergic system on spontaneous activity or excitatory response induced from thalamic mediodorsal nucleus in the rat medial prefrontal cortex. *Brain Res.*, 302: 257–265.

File, S.E., Pellow, S. and Braestrup, C. (1985) Effects of the β-carboline, FG 7142, in the social interaction test of anxiety and the holeboard: correlations between behavior and plasma concentrations. *Pharmacol. Biochem. Behav.*, 22: 941–944.

Fonnum, F., Gottesfeld, Z. and Grofova, I. (1978) Distribution of glutamate decarboxylase, choline acetyltransferase, and aromatic amino acid decarboxylase in the basal ganglia of normal and operated rats. Evidence for striatopallidal, striatoentopeduncular, and striatonigral GABAergic fibers. *Brain Res.*, 143: 125–138.

Freeman, A.S., Meltzer, L.T. and Bunney, B.S. (1985) Firing properties of nigral dopaminergic neurons in freely moving rats. *Life Sci.*, 36: 157–164.

Gallager, D.W. (1978) Benzodiazepines: potentiation of a GABA inhibitory response in the dorsal raphe nucleus. *Eur. J. Pharmacol.*, 49: 133–143.

Galloway, M.P., Wolf, M.E. and Roth, R.H. (1986) Regulation of dopamine synthesis in the medial prefrontal cortex is mediated by release modulating autoreceptors: studies in vivo. *J. Pharmacol. Exp. Ther.*, 236: 689–698.

Gariano, R.F. and Groves, P.M. (1988) Burst firing induced in midbrain dopamine neurons by stimulation of the medial prefrontal and anterior cingulate cortices. *Brain Res.*, 462: 194–198.

Gariano, R.F., Tepper, J.M., Sawyer, S.F., Young, S.J. and Groves, P.M. (1989) Mesocortical dopaminergic neurons. 1. Electrophysiological properties and evidence for somadendritic autoreceptors. *Brain Res. Bull.*, 22: 511–516.

Giorgi, O., Corda, M.G. and Biggio, G. (1988) Ro 15-4513, like anxiogenic β-carbolines, increases dopamine metabolism in the prefrontal cortex of the rat. *Eur. J. Pharmacol.*, 156: 71–75.

Girault, J.A., Barbeito, L., Spampinato, U., Gozlan, H., Glowinski, J. and Besson, M.J. (1986) In vivo release of endogenous amino acids from the rat striatum: further evidence for a role of glutamate and aspartate in corticostriatal transmission. *J. Neurochem.*, 47: 98–106.

Glavin, G.B. (1985) Stress and brain noradrenaline: A review. *Neurosci. Biobehav. Rev.*, 9: 233–243.

Glavin, G.B., Tanaka, M., Tsuda, A., Kohno, Y., Hoaki, Y. and Nagasaki, N. (1983) Regional rat brain noradrenaline turnover in response to restrain stress. *Pharmacol. Biochem. Behav.,* 19: 287–290.

Glowinski, H., Chéramy, A., Romo, R. and Barbeito, L. (1988) Presynaptic regulation of dopaminergic transmission in the striatum. *Cell. Mol. Neurobiol.,* 8: 7–17.

Gonon, F.G. (1988) Nonlinear relationship between impulse flow and dopamine released by rat midbrain dopaminergic neurons as studied by in vivo electrochemistry. *Neuroscience,* 24: 19–28.

Grace, A.A. (1988) In vivo and in vitro intracellular recordings from rat midbrain dopamine neurons. *Ann. N.Y. Acad. Sci.,* 537: 51–76.

Grace, A.A. and Bunney, B.S. (1979) Paradoxical GABA excitation of nigral dopaminergic cells: indirect mediation through reticulata inhibitory neurons. *Eur. J. Pharmacol.,* 59: 211–218.

Grace, A.A. and Bunney, B.S. (1984) The control of firing pattern in nigral dopamine neurons: burst firing. *J. Neurosci.,* 4: 2877–2890.

Grenhoff, J. and Svensson, T.H. (1988) Clonidine regularizes substantia nigra dopamine cell firing. *Life Sci.,* 42: 2003–2009.

Groenewegen, H.J. and Russchen, F.T. (1984) Organization of the efferent projections of the nucleus accumbens to pallidal, hypothalamic, and mesencephalic structures: a tracing and immunohistochemical study in the cat. *J. Comp. Neurol.,* 223: 347–367.

Gysling, K. and Wang, R.Y. (1983) Morphine-induced activation of A10 dopamine neurons in the rat. *Brain Res.,* 277: 119–127.

Haroutunian, V., Knott, P. and Davis, K.L. (1988) Effects of mesocortical dopaminergic lesions upon subcortical dopaminergic function. *Psychopharmacol. Bull.,* 24: 341–344.

Herman, J.P., Guillonneau, D., Dantzer, R., Scatton, B., Semerdjian-Rouquier, L. and Le Moal, M. (1982) Differential effects of inescapable footshocks and of stimuli previously paired with inescapable footshocks on dopamine turnover in cortical and limbic areas of the rat. *Life Sci.,* 30: 2207–2214.

Herman, J.P., Rivet, J.M., Abrous, N. and Le Moal, M. (1988) Intracerebral dopaminergic transplants are not activated by electrical footshock stress activating in situ mesocorticolimbic neurons. *Neurosci. Lett.,* 90: 83–88.

Hervé, J.P., Simon, H., Blanc, G., Lisoprawski, A., Le Moal, M., Glowinski, J. and Tassin, J.P. (1979) Increased utilization of dopamine in the nucleus accumbens but not in the cerebral cortex after dorsal raphe lesion in the rat. *Neurosci. Lett.,* 15: 127–133.

Hervé, J.P., Simon, H., Blanc, G., Le Moal, M., Glowinski, J. and Tassin J.P. (1981) Opposite changes in dopamine utilization in the nucleus accumbens and the frontal cortex after electrolytic lesion of the medial raphe in the rat. *Brain Res.,* 216: 422–428.

Hervé, D., Tassin, J.P., Studler, J.M., Dana, C., Kitabgi, P., Vincent, J.P. and Glowinski, J. (1986) Dopaminergic control of ^{125}I-labeled neurotensin binding site density in corticolimbic structures of the rat brain. *Proc. Natl. Acad. Sci. (U.S.A.),* 83: 6203–6207.

Hirada, K., Deutch, A.Y. and Goldstein, M. (1989) The effects of a benzodiazepine partial agonist, Ro 16-6028, on basal and stress-induced in vivo tyrosine hydroxylation in the mesotelencephalic dopamine system. *Soc. Neurosci. Abstr.,* 15: 1310.

Hjorth, S., Engel, J.A. and Carlsson, A. (1986) Anticonflict effects of low doses of the dopamine agonist apomorphine in the rat. *Pharmacol. Biochem. Behav.,* 24: 237–240.

Hjorth, S., Carlsson, A. and Engel, J.A. (1987a) Anxiolytic-like action of the 3-PPP enantiomers in the Vogel conflict paradigm. *Psychopharmacology,* 92: 371–375.

Hjorth, S., Soderpalm, B. and Engel, J.A. (1987b) Biphasic effect of L-5-HTP in the Vogel conflict model. *Psychopharmacology,* 92: 96–99.

Hökfelt, T., Everitt, B.J., Theodorsson-Norheim, E. and Goldstein, M. (1984) Occurrence of neurotensin-like immunoreactivity in subpopulations of hypothalamic, mesencephalic, and medullary catecholamine neurons. *J. Comp. Neurol.,* 222: 543–549.

Ida, Y. and Roth, R.H. (1987) The activation of mesoprefrontal dopamine neurons by FG-7142 is absent in rats treated chronically with diazepam. *Eur. J. Pharmacol.,* 137: 185–189.

Ida, Y., Tanaka, M., Kohno, Y., Nakagawa, R., Iimori, K., Tsuda, A., Hoaki, A. and Nagasaki, N. (1982) Effects of age and stress on regional noradrenaline metabolism in the rat brain. *Neurobiol. Aging,* 3: 233–236.

Ida, Y., Tanaka, M., Tsuda, A., Tsujimaru, S. and Nagasaki, N. (1985) Attenuating effect of diazepam on stress-induced increases in noradrenaline turnover in specific brain regions of rats: antagonism by Ro 15-1788. *Life Sci.,* 37: 2491–2498.

Ikeda, M. and Nagatsu, T. (1985) Effect of short-term swimming stress and diazepam on 3,4-dihydroxyphenylacetic acid (DOPAC) and 5-hydroxyindolacetic acid (5-HIAA) levels in the caudate nucleus: an in vivo voltammetric study. *N.-S. Arch. Pharmacol.,* 331: 23–26.

Iuvone, P.M. and Dunn, A.J. (1986) Tyrosine hydroxylase activation in mesocortical 3,4-dihydroxyphenylethylamine neurons following footshock. *J. Neurochem.,* 47: 837–844.

Iversen, S.D. (1984) 5-HT and anxiety. *Neuropharmacology,* 23: 1553–1560.

Jhamandas, K. and Marien, M. (1987) Glutamate-evoked release of endogenous brain dopamine: inhibition of an excitatory amino acid antagonist and an enkephalin analogue. *Br. J. Pharmacol.,* 90: 641–650.

Jones, M.W., Kilpatrick, I.C. and Phillipson, O.T. (1988a) Thalamic control of subcortical dopamine function in the rat and the effects of lesions applied to the medial prefrontal cortex. *Brain Res.*, 475: 8 – 20.

Jones, M.W., Kilpatrick, I.C. and Phillipson, O.T. (1988b) Dopamine function in the prefrontal cortex of the rat is sensitive to a reduction of tonic GABA-mediated inhibition in the thalamic mediodorsal nucleus. *Exp. Brain Res.*, 69: 623 – 634.

Joyce, E.M., Stinus, L. and Iversen, S.D. (1983) Effect of injections of 6-hydroxydopamine into either nucleus accumbens septi or frontal cortex on spontaneous and drug-induced activity. *Neuropharmacology*, 9: 1141 – 1145.

Kalin, N.H., Sherman, J.E. and Takahashi, L.K. (1988) Antagonism of endogenous CRF systems attenuates stress-induced freezing behavior in rats. *Brain Res.*, 457: 130 – 135.

Kalivas, P.W. (1985) Sensitization to repeated enkephalin administration into the ventral tegmental area of the rat. II. Involvement of the mesolimbic dopamine system. *J. Pharmacol. Exp. Ther.*, 235: 544 – 550.

Kalivas, P.W. and Abhold, R. (1987) Enkephalin release into the ventral tegmental area in response to stress: modulation of mesocorticolimbic dopamine. *Brain Res.*, 414: 339 – 348.

Kalivas, P.W. and Duffy, P. (1989) Similar effects of daily cocaine and stress on mesocorticolimbic dopamine neurotransmission in the rat. *Biol. Psychiat.*, 25: 913 – 928.

Kalivas, P.W. and Richardson-Carlson, R. (1986) Endogenous enkephalin modulation of dopamine neurons in ventral tegmental area. *Am. J. Physiol.*, 251: R243 – A249.

Kalivas, P.W. and Taylor, S. (1985) Behavioral and neurochemical effect of daily injection with neurotensin into the ventral tegmental area. *Brain Res.*, 358: 70 – 76.

Kalivas, P.W., Widerlov, E., Stanley, D., Breese, G.R. and Prange, A.J., Jr. (1983a) Enkephalin action on the mesolimbic system: a dopamine-dependent and a dopamine independent increase in locomotor activity. *J. Pharmacol. Exp. Ther.*, 227: 229 – 237.

Kalivas, P.W., Burgess, S.K., Nemeroff, C.B. and Prange, A.J., Jr. (1983b) Behavioral and neurochemical effects of neurotensin microinjected into the ventral tegmental area of the rat. *Neuroscience*, 8: 495 – 505.

Kalivas, P.W., Deutch, A.Y., Maggio, J.E., Mantyh, P.W. and Roth, R.H. (1985) Substance K and substance P in the ventral tegmental area. *Neurosci. Lett.*, 57: 241 – 246.

Kalivas, P.W., Richardson-Carlson, R. and Van Orden, G. (1986) Cross-sensitization between foot shock stress and enkephalin-induced motor activity. *Biol. Psychiat.*, 21: 939 – 950.

Kalivas, P.W., Duffy, P. and Latimer, L.G. (1987) Neurochemical and behavioral effects of corticotropin-releasing factor in the ventral tegmental area of the rat. *J. Pharmacol. Exp. Ther.*, 242: 757 – 763.

Kalivas, P.W., Duffy, P., Dilts, R. and Abhold, R. (1988) Enkephalin modulation of A10 dopamine neurons: a role in dopamine sensitization. *Ann. N.Y. Acad. Sci.*, 537: 405 – 414.

Kalivas, P.W., Duffy, P. and Barrow, J. (1989) Regulation of the mesocorticolimbic dopamine system by glutamic acid receptor subtypes. *J. Pharmacol. Exp. Ther.*, 251: 378 – 387.

Kalivas, P.W., Duffy, P. and Eberhart, H. (1990) Modulation of A10 dopamine neurons by γ-aminobutyric acid agonists. *J. Pharmacol. Exp. Ther.*, 253: 858 – 866.

Kant, G.J., Mougey, E.H., Pennington, L.L. and Meyerhoff, J.L. (1983) Graded footshock stress elevates pituitary cyclic AMP and plasma β-endorphin, LPH, corticosterone and prolactin. *Life Sci.*, 33: 2657 – 2663.

Kawano, M., Nozoe, S., Yamanaka, T., Nagata, M., Takayama, I., Kawa, A., Kanahisa, T., Yoshimuta, Y., Ookubo, N. and Sonada, J. (1975) Evaluation of sulpiride effects on neurosis and psychosomatic diseases by double blind method. *Jap. J. Clin. Exp. Med.*, 52: 304 – 316.

Keller, R.W., Stricker, E.M. and Zigmond, M.J. (1983) Environmental stimuli but not homeostatic challenges produce apparent increases in dopaminergic activity in the striatum: an analysis by in vivo voltammetry. *Brain Res.*, 279: 159 – 170.

Kelley, A.E. and Iversen, S.D. (1978) Behavioral effects of substance P injections into zona reticulata of rat substantia nigra. *Brain Res.*, 158: 474 – 478.

Kelley, A.E., Stinus, L. and Iversen, S.D. (1979a) Behavioral activation induced by substance P infusion into ventral tegmental area: implication of dopaminergic A10 neurons. *Neurosci. Lett.*, 11: 335 – 339.

Kelley, A.E., Stinus, L. and Iversen, S.D. (1979b) Substance P infusion into substantia nigra of rat: behavioral analysis and involvement of striatal dopamine. *Eur. J. Pharmacol.*, 60: 171 – 179.

Kelley, P.H. and Moore, K.E. (1976) Mesolimbic dopaminergic neurones in the rotational model of nigrostriatal function. *Nature*, 263: 695 – 696.

Knorr, A.M., Deutch, A.Y. and Roth, R.H. (1989) The anxiogenic β-carboline FG-7142 increases in vivo and in vitro tyrosine hydroxylation in the prefrontal cortex. *Brain Res.*, 495: 355 – 361.

Kocsis, J.D., Sugimori, M. and Kitai, S.T. (1977) Convergence of excitatory synaptic inputs to caudate spiny neurons. *Brain Res.*, 124: 403 – 413.

Kornhuber, J., Kim, J.-S., Kornhuber, M.E. and Kornhuber, H.H. (1984) The cortico-nigral projection: reduced glutamate content in the substantia nigra following frontal cortex ablation in the rat. *Brain Res.*, 322: 124 – 126.

Kramarcy, N.R., Delanoy, R.L. and Dunn, A.J. (1984) Foot-shock treatment activates catecholamine synthesis in slices of mouse brain regions. *Brain Res.*, 290: 311 – 319.

Krettek, J.E. and Price, J.L. (1977) The cortical projections of the mediodorsal nucleus and adjacent thalamic nuclei in the rat. *J. Comp. Neurol.*, 171: 157 – 192.

Kuriyama, K., Kanmori, K., Taguchi, J. and Yoneda, Y. (1984) Stress-induced enhancement of suppression of [^{3}H]GABA release from striatal slices by presynaptic autoreceptor. *J. Neurochem.*, 42: 943 – 950.

Kvetnansky, R., Palkovits, M., Mitro, A., Torda, T. and Mikulaj, L. (1977) Catecholamines in individual hypothalamic nuclei of acutely and repeatedly stressed rats. *Neuroendocrinology,* 23: 257–267.

Lacey, M.G., Mercuri, N.B. and North, R.A. (1988) On the potassium conductance increase activated by $GABA_B$ and dopamine D2 receptors in rat substantia nigra neurones. *J. Physiol.,* 401: 437–453.

LaHoste, G.J., Mormède, P., Rivet, J.M. and Le Moal, M. (1988) Differential sensitization to amphetamine and stress responsivity as a function of inherent laterality. *Brain Res.,* 453: 381–384.

Latimer, L.G., Duffy, P. and Kalivas, P.W. (1987) *Mu* opioid receptor involvement in enkephalin activation of dopamine neurons in the ventral tegmental area. *J. Pharmacol. Exp. Ther.,* 241: 328–337.

Lavielle, S., Tassin, J.-P., Thierry, A.-M., Blanc, G., Hervé, D., Barthelemy, C. and Glowinski, J. (1978) Blockade by benzodiazepines of the selective high increase in dopamine turnover induced by stress in mesocortical dopaminergic neurons of the rat. *Brain Res.,* 168: 585–594.

Leccese, A.P. and Lyness, W.H. (1987) Lesions of dopamine neurons in the medial prefrontal cortex: effects of self-administration of amphetamine and dopamine synthesis in the brain of the rat. *Neuropharmacology,* 26: 1303–1308.

Lee, E.H.Y., Lin, H.H. and Yin, H.M. (1987) Differential influences of different stressors upon midbrain raphe neurons in rats. *Neurosci. Lett.,* 80: 115–119.

Leonard, C.M. (1969) The prefrontal cortex of the rat. I. Cortical projections of the mediodorsal nucleus. II. Efferent connections. *Brain Res.,* 12: 321–343.

Lisaprawski, A., Blanc, G. and Glowinski, J. (1981) Activation by stress of the habenulo-interpeduncular substance P neurons in the rat. *Neurosci. Lett.,* 25: 47–51.

Loughlin, S.E. and Fallon, J.H. (1984) Substantia nigra and ventral tegmental area projections to cortex: topography and collateralization. *Neuroscience,* 11: 425–436.

Louilot, A., Simon, H., Tagzhouti, K. and Le Moal, M. (1985) Modulation of dopaminergic activity in the nucleus accumbens following facilitation or blockade of the dopaminergic transmission in the amygdala: a study by in vivo differential pulse voltammetry. *Brain Res.,* 346: 141–145.

Louilot, A., Tagzhouti, K., Deminière, J.M., Simon, H. and Le Moal, M. (1987) Dopamine and behavior: functional and theoretical considerations. In: M. Sandler, C. Feuerstein and B. Scatton (Eds.), *Neurotransmitter Interactions in the Basal Ganglia,* Raven Press, New York.

Louilot, A., Le Moal, M. and Simon, H. (1989) Opposite influences of dopaminergic pathways to the prefrontal cortex or the septum on the dopaminergic transmission in the nucleus accumbens. An in vivo voltammetric study. *Neuroscience,* 29: 45–56.

Maggio, J.E. (1988) Tachykinins, *Annu. Rev. Neurosci.,* 11: 13–28.

Marcel, D., Weissman, D., Bardelay, C., Meunier, C. and Pujol, J.-F. (1987) Benzodiazepine binding sites in the cingulate cortex after lesion of the noradrenaline and dopamine containing afferents. *Brain Res. Bull.,* 19: 485–494.

Markstein, R. and Emson, P. (1988) Effect of neurotensin and its fragments neurotensin- (1–6) and neurotensin- (8–13) on dopamine release from cat striatum. *Eur. J. Pharmacol.,* 152: 147–152.

Martin-Iversen, M.T., Szostak, C. and Fibiger, H.C. (1986) 6-Hydroxydopamine lesions of the medial prefrontal cortex fail to influence intravenous self-administration of cocaine. *Psychopharmacology,* 88: 310–314.

Matsuzaki, I., Takamatsu, Y. and Moroji, T. (1989) The effects of intracerebroventricularly injected corticotrophin-releasing factor (CRF) on the central nervous system: behavioral and biochemical studies. *Neuropeptides,* 13: 147–155.

Matthews, R.T. and German, D.C. (1984) Electrophysiological evidence for excitation of rat ventral tegmental area dopamine neurons by morphine. *Neuroscience,* 11: 617–625.

McCarty, R. and Kopin, I.J. (1978) Sympatho-adrenal medullary activity and behavior during exposure to footshock stress: a comparison of seven rat strains. *Physiol. Behav.,* 21: 567–572.

Meloni, R., Childs, J., Gerogan, F., Yurkofsky, S. and Gale, K. (1988) Effect of haloperidol on transplants of fetal substantia nigra: evidence for feedback regulation of dopamine turnover in the graft and its projections. *Prog. Brain Res.,* 78: 457–461.

Miller, J.D., Speciale, S.G., McMillen, B.A. and German, D.C. (1984) Naloxone antagonism of stress-induced augmentation of frontal cortex dopamine metabolism. *Eur. J. Pharmacol.,* 98: 437–439.

Mogenson, G.J., Swanson, L.W. and Wu, M. (1983) Neural projections from nucleus accumbens to globus pallidus, substantia innominata, and lateral preoptic-lateral hypothalamic area: an anatomical and electrophysiological investigation in the rat. *J. Neurosci.,* 3: 189–202.

Mount, H., Welner, S., Quirion, R. and Boksa, P. (1989) Glutamate stimulation of [^{3}H]dopamine release from dissociated cell cultures of rat ventral mesencephalon. *J. Neurochem.,* 52: 1300–1310.

Nakane, T., Audhya, T., Kanie, N. and Hollander, C.S. (1985) Evidence for a role of endogenous corticotropin-releasing factor in cold, ether, immobilization, and traumatic stress. *Proc. Natl. Acad. Sci. (U.S.A.),* 82: 1247–1251.

Nauta, W.J.H., Smith, G.P., Faull, R.L.M. and Domesick, V.B. (1978) Efferent connections and nigral afferents of the nucleus accumbens septi in the rat. *Neuroscience,* 3: 385–401.

Nemeroff, C.B. and Cain, S.T. (1985) Neurotensin-dopamine interactions in the CNS. *Trends Pharmacol. Sci.,* 6: 201–205.

Nieoullon, A. and Dusticier, N. (1983) Glutamate uptake, glutamate decarboxylase and choline acetyltransferase in subcortical areas after sensorimotor cortical ablations in the cat. *Brain Res. Bull.,* 10: 287–293.

Nieoullon, A., Scarfone, E., Kerkerian, L., Errami, M. and Dusticier, N. (1985) Changes in choline acetyltransferase, glutamic acid decarboxylase, high-affinity glutamate uptake and dopaminergic activity induced by kainic acid lesion of the thalamostriatal neurons. *Neurosci. Lett.,* 58: 299–304.

Ninan, P.T., Insel, T.M., Cohen, R.M., Cook, J.M., Skolnick, P. and Paul, S.M. (1982) Benzodiazepine receptor-mediated experimental "anxiety" in primates. *Science,* 218: 1332–1334.

Oades, R.D., Tagzhouti, K., Rivet, J.-M., Simon, H. and Le Moal, M. (1986) Locomotor activity in relation to dopamine and noradrenaline in the nucleus accumbens, septal, and frontal areas: a 6-hydroxydopamine study. *Neuropsychobiology,* 16: 37–42.

Ostrowski, N.L., Hatfield, C.B. and Caggiula, A.R. (1982) The effects of low doses of morphine on the activity of dopamine-containing cells and on behavior. *Life Sci.,* 31: 2347–2350.

Palkovits, M., Lang, T., Patthy, A. and Elekes, I. (1986) Distribution and stress-induced increase in glutamate and aspartate levels in discrete brain nuclei of rats. *Brain Res.,* 373: 252–257.

Palmer, M.R. and Hoffer, B.J. (1980) Catecholamine modulation of enkephalin-induced electrophysiological responses in cerebral cortex. *J. Pharmacol. Exp. Ther.,* 213: 205–215.

Paris, J.M., Lorens, S.A., Van de Kar, L.D., Urban, J.H., Richardson-Morton, K.D. and Bethea, C.L. (1987) A comparison of acute stress paradigms: hormonal responses and hypothalamic serotonin. *Physiol. Behav.,* 39: 33–43.

Passingham, R.E., Myers, C., Rawlins, N., Lightfoot, V. and Fearn, S. (1988) Premotor cortex in the rat. *Behav. Neurosci.,* 102: 101–109.

Peterson, S.L., St. Mary, J.S. and Harding, N.R. (1987) Cis-flupentixol antagonism of the rat prefrontal cortex neuronal response to apomorphine and ventral tegmental area input. *Brain Res. Bull.,* 18: 723–729.

Phillipson, O.T. (1979a) The cytoarchitecture of the interfascicular nucleus and ventral tegmental area of Tsai in the rat. *J. Comp. Neurol.,* 187: 85–98.

Phillipson, O.T. (1979b) Afferent projections to the ventral tegmental area of Tsai and interfascicular nucleus: a horseradish peroxidase study in the rat. *J. Comp. Neurol.,* 187: 117–143.

Pinnock, R.D. (1984) Hyperpolarizing action of baclofen on neurons in the rat substantia nigra slice. *Brain Res.,* 332: 337–340.

Pinnock, R.D. (1985) Neurotensin depolarizes substantia nigra dopamine neurones. *Brain Res.,* 338: 151–154.

Pittaluga, A. and Raiteri, M. (1988) Clonidine enhances the release of endogenous γ-aminobutyric acid through alpha-2 and alpha-1 presynaptic adrenoceptors differentially located in rat cerebral cortex subregions. *J. Pharmacol. Exp. Ther.,* 245: 682–686.

Plantjé, J.F., Steinbusch, H.W.M., Schipper, J., Dijcks, F.A., Verheijden, P.F.H.M. and Stoof, J.C. (1987) D-2 dopamine-receptors regulate the release of [^{3}H]dopamine in rat cortical regions showing dopamine immunoreactive fibers. *Neuroscience,* 20: 157–168.

Pycock, C.J., Carter, C.J. and Kerwin, R.W. (1980) Effect of 6-hydroxydopamine lesions of the medial prefrontal cortex on neurotransmitter systems in subcortical sites in the rat. *J. Neurochem.,* 34: 91–99.

Quintin, L., Gonon, F., Buda, M., Ghignone, M., Hilaire, G. and Pujol, J.-F. (1986) Clonidine modulates locus coeruleus metabolic hyperactivity induced by stress in behaving rats. *Brain Res.,* 362: 366–369.

Quirion, R., Chiueh, C.C., Everist, H.D. and Pert, A. (1985) Comparative localization of neurotensin receptors on nigrostriatal and mesolimbic dopaminergic terminals. *Brain Res.,* 327: 385–389.

Reinhard, J.F. Jr., Bannon, M.J. and Roth, R.H. (1982) Acceleration by stress of dopamine synthesis and metabolism in prefrontal cortex: antagonism by diazepam. *N.-S. Arch. Pharmacol.,* 318: 374–377.

Ribak, C.E., Vaughn, J.E. and Roberts, E. (1980) GABAergic nerve terminals decrease in the substantia nigra following hemitransections of the striatonigral and pallidonigral pathways. *Brain Res.,* 192: 413–420.

Robinson, T.E. and Becker, J.B. (1982) Behavioral sensitization is accompanied by an enhancement in amphetamine-stimulated dopamine release from striatal tissue in vitro. *Eur. J. Pharmacol.,* 85: 253–254.

Rose, J.E. and Woolsey, C.N. (1948) The orbitofrontal cortex and its connections with the mediodorsal nucleus in rabbit, sheep, and cat. *Res. Publ. Assoc. Nerv. Ment. Dis.,* 27: 210–232.

Rosin, D.L., Clark, W.A., Goldstein, M., Roth, R.H. and Deutch, A.Y. (1990) Effects of 6-hydroxydopamine lesions of the prefrontal cortex on tyrosine hydroxylase activity in subcortical dopamine systems of the rat. *Neuroscience,* submitted.

Rossier, J., French, E.D., Rivier, C., Ling, N., Guillemin, R. and Bloom, F.E. (1977) Foot-shock induced stress increases β-endorphin levels in blood but not brain. *Nature,* 270: 618–620.

Roth, R.H., Wolf, M.E. and Deutch, A.Y. (1987) The neurochemistry of midbrain dopamine systems. In: H.Y. Meltzer (Ed.), *Psychopharmacology: The Third Generation of Progress,* Raven Press, New York, pp. 81–94.

Roth, R.H., Tam, S.-Y., Ida, Y., Yang, J.-X. and Deutch, A.Y. (1988a) Stress and the mesocorticolimbic dopamine systems. *Ann. N.Y. Acad. Sci.,* 537:138–147.

Roth, R.H., Tam, S.-Y., Bradberry, C.W., Karasic, D.H. and Deutch, A.Y. (1988b) Precursor availability and function of midbrain dopaminergic neurons. In: G. Huether (Ed.),

Amino Acid Availability and Brain Function in Health and Disease, Nato ASI Series, Vol. H20, Springer, Heidelberg, pp. 191–200.

Sanger, D.J. (1985) GABA and the behavioral effects of anxiolytic drugs. *Life Sci.*, 36: 1503–1513.

Sawaguchi, T., Matsumura, M. and Kubota, K. (1988a) Dopamine enhances the neuronal activity of spatial short-term memory task in the primate prefrontal cortex. *Neurosci. Res.*, 5: 465–473.

Sawaguchi, T., Matsumura, M. and Kubota, K. (1988b) Delayed response deficit in monkeys by locally disturbed prefrontal neuronal activity by bicuculline. *Behav. Brain Res.*, 31: 193–198.

Scatton, B., Dennis, T. and Curet, O. (1984) Increase in dopamine and DOPAC levels in noradrenergic terminals after electrical stimulation of the ascending noradrenergic pathways. *Brain Res.*, 298: 193–196.

Scatton, B., D'Angio, M., Driscoll, P. and Serrano, A. (1988) An in vivo voltammetric study of the response of mesocortical and mesoaccumbens dopaminergic neurons to environmental stimuli in strains of rats with different levels of emotionality. *Ann. N.Y. Acad. Sci.*, 537: 124–137.

Scheibner, T. and Tork, I. (1987) Ventromedial mesencephalic tegmental (VMT) projections to ten functionally different cortical areas in the cat: Topographical and quantitative analysis. *J. Comp. Neurol.*, 259: 247–265.

Schwartz, R.D., Wess, M.J., Labarca, R., Skolnick, P. and Paul, S.P. (1987) Acute stress enhances the activity of the GABA receptor-gated chloride ion channel in brain. *Brain Res.*, 411: 151–155.

Sesack, S.R. and Bunney, B.S. (1989) Pharmacological characterization of the receptor mediating electrophysiological responses to dopamine in the rat medial prefrontal cortex: a microiontophoretic study. *J. Pharmacol. Exp. Ther.*, 248: 1323–1333.

Sesack, S.R., Deutch, A.Y., Roth, R.H. and Bunney, B.S. (1989) Topographical organization of the efferent projections of the medial prefrontal cortex in the rat: an anterograde tract-tracing study with *Phaseolus vulgaris* leucoagglutinin. *J. Comp. Neurol.*, 290: 213–242.

Shanks, N. and Anisman, H. (1988) Stressor-provoked behavioral changes in six strains of mice. *Behav. Neurosci.*, 102: 894–905.

Shepard, P.D. and German, D.C. (1984) A subpopulation of mesocortical dopamine neurons possesses autoreceptors. *Eur. J. Pharmacol.*, 98: 455–456.

Shepard, R.A. (1987) Behavioral effects of GABA agonists in relation to anxiety and benzodiazepine action. *Life Sci.*, 40: 2429–2436.

Simon, H. and Le Moal, M. (1984) Mesencephalic dopaminergic neurons: functional role. In: *Catecholamines: Neuropharmacology and Central Nervous System – Theoretical Aspects*, Alan R. Liss, New York, pp. 293–307.

Sivam, S.P., Breese, G.R., Krause, J.E., Napier, T.C., Mueller, R.A. and Hong, J.-S. (1987) Neonatal and adult 6-hydroxydopamine-induced lesions differentially alter tachykinin and enkephalin gene expression. *J. Neurochem.*, 49: 1623–1633.

Sokoloff, P., Giros, B., Martres, M.-P., Bouthenet, M.-L., and Schwartz, J.-C. (1990) Molecular cloning and characterization of a novel dopamine receptor (D_3) as a target for neuroleptics. *Nature*, 347: 146–151.

Sorg, B.A. and Kalivas, P.W. (1989) Measurement of footshock-induced changes in dopamine and metabolites by in vivo dialysis in rat prefrontal cortex. *Soc. Neurosci. Abstr.*, 15: 558.

Speciale, S.G., Miller, J.D., McMillen, B.A. and German, D.C. (1986) Activation of specific central dopamine pathways: locomotion and footshock. *Brain Res. Bull.*, 16: 33–38.

Spencer, H.J. (1976) Antagonism of cortical excitation of striatal neurons by glutamic acid diethylester: evidence for glutamic acid as an excitatory transmitter in rat striatum. *Brain Res.*, 102: 91–101.

Stanford, C., Fillenz, M. and Ryan, E. (1984) The effect of repeated mild stress on cerebral cortical adrenoceptors and noradrenaline synthesis in the rat. *Neurosci. Lett.*, 45: 163–167.

Stone, E.A. (1975) Stress and catecholamines, in: A.J. Friedhoff (Ed.), *Catecholamines and Behavior. Vol. 2. Neuropsychopharmacology*, Plenum Press, New York, pp. 31–72.

Studler, J.-M., Kitabgi, P., Tramu, G., Hervé, D., Glowinski, J. and Tassin, J.-P. (1988) Extensive co-localization of neurotensin with dopamine in rat meso-cortico-frontal dopaminergic neurons. *Neuropeptides*, 11: 95–100.

Sudo, A. (1983) Time course of the changes of catecholamine levels in rat brain during swim stress. *Brain Res.*, 276: 372–374.

Szigethy, E. and Beaudet, A. (1989) Correspondence between high affinity ^{125}I-neurotensin binding sites and dopaminergic neurons in the rat substantia nigra and ventral tegmental area: a combined radioautographic and immunohistochemical light microscopic study. *J. Comp. Neurol.*, 279: 128–137.

Tam, S.-Y. and Roth, R.H. (1985) Selective increase in dopamine metabolism in the prefrontal cortex by the anxiogenic beta-carboline FG 7142. *Biochem. Pharmacol.*, 34: 1595–1598.

Tassin, J.P., Herve, D., Blanc, G. and Glowinski, J. (1980) Differential effects of a two-minute open-field session on dopamine utilization in the frontal cortices of BALB/C and C57 BL/6 mice. *Neurosci. Lett.*, 17: 67–71.

Tassin, J.P., Kitabgi, P., Tramu, G., Studler, J.M., Hervé, D., Trovero, F. and Glowinski, J. (1988) Rat mesocortical dopaminergic neurons are mixed neurotensin/dopamine neurons: immunohistochemical and biochemical evidence. *Ann. N.Y. Acad. Sci.*, 537: 531–533.

Terreberry, R.R. and Neafsey, E.J. (1987) The rat medial frontal cortex projects directly to autonomic regions of the brainstem. *Brain Res. Bull.,* 19: 639–649.

Thiébot, M.-H. and Soubrié, P. (1983) Behavioral pharmacology of the benzodiazepines. In: E. Costa (Ed.), *The Benzodiazepines: From Molecular Biology to Clinical Practice,* Raven Press, New York, pp. 67–92.

Thiébot, M.-H., Hamon, M. and Soubrié, P. (1982) Attenuation of induced anxiety in rats by chlordiazepoxide: role of raphe dorsalis benzodiazepine binding sites and serotonergic neurons. *Neuroscience,* 7: 2287–2294.

Thierry, A.M., Stinus, L., Blanc, G. and Glowinski, J. (1973a) Some evidence for the existence of dopaminergic neurons in the rat cortex. *Brain Res.,* 50: 230–234.

Thierry, A.M., Stinus, L. and Glowinski, J. (1973b) Dopaminergic terminals in the rat cortex. *Science,* 182: 499–501.

Thierry, A.M., Tassin, J.-P., Blanc, G. and Glowinski, J. (1976) Selective activation of the mesocortical dopamine system by stress. *Nature,* 263: 242–243.

Thierry, A.M., Deniau, J.M. and Feger, J. (1979) Effects of stimulation of the frontal cortex on identified output VMT cells in the rat. *Neurosci. Lett.,* 15: 103–107.

Thierry, A.M., Le Douarin, C., Penit, J., Ferron, A. and Glowinski, J. (1986) Variation in the ability of neuroleptics to block the inhibitory influence of dopaminergic neurons on the activity of cells in the rat prefrontal cortex. *Brain Res. Bull.,* 16: 155–160.

Trulson, M.E. and Preussler, D.W. (1984) Dopamine-containing ventral tegmental neurons in freely moving cats: activity during the sleep-waking cycle and effects of stress. *Exp. Neurol.,* 83: 367–377.

VanderMaelen, C.P., Matheson, G.K., Wilderman, R.C. and Patterson, L.A. (1986) Inhibition of serotonergic dorsal raphe neurons by systemic and ionotophoretic administration of buspirone, a non-benzodiazepine anxiolytic drug. *Eur. J. Pharmacol.,* 129: 123–130.

Van Eden, C.G., Hoorneman, E.M.D., Buijs, R.M., Matthijssen, M.A.H., Geffard, M. and Uylings, H.B.M. (1987) Immunocytochemical localization of dopamine in the prefrontal cortex of the rat at the light and electron microscopic level. *Neuroscience,* 22: 849–862.

Walaas, I. and Fonnum, F. (1980) Biochemical evidence for γ-aminobutyrate containing fibres from the nucleus accumbens to the substantia nigra and ventral tegmental area in the rat. *Neuroscience,* 5: 63–72.

Weiss, J.M., Bailey, W.H., Pohorecky, L.A., Korzeniowski, D. and Grillione, G. (1980) Stress-induced depression of motor activity correlates with regional changes in brain norepinephrine but not in dopamine. *Neurochem. Res.,* 5: 9–22.

Wilcox, R.A., Robinson, T.E. and Becker, J.B. (1986) Enduring enhancement in amphetamine-stimulated striatal dopamine release in vitro produced by prior exposure to amphetamine or stress in vivo. *Eur. J. Pharmacol.,* 124: 375–376.

Wolf, M.E. and Roth, R.H. (1987) Dopamine neurons projecting to the prefrontal cortex possess release modulating autoreceptors. *Neuropharmacology,* 26: 1053–1059.

Wolf, M.E., Galloway, M.P. and Roth, R.H. (1986) Regulation of dopamine synthesis in the medial prefrontal cortex: studies in brain slices. *J. Pharmacol. Exp. Ther.,* 236: 699–707.

Wolf, M.E., Deutch, A.Y. and Roth, R.H. (1987) The neuropharmacology of dopamine. In: F.A. Henn and L.E. DeLisi (Eds.), *Handbook of Schizophrenia. Vol. 2. Neurochemistry and Neuropharmacology,* Elsevier Science Publishers, Amsterdam, pp. 101–147.

Wood, P.L. (1983) Opioid regulation of CNS dopaminergic pathways: a review of methodology, receptor types, regional variations, and species differences. *Peptides,* 4: 595–601.

Woulfe, J. and Beaudet, A. (1989) Immunocytochemical evidence for direct connections between neurotensin-containing axons and dopaminergic neurons in the rat ventral midbrain tegmentum. *Brain Res.,* 479: 402–406.

Young III, W.S. and Kuhar, M.J. (1980) Radiohistochemical localization of benzodiazepine receptors in rat brain. *J. Pharmacol. Exp. Ther.,* 212: 337–346.

Young III, W.S., Niehoff, D., Kuhar, M.J., Beer, B. and Lippa, A.S. (1981) Multiple benzodiazepine receptor localization by light microscopic radiohistochemistry. *J. Pharmacol. Exp. Ther.,* 216: 425–430.

Zahm, D.S. (1989) The ventral striatopallidal parts of the basal ganglia in the rat. II. Compartmentalization of ventral pallidal efferents. *Neuroscience,* 30: 33–50.

Discussion

P.R. Lowenstein: How do you explain the effects of the benzodiazepine inverse agonists in slices onto DA terminals, in the absence of GABAergic synapses onto axonal terminals in cortex?

A.Y. Deutch: Studies which have examined GABA neurons in the cortex have not revealed axo-axonic GABAergic synapses, except onto intial segments. GABA receptors may be present on axons as autoreceptors, but these are presumably GABA$_B$ receptors. Since the precise laminar distribution of GABA neurons in the cortex varies across cortical regions, it will be necessary to specifically examine PFC GABA neurons using EM methods. The apparent absence of GABA synapses onto axons may suggest that the effects of GABA on in vivo tyrosine hydroxylation in the PFC may be indirect. Alternatively, one cannot eliminate the actions of GABA occurring in the absence of conventional synapses. We have observed association of the benzodiazepine receptor with DA neurons in the midbrain of the human in light microscopic studies; in the absence of EM studies we cannot determine if the labeling is cytoplasmic. It is clear that it will be necessary to examine the synaptic localization of benzodiazepine receptors (as opposed to GABAergic neurons) in the PFC at the EM level.

C.G. van Eden: If DA activation (stress induced) in PFC lowers activity of DA in other structures, how does this fit in with the theory of Dr. d'Angio which reads that DA activation serves to activate cognition to deal with the stressor?

A.Y. Deutch: The stress-induced activation of PFC DA may in part contribute to the observed pattern of response across the mesotelencephalic DA system by transynaptically modulating DA tone in the subcortical sites. If the increase in PFC DA release subserves the acquisition of a coping strategy in response to stress, then the suppression of subcortical DA systems may act to prevent behaviors which will interfere with the acquisition or execution of the coping behavior. In addition, non-dopaminergic subcortical systems may be involved in the execution of the coping strategy (for example, serotonergic or noradrenergic systems), and the suppression of mesolimbic or striatal DA release may remove these non-DA systems from a constraining influence.

D.A. Powell: Does the mild level of foot shock employed result in operant avoidance or escape behavior?

A.Y. Deutch: We have not tested the current of footshock used in the stress experiments in operant avoidance paradigms. It is important to note that the precise way in which current intensity is measured in various studies (i.e., the current as measured over a given resistance load) is frequently not cited, and it is therefore difficult to equate shock intensities. However, the currents used in most learning paradigms evoke overt escape behavior and vocalization, whereas our relatively mild footshock current does not result in such behavior. Nonetheless, since we can condition the release of DA in the PFC such that a neutral tone previously paired with the mild footshock results in increased PFC DA, it would appear likely that one could employ our footshock current in operant studies, although more extensive US-CS pairings than are usually presented may be required.

J.P.C. de Bruin: Prefrontal DA "dampens" the DA response (higher DOPAC) in n. accumbens. You reported studies on the DA response in slices of PFC and striatum. Did you also measure the response in slices containing the n. accumbens, and if so, was the DA response in the accumbens present, when devoid of prefrontal (input) connections?

A.Y. Deutch: We examined changes in DA synthesis in response to the β-carboline benzodiazepine inverse agonists and GABA in slices of PFC and striatum, in which transynaptic corticostriatal modulation is not operative. We have not yet examined benzodiazepine/GABA mechanisms controlling nucleus accumbens DA function in slices, although we plan to investigate this question.

CHAPTER 20

Involvement of mesocorticolimbic dopaminergic systems in emotional states

M. Bertolucci-D'Angio*, A. Serrano, P. Driscoll** and B. Scatton

Synthélabo Recherche (L.E.R.S.), Biology Department, 92220 Bagneux, France

Introduction

Accumulating evidence indicates that the mesocorticolimbic dopaminergic neurons, which orginate in the A10 dopaminergic cell group in the ventral tegmental area, are involved in the control of cognitive processes and emotional behavior. Indeed, electrolytic or neurotoxic lesions of the ventral tegmental area in the rat induce a characteristic syndrome including hypoemotionality, locomotor hyperactivity with repetitive behavior, hypoexploratory behavior and difficulties in suppressing previously learned responses (Le Moal et al., 1969, 1975; Simon et al., 1979, 1980). Deficits in cognitive functions have also been observed after specific lesions of the dopaminergic terminals in the prefrontal cortex of both rats (Simon et al., 1979, 1980) and monkeys (Brozoski et al., 1979); the animals showed retention impairments in a delayed alternation task that is considered to be sensitive and selective test for frontal cortex function. Moreover, lesions of dopaminergic terminals in the nucleus accumbens provoke impairment in the functional processes leading from motivation to action (Simon, 1981).

It is also well documented that stressful conditions cause an activation of dopaminergic neurons that project to the frontal cortex and/or nucleus accumbens in the rodent (Claustre et al., 1986; D'Angio et al., 1987; Deutch et al., 1985; Dunn and File, 1983; Fadda et al., 1978; Herman et al., 1982; Lavielle et al., 1978; Reinhard et al., 1982; Tassin et al., 1980; Thierry et al., 1976). Moreover, systemic injection of the anxiogenic ω (benzodiazepine) receptor inverse agonists methyl- or ethyl-β-carboline-3-carboxylate or FG-7142 increases cortical dopamine (DA) metabolism in a manner similar to that produced by stress (Claustre et al., 1986; Scatton et al., 1988; Tam and Roth, 1985). Furthermore, anxiolytic agents antagonize the footshock or methyl-β-carboline carboxylate (β-CCM) induced increase in DA metabolism in the frontal cortex and nucleus accumbens in rats (Claustre et al., 1986; Fadda et al., 1978; Ida and Roth, 1987; Lavielle et al., 1978; Reinhard et al., 1982). All of these data thus clearly implicate the mesocorticolimbic dopaminergic neurons in emotional states.

In spite of the extensive studies that have been devoted to the effects of stress on brain DA metabolism, there is still conflicting evidence on

---

Correspondence: Dr. B. Scatton, Synthélabo Recherche (L.E.R.S.), Biology Department, 31 Avenue Paul Vaillant-Couturier, 92220 Bagneux, France.

* Present address: Mental Retardation Research Center, UCLA, 760 Westwood Plaza, Los Angeles, CA 90024, USA.
** Eidgenössische Technische Hochschule Zürich, Institut für Verhaltenswissenschaft, Laboratorium für Vergleichende Physiologie und Verhaltensbiologie, Turnerstrasse 1, ETH-Zentrum, CH-8092 Zurich, Switzerland.

the selectivity of the response of the different ascending dopaminergic neurons to stress. Moreover, the significance of the stress-induced activation of mesocortical dopaminergic neurons (reflection of emotional reaction to, or coping with stress?) is as yet unclear. Finally, it is not yet known whether anticipating appetitive stimuli, which involve emotional and motivational processes, also affect DA metabolism in the prefrontal cortex. We here review our recent attempts to address these various questions in the rat.

Differential effects of anxiogenic stimuli and forced locomotion on extracellular DOPAC levels in the rat striatum, nucleus accumbens and prefrontal cortex

The involvement of mesocorticolimbic dopaminergic neurons in emotional behavior is supported by the demonstration that stressful conditions lead to an increase in dopamine (DA) metabolism in the corresponding DA terminal fields in the rodent (see Introduction). However, the reports on stress-induced changes in DA metabolism in DA projection fields have been inconsistent. In some cases, stress exposure (brief footshocks, conditioned fear, exposure to a novel environment) produced a *selective* increase in DA metabolism in the prefrontal cortex of rats or mice (Thierry et al., 1976; Tassin et al., 1980; Herman et al., 1982; Reinhard et al., 1982; Bannon and Roth, 1983). However, others reported stress-induced increases in DA metabolism also in the nucleus accumbens, olfactory tubercle and even striatum (Keller et al., 1983; Dunn and File, 1983; Ikeda et al., 1984; Watanabe, 1984; Claustre et al., 1986; D'Angio et al., 1987). A possible explanation for these discrepant findings may be that the ascending dopaminergic pathways projecting to these structures are activated differentially according to the nature and/or intensity of the applied aversive stimulus. To address this question, we compared the changes in DA metabolism in the striatum, nucleus accumbens and prefrontal cortex of rats subjected to anxiogenic environmental situations (tail-pinch, immobilization), forced locomotion on a rotarod or receiving the anxiogenic agent β-CCM. These stimuli are sufficiently different in nature to enhance the probability of producing stimulus specific response patterns.

In this study, DA metabolism was assessed in the different DA-rich brain areas of unrestrained Sprague – Dawley rats by measuring extracellular levels of 3,4-dihydroxyphenylacetic acid (DOPAC) using in vivo voltammetry with electrochemically pretreated carbon fiber electrodes (Gonon et al., 1980, 1981; Serrano et al., 1986) positioned into the brain parenchyma with an implantation assembly (Louilot et al., 1987; D'Angio et al., 1987, 1988).

Under our experimental conditions, 2 clearly separated oxidation peaks corresponding to the oxidation of ascorbic acid (peak 1) and DOPAC (peak 2) were recorded at -100 and $+100$ mV, respectively in the striatum, nucleus accumbens and anteromedial prefrontal cortex of freely moving rats (Fig. 1). Previous pharmacological studies performed on the immobilized or anesthetized rat have established the identity of peak 2 with DOPAC in these different brain areas (Gonon et al., 1980; Buda et al., 1981; Ikeda et al., 1984; Serrano et al., 1986).

In control rats, the electrochemical signal recorded at $+100$ mV in the striatum, nucleus accumbens or prefrontal cortex was stable over at least a 4 h period. Forced locomotion of rats on a rotarod for 40 min, increased markedly the amplitude of the DOPAC oxidation peak in the striatum (maximum augmentation $+110\%$) and to a lesser extent in the nucleus accumbens (maximum augmentation $+60\%$) but not in the prefrontal cortex (Fig. 2) (Bertolucci-D'Angio et al., 1990). In the striatum and nucleus accumbens, the DOPAC peak was stable during the entire test period (40 min) and gradually returned to baseline values 30 – 40 min after the rats had been replaced in their home cages.

Mild tail pressure for 10 min induced a long-lasting increase in the height of the DOPAC oxidation peak in the nucleus accumbens (maximum augmentation +60%) (Fig. 2) (D'Angio et al., 1987); this increase progressed for more than 80 min after tail-pinch has ceased. A small, transient increase in this parameter was observed in the striatum (maximum augmentation +19%) whereas no significant change was seen in the prefrontal cortex (Fig. 2).

When rats were restrained for 4 min in the hands of the experimenter, an increase in the amplitude of the DOPAC oxidation peak was observed in the nucleus accumbens (maximum increase +53%) and to a lesser extent in the prefrontal cortex (maximum increase +30%) but not in the striatum (Fig. 2) (Bertolucci-D'Angio et al., 1990). Systemic administration of β-CCM (10 mg/kg s.c.) caused a comparable increase in the height of the DOPAC oxidation peak in nucleus accumbens (maximum augmentation +48%) and prefrontal cortex (maximum augmentation +54%), whereas no significant change was measured in the striatum (Fig. 2) (Bertolucci-D'Angio et al., 1990).

These results show a differential response of the nigrostriatal, mesolimbic and mesocortical dopaminergic neurons to aversive stimuli and motor-related conditions. The marked elevation of DA metabolism in the striatum when the rats are exhibiting locomotor behavior is compatible with the currently held view that the nigrostriatal DA neurons are associated with motor function (see Ungerstedt et al., 1977). A similar increase in endogenous levels of DOPAC and homovanillic acid has previously been reported in rats induced to walk on a rotarod (Speciale et al., 1986). The fact that in our study DA metabolism in the striatum was increased during the period of induced locomotion and rapidly returned to control values after its cessation strongly supports a direct relationship between locomotion and elevated striatal

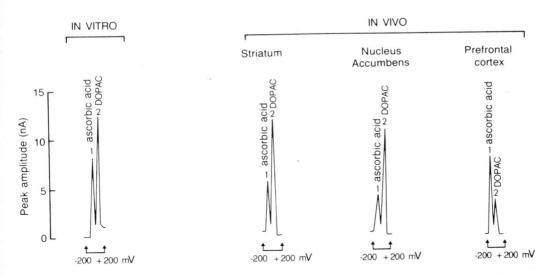

Fig. 1. Typical voltammograms recorded from the striatum, nucleus accumbens and anteromedial prefrontal cortex of unrestrained awake rats. Voltammetric measurements were performed on unrestrained Sprague–Dawley rats using a classical 3-electrode system with working carbon fiber electrode (diameter 8 μm, length 500 μm), reference and auxiliary electrodes and a PRG5 polarograph (Tacussel, France) as described previously (D'Angio et al., 1987, 1988). Working electrodes were electrochemically pretreated before use according to Gonon et al. (1980). The carbon fiber electrode was positioned in the brain areas with the help of an implantation assembly which was implanted stereotaxically 7 days prior to the voltammetric recordings (Louilot et al., 1987; D'Angio et al., 1987, 1988). The following recording parameters were used: ramp potential from −200 mV to +200 mV; scan rate 10 mV/sec; pulse modulation of a square wave form (pulse amplitude 50 mV, pulse period 0.2 sec, pulse duration 48 msec). Before and after each experiment, the response of the carbon fiber electrode was calibrated in a 20 μM DOPAC/200 μM ascorbic acid solution in 0.1 M phosphate buffer, pH 7.4. Electrochemical signals were quantified automatically by measuring the height of the oxidation peak recorded at +100 mV (DOPAC) using an SP 4100 computing integrator (Spectra Physics).

DA metabolism.

There was also an elevation of DA metabolism in the nucleus accumbens during forced locomotion, although less marked than in the striatum as has been previously reported (Speciale et al., 1986). These data are coherent with numerous studies which indicate a locomotor role for those mesolimbic DA neurons projecting to the nucleus accumbens. For instance, the infusion of DA into the nucleus accumbens induces locomotion (Costall et al., 1984); the locomotor stimulant effects of low doses of D-amphetamine or of substance P infusion into the ventral tegmental area are blocked by 6-hydroxydopamine lesions of the nucleus accumbens (Kelly et al., 1975) or by DA receptor blockade in the nucleus accumbens (Kelley et al., 1979).

During locomotion, there was no significant effect on DA metabolism in the frontal cortex. This is in agreement with the results of O'Neill and Fillenz (1985) who reported a good correlation between spontaneous motor activity and extracellular homovanillic acid levels in nucleus accumbens and striatum but not in frontal cortex in the rat.

In contrast to motor behavior, stressful conditions, e.g. tail-pinch, short-lasting immobilization or injection of β-CCM generally failed to affect DA metabolism in the striatum. These results agree

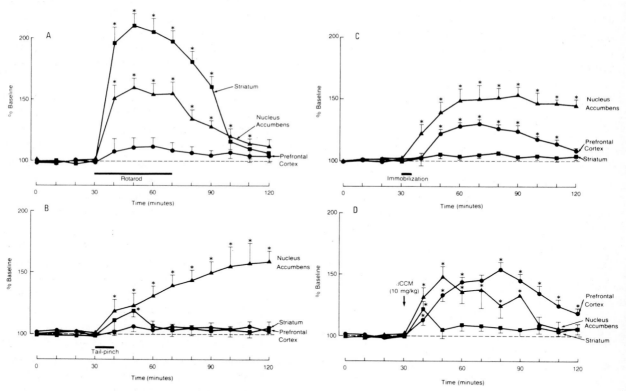

Fig. 2. Effects of forced locomotion, tail-pinch, immobilization and β-CCM on extracellular DOPAC levels in DA-rich brain areas of Sprague–Dawley rats. Stereotaxic coordinates (Paxinos and Watson, 1982): striatum: 1.7 mm anterior to bregma, 2 mm lateral to the midline, 4.5 mm below the cortex surface, nucleus accumbens 2.2, 1.5 and 6 mm, respectively; prefrontal cortex 2.7, 0.7 and 5 mm, respectively. Forced locomotion: rats were forced to walk on a rotating rod (8 rpm) for 40 min daily for 5 days and were recorded on the 5th day so as to eliminate the anxiogenic component (cf. D'Angio et al., 1988). Tail-pinch: a mild tail pressure was administered with sponge-padded forceps for 10 min. Immobilization: animals were restrained for 4 min in the hands of the experimenter. β-CCM: rats received a subcutaneous administration of β-CCM (10 mg/kg). Results are expressed as percentage of prestimulus period and are means with S.E.M. of data obtained on 5 rats per group. *$p < 0.05$ vs. control period.

with previous studies indicating that mild stressors that do not evoke changes in locomotor behavior do not affect nigrostriatal DA neurons (Speciale et al., 1986). In the stressful situations tested, the mesoaccumbens and mesocortical dopaminergic neurons were affected differentially. Thus, injection of β-CCM increased DA metabolism to a similar extent in prefrontal cortex and nucleus accumbens; immobilization increased DA metabolism in the nucleus accumbens and to a lesser extent in the prefrontal cortex; and tail-pinch caused a selective activation of DA metabolism in the nucleus accumbens. This suggests that specific sets of ascending mesencephalic DA systems are activated by aversive stimuli depending on the nature (possibly sensory modality) of the applied stimulus. Supportive evidence fot this view can also be found in previous studies which showed a selective activation of the mesocortical dopaminergic system during certain kinds of stresses e.g. "conditioned fear" or exposure to a novel environment (Tassin et al., 1980; Herman et al., 1982). In contrast, immobilization of the rats caused a selective activation of DA metabolism in the nucleus accumbens (Watanabe, 1984). The amplitude, duration and physical consequences of the stress could also determine the preferential activation of either of these dopaminergic systems. Thus, Speciale et al. (1986) demonstrated that low, but not high, intensity footshocks increase DA metabolism in the nucleus accumbens whereas DA metabolism in the prefrontal cortex was increased under both conditions.

The differential activation of mesocortical and/or mesoaccumbens DA systems in response to stress could be accounted for by the activation of different populations of dopaminergic cells in the ventral tegmental area in response to differing stimuli. Electrophysiological studies (Deniau et al., 1980) indicate that those mesocortical dopaminergic neurons originating in the ventral tegmental area are distinct from those innervating the nucleus accumbens. Moreover, these sets of dopaminergic neurons appear to be modulated by different afferent pathways (Lisoprawski et al., 1980; Hervé et al., 1981, 1982). Neuropeptides located in the ventral tegmental area may play an important role in the differential activation of mesolimbic and mesocortical dopaminergic neurons in response to stress. As specific neuropeptides apparently differentially activate the mesocortical and/or the mesolimbic dopaminergic neurons (Elliot et al., 1986; Deutch et al., 1987; Kalivas and Abhold, 1987), it would be conceivable that various environmental stimuli selectively influence the release of neuropeptides in the ventral tegmental area and thus lead to selective modifications of DA metabolism in the nucleus accumbens and/or the prefrontal cortex.

The tail-pinch- and immobilization-induced increases in extracellular DOPAC levels in the nucleus accumbens were of long duration (more than 2 h). This would suggest that even short-lasting stresses (10 min and 4 min for tail-pinch and immobilization, respectively) are able to induce a sustained activation of DA neurons projecting to the nucleus accumbens. A similar long-lasting augmentation of extracellular DOPAC levels has been observed in the nucleus accumbens after a few minutes of social interaction with an aggressive intruder (Louilot et al., 1986). However, caution should be exercised when interpreting data obtained by measurement of DA metabolites since previous in vivo dialysis studies have revealed that enhanced extracellular DOPAC levels do not necessarily reflect enhanced DA release but may rather index increased DA synthesis (Imperato and DiChiara, 1985; Zetterström et al., 1986). The persistence of the elevation of extracellular DOPAC concentrations long after stress was initiated may thus reflect an increase in tyrosine hydroxylase activity and a subsequent breakdown and spillover of intraneuronal DA rather than sustained DA release.

Response of mesocortical dopaminergic neurons to stressful environmental stimuli in strains of rats with differing levels of emotionallity

As discussed in the previous section, a variety of

anxiogenic environmental or pharmacological stimuli have been shown to increase DA metabolism in the mesocortical dopaminergic system of the rat. However, the significance of the stress-induced activation of the mesocortical dopaminergic neurons is as yet unclear. By analogy with other behavioral and hormonal changes induced by stress (e.g. increases in plasma corticosterone, in heart rate or in blood pressure), the enhancement of cortical DA metabolism could be viewed as a biochemical manifestation of the emotional reaction provoked by the stressor, the extent of which depends on the degree of physical stress. However, since the stress-related behavioral and biochemical changes may be dependent on the possibility of the animal gaining control over the stressor and not on the aversive or noxious nature of the stressor per se (Vogel, 1985) the increase in cortical DA metabolism may also reflect a heightened attention of the animal in an attempt to cope with the stressor. To evaluate the relationship between the emotional status and the response of mesocortical dopaminergic neurons to stress, we have investigated the effects of different stressful environmental situations on extracellular DOPAC levels (as assessed by in vivo voltammetry) in the prefrontal cortex of 2 genetically distinct lines of rats, Roman high- and low-avoidance, which differ markedly in their level of emotionality. The Swiss lines of Roman high-avoidance (RHA/Verh) and Roman low-avoidance (RLA/Verh) rats are selected and bred for the rapid acquisition vs. non-acquisition of a two-way active avoidance behavior. RLA/Verh rats are considered as being more "emotional" or anxious than RHA/Verh rats. Numerous behavioral and physiological studies (for a review see Driscoll and Bättig, 1982) have indicated that this difference in avoidance acquisition is not due to "learning ability" but due, primarily, to emotional factors. RLA/Verh rats, commonly "freeze" in the shuttle box situation (rather than avoid or escape the shock) and as opposed to RHA/Verh, react to repeated handling and injections (as indicated by frequent defecation, urination, "tensing up" and squealing), show greater plasma corticosterone, ACTH and prolactin responses after, and higher defecation scores and lower activity levels during application of various novel environment situations (Gentsch et al., 1982; Driscoll and Battig, 1982).

In this study, various stressful conditions were applied to both lines of rats. These included exposure to an unfamiliar environment, loud noise and immobilization. When chronically implanted freely moving RHA/Verh and RLA/Verh rats were placed in an unfamiliar environment (Y-maze) for 30 min, a rapid increase in the amplitude of the cortical DOPAC oxidation peak was observed in the former but not in the latter rats (Fig. 3)

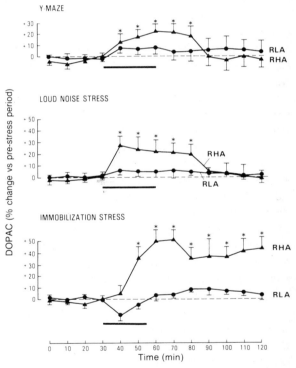

Fig. 3. Effects of various stressful conditions on extracellular DOPAC levels in the anteromedial prefrontal cortex of RHA/Verh and RLA/Verh rats. Y-Maze (unfamiliar environment): rats were placed in a Y-maze and allowed to explore for 30 min before being returned to their home cage. Loud noise stress: rats were subjected to a high-intensity noise for 30 min in their home cage. Immobilization: rats were placed in a restraint box for 20 min and then returned to their home cages. Results are means with S.E.M. of data obtained on 5 rats per group. *$p < 0.05$ vs. pre-stress period.

(D'Angio et al., 1988). Similarly, application of a high-intensity loud noise for 30 min caused an elevation of the amplitude of the cortical DOPAC oxidation peak in RHA/Verh rats but not in RLA/Verh animals (Fig. 3). In both of these experimental situations, cortical DOPAC levels in RHA/Verh rats returned to baseline within 30 min after cessation of the stimulus. Finally, immobilization of the animals for 20 min induced a progressive and marked increase in extracellular DOPAC levels in the frontal cortex of RHA/Verh but not of RLA/Verh rats (Fig. 3) (D'Angio et al., 1988). The DOPAC peak remained elevated for at least 50 min after cessation of the stress in the RHA/Verh rats.

In all stress conditions, RLA/Verh but not RHA/Verh rats showed clear signs of emotional reaction, e.g. defecation, freezing, self-grooming especially when intense stresses such as tail-pinch were applied. Moreover, all the stressful situations tested caused an elevation of heart rate, the amplitude and duration of which were much greater in RLA/Verh than in RHA/Verh rats (D'Angio et al., 1988).

The above results show that a variety of environmental stressors fail to alter extracellular cortical DOPAC levels in a line of rats (RLA/Verh) demonstrating measurable signs of intense anxiety and/or fear but augmented cortical DA metabolism in a line of rats (RHA/Verh) showing a very low emotional response. We have observed recently that placement in a shuttle box and electric footshocks similarly cause an increase in endogenous levels of DOPAC (measured postmortem) in the prefrontal cortex of RHA/Verh but not RLA/Verh rats (Driscoll et al., 1987). These data therefore strongly suggest that the stress-induced increase in cortical DOPAC is unlikely to be a direct reflection of the emotional status of the animals.

Inasmuch as RLA/Verh rats are more emotional than RHA/Verh rats, the lack of stress-induced increase in cortical DA metabolism in this line may have been due to the development of a habituation to stress in these animals (the emotionality of these animals is such that they could be considered as being exposed to stressful conditions even in a normal environment). However, this is highly unlikely as in all stressful situations tested so far, RLA/Verh rats showed clear signs of emotional response (e.g. defecation, freezing, self-grooming, increase in heart rate). Moreover, in contrast to the stressful situations investigated in the present study, social interaction with an intruder (non-aggressive) male Wistar rat for 30 min has been found to elicit an increase in extracellular DOPAC levels in the prefrontal cortex of RLA/Verh rats (M. D'Angio and B. Scatton, unpublished results), which shows that the mesocortical dopaminergic system in these animals can be activated under certain behavioral circumstances. This latter finding and the similar basal levels of extracellular DOPAC (D'Angio et al., 1988) found in both rat lines also exclude the possibility that the differential changes in cortical DA metabolism observed in RLA/Verh and RHA/Verh rats are linked to a difference in the nature of the regulatory mechanisms controlling mesocortical dopaminergic neurons in these two lines.

Since a number of experiments have shown that biochemical, physiological and/or pathological changes in an organism in response to stress do not seem to be caused by the aversive or anxious nature of the stressor but by the ability or inability of the organism to deal with the stressor (see Vogel, 1985), it is tempting to suggest that the increased cortical DA metabolism induced by stress reflects heightened attention of the animal or activation of cognitive processes in an attempt to cope with the stressor. Since in the present study, rats were not permitted to escape from the stress situation, the increased cortical DA metabolism associated with stress may not be related directly to successful control over the stressful event but rather with the preparative cortical processes (avoidance strategy) which lead to an effective control over the stressful situation. That stress-induced increase in cortical DA metabolism is linked to coping with the stressor is strengthened by our recent demonstration that prefrontal cortex endogenous DOPAC levels are increased in RHA/Verh rats exposed to inescapable footshocks, but remain normal in those RHA/Verh rats which were allowed to avoid (about 80–85% of the time) or to terminate (by escaping the other 15–20% of the time) the shocks (and are thus able to cope with the stressor) (Table I) (Driscoll et al., 1989). This hypothesis would also be consistent with the role played by the mesocortical dopaminergic pathway in attentional and cognitive processes (see Introduction) and with some of the functions that have been suggested for the prefrontal cortex. Thus, the prefrontal cortex in the rat serves as the rat's frontal eye field involved in head and eye orientation and, therefore, in attention (Crowne and Pathria, 1982; Neafsey et al., 1986); it is also involved in higher-order functioning, temporal structuring of information (Doar et al., 1987; Kesner and Holbrook, 1987; Groenewegen, 1988), searching for an adequate strategy (Mogensen and Jorgensen, 1987) and provides an excitatory influence on motor function (Morency et al., 1987).

Effects of food presentation or appetitive stimuli on DA metabolism in the prefrontal cortex of food-deprived rats

Food intake is a rather complex process that involves, inter alia, motivated and reward-related behaviors. Inasmuch as the prefrontal cortex dopaminergic system is implicated in cognitive and attentional processes, it is possible that the strong emotional and motivational processes involved in the initiation of eating influence dopaminergic transmission in this system. In order to evaluate this possibility, we have investigated the effects of feeding elicited by a 24 h food deprivation on DA metabolism (as assessed by measuring extracellular DOPAC levels) in the anteromedial prefrontal cortex of Sprague–Dawley rats at various periods following food presentation. Feeding behavior includes preparatory, consummatory and post-ingestive phases. To investigate whether the changes in cortical DA metabolism associated with food intake are linked to the animal's preparation for feeding, cortical DA metabolism was also measured following exposure of the animals to food odors (an appetitive stimulus anticipating the presence of food) but with no opportunity to feed.

In Sprague–Dawley rats that had been food-deprived for a 24 h period, there was a time-lag of about 5 min between food presentation and the onset of meal. Thereafter, animals exhibited consummatory behavior for the entire period of voltammetric recordings, with a decreasing intensity of feeding during the last 30 min. Concurrent to feeding, there was a progressive and marked increase in cortical DOPAC peak height (Fig. 4A) (D'Angio et al., 1989). An enhancement, of a similar magnitude, of the amplitude of the cortical DOPAC oxidation peak was also observed in food-deprived rats exposed to food odors but not allowed to feed (Fig. 4B) (D'Angio et al., 1989).

The above data demonstrate that feeding elicited

TABLE I

Effects of controllable and inescapable electric footshocks on endogenous DOPAC levels in the anteromedial prefrontal cortex of RHA/Verh rats

Condition	DOPAC	
	ng/g	% change
Cage control	62 ± 6	
Shuttle box control	76 ± 8	+22
Inescapable stress	104 ± 12*	+68
Controllable stress	81 ± 14	+30

Groups of RHA/Verh rats were subjected either to normal housing conditions (cage control), or to a 30 min exploratory session in the shuttle box (shuttle box control), to 40 inescapable, scrambled shocks during 30 min while confined to one side of the shuttle box or to 40 avoidable (or escapable, when need be) shocks under two-way avoidance conditions during 30 min. The latter group of rats was subjected to a previous acquisition session, 1 week before the experiment, so that the rats showed a high level of avoidance. Rats were sacrificed immediately following the 30 min session, brains were removed and endogenous DOPAC levels were measured by high-performance liquid chromatopraphy with electrochemical detection according to Semerdjian-Rouquier et al. (1981). Results are means with S.E.M. of data obtained on 14 rats per group. *$p < 0.05$ vs. cage controls.

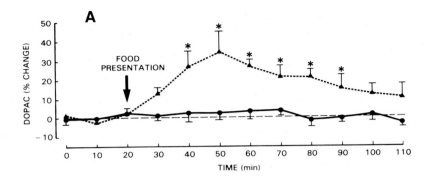

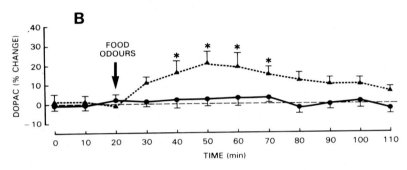

Fig. 4. Effects of food presentation or exposure to food odours on extracellular DOPAC levels in the anteromedial prefrontal cortex of food-deprived Sprague–Dawley rats. Results are means with S.E.M. of data obtained on 4 rats per group. Arrows indicate the onset of food presentation or food odor exposure. *$p < 0.05$ vs. respective controls.

by food deprivation is accompanied by a marked activation of DA metabolism in the rat prefrontal cortex. The time course of the changes in extracellular cortical DOPAC levels shows that the activation of cortical DA metabolism is long-lasting ($\simeq 90$ min), the maximal increase in this biochemical parameter being observed at 30 min following the onset of meal.

At variance with our results, Heffner et al. (1980) failed to observe any change in postmortem levels of DOPAC in the frontal cortex of food-deprived rats given 1 h of access to food though an increase in this biochemical parameter was seen in the nucleus accumbens, hypothalamus and amygdala. This discrepancy could be accounted for by the type of food used, the time of sampling and above all by the dissimilar methods used for measuring DA metabolism in the two studies. In the experiments of Heffner et al. (1980),

postmortem tissue levels of DOPAC were measured in the entire prefrontal cortex whereas in the present study voltammetric recordings were taken specifically from layer V of the anteromedial prefrontal cortex which receives the bulk of dopaminergic afferents originating in the ventral tegmental area. Therefore, the feeding-induced increase in DA metabolism in this cortical subregion may well have been missed in those experiments measuring DOPAC levels in the entire prefrontal cortex.

The increased cortical DA metabolism that accompanied feeding could be related to various components of feeding behavior. The fact that an increased cortical DA metabolism – almost similar in magnitude to that seen following feeding – was observed in rats exposed to food odors (but not allowed to eat) suggests that neither access to food nor the act of feeding per se or the rewarding properties of the food were associated with the increase in mesocortical DA neuron activity. Instead, this study suggests that the olfactory stimulation associated with food presentation, as an appetitive stimulus anticipating the presence of food by hungry rats, was sufficient for altering the activity of the mesocortical dopaminergic system. These results reinforce the currently held view that mesocortical dopaminergic neurons are implicated in cognitive processes.

References

Bannon, M.J. and Roth, R.H. (1983) Pharmacology of mesocortical dopamine neurons. *Pharmacol. Rev.,* 35: 53 – 68.

Bertolucci-D'Angio, M., Serrano, A. and Scatton, B. (1990) Differential effects of forced locomotion, tail-pinch, immobilization and methyl-β-carboline carboxylate on extracellular DOPAC levels in the rat striatum, nucleus accumbens and prefrontal cortex: an in vivo voltammetric study. *J. Neurochem.,* in press.

Brozoski, T.J., Brown, R.M., Rosvold, H.E. and Goldman, P.S. (1979) Cognitive deficit caused by regional depletion of dopamine in prefrontal cortex of rhesus monkey. *Science,* 205: 929 – 932.

Buda, M., Gonon, F., Cespuglio, R., Jouvet, M. and Pujol, J.F. (1981) In vivo electrochemical detection of catechols in several dopaminergic brain regions of anaesthetized rats. *Eur. J. Pharmacol.,* 73: 61 – 68.

Claustre, Y., Rivy, J.P., Dennis, T. and Scatton, B. (1986) Pharmacological studies on stress-induced increase in frontal cortical dopamine metabolism in the rat. *J. Pharmacol. Exp. Ther.,* 238: 693 – 700.

Costall, B., Domeney, A.M. and Naylor, R.J. (1984) Locomotor hyperactivity caused by dopamine infusion into the nucleus accumbens of rat brain: specificity of action. *Psychopharmacology,* 82: 174 – 180.

Crowne, D.P. and Pathria, M.N. (1982) Some attentional effects of unilateral frontal lesions in the rat. *Behav. Brain Res.,* 6: 25 – 39.

D'Angio, M. and Scatton, B. (1989) Feeding or exposure to food odors increases extracellular DOPAC levels (as measured by in vivo voltammetry) in the prefrontal cortex of food-deprived rats. *Neurosci. Lett.,* 96: 223 – 228.

D'Angio, M., Serrano, A., Rivy, J.P. and Scatton, B. (1987) Tail-pinch stress increases extracellular DOPAC levels (as measured by in vivo voltammetry) in the rat nucleus accumbens but not frontal cortex: antagonism by diazepam and zolpidem. *Brain Res.,* 409: 169 – 174.

D'Angio, M., Serrano, A., Driscoll, P. and Scatton, B. (1988) Stressful environmental stimuli increase extracellular DOPAC levels in the prefrontal cortex of hypoemotional (Roman high-avoidance) but not hyperemotional (Roman low-avoidance) rats. An in vivo voltammetric study. *Brain Res.,* 451: 237 – 247.

Deniau, J.M., Thierry, A.M. and Feger, J. (1980) Electrophysiological identification of mesencephalic ventromedial tegmental (VMT) neurons projecting to the frontal cortex, septum and nucleus accumbens. *Brain Res.,* 189: 315 – 326.

Deutch, A.Y., Tam, S.Y. and Roth, R.H. (1985) Footshock and conditioned stress increase 3,4-dihydroxyphenylacetic acid (DOPAC) in the ventral tegmental area but not substantia nigra. *Brain Res.,* 333: 143 – 146.

Deutch, A.Y., Bean, A.J., Bissette, G., Nemeroff, C.B., Robbins, R.J. and Roth, R.H. (1987) Stress-induced alterations in neurotensin, somatostatin and corticotropin-releasing factor in mesotelencephalic dopamine system regions. *Brain Res.,* 417: 350 – 354.

Doar, B., Finger, S., Almli, C.R. (1987) Tactile-visual acquisition and reversal learning deficits in rats with prefrontal cortical lesions. *Exp. Brain Res.,* 66: 432 – 434.

Driscoll, P. and Battig, K. (1982) Behavioral, emotional and neurochemical profiles of rats selected for extreme differences in active, two-way avoidance performance. In I. Lieblich (Ed.), *Genetic of the Brain,* Elsevier, Amsterdam, pp. 95 – 123.

Driscoll, P., Claustre, Y., Fage, D. and Scatton, B. (1987) Recent findings in central dopaminergic and cholinergic neurotransmission of Roman high and low avoidance (RHA/Verh and RLA/Verh) rats. *Behav. Brain Res.,* 26: 213.

Driscoll, P., Dedek, J., D'Angio, M., Claustre, Y. and Scatton, B. (1990) A genetically based model for divergent stress responses: behavioral, neurochemical and hormonal aspects. In V. Pliska and G. Stranzinger (Eds.), *Farm Animals in Biomedical Research, Advances in Animal Breeding and Genetics,* P. Parey Verlag, Hamburg, pp. 97 – 107.

Dunn, A.J. and File, S.E. (1983) Cold restraint alters dopamine metabolism in frontal cortex, nucleus accumbens and neostriatum. *Physiol. Behav.,* 31: 511 – 513.

Elliott, P.J., Alpert, J.E., Bannon, M.J. and Iversen, S.D. (1986) Selective activation of mesolimbic and mesocortical dopamine metabolism in rat brain by infusion of a stable substance P analogue into the ventral tegmental area. *Brain Res.,* 363: 145 – 147.

Fadda, F., Argiolas, A., Melis, M.R., Tissari, A.H., Onali, P.L. and Gessa, G.L. (1978) Stress-induced increase in 3,4-dihydroxyphenylacetic acid (DOPAC) levels in the cerebral cortex and in nucleus accumbens: reversal by diazepam. *Life Sci.,* 23: 2219 – 2224.

Gentsch, C., Lichtsteiner, M., Driscoll, P. and Feer, H. (1982) Differential hormonal and physiological responses to stress in Roman high- and low-avoidance rats. *Physiol. Behav.,* 28:

259–263.

Gonon, F., Buda, M., Cespuglio, R., Jouvet, M. and Pujol, J.F. (1980) In vivo electrochemical detection of catechols in the neostriatum of anaesthetized rats: dopamine or DOPAC? *Nature*, 286: 902–904.

Gonon, F., Buda, M., Cespuglio, R., Jouvet, M. and Pujol, J.F. (1981) Voltammetry in the striatum of chronic freely moving rats: detection of catechols and ascorbic acid. *Brain Res.*, 223: 69–80.

Groenewegen, H.J. (1988) Organization of the afferent connections of the mediodorsal thalamic nucleus in the rat, related to the mediodorsal-prefrontal topography. *Neuroscience*, 24: 379–431.

Heffner, T.G., Hartman, J.A. and Seiden, L.S. (1980) Feeding increases dopamine metabolism in the rat brain. *Science*, 208: 1168–1170.

Herman, J.P., Guillonneau, D., Dantzer, R., Scatton, B., Semerdjian-Rouquier, L. and Le Moal, M. (1982) Differential effects of inescapable footshocks and of stimuli previously paired with inescapable footshocks on dopamine turnover in cortical and limbic areas of the rat. *Life Sci.*, 30: 2207–2214.

Hervé, D., Simon, H., Blanc, G., Le Moal, M., Glowinski, J. and Tassin, J.P. (1981) Opposite changes in dopamine utilization in the nucleus accumbens and the frontal cortex after electrolytic lesion of the median raphe in the rat. *Brain Res.*, 216: 422–428.

Hervé, D., Blanc, G., Glowinski, J. and Tassin, J.P. (1982) Reduction of dopamine utilization in the prefrontal cortex but not in the nucleus accumbens after selective destruction of noradrenergic fibers innervating the ventral tegmental area. *Brain Res.*, 237: 510–516.

Ida, Y. and Roth, R. (1987) The activation of mesoprefrontal dopamine neurons by FG 7142 is absent in rats treated chronically with diazepam. *Eur. J. Pharmacol.*, 137: 185–190.

Ikeda, M. and Nagatsu, T. (1985) Effect of a short-term swimming stress and diazepam on 3,4-dihydroxyphenylacetic acid (DOPAC) and 5-hydroxyindoleacetic acid (5-HIAA) levels in the caudate nucleus: an in vivo voltammetric study. *Naunyn-Schmiedeberg's Arch. Pharmacol.*, 331: 23–26.

Imperato, A. and DiChiara, G. (1985) Dopamine release and metabolism in awake rats after systemic neuroleptics as studied by trans-striatal dialysis. *J. Neurosci.*, 5: 297–306.

Kalivas, P.W. and Abhold, R. (1987) Enkephalin release into the ventral tegmental area in response to stress: modulation of mesocorticolimbic dopamine. *Brain Res.*, 414: 339–348.

Keller, R.W., Stricker, E.M. and Zigmond, M.J. (1983) Environmental stimuli but not homeostatic challenges produce apparent increases in dopaminergic activity in the striatum: an analysis by in vivo voltammetry. *Brain Res.*, 279: 159–170.

Kelly, A.E., Stinus, L. and Iversen, S.D. (1979) Behavioural activation induced in the rat by substance P infusion into ventral tegmental area: implication of dopaminergic A10 neurones. *Neurosci. Lett.*, 11: 335–339.

Kelly, P.H., Seviour, P.W. and Iversen, S.D. (1975) Amphetamine and apomorphine responses in the rat following 6-OHDA lesions of the nucleus accumbens septi and corpus striatum. *Brain Res.*, 94: 507–522.

Kesner, R.P. and Holbrook, T. (1987) Dissociation of item and order spatial memory in rats following medial prefrontal cortex lesions. *Neuropsychologia*, 25: 653–664.

Lavielle, S., Tassin, J.P., Thierry, A.M., Blanc, G., Herve, D., Barthélémy, C. and Glowinski, J. (1978) Blockade by benzodiazepines of the selective high increase in dopamine turnover induced by stress in mesocortical dopaminergic neurons of the rat. *Brain Res.*, 168: 585–594.

Le Moal, M., Cardo, B. and Stinus, L. (1969) Influence of ventral mesencephallic lesion on various spontaneous and conditioned behaviors in the rat. *Physiol. Behav.*, 4: 567–573.

Le Moal, M., Galey, D. and Cardo, B. (1975) Behavioral effects of local injection of 6-hydroxydopamine in the medial ventral tegmentum in the rat. Possible role of the mesolimbic dopaminergic system. *Brain Res.*, 88: 190–194.

Lisoprawski, A., Hervé, D., Blanc, G., Glowinski, J. and Tassin, J.P. (1980) Selective activation of the mesocorticofrontal dopaminergic neurons induced by lesion of the habenula in the rat. *Brain Res.*, 183: 229–234.

Louilot, A., Le Moal, M. and Simon, H. (1986) Differential reactivity of dopaminergic neurons in the nucleus accumbens in response to different behavioral situations. An in vivo voltammetric study in free moving rats. *Brain Res.*, 397: 395–400.

Louilot, A., Serrano, A. and D'Angio, M. (1987) A novel carbon fiber implantation assembly for cerebral voltammetric measurements in freely moving rats. *Physiol. Behav.*, 41: 227–231.

Mogensen, J. and Jorgensen, O.S. (1987) Protein changes in the rat's prefrontal and "inferotemporal" cortex after exposure to visual problems. *Pharmacol. Biochem. Behav.*, 26: 89–94.

Morency, M.A., Stewart, R.J. and Beninger, R.J. (1987) Circling behavior following unilateral microinjections of cocaine into the medial prefrontal cortex: dopaminergic or local anesthetic effect? *J. Neurosci.*, 7: 812–818.

Neafsey, E.J., Hurley-Gius, K.M. and Arvanitis, D. (1986) The topographical organization of neurons in the rat medial frontal, insular and olfactory cortex projecting to the solitary nucleus, olfactory bulb, periaqueductal gray and superior colliculus. *Brain Res.*, 377: 261–270.

O'Neill, R.D. and Fillenz, M. (1985) Simultaneous monitoring of dopamine release in rat frontal cortex, nucleus accumbens and striatum: effect of drugs, circadian changes and correlations with motor activity. *Neuroscience*, 16: 49–55.

Paxinos, G. and Watson, C. (1982) *The Rat Brain in Stereotaxic Coordinates*, Academic Press, New York.

Reinhard, J.F., Bannon, M.J. and Roth, R.H. (1982) Acceleration by stress of dopamine synthesis and metabolism in prefrontal cortex: antagonism by diazepam. *Naunyn-Schmiedeberg's Arch. Pharmacol.*, 318: 374–377.

Scatton, B., D'Angio, M., Driscoll, P. and Serrano, A. (1988) An in vivo voltammetric study of the response of mesocortical and mesoaccumbens dopaminergic neurons to environmental stimuli in strains of rats with differing levels of emotionality. *Ann. N.Y. Acad. Sci.*, 537: 124–137.

Semerdjian-Rouquier, L., Bossi, L. and Scatton, B. (1981) Determination of 5-hydroxytryptophan, serotonin and 5-hydroxyindoleacetic acid in rat and human brain and biological fluids by reversed-phase high-performance liquid chromatography with electrochemical detection. *J. Chromatogr.*, 218: 663–670.

Serrano, A., D'Angio, M. and Scatton, B. (1986) In vivo voltammetric measurement of extracellular DOPAC levels in the anteromedial prefrontal cortex of the rat. *Brain Res.*, 378: 191–196.

Simon, H. (1981) Neurones dopaminergiques A10 et système frontal. *J. Physiol.*, 77: 81–95.

Simon, H., Scatton, B. and Le Moal, M. (1979) Definitive disruption of spatial delayed alternation in rats after lesions in the ventral mesencephalic tegmentum. *Neurosci. Lett.*, 15: 319–324.

Simon, H., Scatton, B. and Le Moal, M. (1980) Dopaminergic A10 neurones are involved in cognitive functions. *Nature*, 286: 150–151.

Speciale, S.G., Miller, J.D., McMillen B.A. and German, D.C. (1986) Activation of specific central dopamine pathways: locomotion and footshock. *Brain Res. Bull.*, 16: 33–38.

Tam, S.Y. and Roth, R.H. (1985) Selective increase in dopamine metabolism in the prefrontal cortex by the anxiogenic beta-carboline FG 7142. *Biochem. Pharmacol.*, 34: 1595–1598.

Tassin, J.P., Hervé, D., Blanc, G. and Glowinski, J. (1980) Differential effects of a two-minute open field on dopamine utilization in the frontal cortices of BALB/C and C57 BL/6 mice. *Neurosci. Lett.*, 17: 67–71.

Thierry, A.M., Tassin, J.P., Blanc, G. and Glowinski, J. (1976) Selective activation of the mesocortical DA system by stress. *Nature*, 263: 242–244.

Vogel, W.H. (1985) Coping, stress, stressors and health consequences. *Neuropsychobiology*, 13: 129–135.

Ungerstedt, U., Ljungberg, T. and Ranje, C. (1977) Dopamine neurotransmission and the control of behaviour. In A.R. Cools, A.H.M. Lohman and J.H.L. Van den Bercken (Eds.), *Psychobiology of the Striatum*, Elsevier, Amsterdam, pp. 85–97.

Watanabe, H. (1984) Activation of dopamine synthesis in mesolimbic dopamine neurons by immobilization stress in the rat. *Neuropharmacology*, 23: 1335–1338.

Zetterström, T., Sharp, T. and Ungerstedt, U. (1986) Effect of neuroleptic drugs on striatal dopamine release and metabolism in the awake rat studied by intracerebral dialysis. *Eur. J. Pharmacol.*, 106: 27–37.

Discussion

P. Goldman-Rakic: Have you compared the response of the RHA and RHA strains to food deprivation or food-odor exposure?

M. d'Angio: No, not so far, but that would be of interest and some differences may be expected.

T. Paus: What is the basis of the better ability to cope with stressors? Might we explain the differences observed to be due not to cognitive (attentional) but to motivational (drive) processes?

M. d'Angio: The possibility that the differences observed between RLA and RHA lines are due to motivational processes cannot be excluded. Further experiments are clearly needed in order to investigate in detail the psychological or behavioral processes involved in the coping response of RHA rats.

J. Scheel-Kruger: A comment — In your very interesting lecture you showed increased DOPAC in PFC in RHA versus RLA rats and your interpretation suggested that dopamine in PFC may be involved in coping behavior of the rats. A similar role has been suggested by T. Robbins and others (Cole and Robbins, 1987) with regard to noradrenaline. Future experiments should deal with the ratio or role of NA versus DA in biochemistry to elucidate the role of motivation, attention and coping behavior. NA in accumbens may switch behavior to dopamine dependent behavior (coping) in striatum (Cools et al., 1990).

A.Y. Deutch: In conditional fear experiments (pairing mild foot shock stress with presentation of a neutral auditory stimulus and exposing the animal for 20 min only to the neutral tone on the test day) we see a significant increase in prefrontal cortical DA metabolism in animals examined after 1 day of pairing (Deutch et al., 1985). In contrast, after presentation of paired stimuli for 4 days, and then exposing the animals to the auditory stimulus no augmentation of PFC DA metabolism was observed. The behavior of the animals exposed to the auditory stimulus on day 5 (after 4 days of CS-UCS pairing) was quite different from animals observed after only 1 pairing session, in that the animals tightly grasped the bars of the shock chamber floor and stretched recumbent upon the bars. In contrast, animals exposed to acute footshock or observed upon CS exposure following only 1 day of CS-UCS pairing explored the cage, rearing on hindlimbs and sniffing. These observations may indicate that the animals had acquired a coping strategy (recumbent position and grasping of the bars through which shock is delivered) aimed at reducing high current density stimulation: in cases when a portion of an animal's body is in close apposition to (but not touching) the bars across which shock is delivered, high density current arcs can result which are quite noxious. The lack of augmentation of PFC DA metabolism under such conditions can be interpreted to be at odds with a role of PFC DA in coping behavior, as suggested by your data.

M. d'Angio: It is possible that DA in the PFC is involved in the acquisition, but not retention or performance of a coping strategy; this hypothesis can be readily tested.

A.Y. Deutch: A second possibility to reconcile our observations and those of you could be based on the temporal resolution afforded by in vivo voltammetry. Our observations were based on ex vivo measurments of DA metabolism in the PFC; the PFC was dissected from animals sacrificed after 20 min of exposure to the CS on day 5. In vivo voltammetry would allow one to determine if a transient increase in DA release occurred, or assess if an increase in DA release was present that was not temporally locked to the CS presentation, i.e., occurred after the discontinuation of the CS. The former possibility would be consistent with DA release in the PFC being part of a coping response.

A.M. Thierry: Is there some difference in the control level of the mesocortical DA activity in RHA and RLA rats.

M. d'Angio: DOPAC control levels were similar in the two strains.

E.J. Neafsey: A comment — The rats may be emitting ultrasonic vocalizations during these different stressors.

References

Cole, B.J. and Robbins, T.W. (1987) Amphetamine impairs the discriminative performance of rats with dorsal noradrenergic bundle lesions on a 5-choice serial reaction time test: New evidence for central dopaminergic-noradrenergic interactions. *Psychopharmacology,* 91, 458–466.

Cools, A.R., Van den Bos, R., Ploeger, G. and Ellenbroek, B.A. (1990) Gating function of noradrenaline in the ventral striatum: its role in behavioural responses to environmental and pharmacological challenges. In P. Willner and J. Scheel-Krüger (Eds.), *The Mesolimbic Dopamine System: From Motivation to Action,* John Wiley, Chichester, in press.

Deutch, A.Y., Tam, S.Y. and Roth, R.H. (1985) Footshock and conditioned stress increase 3,4-dihydroxyphenylacetic acid (DOPAC) in the ventral tegmental area but not substantia nigra. *Brain Res.,* 333: 143–146.

CHAPTER 21

The neurobiological basis of prefrontal cortex self-stimulation: A review and an integrative hypothesis

F. Mora and M. Cobo

Department of Physiology, Faculty of Medicine, University Complutense of Madrid, 28040 Madrid, Spain

Introduction

There can be little doubt that, despite considerable research on the neural basis of rewarding brain stimulation (Olds and Fobes, 1981), an intriguing and apparently simple question remains largely unanswered. That is, when an electrical stimulus is applied at the tip of an electrode that supports self-stimulation, what is activated in that restricted area of the brain to initiate the behavior in the animal? Depending on the site of self-stimulation (and only 3 structures remain within the focus of research in this field, viz. lateral hypothalamus-medial forebrain bundle, ventral tegmental area and prefrontal cortex), fragmentary answer to that central question have been given through investigations done principally during the present decade (see Stellar and Stellar, 1985).

The prefrontal cortex has been an area of intensive research in our laboratory for the past several years. This area is of special interest for studies on self-stimulation not only because it is involved in cognitive, mnemonic and emotional functions but also because, in contrast to other areas of the brain which also support self-stimulation, biochemical and immunohistological studies are revealing its neurochemical map (Mora, 1978; Kolb, 1984; Mora and Ferrer, 1986; Fuster, 1989). In a recent review, we have given an account of the research done up to 1986 on the neuroanatomical and neurochemical basis of self-stimulation of this area of the brain (Mora and Ferrer, 1986). In this chapter, we will summarize those findings and describe recent experiments regarding chemical neurotransmission along with the neural pathways presumably involved. Finally, and as a partial answer to the question posed above, we will elaborate on the role of intrinsic neurons and the interaction of neurotransmitters at the locus where self-stimulation is elicited, and also on the possible circuitry involving other areas of the brain. Implications for further research will also be discussed.

Neuroanatomy and neurophysiology of self-stimulation of the prefrontal cortex

In 1969, 2 areas of the frontal pole in the rat, that of the medial wall of the hemisphere anterior and dorsal to the genu of the corpus callosum and the dorsal bank of the rhinal sulcus were identified as the medial (MPC) and sulcal or orbital (SPC) prefrontal cortex (Leonard, 1969) (see Fig. 1). It was discovered in 1971 that not only these two

Correspondence: Dr. F. Mora, Department of Physiology, Faculty of Mecidine, University Complutense of Madrid, 28040 Madrid, Spain.

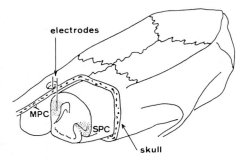

Fig. 1. Artistic view of a coronal section of the rat's brain showing the medial and sulcal prefrontal cortex in one of the hemispheres. The electrodes shown are positioned in the deep layers (V – VI) of the prelimbic area.

regions of the cortex support self-stimulation (SS) (Routtenberg, 1971), but that neurons in the medial prefrontal cortex are activated by self-stimulation at the medial forebrain bundle-lateral hypothalamus (MFB-LH) (Ito and Olds, 1971).

After these initial studies, a series of investigations using a multidisciplinary approach added further evidence for the involvement of the same two areas of the cerebral cortex in the neural substrates of rewarding brain stimulation. Thus, electrophysiological studies showed that neurons in both MPC and SPC were activated either directly or transsynaptically by SS of a number of different brain areas in the rat and in the monkey (Rolls, 1975; Mora et al., 1980a). Reciprocally, SS in the prefrontal cortex of the monkey activated neurons in different brain nuclei, i.e. the lateral hypothalamus, basolateral amygdala, caudate nucleus, nucleus dorsomedialis of the thalamus, nucleus accumbens septi, regions of substantia nigra-ventral tegmental area and locus coeruleus (Rolls et al., 1980; Mora and Ferrer, 1986).

The electrophysiological findings mentioned above have received support using histological techniques. Thus, 2 main studies, using the HRP and the [^{35}S]methionine tracing techniques for anterograde and orthograde neuronal labeling, respectively, showed that neurons in the MPC receive afferent terminals arising from neurons located in numerous thalamic, hypothalamic, mesencephalic and pontine areas (Vives et al., 1983). Reciprocally, neurons in the MPC send efferents to neurons in at least some of the same nuclei (Dalssas et al., 1981), among them the SPC, caudate-putamen, nucleus accumbens, lateral and posterior hypothalamus, basolateral amygdala, locus coeruleus, ventral tegmental area, raphe nuclei and nucleus dorsomedialis of the thalamus (Mora and Ferrer, 1986).

Theoretically, the studies mentioned above give rise to the general idea that at least 3 main neural components shoud be considered in order to understand the neural network responsible for eliciting SS in the MPC: (1) intrinsic neurons located in the MPC, (2) efferents from neurons in the MPC to other neurons located in different nuclei of the brain, (3) fibers arising from the latter neurons terminating in the MPC. Functionally, this type of neural arrangement would imply that the physical stimulation produced through the electrode in the MPC, not only produces localized neuronal activity (the nature of which will be discussed in a next section of this chapter), but also activates neurons in areas of the brain distant from the MPC. Activation of these distant neurons could, through a feedback mechanism, activate terminals in the MPC important for maintaining SS (evidence for this hypothesis will be reviewed below).

If intrinsic neurons in the MPC receiving afferents from and sending efferents to different regions in the brain were the main integrative element in the neuronal local substrate responsible for SS, destruction of those neurons should disrupt the behavior. Two recent studies showed that this is the case (Ferrer et al., 1985; Nassif et al., 1985). Microinjections of either kainic acid or ibotenic acid at the tip of the electrode supporting SS almost completely abolishes the behavior (see Fig. 2). These studies also reported the absence of cell bodies and the loss of normal cytoarchitecture at the tip of the electrode after microinjections of the neurotoxin, the space becoming invaded by glial elements.

Following the same reasoning, lesions of a

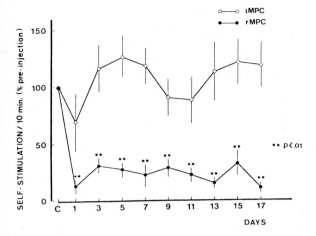

Fig. 2. Mean rate of self-stimulation along time as percentage of the pre-injection control. Mean absolute control rate ± SEM was 252 ± 22 for the right, and 329 ± 40 for the left MPC. Kainic acid was microinjected into the right medial prefrontal cortex (rMPC). The left medial prefrontal cortex (lMPC) was injected with the vehicle. (Taken from Ferrer et al., 1985.)

pathway at the level of the external capsule produced a reduction that lasted for the total duration of the experiment (Cobo et al., 1989a; see also Fig. 3). It is interesting that lesions of the fugal corticocortical prefrontal pathway, before entering the external capsule lead to recovery 22 days later (Corbett et al., 1982; Robertson et al., 1986). It is therefore possible that lesions of the external capsule, through which this corticocortical pathway funnels, could be compensated for after a sufficient recovery time.

Although the full significance of these findings will be discussed later, these results suggest the existence of at least 3 different categories of pathways activated during SS of the MPC (Vives et al., 1986): First, pathways whose activation seems necessary for SS behavior to be sustained (the mesocortical afferent dopaminergic pathway from the ventral tegmental area and a corticocortical fugal pathway through the external capsule) (A.G. Phillips and Fibiger, 1978; Vives et al., 1986; Cobo et al., 1989a); second, pathways whose activation somehow modulates SS (nucleus dorsomedialis of

nucleus that maintains reciprocal input – output connections with MPC (a simple feedback circuit, see Mora and Ferrer, 1986, for further details) should disrupt the behavior. However, as illustrated in Fig. 3, different effects are observed depending on the brain area that has been lesioned. For instance, lesions of the locus coeruleus, nucleus accumbens or SPC have no significant effect on SS (Corbett et al., 1982; Ramírez et al., 1983; Robertson et al., 1986; Cobo et al., 1989a). In contrast, lesions of the caudate-putamen (CP) produced a decrease in SS of up to 3 days after the lesions (Vives et al., 1986), while lesioning the mediodorsal nucleus of the thalamus (MD) or basolateral amygdala (BLA) had effects that lasted up to 6 days (Vives et al., 1986; Ferrer et al., 1987). Lesions of the ventral tegmental area (VTA) or its dopaminergic projections have effects that last as long as 15 – 21 days, but with a trend towards recovery (A.G. Phillips and Fibiger, 1978; Vives et al., 1986). In fact, Simon et al. (1979) reported eventual full recovery after lesions of VTA. Finally, lesions of the corticocortical prefrontal

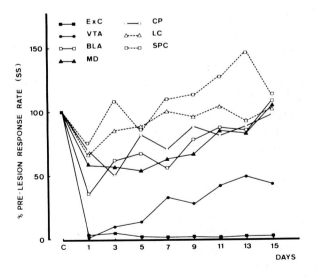

Fig. 3. Effects of electrolytic lesions of SPC (sulcal prefrontal cortex), LC (locus coeruleus), CP (caudate-putamen), MD (nucleus dorsomedialis of the thalamus), BLA (basolateral amygdala), VTA (ventral tegmental area) and ExC (external capsule). Self-stimulation rate is expressed as percentage of the pre-lesion level.

the thalamus, caudate-putamen, basolateral amygdala and others) (Vives et al., 1986; Ferrer et al., 1987); and third, those pathways whose activation by electrical stimulation does not seem to be functionally related to SS of the MPC (locus coeruleus, raphe nuclei, SPC, nucleus accumbens and others) (Corbett et al., 1982; Ramírez et al., 1983; Robertson et al., 1986; Cobo et al., 1989a).

Based on the results of these studies, in which at no single locus a lesion appears to produce a permanent abolition of self-stimulation of the MPC (Mora and Ferrer, 1986), we have proposed the existence of complex feedback circuits as the basis of SS of the MPC (Mora and Ferrer, 1986). That is, electrical self-stimulation of the MPC would involve not only feedback activation of several independent nuclei, but also a functional interaction of those nuclei with still other brain structures. For instance, as depicted in Fig. 4, the MPC, the basolateral nucleus of the amygdala, and the MD nucleus of the thalamus are reciprocally interconnected (Krettek and Price, 1977a, b; Divac et al.,

1978; Beckstead, 1979; Dalssas et al., 1981; Sarter and Markowitsch, 1983a; Vives et al., 1983), and these 3 structures support self-stimulation (Wurtz and Olds, 1963; Routtenberg, 1971; Clavier and Gerfen, 1982). Moreover, high metabolic activity in the amygdala and MD has also been detected by the 2-deoxyglucose method during SS of the MPC (Gallistel, 1983). This "limbic basolateral" circuit which has been proposed to be involved in the physiology of memory related processes (Livingstone and Escobar, 1971; Sarter and Markowitsch, 1983b), has been investigated through simultaneous lesions in our laboratory. We have suggested that this circuit could also play a role in SS of the MPC, and thus be involved in reward related processes (Ferrer et al., 1987).

As depicted in Fig. 4, and based on theoretical and experimental data, other circuits have also been implicated in the neural basis of SS of the MPC. Of those, the MPC-EC-VTA-MPC seem particularly important because these structures support SS and are reciprocally interconnected (Routtenberg, 1971; Crow, 1972; Collier et al., 1977; Beckstead, 1979; Dalssas et al., 1981; Vives et al., 1983). Further, high metabolic activity has been demonstrated to funnel through a corticocortical fugal pathway to the entorhinal cortex during self-stimulation of the MPC (Gallistel, 1983). Moreover, lesions of this corticocortical pathway at the level of the external capsule produce a decrease of SS. Because lesions of the entorhinal cortex reduce SS for only 5 days (Cobo et al., 1989a), however, it is possible that fibers through the external capsule also terminate in other cortical or subcortical areas relevant for SS of the MPC.

Finally, we propose that multiple circuits such as the ones depicted in Fig. 4 are functioning simultaneously during SS of the MPC. This idea has important functional implications for understanding why lesions of one nucleus or even simultaneous lesions of several nuclei do not permanently abolish SS of the MPC (Mora and Ferrer, 1986), since physiological and possibly even anatomical compensation (M.I. Phillips, 1976) could be achieved using the undamaged circuitry.

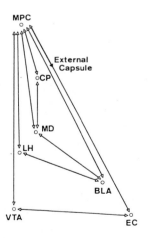

Fig. 4. Circuits presumably involved in SS of the MPC. The circuits depicted are MPC-EC-VTA-MPC; MPC-LH-BLA-MPC; MPC-MD-BLA-MPC and MPC-MD-CP-MPC. VTA = ventral tegmental area; EC = entorhinal cortex; LH = lateral hypothalamus; BLA = basolateral amygdala; MD = mediodorsal nucleus of the thalamus; CP = caudate-putamen.

Neurochemistry and neuropharmacology of self-stimulation of the prefrontal cortex

Throughout the previous section we have seen that intrinsic neurons in the MPC as well as terminals arising from neurons outside the prefrontal cortex and fibers leaving the prefrontal cortex to other areas of the brain, have been proposed as part of the circuitry underlying the local substrates of SS in this area of the brain. However, detailed knowledge of the interconnections among neurons and of the inputs and outputs to those neurons in the MPC, would be essential in providing an explanation of the neural events that initiate electrical SS. At present, however, we have no data indicating how such "wiring" is organized. Nonetheless, knowledge about the neurotransmitter changes occurring "in vivo" at sites were the stimulation is applied, could provide some indication of the neuronal substrate involved.

Mora and Ferrer (1986) have described the neurochemical mapping of the medial prefrontal cortex. Thus, the MPC contains the majority of the neurotransmitters found in the central nervous system. These include monoamines, peptides and amino acids. More recently, evidence has been accumulated on the existence of different peptide containing neurons and terminals and also on the morphology and location of acidic amino acid containing neurons in different layers of the cortex (Dinopoulos et al., 1989; Dori et al., 1989; Kalsbeek, 1989). Moreover, coexistence of several neurotransmitters at the same synaptic terminals in the neocortex, which presumably would apply as well to the prefrontal cortex, have been reported (Docherty et al., 1987). A summary of these neurotransmitter systems in the MPC is presented below.

Both divisions of the prefrontal cortex (MPC and SPC) receive noradrenergic and dopaminergic terminals from the locus coeruleus and the ventral tegmental area in the midbrain (Ungerstedt, 1971; Thierry et al., 1973, 1976; Berger et al., 1976; Kalsbeek, 1989). A very extensive body of literature has been devoted to the study and description of the origin and termination of these catecholaminergic pathways in the cortex, particularly for dopamine (Mora and Ferrer, 1986; Van Eden et al., 1987; Doucet et al., 1988; for a comprehensive review see Kalsbeek, 1989). Also, serotoninergic and acetylcholinergic terminals are present in the MPC. The origins of these monoamines are the dorsal and medial raphe nuclei and the basal forebrain, particularly the Ch4 group (Ungerstedt, 1971; Emson, 1978; Mesulam et al., 1983; Mayo et al., 1984; Wainer et al., 1984).

Inmunohistochemical as well as other types of studies have shown that intrinsic neurons as well as terminals containing substance P, cholecystokinin (CCK), neurophysin, vasoactive intestinal peptide (VIP), somatostatin, enkephalin, bombesin, neuropeptide Y and also multiple opiate receptors and receptors for neurotensin are present in the MPC (Emson, 1978; Emson et al., 1979; Williams and Zieglgansberger, 1981; McGregor et al., 1982; Quirion et al., 1982; Phillipson and Gonzalez, 1983; Sakanaka et al., 1983; Foster and Shultzberg, 1984; Morrison et al., 1984; Schults et al., 1984; De Quidt and Emson, 1986). However, in contrast to the very well known description of the organization of the catecholaminergic projections to the MPC, the origins of the different peptide systems in the MPC are not yet well characterized.

The presumptive putative amino acid neurotransmitters GABA, aspartic acid and glutamic acid are present in the prefrontal cortex (Fonnum et al., 1981a,b; Mora et al., 1986; Esclapez et al., 1987). GABA is located in inhibitory interneurons (Esclapez et al., 1987) while aspartic acid and, particularly, glutamic acid seem to be located in the majority of the large and medium size pyramidal neurons and in a small population of non-pyramidal cells in the cortex (Dinopoulos et al., 1989; Dori et al., 1989). Terminals containing these acidic amino acids have been described in the medial prefrontal cortex mediating ipsi- and contralateral corticocortical connections (Fonnum et al., 1981a; Peinado and Mora, 1986).

It is interesting that, as with subcortical structures, coexistence and interaction of different peptides with dopamine, acetylcholine and amino acids have been described in the neocortex (Stephens and Herberg, 1979; Kelley et al., 1980; Pan et al., 1985; Thal et al., 1986; Bunney, 1987; Voorn et al., 1987; Itoh et al., 1988; Mora et al., 1989). Also of relevance in this chapter is the fact that, although the neurotransmitter systems terminate in different layers of the cortex, terminals containing dopamine, acetylcholine, substance P, as well as glutamergic neurons and NMDA receptors show a similar distribution being concentrated mainly in deep layers (V and VI) of the MPC (Tassin et al., 1978; Cotman et al., 1987; Dinopoulos et al., 1989). The possible functional relevance of this observation will be discussed further.

Throughout the years, some of the neurotransmitters found in the MPC have been suggested to be involved in SS of this same area of the brain (Mora, 1978; Mora and Ferrer, 1986). Among them are dopamine, acetylcholine, substance P and, more recently, glutamic acid (Mora and Ferrer, 1986; Cobo and Mora, 1988; Cobo et al., 1988, 1989b). On the contrary, other neurotransmitters do not seem to participate in SS of the MPC, namely norepinephrine (NE), serotonin (5-HT), CCK and enkephalins (Ramírez et al., 1983; Shaw et al., 1984; Ferrer et al., 1987).

Neuropharmacological studies performed in the orbitofrontal cortex of the rhesus monkey suggested that dopamine might be involved in the neurochemical substrates of SS in this area of the brain (Mora et al., 1976a). Confirmation of these initial findings was done in the rat (Mora, 1978). The strongest evidence in support of a role for dopamine in SS of the MPC comes from a series of studies demonstrating the release of dopamine during SS of this area. (Fig. 5) (Mora, 1978). Dopamine is increased by more than 100% during SS. The same animals, when not receiving electrical stimulation, exhibit a typical radioactive washout curve (Mora and Myers, 1977). Further neuropharmacological experiments also confirm the possible role of dopamine including those suggesting that D1 but not D2 is the dopamine receptor type involved (Ferrer et al., 1983; Mora and Ferrer 1986). In agreement with the hypothesis that dopamine is involved in SS of the MPC, lesions of the ventral tegmental area origin of the dopaminergic terminals in the MPC (Vives et al., 1986), or depletion of dopamine in the MPC by neurotoxic lesions of the mesocortical dopamine system at the level of the MFB-LH (A.G. Phillips and Fibiger, 1978) reduce SS. This reduction of SS is observed immediately after the lesion and lasts for at least 15 – 20 days. Interestingly, full recovery of SS after that time has been reported (Simon et al., 1979) indicating that, although dopamine seems important for SS, recovery of behavior occurs after some time, probably due to a functional reorganization that compensates for the lack of dopamine. Simon et al. (1979) also

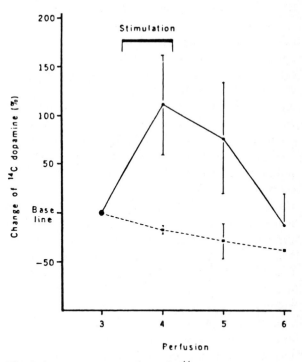

Fig. 5. Average percentage change in [^{14}C]dopamine as verified by chromatographic analysis in samples of perfusates collected from the medial prefrontal cortex during electrical stimulation of this region and the control condition. Control, – – –; stimulation, ———. (Taken from Mora and Myers, 1977.)

demonstrated that, despite full recovery of SS after the lesion, changes in the nature of SS of the MPC persisted. That is, when a rate–intensity curve is performed in the MPC, the typical ceiling effect is found in which the rate of SS does no longer increase when current intensity is increased. This effect disappears after such lesions. This last observation provides evidence for a multifactorial component in SS behavior of the MPC, and suggests that dopamine does not play an exclusive role in mediating SS of this area of the brain.

Acetylcholine has also been implicated in SS of the MPC. Fig. 6 shows one of a series of experiments in which peripheral injections of scopolamine, a muscarinic receptor blocker, produces a dose-related decrease in SS of the MPC. This decrease contrasts with its effects on general arousal (measured by spontaneous motor activity) or an operant motivated behavior, i.e. pressing a bar to obtain water at a rate comparable to the rate to obtain SS (Mora et al., 1980b). These effects were fully replicated with injections of dexetimide, another type of muscarinic antagonist (Mora et al., 1980b), while no effects could be found with nicotinic antagonists (Vives and Mora, 1986). Despite the rather selective effects produced by an acetylcholine receptor blocker on SS, and also the specificity of the type of receptor involved, it is possible (since the injections were made peripherally) that the effects were not produced within the MPC but rather indirectly through some other area. Therefore, it is not possible to draw any definite conclusions about the possible role of acetylcholine in SS of the MPC.

The MPC is probably the cortical area with the highest concentration of substance P, which is found (along with dopamine and acetylcholine) in the deepest layers viz. layers V and VI (Emson, 1978; Sakanaka et al., 1983). It is interesting that intraventricular or intracortical microinjections of substance P produce a decrease in SS (see Fig. 7) (Ferrer et al., 1988). These effects of substance P seem to be selective for SS because neither generalized arousal, measured by spontaneous motor activity, nor SS of the contralateral MPC were affected (Ferrer et al., 1988).

Other peptides microinjected into the MPC have not been shown to produce any effect on SS. Thus, microinjections of CCK-8, at doses previously shown to produce neurochemical and behavioral effects or effects on SS in other areas of the brain (Fekete et al., 1983; Crawley et al., 1985), have no effects on SS in the MPC (Ferrer et al., 1988). Enkephalins also failed to produce any demon-

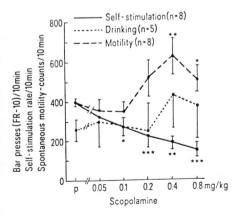

Fig. 6. Effects of intraperitoneal injections of scopolamine on self-stimulation of the medial prefrontal cortex, spontaneous motor activity and operant drinking behavior on an FR-10 schedule. (Taken from Mora et al., 1980.)

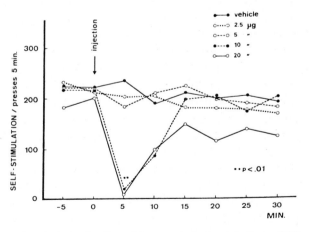

Fig. 7. Effects of unilateral intracortical microinjections of SP (substance P) and vehicle (saline) into the medial prefrontal cortex on self-stimulation of the ipsilateral injected site. Contralateral self-stimulation (not shown) was unaffected. Mean absolute control rate was 216 ± 18. (Taken from Ferrer et al., 1988.)

strable effects (Shaw et al., 1984). The results obtained with substance P suggest a possible role for this peptide in SS of the MPC. Because substance P interacts with dopamine at several brain structures were both neurotransmitters are present (Cheramy et al., 1977; Eison et al., 1982; Starr, 1982) and because the highest concentrations of substance P and dopamine are found in the same layers of the MPC, we have advanced the hypothesis that an interaction between these two neurotransmitters could exist at the tip of the electrode supporting SS (Mora and Ferrer, 1986).

As mentioned previously, GABA as well as aspartic and glutamic acids are present within the prefrontal cortex (Fonnum et al., 1981a, b; Mora et al., 1986; Esclapez et al., 1987). It is noteworthy that, together with the anterior cingulate and pyriform cortices, the prefrontal cortex has the highest density of acidic amino acid (particularly NMDA) receptors in the entire cerebral cortex (Cotman et al., 1987). It should also be noted that dopamine and NMDA receptors coexist in the same cortical layers, and that NMDA receptors have been suggested to play a role in learning, memory and behavior (Koek et al., 1986; Morris et al., 1986). On the basis of these considerations we have recently performed a series of experiments designed to investigate the possible role of these acidic amino acids in SS of the MPC (Cobo et al., 1988, 1989b; Cobo and Mora, 1988).

Thus, we have shown that intraventricular or intracortical microinjections of γ-D-glutamylglycine, a blocker for all 3 known types of glutamatergic receptors (kainate, quisqualate, and N-methyl-D-aspartate (NMDA) receptors), as well as microinjections of DL-2-amino-phosphonovaleric acid (D-AP5), an antagonist of NMDA receptors, produce a dose-related decrease in SS of the MPC. This decrease was selective for SS because neither SS of the contralateral (non-injected) site nor general arousal as measured by an open field test for motor activity were affected. Intracortical microinjections of γ-D-glutamyltaurine, an antagonist of kainate-quisqualate receptors (Watkins and Olverman, 1987) had no effects on SS behavior. Taken together, these results suggest that (a) acidic amino acid could be part of the neurochemical substrates of SS of the MPC, and (b) the receptors involved are of the NMDA subtype (see Fig. 8 for a summary of these experiments).

In a follow up series of neurochemical experiments using the "in vivo" push–pull technique and analysis of the perfusates by HPLC, elec-

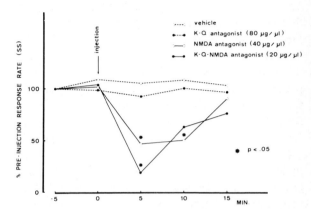

Fig. 8. Effects of unilateral intracortical microinjections of acidic amino acid antagonists and vehicle expressed as percentage of the control pre-injection rate. Mean absolute control rate was 185 ± 18. The doses shown in this figure are the highest of a dose–response curve, performed for each of the 3 antagonists.

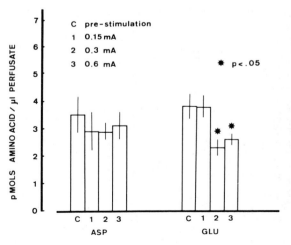

Fig. 9. Endogenous levels of aspartic (ASP) and glutamic (GLU) acids in samples of perfusates collected in the medial prefrontal cortex during electrical stimulation at different intensities and basal conditions (pre-stimulation).

trical stimulation of the MPC, at the same or double the intensity at which the rat presses reliably for SS produces a decrease in the endogenous levels of glutamic but not aspartic acid (Cobo et al., 1989b). GABA was not measured in these experiments (Fig. 9). These results further support the hypothesis that glutamic acid could participate in the neurochemical substrate of SS in the MPC.

Neural events occurring at the tip of the electrode during and after self-stimulation: A hypothesis

It seems obvious that the input that initiates activity in the prefrontal cortex during self-stimulation is not specific, meaning to say that electrical stimulation could initially produce an artificial (and perhaps disorganized) activity of neurons and terminals as a result of which an interaction of excitation and inhibition through the release of different neurotransmitters occurs. This interaction could subserve the local organization of an output from the medial prefrontal cortex.

We propose that this output would activate multiple brain areas "coding" for multiple functions, including those "coding" for natural rewards (Fig. 4). After such activation has occurred, feedback activity from these nuclei (Valenstein, 1969, 1976) would now provide a physiological input to the MPC which would be important for a final organization of a specific activity in this area of the brain. Experimental evidence in support of the theoretical considerations delineated above and reviewed in previous sections is summarized below.

First, activity of neurons located in the MPC as well as activity of neurons in other areas of the brain connected through feedback mechanisms are important elements for maintaining SS of the MPC.

Second, localized activity in the MPC during SS, whether induced directly by physical stimulation or through a feedback circuit produces a release of dopamine and a decrease in the endogenous levels of glutamate. This suggests a possible interaction between these two neurotransmitters within the MPC. Experimental observations in support of this possibility are: (1) dopamine terminals are mainly located in deep layers of the MPC (layers V and VI) and pyramidal cells seem to be the target of these terminals; (2) the vast majority of the large pyramidal cells concentrated in this layer V are probably glutamatergic in nature; and (3) the mesocortical dopaminergic system has been shown to be inhibitory on the spontaneous firing rate of neurons in these deep layers of the MPC as well as on the excitation produced by simultaneous stimulation of the MD (Mora et al., 1976b; Berger et al., 1976; Ferron et al., 1984; Van Eden et al., 1987; Seguela et al., 1988; Dinopoulos et al., 1989).

Third, other neurotransmitters, apart from dopamine and glutamate, could also be involved in SS of the MPC. For instance, GABA, which is located in abundant intrinsic inhibitory neurons in the MPC could also be released by the stimulation and, therefore, be responsible, along with dopamine, for the decrease found in glutamate. Also acetylcholine and substance P could play a role in SS of the MPC and an interaction between these two neurotransmitters and dopamine has been suggested.

Fourth, the data summarized above indicating that intrinsic neurons together with the release and possible interaction of several neurotransmitters are at the basis of SS of the MPC suggest that a complex neuronal interplay (excitation – inhibition) probably subserves the final local organization of an output system from the MPC. Further research is needed in order to determine local activity of different neurotransmitter systems. In this respect, a future direction of research would be not only to analyze neurochemically the dynamics of each of the neurotransmitters which presumably are involved during SS, but also to simultaneously analyze several of them in order to assess their possible interactions.

Fifth, that an output system from the MPC is crucial for SS to be sustained has also been suggested by experiments reviewed in previous sections. Thus, two output systems have been considered to participate in SS of the MPC. (1) A fugal pathway, which courses through the external capsule, could contain an important part of the output set of fibers relevant for processing information to other areas of the brain outside the MPC. This fugal pathway has terminals in entorhinal and perirhinal cortices as well as, via the perforant pathway, in the hippocampus (Gallistel, 1983; Nauta and Domesick, 1983). (2) Another descendent fugal pathway coursing through the internal capsule-MFB, and with possibly a more diffuse set of terminals in different limbic and brainstem structures, could also be involved in SS of the MPC (Clavier and Routtenberg, 1976; Routtenberg and Santos-Anderson, 1977). Both fiber systems would contain, at the same time, the axons through which processed information returns to the MPC (feedback circuits).

In conclusion it can be said that, although it is premature to draw any definitive conclusions about the neural substrates activated and responsible for SS in the MPC, recent experiments have provided evidence suggesting that integrated physiological activity, through the activation of simultaneous circuits and neurotransmitters, are responsible for the rewards elicited by self-stimulation.

Acknowledgements

We thank Prof. Carl V. Gisolfi for his comments on the manuscript and also A. Porras for his help during the preparation of the manuscript.

Supported by grants PB86-0037 from DGYCIT and Ramon Areces Foundation.

References

Beckstead, R.M. (1979) An autoradiographic examination of the cortico-cortical and subcortical projection of the mediodorsal-projection (prefrontal) cortex in the rat. *J. Comp. Neurol.*, 184: 43 – 62.

Berger, B., Thierry, A.M., Tassin, J.P. and Moyne, M.A. (1976) Dopaminergic innervation of the rat prefrontal cortex: a fluorescence histochemical study. *Brain Res.*, 106: 133 – 145.

Bunney, B.S. (1987) Central dopamine – peptide interactions: electrophysiological studies. *Neuropharmacology*, 26: 1003 – 1009.

Cheramy, A., Nieullon, A., Michelot, R. and Glowinski, J. (1977) Effects of intra-nigral application of dopamine and substance P on the 'in vivo' release of newly synthesized [^{3}H] dopamine in the ipsilateral caudate nucleus of the cat. *Neurosci. Lett.*, 4: 105 – 109.

Clavier, R.M. and Gerfen, C.R. (1982) Intracranial self-stimulation in the thalamus of the rat. *Brain Res. Bull.*, 8: 353 – 358.

Clavier, R.M. and Routtenberg, A. (1976) Brainstem self-stimulation attenuated by lesions of medial forebrain bundle but not by lesions of locus coeruleus or caudal ventral norepinephrine bundle. *Brain Res.*, 101: 251 – 271.

Cobo, M. and Mora, F. (1988) Acidic amino acids and self-stimulation of the prefrontal cortex in the rat. *Physiologist*, 31: A225.

Cobo, M., Porras, A. and Mora, F. (1988) Is there a role for NMDA-receptors in mediating self-stimulation of the medial prefrontal.cortex in the rat? *Eur. J. Neurosci.*, Suppl. 1: 225.

Cobo, M., Ferrer, J.M.R. and Mora, F. (1989a) The role of the lateral cortico-cortical prefrontal pathway in self-stimulation of the medial prefrontal cortex in the rat. *Behav. Brain Res.*, 31: 257 – 265.

Cobo, M., Porras, A. and Mora, F. (1989b) In vivo neurochemical analysis of amino acid neurotransmitters during self-stimulation of the prefrontal cortex in the rat. *Eur. J. Neurosci.*, Suppl. 2: 18.

Collier, V.I., Kurtman, S. and Routtenberg, A. (1977) Intracranial self-stimulation derived from entorhinal cortex. *Brain Res.*, 137: 188 – 196.

Corbett, D., Laferriere, A. and Milner, P.M. (1982) Elimination of medial prefrontal cortex self-stimulation following transection of efferents to the sulcal cortex in the rat. *Physiol. Behav.*, 29: 425 – 431.

Cotman, C.W., Monaghan, D.T., Ottersen, O.P. and Storm-Mathisen, J. (1987) Anatomical organization of excitatory amino acid receptors and their pathways. *Trends Neurosci.*, 10: 273 – 280.

Crawley, J.N., Sivers, J.A., Blumstein, L.K. and Paul, S.M. (1985) Cholecystokinin potentiates dopamine-mediated behaviours: evidence for a modulation specific to a site of coexistence. *J. Neurosci.*, 5: 1972 – 1983.

Crow, T.J. (1972) A map of the rat mesencephalon of electrical self-stimulation. *Brain Res.*, 36: 265 – 273.

Dalsass, M., Kiser, S., Menderhausen, M. and German, D.C. (1981) Medial prefrontal cortical projections to the region of the dorsal periventricular catecholamine system. *Neuroscience*, 6: 657 – 665.

De Quidt, M.E. and Emson, P.C. (1986) Distribution of neuropeptide Y-like immunoreactivity in the rat central nervous system. II. Immunocytochemical analysis. *Neuroscience*, 18: 545 – 618.

Dinopoulos, A., Dori, I., Davies, S.W. and Parnavelas, J.G.

(1989) Neurochemical heterogeneity among corticofugal and callosal projections. *Exp. Neurol.*, 105: 36–44.

Divac, I., Kosmal, A., Björklund, A. and Lindvall, O. (1978) Subcortical projection to the prefrontal cortex in the rat as revealed by the horseradish peroxidase technique. *Neuroscience*, 3: 785–796.

Docherty, M., Bradford, H.F. and Wu, J.-Y. (1987) Co-release of glutamate and aspartate from cholinergic and GABAergic synaptosomes. *Nature*, 330: 64–66.

Dori, I., Petrov, M. and Parnavelas, J.G. (1989) Excitatory transmitter amino acid-containing neurons in the rat visual cortex: a light and electron microscopic immunocytochemical study. *J. Comp. Neurol.*, in press.

Doucet, G., Descarries, L., Audet, M.A., García, S. and Berger, B. (1988) Radioautographic method for quantifying regional monoamine innervations in the rat brain. Application to the cerebral cortex. *Brain Res.*, 441: 233–259.

Eison, A.E., Eison, M.S. and Iversen, S.D. (1982) The behavioural effects of a novel substance P analogue following infusion into the ventral tegmental area or substantia nigra of the rat brain. *Brain Res.*, 238: 137–152.

Emson, P.C. (1978) Complementary distribution of dopamine, substance P and acetylcholine in the rat prefrontal cortex and septum. In: P.J. Roberts (Ed.), *Advances in Biochemical Psychopharmacology, Vol. 19,* Raven Press, New York, pp. 397–400.

Emson, P.C., Gilbert, R.F.T., Loren, I., Fahrenkrug, J. Sundler, F. and Schaffalitzky de Muckadell, O.B. (1979) Development of vasoactive intestinal polypeptide (VIP) containing neurones in the rat brain. *Brain Res.*, 177: 437–444.

Esclapez, M., Campistron, G. and Trottier, S. (1987) Immunocytochemical localization and morphology of GABA-containing neurons in the prefrontal and frontoparietal cortex of the rat. *Neurosci. Lett.*, 77: 131–136.

Fekete, M., Lengyel, A., Hegedus, B., Rentzsch, A., Schwarzberg, H. and Telegdy, G. (1983) Effects of cholecystokinin octapeptides on avoidance and self-stimulation behaviour of rats. *Neurosci. Lett.*, 14: S-113.

Ferrer, J.M.R., Sanguinetti, A.M., Vives, F. and Mora, F. (1983) Effects of agonists and antagonists of D1 and D2 dopamine receptors on self-stimulation of the medial prefrontal cortex in the rat. *Pharmacol. Biochem. Behav.*, 19: 211–217.

Ferrer, J.M.R., Myers, R.D. and Mora, F. (1985) Suppression of self-stimulation on the medial prefrontal cortex after local microinjection of kainic acid in the rat. *Brain Res. Bull.*, 15: 225–228.

Ferrer, J.M.R., Cobo, M. and Mora, F. (1987) The basolateral limbic circuit and self-stimulation of the medial prefrontal cortex in the rat. *Physiol. Behav.*, 40: 291–295.

Ferrer, J.M.R., Cobo, M. and Mora, F. (1988) Peptides and self-stimulation of the medial prefrontal cortex in the rat: effects of intracerebral microinjections of substance P and cholecystokinin. *Peptides*, 9: 937–943.

Ferron, A., Thierry, A.M., Ledouarin, C. and Glowinski, J. (1984) Inhibitory influence of the mesocortical dopaminergic system on spontaneous activity or excitatory response induced from the thalamic mediodorsal nucleus in the rat medial prefrontal cortex. *Brain Res.*, 302: 257–265.

Fonnum, F., Soreide, A., Kvala, I., Walker, J. and Walaas, I. (1981a) Glutamate in cortical fibers. *Adv. Biochem. Psychopharmacol.*, 27: 29–42.

Fonnum, F., Storm-Mathisen, J. and Divac, I. (1981b) Biochemical evidence for glutamate as neurotransmitter in corticostriatal and corticothalamic fibers in rat brain. *Neuroscience*, 6: 863–873.

Foster, G.A. and Schultzberg, M. (1984) Immunocytochemical analysis of the ontogeny of neuropeptide Y immunoreactive neurons in foetal rat brain. *Int. J. Dev. Neurosci.*, 2: 387–407.

Foster, G.A., Schultzberg, M., Hokfelt, T., Goldstein, M., Hemmings, H.C., Quimet, C.C., Walaas, S.I. and Greengard, P. (1987) Development of a dopamine and cyclic adenosine $3':5'$-monophosphase-regulated phosphoprotein (DARP-32) in the prenatal rat central nervous system, and its relationship to the arrival of presumptive dopaminergic innervation. *J. Neurosci.*, 7: 1994–2018.

Fuster, J.M. (1989) *The Prefrontal cortex: Anatomy, Physiology and Neuropsychology of the Frontal Lobe,* Raven Press, New York.

Gallistel, C.R. (1983) Self stimulation. In: J.A. Deutsch (Ed.), *The Physiological Basis of Memory,* Academic Press, New York, pp. 269–349.

Ito, M. and Olds, J. (1971) Unit activity during self-stimulation behaviour. *J. Neurophysiol.*, 34: 263–273.

Itoh, S., Katsuura, G. and Takashima, A. (1988) Effects of vasoactive intestinal peptide on dopaminergic system in the rat. *Peptides*, 9: 315–317.

Kalsbeek, A. (1989) *The Role of Dopamine in the Development of the Rat Prefrontal Cortex,* Doctoral thesis, University of Amsterdam, 217 pp.

Kelley, A.E., Stinus, L. and Iversen, S.D. (1980) Interaction between D-ala-met-enkephalin, A-10 dopaminergic neurons and spontaneous behaviour in the rat. *Behav. Brain Res.*, 1: 3–24.

Koek, W., Woods, J.H. and Ornstein, P. (1986) Phencyclidine-like behavioral effects in pigeons induced by systemic administration of the excitatory amino acid antagonist, 2-amino-5-phosphonovalerate. *Life Sci.*, 39: 973–978.

Kolb, B. (1984) Functions of the frontal cortex of the rat. A comparative review. *Brain Res. Rev.*, 8: 65–98.

Krettek, J.E. and Price, J.L. (1977a) The cortical projections of the mediodorsal nucleus and adjacent thalamic nuclei in the rat. *J. Comp. Neurol.*, 171: 157–191.

Krettek, J.E. and Price, J.L. (1977b) Projections from the amygdaloid complex to the cerebral cortex and thalamus in the rat and cat. *J. Comp. Neurol.*, 172: 687–722.

Leonard, C.M. (1969) The prefrontal cortex of the rat. I. Cor-

tical projections of the mediodorsal nucleus. II. Efferent connections. *Brain Res.*, 12: 321–343.
Livingston, K.E. and Escobar, A. (1971) Anatomical bias of the limbic system concept. A proposed reorientation. *Arch. Neurol.*, 24: 17–21.
Mayo, W., Bubois, B., Ploska, A., Javoy-Agid, F., Agid, Y., Le Moal, M. and Simon, H. (1984) Cortical cholinergic projections from the basal forebrain of the rat with special reference to the prefrontal cortex innervation. *Neurosci. Lett.*, 47: 149–154.
McGregor, G.P., Woodhams, P.L., O'Shaughnessy, D.J., Ghatel, M.A., Polak, J.M. and Bloom, S.R. (1982) Developmental changes in bombesin, substance P, somatostatin and vasoactive intestinal polypeptide in the rat brain. *Neurosci. Lett.*, 28: 21–27.
Mesulam, M.M., Mufson, E.J., Wainer, B.H. and Levey, A.I. (1983) Central cholinergic pathway in the rat: an overview based on an alternative nomenclature (Ch1–Ch6). *Neuroscience*, 10: 1185–1201.
Mora, F. (1978) The neurochemical substrates of prefrontal cortex self-stimulation: a review and an interpretation of some recent data. *Life Sci.*, 22: 919–930.
Mora, F. and Ferrer, J.M.R. (1986) Neurotransmitters, pathways and circuits as the neural substrates of self-stimulation of the prefrontal cortex: facts and speculations. *Behav. Brain Res.*, 22: 127–140.
Mora, F. and Myers, R.D.(1977) Brain self-stimulation: direct evidence for the involvement of dopamine in the prefrontal cortex. *Science*, 197: 1387–1389.
Mora, F., Rolls, E.T., Burton, M.J. and Shaw, S.G. (1976a) Effects of dopamine receptor blockade on self-stimulation in the monkey. *Pharmacol. Biochem. Behav.*, 4: 211–216.
Mora, F., Sweeney, F., Rolls, E.T. and Sanguinetti, A.M. (1976b) Spontaneous firing rate of neurones in the prefrontal cortex of the rat: evidence for a dopaminergic inhibition. *Brain Res.*, 116: 516–522.
Mora, F., Avrith, D.B. and Rolls, E.T. (1980a) An electrophysiological and behavioural study of self-stimulation in the orbitofrontal cortex of the rhesus monkey. *Brain Res. Bull.*, 5: 111–115.
Mora, F., Vives, F. and Alba, F. (1980b) Evidence for an involvement of acetylcholine in self-stimulation of the prefrontal cortex in the rat. *Experientia*, 36: 1180–1181.
Mora, F., Peinado, J.M. and Myers, R.D. (1986) Amino acid profiles in cortex of conscious rat: Recent studies and future perspectives. In: R.D. Myers and P.J. Knott (Eds.), *Neurochemical Analysis of the Conscious Brain: Voltammetry and Push–Pull Perfusion, Ann. NY Acad. Sci.*, 473, 461–474.
Mora, F., Peinado, J.M., Ferrer, J.M.R., Cobo, M., Nieto, L. and Saez, J.A. (1989) Peptides–amino acid neurotransmitters interactions in the prefrontal cortex of young and aged rats. *Proc. Int. Un. Physiol. Sci.*, 17: 221.
Morris, R.G.M., Anderson, E., Lynch, G.S. and Baudry, M. (1986) Selective impairment of learning and blockade of long-term potentiation by an N-methyl-D-aspartate antagonist, AP-5. *Nature*, 319: 774–776.
Morrison, J.M., Magistretti, P.J., Benoit, R. and Bloom, F.E. (1984) The distributioin and morphological characteristics of the intracortical VIP-positive cell: an immunohistochemical analysis. *Brain Res.*, 292: 269–282.
Nassif, S., Cardo, B., Labersat, F. and Velley, L. (1985) Comparison of deficits in electrical self-stimulation after ibotenic acid lesion of the lateral hypothalamus and the medial prefrontal cortex. *Brain Res.*, 332: 247–257.
Nauta, W.J.H. and Domesick, V. (1983) Neural associations of the limbic system. In: Beckman (Ed.), *Neural Substrates of Behaviour*, Spectrum, New York.
Olds, M.E. and Fobes, J.J. (1981) The central basis of motivation: intracranial self-stimulation studies. *Annu. Rev. Physiol.*, 32: 523–574.
Pan, H.S., Penney, J.B. and Young, A.B. (1985) Gamma-aminobutyric acid and benzodiazepine receptor changes induced by unilateral 6-hydroxydopamine lesions of the medial forebrain bundle. *J. Neurochem.*, 45: 1396–1404.
Peinado, J.M. and Mora, F. (1986) Glutamic acid as a putative transmitter of the interhemispheric corticocortical connections in the rat. *J. Neurochem.*, 47: 1598–1603.
Phillips, A.G. and Fibiger, H.C. (1978) The role of dopamine in maintaining intracranial self-stimulation in the ventral tegmentum, nucleus accumbens and medial prefrontal cortex. *Can. J. Psychol.*, 32: 58–66.
Phillips, M.I. (1976) Self-stimulation as a model for recovery of function in the brain. In: A. Wauquier and E.T. Rolls (Eds.), *Brain Stimulation Reward*, North-Holland, Amsterdam, pp. 111–115.
Phillipson, O.T. and Gonzalez, C.B. (1983) Distribution of axons showing neurophysin-like immunoreactivity in cortical and anterior basal forebrain sites. *Brain Res.*, 258: 33–44.
Quirion, R., Graudeau, P., St.-Pierre, S., Rioux, F. and Pert, C.B. (1982) Autoradiographic distribution of [^{3}H] neurotensin receptors in rat brain: visualization by tritium-sensitive film. *Peptides*, 3: 757–763.
Ramírez, M., Alba, F., Vives, F., Mora, F. and Osorio, C. (1983) Monoamines and self-stimulation of the medial prefrontal cortex in the rat. *Rev. Esp. Fisiol.*, 39: 351–356.
Robertson, A., Laferriere, A. and Milner, P.M. (1986) The role of corticocortical projections in self-stimulation of the prelimbic and sulcal prefrontal cortex in rats. *Behav. Brain Res.*, 21: 129–142.
Rolls, E.T. (1975) *The Brain and Reward*, Pergamon Press, Oxford.
Rolls, E.T., Burton, J.M. and Mora, F. (1980) Neurophysiological analysis of brain-stimulation reward in the monkey. *Brain Res.*, 194: 339–357.
Routtenberg, A. (1971) Forebrain pathways of reward in *Rattus norvegicus*. *J. Comp. Physiol. Psychol.*, 75: 269–276.
Routtenberg, A. and Santos-Anderson, R. (1977) The central

role of prefrontal cortex in intracranial self-stimulation: a case history of anatomical localization of motivational substrates. In: L.L. Iversen, S.D. Iversen and S.H. Snyder (Eds.), *Handbook of Psychopharmacology. Vol. 8. Drugs, Neurotransmitters and Behaviour,* Plenum, New York, pp. 1–21.

Sakanaka, M., Shiosaka, S., Takatsuki, K. and Tohyama, M. (1983) Evidence for the existence of substance P-containing pathway from the nucleus laterodorsalis tegmenti (castaldi) to the medial frontal cortex in the rat. *Brain Res.,* 259: 123–126.

Sarter, M. and Markowitsch, H.J. (1983a) Convergence of basolateral amygdaloid and mediodorsal thalamic projections in different areas of the frontal cortex in the rat. *Brain Res. Bull.,* 10: 607–622.

Sarter, M. and Markowitsch, H.J. (1983b) Cognitive functions of the basolateral limbic circuit: functional interaction of the basolateral amygdala with prefrontal areas and the mediodorsal thalamic nucleus as studied using different tasks and selective lesions. *Neurosci. Lett.,* 14: S-322.

Seguela, P., Watkins, K.C. and Descarries, L. (1988) Ultrastructural features of dopamine axon terminals in the anteromedial and suprarhinal cortex of the rat. *Brain Res.,* 442: 11–22.

Shaw, S.G., Vives, F. and Mora, F. (1984) Opioid peptides and self-stimulation of the medial prefrontal cortex in the rat. *Psychopharmacology,* 85: 288–292.

Shults, C.W., Quirion, B., Chronwall, B., Chase, T.N. and Obonohva, T.L. (1984) A comparison of the anatomical distribution of substance P receptors in the rat central nervous system. *Peptides,* 5: 1097–1128.

Simon, H., Stinus, L., Tassin, J.P., Lavielle, S., Blanc, G., Thierry, A.M., Glowinski, J. and Le Moal, M. (1979) Is the dopaminergic mesocortico-limbic system necessary for intracranial self-stimulation? *Behav. Neurol. Biol.,* 27: 125–145.

Starr, M.S. (1982) Influence of peptides on [^{3}H] dopamine release from superfused rat striatal slices. *Neurochem. Int.,* 4: 233–240.

Stellar, J.R. and Stellar, E. (1985) *The Neurobiology of Motivation and Reward,* Springer, New York, 255 pp.

Stephens, D.N. and Herberq, L.J. (1979) Dopamine–acetylcholine "balance" in nucleus accumbens and corpus striatum and its effects on hypothalamic self-stimulation. *Eur. J. Pharmacol.,* 54: 331–339.

Tassin, J.P., Bockaert, J., Blanc, G., Stinus, L., Thierry, A.M., Lavielle, S., Premont, J. and Glowinski, J. (1978) Topographical distribution of dopaminergic innervation and dopaminergic receptors of the anterior cerebral cortex of the rat. *Brain Res.,* 154: 241–251.

Thal, L.J., Laing, K., Horowitz, S.G. and Makman, M.H.M. (1986) Dopamine stimulated rat cortical somatostatin release. *Brain Res.,* 372: 205–209.

Thierry, A.M., Blanc, G., Sobel, A., Stinus, L. and Glowinski, J. (1973) Dopaminergic terminals in the rat cortex. *Science,* 182: 499–501.

Thierry, A.M., Tassin, J.P., Blank, G. and Glowinski, J. (1976) Topographic and pharmacological study of the mesocortical dopaminergic system. In: A. Wauquier and E.T. Rolls (Eds.), *Brain Stimulation Reward,* North-Holland, Amsterdam, pp. 290–293.

Ungerstedt, U. (1971) Stereotaxic mapping of the monoamine pathways in the rat. *Acta Physiol. Scand.,* Suppl., 367: 1–122.

Valenstein, E.S. (1969) Behavior elicited by hypothalamic stimulation. A prepotency hypothesis. *Brain Behav. Evol.,* 2: 295–316.

Valenstein, E.S. (1976) The interpretation of behavior evoked by brain stimulation reward. In: A. Wauquier and E.T. Rolls (Eds.), *Brain Stimulation Reward,* North-Holland, Amsterdam, pp. 557–576.

Van Eden, C.G., Hoorneman, E.M.D., Buijs, R.M., Matthijssen, M.A.H., Geffard, M. and Uylings, H.B.M. (1987) Immunocytochemical localization of dopamine in the prefrontal cortex of the rat at the light and electron microscopical level. *Neuroscience,* 22: 849–862.

Vives, F. and Mora, F. (1986) Effects of agonists and antagonists of cholinergic receptors on self-stimulation of the medial prefrontal cortex of the rat. *Gen. Pharmacol.,* 17: 63–67.

Vives, F., Gayoso, M.J., Osorio, C. and Mora, F. (1983) Afferent pathways to points of self-stimulation in the medial prefrontal cortex of the rat as revealed by horseradish peroxidase technique. *Behav. Brain Res.,* 8: 23–32.

Vives, F., Morales, A.B. and Mora, F. (1986) Lesions of connections of the medial prefrontal cortex in rats: differential effects on self-stimulation and spontaneous motor activity. *Physiol. Behav.,* 36: 47–52.

Voorn, P., Roest, G. and Groenewegen, H.J. (1987) Increase of enkephalin and decrease of substance P immunoreactivity in the dorsal and ventral striatum of the rat after midbrain 6-hydroxydopamine lesions. *Brain Res.,* 412: 291–306.

Wainer, B.H., Levey, A.I., Mufson, E.J. and Mesulam, M.M. (1984) Cholinergic systems in mammalian brain identified with antibodies against choline acetyltransferase. *Neurochem. Int.,* 6: 163–182.

Watkins, J.C. and Olverman, H.J. (1987) Agonists and antagonists for excitatory amino acid receptors. *Trends Neurosci.,* 10: 265–272.

Williams, J.T. and Zieglgansberger, W. (1981) Neurons in the frontal cortex of the rat carry multiple opiate receptors. *Brain Res.,* 226: 304–308.

Wurtz, R.H. and Olds, J. (1963) Amygdaloid stimulation and operant reinforcement in the rat. *J. Comp. Physiol. Psychol.,* 56: 941–949.

CHAPTER 22

Role of the prefrontal – thalamic axis in classical conditioning

D.A. Powell[a,b], S.L. Buchanan[a] and C.M. Gibbs[a]

[a] *Neuroscience Laboratory, Wm. Jennings Bryan Dorn VA Medical Center, Columbia, SC 29201 and Department of Psychology, University of South Carolina, Columbia, SC 29208, and* [b] *Department of Neuropsychiatry and Behavioral Science, University of South Carolina School of Medicine, Columbia, SC 29208, USA*

Introduction

Our research on the agranular prefrontal cortex (PFCag) of the rabbit suggests that this area, and its limbic connections (e.g., the amygdala), may provide a CNS substrate for the affective component of associative learning. This conclusion has come from experiments that have investigated classical (Pavlovian) conditioning in rabbits under a variety of circumstances.

This research has been motivated by two major assumptions. The first is that knowledge regarding prefrontal function can be efficaciously obtained by using subprimate animal models of simple kinds of learning phenomena. The extensive use of the Pavlovian eyeblink (EB) and nictitating membrane (NM) conditioning model in the rabbit (e.g. see Gormezano, 1966) has proven useful in a number of laboratories that have focused on the neuroanatomical substrates for the plasticity associated with these responses (e.g. Thompson, 1986; Moore, 1979; Yeo et al., 1984; Schneiderman, 1983; Disterhoff et al., 1988). During Pavlovian or classical conditioning, a signal (or conditioned stimulus, CS) is consistently paired with a reinforcing event (or unconditioned stimulus, UCS). As a result of such pairings, a conditioned response (CR) eventually occurs, during the signal, that in many cases resembles the original unconditioned response (UR) to the reinforcer. This CR is thus a new (or "learned") response, since it did not originally occur to the signal, and one which is elicited by the CS/UCS contingency. In the rabbit EB and NM conditioning model, developed by Gormezano (1966), eyelid closure and nictitating membrane extension occur as CRs to an initially neutral conditioned stimulus, such as a tone, that systematically precedes an airpuff or brief paraorbital electric shock train, which serves as the UCS. Note that occurrence of the learned EB or NM response during the CS has no effect on the later occurrence of the corneal airpuff or eyeshock reinforcer as would be the case in operant conditioning. In the latter paradigm, occurrence of the CR would either prevent UCS occurrence or attenuate its magnitude (Gormezano and Coleman, 1973).

As has been noted by others (e.g., Gantt, 1960; Konorski, 1967; Thompson, 1986; Weinberger and Diamond, 1987; Prokasy, 1984; Schneiderman, 1972), when animals are exposed to classical conditioning contingencies that elicit a specific somatomotor response (e.g., EB or NM responses), a number of nonspecific responses, not usually assessed, are concomitantly elicited. Although nonspecific conditioned responses may also consist

---

Correspondence: D.A. Powell, Ph.D., Neuroscience Laboratory (151A), VA Medical Center, Columbia, SC 29201, USA.

of somatomotor behaviors (Dykman et al., 1965), learned visceral changes have been most often studied, including heart rate (HR), systemic blood pressure (BP), skin conductance, changes in pupil diameter, etc. (e.g., see Obrist, 1981; Cohen and Randall, 1984; Smith and Devito, 1984; Weinberger and Diamond, 1987).

We have previously emphasized that the parametric features of Pavlovian conditioned somatomotor responses, which have been the focus of most research on classical conditioning, and the concomitantly occurring nonspecific visceral responses are very different (Powell and Levine-Bryce, 1988). For example, the temporal parameters that are optimal for somatomotor and visceral conditioning may be quite dissimilar (Powell et al., 1974; Schneiderman, 1972). Autonomic CRs are typically greatest in magnitude when the interval between the onset of the conditioned and unconditioned stimuli (viz. the CS/UCS interval) is fairly long; e.g. from 4 to 8 sec (Powell and Kazis, 1976). However, little or no somatomotor conditioning occurs at these relatively long CS/UCS intervals. Rather the CS/UCS interval that produces the most rapid EB or NM conditioning is 500 msec or less (Schneiderman and Gormezano, 1964; Gormezano, 1972). A second, striking difference between learned somatomotor and visceral responses is that the latter are acquired within just a few trials (e.g., 10 or less), whereas many trials (up to 100) may be required before the first somatomotor CR occurs (e.g., Powell and Levine-Bryce, 1988).

It was while attempting to deal with the discrepancies between what appear to be two very good models of learning and conditioning, i.e. the somatomotor EB and NM responses on the one hand, and the autonomic HR and BP responses on the other, that we came to the second assumption that has guided our research. This assumption is that if the Pavlovian HR and EB CRs both represent valid animal models of associative learning, they must represent different aspects of learning, since the parametric features of the stimuli required to elicit them, and their rate of acquisition,
are different. Single process models of learning and memory thus may be inadequate to explain even simple associative learning phenomena, such as classical conditioning. Consequently, we have employed information processing models to explain differences in classical conditioning of dissimilar response systems (e.g., see Buchanan and Powell, 1988a). We believe that Pavlovian conditioning is an especially good experimental paradigm to assess the usefulness of such models, since classical conditioning almost always involves an explicit stimulus to be processed, presumably according to information processing principles. These paradigms must therefore fulfill all of the properties of information processing if the theoretic information processing models are valid.

A corollary of the second assumption is that if autonomic and somatomotor responses do indeed represent different aspects of information processing, the neuroanatomical substrates which underly them might also differ. This corollary, which was initially stated in a somewhat different form (Powell et al., 1978), is quite compatible, not only with the earlier central processing models of information processing (Anderson, 1983), but even more so with the newer parallel distributed processing models (Rumelhart and McClelland, 1986), since in our experiments different aspects of the learning process, i.e. autonomic and somatomotor responses, are presumably processed at different CNS sites according to principles embodied in parallel distributed processing models. Moreover, since the parallel distributed processing models emphasize connectionistic principles, classical conditioning is probably one of the best experimental approaches for testing these theories of how associative learning takes place (Buchanan and Powell, 1988a; Rescorla, 1988).

To anticipate some of what follows, our experiments suggest that there are indeed different neuroanatomical substrates that mediate these different aspects of learning. The PFCag and its efferent structures appear to be important and perhaps necessary for mediating the early occurring autonomic component of Pavlovian conditioning. In the rabbit, this response consists of cardiac decelerations and slight depressor responses (Powell and Kazis, 1976). We have referred to the cardiac decelerations as "conditioned primary bradycardia" (Powell and Levine-Bryce, 1988). Primary bradycardia has been defined by the Lacey's (Lacey and Lacey, 1980) as HR slowing that is not secondary to some other physiological process, such as baroreceptor activation, respiration, etc. Conditioned primary bradycardia is thus primary bradycardia that is elicited by conditioned stimuli and is not secondary to other physiological processes elicited by these same stimuli. As described in detail elsewhere (e.g., Thompson, 1986), extrapyramidal motor system structures provide a CNS substrate for the plasticity associated with EB and NM conditioning.

In the sections below we will first describe in more detail the kinds of responses that occur during classical conditioning in the rabbit, emphasizing the differential parametric features of EB and HR conditioning in this species. Second, we will describe the afferent and efferent connections of the PFCag that might participate in this process. Third, we will describe the functional studies that provide evidence that the PFCag mediates a mechanism involved in the initial stages of learning and conditioning including: (a) lesion studies, (b) brain stimulation studies, and (c) electrophysiological recording studies. Fourth, we will indicate how the mediodorsal nucleus of the thalamus (MD), which is the thalamic projection nucleus for the PFCag, might also participate in Pavlovian conditioning. Finally, in a last section, we will briefly describe a possible model of the role of the PFCag in learning and memory, based on these studies.

Concomitant Pavlovian conditioning of somatomotor and autonomic responses

As indicated above, our research (e.g., Powell and Kazis, 1976), as well as that of others (e.g., Putnam et al., 1974; Lacey and Lacey, 1980; Pribram and McGuinness, 1975), suggests that classically conditioned cardiovascular responses represent an early component of information processing, which is probably related to initial sensory registration and attention mechanisms. However, the later occurring somatomotor NM or EB response represents the acquisition of a skeletal behavior developed in response to external environmental contingencies. It thus may represent rehearsal and the transfer of information from short-term to more permanent storage.

Fig. 1 illustrates the dramatic differences observed in these two reponse systems in rabbits exposed

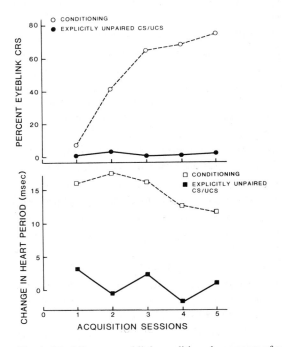

Fig. 1. (Top) Percent eyeblink-conditioned responses of rabbits that received paired CS/UCS presentations (conditioning group) and rabbits that received explicitly unpaired CS/UCS presentations (explicitly unpaired CS/UCS group) as a function of 5 acquisition sessions. The CS was a 1.0-sec, 75-dB, 1216-Hz tone that was followed by a 250-msec, 3-mA paraorbital shock train as the UCS. (Bottom) Mean change in heart period (HP) from pre-CS baseline of the third post-CS interbeat interval (IBI) of the same rabbits that received conditioning and explicitly unpaired CS/UCS presentations over acquisition sessions. The third IBI is shown because it was the last IBI available for analysis in all animals prior to CS offset and is thus usually associated with the largest HP change from pre-CS baseline.

to classical conditioning contingencies. This figure depicts the results of an experiment in which a 1216-Hz, 1.0-sec, 75-dB tone was employed as the CS, and a 250-msec, 3-mA, AC paraorbital electric shock train was the unconditioned stimulus. The top panel of Fig. 1 shows percent EB CRs over 5 acquisition sessions consisting of 60 CS/UCS presentations each (conditioning group). A pseudoconditioning control group received a random sequence of unpaired CS/UCS presentations (explicitly unpaired CS/UCS group). EB CRs increased from a low near-zero rate during session 1 to approximately 80% by session 5, while the explicitly unpaired group showed virtually no EB responses.

The bottom panel of Fig. 1 shows mean change in duration from pre-CS baseline of the third interbeat interval (IBI) of the ECG following tone onset, as a function of acquisition sessions. The third IBI was chosen because it was the last IBI that was available for analysis in all animals before CS offset and the occurrence of the UCS. This IBI usually represents the largest change from pre-CS baseline. The duration of this IBI, which is known as heart period (HP), is the reciprocal of HR; HP was lengthened by 10 – 15 msec as a result of training. Thus, HR CRs consisted of decelerations from pre-CS baseline. The HP changes in the unpaired group were much smaller and more variable. Note however, that no acquisition function is apparent in the conditioning group, as was the case for the EB data above. The reason for this can be seen in Fig. 2. This figure shows mean change in the third IBI over blocks of 2 trials each during the initial session. In Fig. 2, the acquisition function for HP is clearly apparent. Change in heart period decreased in both groups across the first 5 – 10 trials, representing habituation of the cardiac component of the orienting reflex (OR). The OR is the initial response to novel stimulation and normally also consists of bradycardia (Powell and Kazis, 1976). After habituation of the OR, the conditioning group demonstrated a second bradycardiac response of greater than 15 – 20 msec, whereas the unpaired group continued to show small and

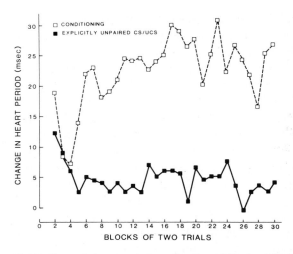

Fig. 2. Change in heart period as a function of blocks of 2 trials each for the conditioning and explicitly unpaired CS/UCS groups illustrated in Fig. 1 for the third interbeat interval during the first session of training.

variable responses of 1 – 3 msec. Thus, the decelerative HR CR appeared by trial 10 and had reached its maximum magnitude by trial 20, well before EB CRs began to occur in any animal.

As noted above, the optimal CS/UCS interval for EB conditioning is relatively short; for example, the EB acquisition function falls off sharply with CS/UCS intervals greater than the 1.0-sec period employed in the experiment described above (Schneiderman and Gormezano, 1964; Gormezano, 1972; Powell et al., 1974). These short intervals are, however, not optimal for autonomic conditioning due to the relatively longer duration and latency of these responses (Schneiderman, 1972; Powell and Levine-Bryce, 1988). We have previously found a 4.0-sec CS/UCS interval to be optimal for HR conditioning (Powell et al., 1974; Powell, 1979; Kazis et al., 1973). Thus the largest magnitude HR CR associated with an eyeshock UCS occurs with CS/UCS intervals that are too long for EB conditioning to occur. The HR acquisition function is shown in Fig. 3 on a trial-by-trial basis using this optimal 4-sec CS/UCS interval. This graph shows differential HP conditioning in one group of animals that received a CS+ paired with eyeshock and a CS− never paired with the UCS. Either a 1216- or 304-Hz, 75-dB (SPL) tone served as the CS+ or CS− for different animals. Data from these animals was compared with a second group that received the same random sequence of 4.0-sec tones and the same number of eyeshocks, but with the latter explicitly never paired with either tone. In this experiment, since CS duration was 4 sec, the last IBI available for analysis in all animals was the 12th IBI. HP change from pre-CS baseline for this IBI is thus plotted in Fig. 3 as a function of consecutive CS+ and CS− trials. The general shape of this acquisition function is similar to that described above for the 1-sec CS/UCS interval. However, the magnitude of the HR changes observed in the conditioning group was considerably greater in this experiment. Responding to both CS+ and CS− was also considerably greater in the conditioning group than in the group that received explicitly unpaired CS/UCS presentations; however, by trial 10, HP magnitude had become somewhat greater in response to the CS+ than to the CS− and for the most part remained greater in response to the CS+ for the remainder of the experiment. Thus, although significant responding occurred to the CS−, probably due to stimulus generalization, the significantly greater response to the CS+ illustrates the specific stimulus control of HP by the reinforced stimulus.

These experiments taken as a whole thus suggest that the initial HR response to tone CSs in the rabbit is in a parasympathetic direction. However, a sympathetic component is also observed when EB conditioning is concomitantly elicited by classical conditioning contingencies. This finding first became apparent in an experiment in which rabbits were studied over a considerable length of time, because the CS involved was a visual stimulus, which is not a salient CS for the albino rabbit (Powell et al., 1971). This experiment showed that, although the HR CR began as a cardiac decelera-

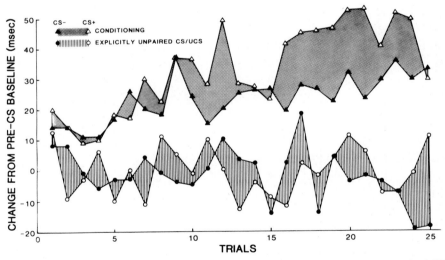

Fig. 3. Change from pre-CS baseline in heart period of the 12th interbeat interval (IBI) of rabbits that received differential conditioning, in which a CS+ that was consistently paired with paraorbital shock and a CS− that was never paired with paraorbital shock, were presented in a predetermined pseudorandom sequence. The 12th interbeat interval is shown because it was the last IBI to occur before CS offset that was available for analysis in all animals and thus most of the time was the IBI associated with the largest HP change from pre-CS baseline. Data are also shown for a group of rabbits that received similar CS/UCS presentations but neither stimulus was paired with the paraorbital shock (explicitly unpaired CS/UCS). The eyeshock in this group was instead randomly interspersed with the 2 tones, using an intertrial interval that was two thirds that of the conditioning group. Data are shown for 25 presentations of each type of CS for each group. (From Powell and Levine-Bryce, 1988.)

tion, as indicated in Figs. 1 – 3, it later developed an accelerative component as the EB CR was acquired.

A similar finding is illustrated in Fig. 4, which shows HR and BP conditioning in rabbits using a 1.0-sec CS/UCS interval that is conducive to somatomotor conditioning. In experiments in which somatomotor conditioning is studied, typically the accompanying autonomic changes are assessed after CS offset, as well as during the CS proper, to determine the full magnitude of these changes in response to the relatively short signal involved. It is thus necessary to assess HR on "test" or "probe" trials on which the UCS is not presented; otherwise the presentation of the UCS interferes with the full expression of the CR. It is after CS offset that the HR accelerations that accompany EB conditioning are observed. Note that the HR changes associated with EB conditioning described in Figs. 1 – 3 above were assessed only during the CS. Fig. 4 shows HR and BP change separately for the first 2 blocks of 10 beats each after CS onset for groups of animals that received 5, 10, 15 or 20 consecutive daily sessions of training. Thus, the cardiovascular changes during, as well as after, CS presentation are shown. These data are the mean values associated with 8 animals per group averaged over 4 CS-alone "test" trials per session. The EB response was also assessed but is not shown in Fig. 4. This experiment was a differential conditioning experiment in which a CS+, reinforced with the eyeshock UCS, and a nonreinforced CS–, were employed (see Powell and Kazis, 1976). However, for purposes of clarity, responses to the unreinforced CS– are also not shown in Fig. 4. As described above, either a 1216-Hz or 304-Hz tone served as CS+ and CS– for different animals.

It is apparent from Fig. 4 that early during training (after 5 sessions) conditioned bradycardia and

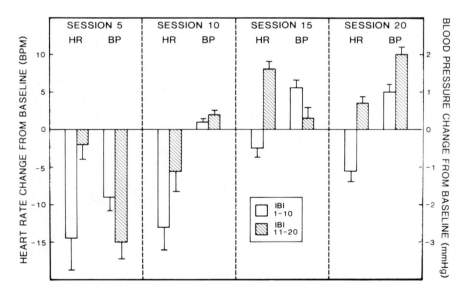

Fig. 4. Heart rate change from pre-CS baseline, in beats/min, and blood pressure change from pre-CS baseline in mm Hg of groups of rabbits that received 5, 10, 15 or 20 consecutive daily sessions of eyeblink conditioning in which a 1-sec duration CS was employed. Eyeblink-conditioned responses are not illustrated. The data shown are conditioned responses to the 1-sec duration CS+, which was consistently paired with paraorbital shock; responding to the CS– is also not shown. The data are shown separately for the first block of 10 interbeat intervals and the second block of 10 beat intervals after CS onset. The first block of 10 interbeat intervals thus includes the initial HR CR that occurred during the CS, while the second 10 intervals represent HR changes that occurred after CS offset on test trials when the UCS was omitted.

depressor responses were obtained. However, after considerable training (i.e. 10, 15, or 20 sessions, after EB acquisition was complete), the initial HR response (during the first 10 IBIs) remained a cardiac deceleration, but the later response (during the second 10 beats, after CS offset and occurrence of the EB CR) consisted of cardiac accelerations. Moreover, the depressor response changed to a pressor response. It has been determined that the initial bradycardia is primarily due to vagal activation, although some sympathetic inhibition is also involved (Kazis et al., 1973; Schneiderman et al., 1969). The peripheral neuronal control of the later occurring HR accelerations and pressor responses has not yet been determined. We have, however, determined, through a series of lesion and brain stimulation experiments, that the PFCag participates in, and possibly provides an essential substrate for, the early occurring bradycardiac response, whereas its thalamic projection nucleus, i.e. the mediodorsal nucleus (MD), and perhaps the closely adjacent thalamic midline nuclei as well, are involved in the sympathetic component.

Prefrontal afferent and efferent connections in the rabbit

The afferent and efferent connections of the prefrontal cortex have been well studied in primates and rodents; for recent reviews see Reep (1984) and Fuster (1989). However, these connections have not been explored to any great extent in lagomorphs. Rose and Woolsey (1948) suggested that the prefrontal area be defined as the frontal projection cortex of the mediodorsal nucleus of the thalamus (MD). Based on neuronal degeneration in MD produced by PFCag lesions, these authors concluded that a small area surrounding the frontal pole comprised in the entire prefrontal area in the rabbit. However, later studies in rats demonstrated that the entire medial prefrontal cortex from a midcallosal level to the frontal pole, and extending laterally down the dorsal lip of the rhinal fissure, receives MD efferents (Domesick, 1969; Leonard, 1969; Krettek and Price, 1977; Groenewegen, 1988).

A similar large projection from MD to the PFCag was also later demonstrated in the rabbit. Benjamin et al. (1978), using horseradish peroxidase (HRP) and autoradiographic techniques, demonstrated that a midline strip of neocortex, beginning at approximately a midcallosal level, received projections from the lateral part of MD, whereas medial MD, which also receives an olfactory projection, projected to the rostral one third of midline PFCag, wrapped around the frontal pole and extended down the dorsal lip of the rhinal fissure into an area comprising the anterior agranular insular cortex (Iag). As Rose and Woolsey (1948) originally noted, much or all of this area in rabbits and rats is agranular and is relatively undifferentiated as well. We will refer to this entire area as the agranular prefrontal cortex (PFCag), and will distinguish between its medial and lateral extent by referring to the former as the medial prefrontal cortex (PFCm) and the latter as the agranular insular prefrontal cortex (Iag).

Arikuni and Ban (1978) using HRP retrograde tracing techniques, reported that the PFCag not only received afferents from MD in the rabbit, but also from the intralaminar nuclei, and the ventromedial/ventroposterior (VM/VP) complex in the thalamus as well. These authors also described labeled cells in the lateral hypothalamus, ventral tegmental area, locus coeruleus, basal forebrain and midbrain raphe subsequent to HRP injections in the PFCag.

Efferents from the PFCag have not been systematically studied in the rabbit. Kapp et al. (1985a, b) examined the efferent projections from the PFCag to the central nucleus of the amygdala and autonomic regulatory nuclei (i.e., the nucleus tractus solitarius (NTS) and dorsal motor nucleus of the vagus (DVM) in the rabbit. These investigators reported that although the Iag projected to the central nucleus, the PFCm did not. Although PFCm was found to project to the basolateral nucleus, it projected only to an area surrounding the central nucleus, i.e., PFCm efferents did not enter the central nucleus proper. A strong projection was also reported from the Iag to the autonomic regulatory nuclei of the dorsal medulla, including both the NTS and DVM. A smaller projection was reported from the more dorsal portions of the PFCm. These efferent connections to the dorsal medulla parallel those reported by other investigators in rodents (Neafsey et al., 1986; van der Kooy et al., 1984; Reep and Winans, 1982; Saper, 1982, 1985). Prefrontal efferents are known to project to other subcortical areas as well in rats and other species (Reep and Winans, 1982; Van Der Kooy et al., 1984; Neafsey et al., 1986; Wyss and Sripanidkulchai, 1984). These efferents have, however, not been verified in lagomorphs. In addition, the exact topography of the PFCag efferents to the thalamus has not been previously investigated in the rabbit. Consequently, several experiments in our laboratory were undertaken to explore the prefrontal afferent and efferent connections in the rabbit, using uncon-

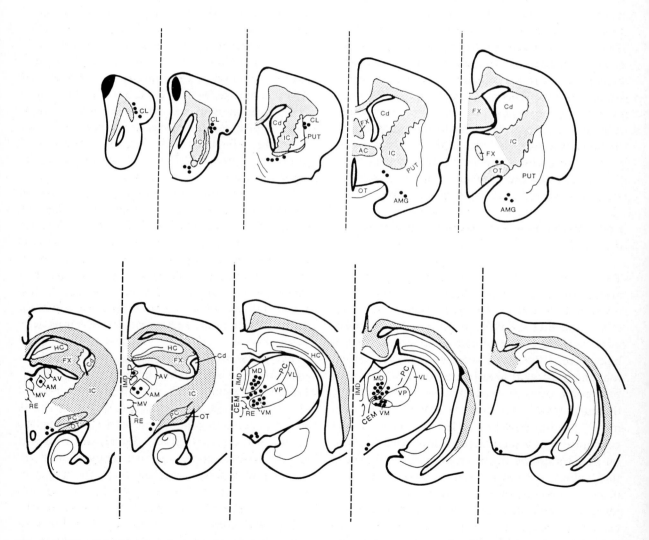

Fig. 5. Cells labeled by retrograde transport of horseradish peroxidase subsequent to injection (solid area) in the midline prefrontal cortex of the rabbit.

jugated HRP as a retrograde neuronal uptake tracer and HRP conjugated to wheat germ agglutinin (WGA-HRP) as an anterograde tracer.

These studies have demonstrated that the afferent connections to the prefrontal cortex of the rabbit are similar but not identical to those previously reported in the rat. A diagram of these connections, as studied by retrograde uptake of unconjugated HRP injected into the prefrontal cortex is shown in Fig. 5. This figure indicates that there is a strong projection from all of the medial nuclei of the thalamus as well as MD. In addition, the paraventricular nucleus of the midline group also projects to the PFCm. A small projection from the anteromedial nucleus was also observed. As might be expected, based on previous results (e.g. Arikuni and Ban, 1978), labeled cells were also found in the amygdala, the raphe nuclei, the locus coerulus, and the ventral tegmentum. A relatively strong projection was also found to arise from the basolateral forebrain, including cells in the nucleus basalis, and extending across the substantia innominata, and into the horizontal diagonal band. A strong projection, not reported by Arikuni and Ban (1978), was also found to emanate from the dorsomedial aspect of the anterior claustrum. Claustral projections to the PFCm have, however, been previously reported in the rat (Minciacchi et al., 1985), cat (Witter et al., 1988), and primate (Pearson et al., 1982).

Since we were specifically interested in the role of the thalamic–prefrontal axis in the control of classical conditioning, a more detailed analysis of the topographical relationships between the efferents of the prefrontal area and the subnuclei of the medial thalamus was undertaken, first by pressure injections and then by iontophoretic application of unconjugated HRP in the medial thalamus, and iontophoretic injections of WGA-HRP in various subregions of the PFCm and Iag. There was a topographical relationship between the medial nuclei of the thalamus and the prefrontal cortex. None of the medial nuclei, including the intralaminar nuclei, MD or the VM/VP complex, received projections from the infralimbic area (Brodman's area 25). The anterior limbic area of the PFCm (approximately Brodman's area 32), on the other hand, was found to provide a strong projection to MD; this field extended up into the ventral parts of the precentral region (approximately Brodman's areas 8 and 24) as well. The intralaminar nuclei, including the centromedian and paracentral nuclei, also received projections from areas 24 and 32. In addition, cells providing efferents to this projection tailed off across the superior aspect of the forceps minor and extended down into the granular and agranular insular cortex. VM and VP were also found to receive PFCag afferents. However, these originated only from the granular area of the insular cortex and from the premotor area overlying the forceps minor. No cells were found in either area 32 or 24 of the PFCag or in the Iag. These topographical relationships are shown in Fig. 6. Anterograde iontophoretic injections of WGA-HRP into the prefrontal cortex revealed that all of these areas received reciprocal projections from the PFCag, and indicated that the previous retrograde HRP projections were not due to labeling of fibers of passage.

WGA-HRP injections in the PFCag also revealed strong bilateral projections to the basolateral nucleus of the amygdala, superior colliculus, the dorsal central gray, and both the dorsal and ventral medulla. A diagram illustrating these PFCag efferents, as well as those to the medial thalamus, is shown in Fig. 7. Retrograde uptake after HRP or fluorescent dye injections into the dorsal, as well as the ventrolateral medulla, including the nucleus ambiguous, indicated that labeled cells were found topographically arranged in the PFCag as had previously been reported by Neafsey et al. (1986) in the rat, and Kapp et al. (1985a) in the rabbit. However, unlike the rat, the majority of these cells were found in the Iag and more dorsolateral aspects of the PFCm, including the dorsal parts of areas 24 and 8 (precentral agranular area). Retrograde uptake following injections in the superior colliculus revealed that PFCm projections to the superior colliculus also originated in these

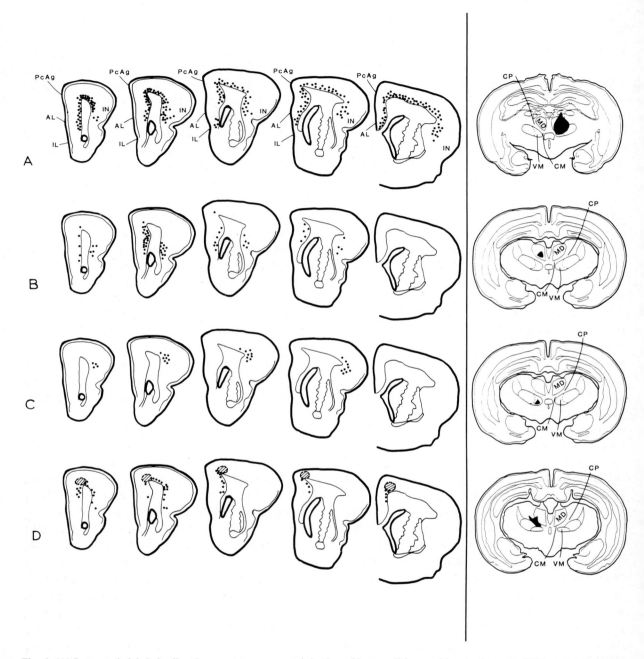

Fig. 6. (A) Retrograde labeled cells subsequent to a pressure injection of horseradish peroxidase in the area of the medial thalamus. (B) Labeled cells subsequent to retrograde iontophoretic injection into the mediodorsal nucleus of the thalamus. (C) Cells retrogradely labeled subsequent to iontophoretic injection of HRP limited to the ventromedial nucleus in the rabbit. (D) Retrogradely labeled cells in the prefrontal cortex subsequent to iontophoretic injection, which was limited primarily to the centromedian and paracentral intralaminar nuclei of the thalamus in the rabbit. The striped area in D is an area of extremely dense concentration in the dorsomedial area, which extended anteriorly from the polar region back to a midcallosal level. The injection site is shown on the right in all cases. (From Buchanan et al., 1989.)

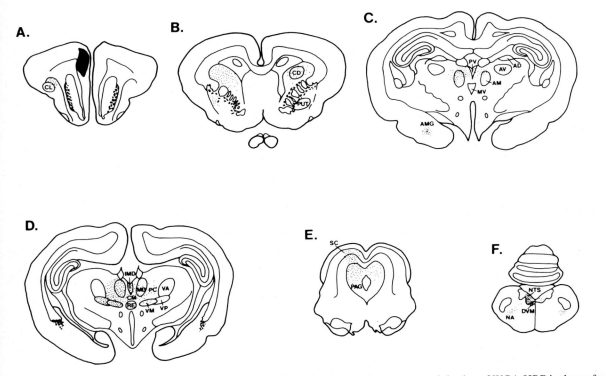

Fig. 7. Diagramatic representation of anterograde labeling in various areas subsequent to an injection of WGA-HRP in the prefrontal cortex.

more dorsal areas. A smaller projection from the infralimbic PFC (area 25) was also found to project to the dorsomedial and ventrolateral medulla as well. Retrograde uptake following central gray injections revealed that cells projecting to the central grey were found throughout the entire PFCm as well as Iag. As reported by Kapp et al. (1985b), we also observed that the projection from the Iag to the medulla was much larger than that from the PFCm.

These studies thus indicate that the connections of the PFCag in the rabbit are very similar to those of the rat, cat, and other subprimate animals. Definite topographical relationships have been established in many cases. However, the extent to which these topographical relationships are related to the functional properties of the PFCag, as demonstrated by behavioral and electrophysiological experiments, remains to be seen. The strong projections from the PFCag to the medulla suggest that the autonomic effects produced by prefrontal stimulation, and possibly the involvement of the PFCag in visceral learning, might be via such monosynaptic connections. However, it should be remembered that the PFCag also has connections with the central grey and hypothalamus, also known to be involved in nociception and other behaviors associated with autonomic adjustments. In addition, there are connections between the PFCag and the amygdala (Beckstead, 1979; Cassell and Wright, 1986; McDonald, 1987); the latter also has monosynaptic relationships with the dorsomedial medulla (Schwaber et al., 1982). This prefrontal system, and its associated limbic structures, thus has many afferent and efferent connections, and its subsystems are also strongly interconnected (Reep and Winans; 1982; Markowitsch and Guldin, 1983; Saper, 1985). The exact role that these afferent and efferent connections might play in mediating various kinds of behaviors, including autonomic changes, as well as learning and memory, has yet to be determined. Nevertheless, it is clear from the experiments described below that the PFCag plays a role in these two closely related phenomena.

Prefrontal control of classical conditioning

Lesion studies

The first evidence suggesting that the PFCm might participate in classical conditioning was discovered serendipitously in an experiment in which HR conditioning of animals with hippocampal lesions, produced by aspiration, was compared with that of control animals which had only the overlying neocortex removed (Buchanan and Powell, 1980). Both types of lesions greatly attenuated the bradycardia normally obtained in response to classical conditioning contingencies. To further explore the reasons for the effects of neocortical lesions on conditioned bradycardia, a second study was done in which lesions of the midline neocortex, comprising both the posterior and anterior portions of the midline region, were compared with animals with more lateral lesions of the isocortex, or animals with lesions of both midline and lateral cortical lesions (Buchanan and Powell, 1982a). The results of this experiment showed that animals with damage to both medial and lateral neocortex combined, and animals with lesions of medial neocortex alone, both showed a dramatic attenuation of conditioned bradycardia, whereas sham-operated animals or animals with lateral lesions alone revealed the normal conditioned response (Buchanan and Powell, 1982a). This experiment thus established that the area of the neocortex involved in attenuating conditioned bradycardia, as a result of cortical damage, was the midline area.

A later experiment (Buchanan and Powell, 1982b) examined whether the anterior or posterior midline regions were involved. In this experiment one group of animals received lesions of the more anterior portions of the prefrontal region including the anterior limbic and agranular precentral areas (approximately Brodmann's areas 8, 24 and 32). However, the infralimbic area (Brodmann's area 25) was not damaged in any animal. A second group of animals received midline damage restricted to the more posterior midline neocortex, in the posterior cingulate area. In some animals this damage extended down into the retrosplenial area as well. As Fig. 8 indicates, this experiment demonstrated quite dramatically that the area of the PFCm involved in the previous experiments, showing a lesion-induced attenuation of conditioned bradycardia, was the anterior PFCm.

Data from this series of experiments also showed that lesions of neither the anterior nor posterior midline prefrontal region produced an attenuation of the cardiac component of the OR. As noted

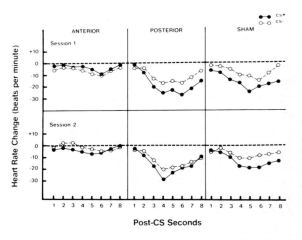

Fig. 8. Heart rate change (beats/min) from pre-CS baseline in response to CS+ and CS− of rabbits which received aspiration lesions of the anterior or posterior midline prefrontal cortex or which received sham lesions. Data are shown for 4 test trials that occurred during 2 consecutive daily sessions of 60 trials each, as a function of 8 sec after CS onset. (From Buchanan and Powell, 1982b.)

above, the OR is also characterized by cardiac decelerations and is the normal response to novel stimulation. Unless reinforced with an appetitive or aversive UCS, this response normally habituates in 10–15 presentations, depending on stimulus intensity, duration, etc. It is thus significant that it was not attenuated by PFCm lesions, since it is basically the unconditioned cardiac response to the CS. Indeed, lesions of the anterior region appeared to enhance this response (Buchanan and Powell, 1982b). The unconditioned HR response as well as EB and general somatomotor thresholds in response to the paraorbital shock were also unaffected by either anterior or posterior lesions. In addition, no general motoric differences were found in animals with lesions of either anterior or posterior PFCm, compared to sham control animals. This series of control experiments thus suggests that the effects of the PFCm lesions were specific to an associative process, since UCRs to neither the conditioned stimulus (i.e. the OR), nor the unconditioned stimulus, were adversely affected by PFCm lesions.

A subsequent experiment in this series of studies examined the effects of anterior and posterior midline cortical lesions on EB conditioning (Buchanan and Powell, 1982b). In this study rabbits received differential classical conditioning, in which 2 tones (304 and 1216 Hz) served as CS+ and CS− and eyeshocks again served as the UCS. The CS/UCS interval was shortened to 1 sec to provide reasonably optimal stimulus parameters for acquisition of the EB response. Separate groups of sham, anterior-lesioned and posterior-lesioned animals were studied. There were no significant group differences in differential acquisition of the EB CR, thus demonstrating that the effects of PFCm lesions on classical conditioning were limited to HR CRs. However, the attenuation of conditioned bradycardia produced in this experiment was somewhat less than that observed in earlier experiments in which a longer (i.e. 4 sec) CS/UCS interval was employed. The reason for this latter finding did not become apparent until a later series of experiments was conducted, in which diencephalic knife cuts were employed to interrupt the efferent fibers of the PFCag in the lateral hypothalamus (LH) and preoptic area (see below).

A subsequent experiment (Powell et al., 1985) examined the effects of lesions of the anterior and posterior Iag, since stimulation of this area was found to produce responses which were very similar and in most cases greater in magnitude and duration than those produced by midline stimulation (see below). However, unlike lesions of the midline area, lesions of the Iag had only minimal effects on conditioned bradycardia. Although a small effect was obtained, differences between a reinforced CS+ and nonreinforced CS− were not significantly different from those of a sham control group. These data therefore suggest that associative HR changes were unaffected by lesions of the Iag.

A series of experiments in which Pavlovian conditioning was studied in rabbits with hypothalamic knife cuts also implicated the PFCag in the acquisition of conditioned bradycardia. The rationale for these experiments was related to the earlier finding by Kapp et al. (1979) that lesions of the amygdaloid central nucleus (ACN) also produced a severe attenuation of conditioned bradycardia in the rabbit. As noted above, these investigators have also reported that, like the PFCm, there is a direct pathway from the ACN to the autonomic regulatory nuclei in the dorsal medulla (Schwaber et al., 1982). However, neither ACN lesions nor lesions of the PFCm completely abolished conditioned bradycardia, although a severe attenuation was produced by both types of lesions (Kapp et al., 1979; Buchanan and Powell, 1982b). In a series of knife cut studies we thus attempted to interrupt the efferents to the medulla from both the ACN and the PFCm, at the level of the lateral preoptic area and hypothalamus (LH) to determine if severing the efferents from both structures would completely abolish conditioned bradycardia.

These experiments also had a second purpose. Previous studies have shown that LH lesions interfere with associative learning of visceral response systems in several species, including primates (Smith et al., 1968), pigeons (Cohen and McDonald, 1976), rabbits (Francis et al., 1981), and rats (Iwata et al., 1986). It was unclear, however, whether the interruption of associative responding in these experiments was due to the destruction of intrinsic neurons in LH or to interruption of fibers, passing through this area, that connect brain stem and forebrain structures. Thus, a second purpose of the experiments utilizing knife cuts was to study the effects of LH coronal knife cuts on HR conditioning and to compare these data with that of animals with ibotenic acid lesions of LH, which destroy intrinsic neurons but leave fibers of passage intact.

A first experiment (Powell and Levine-Bryce, 1989) compared HR conditioning in animals with coronal knife cuts that interrupted the ascending and descending axons which pass through LH, with animals that received sham cuts. It was found that these lesions completely abolished both conditioned bradycardia and the cardiac component of the OR. A second study examined HR conditioning in animals with parasagittal knife cuts that were medial to the temporal lobe but lateral to the major nuclei of the lateral hypothalamus and preoptic area. These lesions also abolished conditioned bradycardia, compared to sham lesions, but, unlike the coronal knife cuts, had no effect on the OR. Histological examination revealed that the parasagittal knife cuts interrupted the sublenticular efferents leaving the ACN, as well as the more ventral portions of the posterior limb of the internal capsule, which carry the efferent fibers from the PFCm (Van Der Kooy et al., 1984). However, the intrinsic cells of LH, damaged by coronal knife cuts in the first experiment described above, and by radio frequency lesions in previous studies (e.g., Francis et al., 1981), were intact. The results of these 2 experiments thus suggested that previous reports of attenuated conditioned bradycardia produced by hypothalamic lesions, were perhaps due to interruption of fibers of passage, rather than to destruction of intrinsic neurons. Results of a third study, in which ibotenic acid lesions were employed to destroy intrinsic neurons in LH in the area damaged by the coronal knife cuts in the previous study supported this conclusion, since these lesions had no effect on conditioned bradycardia (Powell and Levine-Bryce, 1989). However, the OR was attenuated by these lesions.

A subsequent experiment found that parasagittal knife cuts in LH also had no effect on EB conditioning (Powell and Levine-Bryce, 1989). As with the earlier parasagittal knife cut experiment, in which a relatively long CS/UCS interval was employed, tone-evoked HR slowing during the CS was eliminated by the lesion. However, since EB conditioning was of primary interest, a 1.0-sec CS/UCS interval was employed. HR changes were thus also assessed after tone offset to determine the complete expression of the CR, as described above. The later bradycardia that occurred after CS offset on "test" trials was, however, only minimally (although significantly), attenuated by these lesions. This effect is thus similar to that observed earlier in animals with PFCm lesions (Buchanan and Powell, 1982b), in which the attenuation of conditioned bradycardia was also somewhat less with shorter (i.e., 1-sec) than longer (i.e., 4-sec) CS/UCS intervals.

Test trials are, of course, very different from normal acquisition trials, since the UCS is omitted. Presumably the organism might distinguish that such trials are not "normal" and respond differently since the "expected" UCS is not presented. This situation is thus similar to the occurrence of the OR, in which the organism is "surprised" by an event that is "unexpected" and which therefore might be considered novel. Consequently, it may be significant that coronal knife cuts, which interrupt efferents from LH cells anterior to the cuts and also damage many LH cells in the region of the cuts, greatly disrupted the cardiac OR as well as conditioned bradycardia, whereas parasagittal knife cuts, which left these LH cells intact, affected only the learned response, i.e. conditioned bradycardia (see above). Ibotenic acid lesions in LH, in the same area as that in which the coronal knife cuts were previously made, also attenuated the OR, but left conditioned bradycardia intact. These findings thus suggest that LH cells may mediate the cardiac component of the OR, but are not involved in the development of the decelerative HR CR. If the bradycardia occurring after CS offset on test trials is an OR elicited by the absence of the "expected" UCS, the smaller effect of the PFCm lesions in the study by Buchanan and Powell (1982b), and the parasagittal knife cuts in the study by Powell and Levine-Bryce (1989), might be expected, since intrinsic neurons in LH were intact in both cases. On the other hand, both coronal and parasagittal knife cuts interrupted PFCm and ACN connections with midbrain and medullary structures and also abolished the conditioned bradycardia that normally occurs during the CS. Thus, structures rostral to LH, presumably including both the ACN and PFCm are instrumental in producing conditioned bradycardia, but are not necessary for occurrence of the OR.

Brain stimulation studies

Electrical stimulation studies also suggest that PFCm and Iag participate in the control of visceral responses. These stimulation studies were all done in conscious animals with previously implanted bipolar electrodes. We first determined that electrical stimulation of the PFCm, from a midcallosal level caudally to the pregenual area anteriorly produced dramatic bradycardia and depressor responses (Buchanan and Powell, 1982b). However, similar stimulation of the posterior cingulate cortex, in the area in which lesions were previously found to have no effect on conditioned bradycardia or the cardiac OR, produced variable HR and BP changes. It was only at higher stimulus intensities that HR changes were obtained from this area, and these were invariably HR accelerations, accompanied by either pressor responses or a biphasic (pressor/depressor) BP response.

A subsequent experiment examined the effects of medial versus lateral stimulation of the PFCag, beginning at a midcallosal level and extending anteriorly to the frontal pole (Buchanan et al., 1985). Stimulation of the midline prefrontal area throughout its anterior – posterior extent produced dramatic bradycardia, as well as depressor responses in conscious rabbits. As shown in Fig. 9, these cardiovascular changes were invariably accompanied by increases in respiration rate, and decreases in depth. Somatomotor (i.e., EMG) changes also often (but not always) occurred. These cardiovascular responses were found to be greatly attenuated by vagal blockade with methylatropine and were somewhat attenuated by β-adrenergic blockade with propranolol. Further, a double blockade, consisting of both atropine and propranolol, completely abolished the response. Stimulation of the more lateral isocortical areas of the PFC, like posterior cingulate stimulation, produced more variable changes. On occasion depressor or pressor responses were obtained, but these always occurred at relatively high stimulus intensities and were relatively small compared to those obtained from the PFCm. Examples of these responses are also shown in the top panel of Fig. 9.

The Iag was also studied in a series of brain stimulation experiments (Powell et al., 1985). Like stimulation of the PFCm, stimulation of the insular cortex also produced dramatic inhibitory cardiovascular changes. Indeed these stimulus-evoked changes were in most cases much greater in magnitude than those produced by PFCm stimulation and the duration in many instances was considerably longer (see bottom panel, Fig. 9). Like the PFCm, in some cases somatomotor activity was obtained, and increases in respiratory rate and decreases in depth were invariably obtained. Also like the midline region, the use of cholinergic and β-adrenergic blockades was found to produce an attenuation of the response and a double blockade completely abolished stimulation-evoked responses. These findings thus suggest that the Iag is intimately involved in the cortical integration of visceral activities; data from other studies support this conclusion (Saper, 1982; Neafsey et al., 1986;

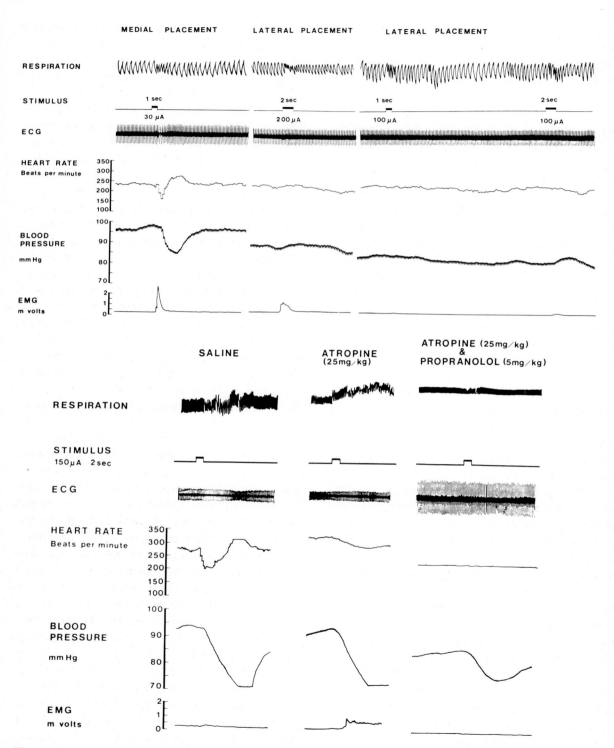

Fig. 9. (Top) Autonomic changes elicited as a result of electrical stimulation in midline prefrontal cortex and lateral isocortical somatomotor cortex in the rabbit. Data are shown for changes in respiration, heart rate, blood pressure and electromyographic activity as indicated. (From Buchanan et al., 1985.) (Bottom) Autonomic changes elicited by electrical stimulation of agranular insular cortex. As indicated data are also shown subsequent to administration of methylatropine (middle tracings) and methylatropine plus propranolol hydrocholoride (tracing on extreme right). (From Powell et al., 1985.)

Cechetto and Saper, 1987). However, the lesion study of the Iag, described above, suggests that this area is only minimally, if at all, involved in learned visceral changes.

In another series of studies, [^{3}H]2-deoxyglucose ([^{3}H]2-DG) autoradiography was combined with brain stimulation to determine the pattern of metabolic activity of MD, PFCm, and Iag when stimulation was presented to MD, PFCm or Iag (Buchanan et al., 1984; Buchanan and Powell, 1987). Stimulation of either the Iag or the PFCm produced increased [^{3}H]2-DG activity in various regions of the thalamus, and, in the case of the Iag of the amygdala. Nuclei which showed increased [^{3}H]2-DG activity in the thalamus included MD, which as noted above has a strong projection to both the Iag and PFCm, and the VM/VP region. Stimulation of the PFCm resulted in increased [^{3}H]2-DG activity only in MD. However, when the stimulated area included the so-called "shoulder" cortex of areas 8 and 24, increases in VM/VP also occurred. As noted above, there is no overlap between the PFCag cells that project to MD from midline PFC and those that project to VM/VP, which include the shoulder cortex as well as more dorsolateral PFC cells. Stimulation of Iag also resulted in VM/VP and MD activity, again compatible with previous retrograde uptake studies (Buchanan et al., 1984).

Stimulation of MD resulted in increased [^{3}H]2-DG activity in both PFCm and Iag, as might be expected (Buchanan and Powell, 1987). Such stimulation also activated certain midline nuclei, as well as contralateral MD. Caudate nucleus and the putamen/globus pallidus were also activated in some animals, but this may have been due to inadvertent stimulation of the adjacent intralaminar nuclei.

In summary, the effects of stimulation of the PFCag are compatible with the lesion studies in suggesting that this area of the brain is important in mediating visceral changes. However, the lesion studies, as reported above, suggest that the PFCm, but not the Iag, participates in learned visceral responses. As described below, the results of electrophysiological recording studies are compatible with this conclusion.

Electrophysiological recording studies

The first study in this series examined multiple unit activity (MUA) in the PFCm of conscious rabbits during orienting and classical HR conditioning (Gibbs and Powell, 1988). Two weeks following implantation of a chronic recording electrode, the animals were initially exposed to 10 trials, during which 4-sec duration tones were presented alone (viz. without the paraorbital shock (UCS), to assess the OR and its habituation. Forty trials then followed in which the 4.0-sec tone was immediately followed by a paraorbital 250-msec, 3-mA electric shock train. A second group of animals, which served as a pseudoconditioning control group, received the same number of tones and paraorbital shocks, except that during training the tones and shocks were explicitly unpaired. A third group continued to receive tone-alone presentations for 40 trials. MUA in the PFCm and HR were recorded in all animals. Peristimulus time histograms were calculated from the MUA data. Standard scores (Z scores) were computed from these histograms, by subtracting the mean number of spikes counted during selected time periods during the tone from the number of spikes counted during a baseline period prior to tone onset, and dividing this difference by the standard deviation of pre-tone neuronal frequency.

Increases in neuronal output from pre-tone baseline occurred during initial nonreinforced trials (viz. during OR assessment), but like evoked HR changes, declined over further nonreinforced trials. However, during conditioning training, in which the CS and UCS were paired, a considerable increase in neuronal activity from this pretraining level occurred. Moreover, this activity was trial-related, reaching its maximum during conditioning trials 11 – 30. Peak activity during the CS was observed to occur with a latency of between 40 and 180 msec. A similar topography was obtained in animals that received tone-alone presentations or explicitly unpaired tone–eyeshock presentations, but tone-evoked MUA declined over trials in these groups and was considerably smaller than that observed in the conditioning group.

Subsequent studies (Gibbs et al., 1989a) compared tone-evoked changes in MUA in the PFCm with those in the Iag, since, as noted above: (a) electrical stimulation of this ventrolateral prefrontal field in rabbits produces a profound, vagally mediated bradycardia (Powell et al., 1985), and (b) the Iag gives rise to a prominent, efferent projection to the dorsal medulla and ACN (Kapp et al., 1985b; Pascoe and Kapp, 1985, 1987). However, also as noted above, lesions of the Iag (Powell et al., 1985), unlike those of the PFCm (Buchanan and Powell, 1982b), have, at best, only a modest decremental effect on HR conditioning. In these studies, MUA in either the PFCm or the Iag was recorded from chronically implanted electrodes during a 2-day training procedure. Day 1 training consisted, as above, of 10 pretraining presentations of a nonreinforced 4-sec, 1216-Hz tone, followed by four 10-trial blocks of either classical conditioning (tones paired with eyeshocks) or pseudoconditioning (explicitly unpaired tone CS/UCS presentations). Day 2 training involved 2 additional blocks of 10 CS/UCS pairings for the conditioning groups and unpaired CS/UCS presentations for the pseudoconditioning groups. These 2 blocks of training trials were followed by extinction training, consisting of three 10-trial blocks of tone-alone presentations. Immediately following extinction, a brief series of unsignalled, 25-msec eyeshocks were delivered to determine their effects on MUA.

Fig. 10 summarizes the day 1 results for these studies. Consistent with the initial findings described above, pretraining presentations of the tone alone elicited increases in MUA in both the PFCm and the Iag; however, the initial, unconditioned tone-evoked increases observed in MUA recorded from the Iag showed a more gradual onset, and were of considerably smaller magnitude, than those recorded from the PFCm. Moreover, tone-evoked MUA in the PFCm was differentially affected by conditioning training whereas evoked activity in the Iag was not. Thus MUA was systematically enhanced by paired training but was progressively attenuated by unpaired CS/UCS stimulation in the PFCm group. While slight enhancements in MUA were observed in the group with Iag placements during conditioning, relative to pretraining response levels, these changes were not significantly different from those observed in animals receiving unpaired CS/UCS presentations and were not systematically trial-related, as they were for the PFCm group. MUA obtained during

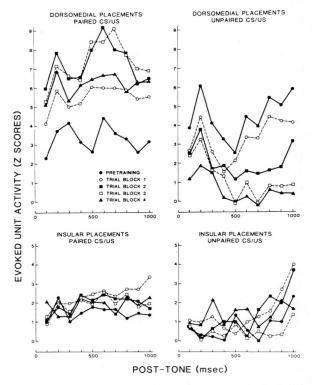

Fig. 10. Evoked unit activity (Z scores) elicited by classical conditioning contingencies in midline prefrontal cortex (dorsomedial placements) and agranular insular cortex (insular placements) of animals that received paired CS/UCS and explicitly unpaired CS/UCS presentations. Data are shown for the last block of 3 trials during pretraining in which tones alone were presented and for four 10-trial blocks, in which tone CSs and paraorbital shock UCSs were presented. The data are shown for 10 bins of 100 msec each.

conditioning and pseudoconditioning training on day 2 was similar to that observed on day 1 for both the PFCm and Iag placement groups. However, MUA declined to pretraining levels during experimental extinction for the PFCm group (not shown in Fig. 10). Further, correlational analyses of tone-evoked MUA and HR CR magnitude were statistically significant only for animals with PFCm electrode placements. During both the first and second conditioning sessions, reliable negative correlations were obtained between training-induced changes in the shortest-latency (i.e. 40–180 msec) component of the evoked MUA response and concomitant bradycardiac changes, when each was normalized relative to their respective pretraining levels. However, no significant correlations were obtained either for the pretraining trial block itself or during extinction, nor were they obtained for PFCm animals subjected to nonassociative unpaired CS/UCS training. These contrasting training effects observed in animals with PFCm vs. Iag placements did not appear to be attributable to differences in regional sensitivity to the UCS, since excitatory or, possibly, disinhibitory patterns of MUA were elicited by unsignaled presentations of eyeshock at all placements within each prefrontal field.

These data are thus consistent with our previous lesion findings (e.g., see above). Electrical activity recorded from neuronal populations in the PFCm appeared to be related to concomitantly occurring conditioned bradycardia, whereas those in Iag were not. Moreover nonassociative MUA (i.e., evoked neuronal activity in response to nonreinforced tones during habituation of the OR and during extinction) was not related to simultaneously occurring HR changes. It may be important, however, that only the early occurring 40–180-msec increase in PFCm MUA was correlated with HR CR magnitude; later occurring MUA was not. It is also significant that animals that received explicitly unpaired CS/UCS presentations, as well as animals in the paired (viz. conditioning) group during orienting and extinction, also showed this short latency MUA increase, although MUA and evoked HR changes were not significantly correlated in this case and the evoked MUA was not nearly as great as that obtained during conditioning. This finding nevertheless suggests that PFCm neurons may be involved in sensory registration, regardless of whether such stimuli are conditioned or not.

Similar studies have been undertaken in which extracellular recordings from single neurons in the PFCm were studied (Gibbs and Powell, 1989). In these experiments, animals were first subjected to surgical procedures that involved (a) removal of a small strip of bone overlying the PFCm and (b) implantation of a bone-fitted adaptor, which was designed to accommodate a miniature hydraulic microdrive assembly and would thus permit acute electrode penetration through the exposed cortical region. Following post-surgical convalescence and adaptation to handling/restraint, a single session of differential aversive conditioning was administered that involved 30 pseudorandom presentations of each of 2 distinctive, 4-sec tones (304 and 1216 Hz), one of which served as CS+ and the other as CS−. On each of the next 3–5 days, the maintained and tone-evoked activities of each of 2–5 individual PFCm cells were sequentially evaluated. As summarized in Table I, each of these test sessions began with a pseudorandom "warm-up" of 5 CS+/UCS pairings interspersed with 5 presentations of the CS− alone. Following this procedure, a microelectrode was slowly advanced via remote control until the activity of a single cell, whose spike conformation was judged to be of somal origin (e.g., Bishop et al., 1962; Fussey et al., 1970), was isolated. Immediately thereafter testing began, which consisted of 5 presentations each of the CS+ and CS− alone, according to one of several pseudorandom sequences. Next, a series of "refresher" trials was administered that, like the warm-up series, involved 5 CS+/UCS pairings interspersed with 5 presentations of the CS− alone; these "refresher" trials were delivered to maintain differential bradycardiac responses to the 2 CSs (cf. Pascoe and Kapp, 1985).

Many neurons were located in the PFCm that

TABLE I

Summary of experimental protocol for single-unit recording studies

Day(s)		
1–4	Adaptation	No stimuli presented
5	Differential conditioning	30 CS+/UCS pairings and 30 CS− presentations
6–10	Testing	
	"Warm-up"	5 CS+/UCS pairings and 5 CS− presentations
	Unit evaluations	5 presentations each of CS+ and CS−
		5 CS+/UCS pairings and 5 CS− presentations (refresher trials)

TABLE II

Functional characteristics of 6 subpopulations of PFCm cells

Cell classification	N	Maintained activity (Hz)	Tone-evoked activity
Type I	37	6.6 ± 1.3	+/+ or +/0
Type II	5	1.1 ± 0.4	0/+
Type III	14	10.1 ± 2.1	+/−
Type IV	16	5.8 ± 1.7	−/−
Type V	2	17.2 ± 6.8	−/+ or −/+/−
Type VI	26	2.3 ± 0.6	0/0

showed changes in firing frequency in response to tone–eyeshock contingencies. Six types of neurons were located (see Table II). Type I neurons consisted of cells that showed tonic increases in activity from pre-CS baseline in response to CS+ with little change in response to CS− ($n = 37$); Fig. 11 shows evoked discharge of 3 of these cells. A second class consisted of neurons that responded with little initial response to either CS+ or CS− but showed later increases in response to CS+ (type II; $n = 5$). A third group consisted of neurons that responded with an initial increase to CS+ and a smaller response to CS−, followed by a later sustained decrease in discharge that was significantly greater in response to CS+ than CS− (type III; $n = 14$). A fourth group consisted of neurons that decreased in activity throughout CS presentation (type IV; $n = 16$). The fifth group (type V) included 2 cells that showed initial decreases in activity followed by increases; finally 26 type VI cells showed no evoked response. Mean changes in evoked discharge during the 4-sec CS+ and CS− of type I–IV cells are shown in Fig. 12. As this figure shows, each of these neuronal subpopulations exhibited differential responsivity to the 2 CSs, in that significantly greater responses were elicited on CS+ trials, irrespective of the direction of the evoked change. Indeed, 85% of the cells from these 4 classes showed significantly greater evoked activity on CS+ trials than on CS− trials.

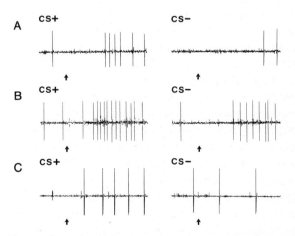

Fig. 11. Examples of discriminative tone(CS)-evoked neuronal discharge in the PFCm during differential aversive Pavlovian conditioning. Shown are the responses of 3 different cells to a single test presentation of the CS+ and the CS−; the arrow below each record indicates tone onset, and the total display times are either 205 (panels A, B) or 410 (panel C) msec. Each of these cells was classified as a type I cell (see Table II), in which short latency increases in activity occurred in response to CS+, but little or no change in activity occurred in response to CS−.

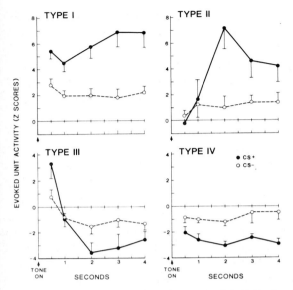

Fig. 12. Evoked unit activity (Z scores) of individual units which showed different types of evoked discharge (see Table II) in response to a CS+, which was paired with paraorbital shock, and a CS− which was not paired with shock. All cells were located in the midline prefrontal cortex.

These findings were not confounded by differences in maintained activity on CS+ and CS− trials, which was generally both irregular and unrelated to concomitant HR variables. These results could also not be attributable to any unconditioned response biases of particular cells towards tones of specific frequencies, since (a) neurons showing greater responsivity to the CS+ were recorded in all animals, and (b) the frequency (i.e., 304 or 1216 Hz) of the tone designated as the CS+ was counterbalanced across animals. Thus, the discriminative activity patterns of these cells appear to reflect the differential Pavlovian contingencies in effect. Moreover, approximately half (52%) of these cells, including 70% of the type I cells, exhibited tone-evoked activity changes that were reliably correlated with concomitant HR changes on a trial-by-trial basis. It should also be noted that the largest number of neurons located were those that consisted of increases in activity, comparable to those that were observed during the multiple unit experiments. As can be seen in Fig. 11, these neurons demonstrated a fairly large initial increase, similar to that seen in the multiple unit experiments. These data thus suggest quite conclusively that neuronal activities of cells in the PFCm are significantly related to Pavlovian conditioned changes in HR activity, whereas the MUA data described above suggests that activity of cells in the Iag are not.

Does the mediodorsal nucleus of the thalamus also participate in Pavlovian heart rate conditioning?

There is substantial evidence relating MD to learning and memory processes, as noted above (e.g. see Markowitsch, 1982, for a recent review). However, MD cells do not appear to play a direct role in mediating learned bradycardia or the bradycardia associated with orienting in spite of its strong PFCm connections. On the other hand, recent evidence, as described below, suggests that MD may be related to other aspects of classical conditioning. Specifically MD function may be related to the selection of relevant adaptive somatomotor responses as a result of exposure to Pavlovian conditioning contingencies.

Our first study of MD function suggested that MD was not involved in cardiovascular conditioning. Parasagittal knife cuts lateral to MD, which interrupt fibers going to, as well as arising from, the PFCag, did not diminish either the cardiac component of the OR or acquisition of conditioned bradycardia (Buchanan and Powell, 1989). Ibotenic acid lesions of MD resulted in similar conclusions with regard to HR CR magnitude (Buchanan, 1988). However, acquisition of differential HR conditioning was impaired in animals with ibotenic acid lesions compared to sham-operated animals, due to an inability of animals with MD lesions to suppress responding to the unreinforced CS−. Moreover, this finding was accompanied by exaggeration of the cardiac OR in the animals with lesions.

Other studies revealed that MD is cardioactive to both electrical and chemical stimulation (Buchanan and Powell, 1986; Powell and Buchanan, 1986; West and Benjamin, 1983). However, the response topography of cardiovascular changes elicited by stimulation of MD is quite different from that elicited from the PFCag. As with PFCag stimulation, HR slowing was obtained subsequent to MD stimulation in almost all experiments; however, with MD stimulation, such bradycardia was usually accompanied by BP pressor as opposed to depressor responses. Moreover, baroreceptor blockade with phentolamine greatly decreased or abolished the magnitude of the HR response, suggesting that it was sympathetically induced (Buchanan and Powell, 1986). Chemical stimulation of MD with either carbachol or glutamate also elicited bradycardia and pressor responses (Powell and Buchanan, 1986), suggesting that the sympathetic responses evoked by electrical stimulation of MD were not due to stimulation of fibers of passage, but to stimulation of intrinsic MD neurons.

The autonomic response constellation elicited by stimulation of MD thus appears similar to that which occurs relatively late during EB conditioning, when skeletal responding has become maximal, suggesting that MD may play a role in skeletal response selection and the concomitant engagement of sympathetic systems to support these responses during learning tasks. Recent studies provide strong support for this interpretation. For example, although parasagittal knife cuts lateral to MD had no effect on learned bradycardia, as noted above, thalamic knife cuts medial to MD had quite dramatic effects on this response (Buchanan and Powell, 1989). Medial thalamic knife cuts, which severed midline connections to and from MD, greatly attenuated conditioned bradycardia and the cardiac OR, although the initial bradycardia that occurred during the first post-CS interbeat interval was considerably greater than that observed in control animals. Interestingly, neither medial nor lateral knife cuts had a major effect on the cardiovascular changes elicited by electrical stimulation of MD, suggesting that these stimulus-evoked cardiovascular changes must be due to efferents located either anterior or posterior to MD in the thalamus (Buchanan and Powell, 1989).

A subsequent experiment in which knife cuts again were made lateral to MD and both HR and EB conditioning assessed, also showed no effect on acquisition of conditioned bradycardia, but EB conditioning was retarded by these lesions (Buchanan and Powell, 1988b). The results of this experiment are illustrated in Fig. 13. As can be seen EB acquisition was significantly delayed in the group with knife cut lesions. The later occurring heart rate accelerations, which as noted above, accompany the acquisition of asymptotic EB responding, and which were obtained in a group of sham animals, were also completely absent in the animals with knife cuts (see the HR changes in session 4, shown in the left panel of Fig. 13). Although the impaired EB acquisition and decreased conditioned tachycardia of the lesioned animals may have been unrelated, it is possible that the ability of animals with damaged MD function to recruit sympathetic mechanisms in support of somatomotor response acquisition may have resulted in the retarded EB conditioning observed. In a recently completed study, we have replicated these findings in animals that received ibotenic acid lesions of MD (Buchanan, 1990). Thus, it appears that although lesions of MD may have little effect on initial HR conditioning, concomitant EB conditioning is impaired, possibly through elimination of sympathetic mechanisms that facilitate somatomotor learning.

In a series of preliminary experiments we have also assessed CS-evoked multiple unit activity from MD during classical HR conditioning (Powell and Watson, 1989). The results of these experiments indicate that, like MUA in the PFCag, CS-evoked neuronal activity in MD showed a short latency (40 – 120 msec) increase that was related to training trials; viz., it increased over consecutive CS/UCS pairings but declined dramatically during OR habituation and extinction. Similar increases in a group of animals that received explicitly unpaired CS/UCS presentations, however, showed trial-related decreases in MUA. Two major differences between tone-evoked PFCm MUA, as described above, and MUA recorded from MD were that (a) the overall magnitude of the MUA increase was initially greater in the unpaired MD group than in the paired group which was not the case for PFCm placements, and (b) the peak increase in MD MUA occurred from 200 to 300 msec after tone onset whereas peak latency was somewhat shorter (100 – 200 msec) in animals with PFCag electrode placements (see above). These differences are no doubt related to the relative roles that these 2 structures might play in associative learning, but further research will be required to relate them to specific aspects of learned behavior.

Taken together the above data thus suggest that MD is responsible for mediating sympathetic mechanisms in support of somatomotor response acquisition during learning tasks. In this regard it appears to function in an opposite manner to its projection cortex in the PFCm. Thus lesions of PFCm produce an attenuation of conditioned bradycardia, whereas lesions of MD produce an exaggeration of this response but interfere with its differential conditioning by increasing responding to the nonrelevant CS – . On the other hand, MD lesions completely abolish the learned tachycardia that accompanies EB conditioning, and retard acquisition of the EB CR. It is thus tempting to speculate that normally the PFCm and its thalamic projection nucleus function in an antagonistic but balanced fashion, such that one (i.e., the PFCm) is active during the early stages of learning, when stimuli are being processed for information, while the other (i.e., MD), as originally suggested by

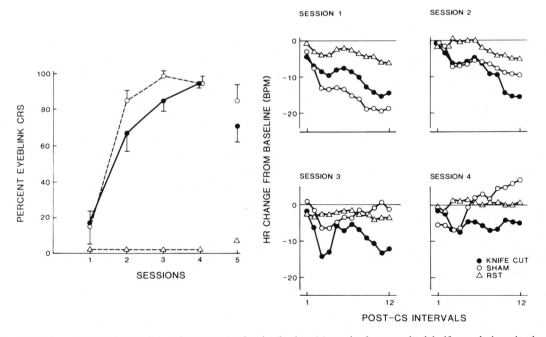

Fig. 13. (Left) Percent eyeblink-conditioned responses of animals that (a) received parasagittal knife cut lesions in the medial thalamus, (b) animals that received sham lesions, or (c) sham-lesioned animals that received explicitly unpaired CS/UCS presentations (random tones and shocks; RST). The data are shown for 4 acquisition sessions in which a 0.5-sec tone CS and a 250-msec paraorbital shock UCS was presented for 60 trials per session and for 1 extinction session consisting of 60 tone-alone trials. (Right) Associated heart rate changes from pre-CS baseline in beats/min of the same groups of animals as a function of 12 post-CS interbeat intervals over the same 4 acquisition sessions. Heart rate data are not shown for experimental extinction. (Adapted from Buchanan and Powell, 1988b.)

Vanderwolf (1971), is active during a response selection process, in which skeletal responses, designed to deal adaptively with environmental contingencies, are selected and executed.

The agranular prefrontal cortex: a possible "reinforcement cue register"

The evidence reviewed above suggests quite strongly that the midline prefrontal region in the rabbit, and perhaps other species of animals as well, provides a CNS substrate that is central to the development of cardiovascular, and perhaps other visceral changes during learning tasks. We have suggested elsewhere (Powell and Levine-Bryce, 1988) that the conditioned bradycardia observed in the rabbit is invariant when sensory processing of significant stimuli occurs. The biobehavioral relevance of these inhibitory cardiovascular changes is at the present time unclear, although circumstantial evidence suggests that general behavioral inhibition during sensory processing enhances the efficiency of such processing (Lacey and Lacey, 1980). It should be noted, however, that as important as the PFCm region appears to be for classically conditioned cardiovascular changes, it is apparently unrelated to skeletal behaviors acquired during exposure to classical conditioning contingencies. No experiments to date have related PFCm or ACN to acquisition of the Pavlovian EB or NM response in the rabbit. Instead, a variety of evidence suggests that a totally separate and nonoverlapping substrate, comprised of structures in the extrapyramidal motor system, provides a necessary substrate for the development of simple classically conditioned somatomotor behaviors (Thompson, 1986; Yeo et al., 1984; Moore, 1979; Kao and Powell, 1988). MD may serve as a link between these 2 substrates, selecting motor programs for execution that are stored in extrapyramidal structures (e.g., cerebellum), based on information processed by PFCm.

The significance of these findings for understanding the theoretical basis of associative learning is obvious; viz. even simple Pavlovian conditioning must involve a multistage process, and these different stages of learning involve different CNS substrates. As noted above, these conclusions are compatible with the currently popular parallel distributed processing models of learning and memory (Rumelhart and McClelland, 1986).

These findings also have important implications for how we view the role of the PFCag in learning and memory. Our working model of this role is, however, somewhat different from that of other investigators, as has been stated elsewhere (Powell and Levine-Bryce, 1988). In two-or-more-stage models of conditioning the autonomic responses that occur early during acquisition have often been thought to reflect the development of "fear" or "emotional" conditioning (e.g. Miller, 1948; Mowrer, 1960; Kapp et al., 1979; Thompson et al., 1983; LeDoux et al., 1984). It is clear, however, that such changes need not reflect simply fear. In appetitive situations certainly they would not (Mowrer, 1960). Even in aversive conditioning paradigms, after considerable exposure to the experimental paradigm, it is difficult to understand these autonomic changes as "fear" responses. Surely after many hundreds of CS/UCS pairings other "subjective states" might be associated with these autonomic changes. For example, if no coping response is generally available, "depression" or "sadness" might be a more appropriate term (Seligman and Weiss, 1980); if coping responses are available, perhaps "anger" or "vigilance" might be more appropriate labels for the subjective state involved. It is possible that only to the extent that ambiguity is involved would "fear" be the subjective emotion experienced. Indeed there are data which show that psychophysiological responses (Brady, 1975; Powell et al., 1974), as well as hormonal changes (Brady, 1975), differ in ambiguous versus clearly unambiguous situations.

In related unpublished studies, we have determined that the bradycardia that accompanies classical appetitive conditioning in many cases is indistinguishable from that which accompanies aversive eyelid or NM conditioning. Thus conditioned bradycardia occurs in response to signals that predict both appetitive and aversive UCSs. Although there is little evidence yet to indicate that the CNS substrates are identical for these 2 situations, we have determined that cells in the PFCm show CS-evoked changes in discharge in response to both aversive and appetitive UCSs (Gibbs et al., 1989b). In any case we believe it is not necessary to specify the subjective states associated with such behaviors for this kind of analysis to be useful. Except for providing heuristic biological models of important human conditions, the use of terms such as "fear", "hope", etc. may be counterproductive, since agreement can seldom be reached on the precise meaning of these terms. On the other hand, if these early stages of the learning process are not described by these (or similar) terms, how should such obviously separable stages of learning be characterized.

We have employed the following as a working model. At its simplest and most empirical level, 3 hierarchically arranged and sequentially identifiable stages of information processing are involved in any task that involves associative learning. These include: (a) nonspecific trace registration associated with attentional processes such as those underlying orienting and nonassociative forms of learning (e.g. see Woody, 1982); (b) the initial registration of associative stimulus significance, as reflected, for example by the development of "nonspecific" (i.e. visceromotor) CRs; and (c) the elaboration of situationally appropriate (i.e., adaptive) somatomotor CRs. The latter would include, for example, EB or NM CRs. According to this model, training-induced changes in the limbic forebrain, including the PFCm, are an integral part of stage 2 memory traces (e.g. Buchanan and Powell, 1982b), whereas extrapyramidal plasticity mediates stage 3 memories (e.g. Thompson et al., 1983). Although it is possible that frontolimbic structures may be at least partially responsible for the mechanisms that underlie orienting and the response to novelty (i.e. stage 1; see above) recent evidence, also as described above (Powell and Levine-Bryce, 1989), suggests that hypothalamic mechanisms may be sufficient for producing the autonomic changes associated with stage 1 ORs and their habituation.

Obviously it is with stage 2 processing that the PFCm appears to be primarily concerned. As noted above, one may view classical conditioning as a simple case of information processing. As such, a typical information processing model, as diagrammed in Fig. 14, can be applied to classical conditioning. This model is adapted from Anderson (1983). Information processing models assume that a series of registers, which operate according to formal control processes, are associated with different stages or kinds of memory formation. According to the model shown in Fig. 14, external stimuli are first processed in a sensory register. Those stimuli that are significant are further processed, via an attention mechanism, and stored in a shortterm, working memory register. The latter has access to at least 2 long-term memory registers; i.e. declarative and procedural memory registers. Declarative memory refers to memory for facts, of either a semantic or episodic nature. It is thus memory that is available to consciousness and which can thereby be "declared". Procedural memory on the other hand is rule-based memory. It contains the habits and skills necessary for deal-

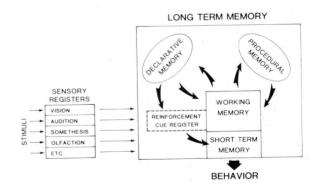

Fig. 14. Diagram of information processing model illustrating the relationship between a postulated "reinforcement cue register", which compares incoming information from sensory registers with elements of working memory, which have been highlighted by current motivational status. We suggest that a possible role for the PFCm may be to associate the signals in such a register with their affective significance. (Adapted from Anderson, 1983.)

ing with objects (both physical and symbolic) in one's environment. We have previously associated conditioning of nonspecific responses (e.g., the HR response) with acquisition of declarative knowledge and conditioning of specific skeletal responses with acquisition of procedural knowledge (Powell and Levine-Bryce, 1988). Not only is this assumption logical, based on the relatively early acquisition of HR CRs and the later acquisition of EB CRs, but neurophysiological data obtained from human clinical studies suggest the participation of frontolimbic structures in acquisition of declarative information and extrapyramidal structures with acquisition of procedural information (Squire, 1986). As discussed at some length above, our studies (e.g., Powell and Levine-Bryce, 1989), as well as as those of others (e.g., Thompson, 1986) suggest that these same CNS substrates underlie acquisition of nonspecific visceral and specific somatomotor responses, respectively.

In any case, based on a model such as that diagrammed in Fig. 14, we suggest that stimuli are selected from the typically defined sensory registers to be placed in working or short-term memory if (a) they are novel, or (b) they serve as signals for reinforcers. The attention process, which, as noted above, is an important control process involved in information processing, is thus governed by these 2 principles, i.e. "novelty" and "reinforcement". However, for this process to function adequately with regard to the second of these principles (i.e. reinforcement), it must sample or in some way come into contact with the current motivational state of the organism. Consequently there must be a register (or another control process) that is: (a) sensitive to the current motivational state of the organism, and (b) compares the events in sensory memory with previously occurring events, presumably in either working or long-term memory, to determine if any are signals for reinforcers that are germane to this motivational state. For this reason, we suggest that an appropriate statement of information processing include such a register, as indicated by the "reinforcement cue register" shown in Fig. 14.

The postulation of such a register has 2 important theoretical consequences. First, as others have noted (e.g., Estes, 1988), a severe shortcoming of most information processing theories is their inability to deal with, or in most cases, their complete omission of the role that motivational mechanisms play in learning and memory. An important aspect of almost all behavioristic accounts of learning and memory, i.e. the role of motivational mechanisms, is thus missing from most information processing models. The major motivational mechanism utilized by behavioristic models is that of reinforcement or feedback (e.g., see Skinner, 1986). Thus, the introduction of a "reinforcement cue register" into an information processing model remedies this omission and makes motivation a central aspect of information processing as it is in behavioristic models of learning and memory. The second important consequence of this new hypothetical register is its association with the PFCag. In this regard, we suggest that the PFCag and possibly its limbic connections, provides a structural entity for such a register, or at the very least for some comparator-like process, in which present sensory events are compared with cues previously associated with reinforcers.

The events to be compared in such a register must come to be there as a consequence of learning. Such learning would consist of classical conditioning, since the events to be compared are all signals. Hence, the rehearsal process is an important mechanism for determining not only the contents of long-term memory, but also the contents of the "reinforcement cue register". In fact the contents of this register might be expected to come directly from working memory, based on the number of previous pairings of the event (i.e., the CS) and a reinforcer (i.e., the UCS), as well as current motivational status of the organism. The operation of this register is thus similar to that of a comparator in which events in the sensory register are compared with those in the "reinforcement cue register". If a match occurs, this event then becomes a part of short-term memory and determines the "procedure" that ultimately produces the desired behavior (see Gray, 1981).

A similar comparator has been proposed to account for orienting and habituation (Sokolov, 1963). According to Sokolov (1963), a neurological model of external ongoing stimulation is associated with neocortical structures. A comparator, possibly associated with hippocampal function (Vinogradova, 1975), continuously compares this cortical model of background stimulation with current sensory input. Any mismatch between current input and background stimulation results in the elicitation of the OR. We wish to utilize a similar mechanism to explain why stimuli become significant, not only as a result of novelty, but also as a result of association with a reinforcer, as in classical conditioning. In the latter case, such a comparator would receive internal input from motivational mechanisms and external input via sensory stimulation. The exact nature of these motivational mechanisms and the way they interact with sensory input at physiological levels is beyond the scope of the present paper; however, others have addressed these issues (e.g. see Bindra, 1974). In any case, the specific needs of the organism would be a major determinant of whether a match or a mismatch occurs. In such a comparison, a stimulus in the sensory register should match one in the reinforcement cue register to be noted as "significant". With novelty, however, the reverse occurs; significance is signaled by a mismatch of the pattern in the cortical model with that in current sensory input. In either case, a major autonomic response to such stimulation consists of primary bradycardia.

At a neurophysiological level, the occurrence of a specific need must allow the "reinforcement cue register" selection process access to those neuronal cells (or cell systems) that have previously signaled the occurrence of a reinforcer that met these particular needs. In essence, then, the "reinforcement cue register" may be conceived as providing momentary access of the comparator to working memory to determine whether the current contents of the sensory register match any stimuli that have been highlighted in long-term memory by current motivational status as being important. The "reinforcement cue register" thus would not consist of reinforcing stimuli per se, but rather signals (viz., "cues") that previously predicted reinforcers associated with current motivational status. It would therefore consist of the representations of "conditional" reinforcers. More importantly for present purposes, we further suggest that the PFCm plays a major role in the establishment of the secondary reinforcing properties of cues associated with reinforcers by providing these signals with their affective qualities.

Two further points should be made regarding the operation of the "reinforcement cue register". First, we do not wish to imply that this register is necessarily a separate memory structure, as is suggested by the model presented in Fig. 14. Obviously this model is only one of many that might be postulated. An equally valid conception of this kind of activity is that stimuli associated with reinforcement in long-term memory become differentially activated or perhaps an "importance" tag becomes attached to such long-term memory elements. Instead, what we would like to emphasize is that a comparison-like process similar to that described above is essential for the operation of any information processing model. Moreover, our data, as described in detail above, suggests that the PFCm is involved in this process in some as yet undetermined fashion. Further, our preliminary research on MD function suggests that it may be involved in selection of an appropriate procedure, or somatomotor behavior, based on such processing of information by the PFCm (again see Fig. 14). Much evidence suggests that the programs for executing these procedures are associated with extrapyramidal structures (see above).

The second point relates to the use of the term "reinforcer". Obviously we do not mean to restrict the use of this term only to biologically relevant events. Many cues that would make up a "reinforcement cue register", at least at the human level, would no doubt be related to social and/or cognitive functioning. Such stimuli include those of conceptual as well as biological importance. Thus, stimulus events that serve as cues for mental operations might also be a part of the "reinforcement cue register". Consequently, such a register might be extremely important in more complex cognitive processes such as thinking, language, problem solving, etc., where chains of stimuli are necessary to accomplish a goal. In this case response-produced stimuli become the cues for further behavior via complicated feedback circuits involving the procedural memory store, as is also illustrated in Fig. 14. In this regard also see Yajeya et al. (1988).

It is unclear whether the affective characteristics of stimuli are represented physically, viz., neurally, in the PFCm or whether the PFCm is merely involved in establishing the importance of such stimuli via the comparison process described above. This is the familiar question of whether acquisition (encoding) or storage of information is involved. A great deal of evidence suggests that although such a process might involve PFCm function, no permanent neural representations are stored in the PFCm. Rather the PFCm may be involved only in increasing the momentary activity of a subset of neural elements that are simultaneously represented in the sensory registers and in the "reinforcement cue register" of working memory. The findings of Orona and Gabriel (1983) for example, that MUA in the PFCm reaches a maximum during differential avoidance conditioning long before the behavioral discrimination is made, supports this hypothesis. Similarly, CS-evoked multiple unit activity of the PFCm increases initially during acquisition of the cardiac changes associated with Pavlovian conditioning but declines during later trials (Gibbs and Powell, 1988), also supporting the hypothesis that the PFCm is involved only in the original acquisition of associative significance.

Summary

The major conclusion to be drawn from the above-described research on the role of the PFCag in classical conditioning is obviously that it plays a primary and perhaps necessary role in the establishment of visceral cues associated with exposure to classical conditioning contingencies. Specifically, these visceral changes appear to be of an inhibitory character. This is significant, since we have postulated that inhibitory cardiac changes invariably accompany initial processing of sensory stimuli for informational value. Such visceral changes are thus not epiphenomena associated with other simultaneously occurring physiological events.

A variety of lesion experiments implicate the PFCm as a central structure in this process, since damage to this area greatly attenuates, and in the case of hypothalamic knife cuts, completely eliminates learned bradycardia. Neuroanatomical tract-tracing experiments revealed that the PFCm and Iag have direct projections to the NTS and DVM in the dorsomedial medulla and the nucleus ambiguous in the ventral medulla, all of which provide medullary output control of visceral activities. The nucleus ambiguous and DVM have been specifically implicated in vagal control in the rabbit (Ellenberger et al., 1983). Electrical stimulation of the PFCm provides additional evidence that this area of the brain participates in parasympathetic activities, including cardiac inhibition, since stimulation of the entire MD projection cortex, including the PFCm, produces HR decelerations accompanied by depressor responses. However, since lesions of the Iag produced relatively little effect on conditioned bradycardia, this part of the PFCag does not appear to play a major role in the development of conditioned bradycardia.

Electrophysiological recording studies, including both multiple unit as well as extracellular single unit studies reinforce these conclusions. A short latency (40–180 msec) CS-evoked increase in MUA was recorded from cells in both the dorsomedial as well as central PFCm. The magnitude of these CS-evoked neuronal changes (a) was correlated with the magnitude of concomitantly occurring conditioned bradycardia; (b) was trial-related; (c) was not obtained in a similar pseudoconditioning group; and (d) declined to pretraining levels during subsequent experimental extinction. Similar, but not identical, CS-evoked changes in neuronal activity were recorded from MD. Although tone-evoked increases in MUA were also obtained from the Iag, this activity did not show the characteristics of associative learning. Single unit analysis also suggests the importance of the PFCm in elicitation of conditioned bradycardia. Although a variety of units were found that responded to CS/UCS contingencies, the most frequent types of cells were those which showed short latency increases in response to the presentation of the CS+ but responded minimally to an unreinforced CS−. Single unit activity was also significantly correlated with concomitantly occurring HR changes in most cases.

The evidence from these experiments taken together thus suggests that the PFCm, but not the Iag, plays a central role in the information processing that accompanies the early stages of classical conditioning. Other data suggest that the thalamic projection nucleus for PFCm may play a role in the later stages of classical conditioning, in which a somatomotor response is selected for execution. The integration of this information into a general model of classical conditioning within an information processing context was suggested. Specifically it was suggested that the PFCm might provide the affective properties associated with cues that predict reinforcers, and therefore might function as a "reinforcement cue register" that is sampled early during information processing.

Acknowledgements

The research reviewed here was supported by VA Institutional Research funds awarded to the William Jennings Bryan Dorn VA Medical Center, Columbia, SC 29201.

The authors thank Dianne Levine-Bryce, Karen Watson, and Richard Thompson for assistance in data collection and analysis. We also thank Elizabeth Hamel for secretarial assistance and the Medical Media Service of the VA Medical Center and the Medical Illustration Department of the University of South Carolina School of Medicine for assistance with the figures.

Abbreviations

AC	Anterior commissure
ACN	Amygdala central nucleus
AL	Anterior limbic area of the prefrontal cortex
AM	Anteromedial nucleus of the thalamus
AMG	Amygdala
AV	Anteroventral nucleus of the thalamus
BP	Blood pressure
Cd	Caudate nucleus
CEM	Centromedian nucleus of the thalamus
Cl	Claustrum
CP	Paracentral nucleus of the thalamus
CR	Conditioned response
CS	Conditioned stimulus
DVM	Dorsal motor nucleus of the vagus
EB	Eyeblink
Fx	Fornix
HP	Heart period
HR	Heart rate
HRP	Horseradish peroxidase
[^{3}H]2-DG	[^{3}H]2-Deoxyglucose
Iag	Agranular insular area of the prefrontal cortex
IC	Internal capsule
IL	Infralimbic area of the prefrontal cortex
IMD	Intermediodorsal nucleus of the thalamus
IN	Insula
LH	Lateral hypothalamus
MD	Mediodorsal nucleus of the thalamus
MUA	Multiple unit activity
MV	Medioventral nucleus of the thalamus
NA	Nucleus ambiguous
NM	Nictitating membrane
NTS	Nucleus of the solitary tract
OR	Orienting reflex
OT	Optic tract
PAG	Periaqueductal grey
PC	Paracentral nucleus of the thalamus
PFCag	Agranular precentral area of the prefrontal cortex
PFCm	Medial area of the agranular prefrontal cortex
PUT	Putamen
Re	Reuniens nucleus of the thalamus
SC	Superior colliculus
UCR	Unconditioned response
UCS	Unconditioned stimulus
VL	Ventrolateral nucleus of the thalamus
VM	Ventromedial nucleus of the thalamus

VP	Ventroposterior nucleus of the thalamus
WGA-HRP	Wheat germ agglutinin conjugated to horseradish peroxidase

References

Anderson, J.R. (1983) *The Architecture of Cognition,* Harvard University Press, Cambridge, MA.

Arikuni, T. and Ban, T., Jr. (1978) Subcortical afferents to the prefrontal cortex in rabbits. *Exp. Brain Res.,* 32: 69–75.

Beckstead, R.M. (1979) An autoradiographic examination of corticocortical and subcortical projections of the mediodorsal-projection (prefrontal) cortex in the rat. *J. Comp. Neurol.,* 184: 43–62.

Benjamin, R.M., Jackson, J.C. and Golden, G.T. (1978) Cortical projections of the thalamic mediodorsal nucleus in the rabbit. *Brain Res.,* 141: 251–265.

Bindra, D. (1974) A motivational view of learning, performance, and behavior modification. *Psychol. Rev.,* 81: 199–213.

Bishop, P.O., Burke, W. and Davies, R. (1962) The identification of single units in central visual pathways. *J. Physiol. (Lond.),* 162: 409–431.

Brady, J.V. (1975) Conditioning and emotion. In L. Levi (Ed.), *Emotions: Their Parameters and Measurement,* Raven Press, New York, pp. 309–340.

Buchanan, S.L. (1988) Mediodorsal thalamic lesions impair differential Pavlovian heart rate conditioning. *Exp. Brain Res.,* 73: 320–328.

Buchanan, S.L. and Thompson, R.H. (1990) Mediodorsal thalamic lesions and Pavlovian conditioning of heart rate and eyeblink responses in the rabbit, *Behav. Neurosci.,* 104: 912–918.

Buchanan, S.L. and Powell, D.A. (1980) Divergencies in Pavlovian conditioned heart rate and eyeblink responses produced by hippocampectomy in the rabbit. *Behav. Neural Biol.,* 30: 20–38.

Buchanan, S.L. and Powell, D.A. (1982a) Cingulate damage attenuates conditioned bradycardia. *Neurosci. Lett.,* 29: 261–268.

Buchanan, S.L. and Powell, D.A. (1982b) Cingulate cortex: its role in Pavlovian conditioning. *J. Comp. Physiol. Psychol.,* 96: 755–774.

Buchanan, S.L. and Powell, D.A. (1986) Electrical stimulation of anteromedial and mediodorsal thalamus elicits differential cardiovascular response patterns from conscious rabbits. *Physiol. Psychol.,* 14: 115–123.

Buchanan, S.L. and Powell, D.A. (1987) ^{3}H-2-Deoxyglucose uptake after electrical stimulation of cardioactive sites in insular cortex and mediodorsal nucleus of the thalamus in rabbits. *Brain Res. Bull.,* 19: 439–452.

Buchanan, S.L. and Powell, D.A. (1988a) Age-related changes in associative learning: studies in rabbits and rats. *Neurobiol. Aging,* 9: 523–534.

Buchanan, S.L. and Powell, D.A. (1988b) Parasagittal thalamic knife cuts retard Pavlovian eyeblink conditioning and abolish the tachycardiac component of the heart rate conditioned response. *Brain Res. Bull.,* 21: 723–729.

Buchanan, S.L. and Powell, D.A. (1989) Parasagittal thalamic knife cuts and cardiac changes. *Behav. Brain Res.,* 32: 241–253.

Buchanan, S.L., Powell, D.A. and Buggy, J. (1984) ^{3}H-2-Deoxyglucose uptake after electrical stimulation of cardioactive sites in anterior medial cortex in rabbits. *Brain Res. Bull.,* 13: 371–382.

Buchanan, S.L., Valentine, J.D. and Powell, D.A. (1985) Autonomic responses are elicited from medial but not lateral frontal cortex in rabbits. *Behav. Brain Res.,* 18: 51–62.

Buchanan, S.L., Powell, D.A. and Thompson, R.H. (1989) Prefrontal projections to the medial nuclei of the dorsal thalamus in the rabbit. *Neurosci. Lett.,* 106: 55–59.

Cassell, M.D. and Wright, D.J. (1986) Topography of projections from the medial prefrontal cortex to the amygdala in the rat. *Brain Res. Bull.,* 17: 321–333.

Cechetto, D.F. and Saper, C.B. (1987) Evidence for a viscerotopic sensory representation in the cortex and thalamus in the rat. *J. Comp. Neurol.,* 262: 27–45.

Cohen, D.H. and Macdonald, R.L. (1976) Involvement of the avian hypothalamus in defensively conditioned heart rate change. *J. Comp. Neurol.,* 167: 465–480.

Cohen, D.H. and Randall, D.C. (1984) Classical conditioning of cardiovascular responses. *Annu. Rev. Physiol.,* 46: 187–197.

Disterhoft, J.F., Golden, D.T., Read, H.L., Coulter, D.A. and Alkon, D.L. (1988) AHP reductions in rabbit hippocampal neurons during conditioning correlate with acquisition of the learned response. *Brain Res.,* 462: 118–125.

Domesick, V.B. (1969) Projections from the cingulate cortex in the rat. *Brain Res.,* 12: 296–320.

Dykman, R.A., Mack, R.L. and Ackerman, P.T. (1965) The evolution of autonomic and motor components of the nonavoidance conditioned response in the dog. *Psychophysiology,* 1: 209–230.

Ellenberger, H., Haselton, J.R., Liskowsky, D.R. and Schneiderman, N. (1983) The location of chronotropic cardioinhibitory vagal motoneurons in the medulla of the rabbit. *J. Auton. Nerv. Syst.,* 9: 513–529.

Estes, W.K. (1988) Human learning and memory. In R.C. Atkinson, R.J. Herrnstein, G. Lindzey and R.D. Luce (Eds.), *Stevens Handbook of Experimental Psychology. Vol. 2. Learning and Cognition,* John Wiley, New York, pp. 351–415.

Francis, J., Hernandez, L.L. and Powell, D.A. (1981) Lateral hypothalamic lesions: effects on Pavlovian conditioning of eyeblink and heart rate responses in the rabbit. *Brain Res.*

Bull., 6: 155–163.

Fussey, I.K., Kidd, C. and Whitwam, J.G. (1970) The differentiation of axonal and soma-dendritic spike activity. *Pflügers Arch.*, 321: 283–292.

Fuster, J.M. (1989) *The Prefrontal Cortex: Anatomy, Physiology and Neuropsychology of the Frontal Lobe*, 2nd Edn., Raven Press, New York.

Gantt, W.H. (1960) Cardiovascular component of the conditional reflex to pain, food, and other stimuli. *Physiol. Rev.*, 40: 266–291.

Gibbs, C.M. and Powell, D.A. (1988) Neuronal correlates of classically conditioned bradycardia in the rabbit: studies of the medial prefrontal cortex. *Brain Res.*, 422: 86–96.

Gibbs, C.M. and Powell, D.A. (1990) Single-unit activity in the dorsomedial prefrontal cortex during the expression of discriminative bradycardia in rabbits. *Behav. Brain Res.*, in press.

Gibbs, C.M., Prescott, L. and Powell, D.A. (1989a) A comparison of CS-evoked multiple unit activity in medial and insular prefrontal cortex during classical heart rate conditioning in the rabbit. Submitted for publication.

Gibbs, C.M., Watson, K.L. and Gibbs, A.W. (1989b) Neuronal activity in the rabbit's medial prefrontal cortex (PFCm) is influenced by aversive and appetitive Pavlovian contingencies. *Soc. Neurosci. Abstr.*, 15: 83.

Gormezano, I. (1966) Classical conditioning. In J.B. Sidowski (Ed.), *Experimental Methods and Instrumentation in Psychology,* McGraw Hill, New York, pp. 385–420.

Gormezano, I. (1972) Investigations of defense and reward conditioning in the rabbit. In A.H. Black and W.F. Prokasy (Eds.), *Classical Conditioning II: Current Research and Theory,* Appleton-Century-Crofts, New York, pp. 151–181.

Gormezano, I. and Coleman, S.R. (1973) The law of effect and CR contingent modification of the UCS. *Condit. Reflex*, 8: 41–56.

Gray, J.A. (1981) *The Neuropsychology of Anxiety,* Oxford University Press, Oxford.

Groenewegen, H. (1988) Organization of the afferent connections of the mediodorsal thalamic nucleus in the rat, related to the mediodorsal-prefrontal topography. *Neuroscience*, 24: 379–431.

Iwata, J., LeDoux, J.E. and Reis, D.J. (1986) Destruction of intrinsic neurons in the lateral hypothalamus disrupts the classical conditioning of autonomic but not behavioral emotional responses in the rat. *Brain Res.*, 368: 161–166.

Kao, K.-T. and Powell, D.A. (1988) Lesions of the substantia nigra retard Pavlovian eyeblink but not heart rate conditioning in the rabbit. *Behav. Neurosci.*, 102: 515–525.

Kapp, B.S., Frysinger, R.C., Gallagher, M. and Haselton, J.R. (1979) Amygdala central nucleus lesions: effect on heart rate conditioning in the rabbit. *Physiol. Behav.*, 23: 1109–1117.

Kapp, B.S., Schwaber, J.S. and Driscoll, P.A. (1985a) Frontal cortex projections to the amygdaloid central nucleus in the rabbit. *Neuroscience,* 15: 327–346.

Kapp, B.S., Schwaber, J. and Driscoll, P. (1985b) The organization of insular cortex projections to the amygdaloid central nucleus and autonomic regulatory nuclei of the dorsal medulla. *Brain Res.*, 360: 355–360.

Kazis, E., Milligan, W.L. and Powell, D.A. (1973) Autonomic-somatic relationships: blockade of heart rate and corneoretinal potential. *J. Comp. Physiol. Psychol.*, 84: 98–110.

Konorski, J. (1967) *Integrative Activity of the Brain,* University of Chicago Press, Chicago, IL.

Krettek, J. and Price, J. (1977) The cortical projections of the mediodorsal nucleus and adjacent thalamic nuclei in the rat. *J. Comp. Neurol.*, 171: 157–192.

Lacey, B.C. and Lacey, J.I. (1980) Cognitive modulation of time-dependent primary bradycardia. *Psychophysiology,* 17: 209–221.

LeDoux, J.E., Sakaguchi, A. and Reis, D.J. (1984) Subcortical efferent projections of the medial geniculate nucleus mediate emotional responses conditioned to acoustic stimuli. *J. Neurosci.*, 4: 683–698.

Leonard, C.M. (1969) The prefrontal cortex of the rat. I. Cortical projection of the mediodorsal nucleus. II. Efferent connections. *Brain Res.*, 12: 321–343.

Markowitsch, H.J. (1982) Thalamic mediodorsal nucleus and memory: a critical evaluation of studies in animals and man. *Neurosci. Biobehav. Rev.*, 6: 351–380.

Markowitsch, H.J. and Guldin, W.O. (1983) Heterotopic interhemispheric cortical connections in the rat. *Brain Res. Bull.*, 10: 805–810.

McDonald, A. (1987) Organization of amygdaloid projections to the mediodorsal thalamus and prefrontal cortex: a fluorescence retrograde transport study in the rat. *J. Comp. Neurol.*, 262: 46–58.

Miller, N.E. (1948) Studies of fear as an acquirable drive. I. Fear as motivation and fear-reduction as reinforcement in the learning of new responses. *J. Exp. Psychol.*, 38: 89–101.

Minciacchi, D., Molinari, M., Bentivoglio, M. and Macchi, G. (1985) The organization of the ipsi- and contralateral claustrocortical system in rat with notes on the bilateral claustrocortical projections in cat. *Neuroscience,* 16: 557–576.

Moore, J.W. (1979) Brain processes and conditioning. In A. Dickinson and R.A. Boakes (Eds.), *Mechanisms of Learning and Motivation: A Memorial Volume to Jerzy Konorski,* Erlbaum, Hillsdale, NJ, pp. 111–142.

Mowrer, O.H. (1960) *Learning Theory and Behavior,* Wiley, New York.

Neafsey, E.J., Hurley-Gius, K.M. and Arvanitis, D. (1986) The topographical organization of neurons in the rat medial frontal, insular and olfactory cortex projecting to the solitary nucleus, olfactory bulb, periaqueductal gray and superior colliculus. *Brain Res.* 377: 261–270.

Obrist, P.A. (1981) *Cardiovascular Psychophysiology: A Perspective,* Plenum Press, New York.

Orona, E. and Gabriel, M. (1983) Multiple-unit activity of the prefrontal cortex and mediodorsal thalamic nucleus during acquisition of discriminative avoidance behavior in rabbits. *Brain Res.,* 263: 295–312.

Pascoe, J.P. and Kapp, B.S. (1985) Electrophysiological characteristics of amygdaloid central nucleus neurons during Pavlovian fear conditioning in the rabbit. *Behav. Brain Res.,* 16: 117–133.

Pascoe, J.P. and Kapp, B.S. (1987) Response of amygdaloid central nucleus neurons to stimulation of the insular cortex in awake rabbits. *Neuroscience,* 21: 471–485.

Pearson, R.C., Brodal, P., Gatter, K.C. and Powell, T.P.S. (1982) The organization of the connections between the cortex and claustrum in the monkey. *Brain Res.,* 234: 435–441.

Powell, D.A. (1979) Peripheral and central muscarinic cholinergic blockade: effects on Pavlovian conditioning. *Bull. Psychonom. Soc.,* 14: 161–164.

Powell, D.A. and Buchanan, S.L. (1986) Autonomic changes elicited by chemical stimulation of the mediodorsal nucleus of the thalamus. *Pharmacol. Biochem. Behav.,* 25: 423–430.

Powell, D.A. and Kazis, E. (1976) Blood pressure and heart rate changes accompanying classical eyeblink conditioning in the rabbit *(Oryctolagus cuniculus). Psychophysiology,* 13: 441–447.

Powell, D.A. and Levine-Bryce, D. (1988) A comparison of two model systems of associative learning: heart rate and eyeblink conditioning in the rabbit. *Psychophysiology,* 25: 672–682.

Powell, D.A. and Levine-Bryce, D. (1989) Conditioned bradycardia in the rabbit: effects of knife cuts and ibotenic acid lesions in the lateral hypothalamus. *Exp. Brain Res.,* 76: 103–121.

Powell, D.A. and Watson, K. (1989) Multiple unit activity in the mediodorsal nucleus of the thalamus evoked by Pavlovian conditioning. *Soc. Neurosci. Abstr.,* 15: 83.

Powell, D.A., Schneiderman, N., Elster, A.J. and Jacobson, A. (1971) Differential classical conditioning in rabbits *(Oryctolagus cuniculus)* to tones and change in illumination. *J. Comp. Physiol. Psychol.,* 76: 267–274.

Powell, D.A., Lipkin, M. and Milligan, W.L. (1974) Concomitant changes in classically conditioned heart rate and corneoretinal potential discrimination in the rabbit *(Oryctolagus cuniculus). Learn. Motivat.,* 5: 532–547.

Powell, D.A., Mankowski, D. and Buchanan, S.L. (1978) Concomitant heart rate and corneoretinal potential conditioning in the rabbit *(Oryctolagus cuniculus):* effects of caudate lesions. *Physiol. Behav.,* 20: 143–150.

Powell, D.A., Hernandez, L.L. and Buchanan, S.L. (1985) Electrical stimulation of insular cortex elicits cardiac inhibition but insular lesions do not abolish conditional bradycardia in rabbits. *Behav. Brain Res.,* 17: 125–144.

Pribram, K.H. and McGuinness, D. (1975) Arousal, activation and effort in the control of attention. *Psychol. Rev.,* 82: 116–149.

Prokasy, W.F. (1984) Acquisition of skeletal conditioned responses in Pavlovian conditioning. *Psychophysiology,* 21: 1–13.

Putnam, L.E., Ross, L.E. and Graham, F.K. (1974) Cardiac orienting during "good" and "poor" differential eyelid conditioning. *J. Exp. Psychol.,* 102: 563–573.

Reep, R. (1984) Relationship between prefrontal and limbic cortex: a comparative anatomical review. *Brain Behav. Evol.,* 25: 5–80.

Reep, R.L. and Winans, S.S. (1982) Efferent connections of dorsal and ventral agranular insular cortex in the hamster *(Mesocricetus auratus). Neuroscience,* 7: 2609–2635.

Rescorla, R.A. (1988) Pavlovian conditioning: it is not what you think it is. *Am. Psychol.,* 43: 151–160.

Rose, J. and Woolsey, C. (1948) Structure and relations of limbic cortex and anterior thalamic nuclei in rabbit and cat. *J. Comp. Neurol.,* 89: 279–348.

Rumelhart, D.E. and McClelland, J.L. (1986) *Parallel Distributed Processing, Vol. 1,* MIT Press, Cambridge, MA.

Saper, C.B. (1982) Convergence of autonomic and limbic connections in the insular cortex of the rat. *J. Comp. Neurol.,* 210: 163–173.

Saper, C.B. (1985) Organization of cerebral cortical afferent systems in the rat. II. Hypothalamocortical projections. *J. Comp. Neurol.,* 237: 21–46.

Schneiderman, N. (1972) Response system divergencies in aversive classical conditioning. In A.H. Black and W.F. Prokasy (Eds.), *Classical Conditioning II: Current Theory and Research,* Appleton-Century-Crofts, New York, pp. 341–376.

Schneiderman, N. (1983) Animal behavior models of coronary heart disease. In D.S. Krantz, A. Baum and J.E. Singer (Eds.), *Handbook of Psychology and Health. Vol. 3. Cardiovascular Disorders and Behavior,* Erlbaum, Hillsdale, NJ, pp. 19–56.

Schneiderman, N. and Gormezano, I. (1964) Conditioning of the nictitating membrane of the rabbit as a function of CS-US interval. *J. Comp. Physiol. Psychol.,* 57: 188–195.

Schneiderman, N., Van Dercar, D.H., Yehle, A.L., Manning, A.A., Golden, T. and Schneiderman, E. (1969) Vagal compensatory adjustment: relationship to heart rate classical conditioning in rabbits. *J. Comp. Physiol. Psychol.,* 68: 175–183.

Schwaber, J.S., Kapp, B.S., Higgins, G.A. and Rapp, P.R. (1982) Amygdaloid and basal forebrain direct connections with the nucleus of the solitary tract and the dorsal motor nucleus. *J. Neurosci.,* 2: 1424–1438.

Seligman, M.E. and Weiss, J.M. (1980) Coping behavior:

learned helplessness, physiological change and learned inactivity. *Behav. Res. Ther.,* 18: 459 – 512.

Skinner, B.F. (1986) What is wrong with daily life in the western world? *Am. Psychol.,* 41: 568 – 574.

Smith, O.A. and DeVito, J.L. (1984) Central neural integration for the control of autonomic responses associated with emotion. *Annu. Rev. Neurosci.,* 7: 43 – 65.

Smith, O.A., Nathan, M.A. and Clarke, N.P. (1968) Central nervous system pathways mediating blood pressure changes. *Hypertension,* 16: 9 – 22.

Sokolov, E.N. (1963) *Perception and the Conditioned Reflex,* McMillan, New York.

Squire, L.R. (1986) Mechanisms of memory. *Science,* 232: 1612 – 1619.

Thompson, R.F. (1986) The neurobiology of learning and memory. *Science,* 233: 941 – 947.

Thompson, R.F., McCormick, D.A., Lavond, D.G, Clark, G.A., Kettner, R.E. and Mauk, M.D. (1983) The engram found? Initial localization of the memory trace for a basic form of associative learning. In J.L. Sprague and A.N. Epstein (Eds.), *Progress in Psychobiology and Physiological Psychology,* Academic Press, New York, pp. 167 – 196.

Van Der Kooy, D., Koda, L.Y., McGinty, J.F., Gerfen, C.R. and Bloom, F.E. (1984) The organization of projections from the cortex, amygdala, and hypothalamus to the nucleus of the solitary tract in rat. *J. Comp. Neurol.,* 224: 1 – 24.

Vanderwolf, C.H. (1971) Limbic-diencephalic mechanisms of voluntary movement. *Psychol. Rev.,* 78: 83 – 113.

Vinogradova, O.S. (1975) The hippocampus and the orienting reflex. In E.N. Sokolov and O.S. Vinogradova (Eds.), *Neuronal Mechanisms of the Orienting Reflex,* Wiley, New York, pp. 128 – 154.

Weinberger, N.M. and Diamond, D.M. (1987) Physiological plasticity in auditory cortex: rapid induction by learning. *Progr. Neurobiol.,* 29: 1 – 55.

West, C.H.K. and Benjamin, R.M. (1983) Effects of stimulation of the mediodorsal nucleus and its projection cortex on heart rate in the rabbit. *J. Autonom. Nerv. Syst.,* 9: 547 – 557.

Witter, M.P., Room, P., Groenewegen, H.J. and Lohman, A.H.M. (1988) Reciprocal connections of the insular and piriform claustrum with limbic cortex: an anatomical study in the cat. *Neuroscience,* 24: 519 – 539.

Woody, C.D. (1982) *Memory, Learning, and Higher Function: A Cellular View,* Springer, New York.

Wyss, J.M. and Sripanidkulchai, K. (1984) The topography of the mesencephalic and pontine projections from the cingulate cortex of the rat. *Brain Res.,* 293: 1 – 15.

Yajeya, J., Quintana, J. and Fuster, J.M. (1988) Prefrontal representation of stimulus attributes during delay tasks. II. The role of behavioral significance. *Brain Res.,* 474: 222 – 230.

Yeo, C.L., Hardiman, M.J. and Glickstein, M. (1984) Discrete lesions of the cerebellar cortex abolish the classically conditioned nictitating membrane response of the rabbit. *Behav. Brain Res.,* 13: 261 – 266.

Discussion

C.G. van Eden: The major input as well as output of MD is through the PFC. Why are MD responses later and through which pathways is the effect achieved?

D.A. Powell: One answer to this question is that information is received and processed by the PFC and the latter, through its efferents, drives the cells of MD causing a later overall increase in neuronal discharge in this nucleus than that originally occurring in the PFC. However, this is contrary to the way one usually thinks of MD – prefrontal relations, since the PFC is usually considered to be the projection cortex for MD, and presumably would receive information through MD efferents. If this is the case, one would expect PFC discharge to be later than that of MD. Most probably MD and PFC are being driven by differential inputs. Although it is true that the major output of MD is to the PFC (its only other efferents being to other thalamic nuclei) it receives a number of afferents from a variety of subcortical nuclei, including the amygdala, ventral tegmental area, basal forebrain, etc. Although some of these nuclei also project to the PFC, I would guess that the differential latency in neuronal discharge by MD and the PFC during classical conditioning are due to differential inputs.

F.H. Lopes da Silva: The system you are investigating here is a control system with a number of feedback loops; one important element of this control system is the input coming from the baroreceptors. I wonder whether sectioning these inputs affects the HR or BP responses elicited by stimulation of the PFC. Another question is, how do you think that the PFC fits into these lower brainstem control systems? Is this a sort of "feedforward" position looking down to the lower control loops or is it itself a part of the feedback loops?

D.A. Powell: To answer your first question, we believe that primary bradycardia occurs somewhat independently of these feedback loops. It has been determined that changing baroreceptor gain, within limits, has no direct effect on the magnitude of the conditioned HR response, and classical conditioning has no effect on baroreceptor sensitivity, although it has been found to increase baroreceptor gain. Also the fact that the conditioned BP response in this paradigm is a depressor rather than a pressor response suggests the lack of baroreceptor participation in classically conditioned bradycardia. We have not studied the effects of eliminating baroreceptor input on the cardiovascular changes elicited by stimulation of the PFC. However, as noted, administration of α-adrenergic receptor antagonists eliminates or greatly diminishes the magnitude of the bradycardia elicited by MD stimulation, suggesting baroreceptor mediation of this response. In answer to the second part of your question, as I indicated, the direct connections between the

PFC and nucleus ambiguous and dorsal motor nucleus of the vagus suggest that this system may operate somewhat independently of the baroreceptor system. However, it is possible that PFC connections with the hypothalamus, or the interaction of baroreceptor input with the medullary nuclei themselves, may indirectly affect such changes. In any case, I feel that the PFC system is in a feedforward position with regard to medullary control loops and is not itself a part of such loops. Indeed primary bradycardia was originally defined by Schneiderman and colleagues and later by the Lacey's as HR slowing that was not secondary to such influences. We believe that conditioned bradycardia meets these criteria and therefore is a type of learned primary bradycardia. It would thus not be directly affected by baroreceptor changes.

H.J. Groenewegen: (1) The injection sites, lesions and recording sites in the medial prefrontal cortex were all situated dorsomedially. Neafsey showed his results in the rat in the more ventral parts (infralimbic). Is this a species difference? (2) To what extent may the effects you described for the MD be attributable to the midline thalamic nuclei, particularly the paraventricular nucleus?

D.A. Powell: To answer the first part of your question, the manipulations which we have made on the PFCm, as you indicated, have for the most part affected areas 8, 24 and 32 while Neafsey's results in the rat, although not limited to area 25 certainly focused on this part of the midline prefrontal region. There are some indications that in cats and primates the more dorsal PFCm subserves parasympathetic cardioinhibitory changes, but more ventral areas are sympathetic in nature. If this is the case in rat and rabbit as well, it is possible that the infralimbic area comprises a sympathetic area, while the more dorsal anterior limbic and precentral agranular areas are parasympathetic. However, there are also species differences between rats and rabbits with regard to their response to stress. For example, the rabbit adapts well to restraint, while rats become extremely sympathetically aroused when restrained. For this reason baseline heart rate in the rat is excessively elevated during restraint compared to rabbits, and conditioned HR decelerations are obtained when rats are restrained. However, the conditioned heart rate changes reported by Neafsey were in the free-moving rat, and thus consisted of heart rate accelerations. Thus not only is the area of the brain which Neafsey studied different from that which we studied but the conditioned response is basically opposite to the conditioned bradycardia which we observed. Thus, further research will be necessary to determine if the differences between our results and those of Neafsey are indeed due to known species differences, or to differential connections of areas 25 and 32 with brainstem cardiovascular control mechanisms.

In answer to the second part of your question, I am positive that the effects which we described for MD can not be attributed to the midline thalamic nuclei. We did find that there are cells in the paraventricular nucleus whose projection overlaps that of MD. In separate experiments, however, we have studied the effects of midline thalamic and MD lesions, and while there are some similarities (for example, midline thalamic lesions also exaggerate conditioned bradycardia), there are also differences. The results of our MD stimulation studies with glutamate and carbachol could not have been due to stimulation of fibers of passage from the midline nuclei since these chemicals could only affect the soma in MD. Likewise, the lesions, which we produced using ibotenic acid, were all limited to MD and since ibotenic acid also affects only the soma, these lesions could not have damaged fibers of passage from the midline nuclei which may pass laterally through MD. Thus, I feel confident that the lesion and stimulation effects which we reported for MD can not be attributed to the action of the midline thalamic nuclei.

CHAPTER 23

Functions of anterior and posterior cingulate cortex during avoidance learning in rabbits

Michael Gabriel

Department of Psychology and Beckman Institute, University of Illinois, Urbana, IL 61801, USA

Introduction

There is currently a great interest in the neural mediation of learning and memory processes. A substantial amount of the research on this issue is being carried out with a model system strategy, wherein neural substrates of learning are studied in specific well-defined classical and instrumental conditioning paradigms (see reviews by Thompson, 1986; Gabriel et al., 1989b; Byrne, 1987). This chapter presents the current status of such a model, concerning analysis of the neural substrates of avoidance learning in rabbits.

The goal of the analysis is to identify the brain processes that mediate avoidance learning. The task is administered as rabbits occupy a running wheel, a replica of an apparatus designed by Brogden and Culler (1936). The rabbits learn to avoid a shock, a 1.5 mA alternating current delivered to the footpads of the rabbits. Avoidance is achieved by performance of a conditioned response (CR) to a tone that precedes the shock by 5 sec. The required CR is simply a step sufficient to turn the wheel 2° or more. The rabbits also learn to ignore a different tone. We refer to the shock-predictive and the non-predictive tones as positive and negative conditional stimuli, CS+ and CS− respectively. Acquisition to asymptotic levels of discriminative CR performance occurs after an average of 3 days of training (120 trials daily, 60 with each tone in an irregular order). The trained rabbits perform CRs, avoiding shock, on an average of 85% of the CS+ trials and they respond on fewer than 10% of the CS− trials. In order to study the neuronal events that support this learning, multi-unit and single-unit activity is recorded simultaneously in several brain areas during acquisition. The analysis of neural circuit interactions is approached by examining the effects of selective deafferenting lesions on the training-related unit activity in various neuronal target populations.

Anatomical substrates

The available data implicate cingulate cortex and the related "limbic" thalamic nuclei in the mediation of CR acquisition and performance. The involved cortical areas, illustrated in Fig. 1, include the anterior cingulate cortex (Brodmann's area 24b) and the posterior cingulate cortex (Brodmann's areas 29b and 29c). Area 24b receives thalamic afferents from the medial dorsal (MD) nucleus and is thus often regarded as functionally analogous to the primate prefrontal cortex (PFC). Neurons in the posterior cingulate cortex are reciprocally interconnected with neurons of the

Correspondence: Dr. M. Gabriel, Department of Psychology and Beckman Institute, University of Illinois, 405 N. Mathews Avenue, Urbana, IL 61801, USA. Tel.: 217-244-3463; Fax: 217-244-8371.

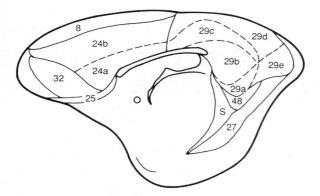

Fig. 1. Schematic drawing of mid-sagittal section of the rabbit brain portraying the cytoarchitectonic areas of cingulate cortex. The diagram is based on original designations of Brodmann recently modified by Vogt et al. (1986). Further explanation is provided in the text.

anterior thalamic nuclei, forming part of the well-known circuit of Papez. Thus, there exists a separate anterior system (the anterior cingulate cortex and the related MD thalamic nuclei), and a posterior system (the posterior cingulate cortex and the anterior thalamic nuclei). This dichotomy is qualified, not absolute, as the two cortical areas in all mammalian species studied are monosynaptically and reciprocally interconnected and they share input from the anteromedial thalamic nucleus (see Vogt, 1985).

Shared functions of the anterior and posterior systems

A theoretical model

The shared functions of these two systems are portrayed in a theoretical model of the neuronal plasticity and dynamic information flow underlying CR acquisition and performance (Gabriel et al., 1980b, 1986; Fig. 2). The model envisions two systems, a GO system and a STOP system. The activation of neurons in the GO system is associated with the output of the avoidance CR, whereas the activation of STOP system neurons is associated with CR inhibition.

The GO system

Neurons of the limbic thalamic (AV and MD) nuclei comprise the core of the GO system. This system's operation is simple. Upon CS+ presentation to a trained rabbit, CR performance is induced by neuronal excitatory volleys that originate in limbic thalamic cells and flow through cingulate cortex to motor system areas. The identity of the specific involved motor structures and certain associative processes of the motor system that contribute to CR acquisition are discussed elsewhere (Gabriel et al., 1986). Plasticity development responsible for the CS-driven thalamic activity occurs at the level of the thalamus and does not require cortical input.

Empirical evidence supporting the model is described in the following paragraphs. All of the neuronal and behavioral findings cited, and references to statistical significance, are based on the occurrence of significant F values ($P < 0.05$) in mixed repeated measures analyses of variance, followed by significant ($P < 0.05$) individual comparisons of relevant treatment means.

Evidence for the GO system

Discriminative neuronal activity

An observation germane to the GO system construct is the development, during behavioral acquisition, of excitatory training-induced multi-unit discharges throughout the two systems. The training-induced discharges result from the associative pairing of the CS+ with the US. Minimal or no training-induced discharges develop in response to the CS−, the CS not paired with the US (e.g., Fig. 3). Thus, an important aspect of the training-induced discharges is their discriminative character, i.e., the greater discharge magnitudes in response to the CS+ than to the CS−. It is this property that indicates that the activity is a reflection of a learning process. Counterbalancing assures that each of two tonal frequencies (8 kHz and 1 kHz) are used as both CS+ and CS−. Thus, the discriminative discharges are not prewired, tone-specific effects.

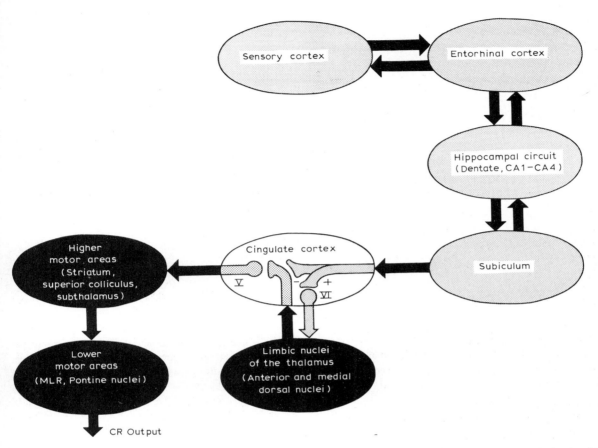

Fig. 2. Theoretical working model of the interactions of cingulate cortex, hippocampal formation and the limbic nuclei of thalamus during learning. The dark and light ovals represent the GO and STOP systems, respectively, as defined in the text. Bidirectional arrows interconnecting the subdivisions of the hippocampal formation denote informational exchanges that support hippocampal perceptual and mnemonic representations of the current environment. The dark arrow connecting the subiculum and the cingulate cortex represents hippocampal formation input to cingulate cortex that is involved in suppressing (limiting) the firing of thalamic neurons, and CR performance. Arrows connecting the dark ovals represent information flow from limbic thalamus through cingulate cortex to the motor system to initiate CR performance. A detailed explanation of this model is provided in the text and in Gabriel et al. (1986, 1987a).

CR predictive neuronal activity

The training-induced discriminative discharges occur in the first 100 – 600 msec after CS onset and have latencies as brief as 15 msec. A different variety of training-induced neuronal activity occurs in the two systems at greater intervals after CS initiation. Substantial proportions of the extracellularly recorded cingulate cortical and limbic thalamic single units in trained rabbits show a firing-frequency build-up in anticipation of CR output (Fig. 4). This pattern of neuronal firing is consistent with the model's hypothesis that CS-related neuronal activities in these areas are involved in the elicitation of CR performance.

Lesions and CR acquisition

Bilateral electrolytic lesions that include both the anterior and posterior cingulate cortex, or lesions that include both the MD and anterior thalamic nuclei essentially block acquisition of the avoi-

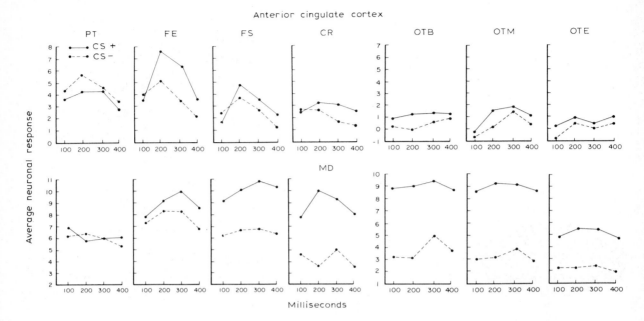

Fig. 3. Average neuronal firing frequency elicited by CS+ (dark lines) and CS− (dashed lines) in the anterior cingulate cortex (upper row) and in the medial dorsal thalamic nucleus (MD, lower row) during pretraining (PT), the first exposure (FE) to conditioning, the session in which the first significant (FS) behavioral discrimination occurred, the session in which the criterion (CR) of discriminative performance was reached and 3 of 6 (overtraining) sessions, the beginning, middle and ending sessions of overtraining (OTB, OTM, and OTE), where overtraining consists of continuing sessions of standard conditioning following the attainment of criterion. Neuronal data are shown for 4 consecutive 100 msec intervals after CS onset. The plotted values are standard scores indicating the neuronal response after CS onset normalized with respect to a 300 msec pre-CS baseline. Note the early development and loss of discriminative activity in PFC and the more gradual and persistent development of discriminative activity in the MD nucleus. However, note also that the MD discriminative discharge is substantially attenuated during the final overtraining session.

dance CR (Lambert and Gabriel, 1982; Gabriel et al., 1983, 1989a). Spontaneous locomotion was normal in the rabbits with these lesions, and the latencies and durations of the unconditioned response (UR) to the US in rabbits with lesions did not differ from control values. Informal observations of the rabbits in these studies and work in other laboratories indicate that other responses (e.g., heart rate change: Buchanan and Powell, 1982; Buchanan, 1988) to auditory stimuli occur at normal or enhanced magnitudes in rabbits with cingulate cortical and limbic thalamic lesions, suggesting that these lesions do not interfere with sensory reception of the CSs. Therefore, the available data are consistent with the hypothesis that the lesion-induced impairments of the stepping CR are due to a disruption of associative processes rather than to sensory, motor or motivational deficits. It should be noted that combined damage of the two systems considered here has yielded severe learning deficits in primates and rats engaged in learning paradigms other than avoidance (Mishkin, 1978; Irle and Markowitsch, 1982).

Lesions and training-induced cortical and thalamic neuronal activity

The combined lesions of the MD and anterior thalamus that abolish CR acquisition also attenuate severely the training-induced excitatory and discriminative unit activity in cingulate cortex (Gabriel et al., 1989a), indicating that thalamic input is essential for the training-induced plasticity.

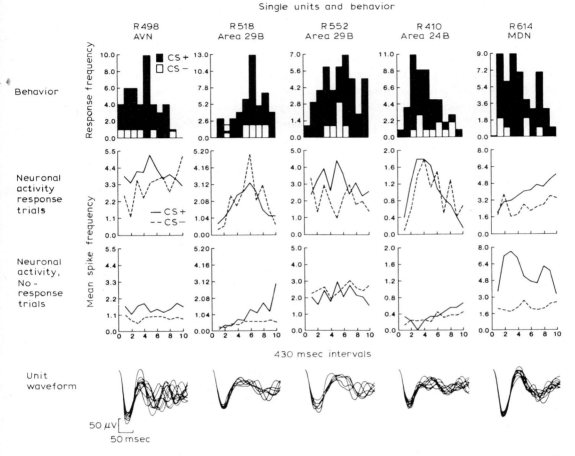

Fig. 4. The frequency of conditioned responses (upper row of panels) and the average firing frequency of single units (second and third rows) during 10 consecutive 430 msec intervals from 700 msec to 5000 msec after CS onset. All data were obtained during the training session in which the first significant behavioral discrimination was attained or during the session in which asymptotic performance occurred. The panels in the top row show the number of times the avoidance response occurred in each of the intervals during the training session. The average single unit firing frequencies are plotted separately for CS+ and CS− for trials in which the avoidance CR occurred (second row) and for trials in which no avoidance CR occurred (third row). The action potential waveforms of the single units are shown in te bottom row. These data illustrate putative premotor activity, i.e., the build-up of unit firing in anticipation of CR output.

These findings provide the basis of the model's hypothesis that the projection of CS-driven thalamic activity to cingulate cortex is essential for CR acquisition and performance. Clearly, cortical CS-driven excitation depends on intact projections from thalamus in this system. On the other hand, the thalamic training-induced excitatory and discriminative discharges are fully spared, indeed they are *enhanced* (as shown below), in animals with the CR-deleting cingulate cortical lesions (Gabriel et al., 1978a, b). Thus, the limbic thalamic nuclei do not appear to require cingulate cortical input for the exhibition of these discharges. These findings provide the basis for the model's hypothesis that the training-induced discharges of the thalamic neurons is elaborated at the diencephalic level and is not dependent on input from cingulate cortex.

Evidence for the STOP system

Subicular lesions and cingulate cortical activity

Neurons in the subiculum of the hippocampal formation send axonal fibers to the cingulate cortex (see Vogt, 1985). Bilateral electrolytic lesions of the dorsal and posterior subicular complex, administered before training, severely attenuated the excitatory and discriminative training-induced discharges in anterior and posterior cingulate cortex (Gabriel et al., 1987a; Fig. 5). These results give rise to two important conclusions. First, information flow from the hippocampal formation to the cingulate cortex contributes importantly to the development of training-induced activity in cingulate cortex. Second, this information flow, and the cingulate training-induced discharges, are not essential for CR acquisition and performance. (The rabbits with subicular lesions acquired and performed CRs as proficiently as intact controls.)

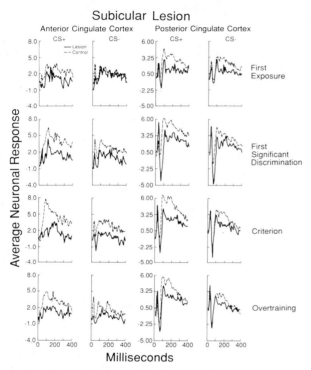

Fig. 5. Average multi-unit firing frequency in anterior cingulate cortex and posterior cingulate cortex in intact rabbits and in rabbits with bilateral electrolytic lesions of the dorsal and posterior subicular complex. The plotted values are standard scores indicating the neuronal response after CS onset normalized with respect to a 300 msec pre-CS baseline. These data show lesion-induced loss of training-induced activity in these cortical areas. In each panel the neuronal discharge magnitude is shown for 40 consecutive 10 msec intervals after CS onset for the stages of behavioral acquisition indicated by the labels in the right margin of the figure. The solid lines in each panel represent the discharges in rabbits with lesions and the dotted lines represent the discharges in intact controls. The 2 left-hand columns portray the discharges in anterior cingulate cortex in response to the CS+ (left column) and in response to the CS− (second column from left). The data in the third and fourth columns from the left represent the posterior cingulate cortical discharges in response to the CS+ and CS−.

Subicular/cingulate cortical lesions and limbic thalamic activity

The training-induced neuronal discharges elicited by the CS+ in the AV thalamic nucleus were *enhanced* in rabbits with bilateral subicular lesions, and in rabbits with posterior cingulate cortical lesions administered before training (Gabriel et al., 1987a; Fig. 6). Also, MD thalamic training-induced activity was enhanced in rabbits with fiber-sparing anterior cingulate ibotenic acid lesions administered before training (Gabriel et al., 1987b; Fig. 6). These results indicate, as mentioned, that cingulate cortical and subicular inputs are *not* necessary for the development of training-induced excitatory and discriminative discharges of AV and MD thalamic neurons. In addition, these data suggest that corticothalamic influences involving the subiculum and cingulate cortex, normally exert a suppressive or inhibitory influence on the firing of limbic thalamic neurons during conditioning.

Subicular/cingulate cortical lesions and CR performance

Rabbits with the subicular lesions performed avoidance CRs more frequently than the intact controls. This commonly reported effect of hippocampal damage (see review by Gabriel et al., 1980b) occurred during novel or "transitional" training stages (the first session of conditioning, the first session of extinction (CS presentation without the US) presented to overtrained rabbits and the first session of re-acquisition presented to extinguished rabbits). These training sessions are transitional in the sense that they are characterized by altered training contingencies relative to the preceding experience of the subjects. For example, the first session of conditioning is the first session in which the CS+ is paired with the US, following a pretraining (PT) session in which the tones and shock are presented in a non-contingent (explicitly

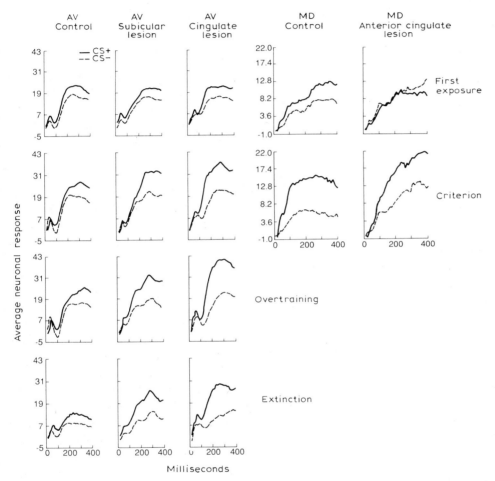

Fig. 6. Average integrated unit discharges elicited by CS+ (dark lines) and CS− (dashed lines) in the anterior ventral (AV) and medial dorsal (MD) thalamic nuclei during the first session of conditioning (top row), the session in which the criterion of behavioral discrimination was reached (second row), postcriterial overtraining (third row) and extinction training (CS but no US presentations, fourth row, AV activity only). AV neuronal discharges are shown for intact controls (left column), in rabbits with bilateral electrolytic lesions of the dorsal/posterior subicular complex (second column from left), and in rabbits with bilateral electrolytic and aspirative lesions of the posterior cingulate cortex (labeled "cingulate cortex", third column from left). MD neuronal discharges are shown for intact controls (fourth column from left) and in rabbits with ibotenic acid lesions of the anterior cingulate cortex (fifth column). In each panel, the average integrated unit response is shown in 40 consecutive 10 msec intervals after CS onset. The plotted values are standard scores indicating the neuronal response after CS onset normalized with respect to a 300 msec pre-CS baseline. The neuronal discharges were enhanced in the AV nucleus during the criterial and overtraining stages and the MD discharges were enhanced during the criterial stage. MD neuronal data for overtraining and extinction were not available. These results form part of the empirical basis for the hypothesis that anterior and posterior cingulate corticothalamic neurons have an inhibitory or "limiting" influence on neurons in limbic thalamic nuclei (see text for further details).

unpaired) manner. The first session of reacquisition is the first session in which the CS+ is paired with the US after a long series of extinction (CS followed by no US) trials. These results suggest that subicular neurons in intact animals are involved in suppressing or inhibiting the avoidance CR, and that this suppression operates during the novel or transitional training sessions.

It should be pointed out that unlike subicular lesions, the bilateral lesions of the posterior cingulate cortex did not increase the frequency of CR performance during the transitional training stages. Quite to the contrary, CRs were performed at a significantly reduced frequency by rabbits with posterior cingulate lesions, relative to controls. This finding, during the late stages of training, is discussed below.

Summary and hypothesis

Lesions of the subiculum *unbalanced* the training-induced discharges in cingulate cortex and limbic thalamus. That is, the discharges in cingulate cortex were attenuated whereas those in thalamus were enhanced in the rabbits with lesions. Also, CR frequency was enhanced in the rabbits with subicular lesions. These facts represent the basis for the model's hypotheses concerning the STOP system. Essentially, the model states that inputs to cingulate cortex originating in the subiculum increase the excitability of the corticothalamic projection neurons in cingulate cortex. The activation of these cells inhibits the firing of cells in their MD and AV thalamic projection targets. This inhibitory effect promotes CR inhibition as the thalamic neurons are viewed as essential for triggering CRs via the operation of the GO system. The activation of the cingulate corticothalamic projections, termed the "limiting process", is especially likely to occur in association with novel or unexpected stimuli. Subicular lesions prevent the activation of the limiting process and, as a consequence, they enhance AV and MD thalamic discharges as well as CR frequency.

It is important to note that both subicular and cingulate cortical lesions enhanced the firing of the limbic thalamic neurons, as expected by the model. However, only subicular lesions enhanced CR frequency. Cingulate lesions reduced CR frequency. This may seem paradoxical. Both lesions enhance thalamic activity. Why then should not both lesions increase CR frequency?

In fact, this result follows directly from the model. Although CS-elicited thalamic activity is enhanced, CRs are reduced after cingulate cortical lesions because the enhanced thalamic activity cannot access the motor system in the rabbits with these lesions. The cingulate lesions abolish the pathway from AV and MD through cingulate cortex to the motor system. Subicular lesions do not block this pathway, and CRs are thus enhanced as a result of the enhanced thalamic activity.

Why a limiting process?

The essence of the avoidance task is that the rabbits perform CRs only on the proper occassions. The stimuli that set the occassion for the CR are the CS+ and the normal (i.e., expected) background contextual stimuli of the experimental environment. It is proposed that purpose of the limiting process is to hold the CR in check until the proper occasion-setting stimuli are detected. In addition, a special purpose of the limiting process is to prevent CR occurrence in response to novel or unexpected stimuli.

If the occasion-setting stimuli are to be detected an analysis of incoming stimuli must be continuously performed. Incoming stimuli must be compared to a mnemonic template derived from past experience in the learning situation. Inputs that are compatible with the mnemonic template generate a match condition, the consequence of which is that the motor program for the CR is executed. Compatible inputs would of necessity be the occasion-setting stimuli themselves. In neurological terms, the occurrence of the occasion-setting stimuli generates a match condition and hippocampal inputs to cingulate cortex cause the withdrawal of CR limiting and thus, CR performance. If there is a mismatch between the actual and expected inputs, the limiting process is maintained and the CR is blocked.

Unique properties of the anterior and posterior systems

Prologue

We have seen that limbic thalamic neurons are involved in circuitry that is essential for acquisition and performance of the avoidance CR. However, the brief-latency excitatory and discriminative neuronal discharges in cingulate cortex do not contribute directly to CR acquisition and performance. Instead, the data indicate that these neuronal plasticities reflect in part, the operation of a limiting process, a product of hippocampal interaction with cingulate cortex. Whereas the limiting process does not bring about CR acquisition, it can exercise an inhibitory influence, a "veto power", over the CR. This prevents CR output as long as uncertainty exists as to whether CR performance is appropriate.

The STOP and GO functions just described appear to be common to both the anterior and posterior cingulate cortices and their related thalamic areas. However, other data indicate that these two systems are functionally distinct. Consideration of the distinctive properties of each system sheds light on the manner in which they function cooperatively to bring about control of the avoidance CR.

Several observations described below provide information about the separate functions of the anterior and posterior systems. Briefly, the data suggest that the anterior system is specialized for the rapid and flexible encoding of significant stimuli. Discriminative discharges and other forms of neuronal plasticity that encode the properties of significant cues are acquired quickly by the neurons in this system, and these plasticities can be quickly abandoned and replaced by new plasticities. This rapid and flexible neuronal encoding in the anterior cingulate suggests the operation of a mnemonic *recency system,* i.e., a system that abandons older stored information in order to capture recent information. This idea is also contained in the idea of a working memory system, a term that has been applied elsewhere in characterizing the functions of the anterior system (e.g., Funahashi et al., 1989). For convenience we shall use the term "recency system" to refer to this configuration of properties.

In contrast, plasticity of posterior system neuronal activity develops more gradually than in the anterior system, and it is not readily abandoned, once acquired. Instead, this system seems to specialize in primacy or the relatively long-term retention of plasticity. We shall use the term "primacy system" to refer to this configuration of properties.

More rapid development and decline of training-induced activity in the anterior than in the posterior system

The anterior cingulate cortex and MD thalamic nuclei represent the "leading edge" circuitry in relation to discriminative CR acquisition. Discriminative activity develops more rapidly in the anterior cingulate than in the posterior cingulate cortex, and it develops more rapidly in the MD than in the AV thalamic nucleus (Gabriel and Orona, 1982). Moreover, the amplitudes of the discriminative discharges in the anterior cingulate and MD drop quite dramatically during "overtraining", i.e., standard training sessions given after the achievement of criterion (Fig. 3). This drop is also present but much less pronounced in the neuronal records of the posterior system. These findings are consistent with the hypothesis that anterior system neurons contribute preferentially to CR acquisition and posterior system neurons contribute preferentially to CR performance in later training stages.

Reversability of discriminative discharges

Anterior cingulate neurons exhibit reversal of the discriminative discharges during reversal training given after the completion of original acquisition (e.g., Orona and Gabriel, 1983). In contrast, neurons in the posterior cingulate cortex exhibit a remarkable retention of the original discriminative discharge during the course of discrimination reversal learning. These neurons never exhibit the reverse discrimination. Instead, the most that is achieved is a neutrality, wherein the CS+ and CS− elicit equal and minimal neuronal firing, even as the rabbits discriminate behaviorally at criterial levels on the reversal problem (Gabriel et al., 1976, 1980c).

Effects of restricted lesions

As mentioned, CR acquisition is virtually abolished after cingulate cortical or limbic thalamic lesions, as long as both anterior and posterior systems are damaged. Lesions in only one of the systems yield only mild performance impairments. However, the nature of the mild impairments is of considerable interest. When the damage is confined to the cortical or thalamic components of the posterior system, CR acquisition is indistinguishable from that of intact controls. However, CR performance dissipates relative to that of controls during continued postasymptotic training (Gabriel et al., 1983, 1987a; Fig. 7). In contrast, damage confined to either the cortical or thalamic component of the anterior system results in a significant retardation of CR acquisition. An average of 6 – 8 daily training sessions is needed for criterion attainment rather than the 3 – 4 sessions taken by intact controls (Gabriel et al., 1987b). This retardation is due to a reduced level of CR performance, not to an inability to withold responding to the CS−. CR performance levels do eventually attain those of controls in these rabbits, given sufficient training. These results suggest, in accord with the different rates of plasticity development in these systems, that the anterior system mediates original acquisition, but this system makes a lessened con-

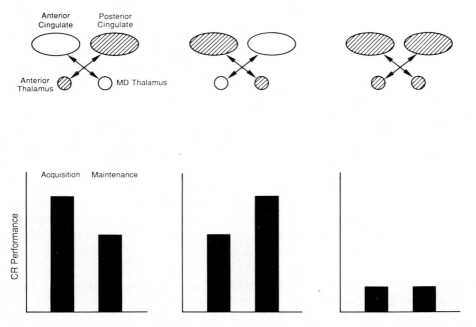

Fig. 7. Schematic illustration of the effects of anterior and posterior system lesions. The left-hand portion of the diagram indicates that lesions in either the posterior cingulate cortex or in the anterior thalamus are associated with normal CR acquisition but impaired CR performance during post-criterial training sessions (maintenance). The central portion of the figure indicates that lesions in the anterior cingulate cortex or the medial dorsal (MD) thalamic nucleus are associated with retarded CR acquisition but normal CR maintenance. The right-hand portion of the figure indicates that lesions of both anterior and posterior cingulate cortex or lesioins of MD and AV thalamus severely impair CR acquisition.

tribution as training is continued after criterion. The more gradual acquisition of plasticity in the posterior system is completed as the anterior system contribution begins to decline. Thus, the posterior system is the principal mediator of CR performance during asymptotic and postasymptotic task performance.

Manipulation of CS duration

Differences between anterior and posterior cingulate cortical neural processing were revealed in a study in which the duration of the conditional stimuli was manipulated (Sparenborg and Gabriel, 1990). Three values of CS duration (200, 500 and 5000msec) were presented in a counterbalanced order, each for 3 consecutive post-asymptotic training sessions after rabbits were trained to criterion using the fixed duration 500msec CS. In all cases the interval from CS onset to US onset was 5000 msec. Neuronal discharges in anterior cingulate cortex were of a significantly greater magnitude in response to the 200 msec CS+ than in response to the 5000 msec CS+. Discharge magnitude in response to the 500 msec CS+ was of intermediate magnitude relative to the other discharges (Fig. 8, left panel). The inverse covariation of CS duration and the neuronal discharge magnitude was due entirely to changes in response to the CS+, as constant discharges of minimal amplitude were elicited by CS−s of varied duration. The inverse relationship suggests that by virtue of anterior cingulate encoding the physiological "weighting" of the CS+ can be enhanced to compensate for minimal physical salience. The finding that anterior cingulate cortical neurons can alter their discharge magnitudes in accordance with day-to-day changes in CS duration is compatible with our hypothesis that these neurons participate in a recency system.

The overall average neuronal records in the posterior cingulate cortex were not significantly affected by CS duration (Fig. 8, right panel). However, effects were observed when the posterior cingulate records were separated into subsets that exhibited early- and late-developing discriminative activity during acquisition (see Foster et al., 1980). The late discriminators, constituting 4 of the 13 multi-unit records analyzed in this study, showed a clear effect of CS duration: the neuronal discrimination between CS+ and CS− increased as CS duration decreased (Fig. 9). This effect is similar to that just described for the anterior cingulate cortex. On the other hand, the average discharges of the 9 posterior cingulate records that exhibited

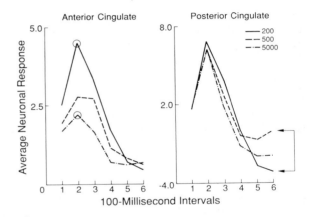

Fig. 8. Average multi-unit discharge profiles of the anterior cingulate cortex (left panel) and in the posterior cingulate cortex (right panel) in response to conditional stimuli (CSs) of varying durations as indicated in milliseconds in the figure legend. The average discharges are those elicited by the CS+, as the CS− elicited discharges were uniformly smaller than those shown here due to the discriminative response, but they did not vary with CS duration. The plotted values shown for 6 consecutive 100 msec intervals after CS onset are standard scores indicating the neuronal response magnitude normalized with respect to a 300 msec pre-CS baseline. The data were obtained during 9 training sessions administered after acquisition to criterion was completed. In these sessions the rabbits experienced the CS durations in a counterbalanced order. Each duration was presented for 3 consecutive sessions. These results demonstrate a significantly greater area 24 discharge to the 200 msec CS than to the 5000 msec CS. This difference was significant only in the interval from 100 to 200 msec after CS onset. Essentially overlapping neuronal discharge profiles in response to the various durations were found in the posterior cingulate cortex (left panel, Area 29) and in the AV and MD nuclei (data not shown). However, the posterior cingulate neurons did exhibit what we interpret as a significant "off response" to the 500 msec CSs, as indicated by the arrows in the lower right-hand area of the figure.

early-developing discrimination were not monotonically related to CS duration. In this instance, the average discharges were affected by CS novelty. Significantly greater discharges were elicited by the novel 200 and 5000 msec durations than by the familiar 500 msec duration, i.e., the duration that was used throughout the course of CR acquisition.

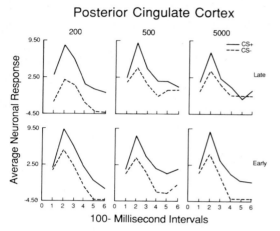

Fig. 9. Average multi-unit discharge profiles recorded in the posterior cingulate cortex (Area 29). The upper row of panels shows data obtained from records that developed discriminative responding in the late training stages, whereas the lower row of panels shows data obtained from early-discriminating records (see text for details). The 3 columns of panels indicate the neuronal discharges in response to conditional stimuli (CSs) of varying durations as indicated in milliseconds above each column. The average discharges are those elicited by the CS+ and CS− for 6 consecutive 100 msec intervals after CS onset. The data are in the form of standard scores indicating the neuronal response magnitude after CS onset normalized with respect to a 300 msec pre-CS baseline. The data were obtained during 9 training sessions administered after the completion of acquisition. During acquisition the CS duration was 500 msec for all rabbits. During the 9 sessions following acquisition, the rabbits experienced the indicated CS durations in a counterbalanced order. Each duration was presented for 3 consecutive sessions. The data indicate a significant inverse relationship between CS duration and discrimination magnitude for the late-discriminating records (upper panels). Significant effects of the CS duration factor were not obtained for the early-discriminating records (lower panels). However, the early-discriminating records did exhibit, within a restricted range of milliseconds after CS onset, an enhanced discharge in response to the novel 200 and 5000 msec CS durations, relative to the standard 500 msec duration (see Fig. 10).

This effect occurred only in the first of the 3 sessions with the same duration, and it occurred only from 70 to 130 msec after CS onset. It can be seen in the lower panels of Fig. 9, but it is better visualized in Fig. 10, in which the discharge magnitudes in 10 msec intervals are plotted for each of the 3 sessions with the same CS duration. This figure shows a significant drop from the first to the third session with the 2 novel CS durations, but no such drop occurred for the familiar duration. The fact that these posterior cingulate records

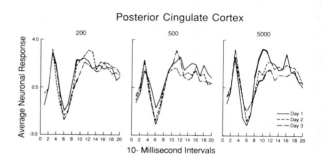

Fig. 10. Average multi-unit firing frequency in posterior cingulate cortex during 9 training sessions presented after the completion of acquisition. A CS duration of 500 msec was used for all rabbits during acquisition. The records shown here all exhibited early-developing discrimination between CS+ and CS− during original acquisition (see text for details). The CS duration during the 9 sessions was either 200, 500 or 5000 msec, each one of these values being presented in a block of 3 consecutive training sessions indicated by the labels, Day 1, Day 2 and Day 3. The 3 durations were presented in a counterbalanced order. Neuronal data are shown for 20 consecutive 10 msec intervals after CS onset. The plotted values are standard scores indicating the neuronal response after CS onset normalized with respect to a 300 msec pre-CS baseline. The activities elicited by CS+ and CS− are pooled in this figure as the CS factor did not yield significant effects in the analysis. In contrast to the results obtained for the late-discriminating records, which exhibited greater discriminative discharges to brief- than to long-duration CSs (Fig. 9), the early-discriminating records were selectively responsive to the novelty of particular CS durations, rather than to duration per se. The novel (200 and 5000 msec) CSs elicited significantly greater neuronal discharges during the first, compared to the last, of the 3 sessions in which a given CS duration was presented. However, the familiar 500 msec CS did not elicit a greater first-session discharge. These results implicate the early-discriminating records of the posterior cingulate cortex in the processing of CS novelty.

exhibited sensitivity to the unexpected durations implicates this region in mismatch functions and in the "tracking" of CS duration for relatively long periods of time (days to weeks). These conclusions are in accord with our hypothesis that posterior cingulate neuronal circuits participate in a primacy system.

Manipulation of CS probability

Clear differences between anterior and posterior cingulate cortical neuronal activity also emerged in relation to the manipulation of the relative incidence or "probability" of the CSs (Stolar et al., 1989). In this study all rabbits were trained to asymptotic levels after which they experienced in a counterbalanced order, training sessions with either frequent or rare CS+ presentations. In the rare sessions the CS+ occurred on 20% (24 of 120) of the trials and the CS- occurred on 80% (96 of 120) of the trials. This relationship was reversed in the frequent CS+ sessions. Neurons in all of the areas showed discrimination between CS+ and CS- (compare the discharge magnitudes in the left versus the right columns of Fig. 11). However, the discrimination as shown by anterior cingulate neurons was subordinated to the probability response, i.e., a significantly greater average discharge occurred in response to the rare CS+ than to the frequent and control CS+s, and a significantly greater response to the rare CS- than to the frequent and control CS-s occurred (Fig. 11, top panels).

Posterior cingulate cortical neurons exhibited marginally significant discharge enhancements in response to the rare and frequent CS+ stimuli, relative to the control CS+ (Fig. 11, second row,

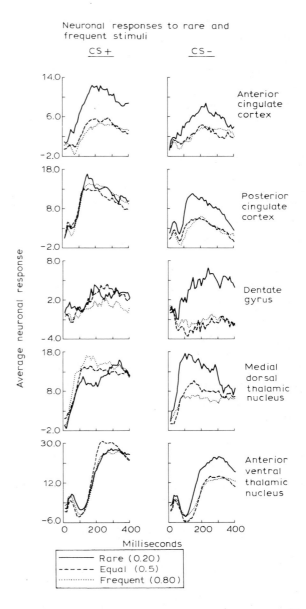

Fig. 11. Average integrated unit firing frequency profiles associated with presentation of the CS+ (left column) and CS- (right column) in the anterior cingulate cortex, posterior cingulate cortex, dentate gyrus of the hippocampal formation, medial dorsal (MD) thalamic nucleus and anteroventral (AV) thalamic nucleus. The plotted values are standard scores indicating the neuronal response after CS onset normalized with respect to a 300 msec pre-CS baseline. Each panel portrays the discharge profile in 40 consecutive 10 msec intervals after CS onset. The profiles in each panel indicate the discharges associated with 3 training sessions, a session in which the CS+

was presented rarely (on 20% of the trials, solid lines), a session in which the CS+ was presented frequently (on 80% of the trials, dotted lines) and a control session in which the CSs were presented on 50% of the trials (dashed lines). The profiles were obtained during postasymptotic training sessions following CR acquisition with CS+ and CS- presented on 50% of the trials.

left panel). Recall that the rare and frequent CS+ stimuli were less familiar than the control CS+ because the control stimulus condition (equiprobability) was present throughout acquisition. These results suggest that whereas CS rareness per se enhances anterior cingulate discharges, the discharges of the neurons in the posterior cingulate in response to the CS+ appear to be governed primarily by CS novelty/familiarity. This outcome is similar to the apparent governance of posterior cingulate activity by familiarity and novelty in the studies of CS duration effects described in the previous section. Discharge enhancement exclusive to the rare stimulus is suggestive, again, of a kind of compensatory encoding to insure that the rare but significant stimulus receives adequate processing. This compensatory encoding can be explained as a response to rare stimulus presentation, i.e., an immediate response to a recent event in the training situation. However, the novelty-driven discharges of the posterior cingulate can only be accounted for as products of a comparison of the rare and frequent stimuli to a primacy-based representation of the equiprobable stimuli formed during acquisition, many days before the rare stimulus sessions were administered.

It is interesting to note that posterior cingulate neurons and indeed, neurons in all of the areas studied, exhibited discharge enhancement in response to the CS− when it was presented rarely (Fig. 11, right-hand column). This finding suggests that rare presentation is an especially effective way to remove the inhibitory properties of a CS−, i.e., the properties that are responsible for the minimal thalamic discharges elicited by the CS− in the standard training paradigm. We are currently carrying out studies to determine whether hippocampal influences are essential for this intriguing effect of rare CS− presentation. Ultimately, we may find that this phenomenon represents a corticothalamic *amplification* process, i.e., a counterpart to the limiting process documented above.

Unlike anterior and posterior cingulate cortical neurons, limbic thalamic (MD and AV) neurons exhibited suppressed rather than enhanced discharges in response to the rare CS+ Fig. 11, fourth and fifth left panels). We interpreted this result as an indication of the limiting process postulated by the theoretical model. Differences in the specific pattern of results on the part of AV and MD thalamic cells were intriguing. The MD average discharge was suppressed in response to the rare CS+ relative to the control CS+, but it was not altered by the frequent CS+ (Fig. 11, fourth left panel). This pattern was precisely the inverse of the pattern exhibited by anterior cingulate neurons which were activated by the rare CS+ but which exhibited equivalent moderate discharges to the frequent and control CS+. In contrast, AV neurons like posterior cingulate neurons were governed by the novelty/familiarity dimension of the CSs. Cells in the AV nucleus were suppressed more by the two categories of novel stimuli, the rare and frequent CS+, than by the familiar control stimulus. These differences reinforce the results of the CS duration manipulation in suggesting that the discharges of anterior and posterior system neurons reflect the operation of recency and primacy systems, respectively.

Conclusions

Diencephalic mnemonic processes centered in the AV and MD thalamic nuclei are critical to the operation of the GO system, which mediates acquisition and performance of the discriminative avoidance CR. Neurons in the anterior and posterior cingulate cortices are also involved in GO system functions, but principally as conduits for discharges of diencephalic origin en route to the motor system. The brief-latency discriminative cingulate cortical discharges elicited by the CS+ and CS− are not at all critical for CR GO system functions, as indicated by the effects of hippocampal lesions. Yet, these discharges do encode the associative properties of the CSs. We propose here that the associative modifications exhibited by these brief-latency cortical discharges represent the operation of mnemonic processes intrinsic to the telencephalon.

It so happens that telencephalic processes are not essential for CR acquisition or performance. However, telencephalic processes govern the CR in an inhibitory fashion. That is, the telencephalic system acts as a STOP system to block CR output continuously, until and unless the criteria for the presence of the occasion-setting stimuli have been met. We hypothesize that this CR prevention results from an inhibitory flow from cingulate cortex to limbic thalamus, termed the limiting process. This flow provides fine cortical control over CR performance, without which the survival of the rabbit in the wild would probably be in jeopardy.

The anterior system and the posterior system share the GO and STOP functions that we have been describing. In this sense, the two systems have redundant functions. Thus, lesions in only one of the systems do not seriously affect discriminative avoidance behavior. We think that this overlapping jurisdiction is the reason why relatively simple behaviors such as the avoidance CR can be as resistant to brain damage as they are. This provides at least a part of the answer to Lashley's conundrum concerning the localization of the "engram". Multiple parallel systems are engaged in mediating this learning and all of the involved systems must be damaged if deficits are to be observed. Nevertheless, redundancy notwithstanding, the two systems differ in the manner in which they exhibit the GO and STOP functions. The anterior system exhibits properties suggesting that it functions as a recency or working memory system, whereas the properties of the posterior system implicate it in primacy, or intermediate/long-term memory functions. It should be noted that information provided by recency and primacy systems is exactly the input needed by the hypothetical comparator, if it is to evaluate the compatibility of past and present environmental circumstances. That is, the recency code conveyed by anterior cingulate neurons may provide the comparator with a representation of the current status of the training environment, whereas the primacy code conveyed by posterior cingulate neurons may provide the comparator with mnemonic data representing environmental features expected on the basis of the animal's accumulation of task-related experience.

Acknowledgements

The preparation of this article and the research reported were supported by grants from the Air Force Office of Scientific Research (89-0046), the National Institute of Health (NS 26736) and the National Institute of Mental Health (MH 37915).

References

Brogden, W.J. and Culler, F.A. (1936) A device for motor conditioning of small animals. *Science*, 83: 269.

Buchanan, S.L. (1988) Mediodorsal thalamic lesions impair differential heart rate conditioning. *Exp. Brain Res.*, 73: 320 – 328.

Buchanan, S.L. and Powell, D.A. (1982) Cingulate cortex: Its role in Pavlovian conditioning. *J. Comp. Physiol.*, 96: 755 – 774.

Byrne, J.H. (1987) Cellular analyses of associative learning. *Physiol. Rev.*, 67: 329 – 439.

Foster, K., Orona, E., Lambert, R.W. and Gabriel, M. (1980) Early and late acquisition of discriminative neuronal activity during differential conditioning in rabbits: specificity within the laminae of cingulate cortex and the anteroventral thalamus. *J. Comp. Physiol. Psychol.*, 94: 1069 – 1086.

Funahashi, S., Bruce, C.J. and Goldman-Rakic, P.S. (1989) Mnemonic coding of visual space in the monkey's dorsolateral prefrontal cortex. *J. Neurophysiol.*, 61: 331 – 349.

Gabriel, M. and Orona, E. (1982) Parallel and serial processes of the prefrontal and cingulate cortical systems during behavioral learning. *Brain Res. Bull.*, 8: 781 – 785.

Gabriel, M., Miller, J.D. and Saltwick, S.E. (1976) Unit activity in cingulate cortex and anteroventral thalamus of the rabbit during differential conditioning, extinction and reversal. *J. Comp. Physiol. Psychol.*, 91: 423 – 433.

Gabriel, M., Miller, J. and Saltwick, S.E. (1977) Unit activity in cingulate cortex and anteroventral thalamus of the rabbit during differential conditioning and reversal. *J. Comp. Physiol. Psychol.*, 91: 423 – 433.

Gabriel, M., Foster, K. and Orona, E. (1980a) Interaction of the laminae of cingulate cortex and the anteroventral thalamus during behavioral learning. *Science*, 208: 1050 – 1052.

Gabriel, M., Foster, K., Orona, E., Saltwick, S. and Stanton, M. (1980b) Neuronal activity of cingulate cortex, anteroventral thalamus and hippocampal formation in discriminative conditioning: Encoding and extraction of the significance of conditional stimuli. In: J. Sprague and A.N. Epstein (Eds.), *Progress in Psychobiology and Physiological Psychology,* Academic Press, New York, pp. 125 – 232.

Gabriel, M., Orona, E., Foster, K. and Lambert, R.W. (1980c) Cingulate cortical and anterior thalamic neuronal correlates of reversal learning in rabbits. *J. Comp. Physiol. Psychol.,* 94: 1087 – 1100.

Gabriel, M., Lambert, R.W., Foster, K., Orona, E., Sparenborg, S. and Maiorca, R.R. (1983) Anterior thalamic lesions and neuronal activity in the cingulate and retrosplenial cortices during discriminative avoidance behavior in rabbits. *Behav. Neurosci.,* 97: 675 – 696.

Gabriel, M., Sparenborg, S. and Stolar, N. (1986) An executive function of the hippocampus: pathway selection for thalamic neuronal significance code. In: L.R. Isaacson and K. Pribram (Eds.), *The Hippocampus, Vol. IV,* Plenum Press, New York, pp. 1 – 39.

Gabriel, M., Sparenborg, S. and Stolar, N. (1987a) Hippocampal control of cingulate cortical and anterior thalamic information processing during learning in rabbits. *Exp. Brain Res.,* 67: 131 – 152.

Gabriel, M., Kubota, Y., Sparenborg, S. and Straube, K. (1987b) Anterior cingulate cortical ibotenic acid lesions enhance conditioning-induced unit activity in the MD thalamic nucleus in rabbits. *Soc. Neurosci. Abstr.,* 305.17: 1104.

Gabriel, M., Kubota, Y. and Sparenborg, S.P. (1989a) Anterior and medial thalamic lesions, discriminative avoidance learning and cingulate cortical neuronal activity in rabbits, *Exp. Brain Res.,* 76: 441 – 457.

Gabriel, M., Cox, A., Ellison-Perrine, C. and Miller, J.D. (1989b) Brainstem mediation of learning and memory. In: W.R. Klemm and R.P. Vertes (Eds.), *Brainstem Mechanisms of Behavior,* John Wiley, New York, in press.

Irle, E. and Markowitsch, H.J. (1982) Single and combined lesions of the cat's thalamic mediodorsal nucleus and the mammillary bodies lead to severe deficits in the acquisition of an alternation task. *Behav. Brain Res.,* 6: 147 – 165.

Lambert, R.W. and Gabriel, M. (1982) Effects of midline limbic cortical lesions on discriminative avoidance behavior and neuronal activity in the neostriatum. *Soc. Neurosci. Abstr.,* 8: 318.

Mishkin, M. (1978) Memory in monkeys severely impaired by combined but not by separate removal of amygdala and hippocampus. *Nature,* 273: 297 – 298.

Orona, E. and Gabriel, M. (1983) Multiple-unit activity of the prefrontal cortex and mediodorsal thalamic nucleus during acquisition of discriminative avoidance behavior in rabbits. *Brain Res.,* 263: 295 – 312.

Sparenborg, S.P. and Gabriel, M. (1990) Neuronal encoding of conditional stimulus duration during avoidance learning of rabbits. *Behav. Neurosci.,* 104: 919 – 933.

Stolar, N., Sparenborg, S.P., Donchin, E. and Gabriel, M. (1989) Conditional stimulus probability and activity of hippocampal, cingulate cortical and limbic thalamic neurons during avoidance conditioning in rabbits. *Behav. Neurosci.,* 103: 919 – 934.

Thompson, R.F. (1986) The neurobiology of learning and memory. *Science,* 223: 941 – 947.

Vogt, B.A. (1985) The cingulate cortex. In: E.G. Jones and A. Peters (Eds.), *Cerebral Cortex,* Plenum Press, New York, pp. 89 – 149.

Vogt, B.A., Sikes, R.W., Swadlow, H.A. and Weyand, T.G. (1986) Rabbit cingulate cortex: cytoarchitecture, physiological border with visual cortex, and afferent cortical connections of visual, motor, postsubicular and intracingulate origin. *J. Comp. Neurol.,* 248: 74 – 94.

Discussion

C.G. van Eden: Concerning the MD nucleus in complete functional recovery of delayed alternation performance after neonatal PFC lesions.

M. Gabriel: Delayed alternation is generally believed to be a "working memory" task that relies on the telencephalic memory system. Thus, it is unlikely, though not impossible, that this task can be mediated entirely by the MD nucleus, without participation of cortex. It seems more likely to me that the recovery in this instance is a result of vicariation. As we have seen today, my work suggests that there is substantial functional redundancy between the anterior and posterior cingulate, and respective thalamic areas, in relation to avoidance learning. The redundancy is inferred from the facts that lesions in either system (at either the cortical or thalamic level) produce modest impairments. Anterior cingulate or MD lesions retard aquisition, whereas posterior cingulate or anterior thalamic lesions spare acquisition but engender poor performance during postasymptotic training sessions. Combined lesions in both systems essentially block acquisition. Thus, whereas the anterior cingulate may be somewhat specialized for acquisition, and the posterior cingulate for habit maintenance, each area can sustain performance "outside of " its specialty. Perhaps developmental processes can endow the posterior cingulate and its anterior thalamic counterparts with the capacity to mediate delayed alternation in the absence of an anterior cingulate.

J. Scheel-Krüger: Concerning the possible contribution of the basolateral amygdala and orbital cortex in relation to MD and anterior thalamic training-induced neuronal activity.

M. Gabriel: First it is necessary to clarify that there is no evidence to my knowledge for the existence of projections from basolateral amygdala to the anterior thalamus. Moreover, orbital cortex is not defined in rabbits, but perhaps insular cortex may be regarded as analogous to the primate orbitofrontal cortex. In any case, insular cortex does not project to the anterior thalamus. However, amygdala as well as insular cortical neurons do project to the MD nucleus. Thus, you raise an extremely important issue in relation to the possible role of the amygdala in the production of MD training-induced neuronal activity. As you point out, it is possible that amygdala projections to MD are involved in sustaining the training-induced activity in the MD in rabbits with anterior cingulate lesions. Graduate student Amy Cox and I are currently studying the activity of the amygdala, and the effects of amygdala lesions on MD and anterior cingulate activity during learning. The preliminary results indicate that the pattern of early-developing training-induced changes in the amygdala is virtually identical to that seen in the anterior cingulate, and just as in the case of anterior cingulate and MD lesions, lesions in the basolateral amygdala produce a retardation of behavioral acquisition in the discriminative avoidance task. We have not yet analyzed the data to discover whether MD training-induced activity is blocked in the rabbits with amygdala lesions. However, preliminary observations suggest that MD discriminative activity is intact in rabbits with these lesions. Thus, as in the anterior thalamic system, the MD nucleus appears to be capable of exhibiting training-induced neuronal plasticity in the absence of forebrain influences. This outcome is consistent with our view that limbic thalamus is a self-contained wellspring of fundamental physiological plasticity underlying learning. We are also looking at the possibility that amygdaloid lesions will produce disinhibitory effects similar to those that we see after subicular lesions. The analysis of the preliminary data has not progressed far enough for me to comment on this.

V.B. Domesick: What type of information do you believe the subiculum is sending to the cingulate cortex? Have you considered the role of direct input from the subiculum to the thalamus?

M. Gabriel: Our data indicate that the subicular inputs to cingulate are essential for the development of training-induced neuronal plasticity in cingulate. This plasticity is, I propose, involved in the encoding of the associative significance of task relevant stimuli, and in suppressing or limiting thalamic activity (and behavioral outputs) until external conditions for the behavior are fully appropriate. These functions could be accomplished by a tonic input to cingulate that continually biases the responsiveness of cingulate cortical neurons as long as the animal is in the training situation. Alternatively, it could be that the subicular input operates within a limited time window during training to bring about biophysical changes supporting relatively permanent, and "local" plasticity of cingulate cortical neurons. This latter mode is reminiscent of theoretical accounts (such as that of Teyler and Discenna) that view the hippocampus as causing "memory storage" in the various cortical areas to which hippocampal formation neurons project. If the latter hypothesis is correct then subicular lesions given after training is completed should not disrupt already established cingulate cortical training-induced plasticity. We are currently examining this possibility in my laboratory. As to the role of subiculo-thalamic projections, we have no information on this issue. Our data show that both subicular and posterior cingulate lesions enhance the firing of AV thalamic neurons during CS presentation. On the basis of these data I proposed the working hypothesis that the subiculum influences the AV nucleus via the posterior cingulate cortex. Hopefully, future work will provide more insight than we have now concerning the specific role of subiculo-thalamic fibers.

S.B. Dunnett: How can you have greater discharges to $CS+$ than to $CS-$ during the "first exposure"? Were the stimuli allocated to $CS+$ and $CS-$ conditions counterbalanced?

M. Gabriel: One of the remarkable properties of the neurons in anterior cingulate cortex is the rapidity with which training-induced excitatory and discriminative discharges develop following the onset of conditioning. Several studies since 1980 in my laboratory have shown that the discriminative activity develops in the very first session of conditioning. No discriminative activity is seen in the immediately preceding "pretraining" sessions in which the tones to be used as CSs, and the US are presented in an explicitly unpaired, non-contingent manner. Of course, the tones assigned as $CS+$ and $CS-$ are counterbalanced. The rapid acquisition of discriminative firing in anterior cingulate is one of the principal items of data underlying my proposal today that the anterior cingulate is part of a neural "recency" system.

B. Kolb: Do you have evidence that your paradigm holds for tasks that are not aversive, such as delayed response?

M. Gabriel: It is almost certain that the circuitry that we are studying is involved in many learning situations, not just aversive conditioning. One basis for this assertion is information concerning the involvement of limbic thalamus and cingulate cortex in human memory processing. Moreover, rats' performance in delayed alternation and spatial mapping tasks, based on appetitive motivation, is disrupted by cingulate cortical and limbic thalamic damage. Indeed, if as I suggested today, the anterior and posterior systems mediate the mnemonic functions of recency and primacy encoding, respectively, then we are dealing with very general learning systems. Yet, is it possible that the principles of operation and the specific information flows in these circuits differ depending on whether appetitive or aversive motivation is operating. We are currently developing an appetitive analog of the aversive conditioning task that will permit us to observe activity of the same neurons in response to the same physical stimuli in both aversive and appetitive settings.

CHAPTER 24

Social behaviour and the prefrontal cortex

J.P.C. de Bruin

Netherlands Institute for Brain Research, 1105 AZ Amsterdam, The Netherlands

Introduction

The high degree of development of the prefrontal cortex (PFC) in *Homo sapiens,* both in terms of size and in the multitude of its fibre connections, has elicited a widespread interest in the functions of this part of the brain, both in terms of its uniqueness in the human species, as well as approached from a comparative, "phylogenetic" view. In non-human primates it is considerably smaller, while in a rodent like the rat only a part of the frontal cortex may be termed truly "prefrontal". In carnivores the PFC is intermediate in size between rodents and non-human primates. However, this view has been challenged by Uylings and Van Eden (1990).

The large size of the human PFC has been linked to the supposedly specific "psychological" characteristics of man. The reasoning for this is as follows: a high degree of development of a certain brain area (in terms of size and differentiation) is accompanied by a high level of functionality associated with that particular brain area. For example, an echolocating bat is extremely good at hearing and has a large auditory cortex. Thus, it should come as no surprise that to the human PFC have been attributed primarily those capacities at which man performs especially well: cognition and language. Indeed, when one reviews the literature on PFC-related behavioural functions, one is struck by the emphasis laid on cognitive aspects of behaviour, learning and memory tasks, attention, perseveration etc., using a great variety of experimental testing paradigms (for a general review, see Kolb, 1984; for a review emphasising primate studies, see Fuster, 1989; human studies have been reviewed by Milner and Petrides, 1984; Stuss and Benson, 1984).

PFC involvement in social aspects of behaviour has been studied to a far lesser extent. This is evident from the contributions presented at two important symposia on the PFC, viz., the Pennsylvania Symposium held in 1963 (Warren and Akert, 1964), and The Frontal Granular Cortex and Behaviour Symposium held in Jablonna, Poland in 1971 (*Acta Neurobiol. Exp.,* 32, 1972). Most of the studies presented at these two symposia addressed research on neuroanatomy, learning and neuropsychological testing. During the second Symposium, Warren (1972) advocated a new line of research directed at studying "species-specific" behaviours in more naturalistic, ethologically inspired settings. In the present paper we will review data on the involvement of PFC in social aspects of behaviour, emphasising social interactions of the agonistic type. We will follow a common practice and deal with 4 mammalian taxa: non-human primates (mainly based on studies using rhesus monkeys and vervet monkeys), carnivores (examplified by studies in cats), rodents (examplified by studies in the laboratory rat and incorporating our own studies in this field), and humans (based on case histories of frontal

Correspondence: Dr. J.P.C. de Bruin, Netherlands Institute for Brain Research, Meibergdreef 33, 1105 AZ Amsterdam, The Netherlands.

lobotomy patients and patients suffering from accidental damage to the frontal lobes).

Primate studies

Of all animals the PFC of apes and monkeys resembles the human PFC most, which is undoubtedly the explanation for the multitude of studies in non-human primates on behavioural changes following PFC damage. A confounding factor in such studies is that both the size and the site of ablation vary greatly, ranging from complete lobotomies to restricted lesions of either dorsolateral or orbitofrontal PFC areas, or even smaller lesions within one of these subareas.

One of the first studies on the effects of frontal lobotomy in primates was conducted by Franz (1907), and one of his observations makes clear that the effects were not nearly as devastating as one might expect when a large part of the cortex was removed: "Among the monkeys there was no marked difference in the emotional response following removal of the frontal lobes".

In view of subsequent studies one might argue that Franz's monkeys were not subjected to stimuli which challenge the brain by presenting them with a novel and demanding situation which requires an appropriate and adaptive response. One way of accomplishing that is a human observer who simply stares impassively at the monkey which suffices to elicit aggressive reactions such as lip smacking, threatening facial gestures, barking or alarm calls (Clarke and Mason, 1988).

Butter and Snyder (1972) have followed this approach in adult rhesus monkeys with orbitofrontal cortex lesions. They witnessed two types of reactions in their monkeys: aggressive responses or aversive ones. When compared with sham-operated controls, the "orbitofrontal" monkeys directed fewer aggressive responses towards the human observers, while aversive reactions (towards an animal-like doll or a model snake) were enhanced. In a study using vervet monkeys, another Old World primate species, a similar approach was used by Raleigh and co-workers (1979), which yielded, however, a diametrically opposed result, viz., an increase in aggressive responses directed at the human observer by monkeys with orbitofrontal cortex damage.

But how would monkeys behave in a more natural testing situation? Butter and Snyder observed their "orbitofrontal" monkeys in the following testing situation. They were introduced singly into an established colony of 4 male rhesus monkeys with a clear linear hierarchy. Being older and heavier than the 4 group members, the intruders, prior to their operation, were always able to quickly achieve the α-position. When reintroduced into this colony following surgery, the sham-operates again were able to achieve the α-position. The "orbitofrontal" monkeys exhibited high levels of aggression when reintroduced postoperatively, and initially achieved the top position as before the operation. The interesting point of this study was that reintroductions were made repeatedly (at bimonthly intervals) and the monkeys' behaviour was recorded for many months. These observations showed that, in contrast to the sham-operates, the "orbitofrontal" monkeys eventually lost their ability to reclaim the α-position. In 1 animal this change occurred already after 1 week, in the other 2 monkeys only after 4 and 6 months, respectively. The conclusion from these observations is that high levels of aggression recorded postoperatively, change in the long run, and eventually lead to the loss of the α-position.

One of the problems with the Butter and Snyder study was that a single experimental animal was facing 4 other monkeys at the same time, which led to his eventual downfall in the hierarchy. But what about a more symmetrical testing situation, one in which each operated monkey meets with just one other monkey? Such an observational approach was used by Deets et al. (1970); in order to record behavioural changes following frontal agranular cortex lesions, the operated animals were removed from their home cage, paired with a stimulus animal in a "neutral" observation cage, and the ensuing dyadic interactions were recorded. This testing situation, unusual in primate studies, resembles the one commonly used in rat studies. Notwithstanding the general observation that "frontal" monkeys were more withdrawn and less explorative, it was found that they directed more challenges and threat gestures toward the stimulus males, but without resulting in higher levels of overt physical aggression.

Since the organisation of behavioural systems varies greatly among primates, it is imperative to study a variety of species. Vervet monkeys have a more loosely organised social system than rhesus monkeys, without the linear hierarchy so characteristic for the latter. Raleigh et al. (1979) have studied the effects of orbitofrontal lesions in a group of vervet monkeys consisting of 21 animals. Nine of these received orbitofrontal PFC lesions, equally divided over sex (male vs. female) and age (adult vs. juvenile). In the social group, neither the rate nor the intensity of aggressive behaviours changed following such operations. However, both operated males and females were less successful in the outcome of their aggressive interactions. Males were more involved than females in dyadic interactions and, following surgery they were more frequently defeated in these interactions. In polyadic aggressive interactions they participated as much as before surgery, and with equal success. In females the reverse was the case: equal success in the winning of dyadic interactions, but a clear-cut drop in participation in polyadic interactions.

Two studies have been conducted to record social behavioural changes following damage to the prefrontal cortex in juvenile monkeys. Bowden et al. (1971) found that "orbitofrontal" rhesus monkeys differed more from controls than did dorsolaterally operated ones. "Orbitofrontals" slept more, as well as being more sedentary even when awake. Since the animals were observed at a juvenile age (starting when they were approximately 10 months old) aggression was at a low intensity level. From the impressive ethogram we may deduce that both fighting and biting another animal were very rarely seen in either "dorsolateral" or "orbitofrontal" monkeys. Although animals of both operated groups huddled more than their controls, there was no sign of social withdrawal, since they sat with other members of the group as often as did control animals and did not retreat any more often from other animals. Miller (1976) observed a group of juvenile stump-tailed macaques following ablation of the dorsolateral PFC. Social interactions were recorded at 3 stages with an increasing number of animals operated upon; by the end of the experiment all members of the group had received dorsolateral PFC lesions. The higher the number of operates, the greater was the increase in aggressive behaviour and the decrease in threat behaviour. Based upon his figure, one can visualise a pandemonium of aggressive and hyperactive macaques by the end of the experiment. Miller explains his findings by hypothesising that (dorsolateral) frontal lobe damage resulted in both a difficulty in stimulus processing, and a failure to "inhibit behavioural mechanisms" while in a hyperactive state.

Laboratory settings, as described above, do not provide evidence as to what the ultimate result of a frontal lobotomy might be in a situation resembling the natural situation as much as possible. An answer to this question came from the study of Myers et al. (1973), who studied the well-known colony of rhesus monkeys on the island of Cayo Santiago (Puerto Rico). A number of animals was removed from the social group and was operated upon (either a frontal lobotomy or a sham operation); following recovery from surgery they were reintroduced into their social group. All "frontal" animals (with the exception of one of the juveniles) failed to rejoin their group, remaining solitary until their premature death. However, the thick undergrowth of the natural environment prevented observations of the interactions which led to this social isolation. Although the general conclusion from this study is clear — PFC regions contribute to the "control and regulation" of social behaviour in the primate — one would like to know more about the processes underlying these dramatic social changes.

The pilot study of Peters and Ploog (1976) with squirrel monkeys, a New World primate, reported some of the changes reported above (such as social withdrawal, inadequacy of responding to threats and dominance gestures) after either dorsolateral or orbitofrontal PFC lesions. Individuals varied greatly in their behavioural alterations, however, with some animals being able to maintain their dominant positions, while others became submissive. This study most of all analysed the problems encountered when social animals are removed from their group, and what the consequences are following reintroduction, in a weakened state shortly after operation as opposed to reintroductions after full recovery from surgical treatment, but creating social tension when encountering the members of its former social group after a long interval.

A primate synthesis?

It is clear that it is difficult to arrive at a general conclusion about the contribution either of the PFC as a whole, or of its dorsolateral and orbitofrontal aspects considered separately, in the regulation of social–agonistic behaviour in non-human primates. There are many factors which have to be taken into account, usually thwartening a common conclusion.
  Species differences. Even related primate species may differ in their social structure, e.g., a strict linear hierarchy in the rhesus monkey, and a loosely organised social system in the vervet monkey.
  Behavioural settings. Usually a social laboratory-kept group, which presents the methodological problems discussed by Peters and Ploog (1976), occasionally dyadic interactions in a neutral cage, or the natural large group situation as used in the study of Myers et al. (1973) with severe problems of being able to just see, let alone accurately observe the animals' behaviour.
  Size and locus of lesion. Reports vary from complete frontal lobotomy to discrete lesions of the two subareas or even part of one subarea.
  Sex and age. Social position within the group is also an important factor influencing the behavioural effects of a lesion.

It is not surprising, therefore, that researchers have often discussed their findings on social behavioural changes in terms of difficulties in sensory processing, deficits in learning to avoid other monkeys, thus coming to the conclusion that a change in aggressive behaviour was secondary to these more general deficits.

Carnivore studies

Quite a few studies exist concerning changes in social behaviour resulting from PFC damage, mostly with primates as their subjects, and a few with rodents, but very few using carnivore species, such as the cat. The reason, I presume, is partly derived from the nature of the feline aggressive display, which is difficult to study in a controlled laboratory setting. One of the first observations on cats subjected to a frontal lobotomy again comes from Franz (1907) who, as with his frontal monkeys (vide supra), expressed great surprise about the absence of gross behavioural changes. The problems of studying aggressive interactions in cats under desired laboratory conditions is obvious from one of the key studies on this topic, the one conducted by Warren (1964), who was able to use male and female cats housed in social isolation for a considerable time. Although one might have expected them to behave in an aggressive way upon encountering each other (isolation-induced aggression), this appeared to be only marginally the case, especially in male cats, prompting Warren to provide extra stimuli in order to elicit such behaviour, stimuli which consisted of food morsels for which the animals had to compete. Summarising his results we may state that
– males exhibited very low scores of aggression, with frontal and normal animals not differing from each other,
– females showed more aggressive interactions than did males, with normals exceeding frontals,
– normals (of both sexes) were more successful in food competition than (matched) operates.

Prefrontal modulation of aggressive behaviour again emerges in the studies of Siegel and co-workers, whose basic interest was in the hypothalamic control of aggression. They investigated the modulatory role of other brain structures in aggression, elicited by hypothalamic stimulation, and their basic findings suggest an inhibitory role for the PFC in hypothalamically elicited aggression (A. Siegel et al., 1974, 1975). When the PFC, either its lateral or medial aspect, was stimulated simultaneously with the hypothalamus, the threshold for attack (measured as latency time) was increased.

Summarising the findings of the two studies, we end up with rather contradictory findings: both stimulation and ablation of the PFC can result in an inhibition of aggression. I will come back to these findings when dealing with the role of the PFC in the aggression of rodents.

Rodent studies

Studies on the behavioural functions of the PFC in rats could be made only after its precise boundaries had been ascertained. Since a granular layer IV is lacking in the frontal cortex of rodents (a criterion used in primates), and since the frontal lobes are devoid of sulcal boundaries, one has had to resort to other criteria in order to establish the precise location of the PFC within the frontal lobe. Species comparisons on the site of the PFC are possible using the definition introduced by Rose and Woolsey (1948): the PFC is defined as the cortical projection area of the nucleus mediodorsalis thalami. The first to determine the location and size of the PFC in rats was Leonard (1969), who made lesions in the nucleus mediodorsalis thalami and used Fink–Heimer degeneration staining for mapping the projection sites within the frontal lobes. Her observations showed that the PFC in the rat consists of two separate areas, a (dorso-)medial part (PFC-M) located anterior to the genu of the corpus callosum, and a sulcal (suprarhinal, orbitofrontal) part (PFC-OF) covering the dorsal bank of the sulcus rhinalis. Subsequent studies have further subdivided these two PFC regions (Krettek and Price, 1977; Van Eden, 1985). Since neuroanatomical precision greatly exceeds the specificity of most neuroethological studies, we will confine ourselves to Leonard's subdivision, and will discuss the behavioural consequences of manipulations of medial and orbital PFC.

Going back into recent history we encounter the study of Lubar et al. (1973), from which the following statement has been taken: "At the present time there is virtually nothing known about the effect of frontal damage in the rat on species-specific behavior". Based upon supposed similarities in the behavioural effects of septal and frontal cortex ablations in monkeys, Lubar and co-workers studied the behaviour in a semi-naturalistic environment of rats that had been subjected to frontal cortex aspiration lesions. The group consisted of controls, rats with frontal lesions and rats with septal lesions, with both sexes being included ($n = 26!$). One of their lesion effects was that frontally operated male rats engaged in more fights and more "homosexual mounting". Whether this augmentation in aggressive behaviour had consequences for a top position in the hierarchy was not reported. Perhaps the group composition of 15 males prevented the establishment of any clear-cut social ranking. The lesions of the frontal cortex in the Lubar et al. study resembled a complete "frontal lobotomy", and the anatomical subdivisions of the PFC described by Leonard (1969) were not taken into account.

The first studies which did incorporate this anatomical dissociation in a functional study of social behaviour were conducted by Kolb and co-workers (Kolb, 1974; Kolb and Nonneman, 1974). They observed rats with either orbital or medial PFC lesions in a variety of testing situations for aggressive behaviour, their choice of testing situations being determined by Moyer's (1968) distinction of various kinds of aggression, each with its own neural and/or hormonal basis. Male Wistar albino rats were used, all of which had participated in various other testing situations at least 6 months earlier, so that the interval between operation and testing was unusually long in this study. The animals were tested in a small open field (floor dimensions 45 cm × 45 cm) for 10 min/day on 4 consecutive days. Dyadic interactions were recorded for like pairs of rats, i.e. with medial PFC lesions, orbital PFC lesions, or sham-operated in both animals. The basic findings of this study were that only rats with orbital PFC lesions showed changes in social behaviour, appearing to be more aggressive in male–male interactions in this testing situation. They also had higher levels of shock-induced aggression, but did not differ from controls in tests of gregariousness, territorial aggression and predatory aggression (mouse killing).

From ethological as well as neurobiological studies we now know that aggression is not a unitary phenomenon. Animals can perform different types of aggressive behaviour, depending on the situation. Moreover, Moyer's (1968) initial distinction between kinds of aggression was mainly based on physiological differences (neural, hormonal, etc.). When we addressed the questions of prefrontal cortical involvement in the expression of aggressive behaviour, it was necessary to carefully select an experimental testing situation in order for aggression to occur and to record the behaviour in a vigorous fashion so as to ensure good reliability with repeated testing. A "neutral" aggression testing cage (80 cm × 38 cm × 75 cm) was opted for, divided into 2 equal compartments by means of a retractable door. Each animal remained in one of the compartments for a 20 h period (adaptation) before the door was raised so that interactions could be started. The animals came from different cages and were totally unfamiliar to each other. In the testing cage they were unable to see each other, but undoubtedly were able to pick up each others smells and sounds during the adaptation period. Thus, this testing situation is basically one for intermale aggression, but elements of territorial aggression cannot be excluded. It resembles the natural situation of 2 rats meeting each other for the first time and becoming engaged in social–agonistic interactions. The main difference is that there is no true escape possibility: they may run away from each other, flee and assume submissive postures, all of which may be quite effective, but they are not able to avoid the presence of the opponent.

In all our studies we have housed the experimental animals together with an ovariectomised female rat, thus, preventing social isolation. At the same time this prevented the establishment of a ranking order among male subjects, a social experience/status which we wanted to eliminate from the experimental paradigm. For this type of behavioural experiment, we chose strains of rats known to exhibit aggressive behaviour in this testing situation, initially the WEzob strain (a non-albino, Wistar-derived strain) and later, randomly bred Long–Evans rats.

With a test duration of 10–20 min male rats usually developed a dominance/submission relationship, with one animal performing most of the agonistic acts and the other one eventually evading the dominant one. Vehement attack behaviour, including bites, was rare in this testing situation. It therefore differed from the other testing situation often used in aggression studies involving rats or mice, viz. the resident–intruder paradigm, which mimicks a pure territorial setting. The outcome in such a situation is highly predictable with the resident being the winner and the intruder often being severely wounded.

PFC proper

We began our studies by following the classical approach on which a large body of data concerning PFC function is based: bilateral damage to either medial or orbital PFC. With bilateral medial PFC (PFC-M) lesions we failed to find any significant changes in social–agonistic behaviour. In terms of frequency, duration and latency of a series of social (e.g. approaching, grooming) and agonistic acts (e.g. keeping down, lateral threat, chasing, freezing) the "medial PFC" rats were indistinguishable from their controls. In addition, no changes were detected in sexual interactions: male sexual behaviour towards a receptive female proceeded in the same way as in control-operated rats. These findings, of course, do not imply that PFC-M does not play a role in other behaviours. Food hoarding and spatial delayed alternation in a T-maze are severely impaired following damage to PFC-M (Kolb, 1984; Kolb and Gibb, 1990; De Brabander et al., 1990). Thus, at that time, we decided to direct our attention to the other subarea of PFC, the orbitofrontal part. Most of the data presented below are based on manipulations of that area.

Contrary to PFC-M damage, bilateral orbital PFC lesions produced clear-cut changes in the aggressive components of social – agonistic behaviour (De Bruin, 1981; De Bruin et al., 1983). In confrontations with control-operated male WEzob rats "orbital PFC" rats were clearly more aggressive, which was apparent not only from the greater incidence of aggressive acts, but also from the longer duration of freezing induced in the opponent. Their behaviour towards humans (the animal caretaker or the experimenter) who handle the animals regularly, remained basically the same: no rage syndrome or attempt to direct bites at their fingers or hands! We have repeated such experiments, also in rats of another strain, Long-Evans, and the outcome was very much the same.

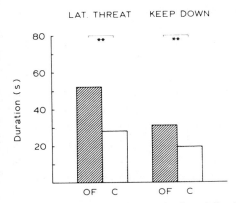

Fig. 1. Changes in aggressive behaviour following bilateral lesions of the orbital PFC in male Long – Evans rats. Mean durations of 2 aggressive acts are portrayed, based on encounters between orbitofrontal (OF) and control-operated (C) rats. Analysis time: 0 – 600 sec. ** $p < 0.01$ (Mann – Whitney U test, two-tailed; S. Siegel, 1956).

An example of the changes in two agonistic acts, lateral threat and keeping down, is presented in Fig. 1.

Based upon this finding two different approaches may be pursued, one of which I would like to term the *ethological* approach, the other the *neurobiological* approach. The former would aim for a further analysis of the stimuli eliciting agonistic responses in rats with orbitofrontal damage, e.g. the stimulus characteristics of the opponent after having been subjected to either defeat or victory (experience of losing or winning a fight). In the second approach, the stimulus properties are more or less kept the same, while the "orbital prefrontal system" is manipulated. I will present some data on both the input and the output side of this "system", using the second approach.

Output: the hypothalamic connection

The findings of increased aggression in "orbitofrontal" rats can be interpreted as being indicative of an inhibitory control exerted by the orbital PFC. The question then becomes which area of the brain is inhibited? The data of A. Siegel and co-workers (1974, 1975) already implicated the hypothalamus, and electrophysiological studies in rats have indeed demonstrated the presence of monosynaptic inhibitory fibres from frontal cortex to hypothalamus (Kita and Oomura, 1981). This led us to examine whether concurrent stimulation of the PFC would affect agressive behaviours elicited by electrical stimulation at hypothalamic sites.

The following experimental paradigm was used: electrodes were implanted in the hypothalamus at sites where electrical stimulation is known to elicit aggressive responses (Kruk et al., 1979). Another electrode which could be moved in a dorsoventral direction by steps of 300 μm (Van der Poel et al., 1983), was implanted in the PFC. Experimental procedures were as follows. First, the hypothalamus was stimulated and, in case of a successful placement, aggression was elicited, which was directed at another rat in the same observation cage; the threshold for triggering such attacks (usually between 40 and 60 µA) was determined using the procedures developed by Kruk et al. (1979). Next, the PFC was stimulated, concurrently with stimulation of the hypothalamus, and the hypothalamic threshold current intensity was again determined. This procedure was repeated, and each time the PFC electrode was lowered to a new, more ventral, position. PFC stimulation current intensity was set at 50 µA, at which intensity no behavioural changes were ever observed when only the PFC was stimulated. Following the dorsoventral axis of the PFC stimulation electrode it can be seen (Fig. 2) that an increase in the hypothalamic threshold current was recorded at some sites, whereas PFC stimulation at other sites failed to result in any noticeable change in the required intensity. Neuroanatomical evaluation of the precise position of the electrode in the frontal cortex revealed that inhibition of the responses had occurred when the electrode was situated in the deeper layers of either medial or orbital PFC, although the effects were larger in the latter area (Fig. 2). Thus, although our measures were somewhat different from those used by Siegel and co-workers in their cat studies, the basic results are similar: PFC exerts an inhibitory control over hypothalamic sites from which aggression can be elicited by electrical stimulation.

Input: the dopaminergic connection

It has been well documented that in the rat the PFC is one of the very few cortical areas to receive a dense dopaminergic projection. This discovery was first made by Thierry et al. (1973), and has elicited a large number of studies on the functional implications of this dopaminergic pathway, some of which are reviewed in this volume (e.g. Kalsbeek et al., 1990; Thierry et al., 1990). Prior to doing any behavioural studies we set out to measure levels of dopamine (DA) in the frontal lobes and to compare medial and orbital PFC in a quantitative way. Basically, this study demonstrated that dopamine is found in both of the prefrontal sub-

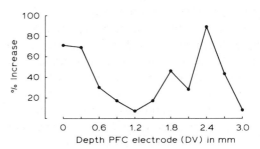

Fig. 2. Inhibition of hypothalamically elicited attack behaviour resulting from concurrent stimulation of both prefrontal cortex and hypothalamus. The various steps of the unilateral PFC electrode are depicted on the *X*-axis, while the increase in current intensity required to elicit attack at threshold value is depicted on the *Y*-axis. Note that there appear to be 2 areas in the frontal cortex where concurrent PFC stimulation results in a clear inhibition. The first one represents the deeper layers of the medial PFC, the latter one the deeper layers of the orbital PFC. Further explanation is given in the text. (Data were collected together with J.S. Slopsema, M.R. Kruk and W. Meelis.)

fields (with levels being somewhat higher in the medial part), but hardly any in the dorsolateral area between them. In contrast, noradrenaline is homogeneously distributed within the rat frontal cortex (Slopsema et al., 1982).

In the experiments to be reported next we have followed two approaches to assess the possible contribution of this dopaminergic pathway in the inhibitory control of the PFC with regard to the control of aggression. The first one was directed at the cells of origin of the mesocortical dopaminergic pathway, the A10 cell group in the ventral tegmental area (VTA), while the second one was directed at the orbital PFC proper.

Damage to the VTA can be accomplished in several ways, e.g., a thermal or electrolytic lesion, an aspecific neurotoxin which acts upon neuronal cell bodies, but leaves passing fibres intact (kainic acid, ibotenic acid), or a specific neurotoxin, 6-hydroxydopamine (6-OHDA), which selectively destroys only dopaminergic cells when used in combination with a noradrenergic receptor reuptake blocker, such as desimipramine. Using the 6-OHDA procedure in male rats of the Long–Evans strain, we found a depletion of DA in the PFC of (approximately) 50% when compared with saline-injected control animals. Despite this (relatively) modest depletion, the behavioural outcome was beyond our expectation: a dramatic increase in offensive behaviour in the experimental animals (Fig. 3).

One disadvantage of this experimental approach is that it is impossible to confine the DA depletion strictly to the area of interest. We have to take into account a depletion not only in the orbital PFC (the area of our specific interest in this study) but also in the medial PFC and other termination areas of the mesocortical dopaminergic pathway. Moreover, there is no clear-cut separation of the VTA cells giving rise to mesocortical and mesolimbic dopaminergic pathways, although some regional differences in their distribution have been suggested (see Oades and Halliday, 1987, and Kalsbeek, 1989, for reviews). Thus, we also have to consider DA depletion in, among other places, the nucleus accumbens and/or septal areas following 6-OHDA lesions of the VTA. One way to approach this problem of specificity is based on the following reasoning. Damage to dopaminergic VTA cells will lead to the degeneration of their axons and presynaptic DA terminals, but postsynaptic elements presumably will still be present and functioning perhaps becoming hypersensitive to DA or its agonists (Woltering, 1988). Thus, following our observations of the dramatic increase in aggression in the rats with 6-OHDA VTA lesions, we decided to continue the experiment by implanting cannulae aimed at the orbitofrontal cortex, either in the left or the right hemisphere. Each animal was again subjected to tests for aggression following infusions into the orbitofrontal cortex of, successively, saline, the DA agonist apomorphine, and the DA antagonist haloperidol, in a random sequence. The results illustrate that apomorphine diminishes lateral threat, thus reducing the behavioural deficit, and resulting in a more normal aggressive display. With haloperidol an expected, but statistically non-significant trend in the opposite direction was observed (Fig. 4).

Since the above results point to an inhibitory involvement of the dopaminergic transmission in the orbital PFC, we have bilaterally injected 6-OHDA directly into the orbital PFC (De Bruin, 1990). With depletion of dopamine limited to orbital PFC, and despite sparing of both noradrenergic and serotonergic innervation (as visualised in immunocytochemical sections) unequivocal behavioural changes in aggression emerged (Fig. 5). Especially lateral threat, but also keeping down, were observed much more frequently in the animals with depleted dopamine in the orbital PFC. The structure of their aggressive behaviour had also changed, as demonstrated by analysing behavioural transitions: the animals with a PFC dopamine depletion were more inclined to react aggressively to noxious stimuli (threats and defensive acts) emanating from their opponents (De Bruin, 1990).

The data presented may be summarised as

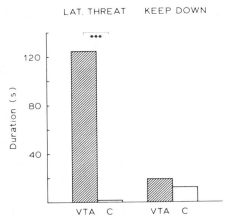

Fig. 3. Changes in aggressive behaviour following bilateral injections of 6-hydroxydopamine (6-OHDA) in the ventral tegmental area (VTA). Mean durations of 2 aggressive acts are portrayed based on encounters between dopamine-depleted (VTA) and control-operated saline-infused (C) rats. The dramatic increase in lateral threat in the VTA animals is especially striking, indicating the asymmetry of the observed interactions. The levels of dopamine in the frontal cortex of the VTA animals had decreased by approximately 50%, based on quantitative measurements using HPLC with electrochemical detection (M. Feenstra et al.). Analysis time: 0 – 1200 sec. *** $p < 0.001$; ⁻$0.05 < p < 0.10$ (Mann – Whitney U test, two-tailed).

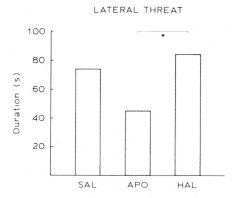

Fig. 4. "Partial reinstatement" of aggressive behaviour following local microinjections of the dopamine agonist apomorphine in animals with a damaged mesotelencephalic dopaminergic system, achieved by bilateral 6-OHDA injections in the ventral tegmental area (cf. Fig. 3). The mean durations of lateral threat are portrayed during conditions in which the experimental animals had been infused with either saline (SAL, 1 μl), the agonist apomorphine (APO, 1 μg in 1 μl) or the antagonist haloperidol (HAL, 1 ng in 1 μl). Microinjections were given approximately 10 min before testing, unilaterally either in the left or right orbital PFC. The order was random with a 1-week interval between tests. Data indicate that even a unilateral infusion of the agonist is sufficient to reduce aggressive behaviour to more normal levels, presumably by stimulating postsynaptic dopamine receptors. Analysis time: 0 – 1200 sec. * $p < 0.05$ (Mann-Whitney U test, two-tailed).

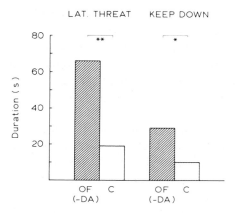

Fig. 5. Changes in aggressive behaviour following a confined dopamine depletion (6-OHDA, 8 μg in 2 μl saline, bilateral injection into the orbital PFC). Mean durations of 2 aggressive acts are portrayed based on encounters between animals with a local dopamine depletion in the orbitofrontal cortex (OF[-DA]) and animals which had been infused with saline (C). The degree of depletion of dopamine was assessed using an immunocytochemical assay with an antibody raised against dopamine (Geffard et al., 1984; Kalsbeek, 1989). The depletion was satisfactory, with few, if any dopaminergic fibres visible in the orbital PFC and sparing of the dopaminergic innervation of adjacent areas (medial PFC and nucleus accumbens). The serotonergic innervation of the orbital PFC had not been affected as evidenced from immunocytochemical staining with an antibody against serotonin. Analysis time: 0 – 600 sec. * $p < 0.05$; ** $p < 0.01$ (Mann – Whitney U test, two-tailed).

follows: The orbital PFC exerts an inhibitory influence on the expression of agonistic behaviours. The effects of thermal lesions can be mimicked by manipulations restricted to depleting dopamine in the orbital PFC, indicating that the dopaminergic input from the A10 area plays a major role. Stimulation studies have shown that this inhibition is directed at hypothalamic sites where aggression can be elicited by electrical stimulation.

Human studies

I have not touched, so far, upon the disturbances in social – agonistic behaviour in humans who have suffered frontal cortex damage, either accidentally or by having received a lobotomy or leucotomy. One of the reported effects of frontal lobotomy is that patients became more docile and were more manageable, reminiscent of Jacobsen's (1936) observations on chimpanzees which lost their temper tantrums after frontal lobotomy. However, such findings fail to account for the occasional reports of such patients suffering from bouts of hyperaggressivity. Contradictory anectodical evidence also stems from one of Egaz Moniz' own patients who, as Valenstein (1986) reported, actually shot the "father" of frontal lobotomy.

I will now mention two recent studies which point towards an increase in aggression following damage to prefrontal cortical areas. The first one (Grafman et al., 1986) is based on observations of Vietnam veterans with PFC damage, often unilateral, and entails behavioural state assessment on the basis of self-reports, i.e., how they would react to certain imaginary situations. Veterans with damage to orbital, and especially frontodorsal PFC, responded that they would be much more inclined to fight than would patients with other cortical damage or "brain-intact" controls. The second example is based on the study of Heinrichs and co-workers (1989), who studied the occurrence of violent incidents in chronic patients. The presence of cerebral damage per se was not a good predictor for violent incidents, but patients with focal frontal cerebral lesions were more frequently involved in such incidents.

Conclusion

In conclusion, experimental data support an inhibitory role of the prefrontal cortex, especially its orbitofrontal part, in the control of aggressive behaviour. In addition, the ascending dopaminergic system strongly modulates this function, presumably by tonically exciting the PFC. Evidence for this conclusion is derived mainly from studies using various strains of laboratory rats, but also receives support from certain studies on cats and primates. It should be emphasized, however, that not all available data permit broad generalisation over the various mammalian taxa. Thus, the "riddle of the frontal lobe" (Teuber, 1972) is still far from being solved, as far as its involvement in "species-specific" behaviours such as social – agonistic interactions is concerned.

Acknowledgements

The author wishes to thank Drs. M.A. Corner and C.G. van Eden for their critical remarks which helped to improve this paper. The figures were drawn by Mr. H. Stoffels, and they were photographed by Mr. G. van der Meulen.

References

Bowden, D.M., Goldman, P.S. and Rosvold, H.E. (1971) Free behavior of rhesus monkeys following lesions of the dorsolateral and orbital prefrontal cortex in infancy. *Exp. Brain Res.*, 12: 265 – 274.

Butter, C.M. and Snyder, D.R. (1972) Alterations in aversive and aggressive behaviors following orbital frontal lesions in rhesus monkeys. *Acta Neurobiol. Exp.*, 32: 525 – 565.

Clarke, A.S. and Mason, W.A. (1988) Differences among three macaque species in responsiveness to an observer. *Int. J. Primatol.*, 9: 347 – 364.

De Brabander, J.M., De Bruin, J.P.C. and Van Eden, C.G. (1990) Comparison of neonatal and adult medial prefrontal cortex lesions on food hoarding and spatial delayed alternation. *Behav. Brain Res.*, submitted.

De Bruin, J.P.C. (1981) Prefrontal cortex lesions and social behavior. In: M.W. van Hof and G. Mohn (Eds.), *Functional Recovery from Brain Damage*, Elsevier, Amsterdam, pp. 239 – 258.

De Bruin, J.P.C. (1990) Orbital prefrontal cortex, dopamine and social agonistic behavior of male Long Evans rats. *Aggr. Behav.*, 16: 231 – 248.

De Bruin, J.P.C., Van Oyen, H.G.M. and Van de Poll, N.E. (1983) Behavioural changes following lesions of the orbital prefrontal cortex in male rats. *Behav. Brain Res.*, 10: 209 – 232.

Deets, A.C., Harlow, H.F., Singh, S.D. and Blomquist, A.J. (1970) Effects of bilateral lesions of the frontal granular cortex on the social behavior of rhesus monkeys. *J. Comp. Physiol. Psychol.*, 72: 452 – 461.

Franz, S.I. (1907) On the functions of the cerebrum: the frontal lobes. *Arch. Psychol.*, 2: 1 – 64.

Fuster, J.M. (1989) *The Prefrontal Cortex*, 2nd Edn., Raven Press, New York.

Geffard, M., Buys, R.M., Sequela, P., Pool, C.W. and LeMoal, M. (1984) First demonstration of highly specific and sensitive antibodies against dopamine. *Brain Res.*, 294: 161 – 165.

Grafman, J., Vance, S.C., Weingartner, H., Salazar, A.M. and Amin, D. (1986) The effects of lateralized frontal lesions on mood regulation. *Brain*, 109: 1127 – 1148.

Heinrichs, R.W. (1989) Frontal cerebral lesions and violent incidents in chronic neuropsychiatric patients. *Biol. Psychiat.*, 25: 174 – 178.

Jacobsen, C.F. (1936) Studies on cerebral functions in primates. *Comp. Psychol. Monogr.*, 13: 1 – 60.

Kalsbeek, A. (1989) *Dopamine and the Development of Rat Prefrontal Cortex*. Doctoral thesis, Krips Repro, Meppel.

Kalsbeek, A., De Bruin, J.P.C., Feenstra, M.G.P. and Uylings, H.B.M. (1990) Age-dependent effects of lesioning the mesocortical dopamine system upon prefrontal cortex morphometry and PFC-related behaviors. *Progress in Brain Research* Vol. 85, Ch. 12, Elsevier, Amsterdam, pp. 257 – 283.

Kita, H. and Oomura, Y. (1981) Reciprocal connections between the lateral hypothalamus and the frontal cortex in the rat: electrophysiological and anatomical observations. *Brain*

Res., 213: 1 – 16.

Kolb, B. (1974) Social behavior of rats with chronic prefrontal lesions. *J. Comp. Physiol. Psychol.,* 87: 466 – 474.

Kolb, B. (1984) Functions of the rat frontal cortex: a comparative review. *Brain Res. Rev.,* 8: 65 – 98.

Kolb, B. and Gibb, R. (1990) Anatomical correlates of behavioral change after neonatal prefrontal lesions in rat. *Progress in Brain Research,* Vol. 85, Ch. 11, Elsevier, Amsterdam, pp. 241 – 256.

Kolb, B. and Nonneman, A.J. (1974) Frontolimbic lesions and social behavior in the rat. *Physiol. Behav.,* 13: 637 – 643.

Krettek, J.E. and Price, J.L. (1976) The cortical projections of the mediodorsal nucleus and adjacent thalamic nuclei in the rat. *J. Comp. Neurol.,* 171: 157 – 192.

Kruk, M.R., Van der Poel, A.M. and De Vos-Frerichs, T.P. (1979) The induction of aggressive behaviour by electrical stimulation in the hypothalamus of male rats. *Behaviour,* 70: 292 – 321.

Lammers, J.H.C.M. (1987) *Behavioural Responses from the Rat Hypothalamus, Origins and Interactions,* Doctoral thesis, Elinkwijk, Utrecht.

Leonard, C.M. (1969) The prefrontal cortex of the rat. I. Cortical projections of the mediodorsal nucleus. II. Efferent connections. *Brain Res.,* 12: 321 – 343.

Lubar, J.F., Herrmann, T.F., Moore, D.R. and Shouse, M.N. (1973) Effect of septal and frontal ablations on species-typical behavior in the rat. *J. Comp. Physiol. Psychol.,* 83: 260 – 270.

Miller, M.H. (1976) Dorsolateral frontal lobe lesions and behavior in the macaque: dissociation of threat and aggression. *Physiol. Behav.,* 17: 209 – 213.

Milner, B. and Petrides, M. (1984) Behavioural effects of frontal-lobe lesions in man. *Trends Neurosci.,* 7: 403 – 407.

Moyer, K.E. (1968) Kinds of aggression and their physiological basis. *Comm. Behav. Biol.,* A2: 65 – 87.

Myers, R.E., Swett, C. and Miller, M. (1973) Loss of social group affinity following prefrontal lesions in free-ranging macaques. *Brain Res.,* 64: 257 – 269.

Oades, R.D. and Halliday, G.M. (1987) Ventral tegmental (A10) system: Neurobiology. 1. Anatomy and connectivity. *Brain Res. Rev.,* 12: 117 – 165.

Peters, M. and Ploog, D. (1976) Frontal lobe lesions and social behavior in the squirrel monkey (Saimiri): a pilot study. *Acta Biol. Med. Germ.,* 35: 1317 – 1326.

Raleigh, M.J., Steklis, H.D., Ervin, F.R., Kling, A.S. and McGuire, M.T. (1979) The effects of orbitofrontal lesions on the aggressive behavior of vervet monkeys *(Cercopithecus aethiops sabaeus). Exp. Neurol.,* 66: 158 – 168.

Rose, J.E. and Woolsey, C.N. (1948) The orbitofrontal cortex and its connections with the mediodorsal nucleus in rabbit, sheep and cat. *Res. Publ. Assoc. Nerv. Ment. Dis.,* 27: 210 – 232.

Siegel, A., Edinger, H. and Lowenthal, H. (1974) Effects of electrical stimulation of the medial aspect of the prefrontal cortex upon attack behavior in cats. *Brain Res.,* 66: 467 – 479.

Siegel, A., Edinger, H. and Dotto, M. (1975) Effects of electrical stimulation of the lateral aspect of the prefrontal cortex upon attack behavior in cats. *Brain Res.,* 93: 473 – 484.

Siegel, S. (1956) *Nonparametric Statistics for the Behavioral Sciences,* McGraw-Hill, Kogakusha.

Slopsema, J.S., Van der Gugten, J. and De Bruin, J.P.C. (1982) Regional concentrations of noradrenaline and dopamine in the frontal cortex of the rat: dopaminergic innervation of the prefrontal subareas and lateralization of prefrontal dopamine. *Brain Res.,* 250: 197 – 200.

Stam, C.J., De Bruin, J.P.C., Van Haelst, A.M., Van der Gugten, J. and Kalsbeek, A. (1989) The influence of the mesocortical dopaminergic system on activity, food hoarding, social-agonistic behavior, and spatial delayed alternation in male rats. *Behav. Neurosci.,* 103: 24 – 35.

Stuss, D.T. and Benson, D.F., (1984) Neuropsychological studies of the frontal lobes. *Psychol. Bull.,* 95: 3 – 28.

Teuber, H.-L. (1972) Unity and diversity of frontal lobe functions. *Acta Neurobiol. Exp.,* 32: 615 – 656.

Thierry, A.M., Blanc, G., Sobel, A., Stinus, L. and Glowinsky, J. (1973) Dopaminergic terminals in the rat cortex. *Science,* 182: 499 – 501.

Thierry, A.-M., Godbout, R., Mantz, J. and Glowinski, J. (1990) Influence of the ascending monoaminergic systems on the activity of the rat prefrontal cortex. *Progress in Brain Research,* Vol. 85, Ch. 18, Elsevier, Amsterdam, pp. 357 – 365.

Uylings, H.B.M. and Van Eden, C.G. (1990) Qualitative and quantitative comparison of the prefrontal cortex in the rat and in primates, including humans. *Progress, in Brain Research,* Vol. 85, Ch. 3, Elsevier, Amsterdam, pp. 31 – 62.

Valenstein, E.S. (1986) *Great and Desperate Cures,* Basic Books, New York.

Van der Poel, A.M., Van der Hoef, H., Meelis, W., Vletter, G., Mos, J. and Kruk, M.R. (1983) A locked, non-rotating, completely embedded, moveable electrode for chronic brain stimulation studies in freely moving, fighting rats. *Physiol. Behav.,* 31: 259 – 263.

Van Eden, C.G. (1985) *Postnatal Development of Rat Prefrontal Cortex.* Doctoral thesis, Rodopi, Amsterdam.

Warren, J.M. (1964) The behavior of carnivores and primates with lesions in the prefrontal cortex. In: J.M. Warren and K. Akert (Eds.), *The Frontal Granular Cortex and Behavior,* McGraw-Hill, New York, pp. 168 – 191.

Warren, J.M. (1972) Evolution, behavior and the prefrontal cortex. *Acta Neurobiol. Exp.,* 32: 581 – 593.

Warren, J.M. and Akert, K. (Eds.) (1964) *The Frontal Granular Cortex and Behavior,* McGraw-Hill, New York.

Woltering, G. (1988) *ACTH-Neuropeptides and Brain Plasticity, Behavioral and Neurotrophic Aspects.* Doctoral thesis, State University of Utrecht.

Discussion

S.B. Dunnett: How specific are your stimulation effects to agonistic behaviors? For example, would stimulation of OF cortex produce a similar rise in threshold of other species-specific behaviours elicited by LH stimulation such as eating or gnawing?

J.P.C. de Bruin: In the experiments reported here we have specifically investigated the modulatory role of prefrontal cortex on hypothalamic sites where electrical stimulation elicited aggressive responses. Of course, it is well known that stimulation of other hypothalamic sites can elicit other, species-typical behaviours (for a recent review, see Lammers, 1987), but they were not investigated in this study. It would certainly be worthwhile to investigate whether prefrontal cortical inhibition can be generalized to other behaviours, such as eating, gnawing, grooming, which can be elicited by electrical stimulation of hypothalamic sites. However, at present we have no data in support of such a generalization.

A.Y. Deutch: Orbital PFC stimulation elevates threshold required to elicit LH-stimulation-evoked aggression, while 6-OHDA lesions of the region are pro-aggression. If DA inhibits neurones in the orbital cortex, then 6-OHDA lesions should remove these neurones from inhibition, i.e., exert effects similar to stimulation. How to reconcile?

J.P.C. de Bruin: The results of our lesion studies of the orbital PFC indicate that this area exerts an inhibiting influence on social–agonistic behaviour. The stimulation studies, concomitant electrical stimulation of orbital PFC and hypothalamus, suggest that this inhibition is exerted through hypothalamic sites, where aggression can be elicited by electrical stimulation (for anatomical support, see Kita and Oomura, 1981). When the dopaminergic input is reduced by 6-OHDA lesions, either directed at the VTA or the orbital PFC proper, then the inhibition is no longer exerted, and opposite behavioural changes are observed: an increase in aggressive behaviour. This, perhaps tentative, explanation reconciles the findings.

B. Kolb: Have you considered the problem of after-discharges induced by the stimulation?

J.P.C. de Bruin: The occurrence of after-discharges can only be reliably assessed with EEG recordings, which were not taken during the present experiments. However, we do not consider after-discharges to be a problem in the present study, because: (1) Current intensity was rather low (never exceeding 50 μA), too low for after-discharges to be expected. (2) At a number of sites in the frontal cortex electrical stimulation did not affect hypothalamic threshold, which is a nice control for the sites where stimulation did affect threshold values for attack.

D.A. Powell: What was the baseline level of aggression shown in this kind of situation in rat studies?

J.P.C. de Bruin: As explained we have used strains of rats which are characterized by their full expression of agonistic behaviour in the testing situation employed, that is, within a testing period of 10–20 min agonistic acts, such as lateral threat, keeping down, chasing, boxing occur at a "reasonable" frequency/duration. Although intensity may be somewhat less than in wild Norway rats, the pattern is basically similar. More detailed descriptions and quantitative data of frequency/duration/latency of agonistic acts recorded in these types of encounters are given in De Bruin et al. (1983) and De Bruin (1990).

References

De Bruin, J.P.C. (1990) Orbital prefrontal cortex, dopamine and social agonistic behavior of male Long Evans rats. *Aggr. Behav.*, 16: 231–248.

De Bruin, J.P.C., Van Oyen, H.G.M. and Van de Poll, N.E. (1983) Behavioural changes following lesions of the orbital prefrontal cortex in male rats. *Behav. Brain Res.*, 10: 209–232.

Kita, H. and Oomura, Y. (1981) Reciprocal connections between the lateral hypothalamus and the frontal cortex in the rat: electrophysiological and anatomical observations. *Brain Res.*, 213: 1–16.

Lammers, J.H.C.M. (1987) *Behavioural Responses from the Rat Hypothalamus, Origins and Interactions,* Doctoral thesis, Elinkwijk, Utrecht.

SECTION IV

Pathology of Prefrontal Cortex

CHAPTER 25

Animal models for human PFC-related disorders

Bryan Kolb

University of Lethbridge, Lethbridge, AB T1K 3M4, Canada

Introduction

The publication of the *Descent of Man* by Darwin in 1871 can probably be taken as the beginning of the general use of nonhuman subjects to study psychological processes in humans. Darwin argued that human and animal minds were similar and that differences were quantitative, not qualitative. Thus, for Darwin there was a continuum of mental complexity; humans differed in degree but not in kind from other animals. As comparative neurology developed in the decades following Darwin's book, anatomists were struck by similarities across mammalian species, with the principal difference appearing to be the volume of neocortex. The implications of this difference remained uncertain, however, and in his discussion of the differences in the brains of rats and humans C. Judson Herrick remarked that "Rats are not men . . . Men are bigger and better than rats" (Herrick, 1926, p. 365)!

Today, we recognize that although the mammalian cortex is remarkably similar in structure across species (e.g. Rockel et al., 1980), there are significant differences both in the details and in the complexity of the organization of mammalian brains. For example, although there are multiple representations of the sensory inputs in the cortex of all mammals the number of cortical regions, as well as the details of their connections and functions, appear to differ in even relatively closely related species such as Old World and New World monkeys (e.g. Kaas, 1987). The problem is to know when it is reasonable to generalize and when it is not. For behavioural neuroscientists this problem concerns not only the issue of structural equivalence across different brains, but also that of behavioural equivalence in different species.

Principles underlying interspecies comparisons

The chief purpose of cross-species comparisons in neuroscience has been to understand the basic mechanisms of brain function. In this type of comparative work, the species chosen for study depend upon the nature of the question asked. For example, neurophysiologists may choose to study the neural activity of giant nerve fibres in the squid because the nerve is so large and accessible. Similarly, the barrel fields of the rat cortex may be chosen as a model of cortical function because of their elegant structural organization, which is closely tied to peripheral structure. In both cases there is a clear rationale for why subjects are being chosen and what basic mechanism is under study. Studies of basic mechanisms of frontal lobe function are of a different class, however, since the research questions are tied to a clearly defined structure, namely the frontal lobe. When looking at the basic mechanisms of frontal lobe function it seems reasonable to compare animals with large frontal lobes, such as nonhuman primates, but it is

Correspondence: Dr. B. Kolb, University of Lethbridge, Lethbridge, AB T1K 3M4, Canada not clear what the rationale would be for using carnivores or rodents. Their frontal areas are far smaller and it is not clear what "basic mechanisms" one would study in nonprimates. Indeed, a perusal of the literature on the effects of frontal lobe lesions in nonhuman primates suggests that those who study primates have this view and find little of interest in the work on other species since nonprimate work is virtually never cited by these researchers.

A second type of comparative study is designed to describe the phylogenetic development of the brain and to demonstrate the phylogenetic development of humanoid intelligence. In these studies, species are chosen because of a presumed phylogenetic relationship. For example, Masterton and Skeen (1972) selected hedgehogs, tree shrews, and bushbabies as experimental animals solely on the basis of paleontological conclusions regarding their successive common ancestry with anthropoids. The authors then chose to study the animals' capacity to perform a delayed alternation task. They suspected that this test would be a behavioural indicator of the function of the prefrontal system because the prefrontal system is necessary for normal performance of delay-type tasks in anthropoids. The results showed a clear relationship between relative volume of prefrontal cortex (and associated regions including nucleus medialis dorsalis and caudate nucleus), performance on the delayed alternation task, and phylogenetic status (Fig. 1). The results suggest that the capacity to perform delayed alternation-type tests has developed in parallel with the prefrontal cortex and could imply that this ability may be fundamental in the development and function of the prefrontal cortex. On the other hand, the improvement in delayed alternation performance also may be related to the general increase in brain size across these species or to some other factor. Hence, it will be necessary to study the performance of these species on other putative tests of anthropoid frontal lobe function but this has not yet been done. In the long run the difficulty, of course, is in deciding what the appropriate behavioural tests should be. One way to approach this problem is to try to establish behavioural homologies but this is fraught with difficulties. Most definitions of homology are versions of Simpson's (1961) definition of structural homology as resemblance due to inheritance from com-

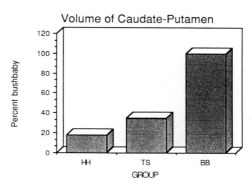

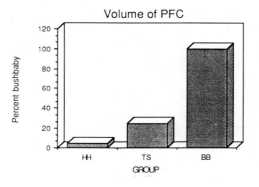

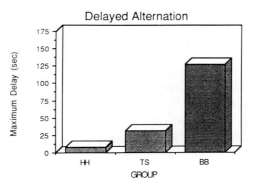

Fig. 1. Comparison of (1) the volume of caudate-putamen and prefrontal cortex and (2) maximum delays on a delayed alternation test in which performance exceeded chance in hedgehogs (HH), tree shrews (TS), and bushbabies (BB). (After Masterton and Skeen, 1972.)

mon ancestry. Atz (1970) applied this to behaviour when he defined behavioural homology by saying that "to be homologous, two behaviours exhibited by two phylogenetically related forms must have been present as a single behaviour in a common ancestor". Such a definition of homology is not too helpful since behaviour leaves no fossil record and it must therefore be tied to decisions regarding structural homology. Furthermore, it is difficult to see how behavioural homology could be established for most behaviours in existing mammals since the common ancestors are so distant. More recent attempts at determining behavioural homology do not really solve this problem (e.g. Hodos, 1976) so it remains difficult to determine the basis for phylogenetic studies of frontal lobe function, except in the limited sense of the Masterton and Skeen type of experiment.

A third approach to comparative studies is to model cerebral organization by placing emphasis upon function rather than structure. The logic of this approach is that although the details of behaviour may differ somewhat, mammals share many similar behavioural traits and capacities that have a similar function. All mammals must detect and interpret sensory stimuli, relate this information to past experience, and act appropriately. Similarly, all mammals appear to be capable of learning complex tasks under various schedules of reinforcement (Warren, 1977) and all mammals are mobile and have developed mechanisms for navigating in space. Although the details of the behaviours vary considerably the general capacities are common to all mammals. Warren and Kolb (1978) proposed that behaviours and behavioural capacities demonstrable in all mammals could be designated as *class-common* behaviours. In contrast, behaviours that are unique to a species and that have presumably been selected to promote survival in a particular niche are designated as *species-typical* (or species-specific) behaviours. Consider the following example.

In our studies of the effects of prefrontal lesions upon the social behaviour of cats we considered the response of cats to species-typical releasers, namely urine and the posture of conspecifics. Urine is used as a territorial marker by cats, who spray distinctive landmarks, such as bushes, large rocks, posts, or walls with a fine mist of urine. Other cats exhibit a striking behavioural sequence, known as flehmen, when they encounter the urine of a conspecific (Kolb and Nonneman, 1975). The most common components of the response pattern include approaching, sniffing, and touching the urine source with the nose, flicking the tip of the tongue repeatedly against the anterior palate behind the upper incisors, withdrawing the head from the urine, and opening the mouth in a gape or "flehmen response", and licking the nose. This behavioural pattern apparently allows olfactory stimuli to reach the secondary olfactory system, which appears to be specialized to analyse odours that are species-relevant. Cats show this response to urine of other cats, and oddly enough to humans, but they do not show it to urine of rhesus monkeys, dogs, rats, or hamsters. They also do not show it to cat fecal matter or cat fur, although they do show it to cat earwax! Like many mammals, cats are also highly responsive to the sight of conspecifics, and their response is strongly affected by the posture and piloerection of other cats. When we studied the effect of prefrontal lesions upon the responses to urine and to silhouettes of cats in different postures we found that although both normal and frontal cats responded to the stimuli, they did so in different ways (Nonneman and Kolb, 1974). The cats with prefrontal lesions sniffed the urine but showed little interest and showed no flehmen response, which means that the accessory olfactory system did not have access to the odour, and although the frontal cats oriented and piloerected to cats in a "halloween" posture, they did not approach the model. Furthermore, when placed in a large room with a novel cat (who was unoperated) the frontal cats were submissive and attempted to escape from the room whereas the normal cats approached the frontal animals and were generally aggressive. Hence, prefrontal lesions in cats altered the normal response to the species-typical social releasers of urine and piloerection. These results led me to the conclusion that prefrontal lesions would have the class common effect of altering the normal response to social releasers, which would have the effect of altering social behaviour. To test this idea my colleagues and I did parallel experiments looking at the response of people with unilateral frontal lobe removals to what appeared to be parallel stimuli in people, namely facial expression and tone of voice (prosody). Our results showed that frontal lobe patients were very poor at the recognition of facial expression and prosody (Kolb and Taylor, 1981, 1988).

The results of our studies on cats and humans showed changes in behaviours unique to each species (urine and piloerection in cats and the facial expression and prosody in humans) as well as a change in a more general class of behaviour, which we can call social behaviour. The details of behavioural change to particular stimuli are *species-typical,* the general function of the behaviour changed is *class-common*. It makes little sense to try to study response to flehmen in humans, since humans do not display it, nor to study facial expression in cats, since cats have little facial expression. These behaviours are species-typical. Similarly, it makes little sense to suggest that the prefrontal cortex has a general function in mammals of controlling facial expression or flehmen. Rather, it is more reasonable to conclude that the general function of the prefrontal cortex is to perceive, and perhaps to produce, species-typical social behaviours, of which flehmen and facial expression are examples. The advantage of this approach to comparative studies is that it is possible to make comparisons in species that are not closely related phylogenetically by focussing upon the function of the behaviour in question rather than upon the details of the behaviours.

The distinction between class-common and species-typical behaviours provides a rationale for making cross-species comparisons but it is not without weaknesses of its own (see Kolb and Whishaw, 1983a). In particular, we must ask whether it is legitimate to assume that since behaviours in different species have the same function, the neural substrates of the behaviours are the same. There is certainly no guarantee that just because mammals have class-common behaviours they have not independently evolved solutions to the class-common problems. Within the class Mammalia this seems unlikely, however, as there is little or no evidence in support of convergent evolution of neural substrates of class common behaviours, or at least within the placental mammals. Neurophysiological stimulation, evoked potential, and lesions studies reveal a similar topography in the motor, somatosensory, visual, and auditory cortices of the mammals, a topography that can provide the basis for the class-common neural organization of fundamental capacities in mammals. Kaas (1987) has suggested that all placental mammalian species appear to have several basic cortical regions in common including primary and secondary visual fields, primary and secondary somatosensory fields, at least a primary auditory cortex, a region of posterior temporal cortex with input primary visual cortex, taste cortex, frontal cortex receiving projections from the medialis dorsalis of the thalamus, and several subregions of limbic cortex related to the anterior and lateral dorsal thalamic nuclei. There are certainly significant species differences in cortical organization, however, particularly in (1) the number of cortical fields beyond the common core in each sensory modality, (2) the size of the class-common subfields, and (3) the details of cortico-cortical connections. These interspecies differences can be presumed to contribute to the unique behavioural repertoire that permits different species to survive in its particular niche. The similarities in organization presumably reflect a class-common solution to the problems of being a mammal. It is likely that the prefrontal cortex as a whole will have class-common functions but that in view of the divergent evolution of existing mammalian forms the organization of the subregions with the prefrontal cortex *may* show significant functional differences, in part because a major source of afferents to the prefrontal cortex is from the various sensory subfields, which themselves show significant cross-species differences in organization.

It will be the goal of the remainder of this chapter to illustrate the class-common functions of the prefrontal cortex of mammals. I shall illustrate the arguments by comparing the anatomical organization of the prefrontal cortex of rodents and primates and then consider the effects of prefrontal lesions in rats, monkeys and humans. I shall argue that there are remarkable parallels in the organization and function of the frontal lobe across the class Mammalia. Further, I will suggest that since species that are markedly divergent both in their level of encephalization and in their ecological adaptations have basic prefrontal functions in common, functions that are likely to be class common. Finally, I will argue that the commonality in function of the prefrontal cortex suggests that much may be learned about the function of the prefrontal cortex by considering it in the context of its biological function in mammals.

Structural comparisons

A primary concern in making cross-species generalizations about prefrontal function is that we are discussing tissue that is structurally equivalent. The traditional way of determining equivalence is to search for homologies. Since brains leave poor fossil records this must be done indirectly. Campbell and Hodos (1970) proposed that similarities in the following criteria could be used to establish homology: (1) connections; (2) topography; (3) position of reliably occurring sulci; (4) embryology; (5) morphology of individual neurones; (6) histochemistry; (7) electrophysiology; and (8) behavioural changes from lesions and stimulation. For the present purposes of comparing rodents and primates I shall focus upon connections.

Thalamo-cortical projections

If we use Rose and Woolsey's (1948) definition of prefrontal cortex as that cortex receiving afferents from medialis dorsalis of the thalamus (MD), then we can identify a region of prefrontal cortex in all mammal brains. In both rodents and primates there is a topographic organization to these projections such that we can identify different cortical subfields (e.g. Divac et al., 1978; Goldman-Rakic and Porrino, 1985). There are difficulties in generalizing from these thalamo-cortical connections, however, since it is now clear that McCulloch's (1944) idea that the principal thalamic nuclei project to separate areas to form a mosaic of functional anatomical areas is in error because thalamic projections are not as discrete as was once believed. For example, in the rat, part of the medial MD projection cortex receives overlapping projections from the medial anterior nucleus (AM), the ventral nucleus (V), or the paratenial nucleus (PT), as well as the lateral posterior nucleus (LP), as illustrated in Fig. 2 (Divac et al., 1978). A parallel pattern of overlapping projections is also seen in the monkey as MD projections overlap with AM and the medial pulvinar nucleus (Goldman-Rakic and Porrino, 1985). Thus, in both species there is frontal tissue receiving only

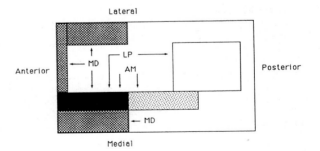

Fig. 2. Schematic illustrations of the types of overlap of thalamic projections to the prefrontal cortex of the rat. The largest rectangle represents the folded-out neocortex of the right hemisphere, with the frontal pole at the top. The checkered area represents the area receiving projections from medialis dorsalis, (MD), ventral tegmentum, and basolateral amygdala; the stippled area represents the area receiving projections from only medialis anterior (AM); the black area represents the area receiving projections from MD, AM, posterior lateralis posterior (LP), basolateral amygdala, and ventral tegmentum. (Adapted from Divac et al., 1978.)

MD projections, tissue receiving MD and AM projections, and tissue receiving projections from the same posterior nucleus that projects to the posterior parietal region. The putative role of posterior parietal cortex in spatial guidance, and the overlapping of posterior parietal thalamic projections (LP/pulvinar) with those of MD and AM in the prefrontal cortex, imply that the prefrontal cortex may be part of a circuit whose function is related to visuospatial guidance.

Cortico-cortical connections

The prefrontal cortex receives two types of cortico-cortical projections: those from sensory regions and those from the posterior parietal area (see Fig. 3). As the prefrontal cortex has expanded in mammalian evolution, the pattern of projections has become more complex: as the sensory areas enlarged and divided into additional sensory fields there has been a corresponding increase in the volume and subregions of the prefrontal cortex (e.g. Petrides and Pandya, 1988). This general pattern of cortico-cortical connections can be seen in both

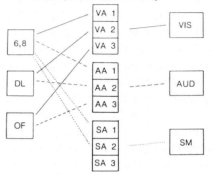

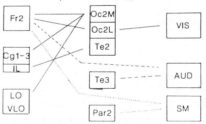

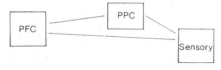

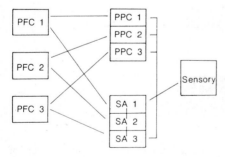

Fig. 3. The putative cortico-cortical connections of the prefrontal cortex of a hypothetical primitive and advanced mammal (e.g. monkey). In the advanced mammal there is an expansion of the posterior parietal region, and the development of the secondary association regions, which replace the primary sensory cortex as the major sensory afferent. Abbreviations: PFC, prefrontal cortex; PPC, posterior parietal cortex. Sensory refers to the visual, auditory and somatosensory cortices. The designations 1, 2 and 3 in the lower panel indicate that there are multiple regions of PFC, PPC and SA cortex.

Fig. 4. A comparison of the cortico-cortical connections of the prefrontal cortex of the rhesus monkey and the rat. The details differ but the general principle is similar, except that the rat still has direct primary sensory projections to the prefrontal cortex. (Note: there is also a direct visual-Fr2 projection in the rat that is not shown.) Nomenclature after Pandya and Yeterian (1985) for the monkey and Zilles (1985) for the rat. Abbreviations: AA, auditory association cortex; AUD, primary auditory cortex; Cg1–3, cingulate regions 1–3; DL, dorsolateral prefrontal cortex; IL, infralimbic cortex; Fr2, frontal cortex, area 2; LO, lateral orbital cortex; Oc2M, medial occipital area 2; Oc2L, lateral occipital area 2; OF, orbital frontal cortex; Par1, 2, parietal areas 1 and 2; SA, somatosensory association cortex; SM, somatosensory cortex; Te1, Te2, Te3, temporal areas 1, 2 and 3; VA, visual association cortex; VIS, visual association cortex; VLO, ventral orbital cortex. The solid, dashed, and dotted lines represent the visual, auditory, and somatosensory regions, respectively.

the rat and monkey, although it is more complex in the monkey. Fig. 4A shows the general organization of sensory projections to the prefrontal regions of the monkey (Pandya and Yeterian, 1985). Each of the visual, auditory and somatosensory regions sends projections from the primary cortex to a series of sensory association zones, which project, in turn, to different prefrontal zones. Fig. 4B summarizes the connections in the rat and it can be seen that although there appear to be fewer secondary sensory zones in the rat, the pattern of connections is strikingly similar to that of the monkey. The one major difference is that there are direct projections from the primary sensory regions, as well as from the secondary regions. Furthermore, the visual inputs appear to be more extensive in the rat than the projections from other cortical areas, whereas in primates there are more equally balanced projections from the visual, auditory, and somatosensory systems.

Fig. 5A summarizes the general pattern of cortico-cortical connections to the posterior parietal cortex in the monkey. This cortex (area PG) receives projections from the visual, auditory and somatosensory regions and, in turn, sends projections to the dorsolateral and periarcuate regions (e.g. Goldman-Rakic, 1987). Again, the pattern of connections in the rat is strikingly similar, although far simpler in detail. There is some debate regarding the nature and position of the posterior parietal cortex of the rat but on the basis of its cytoarchitecture, thalamic connections, cortico-cortical connections, and behavioural changes after lesions, Krieg's area 7 is a likely candidate (Kolb, 1990; Kolb and Walkey, 1987). This region receives projections from visual and somatosensory regions and sends projections to MD projection areas, the heaviest projections being to the orbital region and the frontal eye fields.

In summary, a comparison of the cortico-cortical connections in rodents and nonhuman primates shows a considerable parallel in the general pattern of connections. There are, however, differences in details as the projections are less extensive in the rat.

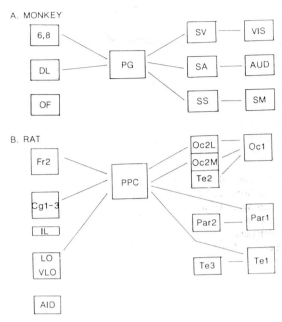

Fig. 5. Connections of the prefrontal and posterior parietal regions of the monkey and rat. Although there a differences, the general pattern is similar. Nomenclature after Pandya and Yeterian (1985) for the monkey and Zilles (1985) for the rat. Abbreviations as in Figs. 3 and 4, and: AID, agranular insular, dorsal, cortex; PG, area PG of the posterior parietal cortex of the monkey; SA, secondary auditory cortex; SS, secondary somatosensory cortex; SV, secondary visual cortex.

Cortico-subcortical connections

In addition to the thalamo-cortical connections there are extensive connections to other subcortical areas in both rodents and monkeys. Table I summarizes these connections and once again, there is a striking parallel. In particular, in both species there is a dopaminergic projection (substantia nigra and ventral tegmental area) and an amygdaloid projection that is coextensive with the MD projections. Every other region that receives prefrontal projections in the monkey also receives projections in the rat, and in addition, the rat receives projections from CA1 of the hippocampus and the nucleus accumbens.

In conclusion, although it is not possible to justify a conclusion that the MD projection cortex

TABLE I

Summary of subcortical connections of prefrontal cortex of monkey and rat

Structure	Monkey	Rat
Amygdala	+	+
Caudate-putamen	+	+
Pretectum, superior colliculus	+	+
Parahippocampal area	+	+
CA1	?	+
Posterior cingulate cortex	+	+
Substantia nigra, VTA	+	+
Nucleus accumbens	?	+
Claustrum	+	+
Hypothalamus	+	+
Mesencephalon	+	+
Pons	+	+

Note: " + " indicates a connection has been demonstrated. For details, see Kolb (1990) and Goldman-Rakic (1987).

of the monkey and the rat is homologous, there can be little doubt that the pattern of cortical and subcortical connections is very similar in general plan in the two species. Although this could be the result of parallel or convergent evolution, it seems more likely that the connections represent a general pattern of prefrontal connections across mammals, possibly resulting from their being present in a common primitive mammalian ancestor. The question we must now address is whether the similarity in structure underlies similarity in function.

Functional comparisons

The oldest and most widely used approach to functional localization in the neocortex is to analyse the behavioural effects of a lesion to circumscribed regions of the neocortex and to compare these effects to those resulting from lesions elsewhere in the cortex. Although the lesion technique is fraught with interpretational problems (see Kolb and Whishaw, 1990, for an extensive discussion), it remains as the major tool of comparative neuropsychology. The major challenge in the current context is to try to identify class-common functions that are disrupted by lesions to the prefrontal cortex of different species. I have grouped the behavioural changes into somewhat arbitrary groups that I believe may measure similar functions in humans, monkeys, and rats. Two difficulties in comparing humans to other species are that (1) the brain lesions in humans are the result of natural events and thus do not correspond to discrete subfields; and (2) the brain lesions in humans are usually unilateral. As we shall see, however, these differences do not appear to lead to major differences in the general pattern of behavioural symptoms (for more details, see Kolb and Whishaw, 1989).

Tables II – IV summarize the general behavioural changes observed after frontal cortex lesions in rodents, monkeys, and humans, respectively. My

TABLE II

Summary of major symptoms of prefrontal cortex damage in rodents

Symptom	Basic reference
1. Poor temporal memory	
a. poor spatial working memory	Kolb et al., 1983
b. poor DNMS	Kolb et al., 1989
c. poor delayed response	Kolb et al., 1974
d. poor habituation	Kolb, 1974b
e. poor associative learning	Passingham et al., 1988
2. Environmental control of behaviour	
a. impaired response inhibition	Kolb et al., 1974
b. alterations in mobility	Kolb, 1974a
c. contralateral neglect	Crowne, 1983
3. Reduced behavioural spontaneity	Kolb and Gorny, 1990
4. Disturbance of motor function	
a. poor execution of chains of movements	Kolb and Whishaw, 1983
b. restricted tongue mobility	Whishaw and Kolb, 1983
5. Impaired spatial orientation	Kolb et al., 1983, 1989
6. Impaired social and sexual behaviour	Kolb, 1974c; Michal, 1973
7. Impaired odour discrimination	Eichenbaum et al., 1980

TABLE III

Summary of major symptoms of prefrontal cortex damage in nonhuman primates

Symptom	Basic reference
1. Poor temporal memory	
a. poor spatial working memory	Passingham, 1985; Funashita et al., 1986
b. poor DNMS	Mishkin and Appenzeller, 1987
c. poor delayed response	Mishkin, 1964
d. poor habituation	Butter, 1964
e. poor associative learning	Petrides, 1982
2. Environmental control of behaviour	
a. impaired response inhibition	Mishkin, 1964
b. alterations in mobility	Gross and Weizkrantz, 1964
c. contralateral neglect	Crowne, 1983
3. Reduced behavioural spontaneity	Myers, 1972
4. Disturbance of motor function	
a. poor execution of chains of movements	Deuel, 1977
b. poor voluntary eye gaze	Latto, 1978
5. Impaired spatial orientation	Mishkin, 1964; Pohl, 1973
6. Impaired social and sexual behaviour	Butter and Snyder, 1972
7. Impaired odour discrimination	Tanabe et al., 1975

review has been selective; the intention was to list representative examples rather than to be comprehensive. I have not concerned myself with lesion locus within the prefrontal cortex, both because of the limited information regarding localization in humans as well as the questions of analogous regions in different species.

Temporal memory

The first evidence that prefrontal lesions might interfere with some type of memory process came from Jacobsen's (1936) discovery of a delayed response deficit in two chimpanzees with frontal lesions. Similar deficits have subsequently been shown in many species ranging from humans (Freedman and Oscar-Berman, 1986) to rats (Kolb et al., 1974). Similarly, there is evidence from studies of both monkeys and rats that prefrontal lesions lead to impairments in "delayed non-matching-to-sample" tests (Kolb et al., 1989; Mishkin and Appenzeller, 1987) in which animals are shown a cue, and then, after a short delay, are given a pair of stimuli, one identical to the first and one different from it. Animals are rewarded for choosing the novel stimulus. Both rats and monkeys with prefrontal lesions perform as well as con-

TABLE IV

Summary of major symptoms of prefrontal cortex damage in humans

Symptom	Basic reference
1. Poor temporal memory	
a. poor spatial working memory	Corkin, 1965
b. poor recency memory	Milner, 1974
c. poor delayed response	Freedman and Oscar-Berman, 1986
d. poor frequency estimate	Smith and Milner, 1985
e. poor habituation	Luria and Homskaya, 1964
f. poor associative learning	Petrides, 1985
2. Environmental control of behaviour	
a. impaired response inhibition	Milner, 1964
b. alterations in mobility	Ackerly, 1964
c. risk taking and rule breaking	Miller, 1985
d. contralateral neglect	Damasio et al., 1980
3. Reduced behavioural spontaneity	Jones-Gotman and Milner, 1977
4. Disturbance of motor function	
a. poor execution of chains of movements	Kolb and Milner, 1981b
b. poor voluntary eye gaze	Guitton et al., 1982
c. Broca's aphasia	Brown, 1972
5. Impaired spatial orientation	Semmes et al., 1963
6. Impaired social and sexual behaviour	Blumer and Benson, 1975
7. Impaired odour discrimination	Potter and Butters, 1980

trols at short delays, but fail to chance at longer delays (Fig. 6).

The source of the deficits on delay tasks has been controversial, but it would appear that successful performance requires that (1) sensory information be received and appropriately processed, (2) the sensory information must be held "on line" for some temporal interval until a behaviour is produced or a decision reached, and (3) the appropriate response needs to be made. It is the second function that is hypothesized to be dependent upon the prefrontal cortex (e.g. Goldman-Rakic, 1987). This process has been given many labels including short-term memory, representational memory, and a temporary memory buffer (e.g. Goldman-Rakic, 1987; Rawlins, 1985), but the essential idea is that there is a memory process that provides a temporary neural record of stimulus or motor events that occur over time. This allows animals to respond to sensory information after some delay and in the absence of the original stimuli. I shall call this temporal memory.

Although deficits in temporal memory are easily demonstrated in delay-type tasks, temporal memory deficits are also apparent in a variety of other behavioural tasks. For example, a series of studies by Brenda Milner and her colleagues have revealed that frontal lobe patients are unable to keep track of their responses on a variety of tasks (e.g. Petrides and Milner, 1982) or to accurately judge which of two stimuli, which were previously presented, was presented most recently (e.g. Milner, 1974). A further example of a temporal memory deficit may be seen in the process of habituation. Rats, monkeys, and humans with prefrontal cortex lesions all show a retarded habituation to novel stimuli as illustrated for rats in Fig. 7 (see Tables II – IV for references). Finally, animals with prefrontal lesions are impaired at tasks that are often described as "associative". For example, Petrides (1982, 1985) found that both monkeys and humans with prefrontal lesions were impaired at learning to make a particular motor response with distinctive stimuli. Although this deficit is often described as one of association memory, it may be reducable to one of temporal memory since successful solution of the task re-

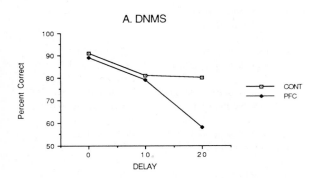

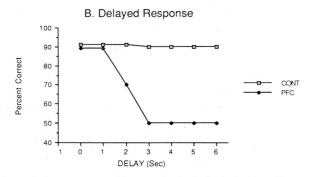

Fig. 6. Summary of delayed nonmatching to sample (DNMS) and delayed response performance of rats with medial prefrontal lesions (PFC). (After Kolb et al., 1974, 1989.)

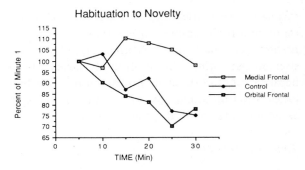

Fig. 7. Change in investigatory behaviour (head poking out of a small hole in a large box) over a 30 min test. Rats with medial frontal lesions did not decline over the test period whereas normal rats and rats with orbital frontal lesions habituated. These data may reflect an absence of normal temporal memory in rats with medial prefrontal lesions. (After Kolb, 1974b.)

quires that the subject keep track of the relationship of a cue and a response, which changes from trial to trial. A failure to keep a temporal record of which combination of response and cue was successful would result in poor performance.

Reduced behavioural spontaneity

One of the common clinical symptoms of patients with frontal lobe lesions is that they seldom spontaneously initiate other than simple behaviours. For example, it is not uncommon for relatives to complain that frontal lobe patients are content to lie in bed, watch television, or just sit. This has been quantified in many ways including tests of verbal fluency (Milner, 1964), facial expression (Kolb and Milner, 1981b), spontaneous speech (Kolb and Taylor, 1981), and doodling (Jones-Gotman and Milner, 1977). With each behaviour, the frontal lobe patients emit fewer behaviours per time unit than do normal controls or patients with lesions elsewhere.

There have not been any systematic attempts to demonstrate similar deficits in other species but at least circumstantial evidence suggests that similar deficits occur. For example, in a recent study of play behaviour in rats with infant frontal lesions Kolb and Gorny (1990) found that although frontal animals were at least as active as control animals, they virtually never initiated play behaviour. Once initiated by a normal animal, however, they did show all the normal components of play behaviour, although again it was less frequent than normal. Similarly, Myers (1972) studied spontaneous vocalizations in monkeys finding a sharp reduction in frontal monkeys. Other lesions did not have this effect. Thus, it appears that the prefrontal cortex plays some role generating novel behaviours. The presence of such a mechanism appears plausible and mammals are characterized by the variability and spontaneity of behaviour. This putative function could also be related to a more general class of behaviours that characterize some forms of "intelligence". Thus, Milner et al. (1985) suggested that the frontal lobe might play some role in what Guilford has termed divergent thinking, which is required to solve open-ended problems such as "How many ways can one use a coat hanger". Frontal lobe patients could be expected to do poorly on such a task, in spite of their normal performance on tests of general intelligence, which tend to test convergent thinking in which there is typically a single answer to the questions posed.

Response to environmental contingencies

A behavioural function that is clearly related to spontaneity is the ability to be flexible when environmental demands change. In contrast to lower vertebrates, mammals have evolved remarkable behavioural plasticity as characterized by the ability to adopt new strategies for problem solving when environmental contingencies change. This flexibility takes many forms including the ability to learn complex associations between seemingly unrelated events (e.g. red means stop), as well as the ability to alter well-learned behaviours when they no longer prove useful. One of the most notorious effects of frontal lobe lesions in mammals is the apparent inability of the subjects to inhibit various types of behaviour. In humans, this can be most clearly shown in tests in which there are specific rules, either implicit or explicit to the task, and failure to respect the rules results in failure at the task. This can be quantified in a number of tests including the Wisconsin Card Sorting Test (e.g. Milner, 1964), and a test of risk taking (Miller, 1985; Miller and Milner, 1985).

Evidence of similar difficulties in nonhuman species comes largely from learning tasks in which the animal is required to learn a particular solution to a problem and then must later make a response that is incompatible with the original learning. Perhaps the best example is spatial reversal learning as rats, cats, dogs and monkeys with frontal lesions are all known to be impaired at such tests.

One of the oldest reported effects of prefrontal lesions in nonhumans is a general increase in activity, which occurs in both monkeys and rats. The reason for this increased activity is uncertain but it could represent a failure to inhibit movement and/or a failure to generate alternate behaviours. Such increased mobility is rare in human subjects but it may require bilateral lesions to be noticeable. Indeed, case reports of people with large bilateral frontal lesions almost uniformly note the increased activity (e.g. Ackerly, 1964).

Although the source of much debate in the literature, there is another peculiar transient behavioural effect of frontal lesions in rats and monkeys, and perhaps in people: unilateral frontal lesions often produce sensory neglect (e.g. Crowne, 1982). The reason for neglect is controversial but it may represent a change in the animal's ability to recognize environmental change, such that the subject continues to engage in the behaviour ongoing at the time of stimulation. Admittedly speculative, this behavioural loss can thus be seen as part of a more general inability to produce flexible behavioural responses to environmental stimulation.

Disturbances of motor function

Damage to prefrontal areas produces disturbances in the execution of complex movements in humans, monkeys and rats. This behavioural change is sometimes described as a deficit in the programming of chains of movements, or possibly in the planning of movements. For example, Kolb and Milner (Kolb and Milner, 1981a, b; Milner and Kolb, in preparation) found a deficit in the copying of series of movements of either the face or the arms by frontal lobe patients (Fig. 8), which was consistent with blood flow studies by Roland et al. (1980) that found increased blood flow in the supplementary motor area during arm movements. Evidence of parallel deficits in nonhuman subjects has been more indirect, although the symptoms are likely comparable. For example, studies using rats and monkeys have shown deficits in the opening of puzzle latches in animals with prefrontal lesions (e.g. Deuel, 1977; Kolb and Whishaw, 1983b). Further, rats and hamsters with MF lesions are impaired at chaining together the behavioural units required for nest building or maternal behavior (e.g. Kolb and Whishaw, 1985). By filming the nest building behaviour of hamsters we were able to show that the animals are capable of performing each of the individual movements but appear unable to reliably execute the behavioural sequence. Similar behavioural disturbances in humans would probably be called apraxias, but the appropriateness of this term for nonhuman species is open to question.

Social and sexual behaviour

One of the most obvious and striking effects of frontal lobe damage in humans is a marked change in social behaviour and personality. Although most reports have been purely descriptive, recent studies have shown an impairment in the perception of affective states in others (Kolb and Taylor, 1981, 1988), reductions in spontaneous facial expressions (Kolb and Milner, 1981) as well as in the intensity of posed expressions (Kolb and Taylor, 1988), and a tendency toward social isolation (Deutsch et al., 1979). For example, Kolb and Taylor (1988) asked subjects to produce facial expressions that were appropriate for particular characters, who had no facial features, in cartoon sketches of everyday situations (e.g. receiving a

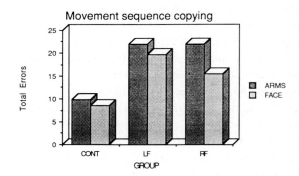

Fig. 8. Summary of arm and facial movement copying deficits in human frontal lobe patients with left or right unilateral frontal excisions. Frontal lobe patients are impaired at both tasks but not at making single movements. (After Kolb and Milner, 1981a.)

traffic ticket). On a subsequent test they were asked to choose an appropriate face for the cartoon character from an array of 6 faces. Frontal lobe patients were impaired at both tests relative to normal controls or patients with temporal lobectomies.

Similarly, monkeys with frontal lobe lesions, especially orbital frontal lesions, have a wide variety of abnormalities in social behaviour. For example, in one study Butter and Snyder (1972) removed the dominant male from each of several groups of monkeys, subsequently removing the frontal lobes from half of the monkeys. When the animals were later returned to their groups they all resumed the position of dominant male, but all of the frontal monkeys were deposed and fell to the bottom of the group hierarchy, although the downfall took varying amounts of time, ranging from a few days to several months. Other studies have shown that like humans with frontal injuries, frontal monkeys do not correctly interpret social displays of others and tend to be social isolates (e.g. Suomi et al., 1970). The social behaviour of both cats and rodents is markedly different from that of primates but we have found that prefrontal lesions in both species disrupt various aspects of social interactions (e.g. Kolb, 1974c; Lubar et al., 1973; Nonneman and Kolb, 1974). As described earlier, cats with frontal lesions do not respond normally to the visual and olfactory cues from other species. Similarly, frontal hamsters do not respond normally to species-typical olfactory cues (Shipley and Kolb, 1977) and rats do not respond normally to tactile cues in social grooming (Kolb, 1974c). Overall, it thus appears that frontal lesions alter the response to species-typical social releasers, which leads to significant changes in social interaction.

Sexual behaviour is presumably social and once again frontal lobe lesions appear disruptive. Complaints about reduced libido are common from relatives of frontal patients, although in some cases right frontal lobe lesions may release sexual activity (e.g. Blumer and Benson, 1975). There are, however, no systematic investigations of sexual behaviour in human frontal lobe patients. Similarly, there are few studies of sexual behaviour in nonhuman subjects with frontal lesions, the principal exceptions being two different reports of altered sexual behaviour in rats with medial frontal lesions (Larsson, 1970; Michal, 1973).

Odour discrimination

Damage to the orbital frontal area of humans, monkeys and rats impairs the ability to discriminate different odours (e.g. Eichenbaum et al., 1980; Potter and Butters, 1980; Tanabe et al., 1975). In addition, the activity of cells in the orbital cortex of the monkey is modulated by specific odours. Taken together, these data suggest that the orbital prefrontal cortex can be considered to be the primary olfactory cortex. The relationship between olfaction and other frontal lobe functions is unclear, however.

Other symptoms of prefrontal lesions

The classical stimulation studies of the 1940s demonstrated that electrical stimulation of at least parts of the prefrontal cortex produced various autonomic effects including changes in respiration, blood pressure, and heart rate (see review by Neafsey, this volume). More recently, there have been several demonstrations of projections from the prefrontal cortex to the vagal solitary nucleus in the medulla of the rat (e.g. Van der Kooy et al., 1982). Similarly, there is evidence that prefrontal lesions reduce food intake in both monkeys and rats, possibly due to drop in "set point" (e.g. Butter and Snyder, 1972; Kolb et al., 1977). The nature of the autonomic functions of the prefrontal cortex is still poorly studied and its relationship to other prefrontal functions is unclear. Nonetheless, there is evidence that such a function exists across mammalian species.

Do frontal lesions produce a deficit in spatial orientation?

On the basis of studies of brain-injured war veterans, Semmes and Teuber (e.g. Semmes et al., 1963; Teuber, 1964) concluded that frontal lobe lesions produced a deficit in personal orientation and that this deficit was dissociable from a deficit in extrapersonal orientation that was characteristic of patients with more posterior lesions. This idea proved extremely influential and there have been claims of a parallel distinction in nonhumans as well (e.g. Pohl, 1973). There are several reasons to doubt the validity of this proposition, however.

First, the deficit on the "personal orientation test" is not necessarily purely spatial. The task requires that patients look at a series of drawings of a man, upon which there are several numbers on different points on the body, and to point to the same locations on their own body. On some of the drawings the back of the man is drawn, and for some of the drawings the front of the man is drawn. The correct solution of the task requires an understanding of the spatial location of the points but it also requires that the patients keep changing the correct rule for solution since the front-sided drawings require a 180° rotation of the figure whereas the back-sided drawings do not. Furthermore, the numbers move all over the body so that the correct answer requires that the subject change the location for each response. As Teuber stated in 1964, "What is not clear is whether the repeated reversal (of a principle) per se produced the difficulty on this task for our frontals, or whether the main difficulty was more specifically related to orientation and reorientation of the body" (Teuber, 1964, p. 438). This question remains today. Second, although subsequent studies with human subjects have shown deficits on various tests requiring the use of spatial information, such as finger and stylus maze tests (Corkin, 1965; Milner, 1965), all of the latter tests confound space and memory. Third, in our own studies on rats with prefrontal lesions we were initially impressed with their deficits on various spatial maze learning tasks but these tasks all confound temporal memory with spatial information. That is, the animal must keep a neural record of both the route it followed to find the platform, as well as the location of various extra-maze cues to the platform location, in addition to being able to recognize the correct configuration of spatial elements that identify the spatial location of the platform. A deficit in either the neural record or in the synthesis of the spatial information would lead to poor performance. Further, on the Morris water task there is only an acquisition deficit, and not a retention deficit (e.g. Sutherland, 1985). If the prefrontal cortex had a significant role in spatial orientation per se, one would expect a deficit on both acquisition and retention of spatial problems. In addition, it is clear that frontal lobe patients are not impaired at most tests of spatial rotation or spatial manipulation (DeRenzi, 1982), again suggesting that there may not be a unique spatial function of the prefrontal cortex. In summary, although frontal lesions impair the performance on tests of spatial orientation and/or navigation, there is no compelling evidence of a unique spatial function of the prefrontal cortex.

Is there a general class-common function of the frontal lobe?

There have been repeated attempts to develop theories of the general role of the prefrontal cortex in the control of behaviour, and although they vary in detail, there is a common theme over the last decade that the prefrontal cortex is involved in the temporal organization of behaviour (e.g. Fuster, 1980; Goldman-Rakic, 1987). The temporal control of movement requires at least 6 components. First, there must be an ongoing record of sensory stimulation that has occurred recently. This allows for behavioural responses to be discontiguous with sensory stimulation. This capacity is taxed in delay-type tasks. Second, there must be an ongoing record of what movements are being produced by the "motor system" at any given moment. It is only with this knowledge that new units or series of units can be initiated. Third, there must be a record of those behaviours that are already executed. This is especially important in situations in which appropriate behaviours are contingent upon previous behaviours, such as in chains of movements. Moreover, a record of past behaviours is needed for strategies of searching since we must keep track of where we have searched. Fourth, there must be inhibition of some motor impulses and excitation of others in order to produce appropriate behavioural sequences. For example, in order to type we need to inhibit certain finger movements and activate others, and of course to do so in the correct order. Fifth, behaviour must be flexible with respect to both internal and external environments. It is one of the characteristics of mammals that we are able to adapt behavioural patterns to changing contexts. The importance of context-dependent behaviour in primate social structure is beautifully illustrated in Jane Goodall's graphic descriptions of the different behavioural patterns exhibited by chimpanzees (Goodall, 1986). Thus, the make-up of the social group at any time dictates the behaviour of each chimpanzee. Given the presence and position of certain animals a given chimp may be bold and relaxed whereas within a different mixture of chimps they are quiet and nervous. Further, it appears that an error in evaluating the context can have grievous consequences. It may be no accident that the frontal lobe has grown so large in primates who are so highly social. Not only must behaviour be regulated by external context but also by internal context, such as autonomic and endocrine status. This presumably accounts for the close relationship between the prefrontal cortex and autonomic systems (see Neafsey, this volume). Sixth, there must be an ongoing monitoring of the consequences of behaviour. If reward is associated with a particular series of movements, but not with another, then the association between movements and consequences needs to be made.

I have argued elsewhere (Kolb, 1984) that whereas the principle of temporal organization of behaviour provides a basis for the unity of prefrontal function, the necessary components for such a function provide the basis for the diversity in the effects of frontal injury. In the absence of an ongoing record of stimulation and behavioural responses, behaviour will appear disorganized. This may be clearly seen in the disruption of various species-typical behaviours such as mating behaviour, maternal behaviour or nest building in rats with prefrontal lesions. The prefrontal cortex is essential for the eventual execution of most of these behaviours (e.g. Whishaw, 1990) but in the absence of prefrontal cortex the complete behavioural action patterns are seldom executed correctly. Furthermore, in the absence of temporal memory, behaviour becomes dependent upon the environmental cues that are present when behaviour is executed. That is, behaviour will not be under the control of internalized knowledge, but of external cues that are present. This may manifest itself in any number of ways, one of which is likely to be a loss of internal inhibition of behaviour.

Conclusion

One of the characteristics of the mammalian cortex is that the sensory systems have evolved multiple representations in each sensory modality. It appears that as this evolution of sensory cortex has occurred there has been a corresponding increase in the volume of prefrontal cortex with which the sensory areas are associated (Fig. 4). It seems likely, therefore, that the prefrontal cortex will have multiple systems designed to provide temporal organization of behaviours related to different types of sensory inputs. It is clear from studies of nonhuman primates that lesions in different prefrontal loci will produce deficits in different delay tasks, each of which may require different types of information (e.g. Mishkin and Appenzeller, 1987; Passingham, 1985). One key to understanding the details of prefrontal organization in mammals may therefore be found in the study of both the cortico-cortical inputs to different prefrontal areas as well as in the study of the differences in behavioural patterns of different mammalian species. These behavioural patterns have evolved in response to specific ecological pressures and the unique pattern of cortical development in different species will likely reflect the different neural requirements for particular species-typical behaviours. The apparent general similarity in prefrontal function across mammals may reflect the general requirements of temporally organizing behaviours in animals with behavioural repertoires that are far more plastic than those of other classes such as birds or reptiles. The interspecies differences in the details of prefrontal organization may reflect the differences in sophistication of the temporal organization of behaviour, or at least of certain classes of behaviours.

As Warren (1972) has pointed out, neuropsychologists have concentrated upon the search for general principles regarding the neural mechanisms of behaviour and have tended to ignore interspecies variability, and the adaptedness of behaviour. It has been the goal of this paper to show that there are general similarities in prefrontal function and organization. I believe, however, that neuropsychologists also can effectively study prefrontal function by taking advantage of the differences in behaviour and prefrontal structure in different mammals, and that this will provide a complementary approach to unravelling prefrontal function. Thus, animal models have provided evidence of unity of prefrontal function in mammals and offer promise of providing evidence of the diversity of prefrontal function as well.

References

Ackerly, S.S. (1964) A case of paranatal bilateral frontal lobe defect observed for thirty years. In: J.M. Warren and K. Akert (Eds.), *The Frontal Granular Cortex and Behavior,* New York, McGraw-Hill, pp. 192 – 218.

Atz, J.W. (1970) The application of the idea of homology to behavior. In: L.R. Aronson, E. Tobach, J.D.S. Lehrman and J.S. Rosenblatt (Eds.), *Development and Evolution of Behavior,* W.H. Freeman, San Francisco, CA.

Blumer, D. and Benson, D.F. (1975) Personality changes with frontal and temporal lobe lesions. In: D.F. Benson and D. Blumer (Eds.), *Psychiatric Aspects of Neurologic Disease,* Grune and Stratton, New York, pp. 151 – 170.

Brown, J. (1970), *Aphasia, Apraxia and Agnosia,* Charles C. Thomas, Springfield, IL.

Butter, C.M. (1964) Habituation of responses to novel stimuli in monkeys with selective frontal lesions. *Science,* 144: 313 – 315.

Butter, C.M. and Snyder, D.R. (1972) Alterations in aversive and aggressive behaviors following orbital frontal lesions in rhesus monkeys. *Acta Neurobiol. Exp.,* 32: 525 – 565.

Campbell, C.B.G. and Hodos, W. (1970) The concept of homology and the evolution of the nervous system. *Brain Behav. Evol.,* 3: 353 – 367.

Corkin, S. (1965) Tactually guided maze-learning in man: Effects of unilateral cortical excisions and bilateral hippocampal lesions. *Neuropsychologia,* 3: 339 – 351.

Crowne, D.P. (1982) The frontal eye field and attention. *Psych. Bull.,* 93: 232 – 260.

Damasio, A.R., Damasio, H. and Chui, H.C. (1980) Neglect following damage to the frontal lobe or basal ganglia. *Neuropsychologia,* 18: 123 – 132.

Darwin, C. (1871) *The Expression of Emotions in Man and Animals,* D. Appleton, London.

DeRenzi, E. (1982) *Disorders of Space Exploration and Cognition,* John Wiley, New York.

Deuel, K.K. (1977) Loss of motor habits after cortical lesions. *Neuropsychologia,* 15: 205 – 215.

Deutsch, R.D., Kling, A. and Steklis, H.D. (1979) Influence of frontal lobe lesions on behavioral interactions in man. *Res. Commun. Psychol. Psychiat. Behav.,* 4: 415 – 431.

Divac, I., Kosmal, A. Bjorklund, A. and Lindvall, O. (1978) Subcortical projections to the prefrontal cortex in the rat as revealed by the horseradish peroxidase technique. *Neuroscience,* 3: 785 – 796.

Eichenbaum, H., Shedlack, K.J. and Eckmann, K.W. (1980) Thalamocortical mechanisms in odor-guided behavior. I. Effects of lesions of the mediodorsal thalamic nucleus and frontal cortex on olfactory discrimination in the rat. *Brain Behav. Evol.,* 17: 255 – 275.

Eichenbaum, H., Clegg, R.A. and Feeley, A. (1983) Reexamination of the functional subdivisions of the rodent prefrontal cortex. *Exp. Neurol.,* 79: 434 – 451.

Freedman, M. and Oscar-Berman, M. (1986) Bilateral frontal lobe disease and selective delayed response deficits in humans. *Behav. Neurosci.,* 100: 337 – 342.

Funahashi, S., Bruce, C.J. and Goldman-Rakic, P.S. (1986) Perimetry of spatial memory representation in primate prefrontal cortex. *Soc. Neurosci. Abstr.,* 12: 554.

Fuster, J.M. (1980) *The Prefrontal Cortex,* Raven Press, New York.

Goldman-Rakic, P.S. (1987) Circuitry of the primate prefrontal cortex and regulation of behavior by representational memory. In: F. Blum (Ed.), *Handbook of Physiology. Vol.*

5. *The Nervous System,* American Physiological Society, Bethesda, MD, pp. 373–418.

Goldman-Rakic, P.S. and Porrino, L.J. (1985) The primate mediodorsal (MD) nucleus and its projections to the frontal lobe. *J. Comp. Neurol.,* 242: 535–560.

Goodall, J. (1986) *The Chimpanzees of Gombe,* Harvard University Press, Cambridge, MA.

Gross, C.G. and Weiskrantz, L. (1964) Some changes in behavior produced by lateral frontal lesions in the macaque. In: J.M. Warren and K. Akert (Eds.), *The Frontal Granular Cortex and Behavior,* McGraw-Hill, New York, pp. 74–101.

Guitton, D., Buchtel, H.A. and Douglas, R.M. (1982) Disturbances of voluntary saccadic eye-movement mechanisms following discrete unilateral frontal-lobe removals. In: G. Lennerstrand, D.S. Lee and E.L. Keller (Eds.), *Functional Basis of Ocular Motility Disorders,* Pergamon Press, Oxford.

Herrick, C.J. (1926) *Brains of Rats and Men,* University of Chicago Press, Chicago, IL.

Hodos, W. (1976) The concept of homology and the evolution of behavior. In R.B. Masteron, W. Hodos and H. Jerison (Eds.), *Evolution, Brain and Behavior: Persistent Problems,* Lawrence Erlbaum, Hillsdale, NJ.

Jacobsen, C.F. (1936) Studies of cerebral function in primates. *Comp. Psychol. Monogr.,* 13: 1–68.

Jones-Gotman, M. and Milner, B. (1977) Design fluency: The invention of nonsense drawings after focal cortical lesions. *Neuropsychologia,* 15: 653–674.

Kaas, J.H. (1987) The organization and evolution of neocortex. In: S.P. Wise (Ed.), *Higher Brain Functions,* Wiley, New York, pp. 347–378.

Kolb, B. (1974a) Dissociation of the effects of lesions of the orbital or medial aspect of the prefrontal cortex of the rat with respect to activity. *Behav. Biol.,* 10: 329–343.

Kolb, B. (1974b) Some tests of response habituation in rats with prefrontal lesions. *Can. J. Psychol.,* 12: 466–474.

Kolb, B. (1974c) Social behavior of rats with chronic prefrontal lesions. *J. Comp. Physiol. Psychol.,* 87: 466–474.

Kolb, B. (1984) Functions of the frontal cortex of the rat: a comparative review. *Brain Res. Rev.,* 8: 65–98.

Kolb, B. (1990) Posterior parietal and temporal association cortex. In: B. Kolb and R. Tees (Eds.), *The cerebral cortex of the rat,* MIT Press, Cambridge, MA.

Kolb, B. and Gorny, B. (1990) Neonatal frontal lesions alter play behavior in the rat. In preparation.

Kolb, B. and Milner, B. (1981a) Observations on spontaneous facial expression after focal cerebral excisions and after intracarotid injection of sodium Amytal. *Neuropsychologia,* 19: 514–515.

Kolb, B. and Milner, B. (1981b) Performance of complex arm and facial movements after focal brain lesions. *Neuropsychologia,* 19: 505–514.

Kolb, B. and Nonneman, A.J. (1975) The development of social responsiveness in kittens. *Anim. Behav.,* 23: 368–374.

Kolb, B. and Taylor, L. (1981) Affective behavior in patients with localized cortical excisions: An analysis of lesion site and side. *Science,* 214: 89–91.

Kolb, B. and Taylor, L. (1988) Facial expression and the neocortex. *Soc. Neurosci. Abstr.,* 14: 219.

Kolb, B. and Walkey, J. (1987) Behavioural and anatomical studies of the posterior parietal cortex in the rat. *Behav. Brain Res.,* 23: 127–145.

Kolb, B. and Whishaw, I.Q. (1983a) Problems and principles underlying interspecies comparisons. In: T.E. Robinson (Ed.), *Behavioral Approaches to Brain Research,* Oxford University Press, New York, pp. 237–264.

Kolb, B. and Whishaw, I.Q. (1983b) Dissociation of the contributions of the prefrontal, motor, and parietal cortex to the control of movement in the rat. *Can. J. Psychol.,* 37: 211–232.

Kolb, B. and Whishaw, I.Q. (1985) Neonatal frontal lesions in hamsters impair species-typical behaviors and reduce brain weight and neocortical thickness. *Behav. Neurosci.,* 99: 691–706.

Kolb, B. and Whishaw, I.Q. (1990) *Fundamentals of Human Neuropsychology,* 3rd Edn., W.H. Freeman, New York.

Kolb, B., Nonneman, A.J. and Singh, R.K. (1974) Double dissociation of spatial impairments and perseveration following selective prefrontal lesions in rats. *J. Comp. Physiol. Psychol.,* 87: 772–780.

Kolb, B., Whishaw, I.Q. and Schallert, T. (1977) Aphagia, behavior sequencing, and body weight set point following orbital frontal lesions in rats. *Physiol. Behav.,* 19: 93–103.

Kolb, B., Sutherland, R.J. and Whishaw, I.Q. (1983) A comparison of the contributions of the frontal and parietal association cortex to spatial localization in rats. *Behav. Neurosci.,* 97: 13–27.

Kolb, B., Buhrmann, K. and McDonald, R. (1989) Dissociation of prefrontal, posterior parietal, and temporal cortical regions to spatial navigation and recognition memory in the rat. *Soc. Neurosci. Abstr.,* 15, 607.

Larsson, K. (1970) Mating behavior in the male rat. In: L.R. Aronson, E. Tobach, D.S. Lehrman and J.S. Rosenblatt (Eds.), *Development and Evolution of Behavior,* W.H. Freeman, San Francisco, CA.

Latto, R. (1978) The effects of bilateral frontal eye-field, posterior parietal or superior collicular lesions on visual search in the rhesus monkey. *Brain Res.,* 146: 35–50.

Lubar, J.F., Herrmann, T.J., Moore, D.R. and Shouse, M.N. (1973) Effect of septal and frontal ablations on species typical behavior in the rat. *J. Comp. Physiol. Psychol.,* 83: 260–270.

Luria, A.R. and Homskaya, E.D. (1964) Disturbances in regulative role of speech with frontal lobe lesions. In: J.M. Warren and K. Akert (Eds.), *The Frontal Granular Cortex and Behavior,* McGraw-Hill, New York.

Masterton, B. and Skeen, L.C. (1972) Origins of anthropoid intelligence: Prefrontal system and delayed alternation in hedgehog, tree shrew, and bush baby. *J. Comp. Physiol. Psychol.,* 81: 423–433.

McCulloch, W.S. (1944) The functional organization of the cerebral cortex. *Physiol. Rev.,* 24: 390–407.

Michal, E.K. (1973) Effects of limbic lesions on behavior sequences and courtship behaviour of male rats *(Rattus norvegicus). Behavior,* 44: 264–285.

Miller, L. (1985) Cognitive risk taking after frontal or temporal lobectomy. I. The synthesis of fragmented visual information. *Neuropsychologia,* 23: 359–369.

Miller, L. and Milner, B. (1985) Cognitive risk taking after frontal or temporal lobectomy. II. The synthesis of phonemic and semantic information. *Neuropsychologia,* 23: 371–379.

Milner, B. (1964) Some effects of frontal lobectomy in man. In: J.M. Warren and K. Akert (Eds.), *The Frontal Granular Cortex and Behavior,* McGraw-Hill, New York, pp. 313–334.

Milner, B. (1965) Visually-guided maze learning in man: Effects of bilateral hippocampal, bilateral frontal, and unilateral cerebral lesions. *Neuropsychologia,* 6: 215–234.

Milner, B. (1974) Hemispheric specialization: Scope and limits. In: F.O. Schmitt and F.G. Worden (Eds.), *The Neurosciences: Third Study Program,* MIT Press, Cambridge, MA.

Milner, B., Petrides, M. and Smith, M.L. (1985) Frontal lobes and the temporal organization of memory. *Hum. Neurobiol.,* 4: 137–142.

Mishkin, M. (1964) Perseveration of central sets after frontal lesions in monkeys. In: J.M. Warren and K. Akert (Eds.), *The Frontal Granular Cortex and Behavior,* McGraw-Hill, New York, pp. 219–241.

Mishkin, M. and Appenzeller, T. (1987) The anatomy of memory. *Sci. Am.,* 256: 80–89.

Myers, R.E. (1972) Role of the prefrontal and anterior temporal cortex in social behavior and affect in monkeys. *Acta Neurobiol. Exp.,* 32: 567–579.

Nonneman, A.J. and Kolb, B. (1974) Lesions of hippocampus or prefrontal cortex alter species-typical behavior in the cat. *Behav. Biol.,* 12: 41–54.

Pandya, D.N. and Yeterian, E.H. (1985) Architecture and connections of cortical association areas. In: A. Peters and E.G. Jones (Eds.), *Cerebral Cortex, Vol. 4,* Plenum, New York, pp. 3–62.

Passingham, R. (1978) Information about movements in monkeys *(Macaca mulatta). Brain Res.,* 152: 313–328.

Passingham, R.E. (1985) Memory of monkeys *(Macaca mulatta)* with lesions in prefrontal cortex. *Behav. Neurosci.,* 99: 3–21.

Petrides, M. (1982) Motor conditional associative learning after selective prefrontal lesions in the monkey. *Behav. Brain Res.,* 5: 407–413.

Petrides, M. (1985) Deficits on conditional associative-learning tasks after frontal- and temporal-lobe lesions in man. *Neuropsychologia,* 23: 601–614.

Petrides, M. and Milner, B. (1982) Deficits on subject-ordered tasks after frontal- and temporal-lobe lesions in man. *Neuropsychologia,* 20: 249–262.

Petrides, M. and Pandya, D. (1988) Association fiber pathways to the frontal cortex from the superior temporal region in the rhesus monkey. *J. Comp. Neurol.,* 273: 52–66.

Pohl, W. (1973) Dissociation of spatial discrimination deficits following frontal and parietal lesions in monkeys. *J. Comp. Phyiol. Psychol.,* 82: 227–239.

Potter, H. and Butters, N. (1980) An assessment of olfactory deficits in patients with damage to prefrontal cortex. *Neuropsychologia,* 18: 621–628.

Rawlins, J.N.P. (1985) Associations across time: The hippocampus as a temporary memory store. *Behav. Brain Sci.,* 8: 479–496.

Rockel, A.J., Hiorns, R.W. and Powell, T.P. (1980) The basic uniformity in structure of the neocortex. *Brain,* 103: 221–224.

Roland, P.E., Larsen, B., Lassen, N.A. and Skinhoj, E. (1980) Supplementary motor area and other cortical areas in organization of voluntary movements in man. *J. Neurophysiol.,* 43: 118–136.

Rose, J.E. and Woolsey, C.N. (1948) The orbitofrontal cortex and its connections with the mediodorsal nucleus in rabbit, sheep and cat. *Res. Publ. Assoc. Nerv. Ment. Dis.,* 27: 210–232.

Semmes, J., Weinstein, S., Ghent, L. and Teuber, H.-L. (1963) Impaired orientation in personal and extrapersonal space. *Brain,* 86: 747–772.

Shipley, J. and Kolb, B. (1977) Neural correlates of species-typical behavior in the Syrian golden hamster. *J. Comp. Physiol. Psychol.,* 91: 1056–1073.

Simpson, G.G. (1961) *Principles of Animal Taxonomy,* Columbia University Press, New York.

Suomi, S.J., Harlow, H.F. and Lewis, J.K. (1970) Effect of bilateral frontal lobectomy on social preferences of rhesus monkeys. *J. Comp. Physiol. Psychol.,* 70: 448–453.

Sutherland, R.J. (1985) The navigating hippocampus: An individual medley of space, memory and movement. In: G. Bujzsaki and C.H. Vanderwolf (Eds.), *Electrical Activity of the Archicortex,* Hungarian Academy of Sciences, Budapest.

Tanabe, T., Yarita, H., Iino, M., Ooshima, Y. and Takagi, S.F. (1975) An olfactory projection area in orbitofrontal cortex of the monkey. *J. Neurophysiol.,* 38: 1269–1283.

Teuber, H.-L. (1964) The riddle of frontal lobe function in man. In: J.M. Warren and K. Akert (Eds.), *The Frontal Granular Cortex and Behavior,* McGraw-Hill, New York.

Ungerleider, L.G. and Mishkin, M. (1982) Two cortical visual systems. In: D.J. Ingle, M.A. Goodale and R.J.W. Mansfield (Eds.), *Analysis of Visual Behavior,* MIT Press, Cambridge, MA.

Van der Kooy, D., McGinty, J.F., Koda, L.Y., Gerfen, C.R. and Bloom, F.E. (1982) Visceral cortex: a direct connection from prefrontal cortex to the solitary nucleus in rat. *Neurosci. Lett.*, 33: 123–127.

Walker, E.A. and Blumer, D. (1975) The localization of sex in the brain. In: K.J. Zulch, O. Creutzfeldt and G.C. Calbraith (Eds.), *Cerebral Localization,* Springer, Berlin.

Warren, J.M. (1972) Evolution, behavior and the prefrontal cortex. *Acta Neurobiol. Exp.*, 32: 580–593.

Warren, J.M. (1977) A phylogenetic approach to learning and intelligence. In: A. Oliverio (Ed.), *Genetics, Environment and Intelligence,* Elsevier, Amsterdam, pp. 37–56.

Warren, J.M. and Kolb, B. (1978) Generalizing in neuropsychology. In: S. Finger (Ed.), *Recovery from Brain Damage,* Plenum Press, New York.

Whishaw, I.Q. (1990) The decorticate rat. In: B. Kolb and R. Tees (Eds.), *The Cerebral Cortex of the Rat,* MIT Press, Cambridge, MA, in press.

Whishaw, I.Q. and Kolb, B. (1983) Stick out your tongue: tongue protrusion in neocortex and hypothalamic damaged rats. *Physiol. Behav.*, 30: 471–480.

Zilles, K. (1985) *The Cortex of the Rat. A Stereotaxic Atlas,* Springer, Berlin.

Discussion

J.P.C. de Bruin: The Masterton and Skeen data show a correlation between size of PFC and delay interval which can be "mastered". You presented in one of your studies the opposum, what about other marsupials like Echidna (cf. Divac et al.). Would its large PFC predict a good performance in tests with a long delay interval?

B. Kolb: This is difficult to know since there are no functional data on the Echidna but the prediction is straightforward: they should be good on tests like delayed alternation. Curiously, however, cats have a rather modest PFC, yet at least wild cats are rather good at delays.

A.Y. Deutch: The social behavior of the lab rat is very "truncated" relative to the social organization of the wild rat. Should we study the *lab* rat, or look as well to wild rats?

B. Kolb: The data from various labs are quite clear on the natural behaviour of the lab rat: it is not qualitatively different from the wild rat, although there are quantitative differences. I thus see no problem with studying lab rats. It would be interesting, however, to see if there is a difference in the size, or dendritic structure, of the PFC in wild rats since there are interspecies differences in social structure and volume of PFC in different species of voles.

CHAPTER 26

The prefrontal cortex in schizophrenia and other neuropsychiatric diseases: in vivo physiological correlates of cognitive deficits

Karen Faith Berman and Daniel R. Weinberger

Clinical Brain Disorders Branch, National Institute of Mental Health, NIMH Neurosciences Center at St. Elizabeths, Washington, DC 20032, USA

Introduction

The prefrontal cortex in human behavior and cognition

The role of the prefrontal cortex has intrigued students of human behavior for many years. Galen, in the second century, believed that the seat of the soul resided in this part of the brain. Albertus Magnus in the twelfth century insightfully foreshadowed modern thinking that the frontal lobes were necessary for problem solving and planning (McHenry, 1969). However, even in more recent times unraveling the functions of the prefrontal cortex remained an elusive goal. In fact, for many years the human prefrontal cortex was viewed as a physiologically "silent area" and was thought to contribute little if anything to human behavior and cognition. This misconception stemmed in part from inadequate definitions of what constitutes cognition and from the relative insensitivity of standard tests of intelligence (on which patients with frontal lobe lesions are often unimpaired) in measuring so-called "executive functions".

Correspondence: K.F. Berman, M.D., Clinical Brain Disorders Branch, National Institute of Mental Health, NIMH Neurosciences Center at St. Elizabeths, Washington, DC 20032, USA.

New approaches to studying the human brain, including neuropsychological tests devised to measure aspects of cognition such as mental flexibility and concept formation, suggest that, far from being a silent, redundant area, the prefrontal cortex may be at the core of that which is most human, shaping our attitudes and organizing our cognitive repertoire to produce the highest order goal-directed behaviors. Through the formation and execution of long- and short-term plans, the internal manipulation of representational systems, and the use of complex sequencing, the prefrontal cortex appears to play a fundamental role in the hierarchical organization of cognitive control and, indeed, in human consciousness (Perecman, 1987).

Methods for studying prefrontal function in man

Implications of prefrontal lesions

The importance of the prefrontal cortex in higher human cognition and behavior has been elucidated by observations of patients with focal cortical lesions. Published case reports of such patients first appeared in the nineteenth century. One of the earliest well-observed and well-documented cases of prefrontal damage was that of Phineas Gage in whom behavioral and personality changes were described following the accidental destruction of a portion of his frontal lobes that occurred while he was employed as a railway worker (Harlow, 1868, as cited in Fuster, 1989). Since then, studies of patients with prefrontal traumatic injury, including several large series of war veterans (Salazar et al., 1986), as well as those with stroke, or surgical resection of tumors or of epileptic foci have firmly established that damage to the prefrontal cortex in humans results in a syndrome of complex and often subtle signs and symptoms including changes in affect, social behavior, and cognition. Some investigators have delineated distinct behavioral syndromes associated with orbitomedial and dorsolateral lesions.

One problem with making general inferences about human frontal lobe function from observations of patients with gross frontal lobe lesions is that naturally or iatrogenically occurring prefrontal lesions are not well controlled in extent or location. In fact, few pathological lesions exclusively damage dorsolateral prefrontal cortex. Another problem is that in patients with tumors or epilepsy it is not clear whether the remaining brain tissue is actually normal. Similar concerns apply to patients with stroke or trauma where there may be sequelae that are distant to the prefrontal lesions themselves.

In vivo brain imaging

An alternative approach has recently become available through the advent of in vivo functional brain imaging techniques that allow the study of the normal living, working, human brain. These include ^{133}Xe inhalation regional cortical blood flow (rCBF) measurements, single photon emission computed tomography (SPECT), and positron emission tomography (PET), all of which involve relatively non-invasive regional imaging and quantification of various parameters of local brain activity such as cerebral blood flow, glucose metabolism, and receptor occupancy and density. These techniques have allowed direct investigation of prefrontal cortical function during various cognitive and other conditions, and they have proven extremely useful for investigating brain function in disease states in which the prefrontal cortex has been implicated.

Observations of similar patterns of cognitive deficits on neuropsychological tests that are sensitive to prefrontal lobe damage and of behavioral abnormalities that are common both in patients with known frontal lobe lesions and those with other neuropsychiatric diseases have raised the possibility that the prefrontal cortex may play a role in a diverse spectrum of illnesses including schizophrenia, Parkinson's disease, Huntington's disease, progressive supranuclear palsy, Alzheimer's disease, and others. Since gross structural frontal lobe pathology is not a prominent feature of the brains of most patients with these illnesses (with the exception of Alzheimer's disease), a mechanism other than a disordered prefrontal cortex per se must be postulated. This paper explores the concept that disruption at any site along the rich and complex circuits interconnecting the prefrontal cortex with other cortical and subcortical structures can result in the clinical "frontal lobe syndrome" and that measuring brain physiology during specific behaviors can help to delineate at what level such a disruption occurs.

In this chapter we describe the use of in vivo functional brain imaging, via the ^{133}Xe inhalation method for measuring rCBF, to study frontal lobe function in a variety of neuropsychiatric disorders that may be related to prefrontal dysfunction. We will focus on schizophrenia, an illness in which prefrontal cortex is of considerable heuristic interest and has long been impugned. Investigations of schizophrenia exemplify the potential power of such studies to elucidate pathophysiological mechanisms underlying diseases in which there are few clues. Despite nearly a century of post-mortem anatomical investigations of the brains of patients with schizophrenia, no consistent pathological lesion has been uncovered in this illness, and the anatomical basis for the brain functional abnormalities has remained unclear (Kirch and Weinberger, 1986). Comparing the results of blood flow studies in schizophrenia with those of patients with diseases such as Parkinson's disease, Huntington's disease, and Alzheimer's disease, all of which involve behavioral alterations thought to reflect dysfunction of prefrontal cortex, but in which the pathological substrates are better understood than are those underlying schizophrenia, may provide clues about the pathophysiological mechanism in the latter illness.

Evidence for prefrontal involvement in schizophrenia

Interest in the prefrontal cortex as a possible locus of pathology in schizophrenia dates back nearly a century. Alzheimer suggested that neuropathology in layers II and III of frontal cortex was responsible for this illness. Other pioneers in the field of schizophrenia research, notably both Kraepelin (1971) and Bleuler (1950), also considered the frontal lobes important in schizophrenia. In the first half of this century prefrontal leucotomies were performed with the notion that isolating a disordered frontal lobe from other brain areas would alleviate some of the symptoms of schizophrenia (Valenstein, 1986). However, although subtle structural changes in schizophrenia have been suggested (Benes et al., 1986), a great deal of neuropathological experimentation has failed to convincingly and consistently demonstrate gross structural damages of the prefrontal cortex.

While further investigations of the gross, cellular, and even synaptic anatomy of the prefrontal cortex in schizophrenia are necessary before the case is closed on the possibility of structural pathology, evidence continues to accrue that a *functional* aberration of prefrontal cortex may explain some of the clinical phenomena of schizophrenia. Hallucinations and delusions may be the most dramatic clinical features of schizophrenia, but the "negative" or "defect" symptoms may be more disabling. These include flat and inappropriate affect, paucity of thought, social withdrawal, lack of initiative and motivation, poor insight and judgement – a clinical constellation reminiscent of patients with frontal lobe lesions, especially lesions of dorsolateral prefrontal cortex.

The pattern of cognitive deficits seen in schizophrenia also most consistently implicates the frontal lobes (Seidman, 1983) as do the minor or "soft" neurological signs common in schizophrenia (Quitkin et al., 1976).

Of course, none of these clinical features is pathognomonic of frontal lobe disease, but methods of imaging brain function in vivo can now be applied to test the hypothesis of a dysfunction of prefrontal cortex more directly. Such studies were pioneered some 15 years ago by Ingvar and Franzen who used a relatively invasive method of measuring rCBF of one hemisphere at a time with carotid injections of ^{133}Xe dissolved in saline. They reported that virtually all normal subjects under virtually all waking conditions exhibited relatively more blood flow to frontal cortex than to other cortical areas visible with this technique. They termed this characteristic normal pattern "hyperfrontality". In contrast they noted that chronic schizophrenic patients were relatively "hypofrontal", lacking the pattern of augmented anterior blood flow common to normal subjects (Ingvar and Franzen, 1974a).

This intriguing finding not only revitalized interest in the role of the frontal lobes in schizophrenia, but also heralded the era of functional brain imaging techniques as applied to neuropsychiatry research. However, in the decade following the publication of this work, a number of rCBF and PET studies appeared in the literature, and the results were extremely inconsistent: while some investigators confirmed Ingvar and Franzen's observation of hypofrontality, others did not (for reviews see Berman and Weinberger, 1986a, and Weinberger and Berman, 1988). There may be many reasons for these inconsistencies, but one of the most important probably relates to the fact that when studying brain physiology, the results depend upon how the brain is engaged while it is being studied. Measurements of cerebral blood flow or metabolism represent the sum of a complex set of cerebral physiological responses. While some of these responses reflect the illness being studied, many reflect the subject's behavior and experience while the measurement is being made. It may seem obvious that the brain's physiology may be expected to reflect and respond to its surround and internal millieu, but the implications of this fact for functional brain imaging studies, i.e. the need to control as completely as possible all sensory inputs and cognitive and motor outputs to reduce the "noise" of such measurements, may not have been fully appreciated by early investigators in this field. In fact, many of the first rCBF and PET schizophrenia studies were carried out while subjects were "at rest", a condition during which sensory, cognitive, and motor inputs and outputs are least controlled. It has been demonstrated that resting state studies are more non-specifically variable than studies carried out during more controlled conditions (Duara et al., 1987), and it has been suggested that there may be no such thing as a "resting state" in the awake human brain (Mazziotta et al., 1982).

In vivo cortical stress tests to assess prefrontal physiology

An alternative approach that we have taken has been to device "cortical stress tests" in which rCBF is measured while subjects receive controlled sensory input and carry out specified cognitive and/or motor tasks (Berman, 1987). For example, cognitive tasks that reliably activate specific cortical areas in normal subjects can be used to test the ability of patients' brains to physiologically respond to various cognitive or other demands. In our studies we typically make 3 rCBF measurements in each subject during 3 different conditions so that each subject can serve as his or her own control. The first of these is always a resting state study, the main purpose of which is to acclimatize subjects to the rCBF measurement procedure. The second and third of this set of measurements, taken together, have proved more informative about neuropsychiatric disorders, and schizophrenia in particular. These two procedures are usually designed such that one is a regionally and cognitively specific task with relevance to the cortical region of interest (the stress test) and the other is a non-specific sensorimotor control task designed to be like the stress test in as many ways as possible except the cognitive processes involved. The order in which these two tasks are presented is counterbalanced across subjects to control for the possibility of an order effect.

To investigate prefrontal cortex we needed a condition or cognitive test that would elevate neuronal function in prefrontal cortex above ambient levels. We chose to measure rCBF while subjects performed the Wisconsin Card Sorting test (WCS), a neuropsychological test that involves the use of feedback and working memory in the formation of conceptual sets and necessitates the shifting of these sets when appropriate. Patients with known frontal lobe pathology do poorly on this test, and Milner (1963, 1964, 1971) has shown that it is a particularly sensitive indicator of the integrity of the dorsolateral aspect of the prefrontal cortex in man. Patients with schizophrenia also consistently do poorly on the WCS (Malmo, 1974; Kolb and Whisman, 1983; Goldberg et al., 1987) and, like patients with gross frontal lobe disease, their errors are characteristically of a perseverative nature; that is, inability to switch the conceptual set even when given feedback to do so.

We also designed a sensorimotor control task during which subjects were shown slides of the numbers 1 through 4, presented in random order, which they were meant to match to 1 of 4 switches also labeled with these numbers. For this Numbers Match test, the method of presenting stimuli to the subject, the mode for the subjects' response, and the feedback mechanism were identical to those for the WCS. Therefore, it controls for the minimal finger movement necessary to make a response, for the visual stimulation, and for the psychological experience of taking a test while having blood flow measured. Thus, blood flow measured while a subject performs the control task can be subtracted from blood flow measured while that subject performs the WCS to highlight the physiological changes related to the abstract reasoning and problem solving aspects of the latter test.

With this approach we showed (Weinberger et al., 1988a) that when normal subjects perform the WCS their dorsolateral prefrontal blood flow values are significantly elevated above levels during the Number Match control task (Figs. 1 and 2). This confirmed that it is an appropriate stress test for studying prefrontal function in neuropsychiatric patient groups.

Prefrontal physiology in schizophrenia

Activation of prefrontal cortex in schizophrenia

Using the paradigm described above we have studied patients with chronic schizophrenia and compared them with age- and sex-matched normal subjects. We demonstrated that, while between-group differences in prefrontal blood flow were inconsistent during resting and non-existent during simple Number Matching, robust differences could be demonstrated during the WCS (Fig. 1). Even when each subject was used as his or her own control and the differences between blood flow during the two tasks were examined, patients showed a striking failure to activate frontal cortex during the WCS, particularly the dorsolateral prefrontal cortex (Fig. 2). We have now demonstrated this basic finding of a behavior- and region-specific dysfunction of prefrontal cortex in a total of 4 different cohorts of patients, 2 who were on neuroleptics (Berman et al., 1986, 1989) and 2 who were medication free (Weinberger et al., 1986, 1988a). A consistent pattern of findings has emerged from these studies and those of other investigators using different methods (Cohen et al., 1987; Volkow et al., 1987); when schizophrenic patients are studied during rest or other non-specific conditions they may or may not appear hypofrontal, but virtually all studies carried out during prefrontal stimulation have demonstrated lower prefrontal blood flow or metabolism in schizophrenia. Thus, the bulk of the available data indicate that physiological dysfunction of the prefrontal cortex in schizophrenia, at least during physiological demand, is a replicable finding. However, its interpretation must be carefully considered.

How prevalent is hypofrontality in schizophrenia?

One question that has not been answered by the body of published studies of frontal lobe function in schizophrenia is: How characteristic of schizophrenic patients is hypofrontality? Is this finding a consistent characteristic of schizophrenia, or does it just affect a subgroup of patients who may have different pathophysiologies and etiologies underlying their illnesses? The approach taken by virtually all studies has been to compare the mean value for a group of patients with schizophrenia to that for an unrelated group of normal subjects. The results of such a comparison often confirm that, on the whole, patients have lower values. However, examination of the individual values typically reveals that there is a great deal of overlap between the two groups and that only a minority of patient values actually fall beyond the lower limit of the relatively wide range of normal values. One interpretation of this observation could be that only a small subgroup of patients with schizophrenia are hypofrontal. However, since we have no way of knowing what a given patient's value would have been if he or she did not have schizophrenia, the true prevalence of hypofrontality in the schizophrenia population cannot be estimated.

We had the opportunity to address this question by studying 10 pairs of monozygotic twins who were discordant for schizophrenia (Berman et al., 1989). Assuming that the rCBF measurements for the well co-twin of a monozygotic pair discordant for schizophrenia may reflect the values that would have characterized the ill twin if he or she did not have schizophrenia, the former can be used as a genetically (as well as socio-economically and environmentally) perfect control to determine the pathophysiological changes that have occurred in each patient.

As in our previous studies of unrelated groups of normals and schizophrenics, when we compared

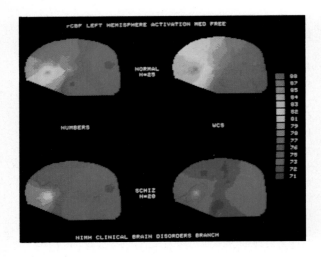
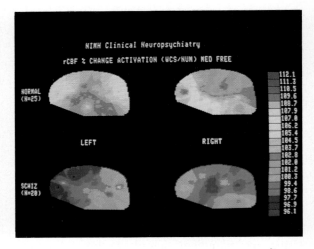
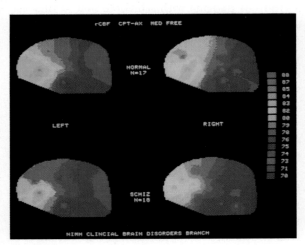
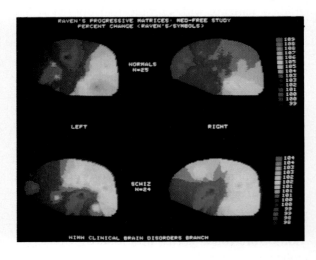
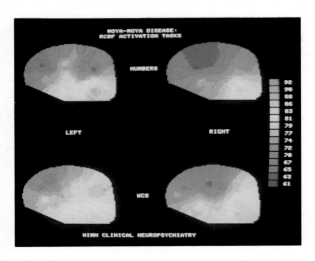

relative prefrontal blood flow values between discordant co-twins, we found no consistent evidence of hypofrontality in the ill twins during resting or Numbers Matching. However, during the WCS in each and every case the twin with schizophrenia was more hypofrontal than his or her well co-twin (Berman et al., 1989). This suggests that in virtually every case something has happened to the brain of the schizophrenic twin to make it more hypofrontal, at least during the prefrontally linked WCS task, than it would have been without schizophrenia.

It is tempting to conclude that these data confirm that there is a primary physiological impairment of prefrontal cortex in schizophrenia, that it is characteristic of the illness, and that it affects most if not all individuals who have schizophrenia. However, an alternative explanation that must be considered is that hypofrontality may not be a primary feature of schizophrenia but may rather be an epiphenomenon due to less important secondary factors. That is, is the finding related to the physiological characteristics of the region subserving the behavior — the prefrontal cortex — or does it, instead, reflect non-specific state variables. Secondary factors that may play a role in studies of schizophrenia include motivation, arousal, effort, and the effects of treatment with neuroleptic medication. Although a growing body of evidence suggests that these secondary factors do not explain hypofrontality in schizophrenia, they are important to consider.

Is hypofrontality in schizophrenia an epiphenomenon?

A medication effect?

Most rCBF and PET studies have been carried out in patients who were either receiving neuroleptics at the time of the study or had been withdrawn from neuroleptics for some period of time (such as the 4-week drug-free interval in our studies). This, along with the fact that hypofrontality could not be demonstrated in several resting state PET studies of small numbers of patients who had never received neuroleptics (Early et al., 1987; Volkow et al., 1986), has suggested to some investigators that neuroleptic treatment is responsible for hypofrontality in schizophrenia. However, a growing body of evidence refutes this notion.

Our own data, demonstrating WCS-related hypofrontality in 24 patients tested while on stable doses of 0.4 mg/kg/day of haloperidol (Berman et al., 1986), suggest that neuroleptic treatment or its withdrawal cannot fully explain hypofrontality. However, these results do not rule out the possibi-

---

Fig. 1. Lateral maps of group mean regional cerebral blood flow (rCBF) values, with the anterior pole at the left, for the left hemisphere in 25 normal controls and 20 medication-free schizophrenics during the Wisconsin Card Sorting test (WCS) and a simple Numbers Matching sensorimotor control task. Note the patients' significantly lower prefrontal flow during the WCS. There were no between-group differences during the control task (from Weinberger et al., 1986).

Fig. 2. rCBF activation maps for the same subjects as in Fig. 1. Data for the WCS are expressed as a percentage of flow during the control task. Note the significant prefrontal activation in the normal subjects and the lack of activation in the patients (from Weinberger et al., 1986).

Fig. 3. rCBF maps for 18 medication-free schizophrenic patients and 17 normal control subjects during a version of the visual Continuous Performance task (CPT). Note the similar pattern in both groups. There were no significant between-group differences (from Berman et al., 1986).

Fig. 4. rCBF maps for 25 medication-free schizophrenic patients and 24 normal controls. Flow values during Raven's Progressive Matrices are expressed as a percentage of those during a simple Symbols Matching sensorimotor control task. Note the pattern of posterior activation in both groups. There were no significant between-group differences (from Berman et al., 1988a).

Fig. 5. rCBF maps for an 18-year-old patient with Moya-Moya disease and intrinsic prefrontal pathology (see text for pertinent history). Note the hypofrontal pattern during both the WCS and the Numbers Matching task.

lity of long-term effects of neuroleptic treatment on prefrontal physiology. We studied 14 patients with Huntington's disease, 5 of whom had been chronically maintained on neuroleptics and found no differences in prefrontal blood flow between these 5 patients and those who were neuroleptic-naive (Weinberger and Berman, 1988).

Moreover, our study of 8 pairs of monozygotic twins who were concordant for schizophrenia, but whose lifetime histories of neuroleptic intake differed greatly, do not support the notion of neuroleptic-mediated hypofrontality. If neuroleptics cause or exacerbate hypofrontality, it would be predicted that within each concordant twin pair the twin with the history of the most neuroleptic intake would be the more abnormal. However, we found that in 6 of the 8 concordant pairs the twin who had been exposed to more neuroleptics was the more hyperfrontal of the pair during the WCS (Berman et al., 1989).

Perhaps the most direct evidence against neuroleptics as a cause of hypofrontality comes from the only study to investigate cerebral metabolism in drug naive patients during prefrontal activation. Using SPECT technology to measure rCBF, Raese et al. (1989) have recently reported that schizophrenic patients never exposed to neuroleptic medication were hypofrontal during the WCS. Taken together, these data do not support neuroleptic treatment as a major factor in hypofrontality.

A role for poor performance, motivation, arousal, or effort?

It is well established that patients with schizophrenia score poorly on the WCS (Goldberg et al., 1987); and we have seen that this poor performance occurs in the context of a failure to activate prefrontal cortex (Weinberger et al., 1986). This raises the possibility that the patients' poor performance is somehow responsible for the prefrontal physiological dysfunction seen during the WCS, rather than that the pathophysiology is responsible for the poor performance. For example, perhaps the patients were not even engaged in the task, but were simply making random responses. The patterns of errors made by patients with schizophrenia does not suggest random, disinterested responses, but rather is typically one of perseveration (i.e., inability to shift the conceptual or cognitive set, even in the face of feedback to do so). Interestingly, this pattern is commonly seen in patients with known frontal lobe lesions (Milner, 1963, 1971).

We have also studied other populations of patients who scored poorly on the WCS. A group of 14 patients with Huntington's disease, who performed as poorly as our patients with schizophrenia, showed no prefrontal blood flow deficit during the WCS (Weinberger et al., 1988b); if anything they tended to have slightly higher blood flow than the normal controls. Thus, a pathophysiological mechanism for the cognitive deficits of Huntington's disease that differs from that operating in schizophrenia is inferred (vide infra). We also used the same WCS-rCBF paradigm to study 10 mildly to moderately mentally retarded patients with Down syndrome whose performance on the WCS was considerably worse than that of either the schizophrenic or Huntington's disease patients. Despite this poor performance, the Down syndrome patients also showed a prefrontal activation in the normal range while attempting to perform the WCS (Berman et al., 1988b). In schizophrenia we have also seen normal dorsolateral prefrontal blood flow in the face of poor performance during other tasks on which the patients perform poorly (see Fig. 3). These include an auditory discrimination task (Berman et al., 1987a) and two versions of a visual Continuous Performance task (CPT) (Berman et al., 1986), all of which require as much attention and effort as the WCS. In fact, the CPT was designed specifically as a test of attention and vigilance rather than abstract reasoning, and patients with schizophrenia characteristically perform poorly on it (Garmezy, 1978). These data suggest that poor performance per se does not indiscriminately produce hypofrontality.

Additional epiphenomenological factors that are

important to consider in studies of neuropsychiatric patients in general, and patients with schizophrenia in particular, center around motivation and interest in the task as well as attention and effort. While these factors are as difficult to measure adequately as they are to define, the fact that patients do not show deficits in dorsolateral prefrontal function during other tasks in which attention is a critical factor, even those such as the CPTs that are effortful and specifically designed to test attention and vigilance, suggests that nonspecific disturbances in attention and/or effort do not account for the WCS findings.

Is hypofrontality in schizophrenia regionally and behaviorally specific?

It is worth noting that tests like the CPT may not be as specifically linked to prefrontal cortex as the WCS. The observation that in schizophrenia dorsolateral prefrontal blood flow is abnormally decreased during the prefrontally linked WCS, but is normal during non-prefrontally linked tasks such as Numbers Matching, CPTs, auditory discrimination, and resting, raises the interesting possibility that hypoprefrontality may only be manifested, or at least may be most apparent, in schizophrenia under conditions that place a regionally and cognitively specific load on the prefrontal cortex. If this conjecture were true, it would have important implications for understanding the pathophysiological mechanism, and ultimately the neural specificity, of WCS-related prefrontal failure in schizophrenia. However, the CPT, Numbers Matching, and resting conditions, unlike the WCS do not involve abstract reasoning and problem solving. An alternative hypothesis could be that schizophrenics appear hypofrontal during any and all abstract reasoning/problem solving tasks, regardless of the cortical region that subserves them. To test this hypothesis we sought to determine whether patients would appear hypoprefrontal during another abstract reasoning task, one that is not specifically linked to prefrontal cortex. Such a paradigm would also demonstrate whether patients, if given a task that requires thought and is linked to an area other than prefrontal cortex, would fail to activate that area.

To address these questions we assessed cortical function while subjects solved Raven's Progressive Matrices (RPM), a non-verbal, abstract reasoning and problem solving test which is a well-accepted indicator of general intelligence (Burke, 1958). Subjects with postrolandic lesions are impaired on this test (Basso et al., 1973), suggesting that posterior cortical areas are critical for the cognitive functions necessary to perform it; however, there is no evidence that patients with dorsolateral prefrontal cortical lesions have particular difficulty with it. This is consistent with results of other tests of intelligence on which patients with dorsolateral prefrontal lesions are also unimpaired, and also agrees with our rCBF data showing that normal subjects activate posterior cortical areas above baseline while performing this task (Fig. 4), suggesting that it is mediated by posterior cortical areas. In contrast to the WCS, no significant activation of frontal cortical areas occurred (Berman et al., 1988a).

These divergent findings in normal subjects during two different abstract reasoning tasks have implications for the role of the prefrontal cortex in normal higher cognitive function; they suggest that not all tests that require problem solving and complex higher order cognitive processing are "frontal" tasks. This notion is consistent with the perspective of many current cognitively based theories. Fuster (1989) has emphasized the role of dorsolateral prefrontal cortex in the crosstemporal integration of information (or spanning time), while Goldman-Rakic (1987) suggests that the dorsolateral prefrontal cortex functions to form "internal representations of situations or stimuli", supporting the "ability to respond to situations on the basis of stored information, rather than on the basis of immediate stimulation". Similarly, Ingvar (1985) proposed a primary role for this brain area in "memory of the future", or "the production of serial action plans . . . used as templates with which is input is compared". The

WCS and RPM differ in several critical aspects that, in light of these cognitive theories, may explain their differential effects on prefrontal cortical physiology (Berman et al., 1988a). For example, determining the proper category by which to sort a particular WCS stimulus requires that the feedback given on previous trials be considered. This may be conceptualized as the use of working memory to span time and make serial judgements. In contrast to the WCS, no previous experience is necessary to complete each RPM trial, and since all the required information is available to the subject throughout the trial, no working memory or "internal representations" are involved. Carrying out prefrontally specific cognitive operations, such as those mentioned above, may result in physiological activation of prefrontal cortex above the ambient, tonal, levels necessary to maintain wakefulness and routine neuronal homeostasis. It appears that successful performance of the prefrontally related WCS requires and results in an augmented prefrontal physiological response, while the RPM does not.

In contrast to the WCS, we found that while performing RPM (an abstract reasoning test that requires at least as much attention, concentration, motivation, and mental effort as any task we have studied with rCBF – including the WCS), patients with schizophrenia did not exhibit significant differences in prefrontal rCBF compared with normal subjects (Fig. 4), and both normal subjects and patients showed similar rCBF patterns with maximal increases in parieto-occipital cortex (Berman et al., 1988a). Similar results were reported by Ingvar and Franzen (1974b). Since performance of RPM involves many of the non-specific cognitive factors also involved in the WCS (e.g. mental effort, attention, etc.), the contrasting results obtained with these two paradigms in schizophrenia probably reflect differences in the functional integrity of the regional neural systems that subserve performance of each test.

These data suggest that since performance of RPM (as well as other non-prefrontally specific tests including Number Matching and the CPT) does not require prefrontally dependent functions such as temporal integration to the same degree as the WCS, augmentation of prefrontal activity levels above baseline may not be required and the RPM challenge is more readily met by the schizophrenic prefrontal cortex than is the WCS challenge. The RPM data also suggest that patients can activate posterior cortical areas when engaged in a task is linked to these areas in normal subjects. Thus, the WCS-related prefrontal pathophysiology does not appear to be due to a generalized inability to activate any cortical area, rather, that it is circumscribed and linked to a specific cognitive behavior.

On the basis of the data outlined above what do we know about the mechanism of the pathophysiology that underlies the prefrontally linked behavioral and cognitive deficits in schizophrenia? A highly simplified scheme might consider 3 possible pathophysiological mechanisms: de-efferentation of prefrontal cortex, deafferentation, and intrinsic prefrontal abnormality. While it has not been definitively delineated which of these mechanisms may play a role in the behaviorally related pathophysiology of the prefrontal cortex in this illness, some clues are offered by the collective results of these blood flow studies carried out in conjunction with a variety of cognitive conditions, and by comparing schizophrenia to other illnesses that have more clearly defined neuropathology. The following sections will consider possible roles of these 3 mechanisms in schizophrenia and other neuropsychiatric diseases.

Regional cerebral blood flow in prefrontal de-efferentation

One possible pathophysiological mechanism for the behavioral abnormalities in schizophrenia could be that, while the prefrontal cortex itself is intact, the pathways carrying information from prefrontal cortex to other brain areas may be aberrant. One illness in which such prefrontal de-efferentation is known to occur is Huntington's disease in which there is degeneration of one of the major outflow connections of the prefrontal cortex, the caudate nucleus. While widespread degenerative changes, involving cortical as well as subcortical areas, do occur in advanced cases, few if any cortical abnormalities are present in early cases (Bryn et al., 1979). However, even in early cases, a constellation of behavioral signs and symptoms as well as neuropsychological deficits characteristics of frontal lobe disease is often present (Caine et al., 1978). Because of the paucity of structural and chemical abnormalities of the cortex in such cases, and because PET studies have consistently found no alterations in cortical metabolism, Huntington's disease is often called a "subcortical dementia".

Consistent with the observations of other investigators, we found no deficits in cortical blood flow at rest or even when patients with Huntington's disease were engaged in cognitive tasks on which they were impaired (Weinberger et al., 1988b). In particular, unlike schizophrenic patients, the Huntington's patients were not hypofrontal during the WCS despite similarly poor performance (Goldberg et al., 1990). This result suggests that the prefrontally related cognitive deficits in Huntington's disease are mediated by a mechanism other than intrinsic prefrontal pathophysiology and that the mechanism is very different from that in schizophrenia. A clue to the nature of the pathophysiology in Huntington's disease is provided by data from our study (Weinberger et al., 1988b) showing a correlation between the degree of caudate atrophy and prefrontal blood flow during the WCS. This correlation was behavior specific, occurring only during the WCS condition; thus, the effect of caudate degeneration on cortical function was most pronounced when there was heightened demand for prefrontal function. The direction of the correlation is also of interest: the smaller the caudate (i.e. the greater the caudate atrophy), the higher the prefrontal blood flow. This finding may demonstrate the normal physiological response of an intact prefrontal cortex attempting to compensate for "downstream" blockage at the head of the caudate, the primary first order corticofugal projection site. The difference between the rCBF pattern in Huntington's disease and that in schizophrenia suggests that prefrontal de-efferentation does not play a major role in schizophrenia.

Regional cerebral blood flow in intrinsic prefrontal pathology

Intrinsic prefrontal pathology, loss of or damage to prefrontal neurons per se, offers a second pathophysiological mechanism by which prefrontal-type cognitive and behavioral deficits could occur. Such pathology is known to be present in ischemic injury, trauma, Pick's disease, and some cases of Alzheimer's disease. Fig. 5 shows the blood flow pattern of an 18-year-old patient with intrinsic pathology restricted mainly to the prefrontal lobe. He had presented with a clinical picture that was indistinguishable from a first episode of schizophrenia, but a CT scan revealed a frontobasilar stroke. Further work-up with cerebral angiography demonstrated vascular anomalies consistent with a diagnosis of Moya-Moya disease, a congenital condition that predisposes to infarction. Unlike patients with schizophrenia, this patient's blood flow studies showed marked hypofrontality under all conditions, both the WCS (on which he performed poorly) and the non-prefrontally specific Numbers Matching test, and even at rest (resting data not shown).

Alzheimer's disease. Another example of intrinsic prefrontal pathology is seen in Alzheimer's disease. Here the blood flow pattern is one of decreased prefrontal blood flow, but in the context of decreases in other cortical areas (particularly parietal and temporal cortex) as well. Like the patient with Moya-Moya disease but unlike patients with schizophrenia, patients with Alzheimer's disease exhibit such decreases during all conditions under which they are studied (Berman and Weinberger, 1986b).

Schizophrenia? As mentioned above there is very little convincing evidence for the existence of structural pathology of the prefrontal cortex in schizophrenia. The fact that in schizophrenia prefrontal blood flow deficits are mainly seen under conditions that impose a regionally specific physiological load or "stress" (like the WCS) — not under all conditions — also tends to weigh against intrinsic prefrontal pathology. However, the possibility of subtle intrinsic pathology that only becomes apparent when the physiological demand placed on the frontal cortex exceeds its capacity to respond cannot be ruled out. Nonetheless, the bulk of available evidence in schizophrenia appears to be more consistent with disordered afferentation than with intrinsic frontal lobe disease or de-efferentation.

Regional cerebral blood flow in prefrontal deafferentation: a possible mechanism for dysfunction in schizophrenia?

Further evidence for deafferentation in schizophrenia: structural – functional relationships and limbic pathology

As reviewed above, the pathophysiological model that appears to best fit the pattern of cortical blood flow abnormalities in schizophrenia is that of a decoupling of prefrontal cortex from the afferenting pathways that provide information to it. While such prefrontal deafferentation has yet to be directly demonstrated in schizophrenia, a number of observations support this concept. One finding that is consistent with this line of thinking concerns a relationship between prefrontal blood flow and subcortical brain structures. One of the most reproducible and time-tested findings reported in the schizophrenia literature is that, on the average, the lateral ventricles of patients with schizophrenia are larger than those of the normal populace (Shelton and Weinberger, 1986). We found a behavior-specific inverse correlation between relative prefrontal blood flow and lateral ventricular size as measured with X-ray computerized tomography (CT); this correlation was seen during the WCS but not during the non-specific Numbers Matching task (Berman et al., 1987b). One possible interpretation of this result is that there is structural pathology of diencephalic and limbic periventricular structures and that this may somehow be responsible for impaired prefrontal activation during the WCS. Since increased lateral ventricle size is a non-specific finding that could reflect pathology anywhere in the brain and since CT is not able to distinguish between soft tissue features of similar densities (such as gray and white matter), the exact structures involved could not be ascertained in this study.

Using magnetic resonance imaging, which offers an anatomically clearer picture of brain structures, Suddath et al. (1989) reported a 20% reduction in temporal lobe gray matter (particularly in the portion of the temporal lobe containing the amygdala and hippocampus) in schizophrenia, a finding that was correlated with ventricular size. Preliminary evidence from our study of monozygotic twins discordant for schizophrenia suggests a further refinement of the structural – functional relationship posited above, that a link between the hippocampus and the prefrontal cortex may play a role in the WCS-related cognitive and physiological deficits in schizophrenia. This observation is particularly intriguing in light of recent demonstrations in the monkey of direct projections from the hippocampus to the prefrontal cortex and metabolic coupling of prefrontal cortex, hippocampus, and medial thalamus during working memory tasks (Friedman and Goldman-Rakic, 1988). While such indirect evidence implicating these structures in the cognitive impairments characteristics of schizophrenia continues to accrue, the exact nature of such a putative deafferentation remains unknown. However, a neurochemical role for dopamine warrants further examination.

The role of dopamine

Although a clear picture of the role of the dopaminergic systems in human higher functions is just beginning to emerge, considerable work in non-human primates has demonstrated that pathways providing dopaminergic innervation to prefrontal cortex are important for cognition. Brozowski et al. (1979) provided compelling evidence that dopaminergic activity in prefrontal cortex is involved in the expression of prefrontally mediated cognition in monkeys. Those investigators found that chemical lesioning of dorsolateral prefrontal cortex (sulcus principalis) with 6-hydroxydopamine, producing depletion of prefrontal dopamine, resulted in a decrement in performance of prefrontally linked cognitive tasks similar to that caused by surgical ablation. It is easy to hypothesize from this work that a disorder of prefrontal dopamine may play a role in prefrontally linked cognitive and physiological deficits in humans. Evidence for such a situation, and particularly for the involvement of the mesocortical dopamine system, comes from clinical and pharmacological studies.

Parkinson's disease. Parkinson's disease is characterized by a loss of nigrostriatal as well as mesocortical dopamine circuits and thus represents a known situation of deafferentation of the prefrontal cortex. In this illness physiological findings that are consistent with such a dopamine deafferentation have been demonstrated and may have a counterpart in schizophrenia. For 10 nondemented patients with mildly to moderately severe Parkinson's disease (Weinberger et al., 1988c) prefrontal blood flow values during the WCS correlated with stage of illness as well as with bradykinesia and rigidity, which are clinical measures traditionally thought to be linked to dopaminergic function. Prefrontal blood flow did not correlate with degree of tremor, which is less associated with the dopamine system. There was also a robust inverse correlation between prefrontal blood flow and performance on the WCS. It is interesting that the direction of this correlation (i.e. poorer performance in the context of lower prefrontal rCBF) is the same as that observed in schizophrenia (Weinberger et al., 1986). In both illnesses the degree to which these patients can activate prefrontal cortex during the card sort relates to how successfully they perform the task. This raises the possibility that there may be some common mechanisms underlying prefrontal defects in schizophrenia and Parkinson's disease.

Schizophrenia. As in Parkinson's disease, the neurochemical system that has traditonally been most implicated in schizophrenia is dopamine. However, the two illnesses are usually thought to have few similarities and, in fact, are often conceptualized as being at opposite extremes of the spectrum of dopamine disorders. Based on the observation that neuroleptics can ameliorate the positive symptoms of schizophrenia while dopamine agonists can exacerbate them, the "dopamine hypothesis" of schizophrenia, in its classical form, postulates a relative overabundance or overactivity of dopamine, while the hallmark of Parkinson's disease is dopamine depletion. Nonetheless, there are a number of clinical and experimental observations suggesting that there may be some common mechanisms in these two patient populations. For example, both groups share the so-called deficit symptoms of flat affect, decreased motivation, amotivation, and specific cognitive deficits (which are not ameliorated by neuroleptics) described above for patients with intrinsic frontal lobe disease. In Parkinson's disease these symptoms have been attributed to decreased dopaminergic activity in prefrontal cortex. There is increasing evidence that a similar mechanism may play a role in the manifestation of these clinical phenomena in schizophrenia.

The first lines of evidence linking Parkinson's disease and schizophrenia are the correlations and clinical features enumerated above. A corollary line of evidence implicating a deficient dopamine system in schizophrenia stems from measurements of cerebrospinal fluid (CSF) monoaminergic metabolites. We found a relationship between CSF levels of the dopamine metabolite homovanillic acid (HVA) and the degree of WCS-related prefrontal activation. Higher HVA levels correlated with higher relative prefrontal flow in the patients (Weinberger et al., 1988a). This relationship was found only for blood flow measured during the WCS and not for that during the non-specific number matching task or other non-prefrontal tasks (Berman et al., 1988a). This pattern of a behavior-specific relationship is remarkably similar to that of the correlations we saw between WCS prefrontal blood flow and lateral ventricular size. Since data from monkeys suggest that CSF HVA levels reflect HVA concentrations in prefrontal cortex (Ellsworth et al., 1987), these observations provide further evidence that the mesocortical dopamine system, perhaps through a structural neuropathological abnormality (reflected by enlargement of the lateral ventricles), is involved in the prefrontal pathophysiology of schizophrenia.

Can the notion of hypodopaminergic function be reconciled with the hypothesis of hyperdopaminergic function that has held sway in schizophrenia for many years? The exact pathways have yet to be completely traced in humans, but it is clear that complex feedback circuits connect the prefrontal and subcortical dopamine systems in the normal brain. For example, mesocortical dopamine neurons appear to affect prefrontal cortical neurons that in turn exert feedback control over mesolimbic activity (Thierry et al., 1984). Pycock et al. (1980) demonstrated in the rat that a disruption of this system at the level of the prefrontal cortex can result in a functional hyperactivity of basal ganglia and limbic dopamine. It is possible that a similar dysregulation of the dopamine systems may exist in schizophrenia (Weinberger, 1987).

If hypodopaminergic function does exist and if it underlies prefrontal pathophysiology in schizophrenia, hypofrontality should be reversible with dopamine agonists. To test this hypothesis we studied the effects of the dopamine agonist apomorphine on WCS-related rCBF in a small group of chronically psychotic subjects. In each of the patients with schizophrenia hypofrontality during the WCS was ameliorated to some degree following a subcutaneous dose of apomorphine (Daniel et al., 1989a). This result is supported by a subsequent study using SPECT technology to measure the effects of amphetamine on rCBF (Daniel et al., 1989b). Additional studies are currently being carried out to to investigate the effects of other agonist agents such as levodopa and SKF38393, a specific dopamine D1 receptor agonist.

Conclusions

Taken together, the studies described above demonstrate that disruption anywhere along the complex circuitry connecting prefrontal cortex with other brain areas can cause a clinically significant syndrome of abnormal behavior suggestive of prefrontal lobe dysfunction. While the final common pathway for the expression of such abnormalities (i.e. the disordered behavior and cognition) may be similar, it is clear that the prefrontal physiological repercussions, and no doubt the neurochemical concomitants, differ and reflect the underlying neural mechanism. In vivo studies of neurophysiology in conjunction with cognitive activation procedures can provide important clues about the neural mechanisms in various illnesses. Such studies may ultimately help to point the way to new treatments.

Schizophrenia is a case in point. Although the agonist treatment studies mentioned above must be considered preliminary, it appears that the mystery of the prefrontal cortex, and its role in disease processes (even the relatively subtle prefrontal pathophysiology in an elusive illness like schizophrenia) is yielding to new technology, lessons from non-human primate models, and new pharmacological agents. A more complete picture of the primate prefrontal cortex may someday lead us to new ways to intervene and correct a primary physiological prefrontal deficit in schizophrenia and other illnesses.

References

Basso, A., DeRenzi, E., Faglioni, P., Scotti, G. and Spinnler, H. (1973) Neuropsychological evidence for the existence of cerebral areas critical to the performance of intelligence tasks. *Brain*, 96: 715 – 728.

Benes, F.M., Davidson, J. and Bird, E.D. (1986) Quantitative cytoarchitectural studies of the cerebral cortex of schizophrenics. *Arch. Gen. Psychiat.,* 43: 31–35.

Berman, K. (1987) Cortical "stress tests" in schizophrenia: regional cerbral blood flow studies. *Biol. Psychiat.,* 22: 1304–1326.

Berman, K.F. and Weinberger, D.R. (1986a) Cerebral blood flow studies in schizophrenia. In H. Nasrallah and D.R. Weinberger (Eds.), *The Neurology of Schizophrenia,* Elsevier North Holland Press, Amsterdam, pp. 227–307.

Berman, K.F. and Weinberger, D.R. (1986b) Cortical physiological activation in Alzheimer's disease: rCBF studies during resting and cognitive states. *Soc. Neurosci. Abstr.,* 12: 1160.

Berman, K.F., Zec, R.F. and Weinberger, D.R. (1986) Physiological dysfunction of dorsolateral prefrontal cortex in schizophrenia. II. Role of neuroleptic treatment, attention, and mental effort. *Arch. Gen. Psychiat.,* 43: 126–135.

Berman, K.F., Rosenbaum, S.C.W., Brasher, C.A., Goldberg, T.E. and Weinberger, D.R. (1987a) Cortical physiology during auditory discrimination in schizophrenia: a regional cerebral blood flow study. *Soc. Neurosci. Abstr.,* 13: 651.

Berman, K.F., Weinberger, D.R., Shelton, R.C. and Zec, R.F. (1987b) A relationship between anatomical and physiological brain pathology in schizophrenia: lateral cerebral ventricular size predicts cortical blood flow. *Am. J. Psychiat.,* 144: 1277–1282.

Berman, K.F., Illovsky, B.P. and Weinberger, D.R. (1988a) Physiological dysfunction of dorsolateral prefrontal cortex in schizophrenia. IV. Further evidence for regional and behavioral specificity. *Arch. Gen. Psychiat.,* 45: 616–622.

Berman, K.F., Schapiro, M.B., Friedland, R.P., Rapoport, S.I. and Weinberger, D.R. (1988b) Cerebral function during cognition in Down syndrome. *Proceedings of the 141st Annual Meeting of the American Psychiatric Association.*

Berman, K.F., Torrey, E.F., Daniel, D.G. and Weinberger, D.R. (1989) Prefrontal cortical blood flow in monozygotic twins concordant and discordant for schizophrenia. *Schizophrenia Res.,* 2: 129.

Bleuler, E. (1950) *Dementia Praecox or the Group of Schizophrenias,* International Press, New York, p. 467.

Brozowski, T.S., Brown, R.M., Rosvold, H.E. and Goldman, P.S. (1979) Cognitive deficits caused by regional depletion of dopamine in prefrontal cortex of rhesus monkeys. *Science,* 929: 929–932.

Bryn, G.W., Bots, G.Th. and Dom, R. (1979) Huntington's chorea: Current neuropathological status. In: T.N. Chase, N.S. Wexler and A. Barbeau (Eds.), *Advances in Neurology,* Raven Press, New York, pp. 83–93.

Burke, H.R. (1958) Raven's progressive matrices: A review and critical evaluation. *J. Genet. Psychol.,* 93: 199–228.

Caine, E.D., Hunt, R.D., Weingarten, H. and Ebert, M.H. (1978) Huntington's dementia: Clinical and neuropsychological features. *Arch. Gen. Psychiat.,* 35: 377–384.

Cohen, R.M., Semple, W.E., Gross, M., Nordahl, T.E., DeLisi, L.E., Holcomb, H.H., Morihisa, J.M. and Pickar, D. (1987) Dysfunction in a prefrontal substrate of sustained attention in schizophrenia. *Life Sci.,* 40: 2031–2039.

Daniel, D.G., Berman, K.F. and Weinberger, D.R. (1989a) The effect of apomorphine on regional cerebral blood flow in schizophrenia. *J. Neuropsychiat. Clin. Neurosci.,* 1: 337–384.

Daniel, D.G., Jones, D.W., Zigun, J.R., Coppola, R., Bigelow, L.B., Berman, K.F., Goldberg, T.E. and Weinberger, D.R. (1989b) Amphetamine and cerebral blood flow (Xe-133 dynamic SPECT) in schizophrenia. *Soc. Neurosci. Abstr.,* 15: 1123.

Duara, R., Gross-Glenn, K., Barker, W.W., Chang, J.Y., Apicella, A., Loewenstein, D. and Boothe, T. (1987) Behavioral activation and the variability of cerebral glucose metabolic measurements. *J. Cerebr. Blood Flow Metab.,* 7: 266–271.

Early, T.S., Reiman, E.M., Raichle, M.E. and Spitznagel, E.L. (1987) Left globus pallidus abnormality in never-medicated patients with schizophrenia. *Proc. Natl. Acad. Sci. (U.S.A.),* 84: 561–563.

Ellsworth, J.D., Leahy, D.J., Roth Jr., R.H. and Redmond Jr., D. (1987) Homovanillic acid concentration in brain, CSF and plasma as indicators of central dopamine function in primates. *J. Neural Transm.,* 68: 51–62.

Friedman, H.R. and Goldman-Rakic, P.S. (1988) Activation of the hippocampus and dentate gyrus by working memory: A 2-deoxyglucose study of behaving rhesus monkeys. *J. Neurosci.,* 8: 4693–4706.

Fuster, J. (1989) *The Prefrontal Cortex,* Raven Press, New York.

Garmezy, N. (1978) Attentional processes in adult schizophrenia and in children at risk. *J. Psychiat. Res.,* 14: 3–34.

Goldberg, T.E., Weinberger, D.R., Berman, K.F., Pliskin, N.H. and Podd, M.H. (1987) Further evidence for dementia of the prefrontal type in schizophrenia. *Arch. Gen. Psychiat.,* 44: 1008–1014.

Goldberg, T.E., Berman, K.F., Mohr, E. and Weinberger, D.R. (1990) Regional cerebral blood flow and neuropsychology in Huntington's disease and schizophrenia: a comparison of patients matched for performance on a prefrontal type task. *Arch. Neurol.,* in press.

Goldman-Rakic, P. (1987) Circuitry of primate prefrontal cortex and regulation of behavior by representational knowledge. In: V. Montcastle (Ed.), *Higher Cortical Function: Handbook of Physiology,* American Physiological Society, Washington, DC, pp. 373–417.

Ingvar, D.H. (1985) "Memory of the future": an essay on the temporal organization of conscious awareness. *Hum. Neurobiol.,* 4: 127–136.

Ingvar, D.H. and Franzen, G. (1974a) Abnormalities of cerebral blood flow distribution in patients with chronic schizophrenia. *Acta Psychiat. Scand.,* 50: 425–462.

Ingvar, D.H. and Franzen, G. (1974b) Distribution of cerebral activity in chronic schizophrenia. *Lancet*, 2: 1484–1486.

Kirch, D.G. and Weinberger, D.R. (1986) Post-mortem histopathological findings in schizophrenia. In: H.A. Nasrallah and D.R. Weinberger (Eds.), *The Neurology of Schizophrenia*, Elsevier/North-Holland, Amsterdam, pp. 325–348.

Kolb, B. and Whisman, I.Q. (1983) Performance of schizophrenia patients on tests sensitive to left or right frontal, temporal, and partietal function in neurologic patients. *J. Nerv. Ment. Dis.*, 171: 435–443.

Kraepelin, E. (1971) *Dementia Praecox and Paraphrenia*. R.E. Krieger, New York.

Malmo, H.P. (1974) On frontal lobe functions: Psychiatric patient controls. *Cortex*, 10: 231–237.

Mazziotta, J.C., Phelps, M.E., Carson, R.E. and Kuhl, D.E. (1982) Tomographic mapping of human cerebral metabolism: Sensory deprivation. *Ann. Neurol.*, 12: 435–444.

McHenry, L.C. (1969) *Garrison's History of Neurology*, Charles C. Thomas, Springfield, IL.

Milner, B. (1963) Effects of different brain lesions on card sorting. *Arch. Neurol.*, 9: 100–110.

Milner, B. (1964) Some effects of frontal lobectomy in man. In: J.M. Warren and K. Akert (Eds.), *The Frontal Granular Cortex and Behavior*, McGraw-Hill, New York, pp. 313–334.

Milner, B. (1971) Interhemispheric differences in the localization of psychological processes in man. *Br. Med. Bull.*, 27: 272–277.

Pereman, E. (1987) *The Frontal Lobes Revisited*, IBRN Press, New York.

Pycock, C.J., Kerwin, R.W. and Carter, C.J. (1980) Effects of lesions of cortical dopamine terminals on subcortical dopamine receptors in rats. *Nature*, 286: 74–77.

Quitkin, F., Rifkin, A. and Klein, D. (1976) Neurologic soft signs in schizophrenia and character disorders. *Arch. Gen. Psychiat.*, 33: 845–853.

Raese, J.D., Paulman, R.G., Steinberg, J.L., Devous, M.D., Judd, C.R. and Gregory, R.R. (1989) Wisconsin Card Sort activated blood flow in dorsolateral frontal cortex in never medicated and previously medicated schizophrenics and normal controls. *Biol. Psychiat.*, 25: 100A.

Salazar, A.M., Grafman, J.H., Vance, S.C., Weingarten, H., Dillon, J.D. and Ludlow, C. (1986) Consciousness and amnesia after penetrating head injury: neurology and anatomy. *Neurology*, 36: 178–187.

Seidman, L.J. (1983) Schizophrenia and brain dysfunction: an integration of recent neurodiagnostic findings. *Psychol. Bull.*, 94: 195–238.

Shelton, R.C. and Weinberger, D.R. (1986) X-Ray computerized tomography studies in schizophrenia: a selective review and synthesis. In: H.A. Nasrallah and D.R. Weinberger (Ed.), *The Neurology of Schizophrenia*, Elsevier/North-Holland, Amsterdam, pp. 207–250.

Suddath, R.L., Casanova, M.F., Goldberg, T.E., Daniel, D.G., Kelsoe, J.R. and Weinberger, D.R. (1989) Temporal lobe pathology in schizophrenia: a quantitative magnetic resonance imaging study. *Am. J. Psychiat.*, 146: 464–472.

Thierry, A.-M., Tassin, J.P. and Glowinski, J. (1984) Biochemical and electrophysical studies of the mesocortical dopamine system. In: L. Descarsies, T.R. Reader and H.H. Jasper (Eds.), *Monoamine Innervation of Cerebral Cortex*, Alan R. Liss, New York, pp. 233–262.

Valenstein, E.S. (1986) *Great and Desperate Cures*, Basic Books, New York.

Volkow, N.D., Brodie, J.D., Wolf, A.P., Angrist, B., Russel, J. and Cancro, R. (1986) Brain metabolism in patients with schizophrenia before and after acute neuroleptic administration. *J. Neurol. Neurosurg. Psychiat.*, 49: 1199–1202.

Volkow, N.D., Wolf, A.P., Van Gelder, P., Brodie, J.D., Overall, J.E., Cancro, R. and Gomez-Mont, F. (1987) Phenomenological correlates of metabolic activity in 18 patients with chronic schizophrenia. *Am. J. Psychiat.*, 144: 151–158.

Weinberger, D.R. (1987) Implications of normal brain development for the pathogenesis of schizophrenia. *Arch. Gen. Psychiat.*, 44: 660–669.

Weinberger, D.R. and Berman, K.F. (1988) Speculation on the meaning of cerebral metabolic hypofrontality in schizophrenia. *Schizophrenia Bull.*, 14: 157–168.

Weinberger, D.R., Berman, K.F. and Zec, R.F. (1986) Physiological dysfunction of dorsolateral prefrontal cortex in schizophrenia. I. Regional cerebral blood flow (rCBF) evidence. *Arch. Gen. Psychiat.*, 43: 114–125.

Weinberger, D.R., Berman, K.F. and Illowsky, B.P. (1988a) Physiological dysfunction of dorsolateral prefrontal cortex in schizophrenia. III. A new cohort and evidence for a monoaminergic mechanism. *Arch. Gen. Psychiat.*, 45: 609–615.

Weinberger, D.R., Berman, K.F., Iadarola, M.K., Driesen, N. and Zec, R.F. (1988b) Prefrontal cortical blood flow and cognitive function in Huntington's disease. *J. Neurol. Neurosurg. Psychiat.*, 51: 94–104.

Weinberger, D.R., Berman, K.F. and Chase, T.N. (1988c) Mesocortical dopaminergic function and human cognition. *Ann. NY Acad. Sci.*, 537: 330–338.

Discussion

J.E. Pisetsky: Schizophrenia is a heterogeneous disorder. Have you differentiated between less regressed and more regressed patients?

K.F. Berman: Patients at the NIMH tend to be a relatively homogenous group and are, therefore, not ideally suited to addressing the question of clinical correlates of hypofrontality. However, other investigators, including Dr. David Ingvar and colleagues over 15 years ago, have shown that those patients who are the most regressed, autistic, and "burnt out", those who have the most severe negative symptoms are the most hypofrontal.

P.R. Lowenstein: A comment – the relation between schizophrenia and dopamine is complex. William Bunney's group showed some years ago that schizophrenic patients respond in the following way after administering low dosis of amphetamine: one-third get better, one-third get worse, one-third do not change. A question: Why does cerebral blood flow increase in all cortical areas?

K.F. Berman: Amphetamine does *not* increase global blood flow. In fact global levels tend to decrease! However, dopamine agonists do appear to increase *frontal* flow. Furthermore, agonists seem to increase the "signal-to-noise" ratio of prefrontal flow relative to other cortical areas during prefrontally linked cognitive tasks, perhaps facilitating a more "normal" physiological response in schizophrenia.

S.B. Dunnett: Is there a contradiction between your data on DA drugs, schizophrenia and parkinsonism: (a) amphetamine can induce schizophrenia-like psychosis in normals but seemed to reduce cerebrovascular pathology in your patients; (b) neuroleptics which reduce schizophrenia symptoms can induce parkinsonian side effects and, yet, your schizophrenia and Parkinson patients appeared to have deficits in frontal cerebral blood flow?

K.F. Berman: A good question. As the question implies, the dopamine theory of schizophrenia in its classical form postulates an overabundance or overactivity of dopamine in this illness. However, new methods of research on primates and new selective pharmacological agents make clear that dopamine in the brain is not a single, unitary system that is either "up" or "down". There are complex interactions between the subcortical and cortical dopamine systems, and it has been shown in the rat that a disruption of the cortical dopamine system at the level of the prefrontal cortex acts to "take the brakes off" the subcortical dopamine system. This is one scenario by which a brain could have too much dopamine and too little dopamine at the same time. A similar scenario could be present in schizophrenia: prefrontal dopamine deficit might underly the negative symptoms of schizophrenia (which agonists help), while subcortical overactivity might explain the positive symptoms (which antagonists ameliorate).

A.Y. Deutch: The convergent data suggest that the deficit may be a DA hypofunction in schizophrenia, matched by the Parkinson's disease DA cortical hypofunction. Is the blood flow altered in progressive supranuclear palsy, in which cortical DA is not depleted?

K.F. Berman: PSP is a group that we are very interested in. We have not yet studied enough subjects to draw firm conclusions.

CHAPTER 27

The prefrontal area and psychosurgery

Elliot S. Valenstein

University of Michigan, Ann Arbor, MI, USA

The origin of psychosurgery

The way it is usually told, the idea to perform psychosurgery suddenly occurred to the Portuguese neurologist Egas Moniz while he was listening to Carlyle Jacobsen and John Fulton describe the effects of frontal lobe damage in chimpanzees. The occasion was the 1935 International Congress of Neurology in London and during the discussion that followed, Moniz is said to have asked: "If frontal lobe removal eliminates frustrational behavior in animals, why would it not be feasible to relieve anxiety states in man by surgical means?" Not long after returning to Lisbon, Moniz reported that he had performed prefrontal leucotomies on 20 seriously ill mental patients, 14 of whom were either "cured" or "significantly improved" (Moniz, 1936). Prefrontal leucotomy was rapidly adopted around much of the world where it was widely praised for its role in helping psychiatry join the ranks of scientific medicine. When Moniz was awarded the Nobel Prize in 1949, the citation described prefrontal leucotomy as "one of the most important discoveries ever made in psychiatric therapy".

In reality, the birth of prefrontal leucotomy was neither an isolated, unanticipated, nor a revolutionary occurrence. It should be recalled that in 1935 there was virtually no effective treatment for mental illness and almost any proposal was considered worthy of exploring. Physical therapies were especially attractive because they offered the possibility of relatively quick and inexpensive help to overcrowded and underfunded mental institutions as well as to desperate patients and their families. Extirpation of endocrine glands, reproductive organs, teeth, tonsils, and other bodily parts had been widely used as treatments for mental illness and some of these "heroic" therapies were still being performed when Moniz started his prefrontal leucotomy series in the fall of 1935. It was less than a decade earlier that the Viennese psychiatrist Wagner-Jauregg received the Nobel Prize for a fever therapy, widely used to treat not only neurosyphilis, but also mental illness. Wagner-Jauregg induced the fever in patients by injecting them with blood infected with malaria. Hydrotherapy, prolonged drug-induced sleep, and the breathing of carbon dioxide were also used extensively at this time and there was little hesitation to explore other physical therapies as possible treatments for mental illness. It was not coincidence that prefrontal leucotomy, insulin coma, metrazol and electroconvulsive "shock" treatments were all introduced during the 1930s.

Moreover, prefrontal leucotomy was not the first attempt to treat mental illness by a brain operation. Psychosurgery can be traced back at least to the 1890s when the Swiss psychiatrist Gotlieb Burckhardt (1891) attempted to treat psychotic patients by destroying parts of the cerebral cortex. This was followed by others, most

Correspondence: Dr. Elliot S. Valenstein, University of Michigan, Ann Arbor, MI, USA.

notably by the Estonian neurosurgeon Lodivicus Puusepp, who tried psychosurgery early in this century. Interest in brain surgery to treat mental illness did not disappear as can be appreciated from the prophetic comment made by the surgeon William Mayo 1 month before Moniz's first prefrontal leucotomy.

> If necessary we shall perform exploratory operations on the head at an early stage of disease and turn the light of day onto many of the pathologic conditions of the brain . . . Are we not in the same position in the treatment of the mentally afflicted that we were in the surgery of the abdomen, when I began fifty years ago, or as we were in the surgery of the chest? Day by day, I can see the extension of remedial measures to conditions of the brain at earlier stages. (Mayo, 1935, as quoted in Woltman, 1941.)

Unquestionably, there had been a long history of interest in all kinds of physical treatments of mental illness, including psychosurgery, at the time Moniz introduced prefrontal leucotomy (Valenstein, 1986; Scull, 1987; Scull and Favreau, 1986).

It was not surprising that the prefrontal area was selected as the brain target for psychosurgery. The association of the prefrontal area of the brain with intellect has a long history dating back at least 1000 years, when the Chinese were depicting the "God of Wisdom and Longevity" with an exaggerated forehead (Fig. 1). During the last century, Paul Broca had concluded that the special majesty of humans over animals resided in "judgment, reflection, invention and abstraction", capacities he, along with many others, assigned to the frontal pole of the brain (Broca, 1861). In the 1920s, Frederick Tilney wrote that it was the prefrontal area which separates man from the apes and he attributed to this region the capacity to act "as the accumulator of experience, the director of behavior, and the instigator of progress" (Tilney, 1928). There was even early speculation that mental illness might be caused by prefrontal area pathology as implied by Franz Nissl's request to have this region amputated to relieve the crippling depression that preceded his death in 1919 (cited in Putman, 1950).

The possibility that some intervention into the prefrontal area might alleviate mental illness had occurred to many people during the 1930s. The French neuropsychiatrist Henri Baruk, for example, wrote in his memoirs that ideas about psychosurgery had "floated around for quite awhile" prior to Moniz's first operation and he reported rejecting a plan to perform prefrontal leucotomy when it was proposed by the neurosurgeon Thierry de Martel (Baruk, 1978, p. 200). During this same period, the American neurologist Richard Brickner and the neurosurgeon Leo Davidoff also had under consideration a plan to perform a prefrontal lobe operation for the relief of depression (see letter dated Oct. 29, 1936 from Richard Brickner to John Fulton, Yale University Library).

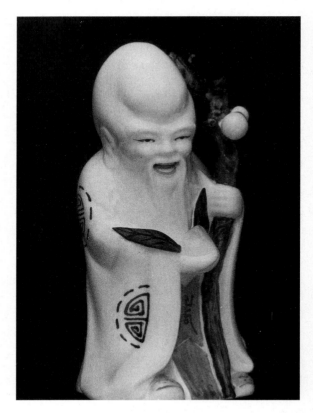

Fig. 1. A contemporary copy of the Chinese statues of the "God of Wisdom and Longevity". The original statues, which are more than 1000 years old, also have exaggerated foreheads.

Rarely remembered today is the fact that numerous frontal lobe operations were performed to treat mental illness during the 1930s, some before Moniz had begun. Maurice Ducosté in France reported that for several years he had been injecting blood drawn from "malarial patients" into the frontal lobes of paretics and mental patients. Whereas Wagner-Jauregg had injected malarial blood systemically, Ducosté injected up to 5 ml of blood directly into the frontal lobes through holes trepanned in the skull. He reported in 1932 that he treated more than 100 patients, repeating injections up to 12 times, and that paretics exhibited "uncontestable mental and physical amelioration" and the results with schizophrenic and manic patients were also "encouraging" (Ducosté, 1932). In 1933, Mariotti and Sciuti in Italy and Ferdière and Coulloudon in France similarly reported injecting malarial blood into the frontal lobes of paretics and the former described withdrawing blood from the arms of schizophrenics and melancholic patients and reinjecting it back into their prefrontal area. During this same period, Chavastelon in France reported 80% "cures" following injections of blood into the prefrontal area of more than 400 demented patients. In 1933, Dogliotti in Italy designed a pointed trocar which could be tapped into the frontal lobes through the orbital plate behind the eyeballs. Dogliotti used this approach mainly to inject opaque substances into the ventricles for ventriculography, but by demonstrating how easy it was to gain access to the frontal lobes, he paved the way for later transorbital lobotomies (Fiamberti, 1947; Freeman, 1948). Ody reported performing a partial frontal lobe resection on a catatonic schizophrenic patient in Switzerland at the time Moniz was just starting. When he published the results several years later he criticized Moniz for his apparent haste in publishing results before determining "if there was a lasting remission" (Ody, 1938; Morel, 1938). Ody's operation was also tried by D. Bagdasar and J. Constantinesco at the Central Hospital in Bucharest. Other early prefrontal interventions to treat psychiatric patients in Europe, Asia, and North America have been described elsewhere (Valenstein, 1986, pp. 118–121).

Prefrontal leucotomy was clearly an idea that had occurred to many and in all probability it would have started even if the chimpanzee study had never been done. Moniz may have been catalyzed into action by the Jacobsen–Fulton report at the London Congress, but he never acknowledged any such influence, claiming always that he had been thinking about the possibility for several years. In any case, the influence of the Fulton–Jacobsen study has been grossly exaggerated and the usual description of the relation of this study to the origin of prefrontal leucotomy must be considered apocryphal.

The purpose of the Fulton–Jacobsen study was to investigate the effects of frontal lobe damage on learning and problem solving, not on emotional behavior. What is usually emphasized in the lobotomy literature, however, is the completely unanticipated observation that the emotional behavior of one animal changed dramatically following frontal lobe ablation. Becky happened to be a highly emotional and volatile chimpanzee and during the initial preoperative testing she usually had temper tantrums if she made a mistake that deprived her of a food treat. Eventually, the temper tantrums became so severe and persistent she could not be tested. After the surgery, Becky was described as appearing as though she had joined a "happiness cult", cheerfully entering the test chamber even though her responses were no more accurate than chance. The conditions that produced Becky's emotional changes, however, were never studied systematically. A review of the methodology described in the initial publications makes it clear that there was no plan to present unsolvable problems in order to induce "experimental neuroses" as later implied, nor was this ever done. Moreover, it seems to have been completely overlooked that Lucy, the other chimpanzee subject, did not exhibit any "neurotic" behavior during the preoperative testing and there was no attempt made to induce such behavior. Lucy was initially a "calm and unexcitable" animal, displaying no temper tantrums prior to the frontal lobe surgery. After the surgery, however, Lucy began to make many errors and apparently in frustation she had frequent temper tantrums and even 2 years later she was still described as responding to failure with "piercing vocalization, drumming with the hands and feet on the floor of the cage and banging on the wire of the apparatus" (Crawford et al., 1948). It should have been apparent to anyone examining the evidence that there was no justification for the conclusion that "experimental neurosis" could not be induced after destruction of the prefrontal area. Fulton, who by this time was far removed from the actual experiments in his large laboratory, consistently maintained, however, that "frustrational behavior" had been eliminated in *both* animals (see, for example, Fulton's discussion in Fulton et al., 1948, p. 209). It is worth noting here that Harlow and Settlage (1948) concluded that monkeys with prefrontal damage often refuse to be tested and are more difficult to tame than normal animals.

In many lectures and writings, Fulton implied that his study with Jacobsen had given Moniz the idea to try prefrontal lobotomy on patients. Pressman, who has closely scrutinized these events, had concluded that "the direct connection between Moniz's decision to attempt leukotomy and the findings of the chimpanzee experiment was largely an invention of Fulton's, made in retrospect" (Pressman, 1986, p. 61; see also Pressman, 1988).* Nevertheless, Fulton's version was so widely accepted that in its award citation to Moniz, the Nobel Committee stated that, "The American physiologist, Fulton, and his collaborators have proved by experiments on anthropoid apes that neuroses caused experimentally disappeared if the frontal lobes were removed and that it was impossible to cause experimental neuroses in animals deprived of their frontal lobes". That same year, while showing a slide of Becky with the caption "The Face that Lopped Ten Thousand Lobes", the Harvard neurologist Stanley Cobb observed that "seldom in the history of medicine has a laboratory observation been so quickly and dramatically translated into a therapeutic procedure" (Cobb, 1949). Fulton, who had an enormous world-wide influence on clinicians as well as neuroscientists, had played a major role in promoting psychosurgery by creating the impression that it was based on a solid scientific foundation of laboratory research.** In reality, Fulton's frontal lobe research prior to Moniz' first leucotomy was mainly concerned with motor control, postural changes, and grasping reflexes

---

* Note that Fulton was not even listed as one of the authors of the original papers describing the chimpanzee frontal lobe study, e.g., see Jacobsen (1934) and Jacobsen et al. (1935). While Fulton was acknowledged for his assistance "on the surgical and physiological aspects of the study" and for making available the facilities of the Laboratory of Physiology, Jacobsen stated that the experiments "very largely owe their inception" to Karl Lashley. Before coming to Yale in 1930, Jacobsen completed his Ph.D. at the University of Minnesota under Lashley's supervision. The thesis was published as *A Study of Cerebral Function in Learning: The Frontal Lobes* (Jacobsen, 1931).

Fulton was included as an author on later papers and some people at the time were convinced that he would share the Nobel Prize with Moniz. Pressman (1988) has cited letters by Yngve Zotterman and the neurosurgeon Olof Sjögvist, both of whom expressed (in letters to Fulton) their dissappointment that he did not share the prize. The Nobel Committee acknowledged the contribution of Fulton, never mentioning Jacobsen, but awarded the prize in Physiology and Medicine jointly to Moniz for prefrontal lobotomy and to Walter R. Hess for neurophysiology.

** In a letter to Richard Brickner (Oct. 28, 1936), Fulton wrote: "On the grounds of your observations and ours in chimps it seems to me that the procedure [Moniz' operation] is warranted and I hope we can stir up a few more courageous neurosurgeons to try it". In a letter written 1 week later (Nov. 4, 1936) Fulton advised the neurosurgeon Paul Bucy to take up the Moniz operation, predicting that "in a year or so this will constitute one of the major phases of neurosurgery, and I think anyone in the field would do well to get in on the ground floor" (Yale University Library). An editorial which appeared in the *New England Journal of Medicine* a few months after the first lobotomy in the United States observed that this procedure was based "on sound physiological observations". Fulton was on the editorial board of the journal at the time.

(Fulton, 1926, 1932) and to a lesser extent with autonomic physiological responses, but not mental or emotional states (Fulton, 1934a, b; see also Watts, 1935).

Although Fulton had not studied the importance of the frontal lobes for mental and emotional states, a large literature on the subject did exist. At the time of the first prefrontal leucotomy there was even some agreement about the extensive impairment that followed damage to the frontal pole of the brain. During the latter part of the last century, David Ferrier in England and Leonardo Bianchi in Italy, had described animals with extensive frontal lobe damage as either apathetic with little interest in the world about them, or as responding purposelessly to momentary stimuli and unable to maintain a consistent goal direction. Bianchi observed that animals with frontal lobe damage were not able to "synthesize" information that occurred serially. The studies during the 1920s and 1930s of humans with frontal lobe damage, most of whom were either tumor patients or soldiers injured during World War I, were consistent with the animal studies. Individual patients were said to exhibit one or several of the following characteristics: exaggerated euphoria (Witzelsucht), restlessness and impulsiveness, unrestrained talkativeness, apathy, loss of motivation, irritability, puerile behavior, and lack of judgment (see review by Rylander, 1939).

In his monograph, Moniz did not dwell on the many impairments that had been associated with frontal lobe damage. Richard Brickner, for example, had reported the results of a thorough study of a patient who had a bilateral frontal lobe tumor removed. Afterwards, the patient was seriously impaired being unable to synthesize information, to behave appropriately, to return to work, or to pursue any but the most immediate goals (Brickner, 1932, 1936). Moniz, however, minimized these deficits by referring only to Brickner's statement that the patient's "psychic functions were altered more quantitatively than qualitatively", a statement clearly misleading when taken out of context.

Moniz's justification for psychosurgery was based only on some general statements about the prefrontal area being most developed in man, the seat of man's highest "psychic activity", and some essentially untestable speculations about the function of this area in mental disorders. Borrowing the concept of "fixed ideas" *(idées fixes)* from French psychiatrists (Janet, 1895), Moniz hypothesized that abnormally stabilized neural activity in the prefrontal area was the root cause of mental illness. He conceived of the pathology in mental illness as functional in nature, existing in abnormally "stabilized" activity in the connections between nerve cells. The purpose of leucotomy, therefore, was to break up the persistent activity in order to free patients of their "fixed ideas".

"In these disorders the psychic disturbances do not affect the intellectual function but rather consist of depressive, hypochondriacial or grandiose ideas, anxiety, delusions of persecutions, and so on which dominate the mental activity of these patients.

These mental disorders arise in our opinion in connection with the formation of cellular connecting groups which have become more or less fixed.

The cell bodies remain altogether normal; their axons may present no anatomic alterations, but their multiple connections, very variable in normal persons, may undergo more or less fixed arrangements which are in relation with the persistent ideas and the delusions of certain morbid psychic states.

... In accordance with the theory which we have just developed, one conclusion is derived: to cure these patients we must destroy the more or less fixed arrangements of cellular connections that exist in the brain, and particularly those which are related to the frontal lobes". (Moniz cited in Freeman and Watts, 1942, p. 12.)

The cores of dead tissue left behind after the initial prefrontal leucotomies were in the centrum semi-ovale, a vaguely defined area consisting of a confluence of fibers located between the genu of the corpus callosum and the frontal neocortex. (See Valenstein, 1986, for a description of Moniz's "core operation" and many of the psychosurgical procedures that followed.) Moniz did not offer any rationale for destroying any particular pathway. His primary concern in inserting the leucotome was to avoid major blood vessels and he varied the location and the number of sites destroyed from patient to patient. Only trial and error guided

Moniz in his attempts to disrupt the "functional fixity of morbid ideas". Nevertheless, despite the lack of any compelling rationale and the crudity of the surgery, prefrontal leucotomy was rapidly adopted. The need for a treatment of mental illness was great, there was a willingness to try any somatic treatment, the idea of performing brain operations to treat mental illness had already occurred to a number of people, and Moniz had acquired considerable credibility based on his earlier pioneering work on cerebral angiography.

Evolution of psychosurgical targets and rationale

In less than 3 months after Moniz had published his monograph, prefrontal leucotomies were being performed in Italy, Rumania, Brazil, Cuba and in the U.S.A. by Walter Freeman and James Watts of George Washington University. Freeman and Watts, who had adopted the Moniz "core operation", were soon reporting that their first 6 "lobotomized" patients had obtained substantial relief from "worry, apprehension, anxiety, insomnia, and nervous tension" (Freeman and Watts, 1937). (They had substituted the term "lobotomy" for "leucotomy" to indicate that nerve cells as well as fibers were inevitably destroyed. Subsequently, the designation "lobotomy" was more commonly used in the U.S.A., while "leucotomy" continued to be used in Europe.) After completing 20 operations, however, it became clear that the relief was not lasting as many of the patients relapsed. When the patients were reoperated the results were often poor as epileptic seizures and other neurological as well as intellectual impairment commonly occurred. Freeman and Watts attributed the poor results to the unreliability of the "core operation" and they proceeded to develop their own psychosurgical operation, which was widely adopted in the U.S.A. and elsewhere. This pattern occurred in most countries where the "core operation" was tried and soon abandoned and replaced by other procedures.

The rationale for psychosurgery, however, did not evolve as rapidly as did the surgery. In 1942 when their influential book on psychosurgery was published, Freeman and Watts were still referring to the need to disrupt the "fixed ideas" of their patients, even though they recognized that there was little to support Moniz's speculation that "fixed ideas" were sustained by activity in fibers in the prefrontal area.

"The theory of Egas Moniz that the operation is successful through the breaking up of constellations of neuron patterns will bear further examination. There is no doubt about the stereotyped thinking and stereotyped activity indulged in by many of the patients. But whether such ideas and activities are actually in relation with abnormal stabilization of synaptic patterns in the brain is another matter". (Freeman and Watts, 1942, p. 18.)

Accepting the then prevalent idea that the thalamus plays a major role in regulating affect, Freeman and Watts speculated that a successful lobotomy might depend on severing the anterior peduncle (the anterior thalamic fasciculus) which connects the thalamus and the prefrontal area. They cited C. Judson Herrick to support this conclusion:

"This region [the extreme prefrontal cortex] is in especially intimate relation with the thalamus through the so-called anterior peduncle of the thalamus, probably by both descending and ascending fibers. Keeping in mind that the thalamus appears to be the organ *par excellence* of affective experience, this thalamo-frontal connection may provide for the addition within the prefrontal field of affective impulses of thalamic origin". (Herrick, 1926.)

Drawing on Earl Walker's anatomical studies of the thalamus in primates (Walker, 1938), Freeman and Watts suggested that the dorsomedial thalamic nucleus, with its connections to both the prefrontal area and the hypothalamus (Le Gros Clark and Boggon, 1935), was ideally positioned to link emotions, autonomic responses, and thought processes:

"It cannot be stated positively that this represents the sole means by which emotion and imagination are linked. Yet the fasciculus of fibers connecting the prefrontal region with the thalamus is unquestionably of great significance, and there is little doubt that its interruption is of major importance in the alteration of the personality seen after frontal lobotomy". (Freeman and Watts, 1942, p. 27.)

Support for this hypothesis came from the brain of a patient who was considered to have had a successful lobotomy before he died from an unrelated cause. Within the thalamus, retrograde degeneration was reported to have been found only in the dorsomedial nucleus. Although Freeman and Watts continued to insists that the leucotome needs to be inserted into all 4 quadrants (upper, lower, medial, and lateral) of the prefrontal area, they implied that destruction in the ventromedial area was most critical (Freeman and Watts, 1942).

By 1947, Freeman and Watts had performed over 400 lobotomies and had the opportunity to examine the brains of 12 patients who had survived long enough for retrograde degeneration studies of the dorsomedial nucleus. They concluded that they had found a "point-to-point relationship between this nucleus and the frontal lobe", the dorsomedial thalamus was critically involved in emotional experience, and (1) the dorsomedial connections to the "frontal pole" needed to be severed for a successful lobotomy; and (2) severing the thalamic connections to the lateral convexity of the prefrontal area seems to complicate the recovery process (Freeman and Watts, 1947).

By 1950, Moniz's speculations about "fixed ideas" and "stabilized neural activity" were totally ignored, except in historical accounts. While it might have been acknowledged that many psychiatric patients have fixed ideas and stereotyped behavior, these were considered symptoms rather than causes. Although Freeman and Watts dedicated the 1950 edition of *Psychosurgery* to Moniz, they clearly dissociated themselves from his speculations about "fixed ideas" and instead placed their emphasis on the emotional changes produced by successful psychosurgery.

"The theory of synaptic stabilization of Egas Moniz loses something of its validity in the consideration of these cases, because in spite of interruption of large numbers of association fibers, the pathologic phenomena persist for a time their patterns unchanged. In our opinion the striking change that develops almost at the moment of sectioning the fibers lies in the reduction of the affect connected with the ideas and acts. The phenomena continue of their own momentum, but, lacking the driving force of emotion, they gradually subside and are displaced to the periphery of consciousness. They are thus compatible with productive activity of the whole individual". (Freeman and Watts, 1950, p. 461.)

While Freeman and Watts had only speculated earlier about the importance of the connections between the prefrontal area and the dorsomedial thalamus, in 1950 they claimed that they had the anatomical support.

"From this study of the retrograde degenerations of the thalamus it would appear that in order to obtain a good clinical result it is not necessary to trespass upon cortical areas receiving their thalamic supply from the lateral or anterior groups of nuclei. In fact, such trespassing would seem to complicate convalescence. This study further points to the importance of the [dorso] medial nucleus as the anatomic substrate for emotion, especially in connection with ideational processing relating to the self. A certain quantitative reduction in the cellular component of this nucleus is compatible with a useful existence freed from abnormal cares. It is our opinion that the localization degeneration of the thalamus has more to do with satisfactory mental results than anything else in the picture". (Freeman and Watts, 1950, pp. 288–289.)

The importance of the ventromedial prefrontal area and its connections to the dorsomedial thalamus was also stressed by others. Earlier, Kleist (1934) had reviewed evidence from brain injured soldiers and concluded that damage to the ventromedial prefrontal area often produced a general disinhibition particularly of the emotions, while damage to the dorsolateral convexity was more likely to produce intellectual deficits (Kleist, 1934). Later, the Swedish psychiatrist Gösta Rylander (1948) reported that lobotomies restricted to the ventromedial prefrontal area attenuate anxieties and obsessive ideation without producing lasting intellectual deficits, but destruction of the lateral convexity of the prefrontal area was often followed by a marked lowering of general intelligence.

During the 1940s, neurosurgeons were also beginning to report that their highest percentage of successful lobotomies were among cases in which the ventromedial prefrontal area had been destroyed. In England, Dax and Radley Smith concluded from a series of 50 leucotomized patients that results were best when damage was restricted to the nerve fibers adjacent to the ventromedial ("orbital") area (Dax and Radley Smith, 1943). Among others in England, Reitman reported good results in 22 patients given an orbital leucotomy (Reitman, 1946); Geoffrey Knight claimed that the post-operative course after an orbital leucotomy was smoother than following the more extensive "major" leucotomies (Knight, 1955); and Ström-Olsen and Northfield (1955) concluded that undercutting restricted to the orbital cortex produced less serious personality defects and other complications.

Similar reports were published in the U.S.A. Leopold Hoffstatter and his colleagues in Saint Louis reported that lobotomies restricted to the orbital area produced improvement that compared favorably with the results following more extensive destruction (Hoffstatter et al., 1945). In Boston, James Poppen modified the dorsal surgical approach introduced earlier by John Lyerly, making it possible to destroy the more medial portion of the prefrontal area. The Lyerly–Poppen procedure soon replaced the Freeman Watts "standard lobotomy" as the most frequently performed psychosurgical operation in the U.S.A. Later, Everett Grantham of the University of Saint Louis further modified the Lyerly–Poppen procedure. By using an X-ray guided, stereotaxic electrode implantation technique and high frequency current to make lesions, Grantham was able to further restrict the damage to the ventromedial quadrant of the prefrontal area (Grantham, 1951). During this same period, William Scoville reported success with a suborbital undercutting procedure that made it possible to elevate the frontal lobes and to apply suction to the orbital prefrontal area (Scoville, 1949). Scoville's approach was later modified and used extensively in Japan (Hirose, 1965, 1972). A more complete discussion of the evolution of psychosurgical procedures has been presented elsewhere (Valenstein, 1980, 1986).

Concurrently, anatomical studies provided more detailed information about the connections between the prefrontal area and the dorsomedial thalamus. Jerzy Rose and Clinton Woolsey at Johns Hopkins University described the connections between the thalamus and the ventromedial prefrontal area in a number of animals, including the rabbit, cat, and sheep (Rose and Woolsey, 1948), while Fred Mettler at Columbia University confirmed and extended Walker's earlier description of these connections in the primate (Mettler, 1947). Studying the autopsied brains of leucotomized patients, Alfred Meyer and his colleagues at the Maudsley Hospital in London found evidence for similar connections in humans (Meyer et al., 1945, 1947; Meyer and Beck, 1954). These anatomical studies served to sharpen the distinction between the ventromedial and dorsolateral prefrontal areas by emphasizing their separate thalamic connections.

By the end of the 1940s, there was much discussion about the possibility of reducing the intellectual deterioration that followed lobotomies by restricting the damage to the more ventromedial aspect of the prefrontal area or to the pathways connecting this region to the dorsomedial thalamus. It was only a small extension of this line of thought to consider that the dorsomedial thalamus itself might be an appropriate target for psychosurgery. When a human model of the Horsley–Clarke stereotactic device was designed in 1947, it was first used to produce dorsomedial lesions in mental patients (Spiegel et al., 1947, 1949). Subsequently, others tried to treat mental illness by destroying parts of the dorsomedial nuclei, but results were generally disappointing and eventually so regarded even by Wycis in a later review (Wycis, 1972).* Stereotaxic instruments of different designs were also introduced in Sweden (Leksell, 1949) and in France (Tailarech et al., 1949;

---

* There were, however, more thalamic operations for intractable pain, the most common thalamic targets to treat pain were the dorsomedial, the centre median, and the parafascicular nuclei. With increasing interest in the limbic system and mental disorders, the anterior thalamic nucleus was sometimes used as a psychosurgical target.

Tailarech, 1952). Tailarech and his colleagues reported using a stereotaxic instrument to make electrolytic lesions in the anterior limb of the internal capsule in order to destroy the fronto-thalamic fibers in psychiatric patients and, as noted above, Grantham (1951) used a stereotaxic instrument to make ventromedial prefrontal lesions primarily in patients suffering from intractable pain. In general, however, stereotaxic instruments were only slowly adopted by those performing psychosurgery.

Thus, during the latter half of the 1940s psychosurgery became increasingly directed toward the ventromedial prefrontal area (including the orbital and rectus gyri) and the anterior thalamic fasciculus. Whereas dorsolateral prefrontal destruction was thought to produce intellectual impairment of "executive" functions such as the planning and evaluating of action and the capacity to change strategies, ventromedial destruction was thought to produce a disinhibition that could have either positive or negative ramifications depending on the degree and on the patients on whom such changes were superimposed. So although, for example, the extraversion, increased motor activity, and euphoria which were produced by ventromedial ablations were claimed to counteract introversion, inertia, indecision, obsessiveness and depression, there was always the danger it might produce restlessness, impulsiveness, and lack of judgement.

Psychosurgery in the 1950s

By 1951, John Fulton had become the major spokesman for the position that the massive destruction of the prefrontal area that characterized much of the early psychosurgery should be discontinued. Fulton (1951) insisted that psychosurgery should be restricted to either the ventromedial area (including its thalamic connections) or to specific limbic system targets, particularly the anterior cingulate gyrus. Fulton cited animal studies, suggesting that animals were tamed following cingulate lesions although it should have been clear that the animals also had cognitive impairment which complicated the interpretation of the emotional changes (Smith, 1945; Ward, 1948a, b). Fulton was also influenced by Paul MacLean's concept of the "visceral brain" (MacLean, 1949). MacLean, who was working at the time in Fulton's laboratory, had revived and elaborated on the heretofore neglected theory of Papez that had identified the limbic lobe as the anatomical substrate of emotions (Papez, 1937). Fulton predicted that psychosurgery in the future would be tailored to a patient's symptoms and he suggested that ventromedial prefrontal targets might be best used on depressed patients with subnormal psychomotor activity, while cingulate lesions were more suitable for aggressive, agitated, and hyperactive patients (Fulton, 1951).

Encouraged by John Fulton, neurosurgeons began to explore the anterior cingulate gyrus as a target for psychosurgery. Successful anterior cingulotomies were reported in England, France and the U.S.A. in the early 1950s (Tow and Whitty, 1953; LeBeau and Petrie, 1953; Livingston, 1953). Acceptance of the theory that the limbic lobe regulated emotionality grew with reports that the temperament of animals could be dramatically changed by stimulation or ablation of selective limbic structures. In the 1940s, the Spanish neurosurgeon Sixto Obrador performed amygdalectomies to treat mental illness after learning that Klüver and Bucy found that monkeys were tamer following temporal lobe excision (Obrador, 1947, 1977, p. xxv). Although one may well wonder why anyone would have been encouraged to perform amygdalectomies on man considering that Klüver and Bucy had reported that hypersexuality, visual agnosia, and other impairment also occurred, amygdalectomies were also performed during this period by Walker (cited in Scoville et al., 1953), Williams and Freeman (1951), and by Scoville et al. (1953). The role of animal experiments in the selection of these and other psychosurgical targets is more thoroughly discussed elsewhere (Valenstein, 1973, pp. 50–63; Valenstein, 1980, pp. 56–59).

The view that gradually emerged was that the prefrontal area, the limbic lobe, the hypothalamus, and interconnected thalamic nuclei constituted a complex neural circuitry modulating emotional expression and "psychic mood". These ideas were vague, however, and had it not been for the animal experiments — which were often misinterpreted — it would have been impossible to predict whether the ablation of any given structure or pathway would exacerbate or alleviate specific psychiatric symptoms. Nevertheless, there seemed to be a wide acceptance of the view that psychosurgery had entered a new phase characterized by a confidence that the anatomy and physiology were (or would soon be) sufficiently understood to justify a more rational selection of brain targets. A 1949 survey in the U.S.A. revealed that most respondents believed that their hospitals would be increasing the amount of psychosurgery performed in the future (Limburg, 1951). Nevertheless, when chlorpromazine was introduced in the mid-1950s there was a precipitous drop in the number of operations. Within a few years after the introduction of chlopromazine and the other psychoactive drugs that became available shortly afterward, the amount of psychosurgery performed worldwide was reduced to between 10 and 20% of what it had been during the 1949–1952 peak period. Interest in psychosurgery dwindled and the annual volumes of advances in psychiatry, neurology, and neurosurgery which had been devoting separate chapters to psychosurgery, rarely even mentioned the topic in the 1960s.

Psychosurgery in the 1960s and 1970s

Despite the precipitous drop in the amount performed after neuroleptic and antidepressant drugs were introduced, some psychosurgery continued to be performed. It had not taken long to realize that many patients were not helped by drugs or by electroconvulsive therapy and some intractably ill patients were still referred for psychosurgery. In general, however, psychosurgery fell from grace and only a few psychiatrists were willing to refer patients for one of these procedures. Starting at the end of the 1960s, however, several articles appeared which suggested that stereotaxic surgical techniques combined with the increased knowledge of the anatomy and function of fronto-limbic-diencephalic pathways justified a "second look" at psychosurgery (Livingston, 1969, 1975). In agreement with John Fulton's earlier recommendation, it was predicted that it would soon be possible to make discretely targeted lesions capable of alleviating different psychiatric symptoms. Although the number of operations did not increase substantially, interest in the future possibilities of modern psychosurgery was clearly on the rise. The First International Congress of Psychosurgery had been held in Lisbon in 1948, but no meeting followed until one was convened in Copenhagen in 1970 (Hitchcock et al., 1972). The Third International Congress was held only 2 years later in England (Laitinen and Livingston, 1973) and this was followed by a fourth in Madrid in 1975 (Sweet et al., 1977).

By the end of the 1970s, stereotaxic instruments were used in most of the psychosurgery being performed and electrode insertion was commonly guided by X-ray monitoring. In some instances, the precise target was selected after observing responses to electrical stimulation. Lesions were usually made with radio frequency current or, to a lesser extent, with cryogenic and electrolytic instruments. A few were using radioactive cobalt or yttrium to make lesions and ultrasound was claimed to have some advantages by one surgeon. While not completely eliminated, psychosurgery performed with hand held leucotomes was generally regarded as a "hangover" from a past era.

Despite the increased precision of the surgery, the prediction that brain targets would be selected based on a patient's predominant symptoms was not fulfilled. A survey found that most neurosurgeons and neuropsychiatrists who performed or recommended psychosurgery did not believe there was any convincing data to justify the target with different patients (Valenstein, 1980, pp. 64–66). While patients who had uncontrollable violent episodes were more likely to be given an amygdalectomy, or in a few instances, a posterior hypothalamic operation (Sano et al., 1972) and some patients exhibiting criminal sexual behavior underwent ventromedial hypothalamic ablations (Roeder et al., 1972) most psychosurgery was targeted either at the ventromedial prefrontal area, the anterior cingulate gyrus, or at specific fronto-limbic fiber tracts. With few exceptions, those performing psychosurgery today use the same procedure on all of their patients. It seems clear from the published record that patients receive a cingulectomy, a bimedial prefrontal leucotomy, or some other procedure depending on the institution to which they are referred rather than on their particular symptoms. As the same success rate is generally claimed regardless of the brain area destroyed, this created an interpretive problem for those arguing for the specificity of targets. It is commonly argued that the targets are all part of the same neural circuit, implying that destruction anywhere in the circuit should have the same effect. The adequacy of this explanation cannot, however, stand up to close scrutiny.

There developed during the 1970s, and there still is, a bewildering array of different brain targets used in psychosurgery, but all of them seem to be based on the experience of the surgeon involved rather than on any explicit rationale. Excluding temporal lobe and diencephalic targets, a partial list of psychosurgical procedures in use today would include: innominate tractotomy, capsulotomy, basofrontal tractotomy, subcaudate tractotomy, limbic leucotomy (combined subcaudate and cingulum lesions), medial mesoloviolotomy (a subrostral cingulotomy that invades the genu of the corpus callosum), bimedial prefrontal leucotomies, cingulotomies, cingulotractomy, and subrostral cingulotomy. More often than not, even when the same name is used to identify a psychosurgical procedure, the area destroyed varies between surgeons. In virtually no instance is there any rationale provided to justify the specific target selected other than to imply that the location is a convenient place to interrupt limbic, or fronto-limbic, or fronto-limbic-diencephalic pathways important for emotions.

In one sense, the rationale for psychosurgery had not changed except that the concept of disrupting "fixed ideas" as hypothesized by Moniz had been replaced by a concept of disrupting fixed emotional patterns. Thus, starting in the 1970s and continuing today, the stated purpose of psychosurgery is to destabilize some circuit presumed to be responsible for maintaining fixed emotional states. An explanation of why "destabilization" is best accomplished by an ablation at the particular point in the brain selected as a target is rarely even attempted. Speculation about what might be inhibited, disinhibited, or excited and how such changes in neural activity might influence emotions is virtually never offered. The arguments justifying psychosurgery are empirical and essentially atheoretical. Reduced to their essence, the arguments are that patients improve, that the incidence of serious complications are minimal and within acceptable limits, that all other appropriate treatments have been tried without success, and that the risk of "doing nothing" is great. Although these arguments and the evidence claimed to support them have been challenged, there is independent support for the claim that modern psychosurgery may produce substantial, and in some cases lasting, improvement in patients and that this may be accomplished without producing serious impairment (Corkin, 1980; Mirky and Orzack, 1980).

The range of patients undergoing psychosurgery became much more restricted during the 1970s. The majority of the patients receiving psychosurgery are now suffering from severe depression, obsessive compulsive disorders, or intractable pain. Although some of the depressed and obsessed patients may exhibit schizophrenic ideation, deteriorated and chronic schizophrenics are no longer considered suitable candidates for psychosurgery.

Although the number of patients that are subjected to psychosurgery to treat violent behavior, sexual deviancy, or drug addiction is a relatively small proportion of the total, the awareness of a renewed interest in psychosurgery and the belief that there were plans a foot to use brain operations more widely to control behavior resulted in a public outcry. The public was particularly aroused by claims that a substantial amount of violence was triggered by brain pathology as it appeared that a biological solution was being proposed for a social problem. The views expressed in the book *Violence and the Brain* (Mark and Ervin, 1970) became a focal point of public reaction as the authors seemed to be suggesting that much more violence than had been suspected — even some of a political or racial origin — can be traced to temporal lobe pathology which was treatable by brain surgery. When it became known that amygdalectomies were being used in many countries to treat violent behavior, the political controversy intensified and legislation restricting psychosurgery was enacted. These developments, which were worldwide, have been reviewed elsewhere (Valenstein, 1980, pp. 39 – 54). It need only to be noted here that despite the renewed interest, the amount of psychosurgery performed actually began to decline by the end of the 1970s (Valenstein, 1980).

Concluding remarks

Few additional changes have occurred during the 1980s. Psychosurgery is still being performed in many countries, despite the enactment of some legislation restricting its use. The procedures used and the brain targets selected are almost identical to what they were at the end of the 1970s. For the most part, the theory has not evolved and the most common rationale offered to justify psychosurgery, if any is given at all, is a general statement implying that the brain targets are at critical crossroads in the neural circuit regulating emotions.

The one new element that is detectable in the psychosurgical literature is the speculation that basal ganglia structures, particularly the striatum, the lentiform nucleus, and their connections, are involved in some emotional disorders. While the basal ganglia have traditionally been viewed as part of a system modulating activity in the motor cortex and these structures have been implicated in Parkinson's disease and other dyskinesias, more recently it has been recognized that basal ganglia also communicate with the prefrontal area, the limbic system, and the sensorimotor cortex. Nauta (1986), for example, has described two major divisions of the striatum: the ventromedial striatum (nucleus accumbens) with connections to the amygdala, hippocampus, cingulate cortex (and other limbic structures) and also to the prefrontal cortex; and an anterodorsolateral portion of the striatum which receives heavy projections from the sensorimotor cortex.

The basal ganglia are now implicated in a number of psychiatric disorders. It has been speculated that prefrontal-striatal connections may play a critical role in the attention disorder seen in schizophrenia (Weinberger et al., 1986) and Laplane and his colleagues (1984, 1989) have described cases where the onset of classical obsessive compulsive symptoms could be traced to basal ganglia damage. It has been argued in a recent review (Modell et al., 1990) that obsessive-compulsive disorder and possibly also substance abuse (including eating disorders), and compulsive gambling may be caused by a disinhibition in a "basal ganglia-limbic striatal-thalamocortical circuit". While it is always heuristically useful to propose a specific mechanism, it has to be recognized that hypotheses about the role of the basal ganglia in mental disorders at this time are highly speculative and that the supporting data are not only limited, but open to many different interpretations. It seems safe to predict, however, that basal ganglia circuitry is likely to be increasingly included in any theoretical model of how psychosurgery might work. Considering the complexity of the basal ganglia circuits themselves, it is not clear whether expanding of the anatomical substrate of theories to include them represents an advance. In the meantime, it should be noted that several of the psychosurgical procedures presently in use disrupt fronto-striatal pathways that traverse through the anterior limb of the internal capsule. It is unlikely, therefore, that the theoretical expansion of the relevant neural substrate will spawn any new psychosurgical targets or significantly increase the number of such operations.

References

Baruk, H. (1978) *Patients are Like Us: The Experience of Half a Century in Neuropsychiatry,* William Morrow, New York.

Brickner, R.M. (1932) An interpretation of function based on the study of a of a case bilateral frontal lobectomy. *Res. Publ. Assoc. Nerv. Ment. Dis.,* 13: 159–351.

Brickner, R.M. (1936) *The Intellectual Functions of the Frontal Lobes,* Macmillan, New York.

Broca, P.B. (1861) *Bull. Soc. Anthropol.,* 2: 301–321.

Burckhardt, G. (1891) Über Rindenexcisionen, als Beitrag zur Operativen Therapie der Psychosen. *Allgem. Z. Psychiat.,* 47: 463–548.

Clark, W.E. le G. and Boggon, R.H. (1935) Thalamic connections of the parietal and frontal lobes of the brain in the monkey. *Phil. Trans. Roy. Soc. B,* 224: 313.

Cobb, S. (1949) "Presidential Address". *Trans Am. Neurol. Assoc.,* p. 3.

Corkin, S. (1980) A prospective study of cingulotomy. In: E.S. Valenstein (Ed.), *The Psychosurgery Debate,* W.H. Freeman, San Francisco, CA, pp. 164–204.

Crawford, M.P., Fulton, J.F., Jacobsen, C.F. and Wolf, J.B. (1948) Frontal lobe ablation in chimpanzees: A resume of "Becky" and "Lucy". *Res. Publ. Assoc. Nerv. Ment. Dis.,* 27: 3–58.

Dax, E.C. and Radley Smith, E.J. (1943) The early effects of prefrontal leucotomy on disturbed patients with mental illness of long duration. *J. Ment. Sci.,* 89: 182–188.

Ducosté, M.L. (1932) Impaludation cérébrale. *Bull. Acad. Med.,* 107: 516–518.

Fiamberti, A.M. (1947) Indicazioni e tecnica della leucotomia prefrontale transorbitaria. *Passegn. Neuropsichiat.,* 1: 3.

Freeman, W.J. (1948) Transorbital lobotomy. *Med. Ann. D.C.,* 17: 257.

Freeman, W.J. and Watts, J.W. (1937) Prefrontal lobotomy in the treatment of mental disorders. *Sthrn Med. J.,* 30: 23–31.

Freeman, W.J. and Watts, J.W. (1942) *Psychosurgery: Intelligence, Emotion and Social Behavior Following Prefrontal Lobotomy for Mental Disorders,* Charles Thomas, Springfield, IL.

Freeman, W.J. and Watts, J.W. (1947) Retrograde degeneration of the thalamus following prefrontal lobotomy. *J. Comp. Neurol.,* 86: 65–93.

Freeman, W.J. and Watts, J.W. (1950) *Psychosurgery: In the Treatment of Mental Disorders and Intractable Pain,* Charles Thomas, Springfield, IL.

Fulton, J.F. (1926) *Muscular Contraction and the Reflex Control of Movement,* Williams and Wilkins, Baltimore, MD.

Fulton, J.F. (1932) *The Sign of Babinski,* Charles Thomas, Springfield, IL.

Fulton, J.F. (1934a) Some functions of the cerebral cortex. Lecture I. Autonomic representation in the cerebral cortex. *Mich. State Med. J.,* 33: 175–182.

Fulton, J.F. (1934b) Some functions of the cerebral cortex. Lecture II. The frontal lobes. *Mich. State Med. J.,* 33: 235–243.

Fulton, J.F. (1951) *Functional Lobotomy and Affective Behavior,* W.W. Norton, New York.

Fulton, J.F., Aring, C.D. and Wortis, S.B. (Eds.) (1948) *The Frontal Lobes, Res. Publ. Assoc. Nerv. Ment. Dis.,* 27: 209.

Grantham, E.G. (1951) Prefrontal lobotomy for relief of pain. With a report of a new operative technique. *J. Neurosurg.,* 8: 405–410.

Harlow, H.F. and Settlage, P.H. (1948) Effect of extirpation of frontal areas upon learning performance of monkeys. *Res. Publ. Assoc. Nerv. Ment. Dis.,* 27: 446–459.

Herrick, C.J. (1926) *Brains of Rats and Men: A Survey of the Origin and Biological Significance of the Cerebral Cortex,* Univ. of Chicago Press, Chicago, IL.

Hirose, S. (1965) Orbito-ventromedial undercutting 1957–1963. *Am. J. Psychiat.,* 121: 1194–1202.

Hirose, S. (1972) The case selection of mental disorder for orbitoventromedial undercutting. In: E. Hitchcock, L. Laitinen and K. Vaernet (Eds.), *Psychosurgery,* Charles C. Thomas, Springfield, IL, pp. 317–330.

Hitchcock, E., Laitiner, L. and Vaernet, K. (Eds.) (1972) *Psychosurgery,* Charles Thomas, Springfield, IL.

Hofstatter, L., Smolik, E.A. and Busch, A.K. (1945) Prefrontal lobotomy in treatment of chronic psychoses. *Arch. Neurol. Psychiat.,* 53: 125–130.

Jacobsen, C.F. (1931) A study of cerebral function in learning the frontal lobes. *J. Comp. Neurol.,* 52: 271–340.

Jacobsen, C.F. (1934) Influence of motor and premotor area lesions upon the retention of acquired skill movements in monkeys and chimpanzees. *Res. Publ. Assoc. Nerv. Ment. Dis.,* 13: 225–247.

Jacobsen, C.F., Wolfe, J.B. and Jackson, T.A. (1935) An experiment analysis of the functions of the frontal association areas in primates. *J. Nerv. Ment. Dis.,* 82: 1–14.

Janet, P. (1895) Névroses et idées fixes. *Presse Méd.,* I, 213, 375.

Laitinen, L. and Livingston, K.E. (Eds.) (1973) *Surgical Approaches in Psychiatry,* University Park Press, Baltimore, MD.

Laplane, D., Baulac, M., Widlocker, D. and Dubois, B. (1984) Pure psychic akinesia with bilateral lesions of basal ganglia. *J. Neurol. Neurosurg. Psychiat.,* 47: 377–385.

Laplane, D., Levasseur, M., Pillon, B., Dubois, B., Baulac, M., Mazoyer, B., Tran Dinh, S., Sette, G., Danze F. and

Baron, J.C. (1989) Obsessive-compulsive and other behavioural changes with bilateral basal ganglia lesions. *Brain,* 112: 699 – 725.

LeBeau, J. and Petri, A. (1953) A comparison of personality changes after (1) prefrontal selective surgery for the relief of intractable pain, and for the treatment of mental cases; and (2) cingulectomy and topectomy. *J. Ment. Sci.,* 99: 53 – 61.

Leksell, L. (1949) A stereotaxic apparatus for intracerebral surgery. *Acta Chir. Scand.,* 99: 209.

Limburg, C.C. (1951) A survey of the use of psychosurgery with mental patients. In: N. Bigelow (Ed.), *Proceedings of the First Research Conference on Psychosurgery, 1949,* National Institutes of Health, Bethesda, MD, U.S. Public Health Service Publ. No. 16, pp. 165 – 173.

Livingston, K.E. (1953) Cingulate cortex isolation for the treatment of psychoses and psychoneurons. *Proc. Assoc. Res. Nerv. Ment. Dis.,* 31: 374 – 378.

Livingston, K.E. (1969) The frontal lobes revisited. The case for a second look. *Arch. Neurol.,* 20: 90 – 95.

Livingston, K.E. (1975) Surgical contributions to psychiatric treatment. In: S. Arieti (Ed.), *American Handbook of Psychiatry, Vol. 5,* Basic Books, New York, pp. 548 – 563.

Kleist, K. (1934) *Kriegverletzungen des Gehirns in ihrer Bedeutung fur Hirnlokalisation und Hirnpathologie,* Barth, Leipzig.

Knight, G.K. (1955) Orbital leucotomy. A review of 52 cases. *Lancet,* 981 – 985.

MacLean, P.D. (1949) Psychosomatic disease and the "visceral brain". Recent developments bearing on the Papez theory of emotion. *Psychosom. Med.,* 11: 338 – 353.

Mark, V.H. and Ervin, F. (1970) *Violence and the Brain,* Harper and Row, New York.

Mettler, F.A. (1947) Extracortical connections of the primate cerebral cortex. I. Thalamo-cortical connections. *J. Comp. Neurol.,* 86: 95 – 118.

Meyer, A. and Beck, E. (1954) *Prefrontal Leucotomy and Related Operations: Anatomical Aspects of Success and Failure,* Oliver and Boyd, London.

Meyer, A., Bonn, M.D. and Beck, E. (1945) Neuropathological problems arising from prefrontal leucotomy. *J. Ment. Sci.,* 41: 385, 411 – 425.

Meyer, A., Beck, E. and McLardy, T. (1947) Prefrontal leucotomy: A neuro-anatomical report. *Brain,* 70: 18 – 49.

Mirsky, A.F. and Orzack, M.H. (1980) Two retrospective studies of psychosurgery. In: E.S. Valenstein (Ed.), *The Psychosurgery Debate,* W.H. Freeman, San Francisco, CA, pp. 205 – 244.

Modell, J., Mountz, J., Curtis, G. and Greden, J. (1990) Neurophysiologic dysfunction in basal ganglia/limbic striatal and thalamocortical circuits as a pathogenetic mechanism of obsessive-compulsive disorder: a new proposal and review of the literature. *J. Neuropsyhiat. Clin. Neurosci.,* in press.

Moniz, E. (1936) *Tentatives Opératoires dans le Traitement de Certaines Psychoses,* Masson, Paris.

Morel, F. (1938) The surgical treatment of dementia praecox. *Am. J. Psychiat.,* 94: 309 – 314.

Nauta, W.J.H. (1986) Circuitous connections linking cerebral cortex, limbic system, and corpus striatum. In: B.K. Doane and K.E. Livingston (Eds.), *The Limbic System: Functional Organization and Clinical Disorders,* Raven Press, New York, pp. 43 – 54.

Obrador, S. (1947) *Las Modernas Intervenciones Quirurgicus en Psiquiatria,* Ediotral Piaz, Montalvo.

Obrador, S. (1977) Opening Remarks. In: W.H. Sweet, S. Obrador and J. Martin-Rodríguez (Eds.), *Neurosurgical Treatment in Psychiatry, Pain, and Epilepsy,* University Park Press, Baltimore, MD. p. xxv.

Ody, F. Le (1938) Traitement de la démence précoce par résection du lobe préfrontal, *Arch. Ital. Chir.,* 53: 321 – 330.

Papez, J.W. (1937) A proposed mechanism of emotion. *Arch. Neurol. Psychiat.,* 38: 725 – 743.

Pressman, J.D. (1986) *Uncertain Promise: Psychosurgery and the Development of Scientific Psychiatry in America, 1935 to 1955.* Ph.D. Dissertation, University of Pennsylvania, Phyladelphia, PA.

Pressman, J.D. (1988) Sufficient promise: John F. Fulton and the origins of psychosurgery. *Bull. Hist. Med.,* 62: 1 – 22.

Putman, T. (1950) Prefrontal lobotomy: its evolution and present status. *Bull. Los Angelos Neurol. Soc.,* 15: 225.

Reitman, F. (1946) Orbital cortex syndrome following leucotomy. *Am. J. Psychiat.,* 103: 238 – 241.

Roeder, F., Orthner, H. and Müller, D. (1972) The stereotaxic treatment of pedophilic homosexuality and other sexual deviations. In: D. Hitchcock, L. Laitinen and K. Vaernet (Eds.), *Psychosurgery,* Charles Thomas, Springfield, IL, pp. 87 – 111.

Rose, J.E. and Woolsey, C.N. (1948) The orbitofrontal cortex and its connections with the mediodorsal nucleus in rabbit, sheep and cat. *Res. Publ. Assoc. Nerv. Ment. Dis.,* 27: 210 – 232.

Rylander, G. (1939) Personality changes after operations on the frontal lobes. *Acta Psychiat. Neurol.,* Suppl. 20.

Rylander, G. (1948) Personality analysis before and after frontal lobotomy. *Res. Publ. Assoc. Nerv. Ment. Dis.,* 27: 691 – 705.

Sano, K., Sekino, H. and Mayanagi, Y. (1972) Results of stimulation and destruction of the posterior hypothalamus in cases with violent, aggressive, or restless behaviors. In: E. Hitchcock, L. Laitinen and K. Vaernet (Eds.), *Psychosurgery,* Charles Thomas, Springfield, IL, pp. 57 – 75.

Scoville, W.C. (1949) Selective undercutting as a means of modifying and studying frontal lobe function in man. *J. Neurosurg.,* 6: 65 – 73.

Scoville, W.B., Dunsmore, R.H., Liberson, W.T., Henry, C.E. and Pepe, A. (1953) Observations on medical temporal lobotomy and uncotomy in the treatment of psychotic states. *Res. Publ. Assoc. Nerv. Ment. Dis.,* 31: 347 – 369.

Scull, A. (1987) Desperate remedies: A Gothic tale of madness and modern medicine. *Psychol. Med.,* 17: 561–577.
Scull, A. and Favreau, D. (1986) "A chance to cut is a chance to cure". Sexual surgery for psychosis in three nineteenth century societies. *Res. Law Deviance Soc. Contr.,* 8: 3–39.
Smith, W.K. (1945) Functional significance of rostral cingular cortex as revealed by its response to electrical excitation. *J. Neurophysiol.,* 8: 241–255.
Spiegel, E.A., Wycis, H.T., Marks, M. and Lee, A.J. (1947) Stereotaxic apparatus for operations on the human brain. *Science,* 106: 349–350.
Spiegel, E.A., Wycis, H.T. and Freed, H. (1949) Thalamotomy in mental disorders. In: *Proc. 1st Int. Conf. Psychosurg.,* Livraria Luso-Espanhola, Lisbon, pp. 91–95.
Ström-Olsen, R. and Northfield, D.W.C. (1955) Undercutting of orbital cortex in chronic neurotic and psychotic tension states. *Lancet,* 986–991.
Sweet, W.H., Obrador, S. and Martin-Rodríguez (Eds.) (1977) *Neurosurgical Treatment in Psychiatry, Pain, and Epilepsy,* University Park Press, Baltimore, MD.
Tailarech, J. (1952) Etude stéréotaxique des structures encéphaliques profundes chez l'homme. *Presse Méd.,* 60: 605.
Tailarech, J., Hecaen, H. and David, M. (1949) Lobotomie préfrontale limitée par électro-coagulation des fibres thalamo-frontales à leur émergence du bras antérieur de la capsule interne. *4th Congr. Neurol. Int. Paris II,* 141.
Tilney, F. (1928) *The Brain from Ape to Man,* Hoeber, New York.
Tow, P.M. and Whitty, C.M.W. (1953) Personality changes after operations on the cingulate gyrus in man. *J. Neurosurg. Psychiat.,* 16: 186–193.
Valenstein, E.S. (1973) *Brain Control,* John Wiley, New York.
Valenstein, E.S. (1980) Rationale and surgical procedures. In: E.S. Valenstein (Ed.), *The Psychosurgery Debate,* W.H. Freeman, San Francisco, CA, pp. 55–75.
Valenstein, E.S. (1986) *Great and Desperate Cures,* Basic Books, New York.
Walker, A.E. (1938) *The Primate Thalamus,* University of Chicago Press, New York.
Ward, A.A. Jr. (1948a) The cingular gyrus, area 24. *J. Neurophysiol.,* 11: 13–23.
Ward, A.A. Jr. (1948b) Anterior cingulate gyrus and personality. *Res. Publ. Assoc. Nerv. Ment. Dis.,* 27: 438–445.
Watts, J.W. (1935) The relation of the frontal lobes to visceral function. *Med. Ann. D.C.,* 4: 99–105.
Weinberger, D.R., Berman, K.F. and Zec, R.F. (1986) Psychologic dysfunction of dorsolateral prefrontal cortex in schizophrenia. *Arch. Gen. Psychiat.,* 43: 114–124.
Williams, J.M. and Freeman, W.F. (1951) Amygdaloidectomy for the suppression of auditory hallucinations. *Med. Ann. Wash. D.C.,* 20: 192–196.
Woltman, H. (1941) *Proc. Staff Meet. Mayo Clin.,* 16: 200.
Wycis, H.T. (1972) The role of stereotaxic surgery in the compulsive state. In: E. Hitchcock, L. Laitinen and K. Vaernet (Eds.), *Psychosurgery,* Charles Thomas, Springfield, IL, pp. 115–116.

Discussion

A. Gevins: What psychosurgery procedures are currently being used in the Soviet Union to treat "psychiatric" patients?

E.S. Valenstein: It is commonly alleged that a large amount of psychosurgery was performed in undemocratic countries, presumably as a means of controlling dissent. That was not the case at all. Actually, no psychosurgery was performed in Germany until 1947 after the Nazi regime had been defeated and replaced. Although psychosurgery, along with a number of other physical treatments of mental illness, had been widely practiced in the Soviet Union, the practice of psychosurgery was discontinued (essentially outlawed) in that country in 1951 at a time when Stalin was still very much in charge of all policy decisions. Psychosurgery in the west on the other hand, continued at a high level for several years after 1951 and was never completely discontinued. To my knowledge, there was no psychosurgery performed in the Soviet Union after 1951. Several articles about lobotomy written by Soviet physicians appeared subsequently, but these were follow-up studies of patients operated prior to 1951. For example, Kornetov (1972) reviewed the results of 40 prefrontal lobotomies, but they had been performed at the 1st Moscow Medical Institute between 1947 and 1950. Although Kornetov concluded that some of the chronic schizophrenics had experienced "excellent recoveries" and stated that he personally believed that there was a place for psychosurgery to treat "pernicious schizophrenia", there have been no documented reports of any resumption of psychosurgery in the Soviet Union. Dr. Bechtereva and her colleagues at the Institute of Experimental Medicine in Leningrad have used "therapeutic electrical stimulation" through chronically implanted electrodes to treat intractable pain (including phantom limb pain) and temporal lobe epilepsy associated with aggression or psychiatric disorders such as paranoia and other psychotic thought processes. Some of these patients had as many as 64 electrodes, assembled in 14 bundles, implanted in the temporal lobes. In a few instances, patients were allowed to engage in self-stimulation, that is, to control the delivery of stimulation themselves. It was reported that stimulation at some brain sites reduced "troublesome symptoms" and elevated mood (Bechtereva et al., 1975). Although brain stimulation is not usually considered psychosurgery, there may be some interest in these reports.

D.H. Linszen: What were the target symptoms of lobotomy in schizophrenia, i.e. positive symptoms, negative symptoms or both?

E.S. Valenstein: Although the distinction between type 1 and type 2 schizophrenia with their associated positive and negative symptoms was not made as sharply in the 1940s and 1950s as it is today, there was considerable agreement that lobotomies were least effective in improving chronic, deteriorated schizophrenics, most of whom today would be considered type 2 schizophrenics with predominately negative symptoms. It was often said during that earlier period that lobotomies were used only as a "last resort" on chronic patients with illnesses of long duration. Actually, however, it was common knowledge among the physicians directly involved that lobotomies were most effective in patients with exaggerated emotional states, not those chronic patients who were flat, emotionally "burnt out", and unresponsive.

With respect to the brain targets, it was true in the 1940s and 1950s (and it is still true today) that almost everyone who performed psychosurgery used the same operation and brain target for all patients, regardless of symptoms. In the 1950s, there was some recognition that the standard prefrontal lobotomies were not effective in eliminating hallucinations (now considered a positive symptom) and as a result Scoville, Obrador, Freeman and Williams, among others, performed amygdalectomies in the hope they would be effective in alleviating hallucinations. In part, the idea to try temporal lobe targets came from the reports that temporal lobe epilepsy was sometimes preceded by hallucinatory experiences.

C.G. van Eden: IQ scores seem to be little affected by even extensive lobotomies. What does this tell about PFC function and/or IQ tests?

E.S. Valenstein: It is certainly true that IQ scores were not dramatically effected by lobotomy. Although Rosvold and Mishkin reported on the basis of a comparison of pre- and post-lobotomy intelligence test scores that IQ decreased following lobotomy, the great majority of the reports indicated that IQ scores were within the normal range and basically unchanged following a lobotomy. The problem with most of the data, however, was that pre-lobotomy IQ scores were seldom available for comparison and when they were available they were hard to interpret because the psychosis usually had depressed the score. It was not uncommon, therefore, to find that IQ score actually improved after a lobotomy, but this could be attributed to the fact that the patient was more tractable and testable.

The IQ score is notoriously insensitive to frontal lobe damage even in patients whose judgement is so seriously impaired that they must be constantly supervised. Much of the IQ score is determined by information acquired earlier in life. It may be recalled that Donald Hebb described a patient who had a major portion of an abscessed frontal lobe removed and when tested 4 years later, his IQ was 152, a score in the highly superior range.

J.E. Pisetsky: Do you believe that psychosurgery may have stimulated serious neurobiological research such as we have heard at this meeting.

E.S. Valenstein: My personal belief is that little serious research was stimulated by lobotomy. Elsewhere, I have made the argument that animal research often stimulated the exploration of a new lobotomy procedure rather than the other way. In general, the data from lobotomy studies were so poorly controlled and the methodology so flawed that few clearly testable hypotheses emerged. It is probably true, however, that the widespread practice of lobotomy created a social climate that generated additional funding for basic research on the frontal lobes, but that is a different issue and probably not what was intended by the question.

References

Bechtereva, N.P. et al. (1975) *Confin. Neurol.,* 37: 136–140.
Bechtereva, N.P. et al. (1976) In: W.H. Sweet et al. (Eds.), *Neurosurgical Treatment in Psychiatry, Pain and Epilepsy,* University Park Press, Baltimore, MD, pp. 581–613.
Kornetov, A.N. (1972) In: I. Pusek and E.Z. Kunc (Eds.), *Present Limits of Neurosurgery,* Avenicum, Prague, pp. 475–476.

CHAPTER 28

Neurometric studies of aging and cognitive impairment

E.R. John[1,2] and L.S. Prichep[1,2]

[1] Brain Research Laboratories, Department of Psychiatry, New York University Medical Center, New York, NY, and [2] Nathan S. Kline Institute for Psychiatric Research, Orangeburg, NY, USA

Introduction

Normal brain electrical activity, whether EEG or ERP, can be precisely characterized by quantitative descriptors. By statistical comparison of the features extracted from an individual against a normative database, clinically significant deviations can be objectively identified.

We have used this method, which is called "neurometrics", to study large populations of patients with different neurological and psychiatric disorders. Each disorder appears to be characterized by a distinctive pattern of deviant neurometric Z values, or standard scores, for particular descriptors.

The anatomical pattern of values of each descriptor can be topographically mapped. The color coding of such maps reflects the probability that the observed value lies within the normal limits for a healthy person the same age as the patient. Our normative age-regression equations describe over 1000 quantitative descriptors from age 6 to 90 (John et al., 1987).

As can be seen in Fig. 1, normal individuals conform well to predicted values, while patients display abnormal values which are different for different disorders (John et al., 1988).

These distinctive patterns of significantly deviant features can be used to construct multiple discriminant functions which accurately classify patients with different diseases. Fig. 2 is an example of a 4-way discriminant function with independent replication, which successfully separated normal from depressed from alcoholic from demented patients. The depressed patients had a mean Hamilton Depression Score of 17, and the demented patients had Global Deterioration Scores which ranged from 3 to 5 (mild to moderate impairment) (John et al., 1988).

We have applied neurometric methods to the study of EEGs and ERPs in a large sample of elderly people with and without cognitive impairment, as well as to a group of normal subjects under age 55 (Prichep et al., 1990). All patients were assessed for the magnitude of cognitive decline stages from 1 to 7 on the Global Deterioration Scale (GDS) for age-associated cognitive decline and primary degenerative dementia (Reisberg et al., 1982)*. Our sample consisted of 279 persons, divided into 7 groups:

Group 0: 35 normal individuals under the age of 55, with a mean age of 30.8.

Group 1: 42 normal individuals over the age of 55, with a mean age of 68.9 and a Global Deterioration Score of 1, with no subjective com-

---

Correspondence: Dr. E.R. John, Brain Research Laboratories, Department of Psychiatry, New York University Medical Center, New York, NY, USA.

* This work was done in collaboration with Drs. Barry Reisberg and Steven Ferris of the Aging and Dementia Studies Program, Department of Psychiatry, New York University Medical Center, New York, NY.

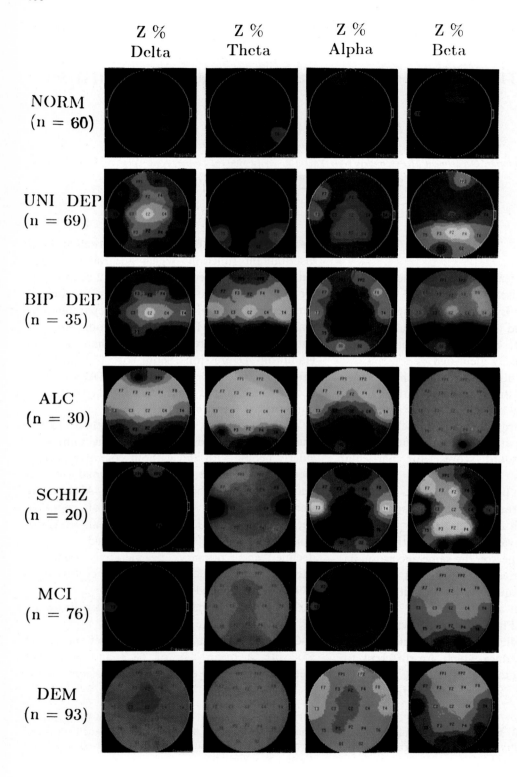

plaints of memory deficit and no memory deficit found in a clinical neurological, psychiatric and psychometric evaluation.

Group 2: 70 individuals with a mean age of 71.9 and a Global Deterioration Score of 2, with subjective complaints of forgetfulness of familiar names or recent placement of objects, but no objective evidence of impairment.

Group 3: 36 individuals with a mean age of 73.5 and a Global Deterioriation Score of 3, showing early confusional signs, decreased work performance and objective evidence of memory deficit.

Group 4: 44 individuals with a mean age of 71.2 and with a Global Deterioration Score of 4, in late confusional phase, showing clear-cut deficits in recent memory and concentration, and diminished work capacity but intact in many areas.

Group 5: 34 individuals with a mean age of 71.4 and with a Global Deterioration Score of 5, with symptoms of early dementia, who can no longer survive without some assistance but retain knowledge of many facts and are capable of basic self-care.

Group 6: 19 individuals with a mean age of 69.0 and with a Global Deterioration Score of 6, with symptoms of middle dementia, including severe memory loss and loss of context and requiring some assistance with activities of daily living.

Table I shows the mean age, standard deviation and age range for each GDS group. The correlation between age and GDS score was 0.005. No significant differences were found by ANOVA.

EEG results

A 1-min sample of eyes closed, resting EEG was collected using an on-line artifact detection algorithm, and artifact-free data was submitted for

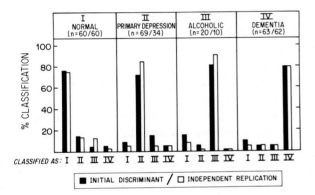

Fig. 2. Percentage of normal (I), and patients with primary depression (II), alcoholic (III) and dementia (IV) classified as I, II, III, or IV by a 4-way discriminant function using neurometric variables. The first and second numbers in parentheses above each panel indicate the sample size of each group used as the training set in the initial discriminant (black bars) and as the test set in the independent replication (open bars). Mean discriminant accuracy of this discriminant was approximately 78% (John et al., 1989).

TABLE I

Mean age, standard deviation, and age range by GDS stage

GDS	n	Mean age	S.D.	Age range
1	41	68.9	6.9	55.6 – 81.8
2	70	71.9	6.7	56.8 – 88.8
3	36	73.5	7.1	55.4 – 85.0
4	44	71.2	6.5	58.8 – 82.8
5	34	71.4	8.3	56.7 – 86.8
6	19	69.0	8.5	56.9 – 84.9
Total	244	71.2		55.4 – 88.8

Fig. 1. Average topographic head maps for Z scores of relative power (percentage) in delta, theta, alpha and beta frequency bands, computed across different groups of individuals. From top to bottom, successive rows represent groups classified as normal, unipolar depressed, bipolar depressed, alcoholic, schizophrenic, mild cognitively impaired and dementia patients. These maps represent the mean relative power difference between each group and the reference group, expressed in standard deviations of the reference (normal) group, not in the figure. Color coding is proportional to the mean Z score for each group, in steps corresponding to those shown on the Z scale. The scale is from +0.7 to −0.7 S.D. for the first 6 rows, and from +1.5 to −1.5 S.D. for the last row. The significance of the Z scale values can be estimated by taking the square root of sample size and the standard deviation of each group into account and ranges from 0.002 to 0.001 (John et al., 1988).

computer analysis by FFT. Absolute and relative power, coherence and symmetry were calculated for the delta, theta, alpha and beta frequency bands. For each of these features for each electrode of the 10–20 system, Z scores were calculated relative to normative age-regression equations, using methods described in detail elsewhere (John et al., 1987). Numerous neurometric variables were significantly correlated with GDS score. Fig. 3 shows the distribution of Z scores for relative power in the theta band for the multivariate measure combining data across all regions (Mahalanobis distance) (John et al., 1987), within each of these groups. The correlations between theta relative power in Fz, Cz and Pz and GDS score were respectively 0.58, 0.56 and 0.56, all significant at P greater than 10^{-4} by ANOVA, but not significantly different from one another by MANOVA.

There is a steady increase in the neurometric abnormality as severity of cognitive impairment increases. The earliest signs are theta increase, already discernible in group 2, followed by delta increase and decrease in the alpha and beta bands in later stages of deterioration. The absolute value of alpha does not decrease, indicating that the theta increase is not simply due to slowing of alpha. These signs of departure from optimum brain state are anatomically diffuse, even though the most extreme abnormality eventually appears frontally.

The correlation coefficients between every neurometric variability and GDS were computed. No significant correlations were found for coherence or symmetry measures. Significant correlations were found between both absolute and relative power measures in different brain regions and cognitive impairment.

Fig. 4 shows topographic maps of correlations between Z scores for absolute (top row) and relative power (bottom row) in each of the 4 frequency bands and cognitive impairment, color coded for the degree of correlation. Neurometric

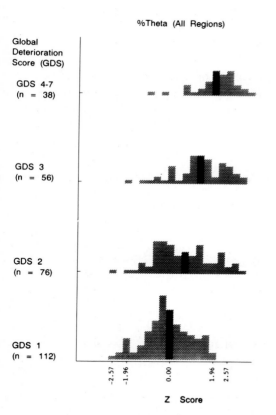

Fig. 3. Distributions of Z scores for the multivariate feature "relative power (%) in theta across all regions" for groups of elderly individuals showing a GDS of 1 (normal), 2 (mild cognitive impairment), and 4 or above (severe impairment). Note that the mean Z score increases with greater impairment (John et al., 1988).

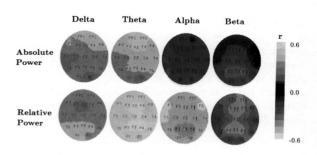

Fig. 4. Topographic maps of the correlations between GDS and Z absolute power (top row) and Z relative power (bottom row) for delta, theta, alpha and beta frequency bands. These maps are color coded for the degree of correlation, in steps corresponding to those shown of the r scale. A correlation of 0.3 with $n = 244$ would have a probability of less than 0.01 (Prichep et al., 1990).

abnormalities in many brain regions were highly correlated with cognitive impairment. Although such correlations were highest in frontal regions for absolute power, significant correlations were found in all brain regions. When relative power measures were considered, the correlations between regional abnormalities and cognitive impairment were much more uniform.

Even though we see a steady progression in the average neurometric abnormality across groups with increasing severity of impairment, these groups are not homogeneous. Cluster analysis reveals at least 3 major subgroups in terms of patterns of EEG abnormality (Prichep et al., 1983). These patterns did not correspond to severity levels, which were approximately the same in each cluster, but seemed to reflect different pathophysiology. Members of different clusters showed differential responsivity to drug treatment, in preliminary studies.

Preliminary results suggest that differential EEG patterns within patients with comparable GDS scores may have predictive utility relative to the subsequent rate of deterioration. A discriminant function was developed and independently replicated, which accurately classified normals versus GDS 4 or greater. The accuracy of this discriminant is shown in the upper half of Fig. 5 and was approximately 95%. Subsequently, a group of 39 patients who had been classified as GDS 2's were recalled for follow-up 3-5 years later, again evaluated clinically, and reclassified into a group which now showed cognitive impairment and a group which showed no further cognitive deterioration. Using the discriminant function described above, the neurometric data initially collected from these patients was used to classify them into those who had resembled 4's more than normals and those who had resembled normals more than 4's. As can be seen in the bottom half of Fig. 5, the majority of those who showed no change were classified as normal on the basis of the earlier data, while the majority of those who had deteriorated were classified as dementias. These preliminary findings are being further explored.

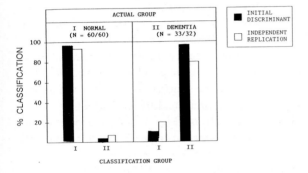

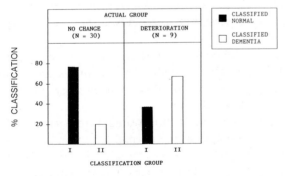

Fig. 5. Top panel shows the percentage of normal subjects (I) and dementia patients (GDS 4 or greater) classified as I or II by a stepwise discriminant function using a small subset of neurometric variables. The first and second numbers in parentheses above each panel indicate the sample size of each group used as the training set (black bars) and as the test set in the independent replication (open bars). The bottom panel of this figure shows the retrospective classification, using the above discriminant, of a group of patients who had been initially evaluated as having mild cognitive impairment with respect to their clinical condition after a 3-5 year follow-up.

ERP results

ERP data from these 7 groups were collected and also subjected to quantitative analysis. Data were gathered using 4 stimuli: (1) a 50% transmission grid flash with 65 lines/inch, perceived as grey, rear projected onto a screen 30 cm in front of the subject; (2) a 50% transmission grid flash with 7 lines/inch similarly rear projected; (3) a regular, predictable slide presented at 45 dB above normal hearing threshold, using a loudspeaker on each side of the subject's chair; and (4) a similar click presented at random, unpredictable temporal intervals. All stimuli were presented while the subject was comfortably seated in a sound-proof darkened chamber, gazing at a defocused television screen to reduce contrast, stabilize direction of gaze and reduce eye movements. One hundred artifact free evoked responses were collected to each stimulus, using an on-line artifact rejection algorithm.

The upper half of Fig. 6 shows the group grand average ERP recorded from 100 normal individuals, including all the members of group 0 and group 1, in response to the 7 line/inch grid flash, obtained from the 19 electrodes of the International 10/20 System referenced to linked earlobes (face upward). The array on the bottom half of Fig. 6 shows the corresponding variance of each group average ERP.

Note that the variance is not flat across the analysis epoch but has peaks, especially in the 100 – 250 msec latency range, indicating some heterogeneity of variance. This demonstrates the variability of the morphology of the ERP waveshape in normal healthy individuals.

The upper half of Fig. 7 shows group grand average ERPs, recorded from 17 severely impaired members of group 6 in response to the 7 line/inch grid flash. The bottom half of Fig. 7 shows the corresponding variances. Inspection of these data shows marked diminution of evoked response amplitude and alterations in waveshape in all leads except the occipital, especially after the primary response.

The variance curves, however, show much larger peaks than the normals. This indicates an even greater diversity of ERP morphology in the demented population.

In view of the demonstrated heterogeneity of ERP waveshapes in the normal group and the patients, and in view of the statistical impropriety of using multiple t tests at successive latency points (as is done by many workers), it was desirable to utilize a method which would provide a quantitative metric of overall abnormal ERP morphology.

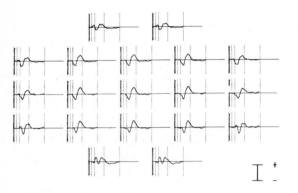

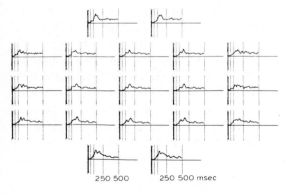

Fig. 6. Group grand average VEPs (top) and variance of grand average VEPs (bottom) computed for each electrode of the 10/20 system and displayed as a topographic array viewed from the top, face upward. Averages from 100 normal individuals, each derived from 100 grid flash stimuli, were combined in this computation. In this and all subsequent EP or factor displays, the vertical lines correspond to 50, 100, 250 and 500 msec. All waves were truncated at 500 msec, although the average inter-stimulus interval was 1000 msec (John et al., 1989).

Factor analysis

For this purpose, we subjected the averaged ERPs recorded from 19 electrodes in each of 75 normal subjects, in each of the 4 conditions, to a principal component analysis, followed by a Varimax rotation. This analysis has been described in detail elsewhere (John et al., 1989).

As illustrated in the schematic shown in Fig. 8, this analysis assumes that the EP in any lead, EP_i, can be described as the sum of the contributions of K independent components or factors $F_{j(t)}$, each multiplied by a weighting coefficient or factor score A_{ij} which describes the unique contribution of factor j to the EP at lead i, plus some residual variance, R.

$$EP_i = \sum_{j=1}^{k} A_{ij} F_j(t) + R$$

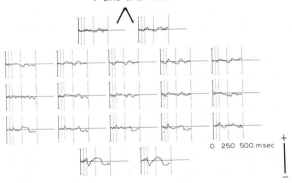

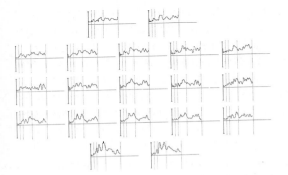

Fig. 7. Group grand average VEPs (top) and variance of grand average VEPs (bottom) as in Fig. 6, but recorded from 17 severely impaired members of the GDS = 6 group.

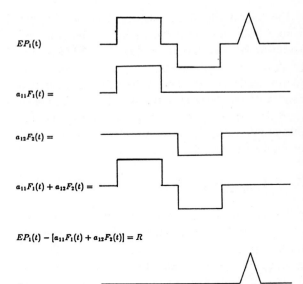

Fig. 8. Schematic discription of the factor analysis procedure: top waveshape = idealized VEP from electrode 1; second waveshape = weighted contribution of 1st factor; third waveshape = weighted contribution of 2nd factor; fourth waveshape = sum of the contributions of the first 2 factors; fifth waveshape = the residual not accounted for by the factors.

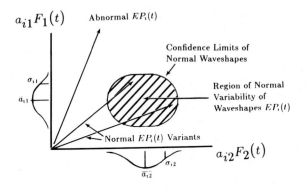

$$EP_i(t) = a_{i1}F_1(t) + a_{i2}F_2(t) + R$$
$$Z_{ij} = [a_{ij} - \bar{a}_{ij}]/\sigma_{ij}$$
$$EP_i(t) = Z_{i1}F_1(t) + Z_{i2}F_2(t)$$

Fig. 9. Schematic representation of the region of normal ERP variability in the factor 1/factor 2 plane. The distribution of the weighting coefficients for the first and second factors is shown on the corresponding axes. Using the mean and standard deviation of each of these distributions, the corresponding weighting coefficients of individual ERPs can be Z transformed.

Further, if all EP_i had identical shapes, each a_{ij} would be a constant. To the extent that the contribution of each component or factor is variable, there will be a distribution of the corresponding factor score A_{ij} in the population, as shown in Fig. 9. If we determine the mean, M_{ij}, and standard deviation, σ_{ij}, of each factor score in the normal population, we can state each A_{ij} as a Z score or standard score Z_{ij}:

$$Z_{ij} = (A_{ij} - M_{ij})/\sigma_{ij}$$

The left side of Fig. 10 shows 6 factors which accounted for 75% of the total variance of the grid flash evoked response normative data. The remaining 25% was estimated to be noise. Each factor was identified by the latency at which it makes its maximum contribution to the signal: 110, 150, 215, 260, 340 and 450 msec. A very similar factor structure was obtained from the click EP data, but with some latency differences for the first 3 factors, as seen on the right side of Fig. 10. It was now possible to reconstruct any visual or auditory ERP as a weighted combination of these factor components.

Each weighting coefficient was Z transformed relative to the normative distribution. Color coded topographic Z score maps were constructed showing the statistical significance of the difference between the contribution made by each factor to the ERP at each electrode position and the normative values. Nine maps quantified each set of EPs: one for each of the 6 factors, the root mean square (RMS) sum of factor Z scores (the total deviation in the normal space), the residual variance (the deviation in "pathological space"), and the total power in the ERP (John et al., 1989).

Fig. 11 shows the group average factor Z score maps for the 7 line/inch grid flash data. Each of the 6 rows of heads represents a different group of subjects: row 1, group 0; rows 2–5, groups 1–4; and row 6 represents groups 5–6 combined. The first 6 columns represent average factor Z scores for factors 1–6, followed by RMS Z scores and Z scores for residual variance and ERP power. Each map represents the 10/20 system seen from above, face upward. Data are color coded in standard deviation units with reds reflecting excessive and blues deficient contributions.

Note that the young normals in group 0 (mean age = 30) deviated from the mean normative value in the opposite direction from the old normals in group 1 (mean age = 70). This confirmed the existence of systematic ERP changes with normal aging, reported by other workers (Dustman and Beck, 1969). It also showed the need to age regress these data, which we have not yet done. This shortcoming can be overcome for our purposes by comparing all GDS groups to group 1, since the correlation of GDS with age was only 0.005.

These data revealed a progressive increase in the abnormality of the visual ERP with increasing cognitive impairment. The primary sensory component at P110 was relatively unchanged. The negativity at N150 (F2) steadily decreased, as did P215 and N260 (F3 and F4), across all leads except O1 and O2. P300 (F5) decreased most in an arc across the posterior temporal and parietal regions.

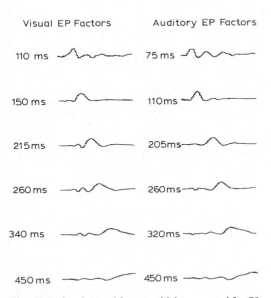

Fig. 10. Left column: 6 factors which accounted for 75 variance of normative grid flash ERP data: Right column: corresponding factor waveshapes which described the normative click ERP data.

N450 (F6) shows a maximum decrement in the frontal and posterior temporal regions, with a marked occipital increase.

Group average factor Z score maps for the random click data (not shown) were similar. The need for age regression was again evident when comparing the young and old normals. Again, the primary component represented by factor 1 was within normal limits. With increasing deterioration, a progressive decrease in N110 and P205 was seen, especially in the central and anterior temporal regions. In contrast to the flash results, little change was seen in N260. The overall abnormality was again diffusely distributed, which could be seen most clearly by the multivariate which estimates overall abnormality (RMS Z scores) in ERP waveshape.

Fig. 12 shows the mean amplitude difference, N1P2, of averaged grid flash and random click ERPs in each GDS group, relative to the amplitude in group 1 taken as 100%, measured for bilateral frontal, central, posterior temporal, parietal and occipital monopolar derivations. The best fit linear regression line across the set of values has also been calculated.

Note that for the visual stimulus, the highest correlation with cognitive decline is equally shown by frontal, central and parietal leads, with

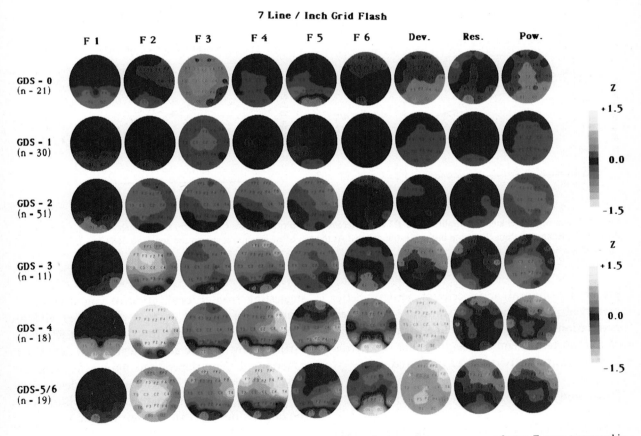

Fig. 11. After Z transformation of the factor scores for each factor for each electrode, group average factor Z score topographic maps were constructed separately for each of the GDS groups, using the same interpolation algorithm as for the EEG data. These data represent the grid flash ERPs and F1 – F6 refer to the factors whose waveshapes are shown on the left side of Fig. 10. All maps are color coded in standard deviation increments as shown on the scale. The significance of Z scores can be estimated as described in the legend of Fig. 1.

Change in Amplitude of N1-P2 with Cognitive Decline

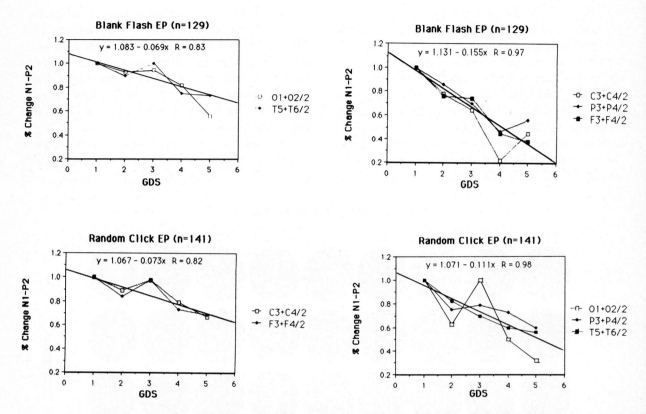

Fig. 12. Mean percent change in amplitude N1-P2 for different leads, plotted versus GDS score. Blank flash EP data are shown in the top graphs and random click data in the bottom graphs. At the top of each graph is shown the equation describing the regression line and the value of the correlation coefficient.

posterior temporal and occipital activity better preserved. For the auditory stimulus, the highest correlation with cognitive decline is shown by occipital, posterior temporal and parietal, with frontal and central activity better preserved.

Discussion

Neither the EEG nor the ERP data provide support for the contention that the frontal cortex is differentially and primarily implicated in the cognitive deterioration of old age. The data support Lashley's Law of Mass Action, rather than suggesting frontal localization of cognitive functions. There appears to be a generalized diffuse departure from optimal function. The ERP data further show that the anatomical loci of decreased function differ, depending upon the sensory modality. The high sensitivity of the parietal lobe, as shown by its similar changes to stimuli of both modalities, might merit further attention.

References

Dustman, R.E. and Beck, E.C. (1969) The effects of maturation and aging on the waveform of visually evoked potentials. *Electroenceph. Clin. Neurophysiol.*, 26: 2-11.

John, E.R., Prichep, L.S. and Easton, P. (1987) Normative data banks and neurometrics: basic concepts, methods and results of norms constructions. In: A. Rémond (Ed.), *Handbook of Electroencephalography and Clinical Neurophysiology. Vol III. Computer Analysis of the EEG and Other Neurophysiological Signals,* Elsevier, Amsterdam, pp. 449-495.

John, E.R., Prichep, L.S. and Easton, P. (1988) Neurometrics: computer assisted differential diagnosis of brain dysfunctions. *Science,* 293: 162-169.

John, E.R., Prichep, L.S., Friedman, J. and Easton, P. (1989) Neurometric topographic mapping of EEG and evoked potential features: Application to clinical diagnosis and cognitive evaluation. In: K. Maurer (Ed.), *Topographic Brain Mapping of EEG and Evoked Potentials,* Springer, Berlin, pp. 90-111.

Prichep, L.S., Gomez-Mont, F., John, E.R. and Ferris, S. (1983) Neurometric electroencephalogram characteristics of dementia. In: B. Reisberg (Ed.), *Alzheimer's Disease: The Standard Reference,* Free Press (MacMillan), New York, pp. 252-257.

Prichep, L.S., John, E.R., Ferris, S.H., Reisberg, B., Alper, K. and Cancro, R. (1990) Neurometric EEG correlates of cognitive deterioration in the elderly. Submitted.

Reisberg, B., Ferris, S.H., de Leon, M.J. and Crook, T.T. (1982) The global deterioration scale for assessment of primary degenerative dementia. *Am. J. Psychiat.,* 139: 165-173.

Discussion

A. Gevins: Would you expect focal changes in cognitively impaired patients performing difficult cognitive tasks?

E.R. John: No, I would not. I might expect an atypical pattern of interactions most deviant for a particular region, as we sometimes see in stroke patients during tasks.

R.W.H. Verwer: I thought that the rhythms you are investigating are global phenomena. Would it be reasonable to expect focal abnormalities in these rhythms due to functional deficits even if these deficits would be focal themselves?

E.J. John: Yes. One commonly sees focal changes in quantitative maps of rhythmic EEG process in patients with tumors, cerebral infarcts or after traumatic head injury.

H. Van de Tweel: Is the GDS score righteously taken as having quantitative significance? I believe you cannot use linear regression.

E.R. John: As usual, I fear you are right. But it may give an approximate measure.

Subject Index

acetylcholine (ACh), 49, 275, 292, 362, 382, 423–425, 427
  activation, 19
  antagonist, 425, 447
  innervation, 19, 46
  neurons, 47, 292
  receptor, 425
acetylcholinesterase (AChE), 42, 97, 225, 227, 230–232, 234–235, 239, 288, 296, 304
aggressive behaviour, 257, 486–497
Alz-50 immunoreactivity, 193, 195–196, 202, 207–211
Alzheimer's disease, 21, 29, 285, 301, 523, 530–531
amygdala, 40–41, 44, 46–49, 51, 95, 132, 133, 252, 422, 439, 441, 443, 449, 465, 483
  basolateral nucleus, 40, 44, 48, 105, 420–422, 482–483
  basomedial, 44
  central nucleus, 44, 105, 445–447, 450
  periamygdaloid cortex, 44, 48
  projections, 106–107, 507
amygdalectomies, 547
amyotrophic lateral sclerosis, 301
anxiety, 545
anxiogenic agents, 371–372, 377, 381, 383, 387, 402, 405, 410
anxiolytic agents, 371–372, 380–383, 387–388, 390, 405
apes, 51–55, 541
apomorphine, 371, 493, 534
archicortex, 43, 64, 74, 87
architectonic trends, 43, 64, 73, 87
aspartate (aspartic acid; Asp), 20–22, 29, 379, 423, 426–427
attention, 48, 88, 133, 156, 371, 411–412, 457, 458, 550

auditory functions, 39, 81
autonomic (cortical) functions, 47, 110, 147–148, 152, 434–436, 439, 443–444, 447, 450, 453–456, 459–460, 466
  parasympathetic, 154
  sympathetic, 152, 154
autonomic responses, 147, 434, 436, 443, 456, 543

basal forebrain, 19, 439, 441, 465
basal ganglia, 40, 46, 48, 95, 119, 550
  caudate nucleus, 47, 48, 127–129, 131, 132, 134, 264, 333, 420, 422, 449, 502, 531
  globus pallidus, 41, 97, 121–123, 127, 129, 131, 134, 135–136, 146, 380
  n. accumbens, 41, 44, 96, 132, 264, 266–267, 269, 276, 283, 360, 363, 368, 389, 403, 405–409, 420–421, 493, 507
  pallidum, 40–41, 44, 100, 110, 132, 134, 136, 380
  putamen, 48, 121–122, 125–126, 131, 134, 422, 449
  striatum, 41, 44, 48, 96–97, 129, 132–134, 136, 190, 196, 202, 252, 262, 265, 268–269, 274, 276, 286, 363, 367–368, 370, 372, 374, 380–381, 388, 389, 406–407, 420–421
bed nucleus of stria terminalis, 41, 378
benzodiazepine, 381–390
  agonist, 371, 383
  antagonist, 371, 383
  inverse agonist, 371–372, 380, 383, 402–403, 405
  receptor (BZDr), 371, 378, 380, 383, 385, 402
blood–brain barrier, 286
brainstem, 45, 47, 49, 357
brain volume, 54
brain weight, 252–253

burst firing, 380, 382

β-carboline, 371–372, 375, 378, 382–383, 387–388, 391, 403, 405, 407–409
cardiovascular responses (*see also* autonomic), 148, 435–436, 447, 454, 456
cell
  death, 180, 233, 249–251
  migration, 300
  number 249, 253, 305
  proliferation, 225, 260
central gray, 49, 441, 443
cerebellum, 41, 122–123, 456
choline acetyltransferase (ChAT), 19–20, 24, 232, 288
cingulate lesions, 483, 547, 549
circuits
  basal ganglia-thalamocortical 102, 119–120
  dorsal agranular insular, 102
  dorsolateral prefrontal, 128–129, 131
  feedback, 346–352, 422
  lateral orbifrontal, 128
  limbic, 47, 97, 119, 547
  motor, 43, 46, 74, 97, 119–122, 124, 126, 268, 313, 319, 322, 407
  neural, 549
  oculomotor, 96, 126
  parallel, 46, 96, 108, 119, 120, 146
  prefrontal circuits, 128
  prelimbic circuit, 102
  thalamocortical, 119, 229
claustrum, 44, 48, 441
cognitive processes, 119, 149, 315, 326, 330, 342, 352, 358, 405, 411–412, 414, 523, 529–531
  deficits, 130–131, 268, 523, 528, 530–533, 558–559, 564
  development, 233, 235, 239
  functions, 119, 130–131, 149, 233, 235, 239, 268, 315, 330, 342, 346, 352, 358, 405, 411, 412, 414, 523, 528–533, 558–559, 564
compulsive disorder, 131, 549–550
conditioned responses (CR), 325, 467–475, 478, 481
  fear, 368–370, 372, 374, 389, 391, 409, 416

heart rate, 152, 438, 444–447, 451, 454–456, 460–461, 465–466
contingent negative variation (CNV), 315, 351
coping behaviour, 370–371, 392, 403, 411–412, 416–417
corollary discharge, 321, 348
cortical areas, 43, 55
  anterior cingulate cortex, *see* prefrontal cortical areas
  auditory areas, 44, 66, 74, 81, 83–85, 131, 227
  entorhinal cortex, 40–43, 85, 87–88, 132, 259, 373, 422, 428
  insula, 39, 78
  motor area, 4–5, 8–9, 38–40, 42–44, 50–51, 72, 76, 119–120, 124, 214, 227, 250–251, 333, 466
  occipital areas, 50, 79, 250–251, 289–290
  (para)limbic areas (*see also* prelimbic and infralimbic under prefrontal cortical areas, 66, 84–85, 89
  parietal areas, 50, 74, 76–78, 86, 88–89, 127, 129, 250, 256, 332, 507
  parinsular cortex, 65–66, 81
  perirhinal areas, 35, 42, 47, 49, 84–85, 88, 132, 428
  posterior cingulate cortex, 38, 444, 447, 467–469, 472, 474–475, 477, 479–481, 483
  prefrontal areas, *see* prefrontal cortical areas
  premotor area, 8–9, 37–38, 40, 44, 47, 50, 66, 69–76, 78, 89, 85–87, 120–123, 126, 227, 331, 333, 507
  pyriform cortices, 40, 43, 373, 426
  retrosplenial area, 43, 87–88, 444
  sensorimotor cortex, 46, 75, 249, 252
  sensory association areas, 40, 87, 89
  somatosensory areas, 9, 43–44, 50–51, 66, 74–75, 77, 87, 121, 123, 210, 230, 245, 250–252, 289
  temporal areas, 40, 49, 74, 78, 81, 84–86, 256
  visual areas, 4, 14, 18, 21–22, 47, 74, 78, 81, 85, 210, 328, 331
  visual association areas, 6, 44, 66, 78–81, 85,

88, 131
visual cortex, 4, 5, 78, 227, 332
cortical connections
   corticocortical, 42–43, 63, 83, 87, 110, 175, 214, 232, 251, 253, 330, 423, 506, 515
   corticostriatal projections, 29, 98, 389
   subcortical connections, 251–253, 507
   thalamo-cortical projections, 32–51, 98–105, 172, 544
cortical layers, 15, 43, 87, 171–172, 191, 196, 229–231, 240
cortical plate, 171, 185, 189, 191, 215, 225, 227, 229, 232, 262–263
cortical thickness, 247, 249–250, 252–253, 269
cortical volume, see volumetry
corticospinal tract, 176
corticosterone, 370–371
cross-temporal contingencies, 313, 315–316, 318–319
cross-temporal integration, 321–322
cuneiform nucleus, 45
cytoarchitectonic parcellation, 4, 31–32, 37, 55, 63–64, 71, 172, 191, 229, 239

decortication, 243, 245, 250, 256
delay, see memory
dementia, 541, 557, 559
dendritic morphology, 186, 189, 194, 250–253, 256, 271, 279, 302–303
2-deoxyglucose, 289, 422
depression, 540
diagonal band of Broca, 19, 41–44, 47–49, 441
DOPAC, 268–269, 371, 389, 406–413, 416–417
dopamine (DA), 15–16, 135, 255–258, 260–276, 286, 288, 358, 361–363, 367–369, 371, 375, 378, 388, 391–392, 402–403, 416, 423–427, 492–495, 507, 532–534, 537
   agonist, 360, 371, 374, 389, 392, 493, 533–534, 537
   antagonist, 260, 360, 371, 374, 389, 548
   autoreceptor, 368, 371–375, 393
   depletion, 131, 271, 274

   fibers, 17, 181, 262–263
   grafts, 286–287
   hypothesis of schizophrenia, 533, 537
   innervation, 15–16, 18, 42
   metabolism, 276, 373, 374, 378, 383, 387–389, 391–393, 405–413, 416–417, 533
   neurons, 261, 266, 272, 357, 367–380, 382, 385, 388–393, 405–409, 414, 534
   receptor, 260, 359–360, 363, 373–374, 424
   release, 367–374, 379–381, 389, 391–392, 403, 409, 417
   synthesis, 368–369, 371–373, 387, 403, 409
   system, 46–47, 261, 286, 421, 427, 492–493
   turnover, 358, 370, 377
   utilization, 368–372, 376–378, 381, 383, 390, 392, 398
dorsal motor nucleus of the vagus, 439, 465
dorsolateral tegmental nucleus, 41–42
Down syndrome, 528

emotion, 63, 88–89, 133, 149, 153, 358, 371, 405, 410–411, 541, 545
endocrine response, 149
epinephrine (Epi), 382, 424
evolution of PFC areas, 3–4, 6, 32, 43, 54, 64, 66, 94, 333
excitatory amino acids, 379, 389
external capsule, 421–422, 428

facial expression, 89, 127, 134, 504, 511, 512
feedback, see circuits
feeding behaviour, 242, 412–413
food hoarding, 242, 245, 258, 272, 273, 276, 491
frontal lobe, 84
   functions, 63, 501
   lesions, 486–489, 494, 523, 529
   pathology, 73, 524, 530, 543
frontal lobotomy (see also psychosurgery), 486–489, 494

G proteins, 374
GABA (γ-aminobutyric acid), 22, 24, 29, 121, 134–135, 211, 269–270, 331, 333, 371, 380–381, 387–388, 402–403, 423,

426–427
  agonist, 136, 381–382
  antagonist, 136, 331
  neurons, 22, 213, 365, 380–381, 402
  receptor, 380–381, 388, 402
  release, 381–382, 388
glutamate (glutamic acid; Glu), 20–22, 29, 362, 379–380, 389, 423–427, 454, 466
  neurons, 21, 424
  receptor, 380, 426
  release, 389
Golgi method, 185–222
gonadal hormones, 247
growth cone, 260–261, 267
gustation, 39, 66, 78

habituation, 242, 436, 449, 454, 459, 510
haloperidol, 271, 274, 360, 367, 493, 527
hedgehog, 6–7, 502
hippocampus, 42–44, 48, 66, 72–73, 132, 153, 332–333, 428, 459, 474–475, 480, 483, 507, 550
histogenesis, 185, 189, 223–224
homologies, 4–6, 31–32, 50–51, 502–503, 505
Huntington's disease, 5, 21, 126, 131, 528, 530, 531
5-hydroxytryptophan (5-HTP), 382
hyperactivity, 133, 272
hyperkinesia, 135
hypoactivity, 272
hypokinesia, 133, 135, 268
hypothalamo-hypophyseal-adrenal axis, 376, 378
hypothalamus, 41, 44, 48, 95, 101, 420, 439, 443, 445–446, 465, 489, 491–492, 494, 497, 544

immobilization, 406, 408, 410, 411
inhibition, 134, 315, 349, 350
initiating movement, 88, 89
intellectual deficits, 545
interpeduncular nucleus, 41, 45, 377
isocortex, 52, 65–66, 85

Katakana, 350

Kennard principle, 243

laminar distribution, 16, 19, 21, 22, 270
Laplacian, 339, 340, 342, 343, 351
Lashley's law, 564
lateralization, 245, 342, 351
learning see memory
limb movements, 134
limbic structures, see (para)limbic structures
locomotor activity, 272, 358, 408

mammillary bodies, 44, 47
marginal zone, 171, 189, 190, 231, 258
maternal behaviour, 242, 512, 515
mediodorsal thalamic nucleus (MD), 31–36, 38–43, 51, 70, 89, 95, 148, 172, 215–216, 230, 239, 256, 257, 362, 392, 420–422, 427, 435, 439, 441, 449, 453–454, 459, 461, 465–467, 469–470, 472, 474–475, 480, 482–483, 489, 502, 544, 546
  central, 35, 40–42
  densocellular division, 37, 39, 41
  lateral, 35, 40–41, 43
  lesions of, 546
  magnocellular division, 37–42, 70, 96, 126, 131–133
  medial 35, 40–42
  multiform division, 39–41, 71, 126–127
  paralamellar, 35, 37, 40–41, 43, 96
  paralaminar division, see multiform division
  parvocellular division, 37–41, 71, 96, 121, 129
  stimulation of, 362
  ventral, 40–41
medulla, 122, 441, 443–447, 450, 465, 466
memory
  alternation learning, 291, 325
  associative learning, 326, 433–434, 445–446, 455, 461, 510
  avoidance learning, 243, 304, 370, 406–409, 456, 460, 467, 475, 483, 481
  classical conditioning, 433–437, 441, 444, 445, 449, 456–461, 465
  context-dependent memory, 319
  declarative memory, 457

delayed alternation, 276, 289–290, 292, 358, 482–483, 491, 502, 519
delayed matching-to-sample, 320, 325, 333
delayed non-matching-to-sample, 325, 509
delayed responding, 130, 243, 313–316, 325, 327–328, 330, 333, 335, 483, 509–510, 514–515
long-term memory, 326, 333, 457–459
memory field, 330, 332, 336
memory maps, 331
mnemonic coding, 332
mnemonic recency system, 475
"mnemonic scotoma", 331
mnemonic template, 474
pattern recognition, 350
primacy encoding, 475, 479–481, 483
procedural memory, 325, 457, 460
"prospective memory", 326
reference memory, 325
representational memory, 326, 510
"retrospective memory", 326
semantic memory, 325
short-term memory, 129, 313, 315–316, 321–322, 326, 458, 510
spatial learning 111, 129–130, 242–243, 247, 257, 273–274, 290, 318, 325–326, 330–333, 335, 483, 511, 514
transient memory, 326
working memory, 242, 325, 457–460, 482
mental illness, 539
α-methyl-p-tyrosine (αMpT), 268, 271, 359, 361
Moniz, 155, 494, 539–542
Morris water maze, 243, 245, 290, 292, 514
motivation, 133, 274
motor functions, 119, 268, 407
motor set, 43, 46, 51, 74, 97, 119–122, 124, 126, 214, 250–251, 268, 313, 319, 321–322, 407, 512
mounting behaviour, 246
movements
    disorders, 122
    execution, 120, 124
    preparation, 120, 124, 126, 315–316, 321
myelinization, 169, 207, 303

nest building behaviour, 242, 245, 512, 515
neural activity, 11, 406
neuroleptics, 360, 373, 525, 527–528, 533, 537
neurons
    Cajal-Retzius, 189–190, 192, 197, 206–207, 209–210
    interneurons, 194, 197–200, 207, 209, 213–214
    interstitial, 202, 205, 207, 211
    pyramidal, 94, 186, 189, 190, 195–197, 202, 205, 207, 209, 212–214, 216–217, 224, 234–235, 240, 270, 287
    subplate neurons, 190, 194, 201–202, 205, 209, 211, 215, 230, 232
neuroses, 542
noradrenaline (norepinephrine; NA; NE), 13–15, 18, 50, 246, 252–253, 256, 258, 261, 263, 265–267, 270, 358, 362–363, 365, 382, 416, 424, 492
    agonist, 358, 382, 388
    antagonist, 447
    fibers, 247
    innervation, 13–15, 18, 493
    neurons, 357, 361, 369
    systems, 371, 403
novelty, 459
nucleus ambiguous, 441, 460, 465
nucleus tractus solitarius, 39, 41, 45–47, 110, 150, 439–440, 460, 513

oculomotor disturbances, 126, 136
oculomotor field (Forel), 45
oculomotor processes, 119, 327
oculomotor task, 331
odour discrimination, 513
olfactory system, 39, 41, 44, 46, 66, 96, 132, 266, 268, 368, 406
open field activity, 272, 276
operculum, 72–73, 76
orang-utan, 52
orienting, 436, 449, 451, 453, 457, 459
overgrowth, 174–175

pain, 227, 547, 549

paleocortex, 65, 74
parabrachial nucleus, 41, 45, 49, 101
parahippocampal gyrus, 74, 78–79, 81, 84, 87–88
(para)limbic structures, 44, 66, 149, 381, 474, 480–483, 547
Parkinson's disease, 126, 131, 135, 268, 269, 530, 533, 537
pedunculopontine tegmental nucleus, 47, 49–50, 122
perception–action cycle, 321–322
periallocortex, 65–66
periaqueductal gray, 41, 45–47, 49
perinatal, 223, 390
perseverative behaviour, 88, 131–132
personality, 63, 88, 543
phylogenetic relationship, 31, 50–51, 79, 502–504
play behaviour, 242–243, 246
pontine nucleus, 45, 49
prefrontal cortical areas, 34, 43–44, 46, 49, 87
   agranular insular cortex (AIdv), 33, 35, 39–40, 43–45, 47, 49, 51, 74–75, 85, 439–440, 443, 445, 447, 449–451, 460–461, 482
   agranular (PFCag), 434–435, 439, 441, 443–445, 449, 454–456, 458, 460
   anterior cingulate (area 24), 33–35, 37–38, 40, 42, 44–51, 67, 69, 72–73, 78, 84–85, 87–88, 119–120, 132–133, 296, 373, 375, 383, 426, 467–472, 474–475, 477, 479–481, 483, 547
   Broca's area, 229, 351
   dorsolateral (areas, 45, 46, 9), 37, 40, 48–51, 67–70, 76, 78–79, 83, 87, 90, 119–120, 127, 129, 139, 188, 326–328, 335, 487–488
   frontal eye field (area 8), 8–9, 37–38, 40, 43–44, 47–50, 67–70, 76, 78, 81, 83–84, 86, 96, 119–120, 127–129, 188, 444
   infralimbic area (area 25), 34, 35, 37, 40, 42, 47, 51, 67, 69, 83–84, 87, 441, 466
   medial, 34, 43–46, 49, 67, 70, 83–84, 87, 241, 268, 427, 439–440, 443, 444, 446–447, 449–451, 453, 455–456, 459–461, 489, 492, 493
   medial orbital cortex (MO), 34
   medial precentral (PrCm, Fr2), 33–36, 40, 42, 44–47, 50
   orbital (areas 11–14, 47), 11, 34, 37, 40, 44, 48–51, 67–70, 80–81, 83–85, 87, 119–120, 129, 131–132, 241, 250, 424, 482, 486, 488–489, 491–494, 497, 513
   prelimbic area (area 32), 34–35, 40, 42, 46–47, 51, 69, 70, 84, 120, 122, 127
   ventral orbital cortex (VO), 34, 36, 40
   ventral PFC, 35, 43, 67, 85
   ventrolateral orbital cortex (VLO), 34, 36, 38, 40, 123
prefrontal deficit, 316
prefrontal lesions, 88–89, 156, 170, 248, 314, 319, 335, 490–491, 541
prefrontal pathology, 530–531, 540
prefrontal size, *see* relative size of PFC
prefrontal volume (*see also* volumetry), 51–55, 492
preoptic area, 40–41, 44, 446
preparatory interval, 350
preparatory process, 126, 313, 348–352
pretectum, 41, 45, 48, 49
principal sulcus, 43, 67, 71, 76, 326, 330, 332
proisocortex, 40, 65, 66, 67
psychiatric therapy, 539
psychosomatic disease, 158
psychosurgery, 486–489, 494, 539, 541, 546–547

rabbits, 467
radial arm maze, 243
radial glia fibers, 185, 191
raphe nuclei (*see also* serotonin), 17, 41, 45–50, 357–358, 363, 365, 381–382, 420, 422–423, 441
reaction time, 341, 349, 352
recency system, 477, 480, 481, 483
regional cortical blood flow (rCBF), 523, 524, 525, 527–531, 533, 534
reinforcement, 458, 459–461
relative size of prefrontal cortex, 31, 51–54
reorganization, 223

reproductive system, 155
response inhibition, 349–350
reticular formation, 41, 45, 49, 126, 357
retrorubral field (A8), 45, 49, 50, 388
reward, 155, 274, 419–420, 422, 427

saccadic eye movements, 38, 127, 328, 332, 335
schizophrenia, 131, 285, 530–537, 541
segregation of cortical regions, 7, 50, 55
self/world model, 352
septum, 44, 48, 296, 378, 493
serotonin (5-hydroxytryptamine; 5HT), 17–18, 256, 258, 260–263, 267, 271, 274–275, 361–363, 381–382, 424
  antagonist, 360
  fibers, 47, 260, 360
  innervation, 17–18, 493
  lesion, 360
  receptor, 360
  systems, 371, 381, 403
sexual behaviour, 275–276, 490, 512, 513
single-unit activity, 467
sleep, 227
somatomotor responses, 434–435, 438, 458–459, 461
somesthetic function, 39
sparing of function, 242–248, 256, 258, 272, 274, 276
spectral intensities, 341, 342
spinal cord, 41, 45, 122, 177–180
spines, 197–198, 202, 207, 213–217, 233
stereotyped behaviour, 545
stress, 155, 268, 358, 368–377, 380–383, 389–393, 402, 405–406, 408–411, 524, 532
striatopallidothalamic systems, 96
subiculum, 40, 44, 111, 472, 483
subplate zone, 172, 189–190, 192–193, 197, 202, 210, 214, 222, 225, 227, 229, 232, 235, 240, 258, 259, 262, 263, 266, 267, 282–283, 378
substantia innominata, 41, 47, 49
substantia nigra (SN), 41–42, 45, 49–50, 101, 121, 127, 129, 131, 134, 136, 252, 261, 264, 282–283, 372, 376–380, 385, 388, 420, 507
subthalamic nucleus, 44, 47–48, 101, 121–123, 128, 132, 134–136, 146
subventricular zone, 189
superior colliculus (SC), 38, 41, 43, 45, 47–49, 126–127, 333, 441, 443
superior temporal sulcus, 79, 81, 89
sympathetic responses, 152, 154
synapse, 15, 17, 22, 223, 227, 258
synaptogenesis, 230, 232, 233, 300

taste aversion, 290
temporal organization, 40, 49, 74, 78, 81, 84–86, 88, 256, 313, 315, 515–516
thalamic connections, 32, 70, 74, 101, 172, 544
thalamic nuclei, 32–33, 422, 505
  area X, 38, 122–123
  anterior nuclei, 37–38, 468–470, 482, 546
    anterodorsal (AD) nucleus, 37–38, 40
    anteromedial, 35, 37–38, 51, 468
    anteroventral, 37, 121, 472, 474–475, 480
    lateral dorsal nucleus (LD), 37
  habenula, 44, 101, 133, 377, 446
  intralaminar nuclei, 36, 38–39, 95, 122, 439, 441, 449, 546
    parafascicular nucleus, 546
  lateral geniculate nucleus, 289
  lateral posterior (LP), 36
  mediodorsal, see mediodorsal thalamic nucleus
  midline nuclei, 33, 439
    paraventricular nucleus, 105, 441, 466
  pulvinar nucleus, 36, 38, 39, 78, 81, 239
  reticular nucleus, 41
  ventral medial complex
    basal ventral medial (VMb), 36, 39
    submedial (gelatinosus), 32, 36
    VA–VL complex, 36
    ventral anterior, 36, 38–39, 122, 126–127, 129, 131
    ventral medial (principal), 12, 33, 36, 39, 74, 439, 449
    ventrolateral (VL), 8, 35–36, 38–39, 120–121, 134–135
tongue use, 110, 245, 246
topographic maps, 349, 555, 557–558

transient
  AChE staining, 42
  collateral projection, 180
  corticospinal projection, 177–180
  neurons, *see* neurons, subplate
  overgrowth, 174–175
transplantation, 255, 274, 285–292, 376
trigeminal nucleus, 45
trophic factors, 292, 296
tyrosine hydroxylase (TH), 16, 262, 368, 375
  activity, 387–388, 402, 409

vagus, n., 152
ventral tegmental area (A10), 15, 41, 46, 49, 133, 252, 257, 261, 264, 266, 267, 273, 282, 357–359, 361, 363–365, 370, 372, 373–389, 391, 405, 419–421, 423, 424, 439, 441, 465, 492, 507
  stimulation of, 359, 360, 362
ventricular zone, 189, 260
verbal fluency, 511
violence, 495, 550
visceral functions, 39, 46, 49, 111, 148–149, 249, 252, 256, 434, 449, 458, 460, 547
visceral responses, 39, 148–149, 434, 443, 449, 458, 460
vocalization, 156
volumetry, 52–54, 174–175, 233

Wisconsin Card Sorting test (WCS), 511, 524–525, 527–534